TABLE OF FORMULA WEIGHTS

No more than three decimals are carried, although in some instances four could be retained according to the rules of significant figures.

AgBr	187.772	$H_2C_2O_4$	90.035	MgO	40.304
AgCl	143.321	$H_2C_2O_4 \cdot 2H_2O$	126.066	$Mg_2P_2O_7$	222.553
$Ag_2C_2O_4$	303.756	HCOOH	46.026	MnO_2	86.937
Ag_2CrO_4	331.730	HCl	36.461	NaBr	102.894
AgI	234.773	$HClO_4$	100.459	$NaCHO_2$	68.007
$AgNO_3$	169.873	HF	20.006	$NaC_2H_3O_2$ (NaOAc)	82.034
AgSCN	165.95	HNO_2	47.013	NaCl	58.443
Al_2O_3	101.961	HNO_3	63.013	NaCN	49.007
As_2O_3	197.841	H_2O	18.015	Na_2CO_3	105.989
$BaCl_2$	208.24	H_3PO_4	97.995	$Na_2C_2O_4$	133.999
$BaCO_3$	197.34	$HgCl_2$	271.50	$NaHCO_3$	84.007
$BaCrO_4$	253.32	Hg_2Cl_2	472.09	NaH_2PO_4	119.977
BaO	153.33	$Hg(NO_3)_2$	324.60	Na_2HPO_4	141.959
$Ba(OH)_2$	171.34	H_2S	34.08	$NaNO_2$	68.995
BaS	169.39	H_2SO_4	98.07	$NaNO_3$	84.995
$BaSO_4$	233.39	KBr	119.002	Na_2O	61.979
$CaCO_3$	100.09	$KBrO_3$	167.000	NaOH	39.997
CaC_2O_4	128.10	KCl	74.551	Na_2SO_4	142.04
CaF_2	78.08	$KClO_3$	122.550	$Na_2S_2O_3 \cdot 5H_2O$	248.17
CaO	56.08	$KClO_4$	138.549	NH_3	17.030
$Ca(OH)_2$	74.09	KCN	65.116	N_2H_4	32.045
$CaSO_4$	136.14	K_2CrO_4	194.190	NH_4Cl	53.491
$Ce(HSO_4)_4$	528.38	$K_2Cr_2O_7$	294.184	N_2H_5Cl	68.506
$Ce(NO_3)_4 \cdot 2NH_4NO_3$	548.23	$KHC_8H_4O_4$ (KHP)	204.223	$(NH_4)_3PO_4 \cdot 12MoO_3$	1876.34
$Ce(SO_4)_2$	332.24	KHC_2O_4	128.126	$(NH_4)_2PtCl_6$	443.88
$Ce(SO_4)_2 \cdot 2(NH_4)_2SO_4 \cdot 2H_2O$	632.53	$KHC_2O_4 \cdot H_2C_2O_4$	218.161	$Ni(C_4H_7O_2N_2)_2$	288.93
CO_2	44.010	KI	166.003	$PbCO_3$	267.2
Cr_2O_3	151.990	KIO_3	214.001	$PbCrO_4$	323.2
CuO	79.545	$KMnO_4$	158.034	$Pb(IO_3)_2$	557.0
$Cu(OH)_2$	97.561	K_2O	94.196	PbO	223.2
CuSCN	121.62	KOH	56.106	$PbSO_4$	303.3
FeO	71.846	K_2PtCl_6	486.00	P_2O_5	141.945
Fe_2O_3	159.692	KSCN	97.18	SO_2	64.059
Fe_3O_4	231.539	K_2SO_4	174.25	SO_3	80.06
FeS_2	119.97	$Mg(C_9H_6ON)_2 \cdot 2H_2O$	348.640	SnO_2	150.69
$FeSO_4$	151.90	$MgCO_3$	84.314	SrC_2O_4	175.64
$HAsO_2$	107.928	$Mg_2C_2O_4$	112.325	U_3O_8	842.082
$HC_2H_3O_2$ (HOAc)	60.052	$MgNH_4PO_4 \cdot 6H_2O$	245.406	$Zn_2P_2O_7$	304.70

QUANTITATIVE ANALYSIS

4TH EDITION

R. A. Day, Jr.
Emory University

A. L. Underwood
Emory University

PRENTICE-HALL, INC., ENGLEWOOD CLIFFS, NEW JERSEY 07632

Library of Congress Cataloging in Publication Data

Day, Reuben Alexander, 1915–
Quantitative analysis.

Bibliography
Includes index.
1. Chemistry, Analytic—Quantitative. I. Underwood, Arthur Louis, 1924– joint author. II. Title.
QD101.2.D37 1980 545 79-21774
ISBN 0-13-746545-9

© 1980, 1974, 1967, 1958 by Prentice-Hall, Inc., Englewood Cliffs, N.J. 07632

All rights reserved. No part of this book may be reproduced in any form or by any means without permission in writing from the publisher.

Printed in the United States of America

10 9 8 7 6 5 4 3 2

Editorial/production supervision and interior design by Ian M. List
Chapter opening design by Linda Conway
Cover design by Linda Conway
Manufacturing buyer: Edmund W. Leone

Prentice-Hall International, Inc., *London*
Prentice-Hall of Australia Pty. Limited, *Sydney*
Prentice-Hall of Canada, Ltd., *Toronto*
Prentice-Hall of India Private Limited, *New Delhi*
Prentice-Hall of Japan, Inc., *Tokyo*
Prentice-Hall of Southeast Asia Pte. Ltd., *Singapore*
Whitehall Books Limited, Wellington, *New Zealand*

CONTENTS

PREFACE

Analytical chemistry has been taught in a variety of ways at different schools. At the present time, many schools offer a one-semester or one-quarter course emphasizing traditional topics following the general chemistry course, and a one-semester course in the senior year on instrumental topics with physical chemistry as a prerequisite. A third course, for students who do not expect to become professional chemists, is also offered at many schools. This course includes instrumental topics, and although it may be considered superficial, it has considerable value in showing students who will never study analytical chemistry again something of the field beyond the classical titrimetric and gravimetric techniques. The majority of the students in this course are premedical and predental; some anticipate graduate work in biomedical fields such as biochemistry, physiology, pharmacology, or microbiology. Biology majors may also take the course. At Emory University, we use this text for the first and third types of courses.

In revising the text, we have kept such a pattern of courses in mind. The major changes, both in reorganization of old material and in the introduction of new topics, are designed to improve the presentation of traditional topics for the introductory course. A clearer separation of gravimetric and titrimetric methods has been made, and a greater emphasis has been given to applications of these methods to practical analyses.

Changes in the later chapters have been fewer, and have been made primarily to update the material and improve the presentation. These chapters, which deal with such topics as electrical and spectroscopic methods of analysis and various aspects of chromatogrphy, are not designed for the senior-level instrumental course for professional chemistry students, although such students may find them useful as a starting point for reading more advanced material.

Specifically, we have made the following major changes. Separate chapters on titrimetric and gravimetric methods, covering mainly stoichiometric calculations, are included. Although the titrimetric chapter is presented first, the order of coverage can be reversed if the instructor desires. A new chapter on review of chemical equilibrium follows the stoichiometry chapters. This chapter covers all the types of equilibrium calculations encountered in the reactions used for analysis. The instructor can thus focus attention on equilibrium calculations before considering their applications to analysis. This chapter, which reviews the equilibrium treatment found in general chemistry texts, may be omitted if the class is well prepared. Chapters 6 through 11 then cover the traditional acid–base, complex formation, precipitation, and oxidation-reduction theory and applications. Chapter 11 is a new one discussing the numerous applications of oxidation-reduction reactions to analysis, and offers a respite from theory.

The chapters on spectrophotometry, flame emission and atomic absorption spectroscopy, solvent extraction, gas-liquid chromatography, and liquid chromatography have undergone less drastic changes. The chapters on electroanalytical chemistry, however, have been extensively rewritten in order to clarify points that students, in our own teaching experience, found troublesome in the earlier edition. A new chapter called "Real Analytical Chemistry" has been added in an effort to give a perspective regarding the importance and utility of the field of analytical chemistry.

Throughout the text, an attempt has been made to improve the questions and problems at the end of each chapter. In many cases involving fundamental concepts, two very similar problems are given consecutively, with answers furnished for the odd-numbered ones. This arrangement allows the student to check himself on one such calculation before attempting a problem for which the answer is not furnished.

One major departure from previous editions is the inclusion of laboratory directions in the last section of the text. This change has been made at the request of many users who find the cost of a separate laboratory manual excessive. In this laboratory section, we have included two chapters on general laboratory techniques and the analytical balance. Directions are then given for a large number of experiments which illustrate a wide variety of analytical applications. Most of the examples are taken from the classical areas of titrimetric and gravimetric methods and should provide ample coverage for the introductory, one-semester course. A few experiments are included in which a simple spectrophotometer or pH meter is employed. Several experiments of a "relevant" nature are included, such as the analysis of commercial antacids and bleaches, the determination of nitrites in water, and the identification of amino acids by potentiometric titration.

A separate laboratory manual will still still be available for those teachers who may use the manual but not the text. This manual will include all of the laboratory experiments given in the text as well as exercises of a more advanced nature.

As with previous editions, teachers may obtain the *Solutions to Problems* manual from the publisher. This manual can also be made available to students through ordinary bookstore orders if the teacher wishes.

We express our sincere appreciation to the many users of the third edition who were kind enough to forward criticisms and suggestions for improvement. These include Professors Roger G. Bates, K. P. Li, and J. D. Winefordner, *University of Florida*; T. Dobbelstein, Frederick W. Koknat, and F. W. Smith, *Youngstown State University*; Herschel Frye, *College of the Pacific*; Robert Glenn, *Hutchinson Community College*; Richard H. Groth, *Central Community College*; Howard L. Hodges, *University of Arkansas, Little Rock*; Gary D. Howard, *University of North Carolina, Charlotte*; Thomas A. Lehman, *Bethel College*; Charles A. Long, *Lake Forest College*; Richard McCreery, *Ohio State University*; David B. Moss, *Hiram College*; Daniel S. Polcyn, *University of Wisconsin, Oshkosh*; Lou R. Raasch, *East Tennessee State University*; Veronica L. Vogel, *Rutgers University*; and Hubert L. Youmans, *Western Carolina University*. Special thanks go to Professors Douglas G. Berge, *University of Wisconsin, Oshkosh*; M. Dale Hawley, *Kansas State University*; and Alfred A. Schilt, *Northern Illinois University*, who have read the entire manuscript and made numerous helpful suggestions.

Atlanta, Georgia

R. A. Day, Jr.
A. L. Underwood

CHAPTER ONE

introduction

ANALYTICAL CHEMISTRY

It used to be possible to subdivide chemistry into several clear and well-defined branches—analytical, inorganic, organic, physical, and biological. Although there was always a certain overlap among these simple categories, it was not difficult to define the branches in terms that were acceptable to most chemists. It was generally fairly clear into which category any particular chemist fitted, and a label such as "organic chemist" usually implied a reasonably clear picture of the sorts of things such a chemist did.

One of the most prominent trends in chemistry in recent years has been a general blurring of the borders of its branches; actually, the boundaries between chemistry itself and other major sciences such as physics and biology are considerably less clear than they used to be. Fields such as chemical physics, biophysical chemistry, physical organic chemistry, geochemistry, and chemical oceanography have achieved recognition in at least a vague way, although precise definitions of these fields are exceedingly difficult to formulate.

During all of this change, chemistry courses in the undergraduate curriculum have largely retained their traditional titles, but they have undergone major changes in content. For example, it is not at all unusual to find such topics as molecular spectroscopy and chemical kinetics in organic chemistry courses. Solution thermodynamics, kinetics of electrode processes, and electronics appear in analytical courses at some schools. In most cases, the college freshman course has undergone drastic change, and interesting changes at the high school level have been initiated.

Analytical chemistry is as old, and as new, as the science of chemistry itself. It may be fairly said that analytical research, as opposed broadly to synthetic, ushered in the change from magic and alchemy to quantitative, scientific chemistry. Analytical work led directly to the revolution that overthrew the phlogiston theory, and rational experiments that would place chemistry on a sound basis of fact and theory became possible with the increasing use of the analytical balance. Careful analyses led to the laws of definite and multiple proportions, and made possible Dalton's great achievement—an atomic hypothesis grounded in fact rather than mystical speculation.

The late nineteenth and early twentieth centuries saw developments in organic and physical chemistry and in physics that were bound to dwarf other fields. During this time many analytical chemists concerned themselves with the chemical composition of various materials which were important in the commerce of a simple industrial society. In many universities, analytical chemistry was taught as a routine, cookbook subject, although a small number of excellent men kept the field alive as a science.

Roughly since World War II, an increasing sophistication of research in all areas of chemistry, physics, and biology and an explosive technological development have combined to create analytical problems which demand increasingly sophisticated knowledge and instrumentation for their solution. Typical examples of such problems are determining traces of impurities at the part per billion level in ultrapure semiconductor materials, deducing the sequence of some 20 different amino acids in a giant protein molecule, detecting traces of unusual molecules in the polluted atmosphere of a smog-bound city, determining pesticide residues at the part per billion level in food products, and determining the nature and concentration of complex organic molecules in, say, the nucleus of a single cell.

The solutions to a host of problems such as these have been developed by research workers of the most diverse backgrounds. For example, a biochemist received a Nobel Prize for working out the amino acid sequence in the protein insulin, and physicists were actively involved in the first semiconductor analyses by mass spectrometry. Research workers in many fields are constantly confronted by analytical problems, and in many cases they work out their own solutions. It is interesting to note that in a recent year, nearly 60% of the papers in the journal *Analytical Chemistry* were authored by people who did not consider themselves to be analytical chemists. The papers originated in a wide variety of laboratories associated with medical schools, hospitals, oceanographic institutes, agricultural experiment stations, physics departments, and many more.

The trends of recent years have drawn analytical chemistry into the forefront of research in many exciting areas, but this very intimacy has blurred the borders of the discipline and made it nearly impossible in many cases to decide what an analytical chemist is. In this connection we may quote from the Fisher Award Address of David N. Hume.[1]

[1] D. N. Hume, *Anal. Chem.*, **35**, 29A (1963).

> One of the most difficult problems facing the analytical chemist today is explaining to others just what analytical chemistry is. Much of the difficulty derives from the changes in the nature of the profession and the fact that a given word may have a whole spectrum of meanings .... The increasing complexity of modern chemistry is to some extent the cause of this confusion, as is the fact that a chemist seldom works in only one branch of the subject, more often combining the techniques and approaches of several.

With this extensive overlap into a variety of fields, what distinguishes the analytical chemist from all others working in these areas? The analytical chemist has, usually, more interest in the methods and techniques in their own right. Physical, organic, and biochemists often need to develop new analytical methods for their own purposes, but their primary interests do not lie in the method itself. To the analytical chemist, developing the methods is the challenging part of the research. Because of his interest in the method per se, the analytical chemist is likely to be skeptical of data presented without a full disclosure of experimental details, and he retains a critical attitude toward results which some workers would like to accept so as to get on with other things. The analytical chemist deals with real, practical systems, and much of his effort is expended in an attempt to apply sound theory to actual chemical situations.

QUANTITATIVE ANALYSIS

Analytical chemistry can be divided into areas called qualitative analysis and quantitative analysis. *Qualitative* analysis deals with the identification of substances. It is concerned with *what* elements or compounds are present in a sample. The student's first encounter with qualitative analysis is often in the general chemistry course, where he separates and identifies a number of elements by precipitation with hydrogen sulfide. In organic chemistry the student may identify his product of synthesis using such instrumental techniques as infrared and nuclear magnetic resonance spectroscopy.

Quantitative analysis is concerned with the determination of *how much* of a particular substance is present in a sample. The substance determined, often referred to as the *desired constituent* or *analyte*, may constitute a small or large part of the sample analyzed. If the analyte constitutes more than about 1% of the sample, it is considered a *major* constituent. It is considered *minor* if it amounts to 0.01 to 1% of the sample. Finally, a substance present to the extent of less than 0.01% is considered a *trace* constituent.

Another classification of quantitative analysis may be based upon the size of the sample which is available for analysis. The subdivisions are not clear-cut, but merge imperceptibly into one another, and are roughly as follows: When a sample weighing more than 0.1 g is available, the analysis is spoken of as *macro*; *semimicro* analyses are performed on samples of perhaps 10 to 100 mg; *micro* analyses deal with samples weighing from 1 to 10 mg; and *ultramicro* analyses involve samples of the order of a microgram ($1\ \mu g = 10^{-6}$ g).

Steps in an Analysis

In the introductory course in quantitative analysis the student will deal mainly with major constituents of macro samples. He will seldom perform a *complete* quantitative analysis of a sample. A chemical analysis actually consists of four main steps: (1) sampling, that is, selecting a representative sample of the material to be analyzed; (2) conversion of the analyte into a form suitable for measurement; (3) measurement; and (4) calculation and interpretation of the measurements. Often the beginner carries out only steps 3 and 4 since these are usually the easiest ones.

In addition to the steps mentioned above, there are other operations that may be required. If the sample is a solid, it may be necessary to dry it before performing the analysis. Solids also need to be dissolved in an appropriate solvent before the measurement step. And an accurate measurement of weight of the sample (volume if it is a gas) must be made since quantitative results are usually reported in relative terms, for example, the number of grams of analyte per 100 g of sample (percent by weight).

At this time we shall make some general comments on the four steps of an analysis as well as the operation of dissolving the sample. In subsequent chapters these topics will be developed in much greater detail.

Sampling

The beginning student seldom encounters the problem of sampling since the sample he is given is homogeneous, or nearly so. Nevertheless, he should be aware of the importance of sampling and he should know where to find proper directions when he is confronted with an unfamiliar problem.[2]

Coal is a particularly difficult material to sample, and we shall use it as an illustration of the problems and methods used to solve them.

The first step in the sampling procedure is to select a large portion of coal, called the *gross* sample, which though not homogeneous itself, represents the average composition of the entire mass. The size of the sample needed depends on such factors as particle size and homogeneity of the particles. In the case of coal the gross sample must be about 1000 lb if the particles are no greater than about 1 inch in any dimension.

There are many techniques used to obtain the gross sample. If the coal is in motion, on a conveyor of some type, a definite fraction may be continuously diverted to give the gross sample. If, on the other hand, the coal were being shoveled from a car, every fiftieth shovelful might be placed aside to form the sample.

After the sample has been selected, it is ground or crushed and systematically mixed and reduced in size. One method used for reducing the sample of coal involves piling the sample into a cone with a shovel, flattening the cone, and

[2] W. W. Walton and J. I. Hoffman, "Principles and Methods of Sampling," Chap. 4, p. 67, of I. M. Kolthoff and P. J. Elving, eds., *Treatise on Analytical Chemistry*, Part I, Vol. 1, Interscience Publishers, Inc., New York, 1959.

dividing it into four equal parts, two of which are discarded. A mechanical device for subdividing the sample is called a *riffle.* The riffle consists of a row of small sloping chutes arranged so that alternate chutes discharge the sample in opposite directions. In this manner the sample is halved automatically.

In the laboratory further grinding of the sample may be done with a mortar and pestle. It is often necessary to grind a sample to pass through a sieve of a certain mesh. The final laboratory sample, 1 g or so, is hopefully representative of the entire gross sample. The analytical data obtained cannot be better than the care exercised in the sampling procedure.

Dissolving the Sample

Many of the samples analyzed in the beginning course in quantitative analysis are soluble in water. Generally speaking, however, naturally occurring materials, such as ores, and metallic products, such as alloys, must be given special treatments to effect their solution. While each material may present a specific problem, the two most common methods employed in dissolving samples are (1) treatment with hydrochloric, nitric, sulfuric, or perchloric acid; and (2) fusion with an acidic or basic flux followed by treatment with water or an acid.

The solvent action of acids depends upon several factors:

1. The reduction of hydrogen ion by metals more active than hydrogen: for example,

$$Zn(s) + 2H^+ \longrightarrow Zn^{2+} + H_2(g)$$

2. The combination of hydrogen ion with the anion of a weak acid: for example,

$$CaCO_3(s) + 2H^+ \longrightarrow Ca^{2+} + H_2O + CO_2(g)$$

3. The oxidizing properties of the anion of the acid: for example,

$$3Cu(s) + 2NO_3^- + 8H^+ \longrightarrow 3Cu^{2+} + 2NO(g) + 4H_2O$$

4. The tendency of the anion of the acid to form soluble complexes with the cation of the substance dissolved: for example,

$$Fe^{3+} + Cl^- \longrightarrow FeCl^{2+}$$

Hydrochloric and nitric acids are most commonly used to dissolve samples. The chloride ion is not an oxidizing agent as is nitrate ion, but it has a strong tendency to form soluble complexes with many elements. A very powerful solvent, aqua regia, is obtained by mixing these two acids.

Many substances that are resistant to attack by water or acids are more soluble after fusion with an appropriate flux. Basic fluxes such as sodium carbonate are used to attack acidic materials such as silicates. Acidic fluxes such as potassium hydrogen sulfate are used with basic materials such as iron ores. Oxidizing or reducing substances can also be used in certain cases. Sodium peroxide, for example, is often employed as a flux.

Conversion of the Analyte to a Measurable Form

Before a physical or chemical measurement can be made to determine the amount of analyte in the solution of the sample, it is usually necessary to solve the problem of "interferences." Suppose, for example, that the analyst wishes to determine the amount of copper in a sample by adding potassium iodide and titrating the liberated iodine with sodium thiosulfate. If the solution also contains iron(III) ion, this ion will interfere since it also oxidizes iodide to iodine. The interference can be prevented by adding sodium fluoride to the solution, converting iron(III) into the stable complex, FeF_6^{3-}. This is an illustration of a general method in which interferences are effectively "immobilized" by alteration of their chemical nature.

A second method involves physical separation of the analyte from the interferences. Suppose that one wishes to determine magnesium in a sample which also contains iron(III) ion and the magnesium is to be precipitated as the oxalate. The iron will interfere since it also forms a precipitate with oxalate. The iron can be precipitated as the hydroxide using ammonia at a *p*H of about 6.5. The magnesium is not precipitated at this *p*H, and hence the interference is removed.

In a *gravimetric* analysis the analyte is physically separated from all other components of the sample as well as from the solvent. For example, the chloride in a sample may be determined by precipitation of silver chloride, which is then filtered, dried, and weighed. Precipitation is one of the more widely used techniques for separating the analyte from interferences. Other important methods include electrolysis, solvent extraction, chromatography, and volatilization.

Measurement

The measurement step in an analysis can be carried out by chemical, physical, or biological means. The laboratory technique which is employed has led to the classification of quantitative methods into the subdivisions *titrimetric* (*volumetric*), *gravimetric*, and *instrumental*. A *titrimetric* analysis involves the measurement of the volume of a solution of known concentration which is required to react with the analyte. In a *gravimetric* method the measurement is one of weight; an example was mentioned above in which chloride is determined by precipitating and weighing silver chloride. The term *instrumental* analysis is used rather loosely, originally referring to the use of a special instrument in the measurement step. Actually, instruments may be used in any or all steps of the analysis, and, strictly speaking, burets and analytical balances are instruments. Spectroscopy, both absorption and emission, is perhaps the most widely used instrumental method and is generally discussed in some detail in introductory texts. Other instrumental methods include potentiometry, polarography, coulometry, conductimetry, polarimetry, refractometry, and mass spectrometry.

Calculation and Interpretation of the Measurements

The final step in an analysis is the calculation of the percentage of analyte in the sample. The principles involved in such calculations are normally straightforward. For example, titrimetric and gravimetric methods are based on the simple stoichiometric relationships of chemical reactions. In spectrophotometric methods the property measured, absorbance, is directly proportional to the concentration of the analyte in the solution. On the other hand, the interpretation of the results obtained by analytical methods is not always simple. Since errors can be made in any measurement, the analytical chemist must consider this possibility in interpreting his results. The methods of statistics are commonly used and are especially useful in expressing the significance of analytical data. We shall devote a full chapter to the topic of errors and the treatment of analytical data.

CHAPTER TWO

errors and the treatment of analytical data

INTRODUCTION

In an experimental science such as chemistry, much effort is expended in gathering data, and as chemistry has developed into a modern science, most of the data have become quantitative, that is, they derive from measurements. When any scientific measurement is performed, it is necessary to consider the fact that an error has been made, and it is important to develop the ability to evaluate data, learning to draw justified conclusions while rejecting interpretations that are unwarranted because of limitations in the measurements. Although analytical chemists in particular like to emphasize the techniques by which data may be evaluated, it is clear that any chemist may enhance his competence by learning methods which are more reliable than intuition alone in assessing the significance of experimental results. The methods which are most suitable for the treatment of analytical data are powerful, general tools which may be used in many other scientific situations.

Most of the techniques which we shall consider are based upon statistical concepts. There is increasing awareness that statistical methods are efficient in planning experiments that will yield the most information from the fewest measurements and in "boiling down" data so that their significance is concisely presented. Statistics, on the other hand, should not be expected to lessen the necessity of obtaining good measurements, and statistical methods are most powerful when applied to good data.

Statistics and the theory of probability represent an important branch of mathematics which possesses a logical and rigorous structure. Although chemists may profit from study in this field, it is impossible in this textbook to examine the foundations of probability theory and to derive their consequences. We must here accept the conclusions of the mathematicians largely on faith, and then attempt to see how they may be useful to chemists. We may hope to learn how our intuitive judgments of data may be validated by quantitative expressions of their probable reliability, and even what the term "reliability" means in connection with measurements of quantities that are actually unknown. We shall see how sets of data may be compared to learn whether they are *really* different or whether an apparent difference could be attributable not to an assignable cause, but to chance alone. We shall see how errors are propagated through a series of experimental steps and calculations. The student should emerge from this study with a heightened skepticism of data which is moderated by an increased confidence in his ability to draw justified conclusions.

ERRORS

The term *error* as used here refers to the numerical difference between a measured value and the true value. The *true value* of any quantity is really a philosophical abstraction, something that man is not destined to know, although scientists generally feel that there is such a thing and believe that they may approach it more and more closely as their measurements become increasingly refined. In analytical chemistry, it is customary to act as though the true value of a quantity were known when it is believed that the uncertainty in the value is less than the uncertainty in something else with which it is being compared. For example, the percentage composition of a standard sample certified by the National Bureau of Standards may be treated as correct in evaluating a new analytical method; differences between the standard values and the results obtained by the new method are then treated as errors in the latter. Values which we are willing to treat as *true* are generally arrived at by a variety of methods whose limitations and pitfalls are sufficiently different that agreement among them cannot reasonably be ascribed to coincidence. Even so, it is well to remain skeptical about standard, accepted, or certified values, because they stem from experimental measurements performed by human, albeit expert, hands.

Determinate Errors

Errors which can, at least in principle, be ascribed to definite causes are termed *determinate* or *systematic* errors. A given determinate error is generally unidirectional with respect to the true value, in contrast to indeterminate errors, discussed below, which lead to both high and low results with equal probability. Determinate errors are often reproducible, and in many cases they can be predicted by a person who thoroughly understands all the aspects of the measurement. Examples of sources of determinate errors are: a corroded weight, a poorly calibrated buret, an impurity in a reagent, an appreciable solubility of a precipitate, a side reaction in a titration, and heating a sample at too high a temperature.

Determinate errors have been classified as *methodic*, *operative*, and *instrumental* in accordance with their origin in (a) the method of analysis as it reflects the properties of the chemical systems involved, (b) ineptitude of the experimenter, and (c) failure of measuring devices to perform in accordance with required standards.[1] Frequently, the source of an error may lie in more than one of these categories. For example, some error may always be expected in weighing a hygroscopic substance, but it may be increased if the analyst has poor balance technique; the environment outside the system may influence the error, as, for example, in the effect of humidity upon the error in weighing a hygroscopic substance.

Constant Errors

Sometimes the magnitude of a determinate error is nearly constant in a series of analyses, regardless of the size of the sample. This may be the case, for example, with an indicator blank that is not corrected for in a series of titrations. Some writers have used the term *additive* for this type of error. The significance of a constant error generally decreases as the size of the sample increases, since usually we are not so interested in the absolute value of an error as in its value relative to the magnitude of the measured quantity. For example, a constant end-point error of 0.1 ml in a series of titrations represents a relative error of 10% for a sample requiring 1 ml of titrant, but only 0.2% if 50 ml of titrant is used.

Proportional Errors

The absolute value of this type of error varies with sample size in such a way that the relative error remains constant. A substance that interferes in an analytical method may lead to such an error if present in the sample. For example, in the iodometric determination of an oxidant like chlorate, another oxidizing agent such as iodate or bromate would cause high results if its presence were unsuspected and not corrected for. Taking a larger sample would increase the absolute error, but the relative error would remain constant provided the sample were homogeneous. Errors may be encountered which vary with the size of the sample but not in a strictly linear fashion. Many writers use the term "proportional" for these also, although of course it is not strictly correct for such cases.

Data obtained a number of years ago by Benedetti-Pichler[2] are often quoted to illustrate the interplay of constant and proportional errors and to suggest how they may be distinguished. The ideas apply as well to modern measurements of a much more sophisticated type. The data are given in Table 2.1. Samples of potassium alum were dissolved and acidified with proper amounts of hydrochloric acid so that the quantity of ammonia required to precipitate hydrous aluminum

[1] E. B. Sandell, "Errors in Chemical Analysis," Chap. 2 of I. M. Kolthoff and P. J. Elving, eds., *Treatise on Analytical Chemistry*, Part I, Vol. 1, Interscience Publishers, Inc., New York, 1959.

[2] A. A. Benedetti-Pichler, *Ind. Eng. Chem., Anal. Ed.*, **8**, 373 (1936).

TABLE 2.1 Determination of Aluminum (as Al_2O_3) in Potassium Alum*

$KAl(SO_4)_2 \cdot 12\ H_2O$ taken, g	Al_2O_3 taken, g	Al_2O_3 found using stock NH_3, g	Difference, g	Al_2O_3 found using distilled NH_3, g	Difference, g
1.0000	0.1077	0.1288	0.0211	0.1087	0.0010
2.0000	0.2154	0.2384	0.0230	0.2178	0.0024
3.0000	0.3231	0.3489	0.0258	0.3258	0.0027
4.0000	0.4308	0.4588	0.0280	0.4352	0.0044

* Data of A. B. Benedetti-Pichler, *Ind. Eng. Chem., Anal. Ed.*, **8**, 373 (1936).

oxide was nearly constant. In one set of experiments, ammonia from a stock bottle was used; in the other set, freshly distilled ammonia. In the former case, it is seen that the errors were nearly constant. This was attributed to the fact that coprecipitation of silicic acid, originating from the attack of the old ammonia solution on the glass bottle, was constant because the same volume of ammonia solution was used in each case. In the latter experiments, silicic acid was absent, and the errors, now much smaller, were much more nearly proportional to sample size. These errors were attributed to the presence of water in the ignited precipitate, the quantity of water retained depending upon the quantity of alumina and hence upon sample size. In evaluating a new analytical method, information about the type of errors present and sometimes clues pointing toward their minimization may be obtained simply by varying the size of the sample.

Indeterminate Errors

If a measurement is sufficiently coarse, repetition will yield exactly the same result each time. For example, in weighing a 50-g object to the nearest gram with a good balance, only by extreme negligence could a person obtain different values or a group of people fail to agree. The only reasonable errors in such a measurement would be determinate ones, such as a seriously defective weight. On the other hand, any measurement can be refined to the point where it is mere coincidence if replicates agree to the last recorded digit. Sooner or later, the point is approached where unpredictable and imperceptible factors introduce what appear to be random fluctuations in the measured quantity. In some cases, it may be possible to specify definite variables that are beyond control near the performance limit of an instrument: noise and drift in an electronic circuit, vibrations in a building caused by passing traffic, temperature variations, and the like. Often the inability of the eye to detect slight changes in a readout device may be invoked as a source of error. To be sure, variations which a slipshod person considers random may appear obvious and controllable to a careful onlooker, but nevertheless the point must be reached where anyone, however meticulous, will encounter random errors which he cannot further reduce. These errors are classified as *indeterminate.*

It is tempting at first glance to retreat from indeterminate errors simply by performing coarser measurements. After backing off to the point where scatter in the data ceases to exist, an observer will obtain exactly the same result each time, and superficially this seems as good as recording an additional digit which varies from one time to the next. But this withdrawal from the challenge to push measurements as far as possible is unacceptable to most scientists. More cogent, however, is the fact that the average of a number of fine observations with random scatter is more precise than coarser data which agree perfectly. Data that exhibit random scatter may be subjected to an analysis that does attach significance to the last recorded digit, as we shall see below.

Accuracy and Precision

The terms *accuracy* and *precision,* often used synonymously in ordinary discourse, should be carefully distinguished in connection with scientific data. An accurate result is one that agrees closely with the true value of a measured quantity. The comparison is usually made on the basis of an inverse measure of the accuracy, that is, the error (the smaller the error, the greater the accuracy). The *absolute error* is the difference between the experimental value and the true value. For example, if an analyst finds a value of 20.44% iron in a sample which actually contains 20.34%, the absolute error is

$$20.44 - 20.34 = 0.10\%$$

The error is most frequently expressed relative to the size of the measured quantity, for example, in percent or in parts per thousand. Here the *relative error* is

$$\frac{0.10}{20.34} \times 100 = 0.5\%$$

or

$$\frac{0.10}{20.34} \times 1000 = 5 \text{ ppt}$$

The term *precision* refers to the agreement among a group of experimental results; it implies nothing about their relation to the true value. Precise values may well be inaccurate, since an error causing deviation from the true value may affect all the measurements equally and hence not impair their precision. A determinate error which leads to inaccuracy may or may not affect precision, depending upon how nearly constant it remains throughout a series of measurements. The precision is commonly stated in terms of the *standard deviation, average deviation,* or *range.* These terms will be defined later. As in the case of error (above), the precision can be expressed on an absolute or a relative basis.

DISTRIBUTION OF RANDOM ERRORS

After the search for determinate errors has been carried as far as possible and all precautions taken and corrections applied, the remaining fluctuations in the data are found to be random in nature. Results that scatter in a random fashion are best treated by the powerful techniques of statistics. It will now be our goal to show how these techniques are applied and what information they furnish beyond what may be seen by simply inspecting the data.

Frequency Distributions

Table 2.2 contains some actual data obtained by a person who prepared sixty replicate colored solutions and measured their absorbance values with a spectrophotometer. (Absorbance is discussed in a later chapter, but the nature of the measured quantity need not concern us here.) The data in Table 2.2 have not been treated in any way, but are simply listed in the order in which they were obtained.

TABLE 2.2 Individual Values, Unorganized

1	0.458	21	0.462	41	0.450
2	0.450	22	0.450	42	0.455
3	0.465	23	0.454	43	0.456
4	0.452	24	0.446	44	0.456
5	0.452	25	0.464	45	0.459
6	0.447	26	0.461	46	0.454
7	0.459	27	0.463	47	0.455
8	0.451	28	0.457	48	0.458
9	0.446	29	0.460	49	0.457
10	0.467	30	0.451	50	0.456
11	0.452	31	0.456	51	0.455
12	0.463	32	0.455	52	0.460
13	0.456	33	0.451	53	0.456
14	0.456	34	0.462	54	0.463
15	0.449	35	0.451	55	0.457
16	0.454	36	0.469	56	0.456
17	0.456	37	0.458	57	0.457
18	0.441	38	0.458	58	0.453
19	0.457	39	0.456	59	0.455
20	0.459	40	0.454	60	0.453

We are here concerned, not with the "correct" result, only with the relationships of the measured values among themselves. It is apparent that the values in Table 2.2 must be treated in some manner before they can be discussed intelligently. A reader with an exceptionally quick eye may notice that the lowest value is 0.441 and the highest 0.469, and perhaps it is apparent that many values are between 0.45 and 0.46, but on the whole the table is relatively uninstructive. Let us now enumerate some steps that will enable us to interpret the data more fully.

First, we arrange the results in order from lowest to highest. This has been done in Table 2.3. This simple operation discloses information not so readily apparent in the raw data, namely the maximum and minimum values, and, by simple counting, the middle or median value. This is still an inadequate presentation of the data, however; the mind does not grasp the meaning of 60 numbers on a piece of paper, regardless of how they are arranged. We need more compactness in order to make practical use of the data.

TABLE 2.3 Individual Values Arranged in Order

1	0.441	21	0.454	41	0.457
2	0.446	22	0.455	42	0.458
3	0.446	23	0.455	43	0.458
4	0.447	24	0.455	44	0.458
5	0.449	25	0.455	45	0.458
6	0.450	26	0.455	46	0.459
7	0.450	27	0.456	47	0.459
8	0.450	28	0.456	48	0.459
9	0.451	29	0.456	49	0.460
10	0.451	30	0.456	50	0.460
11	0.451	31	0.456	51	0.461
12	0.451	32	0.456	52	0.462
13	0.452	33	0.456	53	0.462
14	0.452	34	0.456	54	0.463
15	0.452	35	0.456	55	0.463
16	0.453	36	0.456	56	0.463
17	0.453	37	0.457	57	0.464
18	0.454	38	0.457	58	0.465
19	0.454	39	0.457	59	0.467
20	0.454	40	0.457	60	0.469

The second step involves condensing the data by grouping them into cells. We divide the range from the lowest to the highest value into a convenient number of intervals or *cells* and then count the number of values falling within each cell. Strictly, this process involves some loss of information, but this is more than compensated by the increased efficiency with which the significance of the condensed data may be perceived. In order to proceed, we must first decide upon the number of cells to be used and choose their boundaries. Usually the range is divided into equal intervals, and sometimes confusion is avoided by choosing cell boundaries halfway between possible observed values. In the present case, the absorbance was recorded to three decimal places, and we choose cell boundaries such as 0.4605 so that none of the values coincides with a boundary. Judgment is required in selecting the number of cells: 13 to 20 are sometimes recommended, but 10 or even fewer may be preferable if the number of values to be grouped is small, say, less than 250. A fairly satisfactory grouping of our data into eight cells is shown in Table 2.4.

A glance at Table 2.4 shows that information buried in Tables 2.2 and 2.3 is now obvious. Thus, although the values range from 0.441 to 0.469, we see immediately that very few results are below 0.448 or above 0.464.

TABLE 2.4 Grouping of Individual Values into Cells

Cell midpoint	Cell boundaries	Number of values
	0.4405	
0.4425		1
	0.4445	
0.4465		3
	0.4485	
0.4505		11
	0.4525	
0.4545		21
	0.4565	
0.4585		14
	0.4605	
0.4625		7
	0.4645	
0.4665		2
	0.4685	
0.4705		1
	0.4725	

Next, we may devise a pictorial representation of the frequency distribution. This step is actually unnecessary, and it is rarely performed except for teaching purposes or for popular presentation of what might otherwise be "dry" data to laymen. Two types of graphs are shown in Fig. 2.1: The *histogram* consists of contiguous columns of heights proportional to the frequencies, erected upon the full widths of the cells; the *frequency polygon* is constructed by plotting frequencies at cell midpoints and connecting the points with straight lines.

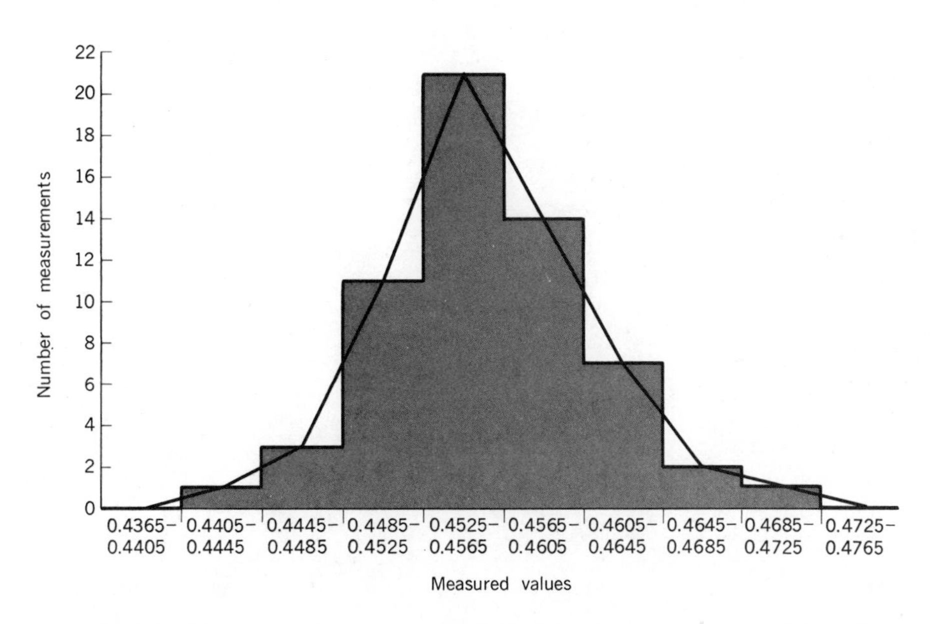

Figure 2.1 Histogram and frequency polygon for absorbance measurements of 60 replicate solutions.

The Normal Error Curve

The limiting case approached by the frequency polygon as more and more replicate measurements are performed is the *normal* or *Gaussian* distribution curve, shown in Fig. 2.2. This curve is the locus of a mathematical function which is well-known, and it is more easily handled than the less ideal and more irregular curves that are often obtained with a smaller number of observations. Data are often treated as though they were normally distributed in order to simplify their analysis, and we may look upon the normal error curve as a model which is approximated more or less closely by real data. It is supposed that there exists a "universe" of data made up of an infinite number of individual measurements, and it is actually this "infinite population" to which the normal error function pertains. A finite number of replicate measurements is considered by statisticians to be a sample drawn in a random fashion from a hypothetical infinite population; thus the sample is at least hopefully a representative one, and fluctuations in its individual values may be considered to be normally distributed, so that the terminology and techniques associated with the normal error function may be employed in their analysis.

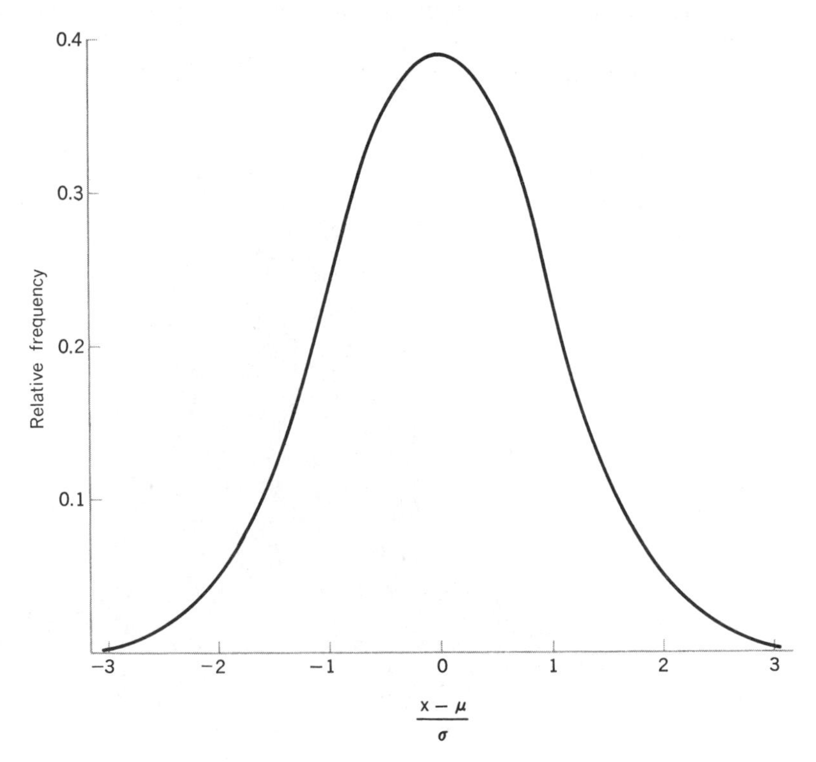

Figure 2.2 Normal distribution curve; relative frequencies of deviations from the mean for a normally-distributed infinite population; deviations $(x-\mu)$ are in units of σ.

The equation of the normal curve may be written for our purposes as follows:

$$y = \frac{1}{\sigma\sqrt{2\pi}} e^{-(x-\mu)^2/2\sigma^2}$$

Here y represents the relative frequency with which random sampling of the infinite population will bring to hand a particular value x. The quantities μ and σ, called the population parameters, specify the distribution. μ is the *mean* of the infinite population, and since we are not here concerned with determinate errors, we may consider that μ gives the correct magnitude of the measured quantity. It is clearly impractical to determine μ by actually averaging an infinite number of measured values, but we shall see below that a statement can be made from a finite series of measurements regarding the probability that μ lies within a certain interval. To the extent of our confidence in having eliminated determinate errors, such a statement approaches an assessment of the true value of the measured quantity. σ, which is called the *standard deviation,* is the distance from the mean to either of the two inflection points of the distribution curve, and may be thought of as a measure of the spread or scatter of the values making up the population; σ thus relates to precision. π has its usual significance, and e is the base of the natural logarithm system. The term $(x - \mu)$ represents simply the extent to which an individual value x deviates from the mean.

The distribution function may be normalized by setting the area under the curve equal to unity, representing a total probability of 1 for the whole population. Since the curve approaches the abscissa asymptotically on either side of the mean, there is a small but finite probability of encountering enormous deviations from the mean. A person who happened to encounter one of these in performing a series of laboratory observations would be unfortunate indeed; some of us who have faith in never obtaining such a "wild" result in our own work are inclined to the view that the normal distribution as a model for real data breaks down, and that only the central region of the distribution curve is pertinent when applied to scientific measurements by competent workers. The area under the curve between any two values of $(x - \mu)$ gives the fraction of the total population having magnitudes between these two values. It may be shown that about two-thirds (actually 68.26%) of all the values in an infinite population fall within the limits $\mu \pm \sigma$, while $\mu \pm 2\sigma$ includes about 95% and $\mu \pm 3\sigma$ practically all (99.74%) of the values. Happily, then, small errors are more probable than large ones. Since the normal curve is symmetrical, high and low results are equally probable once determinate errors have been dismissed.

When a worker goes into the laboratory and measures something, we suppose that his result is one of an infinite population of such values that he might obtain in an eternity of such activity; then the chances are roughly 2:1 that his measured values will be no further than σ from the mean of the infinite population, and about 20:1 that his result will lie in the range $\mu \pm 2\sigma$. In practice, of course, we can never find σ for an infinite population, but the standard deviation of a finite number of observations may be taken as an estimate of σ. Thus we may predict something about the likelihood of occurrence of an error of a certain magnitude in the work of a particular individual once he has performed enough measurements to permit estimation of the characteristics of his particular infinite population.

STATISTICAL TREATMENT OF FINITE SAMPLES

Although there is no doubt as to its mathematical meaning, the normal distribution of an infinite population is a fiction so far as real laboratory work is concerned. We must now turn our attention to techniques for handling scientific data as we obtain them in practice.

Measures of Central Tendency and Variability

The *central tendency* of a group of results is simply that value about which the individual results tend to "cluster." For an infinite population, it is μ, the mean of such a sample. The *mean* of a finite number of measurements, $x_1, x_2, x_3, \ldots, x_n$, is often designated $\bar{x}$ to distinguish it from μ. Of course, $\bar{x}$ approaches μ as a limit when n, the number of measured values, approaches infinity. Calculation of the mean involves simply averaging the individual results:

$$\bar{x} = \frac{x_1 + x_2 + x_3 + \cdots + x_n}{n} = \frac{\sum_{i=1}^{i=n} x_i}{n}$$

The mean is generally the most useful measure of central tendency. It may be shown that the mean of n results is $\sqrt{n}$ times as reliable as any one of the individual results. Thus there is a diminishing return from accumulating more and more replicate measurements: The mean of four results is twice as reliable as one result in measuring central tendency; the mean of nine results is three times as reliable; the mean of 25 results, five times as reliable, etc. Thus, generally speaking, it is inefficient for a careful worker who gets good precision to repeat a measurement more than a few times. Of course the need for increased reliability, and the price to be paid for it, must be decided on the basis of the importance of the results and the use to which they are to be put.

The *median* of an odd number of results is simply the middle value when the results are listed in order; for an even number of results, the median is the average of the two middle ones. In a truly symmetrical distribution, the mean and the median are identical. Generally speaking, the median is a less efficient measure of central tendency than is the mean, but in certain instances it may be useful, particularly in dealing with very small samples.

Since two parameters, μ and σ, are required to specify a frequency distribution, it is clear that two populations may have the same central tendency but differ in "spread" or *variability* (or, as some say, *dispersion*), as suggested in Fig. 2.3. For a finite number of values, the simplest measure of variability is the *range*, which is the difference between the largest and smallest values. Like the median, the range is sometimes useful in small sample statistics, but generally speaking it is an inefficient measure of variability. Note, for example, that one "wild" result

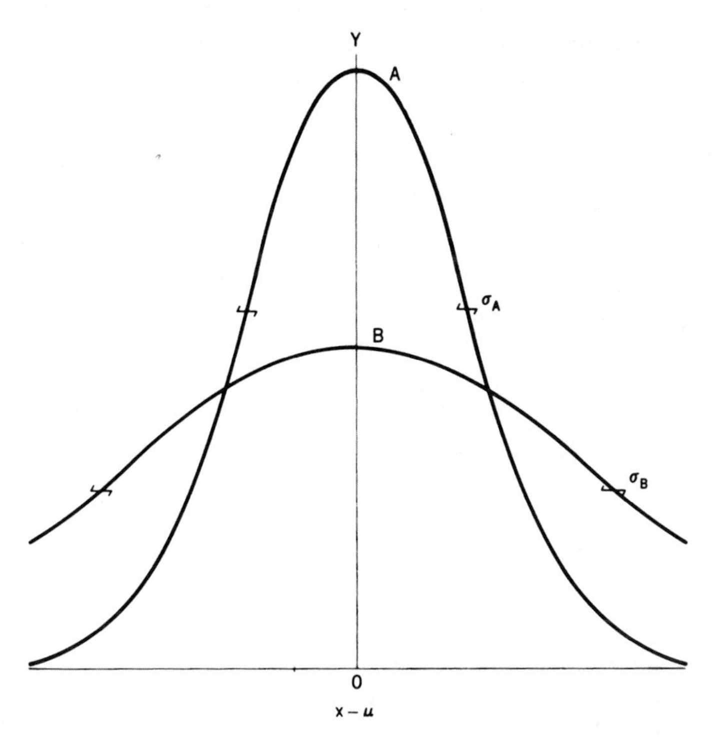

Figure 2.3 Two populations with the same central tendency μ, but different variability.

exerts its full impact upon the range, whereas its effect is diluted by all of the other results in the better measures of variability noted below.

The *average deviation* from the mean is often given in scientific papers as a measure of variability, although strictly it is not very significant from a statistical point of view, particularly for a small number of observations. For a large group of data which are normally distributed, the average deviation approaches 0.8σ. To calculate the average or mean deviation, one simply finds the differences between individual results and the mean, regardless of sign, adds these individual deviations up, and divides by the number of results:

$$\text{Average deviation} = \bar{d} = \frac{\sum_{i=1}^{i=n} |x_i - \bar{x}|}{n}$$

Often the average deviation is expressed relative to the magnitude of the measured quantity, for example as a percentage:

$$\text{Relative average deviation (\%)} = \frac{\bar{d}}{\bar{x}} \times 100 = \frac{\sum_{i=1}^{i=n} |x_i - \bar{x}|/n}{\bar{x}} \times 100$$

Because analytical results are often expressed as percentages (e.g., percent iron in an iron ore sample), it may be confusing to report relative deviations on a percentage basis, and it is preferable to use parts per thousand instead of percent (parts per hundred):

$$\text{Relative average deviation (ppt)} = \frac{\sum_{i=1}^{i=n} |x_i - \bar{x}|/n}{\bar{x}} \times 1000$$

The *standard deviation* is much more meaningful statistically than is the average deviation. The symbol s is used for the standard deviation of a finite number of values; σ is reserved for the population parameter. The standard deviation, which may be thought of as a root mean square deviation of values from their average, is calculated using the formula

$$s = \sqrt{\frac{\sum_{i=1}^{i=n} |x_i - \bar{x}|^2}{n - 1}}$$

If n is large (say 50 or more), then, of course, it is immaterial whether the term in the denominator is $n - 1$ (which is strictly correct) or n. When the standard deviation is expressed as a percentage of the mean, it is called the *coefficient of variation*, v:

$$v = \frac{s}{\bar{x}} \times 100$$

The *variance*, which is s^2, is fundamentally more important in statistics than is s itself, but the latter is much more commonly used in treating chemical data.

For the data in Tables 2.2, 2.3, and 2.4 the following measures of central tendency and variability were calculated:

Mean: $\bar{x} = 0.456$

Median: $M = 0.456$

Range: $R = 0.028$

Average deviation: $\bar{d} = 0.0038$

Relative average deviation: $\frac{\bar{d}}{\bar{x}} \times 1000 = 8.3$ ppt

Standard deviation: $s = 0.0052$

Coefficient of variation: $v = \frac{s}{x} \times 100 = 1.1\%$

The following example illustrates the calculation of the foregoing terms in the case of a determination of the normality of a solution.

Example. The normality of a solution is determined by four separate titrations, the results being 0.2041, 0.2049, 0.2039, and 0.2043. Calculate the mean, median, range, average deviation, relative average deviation, standard deviation, and the coefficient of variation.

$$\text{Mean:}\quad \bar{x} = \frac{0.2041 + 0.2049 + 0.2039 + 0.2043}{4}$$

$$\bar{x} = 0.2043$$

$$\text{Median:}\quad M = \frac{0.2041 + 0.2043}{2}$$

$$M = 0.2042$$

$$\text{Range:}\quad R = 0.2049 - 0.2039$$

$$R = 0.0010$$

$$\text{Average deviation:}\quad \bar{d} = \frac{(0.0002) + (0.0006) + (0.0004) + (0.0000)}{4}$$

$$\bar{d} = 0.0003$$

$$\text{Relative average deviation:}\quad \frac{\bar{d}}{\bar{x}} \times 1000 = \frac{0.0003}{0.2043} \times 1000$$

$$= 1.5 \text{ ppt}$$

$$\text{Standard deviation:}\quad s = \sqrt{\frac{(0.0002)^2 + (0.0006)^2 + (0.0004)^2 + (0.0000)^2}{4 - 1}}$$

$$s = 0.0004$$

$$\text{Coefficient of variation:}\quad v = \frac{0.0004}{0.2043} \times 100$$

$$v = 0.2\%$$

Student's *t*

We have seen that, given μ and σ for the normal distribution of an infinite population, a precise statement can be made regarding the odds of drawing from the population an observation lying outside certain limits. But in practical work, we deal with finite numbers of observations, and we know not μ and σ, but rather $\bar{x}$ and s, which are only estimates of μ and σ. Since these estimates are subject to uncertainty, what we really have is a sort of blurred distribution curve on which to base any predictions we wish to make. This naturally widens the limits corresponding to any given odds that an individual observation will fall outside such limits. An English chemist, W. S. Gosset, writing under the pen name of Student, studied the problem of making predictions based upon a finite sample drawn from an unknown population and published a solution in 1908.[3] The theory of

[3] *Biometrika*, **6**, 1 (1908).

Student's work is beyond the scope of this book, but we may accept it as soundly based and see how it may be used in chemistry. The quantity t (often called *Student's t*) is defined by the expression

$$\pm t = (\bar{x} - \mu)\frac{\sqrt{n}}{s}$$

Tables of t values relating to various odds or probability levels and for varying degrees of freedom may be found in statistical compilations; a portion of such a table is reproduced here in Table 2.5. *Degrees of freedom* in the present connection are one less than n, the number of observations.[4] The t values are calculated to take into account the fact that $\bar{x}$ will not in general be the same as μ and to compensate the uncertainty in using s as an estimate of σ. Values of t such as those in Table 2.5 are used in several statistical methods, some of which are outlined below.

TABLE 2.5 Some Values of Student's *t*

Number of observations, n	Number of degrees of freedom, $n-1$	Probability levels 50%	90%	95%	99%
2	1	1.000	6.314	12.706	63.66
3	2	0.816	2.920	4.303	9.925
4	3	0.765	2.353	3.182	5.841
5	4	0.741	2.132	2.776	4.604
6	5	0.727	2.015	2.571	4.032
7	6	0.718	1.943	2.447	3.707
8	7	0.711	1.895	2.365	3.500
9	8	0.706	1.860	2.306	3.355
10	9	0.703	1.833	2.262	3.250
11	10	0.700	1.812	2.228	3.169
21	20	0.687	1.725	2.086	2.845
∞	∞	0.674	1.645	1.960	2.576

Confidence Interval of the Mean

By rearranging the equation above which defines t, we obtain the so-called *confidence interval* of the mean, or *confidence limits*:

$$\mu = \bar{x} \pm \frac{ts}{\sqrt{n}}$$

[4] Degrees of freedom may be defined as the number of individual observations that could be allowed to vary under the condition that $\bar{x}$ and s, once determined, be held constant. For example, once the mean is obtained and we decide to keep it constant, all but one observation can be varied; the last one is fixed by $\bar{x}$ and all of the other x_i values, and the degrees of freedom then equal $n - 1$. In general, if s is calculated from the same number of observations as were used to calculate $\bar{x}$ (which would normally be the case in treating analytical data), then the degrees of freedom equal $n - 1$.

We might use this to estimate the probability that the population mean, μ, lies within a certain region centered at $\bar{x}$, the experimental mean of our measurements. It is more usual in treating analytical data, however, to adopt an acceptable probability and then find the limits on either side of $\bar{x}$ to which we must go in order to be assured that we have embraced μ. It may be seen in Table 2.5 that t values increase as n, the number of observations, decreases. This is reasonable, since the smaller n becomes, the less information is available for estimating the population parameters. Increases in t exactly compensate for the lessening information.

The following example illustrates the use of Table 2.5.

Example. A chemist determined the percentage iron in an ore, obtaining the following results: $\bar{x} = 15.30$, $s = 0.10$, $n = 4$.

(a) Calculate the 90% confidence interval of the mean.
From Table 2.5, $t = 2.353$ for $n = 4$. Hence

$$\mu = 15.30 \pm \frac{2.353 \times 0.10}{\sqrt{4}}$$

$$\mu = 15.30 \pm 0.12$$

(b) Calculate the 99% confidence interval of the mean.
From Table 2.5, $t = 5.841$ for $n = 4$. Hence

$$\mu = 15.30 \pm \frac{5.841 \times 0.10}{\sqrt{4}}$$

$$\mu = 15.30 \pm 0.29$$

The meaning of confidence intervals is sometimes confused by the beginning student. The correct interpretation, using part (a) of the example above, is as follows: Suppose the chemist repeats the analysis ten times, each time performing four determinations and calculating an interval as illustrated above. He would obtain 10 intervals, such as 15.30 ± 0.12, 15.28 ± 0.14, 15.33 ± 0.11, etc. He could expect nine of these 10 intervals to embrace the population mean, μ. It is a common misconception that 90% of the experimental means would lie within the interval 15.30 ± 0.12. Predicting the interval within which future $\bar{x}$ values will lie is a different statistical problem which can be treated only with another sort of limits which are much wider than the confidence limits discussed here.

In some cases where analyses have been repeated extensively, a chemist may have a reliable estimate of the population standard deviation, σ. In this case there is no uncertainty in the value of σ, and the confidence interval is given by

$$\mu = \bar{x} \pm \frac{Z\sigma}{\sqrt{n}}$$

where Z is simply the value of t at $n = \infty$ (Table 2.5). Note that in the example above the confidence interval for part (a) would be given by

$$15.30 \pm \frac{1.645 \times 0.10}{\sqrt{4}} = 15.30 \pm 0.08$$

The interval is narrower since the uncertainty in σ has been removed.

It is possible to calculate a confidence interval from the range, R, of a series of measurements, using the relationship

$$\mu = \bar{x} \pm c_n R$$

Values of c_n for various numbers of observations and probability levels have been tabulated; some of these are given in Table 2.6. The values of c_n are based upon estimates of s obtained from the range. It should be emphasized that, while it is easy to calculate a confidence interval from the range, an occasional large error will have an undue impact upon the result. The range is normally used in this way only when dealing with a very small number of observations, say, ten or less.

TABLE 2.6 Some Values of c_n for Calculating Confidence Intervals from the Range

Number of observations	Probability levels 95%	99%
2	6.353	31.828
3	1.304	3.008
4	0.717	1.316
5	0.507	0.843
6	0.399	0.628

Testing for Significance

Suppose that a sample is analyzed by two different methods, each repeated several times, and that the mean values obtained are different. Statistics, of course, cannot say which value is "right," but there is a prior question in any case, namely, is the difference between the two values significant? It is possible simply by the influence of random fluctuations to get two different values using two methods; but it is likewise possible that one (or even both) of the methods is subject to a determinate error. There is a test, using Student's t, that will tell (with a given probability) whether it is worthwhile to seek an assignable cause for the difference between the two means. It is clear that the greater the scatter in the two sets of data, the less likely it is that differences between the two means are real.

The statistical approach to this problem is to set up the so-called *null hypothesis*. This hypothesis states, in the present example, that the two means are identical. The t-test gives a *yes* or *no* answer to the correctness of the null hypothesis with a certain confidence such as 95 or 99%. The procedure is as follows: Suppose that a sample has been analyzed by two different methods, yielding means $\bar{x}_1$ and $\bar{x}_2$ and standard deviations s_1 and s_2; n_1 and n_2 are the number of individual results obtained by the two methods. The first step is to calculate a t value using the formula

$$t = \frac{|\bar{x}_1 - \bar{x}_2|}{s}\sqrt{\frac{n_1 n_2}{n_1 + n_2}}$$

(This procedure presupposes that s_1 and s_2 are the same; there is a test for this, noted below.) Second, enter a t table such as Table 2.5 at a degree of freedom given by $(n_1 + n_2 - 2)$ and at the desired probability level. If the value in the table is greater than the t calculated from the data, the null hypothesis is substantiated (i.e., $\bar{x}_1$ and $\bar{x}_2$ are the same with a certain probability. If the t value in the table is less than the calculated t, then by this test the null hypothesis is incorrect and it might be profitable to look for a reason to explain the difference between $\bar{x}_1$ and $\bar{x}_2$.

If s_1 and s_2 are really different, a much more complicated procedure, which is not discussed here, must be used. Usually in analytical work involving methods that would by ordinary common sense be considered comparable, s_1 and s_2 are about the same. A test is available for deciding whether a difference between s_1 and s_2 is significant: This is the *variance-ratio* or *F-test.* The procedure is simple: Find the ratio $F = s_1^2/s_2^2$, placing the larger s value in the numerator so that $F > 1$; then go to a table of F values. If the F value in the table is less than the calculated F value, then the two standard deviations are significantly different; otherwise, they are not. Some sample F values are given in Table 2.7 for a probability level of 95%. The F-test may be used to determine the validity of the simple t-test described here, but it may also be of interest in its own right to determine whether two analytical procedures yield significantly different precision.

TABLE 2.7 *F* Values at the 95% Probability Level

$n - 1$ for smaller s^2	$n - 1$ for larger s^2: 3	4	5	6	10	20
3	9.28	9.12	9.01	8.94	8.79	8.66
4	6.59	6.39	6.26	6.16	5.96	5.80
5	5.41	5.19	5.05	4.95	4.74	4.56
6	4.76	4.53	4.39	4.28	4.06	3.87
10	3.71	3.48	3.33	3.22	2.98	2.77
20	3.10	2.87	2.71	2.60	2.35	2.12

Sometimes it may be of interest to compare two results, one of which is considered a priori to be highly reliable. An example of this might be a comparison of the mean $\bar{x}$ of several analyses of an NBS sample with the value certified by the National Bureau of Standards. The goal would be not to pass judgment upon the Bureau, but to decide whether the method employed gave results that agreed with the Bureau's. In this case, the Bureau's value is taken as μ

in the equation defining Student's t, and a t value is calculated using $\bar{x}$, n, and s for the analytical results at hand. If the calculated t value is greater than that in the t table for $n - 1$ degrees of freedom and the desired probability, then the analytical method in question gave a mean value significantly different from the NBS value; otherwise, differences in the two values would be attributable to chance alone.

The following examples illustrate the foregoing points.

Example 1. A sample of soda ash (Na_2CO_3) is analyzed by two different methods, giving the following results for the percentage of Na_2CO_3:

Method 1	*Method 2*
$\bar{x}_1 = 42.34$	$\bar{x}_2 = 42.44$
$s_1 = 0.10$	$s_2 = 0.12$
$n_1 = 5$	$n_2 = 4$

(a) Are s_1 and s_2 significantly different? Apply the variance-ratio, or F-test:

$$F = \frac{s_2^2}{s_1^2} = 1.44$$

Consult Table 2.7 under column $n - 1 = 3$ (since $s_2 > s_1$) and row $n - 1 = 4$, finding $F = 6.59$. Since $6.59 > 1.44$, the standard deviations are not significantly different.

(b) Are the two means significantly different at the 95% probability level? Calculate a t value (either s_1 or s_2 may be used):

$$t = \frac{|42.34 - 42.44|}{0.10}\sqrt{\frac{5 \times 4}{5 + 4}}$$

$$t = 1.491$$

Consult Table 2.5 at degrees of freedom $n_1 + n_2 - 2 = 7$, finding t for the 95% probability level $= 2.365$. Since $1.149 < 2.365$, the null hypothesis is correct and the difference is not significant.

Example 2. A chemist analyzes a sample of iron ore furnished by the Bureau of Standards and obtains the following results: $\bar{x} = 10.52$, $s = 0.05$, $n = 10$. The Bureau's value for this sample is 10.60% Fe. Are the results significantly different at the 95% probability level?

Calculate t from the equation

$$\mu = \bar{x} \pm \frac{ts}{\sqrt{n}}$$

$$10.60 = 10.52 \pm \frac{t \times 0.05}{\sqrt{10}}$$

$$t = 5.06$$

In Table 2.5, at degrees of freedom $= 9$ and 95% probability level, $t = 2.262$. Since $5.06 > 2.262$ the results are significantly different from the Bureau's value.

Criteria for Rejection of an Observation

Sometimes a person performing measurements is faced with one result in a set of replicates which seems to be out of line with the others, and he then must decide whether to exclude this result from further consideration. This problem is encountered in beginning analytical chemistry courses, later in physical chemistry laboratory work, and even in advanced research, although hopefully with lessening frequency as the student progresses. It is a generally accepted rule in scientific work that a measurement is to be automatically rejected when it is known that an error was made; this is a determinate situation with which we are not concerned here. It should be noted that it is incorrect (but all too human) to reject results which were subject to known errors only when they appear to be discordant. The only way to avoid an unconscious introduction of bias into the measurements is to reject every result where an error was known to be made, regardless of its agreement with the others. The problem to which we address ourselves here is a different one: How do we decide whether to throw out a result which appears discordant when there is no known reason to suspect it?

If the number of replicate values is large, the question of rejecting one value is not an important one; first, a single value will have only a small effect upon the mean, and second, statistical considerations give a clear answer regarding the probability that the suspected result is a member of the same population as the others. On the other hand, a real dilemma arises when the number of replicates is small: The divergent result exerts a significant effect upon the mean while at the same time there are insufficient data to permit a real statistical analysis of the status of the suspected result.

The many different recommendations that have been promulgated by various writers attest to the conclusion that the question of rejecting or retaining one divergent value from a small sample really cannot satisfactorily be answered. Some of the more widely recommended criteria for rejection are considered below, and the student is referred to the excellent discussion by Blaedel et al.,[5] and interesting briefer commentaries by Laitinen[6] and Wilson.[7]

In the first place, it is necessary to decide how large the difference between the suspected result and the other data must be before the result is to be discarded. If the minimum difference is made too small, valid data may be rejected too frequently; this is said to be an "error of the first kind." On the other hand, setting the minimum difference too high leads to "errors of the second kind," that is, too frequent retention of highly erroneous values. The various recommendations for criteria of rejection steer one course or another between the Scylla and Charybdis of these two types of errors, some closer to one and some closer to the other.

The *2.5d rule* is applied as follows:

1. Compute the mean and the average deviation of the "good" results.

[5] W. J. Blaedel, V. W. Meloche, and J. A. Ramsey, *J. Chem. Ed.*, **28**, 643 (1951).

[6] H. A. Laitinen, *Chemical Analysis*, McGraw-Hill Book Company, New York, 1960, p. 574.

[7] E. B. Wilson, Jr., *An Introduction to Scientific Research*, McGraw-Hill Book Company, New York, 1952, p. 256.

2. Find the deviation of the suspected result from the mean of the "good" ones.
3. If the deviation of the suspected result from the mean of the "good" ones is at least 2.5 times the average deviation of the "good" results, then reject the suspected result. Otherwise, retain it.

Strictly, the limit for rejection is too low with the $2.5d$ rule: Valid data are rejected too often (errors of the first kind). The degree of confidence often quoted for the rule is based upon large sample statistics extended to small samples without proper compensation.

The *4d rule* is used in the same manner as the $2.5d$ rule above: This rule likewise leads to errors of the first kind, although obviously not so frequently. There is no statistical justification for using either the $2.5d$ or the $4d$ rule, although both are widely recommended. It should be noted that these rules are meant to apply to the rejection of only one result from a group of four to eight, not to one out of three, two out of five, etc.

The *Q-test*, described by Dean and Dixon,[8] is statistically correct, and it is very easy to apply. When the Q-test calls for rejection, confidence is high (90%) that the suspected result was indeed subject to some special error. Using the Q-test for rejection, errors of the first kind are highly unlikely. However, when applied to small sets of data (say, three to five results), the Q-test allows rejection only of results that deviate widely, and hence leads frequently to errors of the second kind (retention of erroneous results). Thus, the Q-test provides excellent justification for the rejection of grossly erroneous values, but it does not eliminate the dilemma with suspicious but less deviant values. The reason for this, of course, is that with small samples only crude guesses of the real population distribution are possible, and thus sound statistics lends assurance only to the rejection of widely divergent results.

The Q-test is applied as follows:

1. Calculate the range of the results.
2. Find the difference between the suspected result and its nearest neighbor.
3. Divide the difference obtained in step 2 by the range from step 1 to obtain the rejection quotient, Q.
4. Consult a table of Q values. If the computed value of Q is greater than the value in the table, the result can be discarded with 90% confidence that it was indeed subject to some factor which did not operate on the other results.

Some Q values are given in Table 2.8.

The following example illustrates the application of the preceding tests.

Example. Four results obtained for the normality of a solution are 0.1014, 0.1012, 0.1019, and 0.1016. Apply the tests above to see if the 0.1019 result can be discarded.

[8] R. B. Dean and W. J. Dixon, *Anal. Chem.*, **23**, 636 (1951).

TABLE 2.8 Values of Rejection Quotient, Q

Number of observations	$Q_{0.90}$
3	0.90
4	0.76
5	0.64
6	0.56
7	0.51
8	0.47
9	0.44
10	0.41

(a) Compute the mean and average deviation of the three "good" results:

	Results		*Deviations (ppt)*
	0.1014		0.0
	0.1012		2.0
	0.1016		2.0
Average:	0.1014	Average:	1.3

(b) Compute the deviation of the suspected result from the mean of the three "good" results:

$$0.1019 - 0.1014 = 0.0005 \text{ or } 5.0 \text{ ppt}$$

Using the 2.5d rule,

$$2.5 \times 1.3 = 3.3 < 5.0 \quad \text{(discard)}$$

Using the 4.0d rule,

$$4.0 \times 1.3 = 5.2 > 5.0 \quad \text{(do not discard)}$$

Using the Q-test,

$$Q = \frac{0.1019 - 0.1016}{0.1019 - 0.1012}$$

$$Q = \frac{0.0003}{0.0007}$$

$$Q = 0.43$$

Since $Q < 0.76$ (Table 2.8), do not discard.

As noted above, the Q-test affirms the rejection of a value at a confidence level of 90%. Willingness to reject a result with less confidence would make possible a Q-test which allowed retention of fewer deviant values (errors of the second kind). While this appears superficially attractive, there are valid reasons for conservatism in rejecting measurements. Actually, low confidence levels (say 50%) are scarcely meaningful when only a small number of observations is involved. Further, although to many students in introductory courses laboratory measurements are only exercises, it must be remembered that the collection of data is a scientific enterprise with a purpose, and the matter must be discussed as though it were important. The worker who has carefully conceived his measurement and executed it painstakingly, and who has reason to hope that the outcome will be significant, will not quickly throw his work away. He will be more likely to repeat the measurement until the dilemma of the discordant result has evaporated through the operation of two factors: Dilution of any one result by all the others will lessen its significance, and, as the number of observations increases, statistical evaluation of the suspected result will become more meaningful.

A sort of compromise between outright rejection and the retention of a suspected value is sometimes recommended, which is to report the median of all the results rather than a mean either with or without the deviant value. The median is influenced by the *existence* of one discordant result, but it is not affected by the *extent* to which the result differs from the others. For a sample containing three to five values, Blaedel et al. recommend testing the suspected value with the *Q*-test and rejecting it if the test allows this; if not, the median is reported rather than the mean. Some writers (e.g., Wilson) recommend that the highest and lowest values both be rejected and the mean of the others reported: "The best procedure to use depends on what is known about the frequency of occurrence of wild values, on the cost of additional observations, and on the penalties for the various types of error. In the absence of special arguments, the use of the interior average . . . would appear to be good practice."[9] It may be noted that this interior average and the median are necessarily identical in the special case where there are just three results.

PROPAGATION OF ERRORS

Usually, the numerical result of a measurement is not of interest in its own right, but rather is used, sometimes in conjunction with several other measurements, to calculate the quantity which is actually desired. Attention is naturally focused upon the precision and accuracy of the final, computed quantity, but it is instructive to see how errors in the individual measurements are propagated into this result. A rigorous treatment of this problem requires more space than is available and mathematics beyond that presupposed for this book. An interesting elementary approach has been given by Waser,[10] and the interested student may find the elements of a more sophisticated treatment discussed briefly by Wilson[11] and by Shoemaker and Garland.[12] A discussion with particular emphasis on analytical chemistry has been given by Benedetti-Pichler.[13]

[9] Wilson, *loc. cit.*, p. 257.

[10] J. Waser, *Quantitative Chemistry*, rev. ed., W. A. Benjamin, Inc., New York, 1964, p. 371.

[11] Wilson, *loc. cit.*, p. 272.

[12] D. P. Shoemaker and C. W. Garland, *Experiments in Physical Chemistry*, McGraw-Hill Book Company, New York, 1962, p. 30.

[13] A. A. Benedetti-Pichler, *Ind. Eng. Chem., Anal. Ed.*, **8**, 373 (1936).

Determinate Errors

Consider a computed result, R, based upon the measured quantities A, B, and C. Let α, β, and γ represent the absolute determinate errors in A, B, and C, respectively, and let ρ represent the *maximum* resulting error in R. To see how the errors are transmitted through addition and subtraction, suppose that $R = A + B - C$. Changing each quantity by the amount of its error, we may write

$$R + \rho = (A + \alpha) + (B + \beta) - (C - \gamma)$$

or

$$R + \rho = (A + B - C) + (\alpha + \beta + \gamma)$$

Subtracting $R = A + B - C$ gives

$$\rho = \alpha + \beta + \gamma$$

Now, suppose, on the other hand, that multiplication and division are involved; i.e., let $R = AB/C$. Again insert the appropriate errors:

$$R + \rho = \frac{(A + \alpha)(B + \beta)}{C - \gamma} = \frac{AB + \alpha B + \beta A + \alpha\beta}{C - \gamma}$$

Let us neglect $\alpha\beta$, since it may be supposed that the errors are very small compared with the measured values. Then subtracting $R = AB/C$ gives

$$\rho = \frac{AB + \alpha B + \beta A}{C - \gamma} - \frac{AB}{C}$$

Placing the right-hand terms over a common denominator gives

$$\rho = \frac{\alpha BC + \beta AC + \gamma AB}{C(C - \gamma)}$$

It is now convenient to consider the relative error, ρ/R, by dividing by $R = AB/C$, which leads, after appropriate cancellation, to

$$\frac{\rho}{R} = \frac{\alpha BC + \beta AC + \gamma AB}{AB(C - \gamma)}$$

Since γ is very small compared with C, this reduces to

$$\frac{\rho}{R} = \frac{\alpha}{A} + \frac{\beta}{B} + \frac{\gamma}{C}$$

Thus it is found that determinate errors are propagated as follows:

1. Where addition or subtraction is involved, the *absolute* determinate errors are transmitted directly into the result.
2. Where multiplication or division is involved, the *relative* determinate errors are transmitted directly into the result.

Indeterminate Errors

In the case of determinate errors, it was reasonable to assume, at least for the purpose of illustration, that each measurement of some quantity A was attended by a definite error α; we were able to work with the errors of individual measurements. Indeterminate errors, on the other hand, are manifested by scatter in the data when a measurement is performed more than once. In considering the propagation of indeterminate errors, then, we must inquire how scatter in measurements of quantities A, B, C, etc., is translated into random variation in the final result R.

Suppose again that $R = A + B - C$ on the one hand, and that $R = AB/C$ on the other. The result of statistical theory is:

1. In addition or subtraction, the variances (squares of the standard deviations) of the measured values are additive in determining the variance of the result, that is, for $R = A + B - C$, $s_R^2 = s_A^2 + s_B^2 + s_C^2$.
2. With multiplication or division, the squares of the relative standard deviations are transmitted; that is for $R = AB/C$,

$$\left(\frac{s_R}{R}\right)^2 = \left(\frac{s_A}{A}\right)^2 + \left(\frac{s_B}{B}\right)^2 + \left(\frac{s_C}{C}\right)^2$$

Calculation of Analytical Results

In considering the propagation of errors in the calculations involved in analysis, the chemist normally treats the errors as though they were determinate ones. In practice he does not always push his measurements to the point where indeterminate errors are seen, because in many real analyses the precision of the final result is determined by factors such as sample inhomogeneity and losses or contamination. If experience has shown that such errors amount to several parts per thousand, there is no point in weighing a sample to better than 1 part per thousand or so.

In the operation of weighing a sample to the nearest 0.1 mg on an analytical balance, the analyst will ordinarily make no effort to estimate vernier readings to any better than 0.05 mg, regardless of the capability of his particular balance. He rounds off the weight to 0.1 mg and expects to be in error by no more than 0.1 mg unless some special accident occurs. If he records a weight of, let us say, 0.1036 g, the operator is reasonably certain that the correct value to four decimal places is 0.1035, 0.1036, or 0.1037. These, of course, are not randomly distributed numbers. Rather, the maximum error, once decided upon, is treated as a determinate error with respect to its propagation through a series of computations. In treating this propagation, it is usual not to count upon even partial cancellation of the errors in the various measured quantities, but rather to predict the errors in the result on the most pessimistic grounds. For example, in weighing a solid sample on an analytical balance, a sample bottle containing the solid is weighed, some sample poured from the bottle, and the bottle is reweighed. Consider the following example.

Example 1. A weighing bottle containing some pure NaCl is found to weigh 25.8432 g. A portion of the NaCl is removed from the bottle, the bottle reweighed and found to weigh 25.6426 g.

(a) If the uncertainty in each weighing is 0.0001 g (0.1 mg), what is the maximum uncertainty in the weight of the NaCl sample?

The uncertainties are treated as determinate errors. Hence

$$\rho = \alpha + \beta$$

$$\rho = 0.0001 + 0.0001$$

$$\rho = 0.0002$$

(b) What is the relative uncertainty in the weight of the sample?

The sample weighs 25.8432 − 25.6426 = 0.2006 g. Hence

$$\frac{0.0002}{0.2006} \times 1000 \cong 1 \text{ part per thousand}$$

Another common operation in the analytical laboratory is the measurement of volume using a 50-ml buret. The smallest graduation on such a buret is one-tenth of a milliliter and the chemist estimates a reading to one-hundredth of a milliliter by mentally dividing the smallest division into 10 equal parts. He thus obtains a reading such as 34.46 ml. Other observers might obtain 34.45 or 34.47, but it is highly probable that all observers would report a reading in the range 34.45–34.47. The chemist therefore estimates the maximum uncertainty in a single buret reading of 0.02 ml and treats this as a determinate error with respect to its propagation. Consider the following example.

Example 2. What volume of solution should be measured from a buret so that the maximum error is 1 part per thousand? Two readings are required, an initial and final reading, and the uncertainty in each reading is 0.02 ml.

Treating the uncertainties as determinate errors

$$\rho = 0.02 + 0.02$$

$$\rho = 0.04 \text{ ml}$$

Hence

$$\frac{0.04}{V} \times 1000 = 1 \text{ ppt}$$

$$V = 40 \text{ ml}$$

As a general rule, students in the introductory quantitative analysis laboratory are instructed to weigh samples of 0.2 g or more and to use samples large enough to require about 40 ml of solution in a titration. If these suggestions are followed the errors in weighing and measuring volume should be 1 part per thousand or less.

One further point should be emphasized. Suppose that an error of ten parts per thousand is expected in a certain analysis for the percentage of some constituent in a sample because of the known limitations of a certain instrument. Then for this analysis, it would be a waste of time to weigh out a starting sample of 10 g to the nearest 0.0001 g, even though the balance might be capable of this. A weighing to the nearest 0.1 g would represent 10 parts per thousand and would then be adequate, although admittedly a cautious person might well prefer to weigh to the nearest 0.01 g just to be on the safe side. In this example, it is supposed that multiplication and division are involved in calculating the result; for example, a formula of this sort might be used:

$$\% \text{ of constituent} = \frac{\text{instrument reading} \times \text{some factor}}{\text{weight of sample}}$$

Where addition or subtraction is used, the *absolute* rather than the relative errors must be considered. For example, suppose that a 50-g vessel is weighed; then a sample of 0.1 g is added to it, and the weighing is repeated. If the weight of the sample is desired to a part in a thousand, then weighing to the nearest 0.0001 g is required, even though this represents precision of 1 part in 500,000 so far as the weight of the container is concerned.

SIGNIFICANT FIGURES AND COMPUTATION RULES

Significant Figures

When a computation is made from experimental data, the error or uncertainty in the final result can be calculated by the procedures just described. A widely used procedure for making a crude estimate of this uncertainty involves the use of *significant figures.* The principal advantage of this procedure is that it is less laborious than the calculations of actual uncertainties, particularly those based on indeterminate errors. The principal disadvantage is that only a rough estimate of uncertainty is obtained. In most situations encountered in analyses, an estimate is all that is needed, and hence significant figures are widely used.

Most scientists define significant figures as follows: All digits that are certain plus one which contains some uncertainty are said to be significant figures.[14] For example, in weighing an object on an analytical balance, the figures 10.746 can be recorded with certainty. The fourth decimal is estimated by reading a pointer scale or a vernier, and the final weight recorded as 10.7463. The last digit is uncertain, probably to ±1 in a single reading, or ±2 if a difference of two readings is involved. The six digits in this weight are all significant figures.

It is important to use only significant figures in expressing analytical data. The use of too many or too few figures may mislead another person with respect to the precision of the experimental data. If a volume is recorded as 1.234 ml, for example, it would be understood that the graduations on the buret were in

[14] Note that in terms of this definition it is improper to say, as do many authors, that "one should use the appropriate number of significant figures." Rather, one should say that only significant figures should be recorded. One can record too many digits, but not too many significant figures.

0.01-ml intervals and that the third decimal was estimated by reading between the graduations. The same volume read on an ordinary 50-ml buret could be estimated only to the second decimal, since the graduations are in 0.1-ml intervals. Hence the reading should have no more than three figures, say, 1.23 ml.

The digit zero may or may not be a significant figure, depending upon its function in the number. In a buret reading of, say, 10.06 ml, both zeros are measured and are therefore significant figures; the number contains four significant figures. Suppose the foregoing volume is expressed in liters, that is, 0.01006 ℓ. We do not increase the number of significant figures by changing the unit of volume. The function of the initial zero is to locate the decimal point; hence initial zeros are not significant. Usually a zero is also placed before the decimal, as 0.01006, and this also is not significant. Terminal zeros are significant. For example, a weight of 10.2050 g has six significant figures. When it is necessary to use terminal zeros merely to locate the decimal properly, powers of ten may be used to avoid confusion with regard to the number of significant figures. For example, a weight of 24.0 mg expressed as micrograms should not be written as 24,000. The last two zeros are not significant, and this is indicated by writing the numbers as 24.0×10^3, or 2.40×10^4.

Computation Rules

As previously mentioned, the analytical chemist uses his experimental data to compute a final result. We would like to examine now the rules which are suggested to ensure that the final result contains only the number of digits justified by the uncertainties in the data.

Addition and Subtraction

The rule suggested here is to keep only as many decimals in the final result as occur in that one of the numbers which has the fewest decimals. This rule follows from the result we obtained for the propagation of errors in addition and subtraction (i.e., $\rho = \alpha + \beta$). For example, if we are to add the numbers 50.1 ± 0.1, 1.36 ± 0.02, 0.5182 ± 0.0001, and 6.453 ± 0.003, the uncertainty in the sum would be

$$\rho = 0.1 + 0.02 + 0.0001 + 0.003$$

$$\rho = 0.1231$$

Since the final result should contain only one uncertain digit, we should round this off to

$$\rho = 0.1$$

The following example illustrates the use of significant figures in the operation of addition and subtraction.

Example. Add algebraically the following numbers: 14.23 + 8.145 − 3.6750 + 120.4.

The final answer should be expressed to only one decimal place. It is recommended that the numbers be rounded[15] preliminarily to two decimal places, then added. The final result is then rounded to one decimal place.

+14.23	+14.23
+8.145	+8.14
−3.6750	−3.68
+120.4	+120.4
	139.09

Final rounding gives: 139.1

This procedure tends to avoid the accumulation of rounding errors in the final result.

Multiplication and Division

The rule suggested here is to retain in each term and the final result a number of digits which will indicate a relative uncertainty no greater than that of the term with the greatest relative uncertainty. This rule is derived from the fact that relative determinate errors are transmitted directly into the result in the operations of multiplication and division. Consider the following examples.

Example 1. Calculate the following, giving the maximum uncertainty in the final result. Treat uncertainties as determinate errors.

$$\frac{(10.00 \pm 0.02) \times (5.000 \pm 0.001)}{2.50 \pm 0.01}$$

The relative uncertainties are:

10.00	2.0 ppt
5.000	0.2 ppt
2.50	4.0 ppt
Sum =	6.2 ppt

Hence the uncertainty in the final result is 6.2 ppt. The final result, 20.0000 . . . , should be properly rounded off to express this relative uncertainty. If two zeros are retained, 20.00, an uncertainty of at least ±0.01 is implied. The relative uncertainty is 0.5 ppt, less than we actually have. If the number is written 20.0, an uncertainty of at least ±0.1 is indicated. This is a relative uncertainty of 5.0 ppt, close to our value of 6.2. If the number is written 20, the uncertainty is ±1 with a relative value of 50 ppt, obviously too high. Hence the rule says to retain one zero, 20.0, this giving an uncertainty closest to that portion of the data which has the greatest relative uncertainty.

The approximate nature of significant figures can be seen from the above example. In rounding off numbers in this fashion, one obtains only an estimate of

[15] In rounding off numbers, drop the last digit if it is less than 5 : 4.33 becomes 4.3. If the last digit is greater than 5, increase the preceding digit by one: 4.36 becomes 4.4. If the digit to be dropped is 5, round the preceding digit to the nearest even number: 4.35 becomes 4.4; 4.65 becomes 4.6. This procedure avoids a tendency to round in one direction only.

the uncertainty in the final result. An extreme case is that in which all the uncertainties are of equal magnitude. For example, suppose the calculation above had been

$$\frac{(10.00 \pm 0.04) \times (5.00 \pm 0.02)}{2.50 \pm 0.01}$$

All three relative uncertainties are 4.0 ppt, making a total of 12 ppt in the final result. Expressing the number as 20.0, as our rule would suggest, means that the uncertainty in the last digit is as large as ±2.

The following example involves a calculation of percentage purity from typical titrimetric data. It illustrates the fact that the chemist knows some of the uncertainties from his experience with the measurements which were made.

Example 2. The percentage chromium is calculated from a titration as follows:

$$\%\text{Cr} = \frac{40.64 \text{ ml} \times 0.1027 \text{ mmol/ml} \times (51.996 \text{ mg/mmol}/3)}{346.4 \text{ mg}} \times 100$$

$$\%\text{Cr} = 20.883096\ldots$$

How should the final %Cr be expressed?

The chemist would estimate the uncertainties as follows:

40.64 ± 0.04 ml	or	1 ppt
0.1027 ± 0.0001	or	1 ppt
51.996 ± 0.001	or	0.02 ppt
346.4 ± 0.2	or	0.6 ppt

He would round off his answer as 20.88 since this implies a relative uncertainty reasonably close to 1 ppt. A more exact analysis shows that the sum of the uncertainties is actually 3 ppt, indicating that the actual uncertainty in the final digit of our result is about ±6.

The student may have noted that the number of significant figures in the final result of multiplication and division is normally the same as that portion of the data which contains the least number of significant figures. Sometimes such a rule is cited for this operation. The ultimate criterion in rounding off an answer in multiplication and division, however, is the relative uncertainty, not the number of significant figures in the data. For example, consider the following two multiplications:

$$(0.98 \pm 0.01)(1.07 \pm 0.01) = 1.05 \pm 0.02 \quad (1)$$

$$(1.02 \pm 0.01)(1.03 \pm 0.01) = 1.05 \pm 0.02 \quad (2)$$

The uncertainties in the two products are the same. However, one of the numbers in (1), 0.98, contains only two significant figures and if we rigorously followed the rule above we would retain only two digits in (1) and three in (2). Note that there is a relative uncertainty of about 1% in 0.98 and 1.02, even though the first number contains two significant figures and the second contains three. Since the relative uncertainty is the same in each of the two products, each should contain three significant figures.

Logarithms

A logarithm is composed of two parts: a whole number, the characteristic, plus a decimal fraction, the mantissa. The *characteristic* is a function of the position of the decimal in the number whose logarithm is being determined, and therefore is not a significant figure. The *mantissa* is the same regardless of the position of the decimal and all the digits are considered significant.

Consider the following illustration.

Example. (a) Calculate properly the logarithm of 2.4×10^5.

Using a four-place log table we find that the mantissa is 0.3802. The characteristic is 5. The log is expressed as 5.38 since there are two significant figures in the number 2.4×10^5.

(b) Calculate the antilog of 2.974.

If a calculator is used to obtain the antilog the result is 941.8896. The answer should contain only three significant figures and hence should be written 942 or 9.42×10^2.

QUESTIONS

1. *Terms.* Explain clearly the meaning of the following terms: (a) mean; (b) median; (c) standard deviation; (d) range; (e) central tendency; (f) variability; (g) variance; (h) coefficient of variation; (i) relative average deviation; (j) absolute error; (k) relative error; (l) additive error; (m) proportional error.

2. *Errors.* Explain whether the following errors are determinate or indeterminate and whether they affect the accuracy or precision of the measurement. If the error is determinate, tell whether it is methodic, operative, or instrumental, and how it could be eliminated. (a) The analytical weights are corroded. (b) The analyst unknowingly spatters some solution from a flask during a titration. (c) A sample picks up moisture during a weighing. (d) A reagent is used which contains some of the substance being determined. (e) The buret is misread once. (f) A student uses the wrong equivalent weight in his calculations.

3. *Confidence intervals.* Explain clearly the meaning of (a) a 95% confidence interval; (b) a 99% confidence interval of the mean.

4. *Testing for significance.* Explain clearly how to test two sets of results to determine if they differ significantly.

5. *Rejection of a result.* Explain briefly the nature of the statistical problem involved in deciding whether to retain or reject a suspicious value in a small series of replicate measurements. What may be wrong in using the $2.5d$ rule in this connection? With the Q-test?

6. *Weights and volumes.* Explain why it is recommended that students weigh samples of at least 0.2 g and use volumes of solutions of about 40 ml in laboratory procedures.

Multiple-choice: In the following multiple-choice questions, select the *one best* answer.

7. Which of the following statements is true? (a) Precise values may be inaccurate; (b) Determinate errors can be constant or proportional; (c) The variance is the square of the standard deviation; (d) All of the above.

8. The average of 64 results is how many times more reliable than the average of 4 results? (a) 16; (b) 4; (c) 2; (d) 8.

9. Titrator A obtains a mean value of 6.96% and a standard deviation of 0.03 for the purity of a compound. Titrator B obtains corresponding values of 7.10% and 0.05. The true value is 7.02%. Compared to titrator B, titrator A is (a) less accurate but more precise; (b) more accurate but less precise; (c) more accurate and more precise; (d) less accurate and less precise.

10. A student performed an analysis obtaining a percentage purity of 30.68% and a relative standard deviation of 5.0 ppt. He later discovered that he failed to divide the MW by 2 to get the EW, and that his correct percentage purity was 15.34%. What is his correct relative standard deviation in ppt? (a) 5.0; (b) 2.5; (c) 10.0; (d) $\sqrt{5.0}$.

PROBLEMS

(Unless otherwise indicated, it is to be understood that any weight or volume involves the difference of two readings.)

1. *Precision.* Refer to the Table of Atomic Weights on the inside front cover. Assume that the uncertainty in the last digit of each weight is ± 1. (a) Which of the following atomic weights is known with the greatest relative precision, that of iodine, cobalt, or beryllium? (b) Express the precision of each weight in ppt.

2. *Precision.* Repeat problem 1 for the following elements: calcium, sulfur, and lithium.

3. *Mean, median, etc.* Analyst A reported the following percentages of iron in a sample: 19.95, 19.98, 19.90, 20.00, 19.88, and 19.93. For this set of results calculate (a) the mean, median, range, average deviation, relative average deviation (ppt), standard deviation, and coefficient of variation; (b) the 95% confidence interval of the mean, first from the standard deviation, and then from the range.

4. *Mean, median, etc.* Analyst B reported the following percentages of iron in the same sample as in Problem 3: 19.90, 20.04, 19.96, 19.94, 20.12, and 19.80. For this set of results calculate (a) the mean, median, range, average deviation, relative average deviation (ppt), standard deviation, and coefficient of variation; (b) the 95% confidence interval of the mean, first from the standard deviation, and then from the range.

5. *Accuracy.* The Bureau of Standards' value for the percentage of iron in the sample in Problems 3 and 4 is 19.85. (a) Calculate the absolute and relative errors of analysts A and B. (b) What can you say about the work of the two analysts?

6. *Mean and average deviation.* A student standardized a solution of NaOH and obtained the following results: 0.1024, 0.1032, 0.1020, and 0.1018. Calculate the mean and relative average deviation in ppt of his results.

7. *Mean and average deviation.* A student standardized a solution of HCl and obtained the following results: 0.0992, 0.0988, 0.1000, and 0.0986. Calculate the mean and relative average deviation in ppt of his results.

8. *Precision.* The uncertainty in each reading on a semimicro balance is ±0.01 mg. How large a sample should be taken for analysis using this balance so that the maximum uncertainty in the sample weight is 1 ppt?

9. *Precision.* The uncertainty in each reading on a micro balance is ±0.001 mg. How large a sample should be taken for analysis using this balance so that the maximum uncertainty in the sample weight is 1 ppt?

10. *Precision.* The uncertainty in each reading of a microburet is ±0.002 ml. How large a volume should be taken from this buret so that the maximum uncertainty in the volume is 1 ppt?

11. *Precision.* What is the maximum uncertainty in ppt in a volume of 50 ml measured using a graduated cylinder? The uncertainty in each reading on the graduate is ±0.1 ml.

12. *Precision.* The readings on a rough balance can be made to within ±1 g. How large a sample should be taken using this balance so that the maximum uncertainty in the weight is 1%?

13. *Constant error.* In a certain method of determining silica, SiO_2 is precipitated and weighed. It is found that the amount of SiO_2 obtained is always 0.4 mg too high regardless of the weight of sample taken for analysis. Calculate the relative error in ppt in a sample which actually contains 10.0% silica if the size of sample analyzed is (a) 0.100 g; (b) 0.500 g; (c) 1.000 g.

14. *Proportional error.* A sample is to be anlyzed for chloride by titration with silver nitrate. The sample actually contains 0.10% bromide, which reacts exactly like chloride in the titration. Calculate the error in the number of milligrams of chloride an analyst would find in a sample which contains 20.0% chloride if the size of the sample analyzed is (a) 0.100 g; (b) 0.500 g; (c) 1.000 g. Then calculate the relative error (ppt) in the number of milligrams of chloride found.

15. *Relative error.* (a) A relative error of 0.5% is how many ppt? (b) A relative error of 0.4% is how many parts per 500? (c) A relative error of 0.01 ppt is how many ppm? (d) A relative error of 4 parts per 2000 is how many parts per 100?

16. *Relative error.* Assuming an uncertainty of ±1 in the last digit, what is the relative uncertainty in ppt in the following numbers? (a) 50.0; (b) 0.2500; (c) 40; (d) 10.0.

17. *Relative error.* Assuming an uncertainty of ±1 in the last digit, how should the following numbers be expressed to indicate the given relative uncertainty? (a) 50, 2%; (b) 25, 4 ppt; (c) 1 million, 10%; (d) 2000, 0.5 ppt.

18. *Testing for significance.* Two sets of results for the percentage of iron in an ore are obtained using two methods of analysis. These results are as follows:

Method 1	*Method 2*
$x = 10.56\%$	$x = 10.64\%$
$s_1 = 0.10$	$s_2 = 0.12$
$n_1 = 11$	$n_2 = 11$

(a) Are the standard deviations significantly different at the 95% level? (b) Are the two means significantly different (i) at the 90% level; (ii) at the 95% level; (iii) at the 99% level?

19. *Testing for significance.* Repeat Problem 18 for the case where the mean for method 2 is 10.70 rather than 10.64.

20. *Testing for significance.* Repeat Problem 18 for $s_1 = 0.05$ and $s_2 = 0.06$.

21. *Testing for significance.* Repeat Problem 18 for $n_1 = 5$ and $n_2 = 6$.

22. *Rejection of a result.* A student obtained the following results for the percentage of chloride in a sample: 20.38, 20.32, 20.28, 20.40, and 20.68. (a) Can any result be rejected by the *Q*-test? (b) What value should be reported for the percentage of chloride in the sample? (c) Calculate the 99% confidence interval of the mean.

23. *Rejection of a result.* A student obtained the following results for the normality of a solution: 0.1029, 0.1060, 0.1036, 0.1032, 0.1018, and 0.1034. (a) Can any result be rejected by the *Q*-test? (b) What value should be reported for the normality? (c) Calculate the 95% confidence interval of the mean.

24. *Rejection of a result.* A student obtained the following values for the normality of a solution: 0.0990, 0.0991, 0.0992, and 0.0998. (a) Can the last result be rejected according to the *Q*-test? A fifth result was run and a value of 0.0991 obtained. (b) Can 0.0998 now be discarded? Explain.

25. *Rejection of a result.* A student obtained the following values for the percentage of iron in an ore: 11.53, 11.51, and 11.55. What is (a) the highest, and (b) the lowest value a fourth result could be without being discarded by the *Q*-test?

26. *Testing for significance.* An analyst develops a new method for copper in an ore. He analyzes a sample from the Bureau of Standards by his method and obtains the following data: mean of five results = 9.92%; standard deviation = 0.05. The Bureau's value is 9.82%. Are the results significantly different at the 95% level?

27. *Testing for significance.* Repeat Problem 26 for the following data: mean of six results = 9.89; standard deviation = 0.08. The Bureau's value is 9.82%.

28. *F-test.* Two students are given the same sample to analyze. Student A makes seven determinations with a standard deviation of 0.05. Student B makes six determinations with a standard deviation of 0.08. Does the difference in standard deviations imply a significant difference in the technique of the two students?

29. *Confidence interval of the mean.* A chemist analyzes an ore for manganese and obtains a value of 8.42% with a standard deviation of 0.12. Calculate the 95% confidence interval of the mean if the analysis is based on (a) four determinations; (b) nine determinations.

30. *Confidence interval of the mean.* Suppose that the procedure used for determining manganese in Problem 29 had been run many times and it is known that the standard deviation for this method is 0.12. Calculate the confidence interval of the mean if the result of 8.42% is based on (a) four determinations; (b) nine determinations. (*Hint*: use the value of t at $n = \infty$ in Table 2.5.)

31. *Confidence interval of the mean.* The standard deviation of a method for the determination of iron in steel is known from a large number of determinations to be 0.07.

How many determinations must be run by this method if (a) the 99% confidence interval of the mean is to be ±0.10; (b) the 95% confidence interval is to be ±0.10?

32. *Propagation of errors.* Calculate the following properly, giving the maximum uncertainties in the result:

(a) $(8.32 \pm 0.02) + (15.43 \pm 0.04) - (4.82 \pm 0.03)$

(b) $\dfrac{(10.34 \pm 0.02)(5.12 \pm 0.03)}{4.22 \pm 0.01}$

33. *Propagation of errors.* Calculate the following properly, giving the maximum uncertainties in the result:

(a) $(1.213 \pm 0.001) + (5.2 \pm 0.1) - (2.34 \pm 0.01)$

(b) $\dfrac{(42.34 \pm 0.04)(0.1016 \pm 0.0001)(204.2 \pm 0.1)}{41.86 \pm 0.02}$

34. *Significant figures.* How many significant figures does each of the following numbers contain? (a) 0.02030; (b) 6.023×10^{23}; (c) 4.80×10^{-10}; (d) 96,500; (e) 999; (f) 1000.

35. *Relative error.* If the uncertainty in the last digit of each number in Problem 34 is ±1, what is the relative uncertainty in each number in ppt?

36. *Significant figures.* Express the results of the following calculations using only significant figures:

(a) $\dfrac{0.0252 \times 4.11 \times 10^2 \times 1.506}{6.22 \times 10^5}$

(b) $\dfrac{3.08 \times 10^2 \times 21.22 \times 0.0510}{1.122 \times 10^{-4}}$

(c) $124.165 + 8.2 + 1.4250$

37. *Significant figures.* Express the results of the following calculations using only significant figures:

(a) $\dfrac{5.24 \times 10^2 \times 4.12 \times 10^{-4}}{0.02538 \times 2.014 \times 10^3}$

(b) $\dfrac{9.2 \times 10^3 \times 3.12 \times 10^{-2} \times 2.08 \times 10^3}{0.00621}$

(c) $34.2335 + 16.62 - 8.6885$

38. *Significant figures.* How should the percentage of (a) S in $BaSO_4$, and (b) Mg in $Mg_2P_2O_7$ be properly expressed using the atomic weights given in the text and the rules for significant figures?

39. *Significant figures.* Express the following properly: (a) log of 4.68×10^6; (b) log of 2.543×10^{-5}; (c) antilog of 3.450; (d) antilog of 6.26.

40. *Absolute and relative errors.* A student analyzed a sample for chloride by precipitating and weighing AgCl. A 1.184-g sample gave him 0.8152 g of AgCl. (a) Calculate the percentage of chloride in the sample. (See page 72.) (b) If the student by mistake used the atomic weight of Cl as 35.345 instead of 35.453 (both in calculating the MW of AgCl and in calculating the gravimetric factor), what percentage chloride would he find? Calculate his absolute error in the percentage of chloride and the relative error in ppt.

41. *Absolute and relative errors.* A student analyzed a sample of soda ash weighing 0.4134 g which actually contained 40.24% Na_2CO_3. He used 28.56 ml of 0.1106 *N* HCl for the titration. (See page 58. The EW of Na_2CO_3 is 53.00.) Calculate (a) the absolute error in the percentage of Na_2CO_3, and (b) the relative error in ppt.

42. *Sample size.* A sample contains about 10% of the ion B^-. B^- is determined by precipitating the compound AB_2. A has an atomic weight about twice that of B. If the uncertainty in determining the weight of the precipitate is not to exceed 1 ppt on a balance that is sensitive to 0.1 mg, what size sample should be taken for analysis?

43. *Sample size.* A sample containing about 5% sulfur is to be analyzed by oxidizing the S to SO_4^{2-}, precipitating, and weighing the $BaSO_4$. If the uncertainty in determining the weight of $BaSO_4$ is not to exceed 1 ppt on a balance sensitive to 0.1 mg, how large a sample should be taken for analysis?

CHAPTER THREE

titrimetric methods of analysis

We have already mentioned (page 6) that *titrimetric* analysis is one of the major divisions of analytical chemistry and that the calculations involved are based on the simple stoichiometric relationships of chemical reactions. In this chapter we shall discuss the general principles involved in titrimetry and the stoichiometric calculations which are employed.

GENERAL PRINCIPLES

A titrimetric method of analysis is based on a chemical reaction such as

$$aA + tT \longrightarrow \text{products}$$

where a molecules of the analyte, A, react with t molecules of the reagent, T. The reagent T, called the *titrant,* is added incrementally, normally from a buret, in the form of a solution of known concentration. The latter solution is called a *standard* solution and its concentration is determined by a process called *standardization.* The addition of the titrant is continued until an amount of T chemically equivalent to that of A has been added. It is then said that the *equivalence point* of the titration has been reached. In order to know when to stop the addition of titrant the chemist may use a chemical substance, called an *indicator,* which responds to the appearance of excess titrant by changing color. This color change may or may not occur precisely at the equivalence point. The point in the titration where the indicator changes color is termed the *end point.* It is desirable, of course, that the end point be as close as possible to the equivalence point. Choosing indicators to make the two points coincide (or correcting for the difference between the two) is one of the important aspects of titrimetric analysis.

The term *titration* refers to the process of measuring the volume of titrant required to reach the equivalence point. For many years the term *volumetric* analysis was used rather than titrimetric. However, from a rigorous standpoint the term "titrimetric" is preferable because volume measurements need not be confined to titrations. In certain analyses, for example, one might measure the volume of a gas.

Reactions Used for Titrations

The chemical reactions which may serve as the basis for titrimetric determinations are conveniently grouped into four types:

1. *Acid-base.* There are a large number of acids and bases which can be determined by titrimetry. If HA represents the acid to be determined and BOH the base, the reactions are

$$HA + OH^- \longrightarrow A^- + H_2O$$

and

$$BOH + H_3O^+ \longrightarrow B^+ + 2H_2O$$

The titrants are generally standard solutions of strong electrolytes, such as sodium hydroxide and hydrochloric acid.

2. *Oxidation-reduction (redox).* Chemical reactions involving oxidation-reduction are widely used in titrimetric analyses. For example, iron, in the +2 oxidation state can be titrated with a standard solution of cerium(IV) sulfate:

$$Fe^{2+} + Ce^{4+} \longrightarrow Fe^{3+} + Ce^{3+}$$

Another oxidizing agent which is widely used as a titrant is potassium permanganate, $KMnO_4$. Its reaction with iron(II) in acid solution is

$$5Fe^{2+} + MnO_4^- + 8H^+ \longrightarrow 5Fe^{3+} + Mn^{2+} + 4H_2O$$

3. *Precipitation.* The precipitation of silver cation with the halogen anions is a widely used titrimetric procedure. The reaction is

$$Ag^+ + X^- \longrightarrow AgX(s)$$

where X^- can be chloride, bromide, iodide, or thiocyanate (SCN^-) ion.

4. *Complex formation.* An example of a reaction in which a stable complex is formed is that between silver and cyanide ions:

$$Ag^+ + 2CN^- \longrightarrow Ag(CN)_2^-$$

This reaction is the basis of the so-called Liebig method for the determination of cyanide. Certain organic reagents, such as ethylenediaminetetraacetic acid (EDTA), form stable complexes with a number of metal ions, and are widely used for the titrimetric determination of these metals.

Requirements for Reactions Used in Titrimetric Analysis

Of the host of known chemical reactions, relatively few can be used as the basis for titrations. A reaction must satisfy certain requirements before it can be used:

1. The reaction must proceed according to a definite chemical equation. There should be no side reactions.
2. The reaction should proceed to virtual completion at the equivalence point. Another way of saying this is that the equilibrium constant of the reaction should be very large.
3. Some method must be available for determining when the equivalence point is reached. An indicator should be available or some instrumental method may be used to tell the analyst when to stop the addition of titrant.
4. It is desirable that the reaction be rapid, so that the titration can be completed in a few minutes.

Consider, as an example of a reaction well suited for titrations, the determination of the concentration of a hydrochloric acid solution by titration with standard sodium hydroxide. There is only one possible reaction and it is immeasurably fast. The reaction goes to virtual completion; the equilibrium constant is about 10^{14}. At the equivalence point the *p*H of the solution changes by several units, and a number of indicators are available which respond to this *p*H change by changing color.

On the other hand, the reaction between boric acid and sodium hydroxide is not sufficiently complete to satisfy requirement 2; the equilibrium constant is about 6×10^4. The reaction between ethyl alcohol and acetic acid is likewise unsuitable: it is too slow for convenience and does not go to completion. The reaction between tin(II) and potassium permanganate is not satisfactory unless air is excluded, because tin is also readily oxidized by atmospheric oxygen. The precipitation of certain metal ions by sulfide ion satisfies all the requirements except number 3; that is, no suitable indicator is available.

STOICHIOMETRY

The branch of chemistry which deals with the weight relations between elements and compounds in chemical reactions is called *stoichiometry*. Since aqueous solutions are so commonly used in most areas of quantitative analysis, we shall review here methods used to express the amount of solute in solution and stoichiometric calculations involving solutions.

Molecular and Formula Weights

The student should recall that the *mole* is defined as the amount of a substance containing as many entities as there are atoms in 12 g of the carbon-12 isotope,

$^{12}_{6}C$. The entities may be atoms, molecules, ions, or electrons. Since 12 g of carbon-12 contains Avogadro's number of atoms, it follows that 1 mol of any substance contains 6.023×10^{23}elementary particles. If the elementary particles are molecules, the weight in grams of a mole of the substance is called the *gram-molecular weight* (usually shortened to *molecular weight*). Hence the gram-molecular weight of H_2 is 2.016 g and contains 6.023×10^{23} H_2 molecules. If the elementary particles are atoms, the weight in grams of 1 mol of the substance is called the *gram-atomic weight* (usually shortened to *atomic weight*). The gram-atomic weight of copper is 63.54 g and contains 6.023×10^{23} Cu atoms.

The term *gram-formula weight* (or *formula weight*) is the summation of the atomic weights of all the atoms in the chemical formula of a substance and is normally the same as the molecular weight. Some chemists use formula weight rather than molecular weight in cases where it would be inappropriate to talk about "molecules" of a substance, particularly ionic compounds. In sodium chloride (NaCl), for example, the smallest units in the solid are Na^+ and Cl^- ions; a molecule of NaCl does not exist. Since the mole, as defined above, refers to other entities as well as molecules, we shall use the term "molecular weight" as synonomous with "formula weight" in such cases. It is understood that this usage does not imply anything about the structure of the compound. In situations where dissociation or complex formation occurs, resulting in appreciable amounts of both molecules and ions in a solution, we shall use the concentration system *formality* (page 51) to indicate the *total* amount of a substance added to a solution, and *molarity* to indicate the *equilibrium* concentrations of individual species.

Equivalent Weights

The equivalent weight of a substance involved in a reaction used as the basis for a titration is defined as follows:

1. *Acid-base.* The *gram-equivalent weight* is the weight in grams of the substance required to furnish or react with *one mole* (1.008 g) of H^+.
2. *Redox.* The *gram-equivalent weight* is the weight in grams of the substance required to furnish or react with 1 *mol* of electrons.
3. *Precipitation or complex formation.* The *gram-equivalent weight* is the weight in grams of the substance required to furnish or react with 1 *mol* of a univalent cation, $\frac{1}{2}$ *mol* of a divalent cation, $\frac{1}{3}$ *mol* of a trivalent cation,[1] and so on.

The equivalent weight of a substance is called an *equivalent,* just as the molecular weight is called a mole. The equivalent and molecular weights are

[1] The cation referred to is the one directly involved in the titration reaction. For example, the equivalent weight of $AgNO_3$ in the reaction

$$Ag^+ + 2KCN \longrightarrow Ag(CN)_2^- + 2K^+$$

is the molecular weight, 169.87 g, since this amount of the salt furnishes 1 mol of the univalent cation, Ag^+. The equivalent weight of KCN is twice the molecular weight, since 2 mol of KCN react with 1 mol of Ag^+.

related by the equation

$$EW = \frac{MW}{n}$$

where n is the number of moles of hydrogen ion, electrons, or univalent cation furnished or combined with by the reacting substance. There is actually no need to use the term "equivalent"; any stoichiometric calculation can be done in terms of moles. However, the term is widely used and it is unwise to ignore it.

The reason for introducing the term "equivalent" is one of convenience: 1 equivalent of any acid reacts with 1 equivalent of any base; 1 equivalent of any oxidizing agent reacts with 1 equivalent of any reducing agent; and so on. It should be noted, however, that many compounds undergo more than a single reaction and hence can have more than one equivalent weight. For example, the permanganate ion can undergo the following reactions:

$$MnO_4^- + e \longrightarrow MnO_4^{2-} \tag{1}$$

$$MnO_4^- + 4H^+ + 3e \longrightarrow MnO_2 + 2H_2O \tag{2}$$

$$MnO_4^- + 8H^+ + 4e \longrightarrow Mn^{3+} + 4H_2O \tag{3}$$

$$MnO_4^- + 8H^+ + 5e \longrightarrow Mn^{2+} + 4H_2O \tag{4}$$

The equivalent weight of a permanganate salt, such as $KMnO_4$, is its molecular weight divided by 1, 3, 4, or 5, depending upon which of the foregoing reactions occurs.

The reaction of phoshporic acid with a base can be stopped when the following reaction has occurred:

$$H_3PO_4 + OH^- \longrightarrow H_2O + H_2PO_4^-$$

The equivalent weight of the acid is the same as the molecular weight. But the reaction can be carried further:

$$H_3PO_4 + 2OH^- \longrightarrow 2H_2O + HPO_4^{2-}$$

Here the equivalent weight is one-half the molecular weight. In aqueous solutions it is not feasible to titrate the third hydrogen; if the reaction could be carried out, the equivalent weight of the acid would be one-third of the molecular weight.

The following are some examples illustrating the calculation of equivalent weights:

Example 1. Calculate the equivalent weight of SO_3 used as an acid in aqueous solution.

SO_3 is the anhydride of sulfuric acid, H_2SO_4. When the latter acid is titrated with a strong base it furnishes two protons:

$$SO_3 + H_2O \longrightarrow H_2SO_4 \longrightarrow 2H^+ + SO_4^{2-}$$

Hence 1 mol of SO_3 is responsible for furnishing 2 mol of H^+, and

$$EW = \frac{MW}{2} = \frac{80.06}{2}$$

$$EW = 40.03 \text{ g/eq}$$

The equivalent weight of H_2SO_4 is also one-half the molecular weight, 98.07/2 = 49.04 g/eq.

Example 2. Calculate the equivalent weights of $Na_2C_2O_4$, the reducing agent, and $K_2Cr_2O_7$, the oxidizing agent, in the following reaction:

$$3C_2O_4^{2-} + Cr_2O_7^{2-} + 14H^+ \longrightarrow 2Cr^{3+} + 6CO_2 + 7H_2O$$

The number of electrons gained or lost can be determined from the change in oxidation number or the half-reactions (Appendix II). The half-reactions are

$$C_2O_4^{2-} \longrightarrow 2CO_2 + 2e$$

$$Cr_2O_7^{2-} + 14H^+ + 6e \longrightarrow 2Cr^{3+} + 7H_2O$$

Oxalate ion furnishes two electrons and dichromate ion gains six electrons. Hence the equivalent weights are

$$Na_2C_2O_4\text{:} \quad \frac{MW}{2} = \frac{134.0}{2} = 67.00 \text{ g/eq}$$

$$K_2Cr_2O_7\text{:} \quad \frac{MW}{6} = \frac{294.2}{6} = 49.03 \text{ g/eq}$$

Example 3. Calculate the equivalent weights of $AgNO_3$ and $BaCl_2$ in the reaction

$$2Ag^+ + BaCl_2 \longrightarrow 2AgCl(s) + Ba^{2+}$$

One mole of silver nitrate furnishes 1 mol of the univalent cation, Ag^+, 1 mol of $BaCl_2$ reacts with 2 mol of Ag^+. Hence

$$EW\ AgNO_3 = \frac{MW}{1} = \frac{169.9}{1} = 169.9 \text{ g/eq}$$

$$EW\ BaCl_2 = \frac{MW}{2} = \frac{208.2}{2} = 104.1 \text{ g/eq}$$

Concentration Systems

In titrimetric analyses the concentration systems *molarity* and *normality* are most frequently employed. *Formality* and *analytical concentration* are useful in situations where dissociation or complex formation occurs. The *percent by weight* system is commonly employed to express approximate concentrations of laboratory reagents. For very dilute solutions *parts per million* or *parts per billion* units are convenient.

Molarity

This concentration system is based on the volume of *solution* and hence is convenient to use in laboratory procedures where the volume of solution is the quantity measured. It is defined as follows:

$$\text{Molarity} = \text{number of moles of solute per liter of solution}$$

or

$$M = \frac{n}{V}$$

where M is the molarity, n the number of moles of solute, and V the volume of solution in *liters*. Since

$$n = \frac{g}{MW}$$

where g is the grams of solute and MW the molecular weight of solute, it follows that

$$M = \frac{g}{MW \times V}$$

This equation can be solved for grams of solute, giving

$$g = M \times V \times MW$$

The following examples illustrate the molarity system of concentration.

Example 1. Calculate the molarity of a solution which contains 6.00 g of NaCl (MW = 58.44) in 200 ml of solution.

$$M = \frac{g}{\text{MW} \times V}$$

$$M = \frac{6.00\text{ g}}{58.44\text{ g/mol} \times 0.200\text{ liter}}$$

$$M = 0.513\text{ mol/liter}$$

Example 2. Calculate the number of moles and the number of grams of $KMnO_4$ (MW = 158.0) in 3.00 liters of a 0.250 M solution.

$$M = \frac{n}{V}$$

$$n = M \times V$$

$$n = 0.250\text{ mol/liter} \times 3.00\text{ liters}$$

$$n = 0.750\text{ mol}$$

$$g = n \times \text{MW}$$

$$g = 0.750\text{ mol} \times 158.0\text{ g/mol}$$

$$g = 119$$

Formality

This system of concentration is defined as follows:

$$\text{Formality} = \text{number of formula weights of solute per liter of solution}$$

or

$$F = \frac{n_f}{V}$$

where F is the formality, n_f = the number of formula weights, and V the volume of solution in liters. Since

$$n_f = \frac{g}{\text{FW}}$$

where g is the grams of solute and FW the formula weight, it follows that

$$F = \frac{g}{\text{FW} \times V}$$

The following example illustrates the use of the formality system of concentration.

Example 3. A sample of dichloroacetic acid, $Cl_2CHCOOH$ (FW = 128.94), weighing 6.447 g is dissolved in 500 ml of solution. At this concentration the acid is about 45% dissociated:

$$Cl_2CHCOOH \rightleftharpoons H^+ + Cl_2CHCOO^-$$

Calculate the formality of the dichloroacetic acid and the molarities of the two species, $Cl_2CHCOOH$ and Cl_2CHCOO^-.

$$F = \frac{g}{\text{FW} \times V}$$

$$F = \frac{6.447 \text{ g}}{128.94 \text{ g/FW} \times 0.500 \text{ liter}}$$

$$F = 0.100$$

This is the *total* concentration of the species arising from dichloroacetic acid and is also referred to by analytical chemists as the *analytical concentration* of the acid, c_a. The *equilibrium concentrations* of dichloroacetic molecules and dichloroacetate ions are

$$[Cl_2CHCOO^-] = 0.100 \times 0.45 = 0.045\ M$$

$$[Cl_2CHCOOH] = 0.100 \times 0.55 = 0.055\ M$$

Such concentrations are expressed as molarities and are indicated by enclosing the molecule or ion in brackets. Hence

$$F = c_a = [Cl_2CHCOOH] + [Cl_2CHCOO^-]$$

$$F = c_a = 0.055 + 0.045 = 0.100$$

Normality

Like molarity and formality, this concentration system is based on the volume of solution. It is defined as follows:

$$\text{Normality} = \text{number of equivalents of solute per liter of solution}$$

or

$$N = \frac{\text{eq}}{V}$$

where N is the normality, eq the number of equivalents, and V the volume of solution in liters. Since

$$\text{eq} = \frac{g}{\text{EW}}$$

where g is the grams of solute and EW the equivalent weight, it follows that

$$N = \frac{g}{\text{EW} \times V}$$

This equation can be solved for grams of solute, giving

$$g = N \times V \times \text{EW}$$

The relation between normality and molarity is

$$N = nM$$

where n is the number of moles of hydrogen ion, electrons, or univalent cation furnished or combined with by the reacting substance (page 48).

The following examples illustrate the normality system of concentration.

Example 4. Arsenic can be determined by titration with iodine solution. The reaction involved is

$$HAsO_2 + I_2 + 2H_2O \longrightarrow H_3AsO_4 + 2H^+ + 2I^-$$

Pure As_2O_3 can be used for standardizing the iodine solution.

A sample of pure As_2O_3 weighing 4.0136 g is dissolved in 800 ml of solution. Calculate the normality of the solution assuming that it is used in the reaction above. Also calculate the molarity of the solution.

Note that each arsenic atom loses two electrons. Each As_2O_3 contains two arsenic atoms and hence loses four electrons:

$$As_2O_3 \longrightarrow 2HAsO_2 \longrightarrow 2H_3AsO_4 + 4e$$

The equivalent weight of As_2O_3 is one-fourth of the molecular weight. Hence

$$N = \frac{4.0136\text{ g}}{197.84/4\text{ g/eq} \times 0.800\text{ liter}}$$

$$N = 0.1014\text{ eq/liter}$$

$$N = 4 \times M$$

$$M = 0.02535\text{ mol/liter}$$

Example 5. Calculate the number of grams of pure Na_2CO_3 required to prepare 250 ml of a 0.150 N solution. The sodium carbonate is to be titrated with HCl according to the equation

$$CO_3^{2-} + 2H^+ \longrightarrow H_2CO_3$$

Each Na_2CO_3 reacts with $2H^+$; the equivalent weight is therefore one-half the molecular weight or 105.99/2 = 53.00 g/eq. Hence

$$g = 0.250 \text{ liter} \times 0.150 \text{ eq/liter} \times 53.00 \text{ g/eq}$$

$$g = 1.99$$

Example 6. Calculate the normality of a solution of nickel nitrate made by dissolving 2.00 g of pure nickel metal in nitric acid and diluting to 500 ml. The nickel is to be titrated with KCN, the following reaction occurring:

$$Ni^{2+} + 4CN^- \longrightarrow Ni(CN)_4^{2-}$$

Also calculate the molarity.

The equivalent weight of nickel is one-half the atomic weight since nickel is a divalent ion. Hence

$$N = \frac{2.00 \text{ g}}{58.70/2 \text{ g/eq} \times 0.500 \text{ liter}}$$

$$N = 0.136 \text{ eq/liter}$$

$$N = 2 \times M$$

$$M = 0.0680 \text{ mol/liter}$$

Weight Percent

This system specifies the number of grams of solute per 100 g of solution. Mathematically this is expressed as follows:

$$P = \frac{w}{w + w_0} \times 100$$

where P is the percent by weight solute, w the number of grams of solute, and w_0 the number of grams of solvent.

The following examples illustrate the weight percent system of concentration.

Example 7. A sample of NaOH weighing 5.0 g is dissolved in 45 g of water. (1 g of water is approximately 1 ml.) Calculate the weight percent NaOH in the solution.

$$P = \frac{5.0}{5.0 + 45} \times 100$$

$$P = 10\%$$

Example 8. Concentrated HCl (MW = 36.5) has a density of 1.19 g/ml and is 37% by weight HCl. How many milliliters of the concentrated acid should be taken and diluted to 1.00 liter with water to prepare a 0.100 M solution?

$$\text{g HCl needed} = 1.00 \text{ liter} \times 0.100 \text{ mol/liter} \times 36.5 \text{ g/mol}$$

$$\text{g HCl needed} = 3.65$$

$$\text{g HCl per ml} = 1.19 \text{ g/ml} \times 0.37 = 0.44$$

$$\frac{3.65 \text{ g}}{0.44 \text{ g/ml}} = 8.3 \text{ ml}$$

Parts per Million (ppm)

This system specifies the number of parts of one component in 1 million parts of a mixture. We can express this using weight units in a manner similar to weight percent (above):

$$\text{ppm} = \frac{w}{w + w_0} \times 10^6$$

where w is the number of grams of solute and w_0 the number of grams of solvent. Since w is usually very small compared to w_0 this becomes

$$\text{ppm} \cong \frac{w}{w_0} \times 10^6$$

One liter of water at room temperature weighs approximately 10^6 mg. Hence a convenient relationship to remember is that 1 mg of a solute in 1 liter of water is a concentration of about 1 ppm.[2]

The following example illustrates a calculation involving parts per million.

Example 9. If drinking water contains 1.5 parts per million of NaF, how many liters of water can be fluoridated with 1.0 lb (454 g) of NaF?

Let V = liters of water to be fluoridated. Since 1 ppm = 1 mg/liter of H_2O,

$$\frac{454 \times 10^3 \text{ mg NaF}}{V \text{ (liters)}} = 1.5$$

$$V = 3.0 \times 10^5 \text{ liters}$$

For even more dilute solutions the system *parts per billion* (ppb) is employed, where

$$\text{ppb} \cong \frac{w}{w_0} \times 10^9$$

Millequivalents and Millimoles

In titrimetric procedures the volume of titrant employed is usually less than 50 ml and the concentration is about 0.1 to 0.2 N. This means that the number of equivalents of titrant is of the order of

$$0.050 \text{ liter} \times 0.10 \text{ eq/liter} = 0.0050 \text{ eq}$$

[2] The term *milligram percent* is sometimes employed in clinical chemistry. It is the number of milligrams of solute per 100 ml of solution (i.e., 1 mg percent is 1 part per 100,000).

Because this number is so small, it is convenient to adopt a smaller unit than the equivalent. A *milliequivalent* (meq) is one-thousandth of an equivalent, or

$$1000 \text{ meq} = 1 \text{ eq}$$

and the calculation above becomes

$$50 \text{ ml} \times 0.10 \text{ meq/ml} = 5.0 \text{ meq}$$

A *millimole* (mmol) is similarly defined as one-thousandth of a mole.

In practice it is customary to use moles and equivalents when dealing with several liters of solution, and milliequivalents and millimoles when the volume is considerably less than 1 liter. Note that normality and molarity can be expressed in either the large or small units; the numerical value is the same. Thus a solution containing 0.00200 equivalent in 0.00500 liter also contains 2.00 meq in 5.00 ml, and normality is

$$N = \frac{0.00200 \text{ eq}}{0.00500 \text{ liter}} = \frac{2.00 \text{ meq}}{5.00 \text{ ml}} = 0.400 \text{ eq/liter or meq/ml}$$

It should also be noted that the equivalent weight can be expressed with either the large or small units. The equivalent weight of NaOH is 40.00 g/eq or 40.00 mg/meq. It is not necessary to use the term "milliequivalent weight." It should be evident that

$$N = \frac{g}{\text{EW} \times V(\text{liters})} = \frac{\text{mg}}{\text{EW} \times V\,(\text{ml})}$$

Titer

Still another method of expressing concentration that is frequently employed in analytical chemistry is the *titer.* We discuss it here because the units employed are usually milliliters and milliequivalents. The units of titer are weight per volume, but the weight is that of some reagent with which the solution reacts rather than the weight of the solute. For example, if 1.00 ml of a hydrochloric acid solution will exactly neutralize 4.00 mg of sodium hydroxide, the concentration of the acid solution can be expressed as a sodium hydroxide titer of 4.00 mg/ml.

Titer (T) can be easily converted to normality, as seen from the following relations:

$$T = \frac{\text{mg}}{\text{ml}} \qquad N = \frac{\text{mg}}{\text{ml} \times \text{EW}}$$

Thus

$$T = N \times \text{EW}$$

The equivalent weight employed in the transformation is that of the substance with which the solution reacts, not the solute. In the example above, if the titer of the hydrochloric acid solution is 4.000 mg/ml of sodium hydroxide, the normality is obtained upon dividing by 40.00 mg/meq, the equivalent weight of sodium hydroxide, giving a normality of 0.1000 meq/ml.

Example 1. What is (a) the NH_3 titer of a 0.120 N solution of HCl; (b) the BaO titer of the same solution?

(a)
$$T = 0.120\frac{\text{meq}}{\text{ml}} \times 17.0\frac{\text{mg}}{\text{meq}}(NH_3)$$

$$T = 2.04\frac{\text{mg}}{\text{ml}}(NH_3)$$

(b)
$$T = 0.120\frac{\text{meq}}{\text{ml}} \times \frac{153.3}{2}\frac{\text{mg}}{\text{meq}}(\text{BaO})$$

$$T = 9.2\frac{\text{mg}}{\text{ml}}(\text{BaO})$$

This means that 1 ml of the HCl solution will neutralize 2.04 mg of NH_3 or 9.2 mg of BaO.

Example 2. A solution of NaOH has an oxalic acid (MW 126.0) titer of 9.45 mg/ml. Calculate the normality of the NaOH solution. (Oxalic acid furnishes two hydrogen ions.)

$$N = \frac{9.45 \text{ mg/ml}}{63.0 \text{ mg/meq}}$$

$$N = 0.150 \text{ meq/ml}$$

Standardization of Solutions

It has been previously mentioned (page 44) that the process by which the concentration of a solution is accurately ascertained is known as *standardization*. A *standard* solution can sometimes be prepared by dissolving an accurately weighed sample of the desired solute in an accurately measured volume of solution. This method is not generally applicable, however, since relatively few chemical reagents can be obtained in sufficiently pure form to meet the analyst's demand for accuracy. Those few substances which are adequate in this regard are called *primary standards*. More commonly a solution is standardized by a titration in which it reacts with a weighed portion of a primary standard.

The reaction between the titrant and the substance selected as a primary standard should fulfill the requirements for titrimetric analysis (page 46). In addition, the primary standard should have the following characteristics:

1. It should be readily available in a pure form or in a state of known purity at a reasonable cost. In general, the total amount of impurities should not exceed 0.01 to 0.02%, and it should be possible to test for impurities by qualitative tests of known sensitivity.
2. The substance should be stable. It should be easy to dry and should not be so hygroscopic that it takes up water during weighing. It should not lose weight on exposure to air. Salt hydrates are not normally employed as primary standards.

3. It is desirable that the primary standard have a reasonably high equivalent weight in order to minimize the consequences of errors in weighing.

For acid-base titrations it is customary to prepare solutions of an acid and base of approximately the desired concentration and then to standardize one of the solutions against a primary standard. The solution thus standardized can be used as a *secondary standard* to obtain the normality of the other solution. For highly accurate work, it is preferable to standardize both the acid and base independently against primary standards. One of the most widely used primary standards for base solutions is the compound potassium hydrogen phthalate, $KHC_8H_4O_4$, abbreviated KHP. Sulfamic acid, HSO_3NH_2, and potassium hydrogen iodate, $KH(IO_3)_2$, are both strong acids and are excellent primary standards. Sodium carbonate, Na_2CO_3, and tris(hydroxymethyl)aminomethane, $(CH_2OH)_3CNH_2$, known as TRIS or THAM, are common primary standards for strong acids.

Many primary standards are available for redox reagents. Table 3.1 gives a summary of the primary standards commonly used in the laboratory for the reagents listed. The methods used for balancing redox equations are reviewed in Appendix II.

TABLE 3.1 Primary Standards for Redox Reagents

Solution to be standardized	Primary standard	Reaction
$KMnO_4$	As_2O_3	$5H_3AsO_3 + 2MnO_4^- + 6H^+ \longrightarrow 2Mn^{2+} + 5H_3AsO_4 + 3H_2O$
$KMnO_4$	$Na_2C_2O_4$	$5C_2O_4^{2-} + 2MnO_4^- + 16H^+ \longrightarrow 2Mn^{2+} + 10CO_2 + 8H_2O$
$KMnO_4$	Fe	$5Fe^{2+} + MnO_4^- + 8H^+ \longrightarrow 5Fe^{3+} + Mn^{2+} + 4H_2O$
$Ce(SO_4)_2$	Fe	$Fe^{2+} + Ce^{4+} \longrightarrow Fe^{3+} + Ce^{3+}$
$K_2Cr_2O_7$	Fe	$6Fe^{2+} + Cr_2O_7^{2-} + 14H^+ \longrightarrow 6Fe^{3+} + 2Cr^{3+} + 7H_2O$
$Na_2S_2O_3$	$K_2Cr_2O_7$	$Cr_2O_7^{2-} + 6I^- + 14H^+ \longrightarrow 2Cr^{3+} + 3I_2 + 7H_2O$ $I_2 + 2S_2O_3^{2-} \longrightarrow 2I^- + S_4O_6^{2-}$
$Na_2S_2O_3$	Cu	$2Cu^{2+} + 4I^- \longrightarrow 2CuI(s) + I_2$ $I_2 + 2S_2O_3^{2-} \longrightarrow 2I^- + S_4O_6^{2-}$
I_2	As_2O_3	$HAsO_2 + I_2 + 2H_2O \longrightarrow H_3AsO_4 + 2I^- + 2H^+$

For precipitation and complex formation titrations pure salts are usually employed as primary standards. Sodium or potassium chloride can be used to standardize a solution of silver nitrate, the reaction being

$$Ag^+ + Cl^- \longrightarrow AgCl(s)$$

Calcium carbonate, $CaCO_3$, is used as a primary standard for solutions of the complexing agent ethylenediaminetetraacetic acid (EDTA). The reaction is

$$Ca^{2+} + Y^{4-} \longrightarrow CaY^{2-}$$

where Y^{4-} stands for the anion of EDTA.

The following examples illustrate the calculations involved in standardizing a solution. It should be recalled that at the equivalence point of a titration the number of equivalents, or milliequivalents, of analyte equals the number of equivalents, or milliequivalents, of titrant.

Example 1. A sample of pure sodium carbonate, Na_2CO_3, weighing 0.3542 g is dissolved in water and titrated with a solution of hydrochloric acid. A volume of 30.23 ml is required to reach the methyl orange end point, the reaction being

$$Na_2CO_3 + 2HCl \longrightarrow 2NaCl + H_2O + CO_2$$

Calculate the normality of the acid.

We know that

$$\text{meq HCl} = \text{meq } Na_2CO_3$$

The equivalent weight of Na_2CO_3 is one-half the molecular weight, or 106.0/2 = 53.00 mg/meq. Hence

$$V_{HCl} \times N_{HCl} = \frac{\text{mg } Na_2CO_3}{\text{EW } Na_2CO_3}$$

$$30.23 \times N_{HCl} = \frac{354.2}{53.00}$$

$$N_{HCl} = 0.2211 \text{ meq/ml}$$

Frequently, the analyst "overruns" the end point—i.e., adds too much titrant—and then "back-titrates" with a second solution. He must know the normality of this second solution, or the volume relationship between it and the titrant. The following examples illustrate the calculations involved.

Example 2. A sample of pure sodium oxalate, $Na_2C_2O_4$, weighing 0.2856 g is dissolved in water, sulfuric acid is added, and the solution is titrated at 70°C, requiring 45.12 ml of a $KMnO_4$ solution. The end point is overrun and back-titration is carried out with 1.74 ml of a 0.1032 *N* solution of oxalic acid. Calculate the normality of the $KMnO_4$ solution.

The reaction, written ionically, is

$$5C_2O_4^{2-} + 2MnO_4 + 16H^+ \longrightarrow 2Mn^{2+} + 10CO_2 + 8H_2O$$

We know that

$$\text{meq permanganate} = \text{meq oxalate}$$

or

$$\text{meq } KMnO_4 = \text{meq } Na_2C_2O_4 + \text{meq } H_2C_2O_4$$

$$V_{KMnO_4} \times N_{KMnO_4} = \frac{\text{mg } Na_2C_2O_4}{\text{EW } Na_2C_2O_4} + V_{H_2C_2O_4} \times N_{H_2C_2O_4}$$

Since the oxalate ion loses two electrons in the reaction above, the equivalent weight of $Na_2C_2O_4$ is one-half its molecular weight, or 134.0/2 = 67.00. Hence

$$45.12 \times N_{KMnO_4} = \frac{285.6}{67.00} + 1.74 \times 0.1032$$

$$N_{KMnO_4} = 0.0985 \text{ meq/ml}$$

Sometimes it is convenient or necessary to add an excess of titrant and to back-titrate the excess with a solution of known concentration. For example, in the Volhard method (Chapter 9) for chloride, excess silver nitrate is added to precipitate chloride:

$$Ag^+ + Cl^- \longrightarrow AgCl(s)$$

The excess silver is titrated with standard potassium thiocyanate solution:

$$Ag^+ + SCN^- \longrightarrow AgSCN(s)$$

Iron(III) ion serves as the indicator. The following example illustrates the standardization of a silver nitrate solution against pure sodium chloride using the Volhard method.

Example 3. A sample of pure sodium chloride (MW = EW = 58.44) weighing 0.2286 g is dissolved in water and exactly 50.00 ml of silver nitrate solution is added to precipitate AgCl. The excess Ag^+ is titrated with 12.56 ml of a 0.0986 *N* solution of KSCN. Calculate the normality of the $AgNO_3$ solution.

We know that

$$\text{meq } AgNO_3 = \text{meq NaCl} + \text{meq KSCN}$$

$$50.00 \times N_{AgNO_3} = \frac{228.6}{58.44} + 12.56 \times 0.0986$$

$$N_{AgNO_3} = 0.1030 \text{ meq/ml}$$

An example of back-titration in the determination of nitrogen by the Kjeldahl method is given on page 61.

Sometimes the analyst weighs a large sample of the primary standard (or an unknown), dissolves it in a volumetric flask, and withdraws a portion of the solution using a pipet. The portion withdrawn with the pipet is called an *aliquot.* An aliquot is a portion of the whole, usually some simple fraction. This process of dilution to a known volume and removing a portion for titration is called *taking an aliquot.* The following example illustrates this procedure.

Example 4. A sample of pure $CaCO_3$ (MW 100.09) weighing 0.4148 g is dissolved in 1 : 1 hydrochloric acid and the solution is diluted to 500 ml in a volumetric flask. A 50.0-ml aliquot is withdrawn with a pipet and placed in an Erlenmeyer flask. The solution is titrated with 40.34 ml of an EDTA (page 58) solution using Eriochrome Black T indicator. Calculate the molarity of the EDTA solution.

The reaction in the titration is (page 58)

$$Ca^{2+} + Y^{4-} \longrightarrow CaY^{2-}$$

where Y^{4-} stands for the anion of EDTA. At the equivalence point

$$\text{mmol EDTA} = \text{mmol CaCO}_3$$

$$V \times M_{\text{EDTA}} = \frac{\text{mg CaCO}_3}{\text{MW CaCO}_3}$$

The weight of $CaCO_3$ in the aliquot is one-tenth of 0.4148, or 0.04148 g, since 50.0 ml was taken from a volume of 500 ml. Hence

$$40.34 \times M_{\text{EDTA}} = \frac{41.48 \text{ mg}}{100.09 \text{ mg/mmol}}$$

$$M_{\text{EDTA}} = 0.01027 \text{ mmol/ml}$$

Complex-formation titrations involving EDTA are usually 1:1 reactions. In such examples the use of millimoles and molarity rather than milliequivalents and normality is recommended.

Calculation of Percent Purity

To analyze a sample of unknown purity, the analyst weighs accurately a portion of the sample, dissolves it appropriately, and titrates it with a standard solution. He then knows that

$$\text{meq titrant} = \text{meq analyte}$$

If V and N represent the volume (ml) and normality, respectively, of the titrant,

$$V \times N = \text{meq titrant} = \text{meq analyte}$$

To express the results as a percentage, the milliequivalents of analyte are converted to weight and divided by the weight of the sample:

$$\% = \frac{\text{mg of analyte}}{\text{mg of sample}} \times 100$$

$$\% = \frac{V \text{ (ml)} \times N \text{ (meq/ml)} \times \text{EW (mg/meq)}}{\text{weight of sample (mg)}} \times 100$$

Note the cancellation of units to give the percentage, which is dimensionless.

The following examples illustrate the calculation of percent purity.

Example 1. A sample of impure potassium acid phthalate (KHP, page 57) weighing 2.1283 g required 42.58 ml of a 0.1084 N base solution for titration to the phenolphthalein end point. Calculate the percentage of KHP (EW 204.2) in the sample.

$$\% \text{KHP} = \frac{42.58 \text{ ml} \times 0.1084 \text{ meq/ml} \times 204.2 \text{ mg/meq}}{2128.3 \text{ mg}} \times 100$$

$$\% \text{ KHP} = 44.29$$

Example 2. A sample of iron ore weighing 0.6428 g is dissolved in acid. The iron is reduced to Fe^{2+} and titrated with 36.30 ml of a 0.1052 N solution of $K_2Cr_2O_7$. (See

Table 3.1 for this reaction.) (a) Calculate the percentage of iron (Fe) in the sample. (b) Express the percentage as Fe_2O_3 rather than Fe.

(a) The equivalent weight of iron is the atomic weight, 55.847 mg/meq, since each iron loses one electron in the reaction. Hence

$$\% \text{ Fe} = \frac{36.30 \text{ ml} \times 0.1052 \text{ meq/ml} \times 55.847 \text{ mg/meq}}{642.8 \text{ mg}} \times 100$$

$$\% \text{ Fe} = 33.18$$

(b) To express the percentage as Fe_2O_3 the equivalent weight of this oxide rather than that of Fe should be used in the expression above. Since each iron atom loses one electron and each Fe_2O_3 contains two iron atoms, the equivalent weight of Fe_2O_3 is the molecular weight divided by two: $159.69/2 = 79.85$ mg/meq. Then

$$\% \text{ Fe}_2\text{O}_3 = \frac{36.30 \text{ ml} \times 0.1052 \text{ meq/ml} \times 79.85 \text{ mg/meq}}{642.8 \text{ mg}} \times 100$$

$$\% \text{ Fe}_2\text{O}_3 = 47.44$$

Example 3. In the Kjeldahl method for nitrogen, the element is converted into NH_3, which is then distilled into a known volume of standard acid. There is more than enough acid to neutralize the NH_3, and the excess is titrated with standard base.

The ammonia from a 1.325-g sample of fertilizer is distilled into 50.00 ml of 0.2030 *N* H_2SO_4, and 25.32 ml of 0.1980 *N* NaOH is required for back-titration. Calculate the percentage of nitrogen (N) in the sample.

The equivalent weight of N is 14.007 mg/meq, since $1\text{N} = 1\text{NH}_3 = 1\text{OH}^-$. Hence

$$\text{meq NH}_3 + \text{meq NaOH} = \text{meq H}_2\text{SO}_4$$

$$\text{meq NH}_3 + 25.32 \times 0.1980 = 50.00 \times 0.2030$$

$$\text{meq NH}_3 = \text{meq N} = 50.00 \times 0.2030 - 25.32 \times 0.1980$$

$$\text{meq NH}_3 = 5.137$$

$$\% \text{ N} = \frac{5.137 \times 14.007 \text{ mg/meq}}{1325 \text{ mg}} \times 100$$

$$\% \text{ N} = 5.43$$

Example 4. The ore pyrolusite can be analyzed for MnO_2 (and other higher metal oxides) by adding an excess of pure sodium oxalate ($Na_2C_2O_4$) and heating with H_2SO_4. The following reaction occurs:

$$\text{MnO}_2 + \text{C}_2\text{O}_4^{2-} + 4\text{H}^+ \longrightarrow \text{Mn}^{2+} + 2\text{CO}_2 + 2\text{H}_2\text{O}$$

The excess oxalate is titrated with standard $KMnO_4$ in acid solution. The reaction is that given in Table 3.1.

Given the following data: weight of ore = 1.000 g; weight of $Na_2C_2O_4$ = 0.4020 g; volume of 0.1000 N $KMnO_4$ = 20.00 ml. Calculate the percentage oxygen in the sample. (Since oxides besides MnO_2 can contribute to the oxidation, it is convenient to report the percentage as oxygen. The equivalent weight of oxygen is 8.000 mg/meq since an oxygen atom gains two electrons.)

$$\text{meq oxygen} + \text{meq } KMnO_4 = \text{meq } Na_2C_2O_4$$

$$\text{meq oxygen} + 20.00 \times 0.1000 = \frac{402.0}{67.00}$$

$$\text{meq oxygen} = 4.000$$

$$\% \text{ oxygen} = \frac{4.000 \text{ meq} \times 8.000 \text{ mg/meq}}{1000 \text{ mg}} \times 100$$

$$\% \text{ oxygen} = 3.200$$

Factor weight solutions. It is possible to adjust the normality of a standard solution and the weight of sample taken for analysis so that the number of milliliters used in a titration equals the percentage of the analyte (or a factor thereof). This is of particular advantage in laboratories where many determinations of the same constituent are made. The following are examples of the calculations.

Example 5. What weight of sample should be taken for analysis so that the volume of 0.1074 N NaOH used for titration equals the percentage of potassium acid phthalate (KHP, page 57) in the sample. The equivalent weight of KHP is 204.2 mg/meq.

$$\% \text{ KHP} = \frac{\text{ml NaOH} \times 0.1074 \text{ meq/ml} \times 204.2 \text{ mg/meq}}{\text{mg sample}} \times 100$$

Since % KHP = ml NaOH, these terms cancel and

$$\text{mg sample} = 0.1074 \times 204.2 \times 100$$

$$\text{mg sample} = 2193$$

$$\text{g sample} = 2.193$$

Example 6. Samples containing arsenic weighing 0.8000 g are titrated with standard iodine solution. (See Table 3.1 and Example 4, page 52, for reactions.) What should be the normality of the iodine so that each milliliter of titrant represents $\frac{1}{2}$% As_2O_3 in the sample?

The EW of As_2O_3 is 197.84/4 = 49.46 mg/meq.

$$\% \text{ } As_2O_3 = \frac{\text{ml } I_2 \times N \times 49.46}{800} \times 100$$

$$0.500 = \frac{1.00 \times N \times 49.46}{800} \times 100$$

$$N = 0.0809 \text{ meq/ml}$$

Coulometric analysis. A substance can be determined by electrolysis; the procedure can be either gravimetric or titrimetric. In the titrimetric technique, called *coulometry*, the amount of analyte in a solution is determined by measuring the quantity of electricity required to react with it completely. The quantity of electricity which brings about 1 eq of chemical change at an electrode is called a *faraday*. One equivalent of chemical change corresponds to the loss or gain of 1 mol of electrons. In electrical units, the faraday is equal to 96,490 coulombs of charge:

$$6.0229 \times 10^{23} \text{ electrons/faraday} \times 1.60206 \times 10^{-19} \text{ coul/electron}$$

$$= 96{,}490 \text{ coul/faraday}$$

For most purposes, this number is rounded to 96,500. It should also be recalled that the unit of current, the ampere (A), is defined as 1 coulomb per second. By measuring the current and time required for a reaction to go to completion, the chemist can calculate the number of coulombs and hence the number of equivalents of analyte in a solution.

The following example illustrates the calculations involved in a coulometric analysis.

Example 7. A sample of copper ore weighing 2.132 g is dissolved in acid and the copper electrolyzed:

$$Cu^{2+} + 2e \longrightarrow Cu$$

If 8.04 min is required for the electrolysis using a constant current of 2.00 A, calculate the percentage of copper in the ore.

The meq of copper is given by

$$\frac{8.04 \text{ min} \times 60 \text{ sec/min} \times 2.00 \text{ coul/sec}}{96.5 \text{ coul/meq}} = 10.0$$

Since 1 mmol Cu = 2 meq, the sample must contain 5.00 mmol Cu. Hence the percentage is

$$\% \text{ Cu} = \frac{5.00 \text{ mmol} \times 63.55 \text{ mg/mmol}}{2132 \text{ mg}} \times 100$$

$$\% \text{ Cu} = 14.9$$

QUESTIONS

1. *Terms.* Explain clearly the meaning of the following terms: (a) formula weight; (b) molecular weight; (c) equivalent weight; (d) mole; (e) equivalent of an acid-base reaction; (f) equivalent of a redox reagent; (g) equivalent of a precipitation or complex-forming reagent.

2. *Terms.* Explain clearly the meaning of the following terms: (a) titration; (b) standard solution; (c) titrant; (d) equivalence point; (e) end point; (f) standardization; (g) primary standard; (h) analyte; (i) aliquot; (j) milliequivalent.

3. *Types of reaction.* What four types of chemical reactions can be used as the basis of titrimetric analyses? What requirements must be satisfied by these reactions before they can be used successfully?

4. *Concentration systems.* Define clearly the following: (a) molarity; (b) formality; (c) percent by weight; (d) parts per million.

5. *Errors.* Explain the effect that the following errors would have on the normality of a NaOH solution being standardized against pure KHP. Would the error cause the normality to be high or low, or would it have no effect? (a) The buret containing NaOH is read too quickly, not allowing time for drainage. (b) The initial reading of the NaOH buret is recorded as 1.90 when it is actually 2.10. (c) The weight of KHP is recorded as 0.6234 g when it is actually 0.6324 g. (d) The sample is dissolved in 100 ml of water although the directions called for only 50 ml.

6. Repeat Question 5 for the analysis of an impure sample of KHP by titration with standard base. Explain the effect of the errors on the percentage of KHP in the sample.

Multiple-choice: In the following multiple-choice questions, select the *one best* answer.

7. How many equivalents of $KMnO_4$ are there in 1 mol of $KMnO_4$? (a) $\frac{1}{5}$; (b) 5; (c) 1; (d) depends upon the chemical reaction.

8. The MW of a substance is 128. Its EW in a particular reaction is 32. A 1.0 M solution of this substance is (a) 2.0 N, (b) 0.50 N, (c) 4.0 N, (d) 0.25 N.

9. Suppose that substance A, molecular weight MW, reacts with permanganate ion as follows:

$$5A + 2MnO_4^- + \cdots \longrightarrow 2Mn^{2+} + \cdots$$

The equivalent weight of A in this reaction is (a) $\frac{MW}{2}$; (b) $2 \times MW$; (c) $\frac{2 \times MW}{5}$; (d) $\frac{5 \times MW}{2}$.

10. The KHP (MW 204.2) titer of a NaOH solution is T mg/ml. If a sample containing KHP is titrated with NaOH, the volume of titrant equals the percentage KHP in the sample if the weight in milligrams is (a) T; (b) $10T$; (c) $100T$; (d) $204.2T$.

PROBLEMS

1. *Equivalent weights.* Given the following unbalanced equations:
   (i) $NaOH + H_3PO_4 \longrightarrow Na_2HPO_4 + H_2O$
   (ii) $Cr_2O_7^{2-} + Fe^{2+} + H^+ \longrightarrow Cr^{3+} + Fe^{3+} + H_2O$
   (iii) $Ag^+ + Br^- \longrightarrow AgBr(s)$
   (iv) $Hg^{2+} + Cl^- \longrightarrow HgCl_2$
   Balance each equation and calculate the EW's of the following substances when used in these reactions: (a) NaOH; (b) H_3PO_4; (c) $K_2Cr_2O_7$; (d) $FeSO_4$; (e) $AgNO_3$; (f) KBr; (g) $Hg(NO_3)_2$; (h) KCl.

2. *Equivalent weights.* Given the following unbalanced equations:
   (i) $Ca(OH)_2 + HCl \longrightarrow CaCl_2 + H_2O$
   (ii) $MnO_4^- + CN^- + H_2O \longrightarrow MnO_2(s) + CNO^- + OH^-$
   (iii) $BaCl_2 + H_2SO_4 \longrightarrow BaSO_4(s) + HCl$
   (iv) $Ag^+ + CN^- \longrightarrow Ag(CN)_2^-$

Balance each equation and calculate the EW's of the following substances when used in these reactions: (a) $Ca(OH)_2$; (b) HCl; (c) $KMnO_4$; (d) KCN in reaction (ii); (e) $BaCl_2$; (g) H_2SO_4; (g) $AgNO_3$; (h) KCN in reaction (iv).

3. *Molarity.* Calculate the molarity of each of the following solutions: (a) 4.00 g of NaOH in 0.500 liter of solution; (b) 5.30 mg of Na_2CO_3 in 5.00 ml of solution; (c) 8.20 g of CaO in 2.20 liters of solution.

4. *Molarity.* Calculate the molarity of each of the following solutions: (a) 2.00 g of NH_3 in 500 ml of solution; (b) 4.00 g of HCl in 100 ml of solution; (c) 40.0 mg of H_2SO_4 in 2.00 ml of solution.

5. *Normality.* Calculate the normality of each solution in Problem 3, assuming that $CaO = 2H^+$ and $Na_2CO_3 = 2H^+$.

6. *Normality.* Calculate the normality of each solution in Problem 4, assuming that $H_2SO_4 = 2H^+$.

7. *Molarity.* Calculate the molarity of each of the following aqueous solutions: (a) 3.20 g of $KMnO_4$ in 0.800 liter of solution; (b) 40.0 mg of $K_2Cr_2O_7$ in 10.0 ml of solution; (c) 25.0 g of $Na_2S_2O_3 \cdot 5H_2O$ in 1.00 liter of solution.

8. *Molarity.* Calculate the molarity of each of the following aqueous solutions: (a) 7.50 g of $H_2C_2O_4 \cdot 2H_2O$ in 100 ml of solution; (b) 10.0 g of I_2 in 0.600 liter of solution; (c) 2.52 g of As_2O_3 in 500 ml of solution.

9. *Normality.* Calculate the normality of each solution in Problem 7, assuming that the reagents undergo the reactions listed in Table 3.1.

10. *Normality.* Calculate the normality of each solution in Problem 8, assuming that the reagents undergo the reactions listed in Table 3.1. (The reaction of $H_2C_2O_4$ is the same as that of $Na_2C_2O_4$.)

11. *Molarity and normality.* Calculate the molarities and normalities of the aqueous solutions listed below. The reactions are the ones given in Problems 1 and 2. (a) 3.40 g of $Hg(NO_3)_2$ in 100 ml of solution; (b) 150 mg of $AgNO_3$ in 10.0 ml of solution; (c) 25.0 g of KCN in 0.500 liter of solution.

12. *Formality.* A solution is made by dissolving 14.82 g of trichloroacetic acid, Cl_3CCOOH (FW 163.39), in water and diluting to a volume of 750.0 ml. At this concentration the acid is about 70% dissociated. Calculate (a) the formality of the trichloroacetic acid; (b) the molarities of the species Cl_3CCOOH and Cl_3CCOO^-.

13. *Titer.* Calculate the following titers in mg/ml: (a) the CaO titer of a 0.120 *N* HCl solution; (b) the HNO_3 titer of a 0.210 *N* solution of NaOH; (c) the $Na_2C_2O_4$ titer of a 0.150 *N* solution of $KMnO_4$; (d) the FeO titer of a 0.0900 *N* solution of $K_2Cr_2O_7$. See Table 3.1 for the reactions in parts (c) and (d).

14. *Titer.* Calculate the following titers in mg/ml: (a) the As_2O_3 titer of a 0.100 *N* solution of I_2; (b) the Fe_2O_3 titer of a 0.0500 *N* solution of $KMnO_4$. See Table 3.1 for the reactions. (c) An HCl solution has a NaOH titer of 4.80 mg/ml. Calculate the normality of the HCl. (d) Calculate the KOH titer of the HCl solution in part (c).

15. *Density and molarity.* (a) Calculate the molarity of concentrated HCl given that the density is 1.18 g/ml and the percent by weight HCl is 36.0. (b) Calculate the molarity of a solution of H_2SO_4 given that the density is 1.30 g/ml and the solution contains 32.6% SO_3 by weight.

16. *Density and molarity.* (a) Calculate the molarity of concentrated aqueous ammonia given that the density is 0.954 g/ml and the percent by weight NH_3 is 11.6. (b) Calculate the molarity of a solution of perchloric acid given that the density is 1.242 g/ml and that the solution contains 34.0% $HClO_4$ by weight.

17. *Parts per million.* (a) A certain solution contains 0.00300 g of $CaCO_3$ in 200 ml of water. Calculate the number of parts per million of $CaCO_3$ in the water. (b) A certain sample of water contains 20.0 ppm of $MgCO_3$. Calculate the molarity of $MgCO_3$ in the solution.

18. *Parts per million.* (a) How many pounds of NaF would be required to fluoridate 6.0 million liters of water to a level of 1.4 ppm? (1 lb = 454 g.) (b) If a certain water contains 2.1 parts per billion of phosphate (PO_4), what is the molarity of the solution?

19. *Dilution of solutions.* (a) 175 ml of a 0.180 *M* solution is diluted to 400 ml with water. What is the molarity of the final solution? (b) What volume of water should be added to 300 ml of a 0.600 *M* solution to make the molarity 0.200? Assume that the volumes are additive.

20. *Dilution of solutions.* A 10.0-ml sample of a 0.250 *M* solution is placed in a volumetric flask and diluted to exactly 1.00 liter. A 1.00-ml sample of this solution is withdrawn from the volumetric flask and diluted to exactly 250 ml in another volumetric flask. Calculate the molarity of the final solution.

21. *Mixing solutions.* A 2.00-ml sample of a 0.100 *M* solution of $KMnO_4$ is mixed with 3.00 ml of a 0.0800 *M* solution of $KMnO_4$. Calculate the molarity of the final solution, assuming that the volumes are additive.

22. *Mixing solutions.* 40.0 ml of a 0.150 *M* HCl solution is mixed with 60.0 ml of 0.200 *M* NaOH. Is the resulting solution acidic, basic, or neutral? Calculate the molarity of the reactant which is in excess.

23. *Standardization.* A hydrochloric acid solution is standardized using pure Na_2CO_3 as a primary standard. The reaction is

$$Na_2CO_3 + 2HCl \longrightarrow 2NaCl + H_2O + CO_2$$

A sample of Na_2CO_3 weighing 0.2568 g required 38.72 ml of HCl for titration. Calculate the normality of the HCl solution.

24. *Standardization.* A sodium hydroxide solution is standardized using potassium hydrogen phthalate (KHP, page 57) as a primary standard. A sample of KHP weighing 0.8426 g required 42.14 ml of NaOH for titration. Calculate the normality of the NaOH solution.

25. *Standardization.* A potassium permanganate solution is standardized against pure sodium oxalate as a primary standard. (See Table 3.1 for the reaction.) A sample of $Na_2C_2O_4$ weighing 0.3148 g required 44.86 ml of $KMnO_4$ for titration. (a) Calculate the normality of the $KMnO_4$ solution. (b) What is the molarity of the solution?

26. *Standardization.* It is found that 16.34 ml of a solution containing Ni^{2+} is required to titrate a sample of pure KCN weighing 0.5024 g. The reaction is

$$Ni^{2+} + 4CN^- \longrightarrow Ni(CN)_4{}^{2-}$$

Calculate (a) the molarity, and (b) the normality of the nickel solution.

27. *Volume relations.* It is known that 35.24 ml of an acid solution is needed to titrate 38.46 ml of a base solution. (a) Calculate the volume relationship (or volume ratio) of the two solutions (i.e., 1.000 ml of the acid neutralizes how many milliliters of base)? (b) If the normality of the acid is 0.1186, what is the normality of the base?

28. *Standardization with back-titration.* From the following data calculate the normalities of the acid and base solutions: weight of pure Na_2CO_3 = 0.2448 g; volume of acid = 43.65 ml; volume of base used in back-titration = 0.84 ml; 1.000 ml of base = 0.982 ml of acid. Reaction: $CO_3^{2-} + 2H^+ \longrightarrow H_2CO_3$.

29. *Titrimetric analysis.* A 0.6234-g sample containing chloride ion is titrated with 34.68 ml of 0.1156 *N* $AgNO_3$ solution. (a) Calculate the percentage of Cl^- in the sample. (b) Suppose that the chloride is present entirely as NaCl. Calculate the percentage of NaCl in the sample.

30. *Titrimetric analysis with back-titration.* A 1.984-g sample of impure oxalic acid ($H_2C_2O_4$) required 39.26 ml of 0.0988 *N* base for titration; 1.24 ml of an acid solution was used in back titration. Calculate the percentage of $H_2C_2O_4$ in the sample. $H_2C_2O_4 = 2H^+$; 1.000 ml of base = 1.122 ml of acid.

31. *Titrimetric analysis—aliquot portions.* A sample containing arsenic and weighing 3.523 g is dissolved and diluted to exactly 250.0 ml in a volumetric flask. A 50-ml aliquot (page 59) is withdrawn and titrated with 33.84 ml of 0.1066 *N* I_2 solution. (Reaction in Table 3.1.) Calculate (a) the percentage of As in the sample; (b) the percentage of As_2O_3 in the sample.

32. *Titrimetric analysis.* A 1.000-g sample of iron ore required 32.46 ml of 0.1043 *N* oxidizing agent for titration. The iron is oxidized from the +2 to the +3 oxidation state. Calculate the percentage purity as (a) Fe; (b) FeO; (c) Fe_2O_3; (d) Fe_3O_4.

33. *Analysis of pyrolusite.* Pyrolusite ores contain MnO_2 and can be analyzed for their oxidizing capacity by heating the ore with excess pure sodium oxalate, $Na_2C_2O_4$, in acid solution. The following reaction occurs:

$$MnO_2(s) + C_2O_4^{2-} + 4H^+ \longrightarrow Mn^{2+} + 2CO_2(g) + 2H_2O$$

The excess oxalate is titrated with standard $KMnO_4$, the oxalate being oxidized to CO_2 and the permanganate reduced to Mn^{2+}. Given the following data: weight of ore = 0.9432 g; weight of pure $Na_2C_2O_4$ = 0.3948 g; volume of 0.1010 *N* $KMnO_4$ = 19.74 ml, (a) calculate the percentage of MnO_2 in the sample. (b) Since higher oxides of other metals, such as PbO_2, may be present and also oxidize the oxalate, the results are sometimes expressed as percentage of oxygen, where the EW of oxygen is 8.000 g/eq. Express the results given above as percentage oxygen.

34. *Kjeldahl analysis.* In the Kjeldahl determination of nitrogen the element is converted into NH_3, which is then distilled into a measured excess of standard acid. The excess acid is titrated with standard base. The ammonia from a 1.235-g sample of fertilizer is distilled into 50.00 ml of 0.1988 *N* H_2SO_4, and 24.62 ml of 0.1944 *N* NaOH is required for back-titration. Calculate the percentage of nitrogen (N) in the sample.

35. *Analysis of iron ore.* In the titrimetric analysis of iron in an ore the sample is dissolved in acid and all the iron reduced to Fe^{2+}. The Fe^{2+} is then oxidized to Fe^{3+} by titration with a standard solution of an oxidizing agent. A certain iron ore contains 24.4%

Fe_2O_3 and 14.6% Fe_3O_4. What volume (ml) of a 0.124 N oxidizing agent is required to titrate a 0.615-g sample of the ore?

36. *Determination of sulfur in steel.* The sulfur in steel can be determined by conversion of the element to H_2S, absorption of the H_2S in water, and titration with standard I_2 solution. The H_2S is oxidized to S and I_2 is reduced to I^-. The sulfur in a 1.000-g sample of steel is converted to H_2S and titrated with 19.86 ml of 0.0502 N I_2. Calculate the percentage of sulfur in the steel.

37. *Analysis of vinegar.* Ordinary vinegar contains acetic acid, $HC_2H_3O_2$, and this acid can be determined by titration with standard base. A 20.00-ml sample of vinegar having a density of 1.055 g/ml requires 39.26 ml of 0.3108 N base for titration. Calculate the percentage of acetic acid in the sample.

38. *Analysis of an oxalate mixture.* A 0.2586-g sample consists of only $Na_2C_2O_4$ and KHC_2O_4. It requires 48.39 ml of a 0.0830 N $KMnO_4$ solution for titration in acid solution. The oxalate is oxidized to CO_2 and the MnO_4^- is reduced to Mn^{2+}. How many ml of a 0.1622 N solution of NaOH will be required to titrate another 0.2856-g sample of the same mixture in an acid-base titration?

39. *Determination of calcium in blood.* Calcium can be determined in blood by precipitating CaC_2O_4, dissolving the precipitate in H_2SO_4, and titrating the oxalate with standard permanganate. The oxalate is oxidized to CO_2 and the $KMnO_4$ reduced to Mn^{2+}. A 10.0-ml blood sample from a patient is diluted to 50.0 ml in a volumetric flask. A 20.0-ml aliquot from the flask is treated with excess oxalate to precipitate CaC_2O_4. The precipitate is redissolved and titrated with 1.32 ml of 0.00410 N $KMnO_4$. Calculate the number of milligrams of Ca^{2+} per 10.0 ml of blood.

40. *Titrimetric determination of silver.* The silver in a 1.000-g sample is determined by first precipitating the silver as Ag_2CrO_4. The precipitate is redissolved in acid and treated with excess KI. The chromate is reduced to Cr^{3+} and the iodide is oxidized to I_2. The I_2 requires 31.82 ml of 0.1016 N sodium thiosulfate, $Na_2S_2O_3$, for titration. The thiosulfate is oxidized to $Na_2S_4O_6$ and the I_2 is reduced to I^-. Calculate the percentage of silver in the sample.

41. *Titrimetric analysis.* A 2.000-g sample of a substance having an equivalent weight of 50.0 is titrated with a standard solution of titrant. It is found that the percentage purity is exactly 60.0 times the normality of the titrant. What volume (ml) of titrant was used?

42. *Factor weight solution.* What should be the normality of a solution of an oxidizing agent so that the milliliters of titrant multiplied by 2 gives the percentage of Fe_2O_3 in a 0.6000-g sample of iron ore? All the iron is first reduced to Fe^{2+} and oxidized to Fe^{3+} by the titrant.

43. *Factor weight solution.* How many milligrams of sample should be taken for analysis so that the milliliters of 0.1200 N HCl titrant divided by 2 equals the percentage of NaOH in the sample?

44. *Sample size.* A student wishes to weigh a sufficiently large sample so that he will use about 40 ml of 0.10 N reagent for titration. How many grams should he take if the sample (a) is pure potassium hydrogen phthalate; (b) contains 25% KHP; (c) contains 30% FeO (the iron is oxidized from +2 to +3); (d) contains 15% As_2O_3 (the arsenic is oxidized from +3 to +5)?

CHAPTER FOUR

gravimetric methods of analysis

It has been previously mentioned that *gravimetric* analysis is one of the major divisions of analytical chemistry. The measurement step in a gravimetric method is one of weight. The analyte is physically separated from all other components of the sample as well as from the solvent. Precipitation is a widely used technique for separating the analyte from interferences; electrolysis, solvent extraction, chromatography, and volatilization are other important methods of separation.

In this chapter we shall discuss the general principles involved in gravimetric analysis, including stoichiometric calculations. We shall also examine the topic of formation and properties of precipitates, as this relates to the use of precipitation in gravimetric analysis. Other methods of separation will be discussed in later chapters.

GENERAL PRINCIPLES

A gravimetric method of analysis is usually based on a chemical reaction such as

$$aA + rR \longrightarrow A_aR_r$$

where a molecules of the analyte, A, react with r molecules of the reagent, R. The product, A_aR_r, is usually a slightly soluble substance which can be weighed as such after drying, or which can be ignited to another compound of known composition and then weighed. For example, calcium can be determined gravimetrically by

45. *Titrant concentration.* (a) If a 1.0 N base solution is employed as a titrant, how many grams of sample containing 25% KHP should be taken so that the volume of titrant used will be about 40 ml? (b) Repeat the calculation for 0.010 N base. (c) Suggest why titrants are usually about 0.1 to 0.2 N.

46. *Coulometric analysis.* A 1.00-g sample of an alloy containing copper is dissolved and the copper deposited by electrolysis. It requires 12.4 min using a constant current of 4.00 A to complete the electrolysis. Calculate the percentage of copper in the sample.

47. *Coulometric analysis.* How many grams of a sample containing 25.0% silver should be taken for coulometric analysis so that the electrolysis will be complete in 10.0 min using a constant current of 3.00 A?

precipitation of calcium oxalate and igniting the oxalate to calcium oxide:

$$Ca^{2+} + C_2O_4^{2-} \longrightarrow CaC_2O_4(s)$$

$$CaC_2O_4(s) \longrightarrow CaO(s) + CO_2(g) + CO(g)$$

An excess of reagent R is normally added to repress the solubility of the precipitate.

The following requirements should be met in order that a gravimetric method be successful:

1. The separation process should be sufficiently complete so that the quantity of analyte left unprecipitated is analytically undetectable (usually 0.1 mg or less in determining a major constituent of a macro sample).
2. The substance weighed should have a definite composition and should be pure, or very nearly so. Otherwise, erroneous results may be obtained.

The second requirement is the more difficult for the analyst to meet. Errors due to such factors as solubility of the precipitate can generally be minimized and seldom cause significant errors. It is the problem of obtaining pure and filterable precipitates which is of major importance. Much research has been done on the formation and properties of precipitates, and considerable knowledge has been gained which enables the analyst to minimize the problem of contamination of precipitates.

STOICHIOMETRY

In the usual gravimetric procedure a precipitate is weighed, and from this value the weight of analyte in the sample is calculated. The percentage of analyte, A, is then

$$\% \text{ A} = \frac{\text{weight of A}}{\text{weight of sample}} \times 100$$

To calculate the weight of analyte from the weight of the precipitate a *gravimetric factor* is often employed. This factor is defined as the number of grams of analyte in 1 g (or the equivalent of 1 g) of the precipitate. Multiplication of the weight of precipitate, P, by the gravimetric factor gives the number of grams of analyte in the sample:

$$\text{Weight of A} = \text{weight of P} \times \text{gravimetric factor}$$

Then

$$\% \text{ A} = \frac{\text{weight of P} \times \text{gravimetric factor}}{\text{weight of sample}} \times 100$$

The gravimetric factor arises naturally if the mole method is used to solve a stoichiometric problem. Consider the following examples.

Example 1. A 0.6025-g sample of a chloride salt was dissolved in water and the chloride precipitated by adding excess silver nitrate. The precipitate of silver chloride was filtered, washed, dried, and found to weigh 0.7134 g. Calculate the percentage of chloride (Cl) in the sample.

Let g = grams of Cl in the sample. The reaction is

$$Ag^+ + Cl^- \longrightarrow AgCl(s)$$

Since 1 mol of Cl^- gives 1 mol of AgCl,

$$\text{moles Cl} = \text{moles AgCl}$$

$$\frac{g}{35.45} = \frac{0.7134}{143.32}$$

$$g = 0.7134 \times \frac{35.45}{143.32}$$

and

$$\% \text{ Cl} = \frac{\text{weight Cl}}{\text{weight sample}} \times 100$$

$$\% \text{ Cl} = \frac{0.7134 \times (35.45/143.32)}{0.6025} \times 100$$

$$\% \text{ Cl} = 29.29$$

The ratio of the atomic weight of Cl to the molecular weight of AgCl, 35.45/143.32, is the gravimetric factor, the weight of Cl in 1 g of AgCl. Such a factor is often written Cl/AgCl, where Cl stands for the atomic weight of chlorine and AgCl stands for the molecular weight of silver chloride.

Example 2. A 0.4852-g sample of iron ore is dissolved in acid, the iron oxidized to the +3 state, and then precipitated as the hydrous oxide, $Fe_2O_3 \cdot xH_2O$. The precipitate is filtered, washed, and ignited to Fe_2O_3, which is found to weigh 0.2481 g. Calculate the percentage of iron (Fe) in the sample.

Let g be the grams of Fe in the sample. The reactions are

$$2Fe^{3+} \longrightarrow Fe_2O_3 \cdot xH_2O \longrightarrow Fe_2O_3(s)$$

Since 2 mol of Fe^{3+} produce 1 mol of Fe_2O_3,

$$\text{moles Fe} = 2 \times \text{moles } Fe_2O_3$$

$$\frac{g}{55.85} = 2 \times \frac{0.2481}{159.69}$$

$$g = 0.2481 \times \frac{2 \times 55.85}{159.69}$$

$$\% \text{ Fe} = \frac{0.2481 \times [(2 \times 55.85)/159.69]}{0.4852} \times 100$$

$$\% \text{ Fe} = 35.77$$

In this example the gravimetric factor is $2Fe/Fe_2O_3$ since there are two Fe atoms in one Fe_2O_3 molecule.

In general, two points should be noted in setting up a gravimetric factor. First, the molecular weight (or atomic weight) of the analyte is in the numerator; that of the substance weighed is in the denominator. Second, the number of molecules or atoms appearing in the numerator and denominator must be chemically equivalent.

Some additional examples of gravimetric factors are given in Table 4.1. Handbooks of chemistry and physics usually contain rather lengthy lists of these factors and their logarithms.

TABLE 4.1 Some Gravimetric Factors.

Substance weighed	Substance sought	Factor
AgCl	Cl	$\frac{Cl}{AgCl}$
$BaSO_4$	S	$\frac{S}{BaSO_4}$
$BaSO_4$	SO_3	$\frac{SO_3}{BaSO_4}$
Fe_2O_3	Fe	$\frac{2Fe}{Fe_2O_3}$
Fe_2O_3	FeO	$\frac{2FeO}{Fe_2O_3}$
Fe_2O_3	Fe_3O_4	$\frac{2Fe_3O_4}{3Fe_2O_3}$
$Mg_2P_2O_7$	MgO	$\frac{2MgO}{Mg_2P_2O_7}$
$Mg_2P_2O_7$	P_2O_5	$\frac{P_2O_5}{Mg_2P_2O_7}$
$PbCrO_4$	Cr_2O_3	$\frac{Cr_2O_3}{2PbCrO_4}$
K_2PtCl_6	K	$\frac{2K}{K_2PtCl_6}$

The following examples illustrate some applications of stoichiometric calculations in gravimetric analysis.

Example 3. The phosphorus in a sample of phosphate rock weighing 0.5428 g is precipitated as $MgNH_4PO_4 \cdot 6H_2O$ and ignited to $Mg_2P_2O_7$. If the ignited precipitate weighs 0.2234 g, calculate (a) the percentage of P_2O_5 in the sample; (b) the weight of the precipitate of $MgNH_4PO_4 \cdot 6H_2O$.

(a) The percentage of P_2O_5 is given by

$$\% \ P_2O_5 = \frac{\text{weight of precipitate} \times (P_2O_5/Mg_2P_2O_7)}{\text{weight of sample}} \times 100$$

$$\% \ P_2O_5 = \frac{0.2234 \times (141.95/222.55)}{0.5428} \times 100$$

$$\% \ P_2O_5 = 26.25$$

(b) Since 2 moles of $MgNH_4PO_4 \cdot 6H_2O$ produce 1 mol of $Mg_2P_2O_7$,

$$\text{g } MgNH_4PO_4 \cdot 6H_2O = 0.2234 \times \frac{2 \times 245.40}{222.55}$$

$$\text{g } MgNH_4PO_4 \cdot 6H_2O = 0.4927$$

Example 4. A sample containing only $CaCO_3$ and $MgCO_3$ is ignited to CaO and MgO. The mixture of oxides weighs exactly half as much as the original sample. Calculate the percentages of $CaCO_3$ and $MgCO_3$ in the sample.

The problem is independent of the size of the sample. Let us assume a 1.000-g sample. Then the mixture of oxides weighs 0.5000 g. If w = grams of $CaCO_3$, then $1.000 - w$ = grams of $MgCO_3$.

$$\text{Grams of CaO} + \text{grams of MgO} = 0.5000$$

$$w \times \frac{CaO}{CaCO_3} + (1.000 - w)\frac{MgO}{MgCO_3} = 0.5000$$

$$w \times \frac{56.08}{100.09} + (1.000 - w)\frac{40.304}{84.314} = 0.5000$$

$$w = 0.2673 \text{ g } CaCO_3$$

Since the sample weighs 1 g, the % $CaCO_3$ = 26.73 and the % $MgCO_3$ = 100.00 − 26.73 = 73.27.

Example 5. How many grams of a sample containing chloride should be taken for analysis so that the percentage of chloride in the sample can be obtained by multiplying the weight of the silver chloride sample by 10?

Let w_p be the weight in grams of the AgCl precipitate and w_s the weight in grams of the sample. Then

$$\frac{w_p \times (Cl/AgCl)}{w_s} \times 100 = 10w_p$$

The weight of the precipitate cancels and

$$10w_s = \frac{35.453}{143.321} \times 100$$

$$w_s = 2.474 \text{ g}$$

Example 6. *Indirect Analysis.* Two components in a mixture can be determined from two sets of independent analytical data. Two equations containing the two unknowns are set up and the equations solved simultaneously. The following is an illustration in which one set of data is gravimetric and the other titrimetric.

A 0.7500-g sample containing both NaCl and NaBr is titrated with 0.1043 *M* $AgNO_3$, using 42.23 ml. A second sample of the same weight is treated with excess silver nitrate, and the mixture of AgCl and AgBr is filtered, dried, and found to weigh 0.8042 g. Calculate the percentages of NaCl and of NaBr in the sample.

Let x be the millimoles of NaCl and y the millimoles of NaBr. Then

$$x + y = \text{total mmol} = 42.23 \text{ ml} \times 0.1043 \text{ mmol/ml}$$

$$x + y = 4.405$$

Also x = mmol AgCl and y = mmol AgBr produced. Hence

$$\text{AgCl}\, x + \text{AgBr}\, y = 804.2$$

$$143.32\, x + 187.77\, y = 804.2$$

Solving gives

$$x = 0.516 \quad \text{and} \quad y = 3.889$$

Then

$$\%\ \text{NaCl} = \frac{0.516 \text{ mmol} \times 58.443 \text{ mg/mmol}}{750.0 \text{ mg}} \times 100 = 4.02$$

$$\%\ \text{NaBr} = \frac{3.889 \text{ mmol} \times 102.89 \text{ mg/mmol}}{750.0 \text{ mg}} \times 100 = 53.35$$

It should be noted that in solving this expression for y, it is necessary to divide by the difference in molecular weights of AgCl and AgBr. The closer these two weights are to each other, the greater effect an error in the experimental data (say in the weight of the combined precipitates) will have on the value of y and correspondingly of x. In other words, the reliability of the procedure is greatly reduced if the two molecular weights are very close to the same value.

FORMATION AND PROPERTIES OF PRECIPITATES[1]

We have previously mentioned that a problem of major importance in gravimetric analysis is the formation of pure and filterable precipitates. Insight into this problem can be gained by studying the rate at which particles are built up into solid aggregates sufficiently large to settle from the solution as a precipitate. It is this aspect of precipitation that we shall now discuss.

Colloids

Let us consider the process of precipitation of a salt AB, starting with the ions A^+ and B^- in aqueous solution. The ions are of the order of a few angstrom units (10^{-8} cm) in diameter. When the solubility product is surpassed, A^+ and B^- ions

[1] An excellent summary of this topic at a more advanced level is given by H. A. Laitinen and W. E. Harris, *Chemical Analysis*, 2nd ed., McGraw-Hill Book Company, New York, 1975, p. 142.

begin clinging together, forming a crystal lattice and growing sufficiently large to be pulled to the bottom of the container by the force of gravity. As a general rule, it is said that a particle (spherical) must have a diameter greater than roughly 10^{-4} cm before it will settle from solution as a precipitate. During the growth process the particle passes through the colloidal range. Particles with diameters of about 10^{-4} to 10^{-7} cm are said to be *colloids*. We can represent the precipitation process as

$$\underset{(10^{-8}\text{ cm})}{\text{ions in solution}} \longrightarrow \underset{(10^{-7} - 10^{-4}\text{ cm})}{\text{colloidal particles}} \longrightarrow \underset{(>10^{-4}\text{ cm})}{\text{precipitate}}$$

Colloidal particles are electrically charged and resist combining to form larger particles which will settle from the solution. The electrical charge results from the *adsorption* of ions to the surface of the particles. Small particles have a large surface-to-mass ratio and the surface ions attract ions of opposite charge from the solution. For example, suppose that a drop of silver nitrate is added to a sodium chloride solution and the solubility product constant of AgCl is surpassed. When the first particles grow to colloidal size, there are a large number of Ag^+ and Cl^- ions on the surfaces. In the solution are Na^+, Cl^-, and NO_3^- ions. The surface Ag^+ ions attract Cl^- and NO_3^- ions from the solution, and the surface Cl^- ions attract Na^+ ions. As a general rule (Paneth-Fajans-Hahn), the ion in the solution which is more strongly adsorbed is the one in common to the lattice, in this case the chloride ion.[2] Thus the surface of the particle acquires a layer of chloride ions, and the particles become negatively charged. The process is represented schematically in Fig. 4.1. The chloride ions are said to form the *primary layer*; they in turn attract sodium ions, forming a secondary layer. The secondary layer is held more loosely than is the primary layer.

The primary and secondary layers are considered to constitute an *electrical double layer* which imparts a degree of stability to the colloidal dispersion. These

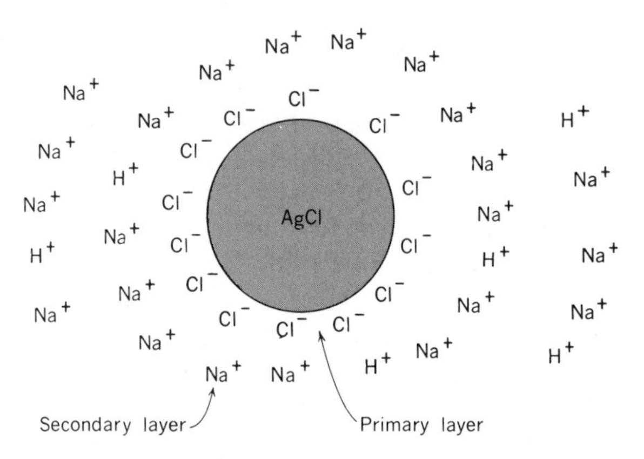

Figure 4.1 Schematic picture of colloidal particle.

[2] If no common ion is present, this rule says that the ion in solution that forms the least soluble compound with one of the lattice ions is the most strongly adsorbed.

layers cause colloidal particles to repel one another and the particles therefore resist combination to form larger particles which will settle from the solution. The particles can be made to *coagulate* (or *flocculate*), that is, to cohere and form larger clumps of material that will settle from the solution, by removal of the charge contributed by the primary layer. In the example of silver chloride, which was cited above, coagulation can be achieved by further addition of silver nitrate until equivalent amounts of silver and chloride ions are present. Since silver ions are more strongly attracted to the primary layer of chloride ions than are sodium ions, they replace sodium ions in the secondary layer and then "neutralize" the negative charge contributed by the primary layer. Stripped of their charge, the particles immediately cohere and form clumps of material which are sufficiently large to settle from the solution. Some colloids when coagulated, carry down large quantities of water, giving a jelly-like precipitate. Such materials are termed *gels* or *hydrogels* if water is the solvent. The solid material is also referred to as an *emulsoid*, or said to be *lyophilic*, meaning that it has a strong affinity for the solvent. (If water is the solvent, the term *hydrophilic* is used.) Iron(III) and aluminum hydroxides and silicic acid are familiar examples of emulsoids. A colloid that has only a small affinity for water is called a *suspensoid*, or said to be *lyophobic*, and when coagulation takes place very little solvent is retained. Silver chloride is this type of material, and the small amount of water that is retained upon coagulation of silver chloride is easily removed by drying above 100°C. The water retained by an emulsoid, such as iron(III) hydroxide, is much more strongly held, and high temperatures are required for complete dehydration.

Coagulation of colloidal dispersions can be brought about by ions other than those of the precipitate itself. When coagulation of a colloid occurs, the coagulating ions may be dragged down with the precipitate. If these ions are dissolved when a precipitate is washed, the solid particles will go back into a colloidal dispersion and pass through the filter. Such a process of dispersing an insoluble material into a liquid as a colloid is termed *peptization* and must be avoided in quantitative procedures. When peptization may occur, an electrolyte is dissolved in the wash water to replace the ions which are washed away. Dilute nitric acid is added for this reason to the water used to wash a silver chloride precipitate. When the precipitate is dried, any nitric acid retained by the silver chloride is volatilized and does not interfere in the analysis.

The Precipitation Process and Particle Size

When the solubility product constant of a compound is exceeded and precipitation begins, a number of small particles, called nuclei, are formed. Subsequent precipitation can take place on these initially formed particles, with the particles growing in size until large enough to settle from the solution. The particle size distribution of the precipitate is determined by the relative rates of the two processes, the formation of nuclei, called *nucleation*, and the growth of nuclei. It is apparent that, if the rate of nucleation is small compared to the rate of growth of nuclei, fewer particles are finally produced, and these particles are of relatively large particle size. Such a material is more easily filterable and frequently purer than is the case with small particles. Hence the analyst tries to adjust conditions during precipitation so that the rate of nucleation is relatively small in order that the particle size will be large.

The first systematic study of the effect of conditions such as the concentrations of the reagents on the rate of precipitation and particle size was made by von Weimarn.[3] He found that, in general, the particle size increases, passes through a maximum, and then decreases as the concentration of reagents increases. He also noted that generally the smaller the solubility of a precipitate, the smaller are its particles. This is only a rough rule, however. Silver chloride and barium sulfate have about the same molar solubilities (K_{sp}'s of about 10^{-10}), but barium sulfate forms much larger particles of precipitate than silver chloride when similar conditions of precipitation are employed.

Von Weimarn introduced the concept of *relative supersaturation*, where

$$\text{relative supersaturation} = \frac{Q - S}{S}$$

Here Q is the total concentration of the substance momentarily produced in the solution by mixing the reagents, and S is the equilibrium solubility. The term $Q - S$ represents the degree of supersaturation at the moment precipitation begins. The larger this term, the greater is the number of nuclei and the smaller are the particles of precipitate. The term S in the denominator represents the force resisting precipitation, or causing the precipitate to redissolve. The greater the value of S, the smaller will be the ratio and the smaller will be the number of nuclei formed. Since the analyst is interested in obtaining large particles, he should try to adjust conditions to make the ratio $Q - S/S$ as small as possible.

Actually, the von Weimarn expression has only qualitative significance; it does not, for example, express the increase in particle size with increasing concentration which is found at low concentration. Extensive studies have been made of the precipitation process since the work of von Weimarn and numerous theories have been proposed.[4] In the precipitation of $BaSO_4$, for example, the process of *heterogeneous* nucleation (the formation of a nucleus on a second phase, such as a particle of dust) is thought to involve the formation of a "critical" cluster, $(Ba^{2+}SO_4^{2-})_3$. The addition of a seventh ion, either Ba^{2+} or SO_4^{2-}, to the cluster constitutes the final step of nucleation. Johnson and O'Rourke[4] concluded that the rate of nucleation of $BaSO_4$ is proportional to the fourth power of the ionic concentration. The rate of crystal growth at concentrations above 10^{-3} M was observed by Nielsen[4] to depend on the first power of the ionic concentration. On the other hand, the Becker-Döring nucleation hypothesis[4] indicates that a much larger number of ions, perhaps 100, is required to form the critical cluster.

In spite of the complexity of the precipitation process and the lack of a generally accepted theory, the analyst can rely on the experimental fact that one does obtain relatively large particles of precipitate by keeping the degree of

[3] P. P. von Weimarn, *Chem. Rev.*, **2**, 217 (1925); *Kolloid-Beihefte*, **18**, 44 (1923).

[4] See A. E. Nielsen, *Acta Chem. Scand.*, **14**, 1654 (1960); R. A. Johnson and J. D. O'Rourke, *J. Am. Chem. Soc.*, **76**, 2124 (1954); R. Becker and W. Döring, *Ann. Phys. (Leipzig)*, **24**, 719 (1935).

supersaturation low. In practice, one routinely brings about a moderate decrease in Q by using reasonably dilute solutions and adding the precipitating agent slowly. It is frequently possible to increase the value of S markedly and thus effect a large decrease in the ratio. This can be done by taking advantage of the factors that may increase solubility: temperature, *p*H, or the use of complexing agents. Precipitations are quite commonly carried out at elevated temperatures for this reason. Salts of weak acids, such as calcium oxalate and zinc sulfide, are better precipitated in weakly acidic, rather than in alkaline, solution.[5] Barium sulfate is better precipitated in 0.01 to 0.05 M hydrochloric acid solution since the solubility is increased by formation of the bisulfate ion. A compound such as iron(III) hydroxide is so insoluble that even in acidic solution the value of $(Q - S)/S$ is still so large that a gelatinous precipitate results. However, a dense precipitate of iron as the basic formate can be obtained by homogeneous precipitation (page 83).

In addition to controlling the conditions during the actual precipitation process, the analyst has one other recourse after the precipitate is formed. This is to *digest*, or *age*, the precipitate, that is, to allow the precipitate to stand in contact with the mother liquor, frequently at elevated temperature, for some time before filtration. The small particles of a crystalline substance, such as barium sulfate, being more soluble than the larger ones, dissolve more readily making the solution supersaturated with respect to the larger particles. In order to establish equilibrium with respect to the larger particles, additional material must leave the solution and enter the solid phase. The ions now deposit on the larger particles, causing these particles to grow even larger. Thus the larger particles grow at the expense of the smaller ones. This process, sometimes called "Ostwald ripening," is useful for increasing the particle size of crystalline precipitates, such as barium sulfate and calcium oxalate. Curdy precipitates, such as silver chloride, and gelatinous precipitates, such as iron(III) hydroxide, are not digested. The particles of the latter compound are either so insoluble, or the small particles do not differ sufficiently from the larger ones, that no appreciable growth in size occurs. Even with crystalline precipitates it is necessary to employ conditions which increase solubility if a beneficial effect is to be attained in a reasonable time. This is the reason for the frequent use of elevated temperatures during digestion.

Thus to obtain a precipitate of large particle size, precipitation is carried out by slow mixing of dilute solutions under conditions of increased solubility of the precipitate. Crystalline precipitates are normally digested at elevated temperature before filtration to further increase particle size.

Purity of Precipitates

One of the most difficult problems that faces the analyst in employing precipitation as a means of separation and gravimetric determination is obtaining the precipitate in a high degree of purity. We wish to look now at the ways in

[5] The solubility should not be increased to such an extent that precipitation is incomplete, of course. Frequently, precipitation is started under conditions of increased solubility. After most of the precipitate is formed, the solubility is lowered to ensure complete precipitation.

which a precipitate can become contaminated, and to see what conditions the analyst can employ to minimize contamination during the precipitation process. We shall also examine methods that can be employed to increase the purity of the precipitate after precipitation has been carried out.

Coprecipitation

The process by which a normally soluble substance is carried down during the precipitation of the desired precipitate is called *coprecipitation*. For example, when sulfuric acid is added to a solution of barium chloride containing a small amount of nitrate ions, the precipitate of barium sulfate is found to contain barium nitrate. It is said that the nitrate is coprecipitated with the sulfate.

Coprecipitation may occur by the formation of *mixed crystals* or by the adsorption of ions during the precipitation process. In the former case, which occurs only infrequently, the impurity actually enters the crystal lattice of the precipitate. In the latter case, adsorbed ions are dragged down with the precipitate during the process of coagulation. A better discussion of coprecipitation which results from adsorption can be given in terms of the three types of precipitates previously described: crystalline, curdy, and gelatinous.

Crystalline Precipitates

A crystalline precipitate, such as barium sulfate, sometimes adsorbs impurities when the particles are small. As the particles grow in size, the impurity may become enclosed in the crystal. This type of contamination is called *occlusion* to distinguish it from the case where the solid does not grow around the impurity. Occluded impurities cannot be removed by washing the precipitate.

There are several things that the analyst can do to minimize coprecipitation with crystalline precipitates. If he is aware of the presence of an ion that readily coprecipitates, he can decrease (but not completely eliminate) the amount of coprecipitation by the method of addition of the two reagents. If it is known that either the sample or the precipitant contains a contaminating ion, the solution containing this ion can be added to the other solution. In this way the concentration of the contaminant is kept at a minimum during the early stages of precipitation.

After a crystalline precipitate is formed, the analyst can still increase the purity. If the substance can be readily redissolved (as salts of weak acids in stronger acids), it can be filtered, redissolved, and reprecipitated. The contaminating ion will be present in a lower concentration during the second precipitation, and consequently a smaller amount will be coprecipitated.

A substance such as barium sulfate is not readily redissolved, but its purity can be improved by the process of aging or digestion. We have already seen that during aging the particle size is increased. At the same time impurities held by these small particles are redissolved and are not readsorbed appreciably by the larger particles. Also, during aging the lattice becomes more compact, probably by ions dissolving at the corners and edges and redepositing in an orderly fashion.

During this perfection process, occluded impurities may be expelled, and since the number of surface ions has decreased, very little impurity is readsorbed.

Curdy Precipitates

Impurities are adsorbed by the primary particles of a substance such as silver chloride in the same manner as by particles of barium sulfate. However, silver chloride particles do not grow beyond colloidal dimensions, and they finally precipitate as a coagulated colloid. The resulting curd is still made up of fine particles that have not grown together to form an extensive lattice structure. Thus curdy precipitates do not enclose, or occlude, foreign ions as do crystalline precipitates. The impurities on the surfaces of the tiny particles can normally be washed off, since the particles are not firmly bound to one another, and the wash liquid can penetrate to all parts of the curd. As previously mentioned, peptization of the particles must be avoided, and hence the wash liquid must contain a volatile electrolyte.

Digestion of a curdy precipitate is not normally employed for purification because there are no occluded impurities. However, a silver chloride precipitate is usually heated and allowed to stand for 1 or 2 hr in contact with the mother liquor containing nitric acid in order to promote coagulation of the colloidal particles. The primary particles do not grow larger during digestion, although it is thought that in some cases colloids form loose agglomerates by sharing the "jackets" of water by which they are surrounded.

Gelatinous Precipitates

The primary particles of a gelatinous precipitate are much larger in number and of much smaller dimensions than those of crystalline or curdy precipitates. The surface area exposed to the solution by such a precipitate is extremely large. A large quantity of water is adsorbed, of course, rendering the precipitate gelatinous, and also the adsorption of foreign ions can be quite extensive. Since the flocculated primary particles do not readily grow into larger crystals, the impurities are not occluded, as with barium sulfate, but are held by adsorption on the surface of the tiny particles.

The electric charge of the primary particles of substances such as iron(III) and aluminum hydroxides is primarily a function of the *p*H of the solution, since hydrogen and hydroxide ions are readily adsorbed by such precipitates. Iron(III) hydroxide is positively charged at *p*H values less than about 8.5 and negatively charged at higher *p*H values. Thus anions tend to be coprecipitated by secondary adsorption at low *p*H, cations at high *p*H. This point is important in processes involving the separation of iron from other cations by precipitation of the hydrous oxide. In analyses of minerals such as limestone, where calcium and magnesium are present, iron is precipitated at as low a *p*H as possible to avoid coprecipitation of these cations.

Washing and reprecipitation can be employed to increase the purity of a gelatinous precipitate once it has been formed. Digestion is not beneficial, since the precipitate is so slightly soluble that the particles have little tendency to grow in size. Washing is normally employed, and, as with curdy precipitates, a volatile electrolyte must be present to avoid peptization. Reprecipitation is employed with a precipitate such as iron(III) hydroxide. The hydroxide is filtered, washed, and then redissolved in dilute hydrochloric acid. The concentration of impurities is lower in the new solution, and when the precipitate is reformed by raising the *p*H, a smaller degree of contamination results. The dissolving and reprecipitation, with intervening filtration, can be repeated several times, but usually one reprecipitation insures a sufficiently pure precipitate. This procedure, as previously mentioned, can also be employed for purification of crystalline precipitates that can be readily dissolved in acids. Calcium oxalate and magnesium ammonium phosphate are usually treated in this manner in the limestone analysis.

Summary

We can now summarize the procedures that can be employed to minimize coprecipitation.

1. *Method of addition of the two reagents.* This can be used to control the concentration of impurity and the electric charge carried by the primary particles of precipitate. In the case of hydrous oxides the charge can be controlled by using the proper *p*H.
2. *Washing.* With curdy and gelatinous precipitates one must have an electrolyte in the wash solution to avoid peptization.
3. *Digestion.* This is of considerable benefit to crystalline precipitates, of some benefit to curdy precipitates, but not used for gelatinous precipitates.
4. *Reprecipitation.* This is used where the precipitate is readily redissolved, primarily for hydrous oxides and crystalline salts of weak acids.
5. *Separation.* The impurity may be separated or its chemical nature changed by some reaction before the precipitate is formed.
6. *Use of conditions that lead to large particle size* (page 77). This point needs further clarification. It would be expected that conditions which lead to large particles would result in a purer precipitate since the surface area of the particles is then relatively small. This is found to be true if the precipitation is sufficiently slow (next section), but it is not true under the usual conditions of precipitation. We have seen how crystalline precipitates, formed under conditions leading to large particles, can occlude impurities during their growth. Such impurities are not readily removed by washing, but digestion is beneficial. Therefore, it is sometimes recommended that a crystalline substance be precipitated from reasonably concentrated solutions at room temperature, the solution then diluted and the precipitate digested under conditions of increased solubility. An electrolyte may be added and the digestion performed at elevated temperature for this purpose. The precipitate is then first formed in a finely divided state, with impurities held on the surfaces of the small particles, not occluded. Upon digestion, Ostwald ripening occurs, leading finally to both large and pure particles. Kolthoff and Sandell[6] obtained

[6] I. M. Kolthoff and E. B. Sandell, *J. Phys. Chem.*, **37**, 443, 459 (1953).

large particles of calcium oxalate in this manner in the presence of iodate ions. Less iodate was coprecipitated using this procedure than when the compound was precipitated under conditions of increased solubility and then digested.

Precipitation from Homogeneous Solution

When a precipitant is added to a solution, even when the solution is dilute and well stirred, there will always be some local regions of high concentration. However, by using a procedure in which the precipitant is produced as the result of the reaction *taking place in the solution*, such local effects can be avoided. This technique is usually called *precipitation from homogeneous solution*, and it can lead to both large and pure particles of a precipitate. The best-known example of this method is the use of the hydrolysis of urea to increase the *p*H and precipitate hydrous oxides, or salts of weak acids. Urea hydrolyzes according to the equation

$$CO(NH_2)_2 + H_2O \longrightarrow CO_2 + 2NH_3$$

The hydrolysis is slow at room temperature but is fairly rapid at 100°C. Thus the *p*H can be well controlled in effecting separation by controlling the temperature and duration of heating. Also, the carbon dioxide liberated as bubbles prevents "bumping." Precipitation is usually complete in one to two hours. During this slow growth the particles have time to attain a large size without imperfections occurring in the lattice structure, and therefore the amount of occluded impurity is minimized.

The precipitation of a number of compounds used for gravimetric analysis has been carried out by this technique. Willard and Chan[7] recommend the precipitation of calcium oxalate by neutralizing an acid solution of calcium, containing excess oxalate, by hydrolysis of urea. The precipitate is not contaminated by magnesium or phosphate after only one precipitation. Barium sulfate can be precipitated in this manner by hydrolyzing dimethyl sulfate[8] or sulfamic acid to generate sulfate ions. The data in Table 4.2 show that much less calcium is coprecipitated with barium sulfate when the latter is precipitated homogeneously.

TABLE 4.2 Coprecipitation of Calcium with Barium Sulfate*

Precipitant	Ca added, mg	Ca in precipitate, mg
Dilute H_2SO_4	5.4	3.4
Sulfamic acid	100	0.4
Ethyl sulfate	100	0.6

* Reprinted by permission from H. F. Walton, *Principles and Methods of Chemical Analysis*, 2nd ed., Prentice-Hall, Inc., Englewood Cliffs, N.J., 1964.

[7] H. H. Willard and F. L. Chan; see *Elementary Quantitative Analysis*, by H. H. Willard and N. F. Furman, D. Van Nostrand Co., New York, 1940, p. 344.

[8] P. J. Elving and R. E. Van Atta, *Anal. Chem.*, **22**, 1375 (1950).

Other ions that have been generated homogeneously include: chloride, from chlorohydrin; phosphate, from ethyl phosphate; and oxalate, from ethyl oxalate.

Hydrous oxides are gelatinous whether formed under ordinary analytical conditions or homogeneously. However, Willard and his co-workers[9] have obtained dense precipitates of iron and aluminum by precipitation with urea in the presence of certain anions. The succinate ion is best for aluminum, and formate is best for iron. The precipitates are of indefinite composition but contain basic salts of aluminum and succinate or of iron and formate. Coprecipitation of foreign ions is less than when the hydrous oxides are precipitated by addition of ammonia. The aluminum precipitate has been found to lose water more readily than the hydrous oxide.[10] Normally, a temperature of about 1100°C is required for ignition of hydrous alumina, but the precipitate obtained using urea-succinate reaches constant weight at about 650°C.

Postprecipitation

The process by which an impurity is deposited *after* the precipitation of the desired substance is termed *postprecipitation.* This process differs from coprecipitation principally in the fact that the amount of contamination increases, the longer the desired precipitate is left in contact with the mother liquor. When there is a possibility that postprecipitation may occur, directions call for filtration to be made shortly after the desired precipitate is formed.

Postprecipitation occurs when the solution is supersaturated with a foreign substance that precipitates very slowly. For example, zinc sulfide does not readily precipitate from solutions containing zinc ion, hydrogen ion (0.1 to 0.2 *M*), and saturated with hydrogen sulfide. However, if mercury(II) sulfide is precipitated under the same conditions in the presence of zinc, over 90% of the zinc comes down as the sulfide within 20 minutes. Apparently, zinc sulfide forms very stable supersaturated solutions. When mercury(II) sulfide is present, sulfide ions are strongly adsorbed at the interface of the solid and solution. The solubility product constant of zinc sulfide is exceeded to an even greater extent at the interface than in the bulk of the solution, and the rate of precipitation is increased.

Magnesium oxalate forms stable, supersaturated solutions, and unless precautions are taken, postprecipitates on calcium oxalate when calcium and magnesium are separated by precipitation of the latter compound. Postprecipitation can be avoided by using as high acidity as possible and filtering off the calcium precipitate within 1 or 2 hr after precipitation.

IGNITION OF PRECIPITATES

In any gravimetric procedure involving precipitation, one must finally convert the separated substance into a form suitable for weighing. It is necessary that the

[9] H. H. Willard, *Anal. Chem.*, **22**, 1372 (1950).
[10] T. Dupuis and C. Duval, *Anal. Chem. Acta*, **3**, 191 (1949).

substance weighed be pure, stable, and of definite composition for the results of the analysis to be accurate. Even if coprecipitation has been minimized, there still remains the problem of complete removal of water and of any electrolytes added to the wash water. Some precipitates are weighed in the same chemical form as that in which they precipitate. Others undergo chemical changes during ignition, and these reactions must go to completion for correct results. The procedure used in this final step depends both upon the chemical properties of the precipitate and upon the tenacity with which water is held by the solid.

Some precipitates can be dried sufficiently for analytical determination without resort to high temperature. For example, magnesium ammonium phosphate hexahydrate, $MgNH_4PO_4 \cdot 6H_2O$, is sometimes dried by washing with a mixture of alcohol and ether and drawing air over the precipitate for a few minutes.[11] Generally, however, such a procedure is used only when considerable difficulty is encountered upon ignition of the precipitate. It is not usually recommended because of the danger of incomplete removal of water by washing. Water that merely *adheres* to the precipitate is removed, but water that is adsorbed or occluded (or water of hydration) is not removed by washing.

Some precipitates lose water readily in an oven at temperatures of 100 to 130°C. Silver chloride does not adsorb water strongly and is normally dried in this manner for ordinary analytical work. In the determination of atomic weights, however, it has been found necessary to fuse silver chloride to remove the last traces of water.

Ignition at high temperature is required for complete removal of water that is occluded or very strongly adsorbed, and for complete conversion of some precipitates to the desired compound. Water can become enclosed within a particle during crystal growth and is then expelled only at high temperatures, probably by the crystal's bursting from the steam pressure generated. Gelatinous precipitates, such as the hydrous oxides, adsorb water quite strongly and must be heated to very high temperatures to remove water completely. Hydrous silica and alumina are well-known examples of precipitates that require very high ignition temperatures. The ignition of calcium oxalate to calcium oxide involves an example of a chemical change that requires a high temperature for complete reaction. At about 880°C the dissociation pressure of calcium carbonate reaches one atmosphere, but the rate of decomposition is rather slow at this temperature. Therefore, it is usually recommended that temperatures in the range of 1100°C be employed.

Errors other than incomplete removal of water or volatile electrolytes can occur during ignition. One of the most serious is reduction of the precipitate by carbon when filter paper is employed. Substances that are very easily reduced, such as silver chloride, are never filtered on paper; filtering crucibles are always employed. Students frequently encounter trouble with precipitates of barium sulfate and iron(III) oxide. Unless the paper is burned off with a plentiful supply of air, these precipitates will be reduced. Magnesium ammonium phosphate is

[11] H. A. Fales, *Inorganic Quantitative Analysis*, Century Co., New York, 1925, p. 22; J. P. Mehlig, *J. Chem. Ed.*, **12,** 288 (1935).

also easily reduced when ignited to the pyrophosphate. This substance is frequently collected on a porcelain filter crucible to avoid using filter paper.

Precipitates can be overignited, leading to decomposition and to substances of indefinite composition. Errors can also result from an ignited precipitate's reabsorbing water or carbon dioxide upon cooling. Crucibles should be properly covered and kept in a desiccator while cooling.

The Thermobalance

Until recent years very few careful studies had been made of the ignition temperatures required for different precipitates. In 1944 Chevenard[12] designed a balance, called a *thermobalance*, which allows a sample to be weighed while it is actually in a furnace. The balance is sensitive to 0.2 mg, and the temperature of the furnace can be measured to within about 1° between room temperature and 1100°C. Duval[13] has used this balance to study the ignition of a large number of precipitates of analytical interest. The data are recorded in the form of a graph of weight of the precipitate against temperature. Such a graph is called a pyrolysis curve. It is evident that one should ignite a sample in a temperature range where the curve is flat, that is, where the weight is constant over a wide temperature range. The pyrolysis curves for a few substances are shown in Fig. 4.2. The curves for calcium and magnesium oxalates are particularly interesting. The monohydrate $CaC_2O_4 \cdot H_2O$ is stable at 100°C and then loses water up to about 226°C. Up to 398°C the form CaC_2O_4 is stable, and then the oxalate loses carbon monoxide abruptly to form $CaCO_3$. The carbonate is stable in the range of about 420 to 600°C, and then the dissociation to calcium oxide commences. The weight finally becomes constant at about 850°C. Magnesium oxalate differs in its behavior in that is loses carbon monoxide and dioxide simultaneously, forming magnesium oxide directly with no intermediate carbonate.

ORGANIC PRECIPITANTS

Many inorganic ions can be precipitated with certain organic reagents called "organic precipitants." A number of these reagents are useful not only for separations by precipitation, but also by solvent extraction. This topic will be discussed in some detail in Chapter 16.

Most of the organic precipitants, about which our discussion will be centered, combine with cations to form *chelate* rings. We shall be concerned here with neutral metal chelate compounds which are insoluble in water. In Chapter 8 we will discuss reagents which form soluble 1 : 1 complex ions with various cations and which can be employed as titrants for metals. In this chapter we shall also

[12] P. Chevenard, X. Wache, and R. de la Tullaye, *Bull. Soc. Chim.* (5), **11**, 41 (1944).

[13] C. Duval, *Inorganic Thermogravimetric Analysis*, 2nd ed., Elsevier Publishing Co., New York, 1963.

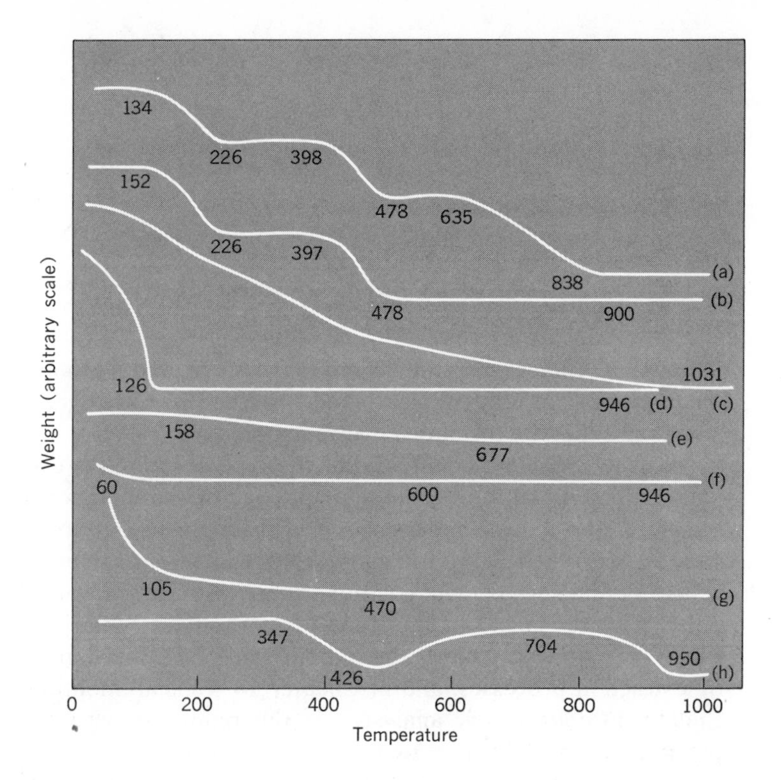

Figure 4.2 Pyrolysis curves: (a) CaC_2O_4, (b) MgC_2O_4, (c) Al_2O_3, precipitated by aqueous NH_3, (d) Al_2O_3, precipitated by urea, (e) $BaSO_4$, (f) AgCl, (g) Fe_2O_3, (h) CuSCN. (Taken by permission from C. Duval, *Inorganic Thermogravimetric Analysis*, 2nd ed., Elsevier Publishing Co., New York, 1963.)

mention a few examples of organic precipitants which form saltlike precipitates with metal ions.

Reagents Forming Chelate Compounds

Generally speaking, most of the better-known organic precipitants which form chelate compounds with cations contain both an acidic and a basic (electron-donating) functional group. The metal, interacting with both of these groups, becomes itself one member of a heterocyclic ring. From the strain theory of organic chemistry it is expected that rings of this type would be mainly five- and six-membered. Hence the acidic and basic functional groups in the organic molecule must be situated in position with respect to each other which permits the closure of such rings.

8-Hydroxyquinoline (often called 8-quinolinol, or "oxine") forms insoluble compounds with a number of metal ions, aluminum for one. The formation of this compound may be formulated as follows:

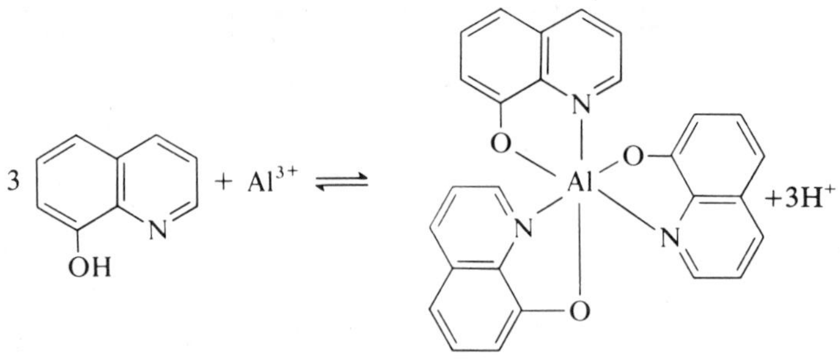

$$3\ \text{(8-hydroxyquinoline: N, OH)} + Al^{3+} \rightleftharpoons \text{(Al chelate: Al, 3 O, 3 N)} + 3H^{+}$$

Aluminum replaces the acidic hydrogen of the hydroxyl group. At the same time, the previously unshared pair of electrons on the nitrogen is donated to the aluminum, thereby forming a five-membered ring.

A neutral chelate compound of the type described is essentially organic in nature. The metal ion becomes simply one of the members of an organic ring structure, and its usual properties and reactions are no longer readily demonstrable. With the reservation in mind that exceptions can be found, we may state generally that such chelate compounds are insoluble in water but soluble in less polar solvents such as chloroform or carbon tetrachloride. We shall see in Chapter 16 that this differential solubility may be utilized in effecting separations by extraction processes, and in Chapter 14 we shall mention briefly the use of chelates in colorimetric analysis. At this point, we wish to consider only the precipitation of metal ions by these organic reagents.

Let us consider first the advantages offered by organic precipitants.

1. Many of the chelate compounds are very insoluble in water, as noted above, so that metal ions may be quantitatively precipitated.
2. The organic precipitant often has a large molecular weight. Thus a small amount of metal may yield a large weight of precipitate.
3. Some of the organic reagents are fairly selective, yielding precipitates with only a limited number of cations. Certain people once thought that we should ultimately have available an absolutely specific reagent for each cation. Although there is little optimism in this regard today, modern research has demonstrated that a sound knowledge of the chemistry of ions in solution makes such specificity seem less necessary. By controlling such factors as *p*H and the concentration of masking reagents, the selectivity of an organic reagent can often be greatly enhanced.
4. The precipitates obtained with organic reagents are often coarse and bulky, and hence easily handled.
5. In some cases a metal can be precipitated with an organic reagent, the precipitate collected and dissolved, and the organic molecule titrated, furnishing an indirect volumetric method for the metal.

8-Hydroxyquinoline can be quantitatively brominated with a bromate-bromide mixture. Methods of this type are available for only a few organic precipitants because of the difficulty in finding organic oxidations which satisfy the general requirements for titrations. Since most organic precipitants are weak acids and bases (too weak to be titrated in water), it is sometimes possible to titrate the organic molecule in nonaqueous media.[14]

We must also consider certain disadvantages in the use of organic precipitants.

1. The low solubility of the metal chelate compounds was listed as an advantage. However, the very limited aqueous solubility of most organic reagents themselves is often troublesome. It is generally necessary to add at least a slight excess of the precipitant, and thus the danger of contaminating the precipitate with excess reagent is often a real one. Occasionally, but not always, the excess reagent can be washed out of the precipitate with a solvent such as hot water or alcohol.
2. Many of the organic precipitates do not have good weighing forms, largely because of uncertainty in the drying process. Some of the most attractive of the organic precipitants can be used only for separations, not for determinations, because of this difficulty in drying to a product of definite composition. Some of the metal chelates tend to volatilize at the temperatures required to remove water. In other cases, decomposition of the organic molecule sets in before drying to constant weight has been assured.
3. A minor disadvantage is the fact that the precipitates are not easily wet by water, and hence tend to float on the surface of the solution and to creep up the sides of glass vessels. This trouble can be alleviated by addition of a small amount of wetting agent to the solution before filtration.

A list showing a few of the more widely used organic precipitants is given in Table 4.3.

Reagents Forming Saltlike Precipitates

Some organic precipitants form salts rather than chelate complexes with inorganic ions. Oxalic acid is well known in analytical processes for its use in the precipitation of calcium; calcium oxalate is a typical insoluble salt. There are a number of such organic compounds which form precipitates with both cations and anions. A few of these are listed in Table 4.4.

APPLICATIONS OF GRAVIMETRIC ANALYSIS

A gravimetric method for almost every element in the periodic table has been reported in the chemical literature.[15] We have listed in Table 4.5 some examples where precipitation reactions have been used for the elements in various groups of the periodic table. The substances precipitated and the form which is finally

[14] C. H. Hill, Han Tai, A. L. Underwood, and R. A. Day, Jr., *Anal. Chem.*, **28**, 1688 (1956).

[15] See, for example, C. L. Wilson and D. W. Wilson, *Comprehensive Analytical Chemistry*, Vol. IC, Elsevier Publishing Co., New York, 1962.

TABLE 4.3 Some Common Organic Precipitants

Compound	Chelate with metal of valence *n*	Comments
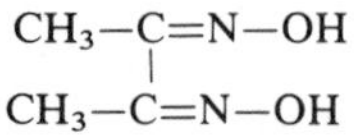 $CH_3-C{=}N-OH$ / $CH_3-C{=}N-OH$ Dimethyl glyoxime	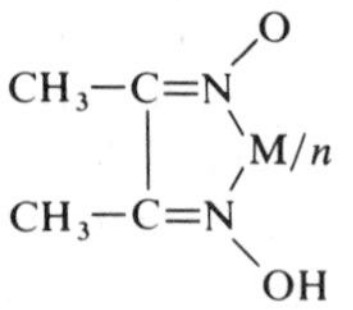 $CH_3-C{=}N(\to O)$, $CH_3-C{=}N(-OH)$, M/*n*	Principally used for determination of nickel
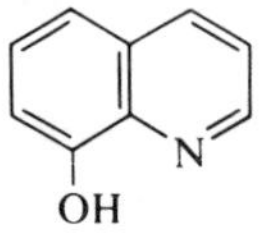 8-Hydroxyquinoline	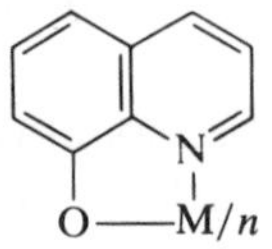 N—M/*n*—O	Precipitates many elements but can be used for group separations by controlling *p*H
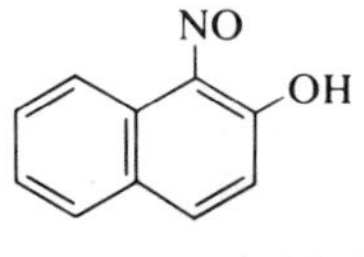 NO, OH α-Nitroso-β-naphthol	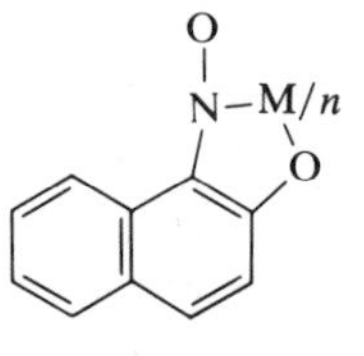 O, N—M/*n*, O	Principally used for precipitation of cobalt in presence of large amounts of nickel
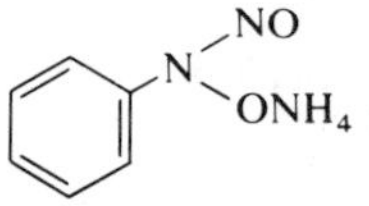 NO, N, ONH_4 Cupferron	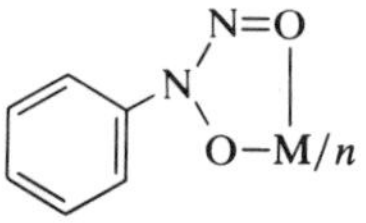 N=O, N, O—M/*n*	Mainly used for separations, such as iron and titanium from aluminum
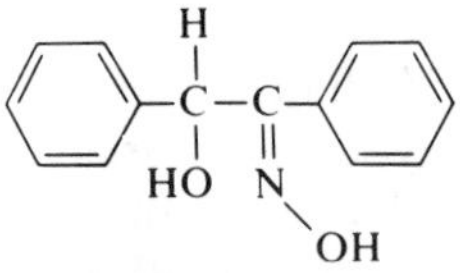 H, C—C, HO, N, OH α-Benzoin oxime	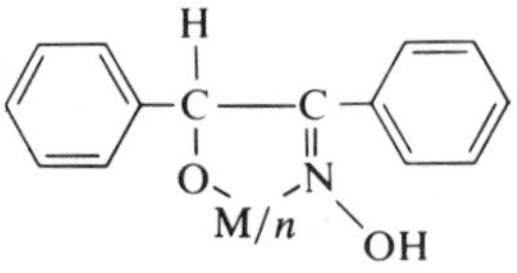 H, C—C, O, N, M/*n*, OH	Good reagent for copper; also precipitates bismuth and zinc
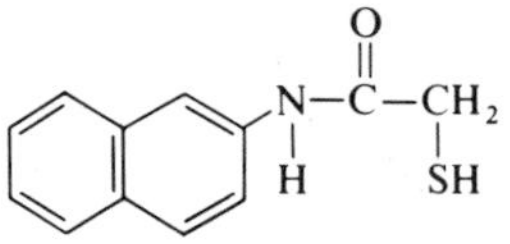 O, N—C—CH_2, H, SH Thionalide	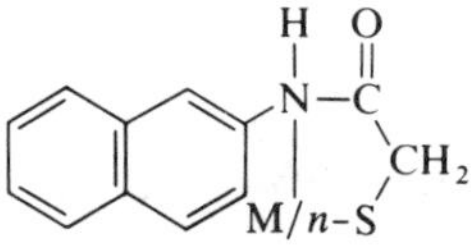 H O, N—C, CH_2, M/*n*-S	Used for precipitation and determination of elements of H_2S group
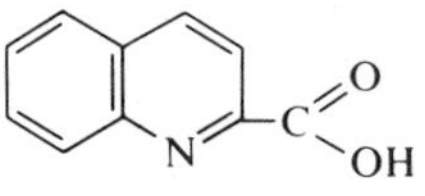 N, C, O, OH Quinaldic acid	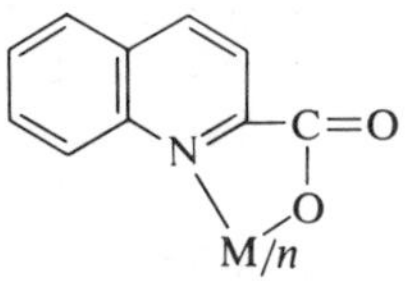 N, C=O, O, M/*n*	Used for determination of cadmium, copper and zinc

TABLE 4.4 Reagents Forming Saltlike Precipitates

Reagent	Comments
$Na^+B(C_6H_5)_4^-$ Sodium tetraphenyl boron	Used principally for K^+; in 0.1 *M* HCl only NH_4^+, Hg^{2+}, Rb^+, and Cs^+ interfere
$H_2N-C_6H_4-C_6H_4-NH_2$ Benzidine	Used principally for SO_4^{2-}
$(C_6H_5)_4As^+Cl^-$ Tetraphenylarsonium chloride	Used for Tl^+; other metals which form precipitates include tin, gold, zinc, mercury, and cadmium
$R-As(=O)(OH)_2$ Arsonic acids; R = phenyl, *n*-propyl, etc.	Acids precipitate quadrivalent metal ions such as tin, thorium, and zirconium from acid media

TABLE 4.5 Gravimetric Determinations of Various Elements

Group in periodic table	Element	Precipitate	Substance weighed
IA	Potassium	$KClO_4$	$KClO_4$
IIA	Calcium	CaC_2O_4	CaO
IIIA	Aluminum	$Al_2O_3 \cdot xH_2O$	Al_2O_3
IVA	Silicon	$SiO_2 \cdot xH_2O$	SiO_2
VA	Phosphorus	$MgNH_4PO_4 \cdot 6H_2O$	$Mg_2P_2O_7$
VIA	Sulfur	$BaSO_4$	$BaSO_4$
VIIA	Chlorine	$AgCl$	$AgCl$
IB	Silver	$AgCl$	$AgCl$
IIB	Zinc	$ZnNH_4PO_4$	$Zn_2P_2O_7$
IIIB	Scandium	Scandium oxinate	Scandium oxinate
IVB	Titanium	Titanium cupferrate	TiO_2
VB	Vanadium	$HgVO_3$	V_2O_5
VIB	Chromium	$Cr_2O_3 \cdot xH_2O$	Cr_2O_3
VIIB	Manganese	MnO_2	Mn_2O_3
VIII	Iron	$Fe_2O_3 \cdot xH_2O$	Fe_2O_3
	Cobalt	CoS	$CoSO_4$
	Nickel	Nickel dimethylglyoxime	Nickel dimethylglyoxime

weighed are given in the table. In addition to inorganic substances, organic compounds have been analyzed by gravimetric techniques. Two examples are the determinations of cholesterol in cereals and lactose in milk products. Organic compounds are also used widely as precipitating agents (page 86), and some of the examples in Table 4.5 involve such precipitants. The elemental analysis of organic compounds is generally done by gravimetric techniques. For example, carbon and hydrogen in organic compounds can be determined by burning the sample in a stream of oxygen and absorbing the CO_2 and H_2O produced on appropriate absorbents. The absorption tubes are weighed before and after the combustion to obtain the weights of CO_2 and H_2O produced.

The student may well have heard the argument that modern instrumental methods have taken the place of classical gravimetric techniques. Although it is true that the gravimetric technique has been displaced in its routine aspects by instrumental methods, gravimetric analysis is still of great importance in the field of analytical chemistry. There are still many instances where it represents the best choice for solving a particular analytical problem. In general, where only a few determinations are required, a gravimetric procedure may actually be faster and more accurate than an instrumental method which requires extensive calibration or standardization. Instruments generally furnish only relative measurements and must be calibrated on the basis of a classical gravimetric or titrimetric method. In the provision of the standards which are required to check the performance of an instrumental method, the gravimetric technique provides a direct and comparatively simple approach.

Gravimetric methods compare favorably with other analytical techniques in terms of the accuracy attainable. If the analyte is a major constituent (>1% of the sample), accuracy of a few parts per thousand can be expected if the sample is not too complex. If the analyte is present in minor or trace amounts (less than 1%), a gravimetric method is generally not employed.

In general, gravimetric methods are not very *specific*. As previously mentioned (page 88) certain chemists once thought that we should ultimately have a specific precipitant for every cation. While this is no longer expected, gravimetric reagents are *selective* in that they form precipitates only with certain groups of ions. Then, too, the selectivity of precipitating agents can often be improved by controlling such factors as *p*H and the concentration of certain masking agents.

As illustrations of gravimetric procedures we shall discuss briefly the determinations commonly used as student exercises. These involve the precipitation of silver chloride, barium sulfate, and iron(III) hydroxide.

Silver Chloride

Silver chloride precipitates in curds or lumps resulting from the coagulation of colloidal material (page 76). The precipitate is easily filtered, and is washed with water containing a little nitric acid. The acid prevents peptization of the precipitate and is volatilized when the precipitate is dried. Silver chloride is normally filtered through a sintered glass or porous porcelain crucible and is dried at 110 to 130°C.

Errors

The precipitation of silver chloride generally gives excellent analytical results. The principal error arises from the decomposition of the precipitate by sunlight:

$$2AgCl(s) \longrightarrow 2Ag(s) + Cl_2(g)$$

The extent of this reaction is negligible unless the precipitate is exposed to direct sunlight.

The solubility of silver chloride in water is slight and losses due to solubility are negligible. However, alkali and ammonium salts, as well as large concentrations of acids, should be avoided since they increase the solubility.

Other Applications

In addition to the determination of silver and chloride, the precipitation of silver chloride can be used to determine chlorine in oxidation states other than -1. Hypochlorites (ClO^-), chlorites (ClO_2^-), and chlorates (ClO_3^-) may be determined by first reducing these ions to chloride, and then precipitating silver chloride. Chlorine in organic compounds is often determined by this precipitation after the organic chlorine is converted to sodium chloride by fusion with sodium peroxide.

Bromide and iodide may be determined by precipitation of their silver salts. Also oxygen-containing anions, such as hypobromite (BrO^-), bromate (BrO_3^-), hypoiodite (IO^-), iodate (IO_3^-), and periodate (IO_4^-) can first be reduced to bromide or iodide and then precipitated as the silver salts.

Barium Sulfate

Barium sulfate is a crystalline precipitate. It is only slightly soluble in water and losses due to solubility are small. The precipitation is carried out in about 0.01 M hydrochloric acid for the purpose of obtaining larger particles, a purer precipitate (page 79), and to prevent the precipitation of such salts as $BaCO_3$.

Errors

Coprecipitation of foreign substances with barium sulfate is very pronounced. The anions most strongly coprecipitated are nitrate and chlorate. Cations, particularly divalent and trivalent ones that form slightly soluble sulfates, are strongly coprecipitated, iron(III) being one of the most prominent examples. The procedures summarized on page 82 are employed where practical to minimize coprecipitation. The digestion process employed to increase particle size also leads to some purification. Reprecipitation is not employed since a suitable solvent is not available.

Barium sulfate is normally filtered with filter paper and washed with hot water. The filter paper must be burned off carefully with a plentiful supply of air. The sulfate is reduced rather readily by carbon of the paper:

$$BaSO_4(s) + 4C(s) \longrightarrow BaS(s) + 4CO(g)$$

If this reduction occurs, the results are low and usually the precision is poor. The precipitate can be reconverted to the sulfate by moistening it with sulfuric acid and reigniting. A porous porcelain crucible can be employed instead of paper.

Other Applications

The sulfur in sulfides, sulfites, thiosulfates, and tetrathionates can be determined by oxidizing the sulfur to sulfate and then precipitating barium sulfate. Permanganate is often used to effect the oxidation. Sulfur in organic compounds is determined by oxidizing the element to sulfate with sodium peroxide. Ores of sulfur, such as pyrites, FeS_2, and chalcopyrite, $CuFeS_2$, may be fused with sodium peroxide to oxidize the sulfur to sulfate.

Other cations often precipitated as sulfates are lead and strontium. Both of these sulfates are more soluble than barium sulfate. Alcohol is added in the determination of strontium to decrease the solubility of the sulfate. The determination of lead in brass can be done by the precipitation of lead sulfate.

Iron(III) Hydroxide

The gravimetric determination of iron involves the precipitation of iron(III) hydroxide (actually $Fe_2O_3 \cdot xH_2O$, called a *hydrous oxide*), followed by ignition at high temperature to Fe_2O_3. The method is used in rock analyses, where iron is separated from elements such as calcium and magnesium by the precipitation. Iron ores are usually dissolved in hydrochloric acid and nitric acid or bromine is used to oxidize the iron to the +3 oxidation state.

The hydrous oxide of iron is a gelatinous precipitate which is very insoluble in water (K_{sp} of 1×10^{-36}). Coagulation of the colloidal material is aided by precipitation from a hot solution. The precipitate is washed with water containing a small amount of ammonium nitrate to prevent peptization. Filtration is carried out using rapid filter paper. The paper is burned off and the precipitate ignited at fairly high temperatures to remove all the water.

Errors

Coprecipitation by adsorption of foreign ions during precipitation can cause serious errors. By using some of the procedures described on page 82, however, coprecipitation can be held to a minimum. Precipitation is generally made from acid solution so that the colloidal particles are positively charged and cations are less strongly adsorbed. Since the oxide can be readily dissolved in acids, reprecipitation is used to advantage to rid the precipitate of adsorbed impurities.

Iron(III) oxide, Fe_2O_3, is fairly easily reduced to either Fe_3O_4 or Fe by carbon of the filter paper. The ignited precipitate can be treated with concentrated nitric acid and reignited to form Fe_2O_3 again.

Other Applications

Several other metals are precipitated as hydrous oxides by ammonia. Among these are aluminum, chromium(III), titanium(IV), and manganese(IV). The hydroxides of aluminum and chromium are amphoteric, and a large excess of ammonia must be avoided because these substances redissolve. Manganese(II) is incompletely precipitated, but if an oxidizing agent, such as bromine, is added, hydrous manganese dioxide is precipitated. All of these hydrous oxides can be ignited to the oxides: Al_2O_3, Cr_2O_3, TiO_2, and Mn_3O_4. Thus the precipitation with ammonia can be used for determining each element provided the others are absent. The method is seldom used for manganese since the conversion of MnO_2 to Mn_3O_4 is not quantitative.

Precipitation with ammonia is used for the quantitative separation of iron and the foregoing elements from the alkali and alkaline earth cations in the analysis of rocks, such as limestone.

QUESTIONS

1. *Requirements.* What are the requirements that must be met for a reaction to serve as the basis of a gravimetric method?

2. *Properties of precipitates.* Explain the reasons for the following procedures: (a) $BaSO_4$ is washed with water, whereas the wash water for AgCl contains a little HNO_3; (b) $BaSO_4$ is digested following precipitation, but this procedure is not used for AgCl or $Fe(OH)_3$.

3. *Coprecipitation.* The sulfate ion in a solution is to be precipitated by the addition of lead(II) ion. The solution also contains small amounts of NO_3^-, Cl^-, and ClO_4^- ions. (a) Which of these ions is most likely to be coprecipitated? (b) How could you minimize this coprecipitation?

4. *Coprecipitation.* The $C_2O_4^{2-}$ ion in a solution which contains a small amount of SO_4^{2-} is to be precipitated as CaC_2O_4. What conditions of precipitation would you use to minimize coprecipitation?

5. *Particle size.* What conditions should be used to precipitate a sulfide, such as ZnS, in order to obtain as large particles as possible?

6. *Postprecipitation.* When is postprecipitation likely to occur? Will ZnS be more likely to postprecipitate on HgS or $BaSO_4$? Explain.

7. *Ignition of a precipitate.* Point out the possible errors in the final step of igniting and weighing a precipitate.

Multiple-choice: In the following multiple-choice questions, select the *one best* answer.

8. A precipitate of $Fe(OH)_3$ is contaminated with $Mg(OH)_2$. The best way to get rid of the impurity is (a) digestion; (b) washing; (c) reprecipitation; (d) ignition.

9. Colloids which carry down large quantities of water when coagulated are said to be (a) gels; (b) lyophilic; (c) emulsoids; (d) all of the above; (e) none of the above.

10. The process of dispersing an insoluble material into a liquid as a colloid is called (a) nucleation; (b) peptization; (c) occlusion; (d) coagulation.

11. Which of the following *does not* promote the formation of large crystals of CaC_2O_4? $H_2C_2O_4$ is a weak acid. (a) Slow mixing of dilute solutions; (b) decreasing $Q - S/S$; (c) digestion; (d) precipitation at high *p*H rather than low *p*H.

12. AgCl is precipitated by adding $AgNO_3$ to an aqueous NaCl solution. The ion most strongly adsorbed to the surface of the colloidal particles before the equivalence point is (a) Na^+; (b) Cl^-; (c) Ag^+; (d) H_3O^+; (e) OH^-.

PROBLEMS

1. *Gravimetric factors.* Calculate the gravimetric factors for the following. The substance weighed is given first, then the substance sought. (a) U_3O_8, U; (b) $KClO_4$, K_2O; (c) $PbCrO_4$, Cr_2O_3; (d) AgCl, $KClO_4$; (e) $(NH_4)_2PtCl_6$, NH_3.

2. *Gravimetric factors.* Calculate the gravimetric factors for the following. The substance weighed is given first, then the substance sought. (a) $Zn_2P_2O_7$, Zn; (b) $BaSO_4$, FeS_2; (c) $CaCO_3$, CO_2; (d) $(NH_4)_3PO_4 \cdot 12MoO_3$, P_2O_5; (e) K_2PtCl_6, K_2O.

3. *Gravimetric determination of chloride.* A 0.7168-g sample containing chloride is dissolved and the chloride precipitated as AgCl. The precipitate is washed, dried, and found to weigh 0.3964 g. Calculate the percentage of chloride in the sample.

4. *Gravimetric determination of sulfur.* The sulfur in a 0.8423-g sample is converted to sulfate and the sulfate precipitated as $BaSO_4$. The precipitate is found to weigh 0.3148 g after it is washed and ignited. (a) Calculate the percentage of sulfur in the sample. (b) Calculate the percentage expressed as SO_3.

5. *Gravimetric determination of lead.* The lead in a 0.6342-g sample of an ore is precipitated as $PbSO_4$, The precipitate is dried and found to weigh 0.4381 g. Calculate (a) the percentage of Pb in the ore; (b) the percentage expressed as PbO.

6. *Precipitation of iron.* Calculate the number of milliliters of ammonia, density of 0.99 g/ml, 2.3% by weight NH_3, which will be required to precipitate the iron as $Fe(OH)_3$ in a 0.70-g sample that contains 25% Fe_2O_3.

7. *Precipitation of barium sulfate.* Calculate the number of milliliters of a solution containing 20 g of $BaCl_2$/liter which will be required to precipitate the sulfur as $BaSO_4$ in a 0.50-g sample which contains 12% S.

8. *Sample size.* What size sample which contains 15.0% Cl^- should be taken for analysis if it is desired to obtain a precipitate of AgCl which weighs 0.400 g?

9. *Sample size.* What size sample which contains 12.5% FeO should be taken for analysis if it is desired to obtain a precipitate of Fe_2O_3 weighing 0.3600 g?

10. *Sample size.* Iron is determined in a sample by precipitating the hydrous oxide and igniting it to Fe_2O_3. What weight of sample (mg) should be taken for analysis so that each mg of Fe_2O_3 represents 0.100% Fe in the sample?

11. *Sample size.* Calcium is determined in a rock sample by precipitating $CaCO_3$ and igniting the precipitate to CaO. What weight of sample should be taken so that each 2.00 mg of the ignited precipitate represents 0.500% CaO in the sample?

12. *Determination of magnesium.* A 0.5000-g sample of a magnesium salt yields a precipitate with 8-hydroxyquinoline which weighs 0.4523 g. Calculate the percentage of MgO in the sample, assuming that the precipitate is magnesium oxinate dihydrate.

13. *Atomic weight.* A sample of pure sodium chloride weighing 0.65310 g is dissolved in water and the chloride precipitated as AgCl. If the AgCl precipitate weighs 1.6016 g, calculate the atomic weight of sodium. Assume that the atomic weights of chlorine and silver are 35.453 and 107.868, respectively.

14. *Organic analysis.* A 1.000-g sample of a pure organic compound containing chlorine is fused with Na_2O_2 to convert the chlorine to NaCl. The sample is then dissolved in water and the chloride precipitated with $AgNO_3$, giving 1.950 g of AgCl. If the MW of the organic compound is about 147, how many chlorine atoms does each molecule contain?

15. *Analysis of brass.* A brass sample contains 9.20% tin, 5.40% lead, 4.30% zinc, and 81.10% copper. The elements are determined gravimetrically by weighing the following precipitates: SnO_2, $PbSO_4$, CuSCN, and $Zn_2P_2O_7$. If a 0.600-g sample of brass is analyzed, what weights of each of these precipitates will be obtained?

16. *Stoichiometry.* A 0.6000-g sample consisting of only CaC_2O_4 and MgC_2O_4 is heated at 500°C, converting the two salts to $CaCO_3$ and $MgCO_3$. The sample then weighs 0.4650 g. If the sample had been heated at 900°C, where the products are CaO and MgO, what would the mixture of oxides weigh?

17. *Stoichiometry.* A mixture which contains only Fe_2O_3 and Al_2O_3 weighs 0.6432 g. It is heated with H_2, reducing the Fe_2O_3 to Fe, leaving the Al_2O_3 unchanged. The mixture now weighs 0.5448 g. Calculate the percentage of Al in the sample.

18. *Stoichiometry.* A 0.4531-g sample containing only $CaCO_3$ and $MgCO_3$ is treated with HCl and 112.0 ml (STP) of CO_2 is produced. Assuming that the reactions went to completion, what percentage of the sample is $CaCO_3$?

19. *Stoichiometry.* A mixture containing only AgCl and AgBr weighs 0.4628 g. It is treated with Cl_2, converting the AgBr to AgCl. The sample now weighs 0.3604 g. Calculate the percentage of AgBr in the original sample.

20. *Stoichiometry.* A 1.000-g sample containing 75.0% K_2SO_4 and 25.0% of another metal sulfate, MSO_4, is dissolved and the sulfate precipitated as $BaSO_4$. If the $BaSO_4$ precipitate weighs 1.490 g, what is the MW of MSO_4?

21. *Determination of sodium and potassium.* Sodium and potassium in a rock were determined as follows: The elements were converted to the chlorides and the mixture of NaCl and KCl was found to weigh 0.6648 g. The chlorides were then converted to sulfates and the mixture of Na_2SO_4 and K_2SO_4 was found to weigh 0.7849 g. If the original sample of rock weighed 0.8792 g, calculate the percentages of Na_2O and K_2O in sample.

22. *Stoichiometry.* A 0.5000-g sample which contains both barium and lead is analyzed as follows. The metals are precipitated as sulfates, giving a precipitate weighing 0.1950 g. The metals in another 0.5000-g sample are precipitated as chromates, giving a precipitate weighing 0.2093 g. Calculate the percentages of barium and lead in the sample.

23. *Errors.* A student analyzed a limestone sample but forgot to dry it. He found 31.20% CaO, whereas the correct value was 31.60. If the error was caused by the failure to dry the sample, what percentage moisture did the sample contain?

24. *Errors.* A 0.8000-g sample of iron ore containing 21.00% Fe is analyzed gravimetrically. If the final precipitate (assumed to be Fe_2O_3) contains 20.0 mg of Al_2O_3 by mistake, what percentage Fe does the analyst find?

25. *Errors.* A student analyzed a 1.000-g sample for sulfur by precipitating $BaSO_4$. The sample actually contained 30.00% SO_3 and the student obtained a precipitate weighing 0.8505 g. (a) Calculate the % SO_3 the student would obtain. (b) If his error was caused by reduction of some $BaSO_4$ to BaS during the ignition process, what percentage of the precipitate he obtained was BaS?

26. *Errors.* A student determined calcium in a limestone sample by precipitating CaC_2O_4 and igniting it to CaO. He found 40.00% CaO where the correct value was 39.12%. If the error was caused by insufficient ignition, leaving some $CaCO_3$ in the final precipitate, what percentage of this final precipitate was $CaCO_3$?

27. *Indirect analysis.* A 0.6400-g sample that contains both NaCl and NaBr gives a precipitate of AgCl and AgBr that weighs 0.5269 g. Another 0.6400-g sample is titrated with 0.1084-*M* $AgNO_3$, requiring 27.52 ml. Calculate the percentages of NaCl and NaBr in the sample.

28. *Indirect analysis.* A 0.5000-g sample that contains KCl and KI gives a precipitate of AgCl and AgI that weighs 0.2818 g. A 0.8000-g sample is titrated with 0.1053 *M* $AgNO_3$, requiring 26.74 ml. Calculate the percentages of KCl and KI in the sample.

CHAPTER FIVE

review of chemical equilibrium

INTRODUCTION

It will be recalled that one of the conditions which must be satisfied by a chemical reaction used for a titrimetric analysis is that the reaction go essentially to completion at the equivalence point. Similarly, in a gravimetric analysis the reaction used for separation must be so complete that the amount of analyte left behind is analytically undetectable. Stoichiometric calculations, treated in Chapters 3 and 4, tell us nothing about the completeness of a reaction. We need to know the magnitude of the *equilibrium constant* of a reaction to determine the degree to which a reaction goes to completion under a given set of conditions.

Equilibrium calculations are introduced in all standard texts in general chemistry. We wish to review here how equilibrium constants are formulated for different reactions, especially those which are used in analytical procedures: acid-base, precipitation, complex formation, and oxidation-reduction. In subsequent chapters we shall apply the principles of equilibrium to answer such a question as whether a reaction goes sufficiently to completion for a titration to be feasible.

THE EQUILIBRIUM CONSTANT

Chemical reactions, such as the formation of hydrogen iodide from hydrogen and iodine in the gaseous phase,

$$H_2(g) + I_2(g) \rightleftharpoons 2HI(g)$$

are generally *reversible*, and when the rates of the forward and reverse reactions are equal, the concentrations of the reactants and products remain constant with time. We say that the reaction has reached a state of *equilibrium*.

It is found experimentally that the extent to which reactions are complete when equilibrium is reached varies tremendously. In some cases the concentrations of products are much larger than those of reactants; in other cases the reverse is true. The equilibrium concentrations reflect the intrinsic tendencies of the atoms to exist as molecules of reactant or product. Although there may be an infinite number of concentrations which satisfy the equilibrium condition, there is only one general expression which is found to be constant at a given temperature for a reaction at equilibrium. For the general reaction in aqueous solution,

$$\text{A(aq)} + \text{B(aq)} \rightleftharpoons \text{C(aq)} + \text{D(aq)}$$

this expression is

$$\frac{[\text{C}]_e[\text{D}]_e}{[\text{A}]_e[\text{B}]_e} = K$$

and is called the *equilibrium constant*. Here the brackets designate concentrations in moles per liter (molarity) at equilibrium. The fraction is often called the *mass action* expression, from the work of Guldberg and Waage (1864), who used the term "active mass" to denote the amount of chemical reactivity. These workers were not clear about the meaning of active masses. Chemists today use the term *activity* as the best meaning of active mass and *concentration* as a good approximation when activity is not known. The concentrations employed are usually molality or molarity for reactions in aqueous solution. For gases molarity or partial pressure is employed. We shall discuss activity in more detail later.

For the completely general reaction

$$a\text{A(aq)} + b\text{B(aq)} \rightleftharpoons c\text{C(aq)} + d\text{D(aq)}$$

the equilibrium constant is

$$\frac{[\text{C}]_e^c[\text{D}]_e^d}{[\text{A}]_e^a[\text{B}]_e^b} = K$$

The exponents in this expression are the coefficients of the reactants and products in the balanced equation for the reaction. Table 5.1 contains the equilibrium expressions for several reactions of the type we will encounter later in our discussions. Note that if a reactant is a solid or liquid, its concentration does not appear in the equilibrium expression. The reason is that the concentration of a solid or liquid is constant. Increasing the amount of a solid or liquid in a reacting system does not change its concentration. The number of moles increases, but the volume also increases, and the number of moles per liter is unchanged.

AQUEOUS SOLUTIONS

Water is one of the most plentiful compounds in nature and is essential to life processes. It dissolves many substances and serves as the medium in which a wide

TABLE 5.1 Equilibrium Constants for Various Reactions

Reaction	Equilibrium expression	Numerical value at 25°C
$HOAc(aq) + H_2O \rightleftharpoons H_3O^+ + OAc^-$	$\frac{[H_3O^+][OAc^-]}{[HOAc]}$	1.8×10^{-5}
$Al(OH)_3(s) \rightleftharpoons Al^{3+} + 3OH^-$	$[Al^{3+}][OH^-]^3$	5×10^{-33}
$Ag^+ + 2NH_3(aq) \rightleftharpoons Ag(NH_3)_2^+$	$\frac{[Ag(NH_3)_2^+]}{[Ag^+][NH_3]^2}$	1.4×10^7
$Sn^{2+} + 2Ce^{4+} \rightleftharpoons Sn^{4+} + 2Ce^{3+}$	$\frac{[Sn^{4+}][Ce^{3+}]^2}{[Sn^{2+}][Ce^{4+}]^2}$	5.4×10^{43}
$HOAc + OH^- \rightleftharpoons OAc^- + H_2O$	$\frac{[OAc^-]}{[HOAc][OH^-]}$	1.8×10^9

variety of chemical reactions take place. Most of the analytical processes we will discuss involve reactions in aqueous solutions.

Weak and Strong Electrolytes

Aqueous solutions of certain compounds are good conductors of an electric current because of the presence of positive and negative ions. Such compounds are called *electrolytes*, whereas compounds whose aqueous solutions do not conduct current are called *nonelectrolytes*. Sodium chloride, NaCl, is completely dissociated into Na^+ and Cl^- ions in aqueous solution and is an electrolyte. Ethylene glycol, $CH_2OH—CH_2OH$, a common antifreeze, is not dissociated in aqueous solution and is an example of a nonelectrolyte. Most *ionic* compounds are completely dissociated in water and are called *strong* electrolytes. Many *covalent* compounds dissociate to only a slight degree when dissolved in water and are called *weak* electrolytes. Acetic acid (CH_3COOH, abbreviated HOAc), ammonia, and water are examples of weak electrolytes. Table 5.2 gives a short list of the most common weak and strong acids and bases encountered in the introductory course in quantitative analysis.

TABLE 5.2 Strong and Weak Acids and Bases in Aqueous Solution

ACIDS		BASES	
Strong	Weak	Strong	Weak
HCl	HOAc	NaOH	$NH_3(NH_4OH)$
HNO_3	HF	KOH	$CH_3NH_2(CH_3NH_3OH)$
$HClO_4$	H_2CO_3	$Ba(OH)_2$	$C_6H_5NH_2$ (aniline)
H_2SO_4	H_2S	$(CH_3)_4NOH$	

We write equilibrium constants for the dissociation of weak electrolytes into ions and call them *dissociation* constants. For example, the dissociation of water is written

$$2H_2O \rightleftharpoons H_3O^+ + OH^- \qquad K_w = [H_3O^+][OH^-]$$

For acetic acid and ammonia we write

$$HOAc + H_2O \rightleftharpoons H_3O^+ + OAc^- \qquad K_a = \frac{[H_3O^+][OAc^-]}{[HOAc]}$$

$$NH_3 + H_2O \rightleftharpoons NH_4^+ + OH^- \qquad K_b = \frac{[NH_4^+][OH^-]}{[NH_3]}$$

The subscripts are used for convenience. K_a is used as the symbol for the dissociation constant of a weak acid; K_b is used for a weak base. K_w refers to the dissociation of water and is sometimes called the *autoprotolysis* constant of water.

The term *salt* deserves special comment. A salt is the product other than water which is formed when an acid reacts with a base. For example, when hydrochloric acid and sodium hydroxide react, the products are a salt, sodium chloride, and water. Written molecularly, this is

$$HCl + NaOH \longrightarrow NaCl + H_2O$$

However, we realize that the net ionic reaction which occurs is simply

$$H_3O^+ + OH^- \longrightarrow 2H_2O$$

and that the salt is not an NaCl molecule, but Na^+ and Cl^- ions; that is, sodium chloride is a strong electrolyte, completely dissociated. Most salts are strong electrolytes, although for convenience we write their formulas molecularly, as Na_2SO_4, KNO_3, $CaCl_2$, etc. A few salts are weak electrolytes; one we will encounter later is mercury(II) chloride, $HgCl_2$. The reaction

$$Hg^{2+} + 2Cl^- \longrightarrow HgCl_2$$

is sufficiently complete at the equivalence point that it can be used for the titration of chloride ions.

Salts of weak electrolytes react with water to produce either hydrogen or hydroxide ions. For example, an aqueous solution of the salt sodium acetate, NaOAc, is basic because the acetate ion reacts with water to produce hydroxide ions:

$$OAc^- + H_2O \rightleftharpoons HOAc + OH^-$$

The reason that this reaction occurs to an appreciable extent is that HOAc is a weak acid and prefers to remain as molecules. The Na^+ ion does not react with water to produce NaOH molecules and H_3O^+ ions since NaOH is a strong base and prefers to remain as Na^+ and OH^- ions. The salt ammonium chloride, NH_4Cl, is acidic since ammonium ions react with water to produce H_3O^+ ions:

$$NH_4^+ + H_2O \rightleftharpoons NH_3 + H_3O^+$$

The Cl^- ion does not react with water to produce HCl and OH^- ions, since HCl is a strong acid.

The equilibrium constant for the reaction of acetate ion with water is

$$\frac{[HOAc][OH^-]}{[OAc^-]} = K_b$$

The symbol K_b is used to indicate that the acetate ion is a base; it is said that the acetate ion is the *conjugate* base of acetic acid. (See discussion of Brønsted theory, page 107.) There is a simple relation between the K_a of an acid and the K_b of its conjugate base. This can be seen by multiplying the two expressions:

$$\frac{[H_3O^+][\cancel{OAc^-}]}{[\cancel{HOAc}]} \times \frac{[\cancel{HOAc}][OH^-]}{[\cancel{OAc^-}]} = [H_3O^+][OH^-] = K_w$$

or

$$K_a \times K_b = K_w$$

The same relation holds for the weak base, NH_3, and its conjugate acid, NH_4^+.

In discussing solutions containing salts, such as sodium acetate or ammonium chloride, chemists sometimes refer to the ion, OAc^- or NH_4^+, as the salt. It is understood that the salt is NaOAc or NH_4Cl. Since the acetate or ammonium ions undergo reaction with water molecules, they are the species of interest in equilibrium calculations of the hydrogen ion concentration.

Activity

The concept of activity may be explained to the student in a rigorous fashion only after he has encountered partial molal free energy or chemical potential in thermodynamics. However, he is probably acquainted with the concept of free energy, the measure of the driving force of a chemical reaction and the maximum work that can be obtained at constant temperature and pressure. This change in free energy for the transfer of 1 mol of a given substance from a state of activity a_1 to activity a_2 is given by

$$\Delta G = 2.3RT \log \frac{a_2}{a_1}$$

In the case of solutions, the volumes must be so large that the transfer does not change the concentrations.

It is evident that the change in free energy is determined by the *ratio* of the two activities. Hence, to define an individual activity it is customary to adopt an *arbitrary* reference or *standard state* and to assign to it an activity of unity at any given temperature and pressure. The customary choices are as follows:

1. For a perfect gas the standard state is 1 atm, and the activity is then the same as the pressure of the gas. For a real gas the standard state is that in which the so-called fugacity is unity.[1] Since at low pressures the real gas approaches ideal

[1] The fugacity is the same as the vapor pressure when the vapor is a perfect gas, and it may be regarded as an "ideal" or "corrected" vapor pressure.

behavior, making fugacity and pressure approximately equal, we will take the pressure of a gas as its activity. Thus

$$\frac{a}{P} = 1 \qquad \text{when } P \longrightarrow 0$$

2. The activity of a pure liquid or solid (in its most stable crystalline state) acting as a solvent for other substances is unity. That is, the standard state is a mole fraction of unity, where X is the mole fraction of the solvent.

$$\frac{a}{X} = 1 \qquad \text{when } X \longrightarrow 1$$

   If the activity of the liquid or solid is changed by dissolving in it a solute, the activity of the solvent is still given by the mole fraction. In most examples that we shall encounter, it will still be acceptable to take a value of unity as the activity of the solvent. For example, a liter of a 0.1 M aqueous solution of a solute contains 0.1 mol of that solute and about 55.3 mol of water. The mole fraction of water is thus about $55.3/55.4 \cong 1$. The possible effect of the solute upon the activity of water will be ignored in our calculations.
3. The activity of a solute is the same as its molality in very dilute solution, where ideal behavior may be assumed. That is,

$$\frac{a}{m} = 1 \qquad \text{when } m \longrightarrow 0$$

   where m is the molality.[2] Here, as for a real gas, the standard state is a hypothetical one in which the solute is at 1 molal concentration (1 atm pressure), but the environment about the solute would be the same as that of an ideal solution. In dilute solutions the behavior of the solute does approach ideal, and we use molality for activity for such solutions.

Activity Coefficient

In solutions where the activity of a solute is expressed as concentration, deviations from ideal solution behavior are generally expressed in terms of the *activity coefficient*, γ, defined as

$$a = \gamma m$$

The more nearly ideal a solute behaves, the closer to unity is γ, and hence the closer activity is to concentration. According to our chosen standard state, as $m \rightarrow 0$, $\gamma \rightarrow 1$ and $a \rightarrow m$. For solutions of electrolytes the activity coefficient is a measure of the deviation from ideality because of ion–ion interactions. Such interactions are general, not just between the specific ions undergoing chemical reaction. For instance, the activities of ions such as Ag^+ and Cl^- in the formation

[2] This is also defined in terms of mole fraction, i.e., $a/X = 1$ when $x \rightarrow 0$, where X is the mole fraction solute. Since in very dilute solution molality becomes proportional to mole fraction, either definition can be used.

of a precipitate of AgCl are lowered by the addition of a nonreacting electrolyte such as potassium nitrate. Debye and Hückel interpreted this diminished activity in terms of electrostatic interactions of the ions: Clustering of NO_3^- about the Ag^+ and of K^+ around the Cl^- tends to shield the Ag^+ and Cl^- from each other and thus hampers the effectiveness of these ions in forming AgCl.

Either the activity or the activity coefficient could be made dimensionless, but generally it is the activity coefficient which is so treated. Since our limiting definition is $a/m = 1$, it seems logical to make the units of activity the same as those of molality, thereby making the activity coefficient dimensionless.

It should also be noted that in analytical chemistry most concentrations are expressed in the molarity system rather than molality. Since these two systems are very nearly the same in dilute aqueous solution, we shall use molarity in place of molality.

In 1923 Peter Debye and Erich Hückel derived an equation which enables one to calculate activity coefficients theoretically. The equation for aqueous solutions at room temperature is approximately

$$-\log \gamma_i = 0.5Z_i^2 \sqrt{\mu}$$

where γ_i is the activity coefficient, Z_i is the charge of the ion of interest, and μ is the *ionic strength* of the solution. The latter term is defined by the equation

$$\mu = \frac{1}{2}\sum_i C_i Z_i^2$$

where C_i is the molar concentration and Z_i is the charge of each ionic species in the solution.

It is not possible to test the preceding equation experimentally since it is impossible to prepare a solution which contains only a single ionic species, such as K^+. The assumption is made that the activity coefficients of all singly charged ions in the solution are mutually equal and equal in turn to the *mean activity coefficient*, defined for a 1:1 type electrolyte as

$$\gamma_\pm = \sqrt{\gamma_+\gamma_-}$$

Mean activity coefficients can be measured by various physiochemical techniques, and some experimental values are given in Table 5.3. It may be seen in this table that the activity of an electrolyte such as HCl depends not only on the concentration of this solute itself but also upon the presence of other ions such as those of NaCl. It may also be noted that activity coefficients are lowered when the electrical charges of the ions are increased; compare, for example, the values for comparable concentrations of NaCl and $CuSO_4$.

In general, the presence of ions will have a lesser effect upon the activity of a neutral molecule than upon that of another electrolyte. However, ions do influence molecules to some degree by interacting with existing dipoles or even inducing them.

For a binary electrolyte A_mB_n the Debye-Hückel equation is written

$$\log \gamma_\pm = 0.5Z_A Z_B \sqrt{\mu}$$

TABLE 5.3 Mean Molal Activity Coefficients of Electrolytes at 25°C

	Molality					
	0.001	0.005	0.01	0.05	0.1	0.5
HCl	0.97	0.93	0.90	0.83	0.80	0.76
HCl (0.01 *M*) (in NaCl)	—	—	0.87	0.82	0.78	0.76
HNO_3	0.97	0.93	0.90	0.82	0.79	0.72
H_2SO_4	0.83	0.64	0.54	0.34	0.27	0.15
KOH	—	0.93	0.90	0.81	0.76	0.67
NaCl	0.97	0.93	0.90	0.82	0.78	0.68
$CaCl_2$	0.89	0.79	0.73	0.57	0.52	0.52
K_2SO_4	—	0.78	0.71	0.53	0.44	0.26
$CuSO_4$	0.74	0.53	0.41	0.21	0.16	0.07

where Z_A and Z_B are charges on the cation and anion taken without regard to sign. Activity coefficients calculated by this equation are satisfactory up to ionic strengths of about 0.1. In more concentrated solutions experimentally determined values should be used.

The following calculation illustrates a simple application of the Debye-Hückel equation.

Example. Calculate the mean activity coefficient for a 0.01 *M* solution of HCl and compare it with the value in Table 5.3.

The ionic strength is

$$\mu = \tfrac{1}{2}\sum_i C_i Z_i^2$$

$$\mu = \tfrac{1}{2}[0.01(1)^2 + 0.01(1)^2]$$

$$\mu = 0.01$$

$$-\log \gamma_\pm = 0.5 Z_A Z_B \sqrt{\mu}$$

$$-\log \gamma_\pm = 0.5(1)(1)\sqrt{0.01}$$

$$-\log \gamma_\pm = 0.05$$

$$\gamma_\pm = 0.89.$$

This is in close agreement with the experimental value 0.90 in Table 5.3. A similar calculation for 0.1 *M* HCl will give a value of 0.69, where the experimental value is 0.80.

Throughout this text we shall use molar concentrations as though they were activities in most of the calculations we make. It is rare that activity coefficients are known in the complex, concentrated solutions frequently encountered in analytical chemistry. Furthermore, many of the answers we seek regarding, for example, the feasibility of a titration, can be obtained by approximate calculations. We shall make a practice, however, of reminding the student that activities should be used in equilibrium calculations, and we shall point out instances where activity effects may appreciably affect the answer we are seeking.

BRØNSTED TREATMENT OF ACIDS AND BASES

Although substances with acidic and basic properties had been known for hundreds of years, the quantitative treatment of acid–base equilibria became possible after 1887, when Arrhenius presented his theory of electrolytic dissociation. In water solution, according to Arrhenius, acids dissociate into hydrogen ions and anions, and bases dissociate into hydroxide ions and cations:

$$\text{Acid:} \quad HX \rightleftharpoons H^+ + X^-$$

$$\text{Base:} \quad BOH \rightleftharpoons OH^- + B^+$$

By applying to these dissociations the principles of chemical equilibrium which had been well systematized before the turn of the century, the behavior of acids and bases in aqueous solution could be quantitatively described, at least approximately. The Debye-Hückel theory (1923) permitted a refined treatment that was even better.

In 1923, Brønsted presented a new view of acid-base behavior which retained the soundness of the Arrhenius equilibrium treatment but which was conceptually broader and facilitated the correlation of a much larger body of information.[3] In Brønsted terms, an acid is any substance that can give up a proton, and a base is a substance that can accept a proton. The hydroxide ion, to be sure, is such a proton acceptor and hence a Brønsted base, but it is not unique; it is one of many species that can exhibit basic behavior. When an acid yields a proton, the deficient species must have some proton affinity, and hence it is a base. Thus in the Brønsted treatment we encounter "conjugate" acid-base pairs:

$$\underset{\text{Acid}}{HB} \rightleftharpoons H^+ + \underset{\text{Base}}{B}$$

The acid HB may be electrically neutral, anionic, or cationic (e.g., HCl, HSO_4^-, NH_4^+), and thus we have not specified the charges on either HB or B.

As the elemental unit of positive charge, the proton possesses a charge density which makes its independent existence in a solution extremely unlikely. Thus, in order to transform HB into B, a proton acceptor (i.e., another base) must be present. Often, as in the dissociation of acetic acid in water, this base may be the solvent itself:

$$HOAc \rightleftharpoons H^+ + OAc^-$$

$$H_2O + H^+ \rightleftharpoons H_3O^+$$

$$\underset{\text{Acid}_1}{HOAc} + \underset{\text{Base}_2}{H_2O} \rightleftharpoons \underset{\text{Acid}_2}{H_3O^+} + \underset{\text{Base}_1}{OAc^-}$$

[3] The same ideas were proposed independently by Lowry in 1924; some writers speak of the Brønsted-Lowry theory.

The interaction of the two conjugate acid–base pairs (designated by subscripts 1 and 2) leads to an equilibrium in which some of the acetic acid molecules have transferred their protons to water. The protonated water molecule or hydrated proton, H_3O^+, may be called a "hydronium ion," but it is usually designated simply "hydrogen ion" and often written "H^+."[4]

Water is not the only solvent to which acids can transfer their protons, and we may write a general dissociation equation, where S is any solvent capable of accepting a proton:

$$HB + S \rightleftharpoons HS^+ + B$$

The species HS^+ is the solvated proton (H_3O^+ in water solution, H_2OAc^+ in glacial acetic acid, $H_3SO_4^+$ in sulfuric acid, $C_2H_5OH_2^+$ in ethanol, etc.). One of the important contributions of Brønsted theory is its emphasis on the role of the solvent in the dissociation of acids and bases. We may suppose that an acid has a certain intrinsic "acidity" if we wish, but the Brønsted treatment makes clear that the extent to which such an acid is dissociated in solution depends importantly upon the basicity of the solvent. Thus perchloric acid, $HClO_4$, is a strong acid, completely dissociated in water solution, but it is only slightly dissociated in nonaqueous sulfuric acid.

If HB is inherently a stronger acid than HS^+, it will transfer its proton to the solvent; in other words, the position of the equilibrium in the reaction $HB + S \rightleftharpoons HS^+ + B$ will lie toward the right. If HB is very much stronger than HS^+, the equilibrium will lie far to the right, and HB will be essentially 100% dissociated. A series of different acids, all of which are very much stronger than the solvated proton, will dissociate completely; such solutions will be brought to a level of acidity governed by the acid strength of HS^+. This is known as the *leveling effect.* Thus in aqueous solution the acids perchloric, nitric, and hydrochloric are equally strong, whereas in the less basic solvent, glacial acetic acid, the three acids are not leveled, and perchloric is stronger than the other two.

In Brønsted terms, the dissociation of bases is treated in a similar fashion except that here the process is promoted by the *acidity* of the solvent. Again, the general case may be formulated as the interaction of two conjugate pairs:

$$SH \rightleftharpoons S^- + H^+$$

$$B + H^+ \rightleftharpoons BH^+$$

$$\underset{\text{Base}_1}{B} + \underset{\text{Acid}_2}{SH} \rightleftharpoons \underset{\text{Acid}_1}{BH^+} + \underset{\text{Base}_2}{S^-}$$

An example is $NH_3 + H_2O \rightleftharpoons NH_4^+ + OH^-$. As with acids, bases may be of any charge type (neutral, cationic, or anionic). The charges have been placed in the foregoing equations simply to show that the base and its conjugate acid differ by

[4] The proton in aqueous solution may actually be more heavily hydrated than H_3O^+. For example, a species $H_9O_4^+$ ($H_3O^+ \cdot 3H_2O$ or $H^+ \cdot 4H_2O$) has been postulated on the basis of the infrared spectra of strong acid solutions, studies of the extraction of strong acids from water into certain organic solvents, and other experimental evidence. For an interesting review, see H. L. Clever, *J. Chem. Educ.*, **40**, 637 (1963).

one. If the solvent is sufficiently acidic, we may again encounter a leveling effect in which a series of bases are brought to a level of basicity in solution determined by the species S^-. In water, for example, so-called basic anhydrides like CaO yield OH^- by a process which may be written

$$O^{2-} + H_2O \rightleftharpoons 2OH^-$$

In anhydrous sulfuric acid sulfates are analogous to the basic anhydrides in the aqueous system:

$$SO_4^{2-} + H_2SO_4 \rightleftharpoons 2HSO_4^-$$

Neutralization reactions involving strong acids and bases in the various solvents become, in Brønsted terms, simply reactions between the cation and the anion of the solvent because of the leveling effect. Water, for example, dissociates as follows:

$$2H_2O \rightleftharpoons H_3O^+ + OH^-$$

One of the two water molecules in the equation acts as an acid, the other as a base, which is to say that water is *amphoteric*. Neutralization of strong acids and bases is simply the reverse of this self-dissociation or autoprotolysis reaction:

$$H_3O^+ + OH^- \rightleftharpoons 2H_2O$$

Likewise, in liquid ammonia solution, strong acids and bases are leveled to NH_4^+ and NH_2^-, respectively, and neutralization may be written

$$NH_4^+ + NH_2^- \rightleftharpoons 2NH_3$$

In sulfuric acid as a solvent, the reaction becomes

$$H_3SO_4^+ + HSO_4^- \rightleftharpoons 2H_2SO_4$$

The Brønsted treatment offers the conceptual advantage of unifying a number of acid-base processes which, in other terms, may appear different. Hydrolysis, for example, need no longer be distinguished as a special process. The hydrolysis of a salt, like sodium acetate, is simply the dissociation reaction of the acetate ion as a base:

$$OAc^- + H_2O \rightleftharpoons HOAc + OH^-$$

It may be seen that it will be a property of a conjugate acid-base pair that a strong acid has a weak conjugate base. Thus chloride ion, the conjugate base of the strong acid hydrochloric, is too weak a base to abstract protons from water, and hydrolysis of the chloride ion is negligible.

ANALYTICAL APPLICATIONS

We shall now review briefly the formulation of equilibrium constants for the four types of chemical reactions which are used in titrimetric analyses. We shall also examine some typical equilibrium calculations which are of interest in analytical chemistry.

We have already mentioned that water is a weak electrolyte, dissociating into H_3O^+ and OH^- ions:

$$2H_2O \rightleftharpoons H_3O^+ + OH^-$$

The extent of dissociation of water has been measured experimentally and at 25°C the hydronium and hydroxide ion concentrations have been found to be 1.0×10^{-7} *M*. This means that the value of K_w, the autoprotolysis constant of water, is 1.0×10^{-14} at 25°C:

$$K_w = [H_3O^+][OH^-]$$

$$K_w = (1.0 \times 10^{-7})(1.0 \times 10^{-7})$$

$$K_w = 1.0 \times 10^{-14}$$

The value of K_w at several other temperatures is shown in Table 5.4.

TABLE 5.4 Ion Product Constant, K_w, of Water at Different Temperatures

Temperature, °C	K_w	pK_w
0	1.14×10^{-15}	14.943
15	4.51×10^{-15}	14.346
25	1.01×10^{-14}	13.996
35	2.09×10^{-14}	13.680
50	5.47×10^{-14}	13.262
100	4.9×10^{-13}	12.31

The term *p*H is convenient to use to express hydrogen ion concentrations since the latter values are very small numbers and may vary over many orders of magnitude during titrations. The term *p*H is defined in such a way as to convert a negative power of ten into a small positive number. The definition is

$$pH = -\log [H_3O^+] = \log \frac{1}{[H_3O^+]}$$

Thus, a hydrogen ion concentration of 1.0×10^{-1} corresponds to a *p*H value of 1.00, and a hydrogen ion concentration of 1.0×10^{-13} becomes $pH = 13.00$. Such numbers, ranging from, say, 0 or 1 up to perhaps 13 or 14 are conveniently plotted on titration curves, as will be seen later. Also, in a later chapter it is seen that the electromotive force (voltage) developed by certain galvanic cells is more directly related to *p*H than to the hydrogen ion concentration itself.

The following examples illustrate the conversion of hydrogen ion concentration to *p*H, and vice versa.

Example 1. The hydrogen ion concentration of a solution is 5.0×10^{-7}. Calculate the pH.

$$pH = -\log (5.0 \times 10^{-7})$$
$$pH = -(\log 5.0 + \log 10^{-7})$$
$$pH = -(0.70 - 7.00)$$
$$pH = 7.00 - 0.70$$
$$pH = 6.30$$

Example 2. The pH of a solution is 10.70. Calculate the hydrogen ion concentration.

$$pH = 10.70 = -\log [H_3O^+]$$
$$\log [H_3O^+] = -10.70 = 0.30 - 11.00$$
$$[H_3O^+] = \text{antilog}\, 0.30 \times \text{antilog}\, (-11)$$
$$[H_3O^+] = 2.0 \times 10^{-11}$$

It is often convenient to use other p-functions analogous to pH. For example, the following functions are frequently used:

$$pOH = -\log [OH^-]$$
$$pK_a = -\log K_a$$
$$pK_b = -\log K_b$$
$$pK_w = -\log K_w$$

Note that since $[H_3O^+][OH^-] = K_w = 1.0 \times 10^{-14}$ at 25°C, then

$$pH + pOH = pK_w = 14.00$$

Also note that in pure water (neutral), $[H_3O^+] = [OH^-]$ and

$$pH = pOH = \frac{pK_w}{2}$$

At 25°C the values of the pH and pOH are 7.00 in neutral solutions. In acidic solutions $[H_3O^+] > 1.0 \times 10^{-7}$ and $pH < 7.00$. In basic solutions $[H_3O^+] < 1.0 \times 10^{-7}$ and $pH > 7.00$.

The following examples illustrate calculations involving acid–base equilibria.

Example 3. In 0.100 M solution at 25°C acetic acid is found to be 1.34% dissociated. Calculate the value of K_a for HOAc.

The dissociation reaction is

$$HOAc + H_2O \rightleftharpoons H_3O^+ + OAc^-$$

The concentrations of hydrogen and acetate ions are approximately equal since these ions are formed in equimolar amounts. (We neglect the small amount of hydrogen ion formed by the dissociation of water.) Hence

$$[H_3O^+] \cong [OAc^-] = 0.100 \text{ mmol/ml} \times 0.0134$$

$$[H_3O^+] \cong [OAc^-] = 0.00134 \text{ mmol/ml}$$

The concentration of undissociated acetic acid molecules is the original concentration minus the amount dissociated:

$$[HOAc] = 0.100 - 0.00134$$

$$[HOAc] \cong 0.0987 \cong 0.10 \text{ mmol/ml}$$

Hence

$$K_a = \frac{[H_3O^+][OAc^-]}{[HOAc]}$$

$$K_a = \frac{(0.00134)(0.00134)}{0.10}$$

$$K_a = 1.8 \times 10^{-5}$$

Example 4. Calculate the pH and pOH of a 0.050 M solution of HCl.
HCl is a strong acid, completely dissociated. Hence

$$[H_3O^+] = 0.050 \quad \text{or} \quad 5.0 \times 10^{-2}$$

$$pH = 2.00 - \log 5.0$$

$$pH = 1.30$$

Since $p\text{H} + p\text{OH} = 14.00$, $p\text{OH} = 12.70$.

Example 5. Calculate the pH of a 0.050 M solution of acetic acid. K_a of HOAc is 1.8×10^{-5}.

This is a weak acid and is only slightly dissociated. We can calculate the hydrogen ion concentration using the dissociation constant.

$$HOAc + H_2O \rightleftharpoons H_3O^+ + OAc^-$$

$$[H_3O^+] \cong [OAc^-]$$

$$[HOAc] = 0.050 - [H_3O^+]$$

Substituting in the equilibrium expression,

$$\frac{[H_3O^+][OAc^-]}{[HOAc]} = K_a$$

$$\frac{[H_3O^+]^2}{0.050 - [H_3O^+]} = 1.8 \times 10^{-5}$$

This is a quadratic equation and can be solved by using the quadratic formula. The calculation can be simplified by recognizing that the acid is weak, and the amount of acid that dissociates will be small with respect to the total amount present (the analytical concentration or formality, page 51). Therefore, a good approximation is

$$[\text{HOAc}] = 0.050 - [\text{H}_3\text{O}^+]$$

$$[\text{HOAc}] \cong 0.050\ M$$

Then

$$\frac{[\text{H}_3\text{O}^+]^2}{0.050} = 1.8 \times 10^{-5}$$

$$[\text{H}_3\text{O}^+] = 9.5 \times 10^{-4}$$

$$p\text{H} = 3.02$$

The value obtained by not making the approximation and solving the complete equation is 9.4×10^{-4} M. The percentage error is therefore

$$\frac{(9.5 - 9.4) \times 10^{-4}}{9.4 \times 10^{-4}} \times 100 = 1.1\%$$

Table 5.5 lists the error made by such an approximation for different values of K_a and concentrations of weak acid. Note that the error increases as the value of K_a increases and as the concentration of acid decreases. We will rarely deal with concentrations below 0.0100 M, but we do occasionally encounter an acid with a K_a as large as 10^{-3} or 10^{-2}. A complete solution of the quadratic is advised when the weak electrolyte has such a large dissociation constant.

TABLE 5.5 Error Introduced by Approximation* in Calculation of Hydrogen ion Concentration

Value of K_a	Analytical concentration, c_a	Approximate $[\text{H}_3\text{O}^+]$	Exact† $[\text{H}_3\text{O}^+]$	Percent error
1.00×10^{-6}	0.100	3.16×10^{-4}	3.16×10^{-4}	0.0
	0.0500	2.24×10^{-4}	2.23×10^{-4}	0.4
	0.0100	1.00×10^{-4}	0.995×10^{-4}	0.5
1.00×10^{-5}	0.100	1.00×10^{-3}	0.995×10^{-3}	0.5
	0.0500	7.07×10^{-4}	7.02×10^{-4}	0.7
	0.0100	3.16×10^{-4}	3.11×10^{-4}	1.6
1.00×10^{-4}	0.100	3.16×10^{-3}	3.11×10^{-3}	1.6
	0.0500	2.24×10^{-3}	2.19×10^{-3}	2.3
	0.0100	1.00×10^{-3}	0.95×10^{-3}	5.3
1.00×10^{-3}	0.100	1.00×10^{-2}	0.95×10^{-2}	5.3
	0.0500	7.07×10^{-3}	6.55×10^{-3}	7.9
	0.0100	3.16×10^{-3}	2.70×10^{-3}	17

* $c_a - [\text{H}_3\text{O}^+] \cong c_a$.

† Solving the equation $\dfrac{[\text{H}_3\text{O}^+]^2}{c_a - [\text{H}_3\text{O}^+]} = K_a$

Example 6. Calculate the pH of a solution which is 0.050 M in acetic acid and 0.10 M in sodium acetate.

This is sometimes called a "common-ion effect" problem. The dissociation of acetic acid

$$HOAc + H_2O \rightleftharpoons H_3O^+ + OAc^-$$

is decreased by the presence of the common ion, acetate, from the completely dissociated salt, sodium acetate:

$$NaOAc \longrightarrow Na^+ + OAc^-$$

According to the *Principle of Le Châtelier*, increasing the acetate ion concentration by adding sodium acetate to the acetic acid solution shifts the equilibrium to the left, thereby reducing the hydrogen ion concentration. The concentrations are as follows:

$$[HOAc] = 0.050 - [H_3O^+]$$

$$[OAc^-] = 0.10 + [H_3O^+]$$

Note that the concentrations of H_3O^+ and OAc^- are not equal as in Example 5 because of the additional acetate ion from the sodium acetate.

Noting again that the dissociation of acetic acid is small, we can make the approximations

$$[HOAc] \cong 0.050\,M$$

$$[OAc^-] \cong 0.10\,M$$

Then, substituting in the equilibrium expression,

$$\frac{[H_3O^+](0.10)}{0.050} = 1.8 \times 10^{-5}$$

$$[H_3O^+] = 9.0 \times 10^{-6}$$

$$p\text{H} = 5.05$$

The calculation given above is sometimes set up in a manner that seems different at first glance, but really amounts to the same thing. Solve the K_a expression for $[H_3O^+]$, take logarithms of both sides of the equation, and multiply by -1:

$$\frac{[H_3O^+][OAc^-]}{[HOAc]} = K_a$$

$$[H_3O^+] = K_a \times \frac{[HOAc]}{[OAc^-]}$$

$$-\log [H_3O^+] = -\log K_a - \log \frac{[HOAc]}{[OAc^-]}$$

$$p\text{H} = pK_a - \log \frac{[HOAc]}{[OAc^-]}$$

The equation in this form shows that the pH is a function of pK_a and the ratio of acid concentration to that of the salt, or conjugate base. Since HOAc and OAc^- are in the same solution, the volume cancels and the ratio of millimoles is the same as the ratio of molar concentrations. This logarithmic form of the dissociation constant frequently appears in biochemistry or physiology textbooks under the designation *Henderson-Hasselbalch equation.* In these texts the log term is usually inverted and the equation written

$$pH = pK_a + \log\frac{[OAc^-]}{[HOAc]}$$

For a weak base B and its conjugate acid HB^+, the equation is

$$pOH = pK_b - \log\frac{[B]}{[BH^+]}$$

Example 7. Calculate the pH of a solution made by mixing 50 ml of 0.10 M NH_3 and 50 ml of 0.040 M HCl. The K_b of NH_3 is 1.8×10^{-5}.

This is a problem involving a common ion, as in Problem 6. Here the ion, NH_4^+, is formed by the reaction of the acid and base. A stoichiometric calculation can be done as follows:

	NH_3	+ H_3O^+	$\longrightarrow$ NH_4^+	+ H_2O
mmol originally:	5.0	2.0	—	
change:	−2.0	−2.0	+2.0	
mmol at equilibrium:	3.0	—	2.0	

The concentrations are:

$$[NH_3] = 0.030 - [OH^-] \cong 0.030$$

$$[NH_4^+] = 0.020 + [OH^-] \cong 0.020$$

Using the logarithmic expression from Example 6,

$$pOH = pK_b - \log\frac{[NH_3]}{[NH_4^+]}$$

$$pOH = 4.74 - \log\frac{0.030}{0.020}$$

$$pOH = 4.56$$

$$pH = 9.44$$

Example 8. Calculate the pH of a 0.10 M solution of sodium acetate.

We have previously noted that the acetate ion is a base, reacting with water to produce hydroxide ions:

$$OAc^- + H_2O \rightleftharpoons HOAc + OH^-$$

The equilibrium expression is

$$\frac{[HOAc][OH^-]}{[OAc^-]} = K_b$$

Note that since $K_a \times K_b = K_w$ and $K_a = 1.8 \times 10^{-5}$, $K_b = 5.6 \times 10^{-10}$. The concentrations are

$$[HOAc] \cong [OH^-]$$

$$[OAc^-] = 0.10 - [OH^-]$$

Substituting in the expression for K_b,

$$\frac{[OH^-]^2}{0.10 - [OH^-]} = 5.6 \times 10^{-10}$$

We make the approximation

$$0.10 - [OH^-] \cong 0.10$$

since OAc^- is a weak base. Hence

$$\frac{[OH^-]^2}{0.10} = 5.6 \times 10^{-10}$$

$$[OH^-] = 7.5 \times 10^{-6}$$

$$pOH = 5.12$$

$$pH = 8.88$$

Solubility Equilibria

The equilibrium constant expressing the solubility of a precipitate in water is the familiar *solubility product constant.* For a precipitate of silver chloride, the equilibrium constant of the reaction

$$AgCl(s) \rightleftharpoons Ag^+(aq) + Cl^-(aq)$$

is

$$K = \frac{a_{Ag^+}a_{Cl^-}}{a_{AgCl}}$$

The activity of solid AgCl is constant and by convention we take it to be unity (page 104). The solid is only slightly soluble; hence the concentrations of Ag^+ and Cl^- ions are small and, unless large concentrations of other ions are present, activities can be approximated by molarities, giving

$$K_{sp} = [Ag^+][Cl^-]$$

The constant K_{sp} is called the solubility product constant.

The proper equilibrium expressions for a few other salts are given below:

Salt	K_{sp}
$BaSO_4$	$[Ba^{2+}][SO_4^{2-}]$
Ag_2CrO_4	$[Ag^+]^2[CrO_4^{2-}]$
CaF_2	$[Ca^{2+}][F^-]^2$
$Al(OH)_3$	$[Al^{3+}][OH^-]^3$

A general expression for the salt A_xB_y dissociating as follows:

$$A_xB_y = xA^{y+} + yB^{x-}$$

is

$$K_{sp} = [A^{y+}]^x[B^{x-}]^y$$

The numerical value of a solubility product constant can be readily calculated from the solubility of the compound. The calculation can be reversed, of course, and the solubility calculated from the K_{sp}. If the ions of the precipitate undergo reactions such as hydrolysis or complex formation, the calculations are more complicated. Such cases will be considered in a later chapter. Typical computations are illustrated in the following examples:

Example 1 The solubility of barium sulfate (MW = 233) at 25°C is 0.00023 g per 100 ml of solution. Calculate the value of K_{sp}.

The soublity is 0.23 mg/100 ml or 0.0023 mg/ml. The molarity is

$$\frac{0.0023\ \text{mg/ml}}{233\ \text{mg/mmol}} = 1.0 \times 10^{-5}\ \text{mmol/ml}$$

Since each mmol of $BaSO_4$ yields 1 mmol of Ba^{2+} and 1 mmol of SO_4^{2-},

$$[Ba^{2+}] = [SO_4^{2-}] = 1.0 \times 10^{-5}$$

$$K_{sp} = [Ba^{2+}][SO_4^{2-}] = [1.0 \times 10^{-5}]^2 = 1.0 \times 10^{-10}$$

Example 2. The solubility of silver chromate (MW = 332) is 0.0279 g/liter at 25°C. Calculate K_{sp}, neglecting hydrolysis of the chromate ion.

The molarity of Ag_2CrO_4 is

$$\frac{0.0279\ \text{g/liter}}{332\ \text{g/mol}} = 8.4 \times 10^{-5}\ \text{mol/liter}$$

Since each Ag_2CrO_4 yields two Ag^+ ions and one CrO_4^{2-} ion,

$$[Ag^+] = 2 \times 8.4 \times 10^{-5} = 1.7 \times 10^{-4}$$

$$[CrO_4^{2-}] = 8.4 \times 10^{-5}$$

Therefore,

$$K_{sp} = [Ag^+]^2[CrO_4^{2-}] = [1.7 \times 10^{-4}]^2[8.4 \times 10^{-5}]$$

$$K_{sp} = 2.4 \times 10^{-12}$$

It should be noted that one can judge on inspection the relative molar solubilities of two compounds from their solubility product constants only if they are the same type of compounds, i.e., both type AB or AB_2, etc. The solubility product constants of both AgCl and $BaSO_4$ are about 1×10^{-10}, and hence both are soluble to the extent of 1×10^{-5} mol/liter. However, a compound of the type AB_2 with the same molar solubility would have a small solubility product constant, 4×10^{-15}.

Example 3. The silver in a solution is precipitated by the addition of chloride ion. The final volume of the solution is 500 ml. What should be the concentration of Cl^- if no more than 0.10 mg of Ag^+ remains unprecipitated?

The concentration of Ag^+ is

$$[Ag^+] = \frac{0.10\ \text{mg}}{107.9\ \text{mg/mmol} \times 500\ \text{ml}}$$

$$[Ag^+] = 1.9 \times 10^{-6}\ M$$

$$[Ag^+][Cl^-] = K_{sp} = 1.0 \times 10^{-10}$$

$$(1.9 \times 10^{-6})[Cl^-] = 1.0 \times 10^{-10}$$

$$[Cl^-] = 5.3 \times 10^{-5}$$

Example 4. Calcium fluoride, CaF_2, has a K_{sp} of 4×10^{-11}. Predict whether a precipitate forms or not when the following solutions are mixed: (a) 100 ml of $2.0 \times 10^{-4}\ M\ Ca^{2+}$ plus 100 ml of $2.0 \times 10^{-4}\ M\ F^-$; (b) 100 ml of $2.0 \times 10^{-2}\ M\ Ca^{2+}$ plus 100 ml of $6.0 \times 10^{-3}\ M\ F^-$.

If the product $[Ca^{2+}][F^-]^2$ exceeds K_{sp}, precipitation will occur; if not, precipitation will not occur. Note that the concentrations are halved when the solutions are mixed.

(a) $[Ca^{2+}] = 1.0 \times 10^{-4}$ and $[F^-] = 1.0 \times 10^{-4}$

$[Ca^{2+}][F^-]^2 = (1.0 \times 10^{-4})(1.0 \times 10^{-4})^2 = 1.0 \times 10^{-12}$.

Since $1.0 \times 10^{-12} < 4 \times 10^{-11}$, precipitation does not occur.

(b) $[Ca^{2+}] = 1.0 \times 10^{-2}$ and $[F^-] = 3.0 \times 10^{-3}$

$[Ca^{2+}][F^-]^2 = (1.0 \times 10^{-2})(3.0 \times 10^{-3})^2 = 9.0 \times 10^{-8}$.

Since $9.0 \times 10^{-8} > 4 \times 10^{-11}$, precipitation does occur.

Complex Formation Equilibria

Reactions involving complex formation are utilized by the chemist in both titrimetric and gravimetric procedures. We shall discuss the topic in detail in a later chapter. At this point we shall use the complex formed by silver and ammonia to illustrate the formulation of the equilibrium constant and its use in a calculation.

A student of chemistry learns early in his career that solid silver chloride will dissolve in a solution of ammonia. The equation can be written molecularly as

$$AgCl(s) + 2NH_3(aq) \longrightarrow Ag(NH_3)_2Cl$$

The compound $Ag(NH_3)_2Cl$ is called a "complex." Actually, the compound is ionic, dissociating into $Ag(NH_3)_2{}^+$ and Cl^- ions, and the species $Ag(NH_3)_2{}^+$ is called a "complex ion."

The silver–ammonia complex ion is formed in steps by the addition of ammonia molecules, called the *ligand*, to silver ion, called the *central metal atom*:

$$Ag^+ + NH_3 \rightleftharpoons AgNH_3{}^+ \tag{1}$$

$$AgNH_3{}^+ + NH_3 \rightleftharpoons Ag(NH_3)_2{}^+ \tag{2}$$

The equilibrium constants for the two reactions are

$$\frac{[AgNH_3{}^+]}{[Ag^+][NH_3]} = K_1$$

and

$$\frac{[Ag(NH_3)_2^+]}{[AgNH_3^+][NH_3]} = K_2$$

They are called *stability* or *formation* constants of the complexes. Sometimes the reactions are written as the dissociation of the complex:

$$AgNH_3^+ \rightleftharpoons Ag^+ + NH_3$$

and the equilibrium constant is called the *instability* or *dissociation* constant. The stability and instability constants are simply reciprocals of one another. We shall write only stability constants in our discussion of this topic.

Note that if the two "stepwise" constants are multiplied together, we obtain the formation constant for the complex $Ag(NH_3)_2^+$:

$$\frac{[\cancel{AgNH_3^+}]}{[Ag^+][NH_3]} \times \frac{[Ag(NH_3)_2^+]}{[\cancel{AgNH_3^+}][NH_3]} = K_1 \times K_2$$

$$\frac{[Ag(NH_3)_2^+]}{[Ag^+][NH_3]^2} = \beta_2$$

The *overall formation constant* for the general reaction

$$M + nL \rightleftharpoons ML_n$$

is usually given the symbol β_n. Here M is the central metal ion and L is the ligand. The relations between stepwise and overall formation constants are: $\beta_1 = K_1$, $\beta_2 = K_1K_2$, $\beta_3 = K_1K_2K_3$, $\beta_4 = K_1K_2K_3K_4$, and so on. Here β_2 for $Ag(NH_3)_2^+$ is $2.3 \times 10^3 \times 6.0 \times 10^3 = 1.4 \times 10^7$. The symbol β is also commonly used to represent the fraction of a metal ion in the uncomplexed form (page 198). To avoid confusion, we shall use the symbol K rather than β for the overall formation constant, and we shall indicate which stepwise constants are involved.

The following example illustrates the use of the stability constant in a calculation.

Example. 1.0 mmol of AgCl is dissolved in 500 ml of ammonia. The final concentration of NH_3 (uncomplexed) is 0.10 *M*. Calculate the concentration of uncomplexed Ag^+ in the solution. The stability constants are: $K_1 = 2.3 \times 10^3$, $K_2 = 6.0 \times 10^3$.

Note that K_1 is relatively large, meaning that most of the Ag^+ will be converted to $AgNH_3^+$. But K_2 is also large, so that most of the $AgNH_3^+$ will be converted to $Ag(NH_3)_2^+$. Therefore, we shall assume most of the Ag^+ is converted to $Ag(NH_3)_2^+$, or

$$[Ag(NH_3)_2^+] = \frac{1.0 \text{ mmol}}{500 \text{ ml}} - [Ag^+]$$

$$[Ag(NH_3)_2^+] \cong 2.0 \times 10^{-3}$$

Since $[NH_3] = 0.10\ M$, we can solve for $[Ag^+]$:

$$\frac{[Ag(NH_3)_2^+]}{[Ag^+][NH_3]^2} = K_1 \times K_2$$

$$\frac{2.0 \times 10^{-3}}{[Ag^+](0.10)^2} = 2.3 \times 10^3 \times 6.0 \times 10^3$$

$$[Ag^+] = 1.4 \times 10^{-8}\ M$$

Oxidation-Reduction Equilibria

The equilibrium constant of an oxidation-reduction reaction is obtained from the potential of an appropriate galvanic cell. We shall consider this topic in detail in Chapter 10. The equilibrium expressions for such reactions are formulated in the usual way. For example, when iron(II) is titrated with cerium(IV) the reaction is

$$Fe^{2+} + Ce^{4+} \rightleftharpoons Fe^{3+} + Ce^{3+}$$

The equilibrium constant for this reaction is

$$\frac{[Fe^{3+}][Ce^{3+}]}{[Fe^{2+}][Ce^{4+}]} = K$$

The following example illustrates a calculation using this equilibrium constant.

Example. 5.00 mmol of Fe^{2+} is titrated with Ce^{4+} in sulfuric acid solution. Calculate the concentration of Fe^{2+} at the equivalence point. The volume of solution is 100 ml and the K for the reaction is 7.6×10^{12}.

Since the K is large we shall assume that the reaction goes essentially to completion. Then at the equivalence point

$$[Fe^{2+}] = [Ce^{4+}]$$

and

$$[Fe^{3+}] = [Ce^{3+}] = 0.050 - [Fe^{2+}]$$

Since $[Fe^{2+}]$ is small,

$$[Fe^{3+}] = [Ce^{3+}] \cong 0.050$$

Hence

$$\frac{(0.050)(0.050)}{[Fe^{2+}]^2} = 7.6 \times 10^{12}$$

$$[Fe^{2+}] = 1.8 \times 10^{-8}\ M$$

Simultaneous Equilibria

We have now reviewed the formulation of equilibrium constants for the four types of reactions used in titrimetric analyses. It is quite common for two or more of these equilibria to be established in the same solution. If any one ion is a participant in more than one of the equilibria, the reactions are said to "interact," or it is said that the equilibria are established simultaneously. The Principle of Le Châtelier can be used to make qualitative predictions of the result of such interactions. Quantitative calculations can be fairly complex.

As an example, consider the effect of *p*H on the solubility of the salt of a weak acid, say calcium carbonate, $CaCO_3$. The solubility equilibrium is

$$CaCO_3(s) \rightleftharpoons Ca^{2+} + CO_3^{2-} \quad (1)$$

The carbonate ion is a base and reacts with hydrogen ions in two steps:

$$CO_3^{2-} + H_3O^+ \rightleftharpoons HCO_3^- + H_2O \quad (2)$$

$$HCO_3^- + H_3O^+ \rightleftharpoons H_2CO_3 + H_2O \quad (3)$$

According to Le Châtelier's Principle increasing the concentration of H_3O^+ (lowering the *p*H) shifts equilibrium (2) to the right, lowering the CO_3^{2-} concentration. This in turn shifts equilibrium (1) to the right, causing more solid $CaCO_3$ to dissolve. It is a general principle that salts of weak acids, such as oxalates, sulfides, carbonates, and hydroxides, are more soluble in acidic than basic solutions.

The dissolving of AgCl in aqueous ammonia is an example of the interaction of solubility and complex formation equilibria. The solubility reaction is

$$AgCl(s) \rightleftharpoons Ag^+ + Cl^- \quad (1)$$

and the reactions forming complexes are

$$Ag^+ + NH_3 \rightleftharpoons AgNH_3^+ \quad (2)$$

$$AgNH_3^+ + NH_3 \rightleftharpoons Ag(NH_3)_2^+ \quad (3)$$

The addition of ammonia to a saturated solution of silver chloride lowers the concentration of Ag^+ by shifting equilibria (2) and (3) to the right. This, in turn, shifts equilibrium (1) to the right, causing more AgCl to dissolve.

We shall consider quantitative calculations involving simultaneous equilibria later in the text.

Completeness of Reactions Used for Analyses

We have said several times that a reaction used for a titration should be "complete" or nearly so at the equivalence point. We have also said that a reaction used for a separation by, say, a precipitation procedure should be "complete." In the latter case an excess of precipitating agent can be added to take advantage of the common-ion effect in forcing the reaction to completion. In titrations, however, we shall need to decide how large the constant must be for the process to be feasible. At this time we can answer the question partially; for a complete answer we shall need to know something about the indicator used to detect the end point of the titration.

Let us consider the following general reaction in aqueous solution:

$$A(aq) + B(aq) \rightleftharpoons C(aq) + D(aq)$$

where the equilibrium constant is

$$\frac{[C][D]}{[A][B]} = K$$

Suppose that we titrate 5.000 mmol of A with 5.000 mmol of B and the volume of the solution at the equivalence point is 100 ml. Let us calculate how large K must be for 99.9% of A and B to be converted to B and C at the equivalence point.

$$[C] = [D] = \frac{5.000 \times 0.999 \text{ mmol}}{100 \text{ ml}} \cong 0.050$$

$$[A] = [B] = \frac{5.000 \times 0.001 \text{ mmol}}{100 \text{ ml}} = 5.0 \times 10^{-5}$$

$$K = \frac{(0.050)(0.050)}{(5.0 \times 10^{-5})(5.0 \times 10^{-5})} = 1.0 \times 10^{6}$$

Had we asked that 99.99% of A and B be converted to C and D, the value of K would be 1.0×10^{8}. Table 5.6 lists examples of the four types of reactions used in titrations with the values of K required for 99.9% and 99.99% completion. We shall discuss the question of feasibility of titrations in detail when we consider these reactions in later chapters.

TABLE 5.6 Values of Equilibrium Constants for Titrations

Reaction	K	Value required for* 99.9%	99.99%
Acid–base:			
$HOAc + OH^- \rightleftharpoons OAc^- + H_2O$	$\frac{[OAc^-]}{[HOAc][OH^-]}$	2×10^7	2×10^9
Complex formation:			
$Ca^{2+} + EDTA^{4-} \rightleftharpoons CaEDTA^{2-}$	$\frac{[CaEDTA^{2-}]}{[Ca^{2+}][EDTA^{4-}]}$	2×10^7	2×10^9
Precipitation:			
$Ag^+ + Cl^- \rightleftharpoons AgCl(s)$	$\frac{1}{[Ag^+][Cl^-]}$	4×10^8	4×10^{10}
Oxidation–Reduction:			
$Fe^{2+} + Ce^{4+} \rightleftharpoons Fe^{3+} + Ce^{3+}$	$\frac{[Fe^{3+}][Ce^{3+}]}{[Fe^{2+}][Ce^{4+}]}$	1×10^6	1×10^8

* Calculated for the titration of 5.000 mmol of analyte, 100 ml volume at the equivalence point.

QUESTIONS

1. *Equilibrium expressions.* Write the equilibrium constant expressions for the following reactions:
   (a) $Ag_2CrO_4(s) \rightleftharpoons 2Ag^+ + CrO_4^{2-}$
   (b) $H_2CO_3 + H_2O \rightleftharpoons H_3O^+ + HCO_3^-$
   (c) $Cu(s) + 2Ag^+ \rightleftharpoons 2Ag(s) + Cu^{2+}$
   (d) $Cu^{2+} + 4NH_3 \rightleftharpoons Cu(NH_3)_4^{2+}$
   (e) $5Fe^{2+} + MnO_4^- + 8H^+ \rightleftharpoons 5Fe^{3+} + Mn^{2+} + 4H_2O$

2. *Terms.* Explain the meaning of the following terms: (a) autoprotolysis constant; (b) salt; (c) standard state; (d) leveling effect; (e) Le Châtelier's Principle; (f) common-ion effect; (g) ligand; (h) conjugate base.

3. *Weak electrolytes.* (a) What is meant by a weak electrolyte? (b) Explain how ions of a weak electrolyte affect the *p*H of an aqueous solution.

4. *Principle of Le Châtelier.* In which direction will the following equilibrium be shifted by making the indicated changes in the solution?

   $$CaF_2(s) \rightleftharpoons Ca^{2+} + 2F^-$$

   (a) Increase $[H_3O^+]$; (b) increase $[F^-]$; (c) increase the amount of solid CaF_2; (d) add more water.

5. *Brønsted theory.* HA and HB are both weak acids, but HA is stronger than HB. (a) Write equations for the reactions of the conjugate bases, A^- and B^-, with water. (b) Write the equilibrium constants for the two reactions. Which equilibrium constant is larger?

6. *Acids and bases.* Tell whether each of the following aqueous solutions is acidic, basic, or neutral. If acidic or basic, write an equation to show the reaction which produces hydrogen or hydroxide ions. (a) Sodium acetate; (b) ammonium chloride; (c) potassium nitrate; (d) barium sulfide; (e) ammonium perchlorate.

Multiple-choice: In the following multiple-choice questions, select the *one best* answer.

7. Which of the following statements is/are always true? (a) An acid and its conjugate base react to form a salt and water; (b) The acid H_2O is its own conjugate base; (c) The conjugate base of any weak acid is a strong base; (d) A base and its conjugate acid react to form a neutral solution; (e) all of the above; (f) none of the above.

8. Which of the following is the most basic solution? (a) *p*H = 5; (b) $[H_3O^+] = 10^{-9}$; (c) $[OH^-] = 10^{-4}$; (d) *p*H = 8; (e) $[H_3O^+] = 10^{-6}$.

9. Which of the following statements is/are true? (a) The solubility of $CaCO_3$ is increased by the addition of HCl; (b) The solubility of $BaSO_4$ is increased by the addition of H_2SO_4; (c) Addition of HCl to a saturated solution of AgCl lowers the solubility product constant of AgCl; (d) more than one of the above.

10. Which of the following statements is/are true? (a) Addition of NH_3 to a saturated solution of AgCl causes more AgCl to dissolve; (b) Addition of NaF to a solution of HF increases the *p*H; (c) Addition of NaCl to a solution of HCl increases the *p*H; (d) all of the above; (e) more than one but not all of the above.

11. In a 0.10 M solution of the salt NaF (a) $[HF] \cong [H_3O^+]$; (b) $[HF] \cong [OH^-]$; (c) $[H_3O^+] \cong [OH^-]$; (d) all of the above.

12. According to the Brønsted theory, (a) The stronger an acid the weaker is its conjugate base; (b) The strongest acid that exists in water is H_3O^+; (c) The conjugate base of H_3O^+ is OH^-; (d) all of the above; (e) two of the above.

PROBLEMS

1. *pH and pOH.* Convert the following hydrogen ion concentrations to pH, and hydroxide ion concentrations to pOH: (a) $[H_3O^+] = 0.0060$; (b) $[OH^-] = 0.10$; (c) $[H_3O^+] = 10$; (d) $[OH^-] = 4.0 \times 10^{-10}$.

2. *pH and pOH.* Convert the following hydrogen ion concentrations to pH and hydroxide ion concentrations to pOH: (a) $[H_3O^+] = 1.8 \times 10^{-5}$; (b) $[OH^-] = 3.0$; (c) $[H_3O^+] = 8.0 \times 10^{-13}$; (d) $[OH^-] = 2.8 \times 10^{-15}$.

3. *pH.* Convert the following pH values to hydrogen ion concentration: (a) $p\text{H} = -0.30$; (b) $p\text{H} = +0.30$; (c) $p\text{H} = 4.74$; (d) $p\text{H} = 11.52$; (e) $p\text{H} = 14.40$.

4. *pOH.* Convert the following pH values to hydroxide ion concentration: (a) $p\text{OH} = -0.70$; (b) $p\text{OH} = +0.70$; (c) $p\text{OH} = 5.26$; (d) $p\text{OH} = 10.84$; (e) $p\text{OH} = 14.22$.

5. *p-functions.* Convert the following to the corresponding p-functions: (a) $K_a = 3.4 \times 10^{-6}$; (b) $K_b = 5.6 \times 10^{-9}$; (c) $K_{sp} = 2 \times 10^{-11}$; *(d)* $[Cl^-] = 0.032\ M$.

6. *Dissociation constant.* A weak acid, HX, is 2.0% dissociated in a 0.080 M solution. (a) Calculate the value of K_a for HX. (b) Calculate the percent dissociation in a 0.040 M solution of HX. (c) At what concentration is the acid 1.0% dissociated?

7. *Dissociation constant.* A weak base, BOH, has a MW of 125 g/mol. A solution prepared by dissolving 0.625 g of BOH in 50.0 ml of solution has a pH of 11.30. Calculate the dissociation constant of BOH.

8. *pH calculations.* Calculate the pH of the following solutions: (a) 5.0 g of NaOH in 150 ml of solution; (b) 0.0500 g of HCl in 3.25 liters of solution; (c) 600 mg of HOAc in 100 ml of solution; (d) 3.0 g of NH_3 in 0.20 liter of solution.

9. *pH calculations.* Calculate the pH of the following solutions: (a) 10.0 g of NaOH in 200 ml of solution; (b) 0.365 g of HCl in 2.50 liters of solution; (c) 500 mg of HOAc in 60.0 ml of solution; (d) 2.7 g of NH_3 in 0.30 liter of solution.

10. *Common-ion effect.* Calculate the pH of a solution which is (a) 0.10 M in formic acid and 0.25 M in sodium formate; (b) 0.26 M in NH_3 and 0.12 M in NH_4Cl; (c) 0.050 M in HCl and 0.10 M in NaCl.

11. *Common-ion effect.* Calculate the number of grams of sodium formate, $NaHCO_2$, which should be added to 400 ml of a 0.10 M solution of formic acid to make the $p\text{H} = 4.00$.

12. *Salt solutions.* Calculate the pH of the following aqueous salt solutions: (a) 0.10 M NaCN; (b) 0.20 M NH_4Cl; (c) 0.50 M KCl.

13. *Mixtures of solutions.* Calculate the pH of the solution made as indicated. Assume that the volumes are additive. (a) 40 ml of 0.20 M HF + 60 ml of 0.080 M NaOH; (b)

50 ml of 0.12 M NH_3 + 50 ml of 0.060 M HCl; (c) 75 ml of 0.15 M NH_3 + 125 ml of 0.14 M HCl.

14. *Mixtures of solutions.* Calculate the pH of the solution made as indicated. Assume that the volumes are additive. (a) 150 ml of 0.095 M HCl + 100 ml of 0.20 M NaOH; (b) 60 ml of 0.12 M HOAc + 40 ml of 0.090 M NaOH; (c) 100 ml of 0.14 M NH_3 + 50 ml of 0.10 M HCl.

15. *Mixtures of solutions.* Calculate the pH of the solution made as indicated. Assume that the volumes are additive. (a) 60 ml of 0.10 M HCN + 40 ml of 0.15 M NaOH; (b) 100 ml of 0.14 M NH_3 + 70 ml of 0.20 M HCl; (c) 40 ml of 0.15 M HCl + 60 ml of 0.10 M NaOH.

16. *Strong and weak acids.* A chemist wishes to prepare 400 ml of a solution of pH 2.30 by dissolving an acid in water. Calculate the number of grams required if the acid is (a) $HClO_4$; (b) HF; (c) HOAc.

17. *Dissociation constant.* A weak acid, HX, has a MW of 125 g/mol. A solution prepared by dissolving 0.500 g of HX in 50.0 ml of solution has a pH of 2.70. Calculate the dissociation constant of HX.

18. *Dissociation constant.* The pH of a 0.10 M solution of the salt BX is 5.70. HX is a strong acid and BOH is a weak base. Calculate the dissociation constant of BOH.

19. *Mixtures of solutions.* Calculate the pH of solutions resulting from mixing equal volumes of the following solutions of strong acids or bases: (a) pH 1.00 + pH 2.00; (b) pH 1.00 + pH 5.00; (c) pH 1.00 + pH 13.00; (d) pH 1.00 + pH 14.00; (e) pH 5.00 + pH 9.00.

20. *Mixtures of solutions.* A chemist wishes to prepare 100 ml of a solution of pH 1.40 from solutions of HCl, pH 0.40, and NaOH, pH 13.70. How many milliliters of each solution should be mixed to give the desired solution? Assume that the volumes are additive.

21. *Approximations.* Calculate the hydrogen ion concentration of the following solutions in two ways: (1) Neglect the anion concentration, as was done on page 113; (2) solve the complete quadratic equation. (a) 0.10 M dichloroacetic acid; (b) 0.10 M $NaHSO_4$; (c) 0.10 M chloroacetic acid; (d) 0.10 M formic acid; (e) 0.10 M acetic acid. Calculate the percent error in the hydrogen ion concentration in each case.

22. *Solubility product constant.* Calculate the solubility product constants from the given solubilities: (a) AgI, 0.00235 mg/liter; (b) $Mg(OH)_2$, 0.000793 g/100 ml; (c) $Ag_2C_2O_4$, 3.28 mg/100 ml.

23. *Solubility.* From the solubility product constants listed in Table 3, Appendix I, calculate the following solubilities in water, neglecting effects such as hydrolysis: (a) $PbCO_3$ in mg/ml; (b) CaF_2 in g/200 ml; (c) $Pb(IO_3)_2$ in mg/100 ml.

24. *Solubility-common-ion effect.* Calculate the following molar solubilities, neglecting such effects as hydrolysis: (a) $BaSO_4$ in 0.05 M K_2SO_4; (b) CaF_2 in 0.10 M NaF; (c) Ag_2CrO_4 in 0.010 M $AgNO_3$.

25. *Precipitation.* Silver chromate, Ag_2CrO_4, has a K_{sp} of 2×10^{-12}. Predict whether a precipitate forms or not when the following solutions are mixed: (a) 100 ml of 0.020 M $AgNO_3$ + 100 ml of 2.0×10^{-5} M K_2CrO_4; (b) 100 ml of 0.060 M $AgNO_3$ + 200 ml of 3.0×10^{-4} M K_2CrO_4; (c) 50 ml of 4.0×10^{-5} M $AgNO_3$ + 150 ml of 4.0×10^{-4} M K_2CrO_4.

26. *Solubility.* 50 ml of 0.060 M K_2CrO_4 is mixed with 50 ml of 0.080 M $AgNO_3$. Calculate the following: (a) the solubility of Ag_2CrO_4 in the solution in moles per liter; (b) the concentrations of the following ions: Ag^+, CrO_4^{2-}, K^+, and NO_3^-.

27. *Solubility.* To 50 ml of 0.10 M NaCl is added 50 ml of 0.12 M $AgNO_3$. (a) Calculate the number of milligrams of Cl^- not precipitated. (b) If the precipitate is washed with 100 ml of water at 25°C, what is the maximum number of milligrams of AgCl that could be lost by solubility?

28. *Precipitation.* To 100 ml of a solution containing 1.0 g of $BaCl_2$ is added 1.0 g of Na_2SO_4. (a) Calculate the number of grams of $BaSO_4$ formed. (b) How many milligrams of barium remains unprecipitated?

29. *Precipitation.* The silver in a solution is precipitated by the addition of a chromate solution, the final volume being 200 ml. What must be the concentration of CrO_4^{2-} in the final solution in order to precipitate all but 0.10 mg of Ag^+?

30. *Complex formation.* Calculate the solubilities in g/100 ml of the following: (a) AgCl in 0.50 M NH_3; (b) AgBr in 2.0 M NH_3; (c) AgI in 0.10 M NH_3. The NH_3 concentrations are those of uncomplexed NH_3 at equilibrium.

31. *Complex formation.* If exactly 1.00 liter of 2.0 M NH_3 (final concentration) is required to dissolve 5.0 mmol of the silver salt AgX, what is the K_{sp} of AgX?

32. *Oxidation-reduction equilibrium.* The equilibrium constant of the reaction

$$Sn^{2+} + 2Ce^{4+} \rightleftharpoons Sn^{4+} + 2Ce^{3+}$$

is 5.4×10^{43} in sulfuric acid solution at 25°C. 4.00 mmol of Sn^{2+} is titrated with Ce^{4+} in sulfuric acid medium. Calculate the concentration of Sn^{2+}(a) at the equivalence point, and (b) after the addition of 12.0 mmol of Ce^{4+}. Assume that the volume is 100 ml in each case.

33. *Oxidation-reduction equilibrium.* The equilibrium constant of the reaction

$$2Ag^+ + Cu(s) \rightleftharpoons 2Ag(s) + Cu^{2+}$$

is 2.1×10^{15} at 25°C. An excess of copper turnings is added to a 0.10 M solution of Ag^+. Calculate the final concentration of Ag^+ after equilibrium is established.

34. *Oxidation-reduction equilibrium.* The equilibrium constant of the reaction

$$2H^+ + Cu(s) \rightleftharpoons H_2(g) + Cu^{2+}$$

is 4.7×10^{-12} at 25°C. Suppose that a strip of pure copper is placed in 100 ml of 1.0 M HCl which contains no Cu^{2+} ions. Assuming that the reaction above actually reaches equilibrium, calculate the number of grams of copper which dissolve. Assume that the pressure of H_2 is 1 atm.

35. *Completeness of reactions.* For the reaction

$$A + B \rightleftharpoons C + D$$

(all reactants in solution) calculate the value of the equilibrium constant for the following percent conversions of A and B into products. (The initial concentrations of A and B are 1.0 M.) (a) 67%; (b) 95%; (c) 99%.

36. *Completeness of reactions.* Repeat Problem 35 for the all-solution reaction

$$\mathrm{A + B \rightleftharpoons C}$$

37. *Equilibrium constant.* How large should the equilibrium constant be for the all-solution reaction

$$\mathrm{A + 2B \rightleftharpoons C + 2D}$$

so that all but 0.10% of A is converted into products when A and B are reacted in an initial molecular ratio 1:2?

CHAPTER SIX

acid-base equilibria

INTRODUCTION

Acid-base equilibrium is an extremely important topic throughout chemistry and in other fields, like agriculture, biology, and medicine, which utilize chemistry. Titrations involving acids and bases are widely employed in the analytical control of many products of commerce, and the dissociation of acids and bases exerts an important influence upon metabolic processes in the living cell. Acid-base equilibrium, as it is taught in analytical chemistry courses, offers the inexperienced student opportunity to broaden his understanding of chemical equilibrium and to gain the confidence to apply this understanding to a wide variety of problems.

In the evaluation of a reaction which is to serve as the basis for a titration, one of the most important aspects is the extent to which the reaction proceeds toward completion near the equivalence point. Stoichiometric calculations do not take into account the position of equilibrium toward which a chemical reaction tends. In stoichiometry, one calculates maximal yields of products (or consumption of reactants) with the implicit assumption that the reaction proceeds to completion, whereas in actuality the realization of completeness may require that one of the reactants be present in large excess or that a reaction product be removed from the mixture. Titrimetry by its very nature generally precludes forcing a reaction to completion by a large excess of reactant, and we shall see that the feasibility of a titration depends, at least in part, upon the position of equilibrium established when equivalent quantities of reactants have been mixed. Although our main goal in this chapter will be the understanding of acid-base titrations, other important aspects of acid-base chemistry will be discussed at appropriate points.

TITRATION CURVES

In examining a reaction to determine whether it can be used for a titration it is instructive to construct a *titration curve*. For acid-base titrations a titration curve consists of a plot of *p*H (or *p*OH) vs. the milliliters of titrant. Such curves are helpful in judging the feasibility of a titration and in selecting the proper indicator. We shall examine two cases, the titration of a strong acid with a strong base and the titration of a weak acid with a strong base.

Strong Acid–Strong Base Titration

Strong acids and bases are completely dissociated in aqueous solution (page 101). Hence the hydrogen or hydroxide ion concentration can be calculated directly from the stoichiometric amounts of acid and base that have been mixed. At the equivalence point the pH is determined by the extent to which water dissociates; at 25°C the *p*H of pure water is 7.00.

The following example illustrates the calculations involved to obtain the data needed to construct a titration curve.

Example. 50.0 ml of 0.100 *M* HCl is titrated with 0.100 *M* NaOH. Calculate the *p*H at the start of the titration and after the addition of 10.0, 50.0, and 60.0 ml of titrant.

(a) *Initial p*H. HCl is a strong acid and is completely dissociated. Hence

$$[H_3O^+] = 0.100$$

$$pH = 1.00$$

In making the calculations above we ignored the H_3O^+ contributed by the dissociation of water molecules. This is a valid assumption, of course, but it is worthwhile to examine more critically what we have done. The general approach is to find a set of equations sufficient in principle to permit calculation of the concentration of each chemical species in the solution. This means as many equations as unknowns. Here there are three species: H_3O^+, OH^-, and Cl^-. We need three independent equations in order to determine the concentrations of these ions. We always have available the autoprotolysis constant of water,

$$[H_3O^+][OH^-] = K_w \tag{1}$$

Second, the *electroneutrality condition* must be satisfied; that is, there must be an equal number of positive and negative charges in any solution. The *charge balance* equation is

$$[H_3O^+] = [OH^-] + [Cl^-] \tag{2}$$

Since HCl is completely dissociated, our third equation, called a *mass balance* on chloride or *analytical concentration*, is

$$[Cl^-] = 0.10 \tag{3}$$

We recognize that water is a very weak acid compared with HCl and thus $[OH^-]$ is very small, smaller, indeed, than the 10^{-7} *M* found in pure water, because the dissociation of water is repressed by the H_3O^+ from the strong acid HCl. Thus we neglect $[OH^-]$ as compared with $[Cl^-] = 0.10\,M$ and obtain

$$[H_3O^+] = [OH^-] + 0.10 \cong 0.10$$

From equation (1) it is now readily obtained that

$$[OH^-] = \frac{K_w}{[H_3O^+]} = \frac{1.0 \times 10^{-14}}{1.0 \times 10^{-1}} = 1.0 \times 10^{-13}$$

Thus the pH of the solution is 1.00 and its pOH is 13.00. Note that the approximation $[H_3O^+] \cong [Cl^-]$ is obviously all right in fairly concentrated HCl solutions. However, in very dilute HCl, say $1.0 \times 10^{-7}\,M$, a large error would result if the H_3O^+ produced by water were neglected. In such a case one would need to substitute for $[OH^-]$ in equation (2), giving

$$[H_3O^+] = \frac{K_w}{[H_3O^+]} + [Cl^-]$$

The quadratic equation can be solved for $[H_3O^+]$.

(b) pH *after the addition of 10.00 ml of base.* The reaction which occurs during the titration is

$$H_3O^+ + OH^- \rightleftharpoons 2H_2O$$

The equilibrium constant, K, is $1/K_w$ or 1.0×10^{14}. This is a very large constant, meaning that the reaction goes well to completion. We start with 50.0 ml × 0.100 mmol/ml = 5.00 mmol of H_3O^+ and add 10.0 ml × 0.100 mmol/ml = 1.00 mmol OH^-. Assuming that the reaction goes to completion, we have 5.00 − 1.00 = 4.00 mmol excess H_3O^+ in 60.0 ml of solution. Hence

$$[H_3O^+] = \frac{4.00\ \text{mmol}}{60.0\ \text{ml}} = 6.67 \times 10^{-2}\ \text{mmol/ml}$$

$$pH = 2 - \log 6.67 = 1.18$$

The pH values for other volumes of titrant can be calculated in a similar fashion.

(c) pH *at the equivalence point.* The equivalence point is reached when 50.0 ml of NaOH has been added. Because the salt formed in the reaction (NaCl) is neither acidic nor basic in water solution (not hydrolyzed), the solution is neutral: $[H_3O^+] = [OH^-] = 1.0 \times 10^{-7}$. Hence the pH is 7.00, as in pure water.

(d) pH *after addition of 60.0 ml of base.* At this point 60.0 ml × 0.100 mmol/ml = 6.00 mmol of OH^- has been added. We have 6.00 − 5.00 = 1.00 mmol excess OH^- in 110 ml of solution. Hence

$$[OH^-] = \frac{1.00\ \text{mmol}}{110\ \text{ml}} = 9.1 \times 10^{-3}\ \text{mmol/ml}$$

$$pOH = 3 - \log 9.1 = 2.04$$

$$pH = 14.00 - 2.04 = 11.96$$

The pH values at other points in the titration are given in Table 6.1 and the titration curve is shown in Fig. 6.1. Note that initially the pH rises only gradually as the titrant is added, rises more rapidly as the equivalence point is approached, and then increases by about 5.20 units for the addition of only 0.10 ml of base at

TABLE 6.1 Titration of a Strong Acid and a Weak Acid with NaOH (50.0 ml of 0.100 *M* Acid Titrated with 0.100 *M* NaOH)

NaOH, ml	Volume of solution	*p*H, HCl	*p*H, HB*
0.00	50.0	1.00	3.00
10.00	60.0	1.18	4.40
20.00	70.0	1.37	4.82
25.00	75.0	1.48	5.00
30.00	80.0	1.60	5.18
40.00	90.0	1.95	5.60
49.00	99.0	3.00	6.69
49.90	99.9	4.00	7.70
†49.95	99.95	4.30	8.00
50.00	100.0	7.00	8.85
†50.05	100.05	9.70	9.70
50.10	100.10	10.00	10.00
51.00	101.0	11.00	11.00
60.00	110.0	11.96	11.96
70.00	120.0	12.23	12.23

* Assuming 20 drops per milliliter, these values are 1 drop before and 1 drop after the equivalence point.

† $K_a = 1.0 \times 10^{-5}$

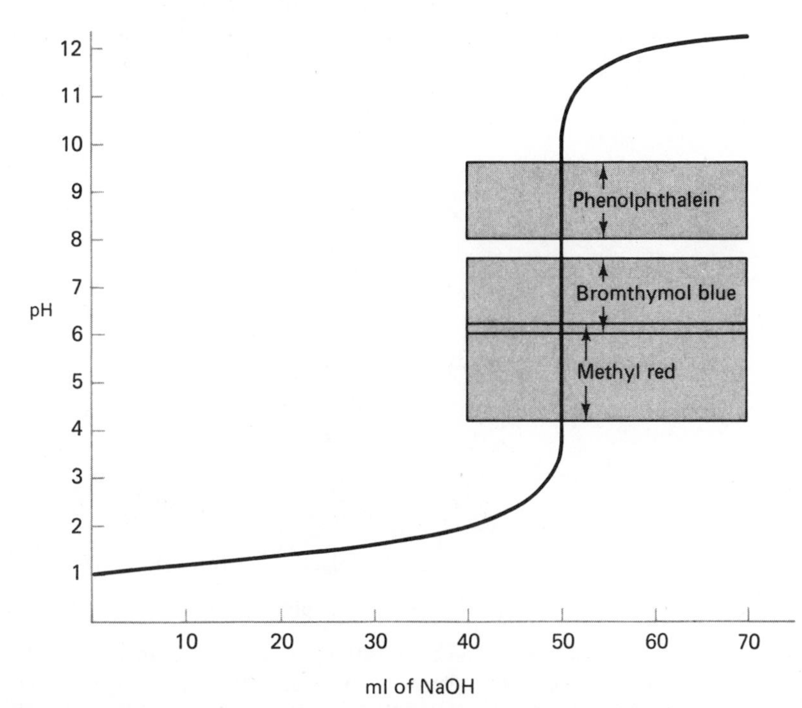

Figure 6.1 Strong acid–strong base titration curve: 50 ml of 0.10 *M* HCl titrated with 0.10 *M* NaOH.

the equivalence point. Beyond the equivalence point the pH again increases only slowly as titrant is added.

The shaded areas in Fig. 6.1 are the "ranges" over which three visual indicators (see next section for discussion of indicators) change color. It is apparent that as the strong acid is titrated, the large increase in pH at the equivalence point is sufficient to span the ranges of all three indicators. Hence any one of these indicators would change color within 1 or 2 drops of the equivalence point.

The titration curve for a strong base titrated with a strong acid, say NaOH with HCl, would be identical to the curve in Fig. 6.1 if pOH were plotted vs. volume of HCl. If pH is plotted the curve in Fig. 6.1 is simply inverted, starting at a high value and dropping to a low pH after the equivalence point.

Weak Acid–Strong Base Titration

Let us now consider the titration of a weak acid with a strong base. We reviewed the principles involved in the dissociation of weak acids in Chapter 5. The following example illustrates the calculations needed to obtain the data for a titration curve.

Example. 50.0 ml of a 0.100 *M* solution of a weak acid, HB, $K_a = 1.0 \times 10^{-5}$, is titrated with 0.100 *M* NaOH. Calculate the pH at the start of the titration and after the addition of 10.0, 50.0, and 60.0 ml of titrant.

(a) *Initial pH.* Since HB is weakly dissociated and produces one B^- and one H_3O^+ on dissociation,

$$HB + H_2O \rightleftharpoons H_3O^+ + B^-$$

we assume that

$$[H_3O^+] \cong [B^-]$$

and

$$[HB] = 0.100 - [H_3O^+] \cong 0.100$$

Substituting in the expression for K_a:

$$\frac{[H_3O^+][B^-]}{[HB]} = K_a$$

$$\frac{[H_3O^+]^2}{0.10} = 1.0 \times 10^{-5}$$

$$[H_3O^+] = 1.0 \times 10^{-3}$$

$$pH = 3.00$$

Let us now examine the assumptions we made to simplify the calculations. We will proceed as we did previously (page 129). There are four species in the solution: H_3O^+, OH^-, HB, and B^-. We need four equations. Three are already familiar: the autoprotolysis constant of water, K_w, the dissociation constant of the weak acid, K_a, and the charge balance equation. The latter is

$$[H_3O^+] = [B^-] + [OH^-] \quad (1)$$

The fourth equation is the mass balance on B, that is the *analytical concentration* (page 51),

$$[HB] + [B^-] = 0.10 \quad (2)$$

This equation says that all of the weak acid introduced into the solution consists of HB molecules or B^- ions.

In solving these equations we first note that the solution is acidic, and the concentration of OH^- ions is small. We therefore neglect $[OH^-]$ in equation (1):

$$[H_3O^+] \cong [B^-]$$

Also, since HB is weak $[B^-]$ is small compared with [HB]. Thus equation (2) becomes

$$[HB] \cong 0.10$$

Substitution into the expression for K_a gives

$$\frac{[H_3O^+]^2}{0.10} = 1.0 \times 10^{-5}$$

$$[H_3O^+] = 1.0 \times 10^{-3} \qquad \text{and} \qquad [OH^-] = 1.0 \times 10^{-11}$$

Note that the assumption that $[OH^-]$ can be neglected in equation (1) is a good one since

$$1.0 \times 10^{-3} \cong 1.0 \times 10^{-3} + 1.0 \times 10^{-11}$$

Checking the second assumption in equation (2) gives

$$0.10 + 0.001 \cong 0.10$$

Here the error is larger. The relative error (disregarding significant figures) is

$$\frac{0.001}{0.10} \times 100 = 1.0\%$$

(b) *pH after addition of 10.0 ml of base.* We started with $50.0 \times 0.100 = 5.00$ mmol of HB and have added $10.0 \times 0.100 = 1.00$ mmol of OH^-. Assuming the reaction is complete, we have 4.00 mmol of HB remaining and have produced 1.00 mmol of B^-. Hence

$$[HB] = \frac{4.00}{60.0} - [H_3O^+] \cong \frac{4.00}{60.0}$$

$$[B^-] = \frac{1.00}{60.0} + [H_3O^+] \cong \frac{1.00}{60.0}$$

Substituting in the expression for K_a,

$$\frac{[H_3O^+](1.00/60.0)}{4.00/60.0} = 1.0 \times 10^{-5}$$

Note that the volume cancels. Hence

$$[H_3O^+] = 4.0 \times 10^{-5}$$

$$pH = 5 - \log 4.0 = 4.40$$

Let us now check the assumptions we made above. We have five species in solution: H_3O^+, OH^-, HB, B^-, and Na^+. Our five equations are K_a, K_w, the charge balance equation,

$$[H_3O^+] + [Na^+] = [B^-] + [OH^-] \tag{1}$$

and the mass balance equations,

$$[HB] + [B^-] = \frac{5.00 \text{ mmol}}{60.0 \text{ ml}} = 0.0833 \tag{2}$$

$$[Na^+] = \frac{1.00 \text{ mmol}}{60.0 \text{ ml}} = 0.0167 \tag{3}$$

Since the solution is acidic we assume that $[OH^-]$ is small; equation (1) becomes

$$[H_3O^+] + 0.0167 \cong [B^-]$$

But since HB is a weak acid, $[H_3O^+]$ is probably much smaller than 0.0167. Then

$$[B^-] \cong 0.0167$$

Substituting this into equation (2) gives

$$[HB] + 0.0167 = 0.0833$$

$$[HB] = 0.0666$$

These values of [HB] and $[B^-]$ can be substituted into the expression for K_a, giving

$$\frac{[H_3O^+] \times 0.0167}{0.0666} = 1.0 \times 10^{-5}$$

$$[H_3O^+] = 4.0 \times 10^{-5} \quad \text{and} \quad [OH^-] = 2.5 \times 10^{-10}$$

Checking the assumption in equation (1) gives

$$4.0 \times 10^{-5} + 0{:}0167 = 0.0167 + 2.5 \times 10^{-10}$$

The left-hand term differs from the right by only about 0.2%.

(c) *pH at the equivalence point.* 50.0 ml of NaOH has been added and 5.00 mmol of B^- has been formed (i.e., $[B^-] = 0.050\ M$). B^- is a base and its reaction with water is

$$B^- + H_2O \rightleftharpoons HB + OH^-$$

We make the following assumption:

$$[HB] \cong [OH^-]$$

Substituting in the expression for K_b,

$$\frac{[HB][OH^-]}{[B^-]} = K_b = 1.0 \times 10^{-9}$$

$$\frac{[OH^-]^2}{0.050} = 1.0 \times 10^{-9}$$

$$[OH^-] = 7.1 \times 10^{-6}$$

$$pOH = 5.15 \quad \text{and} \quad pH = 8.85$$

We can now check the assumption made above. The solution contains the same five species as in part (b). The five equations are K_b, K_w, the charge balance equation,

$$[H_3O^+] + [Na^+] = [B^-] + [OH^-] \tag{1}$$

and the mass balance equations,

$$[HB] + [B^-] = \frac{5.00 \text{ mmol}}{100 \text{ ml}} = 0.050 \tag{2}$$

$$[Na^+] = 0.050 \tag{3}$$

In this situation $[H_3O^+]$, $[OH^-]$, and $[HB]$ are all small terms, and in order to obtain a useful expression involving them equations (1) and (2) should be added, giving

$$[H_3O^+] + [Na^+] + [HB] = [OH^-] + 0.050$$

Since $[Na^+] = 0.050$, this becomes

$$[H_3O^+] + [HB] = [OH^-] \tag{4}$$

Since the solution is basic, $[H_3O^+]$ is small and we can assume it is negligible, giving

$$[HB] \cong [OH^-]$$

If [HB] is neglected in equation (2), we have

$$[B^-] \cong 0.050$$

We can substitute in the expression for K_b, giving

$$\frac{[OH^-]^2}{0.050} = 1.0 \times 10^{-9}$$

$$[OH^-] = [HB] = 7.1 \times 10^{-6} \qquad \text{and} \qquad [H_3O^+] = 1.4 \times 10^{-9}$$

Checking the assumption in equation (4),

$$1.4 \times 10^{-9} + 7.1 \times 10^{-6} = 7.1 \times 10^{-6}$$

The left-hand term differs from the right by only about 0.02%. Checking equation (2),

$$7.1 \times 10^{-6} + 0.050 = 0.050$$

The two terms differ by only about 0.01%.

(d) *pH after the addition of 60.0 ml of base.* This is 10.0 ml, or 1.00 mmol, past the equivalence point. The OH^- ion produced by B^- in the reaction

$$B^- + H_2O \rightleftharpoons HB + OH^-$$

can be neglected since excess OH^- shifts the equilibrium to the left. The *p*H is calculated from the excess strong base, that is,

$$[OH^-] \cong \frac{1.00 \text{ mmol}}{110 \text{ ml}} = 9.1 \times 10^{-3}$$

$$pOH = 2.04$$

$$pH = 11.96$$

We can check this assumption as before. We have the same five species as above.

The five equations are: K_w, K_a (or K_b), the charge balance equation

$$[H_3O^+] + [Na^+] = [OH^-] + [B^-] \tag{1}$$

and the mass balance equations

$$[HB] + [B^-] = \frac{5.0 \text{ mmol}}{110 \text{ ml}} = 0.045 \tag{2}$$

$$[Na^+] = \frac{6.0 \text{ mmol}}{110 \text{ ml}} = 0.055 \tag{3}$$

We assume that $[H_3O^+]$ and [HB] are negligible, giving

$$[B^-] = 0.045$$

and

$$0.055 = [OH^-] + 0.045$$

$$[OH^-] = 1.0 \times 10^{-2} \qquad \text{and} \qquad [H_3O^+] = 1.0 \times 10^{-12}$$

[HB] can be obtained from K_a:

$$\frac{(1.0 \times 10^{-12})(0.045)}{[HB]} = 1.0 \times 10^{-5}$$

$$[HB] = 4.5 \times 10^{-9}$$

Checking equation (2) gives

$$4.5 \times 10^{-9} + 0.045 = 0.045$$

The left-hand side differs from the right by about 1×10^{-5}%. Substitution in equation (1) gives

$$1.0 \times 10^{-12} + 0.055 = 0.010 + 0.045$$

The two sides differ by only about 2×10^{-9}%.

The *p*H values at other points in the titration are given in Table 6.1 and the titration curve is shown in Fig. 6.2. The curve for the titration of a strong acid is included in this figure for comparison. Note that the curve for a weak acid begins to rise rapidly as base is first added; the rate of increase slows down as the concentration of B^- increases. The solution is said to be *buffered* (see page 145) in this region where the rate of increase in *p*H is slow. Note that when half of the acid is neutralized, $[HB] \cong [B^-] = 2.5$ mmol/125 ml. Since

$$pH = pK_a - \log \frac{[HB]}{[B^-]}$$

$$pH \cong pK_a = 5.00$$

After the halfway point, the *p*H slowly increases again until the large change occurs at the equivalence point.

Titration curves for weak acids of pK_a values 3.00, 5.00, 7.00, and 9.00 are included in Fig. 6.2. The value for the K of the reaction occurring during titration

$$HB + OH^- \rightleftharpoons B^- + H_2O$$

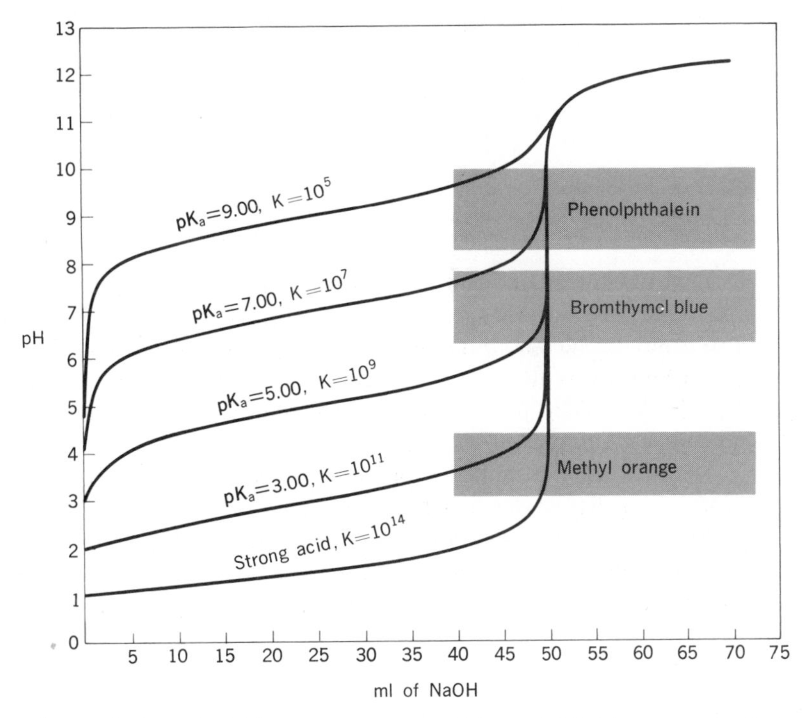

Figure 6.2 Typical acid-base titration curves: 50.0 ml of 0.100 *M* monoprotic acid titrated with 0.100 *M* NaOH; pK_a values of the acids are shown on the curves.

is K_a/K_w, or 10^{11}, 10^9, 10^7, and 10^5, respectively, for acids with the above pK_a values. It is obvious that the weaker the acid, the smaller the value of K and the smaller the change in pH at the equivalence point.

It should also be noted that the pH at the equivalence point for an acid of $K_a = 1.0 \times 10^{-5}$ is 8.85. For the addition of 0.10 ml of titrant at the equivalence point the pH changes from 8.00 to 9.70. The pH range over which the indicator phenolphthalein changes color is about 8.0 to 9.6, and this indicator will change color at the equivalence point of this titration.

ACID-BASE INDICATORS

The analyst takes advantage of the large change in pH that occurs in titrations in order to determine when the equivalence point is reached. There are many weak organic acids and bases in which the undissociated and ionic forms show different colors. Such molecules may be used to determine when sufficient titrant has been added and are termed *visual indicators*. A simple example is *para*-nitrophenol, which is a weak acid, dissociating as follows:

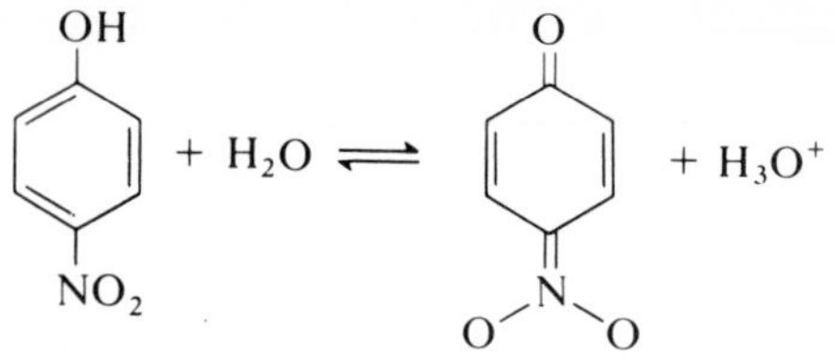

$$p\text{-}HOC_6H_4NO_2 + H_2O \rightleftharpoons O{=}C_6H_4{=}NO_2^- + H_3O^+$$

Colorless   Yellow

The undissociated form is colorless, but the anion, which has a system of alternating single and double bonds (a conjugated system), is yellow. Molecules or ions having such conjugated systems absorb light of longer wavelengths than comparable molecules in which no conjugated system exists. The light absorbed is often in the visible portion of the spectrum, and hence the molecule or ion is colored (see Chapter 14).

The well-known indicator phenolphthalein (below) is a diprotic acid and is colorless. It dissociates first to a colorless form and then, on losing the second hydrogen, to an ion with a conjugated system; a red color results. Methyl orange, another widely used indicator, is a base and is yellow in the molecular form. Addition of a hydrogen ion gives a cation which is pink in color.

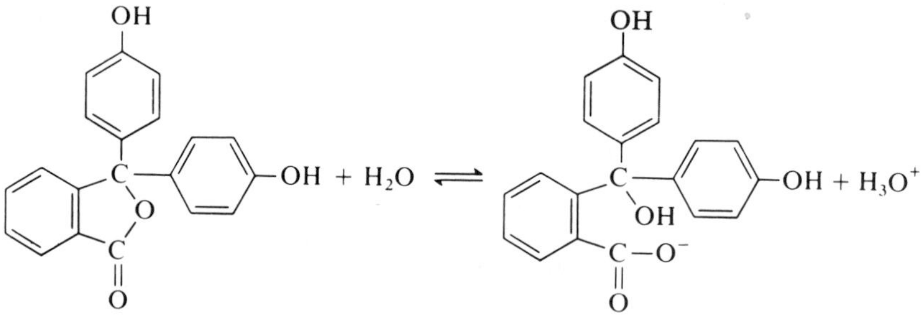

$$H_2In + H_2O \rightleftharpoons HIn^- + H_3O^+$$

H_2In, colorless
Phenolphthalein

HIn^-, colorless

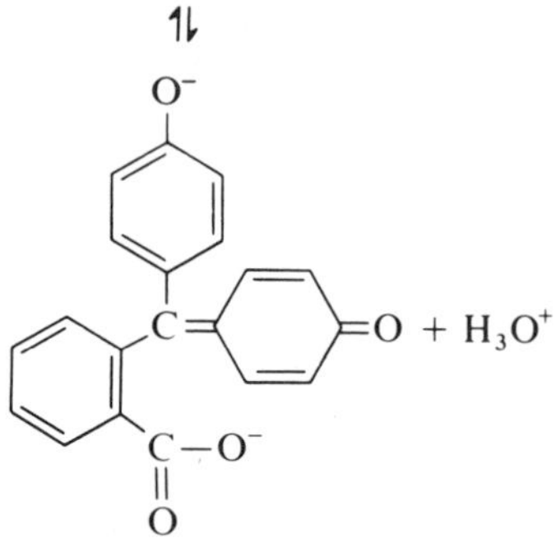

$$HIn^- \rightleftharpoons In^{2-} + H_3O^+$$

In^{2-}, red

$$Na^+\ {}^-O_3S{-}C_6H_4{-}N{=}N{-}C_6H_4{-}N(CH_3)_2 + H_3O^+ \rightleftharpoons Na^+\ {}^-O_3S{-}C_6H_4{-}NH{-}N{=}C_6H_4{=}\overset{+}{N}(CH_3)_2 + H_2O$$

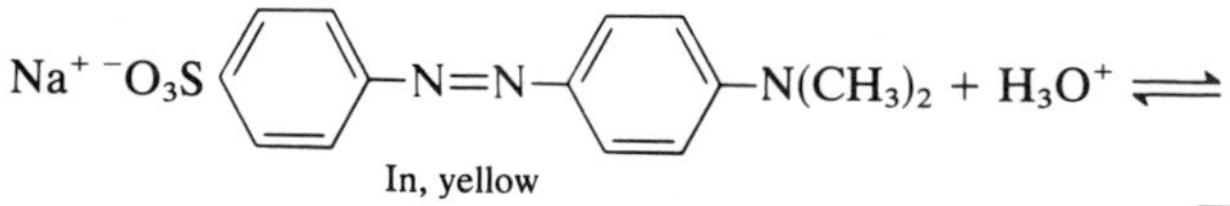

In, yellow
Methyl orange

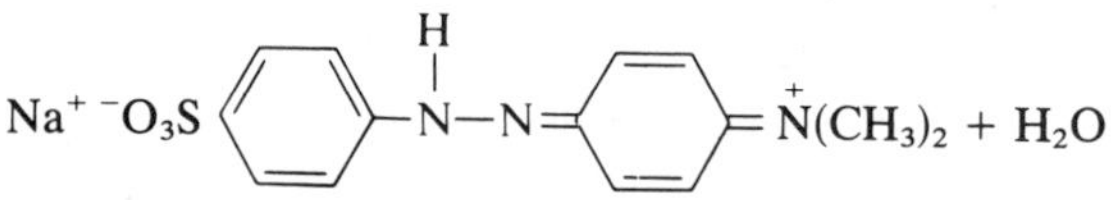

In^+, pink

For simplicity, let us designate an acid indicator as HIn, a basic indicator as InOH. The dissociation expressions are

$$HIn + H_2O \rightleftharpoons H_3O^+ + In^-$$

$$InOH \rightleftharpoons In^+ + OH^-$$

The dissociation constant of the acid is

$$K_a = \frac{[H_3O^+][In^-]}{[HIn)}$$

In the logarithmic form, this becomes

$$pH = pK_a - \log\frac{[HIn]}{[In^-]}$$

Let us for illustration assume that the molecule HIn is red in color and the ion In^- is yellow. Both forms are present, of course, in a solution of the indicator, their relative concentrations depending upon the *p*H. The color that the human eye detects depends upon the relative amounts of the two forms. Obviously, in solutions of low *p*H, the acid HIn predominates and we would expect to see only a red color. In solutions of high *p*H, In^- should predominate and the color should be yellow. At intermediate *p*H values, where the two forms are in about equal concentrations, the color might be orange.

Suppose that the pK_a of HIn is 5.00 and that a few drops of HIn are added to a solution of a strong acid which is being titrated with a strong base. The quantity of HIn added is so small that the amount of titrant used by HIn can be considered negligible. Now let us follow the ratio of the two colored forms as the *p*H changes during the titration. This is shown in Table 6.2. Let us also assume that the solution appears red to the eye when the ratio of $[HIn]/[In^-]$ is as large as 10:1,

TABLE 6.2 Ratio of Colored Forms of Indicator at Various *p*H Values

*p*H solution	$[HIn]/[In^-]$ ratio	Color	
1	10,000:1	Red	
2	1000:1	Red	
3	100:1	Red	
4	10:1	Red	Range
5	1:1	Orange	Range
6	1:10	Yellow	Range
7	1:100	Yellow	
8	1:1000	Yellow	

and yellow when this ratio is 1:10 or less. In such a case, the minimum change in *p*H, designated ΔpH, required to cause a color change from red to yellow is 2 units:

$$\text{Red:} \quad pH_r = pK_a - \log 10/1 = 5 - 1$$

$$\text{Yellow:} \quad pH_y = pK_a - \log 1/10 = 5 + 1$$

$$\Delta pH = pH_r - pH_y = (5 - 1) - (5 + 1) = -2$$

This minimum change in *p*H required for a color change is referred to as the "indicator range." In our example, the range is 4 to 6. At intermediate *p*H values, the color shown by the indicator is not red or yellow but some shade of orange. At *p*H 5, the pK_a of HIn, the two colored forms are in equal concentrations, that is, HIn is half-neutralized. Frequently, one hears terminology such as "An indicator which changed color at *p*H 5 was employed." This means that the pK_a of the indicator is 5, and the range is approximately *p*H 4 to 6.

Table 6.3 lists some acid-base indicators together with their approximate ranges. Note that the ranges are roughly 1 to 2 *p*H units, in general agreement with the assumption we made above. Actually, the range may not be symmetrical about the *pK* of the indicator, since a higher ratio may be required for the observer to see one form than is required to see the other. It should also be noted that various indicators change color at widely different *p*H values. It is necessary for the analyst to select the proper indicator for his titration.

TABLE 6.3 Some Acid–Base Indicators

Indicator	Color change with increasing *p*H	*p*H range
Picric acid	Colorless to yellow	0.1–0.8
Thymol blue	Red to yellow	1.2–2.8
2,6-Dinitrophenol	Colorless to yellow	2.0–4.0
Methyl yellow	Red to yellow	2.9–4.0
Bromphenol blue	Yellow to blue	3.0–4.6
Methyl orange	Red to yellow	3.1–4.4
Bromcresol green	Yellow to blue	3.8–5.4
Methyl red	Red to yellow	4.2–6.2
Litmus	Red to blue	4.5–8.3
Methyl purple	Purple to green	4.8–5.4
p-Nitrophenol	Colorless to yellow	5.0–7.0
Bromcresol purple	Yellow to purple	5.2–6.8
Bromthymol blue	Yellow to blue	6.0–7.6
Neutral red	Red to yellow	6.8–8.0
Phenol red	Yellow to red	6.8–8.4
p-α-Naphtholphthalein	Yellow to blue	7.0–9.0
Phenolphthalein	Colorless to red	8.0–9.6
Thymolphthalein	Colorless to blue	9.3–10.6
Alizarin yellow R	Yellow to violet	10.1–12.0
1,3,5-Trinitrobenzene	Colorless to orange	12.0–14.0

Selection of Proper Indicator

In Fig. 6.2 the shaded areas are the indicator ranges of methyl orange (3.1 to 4.4), bromthymol blue (6.0 to 7.6), and phenolphthalein (8.0 to 9.6). It is apparent that as a strong acid is titrated, the large change in pH at the equivalence point is sufficient to span the ranges of all three indicators. Hence any one of these indicators would change color within one or two drops of the equivalence point, as would any other indicator changing color between pH 4 and 10.

In the titration of weaker acids, the choice of indicators is much more limited. For an acid of pK_a 5, approximately that of acetic acid, the pH is higher than 7 at the equivalence point and the change in pH is relatively small. Phenolphthalein changes color at approximately the equivalence point and is a suitable indicator.

In the case of a very weak acid, for example, $pK_a = 9$, no large change in pH occurs in the vicinity of the equivalence point. Hence a large volume of base would be required to change the color of an indicator, and the equivalence point could not be detected with the usually desired precision.

As a general rule, then, one should select an indicator which changes color at approximately the pH at the equivalence point of the titration. For weak acids, the pH at the equivalence point is above 7 and phenolphthalein is the usual choice. For weak bases, where the pH is below 7, methyl red (4.2 to 6.2) or methyl orange is widely used. For strong acids and strong bases, methyl red, bromthymol blue, and phenolphthalein are suitable.

Indicator Errors

There are at least two sources of errors in the determination of the end point of a titration using visual indicators. One occurs when the indicator employed does not change color at the proper pH. This is a determinate error and can be corrected by the determination of an *indicator blank.* The latter is simply the volume of acid or base required to change the pH from that at the equivalence point to the pH at which the indicator changes color. The indicator blank is usually determined experimentally.

A second error occurs in the case of very weak acids (or bases) where the slope of the titration curve is not great and hence the color change at the end point is not sharp. Even if the proper indicator is employed, an indeterminate error occurs and is reflected in a lack of precision in deciding exactly when the color change occurs. The use of a nonaqueous solvent (page 154) may improve the sharpness of the end point in such cases.

In order to sharpen the color change shown by some indicators, mixtures of two indicators, or of an indicator and an indifferent dye, are sometimes used. The familiar "modified methyl orange" for carbonate titrations is a mixture of methyl orange and the dye xylene cyanole FF. The dye absorbs some of the wavelengths of light that are transmitted by both colored forms, thus cutting down on the overlapping of the two colors. At an intermediate pH, the methyl orange assumes a color which is almost complementary to that of xylene cyanole FF, and the solution thus appears gray. This color change is more easily detected than the gradual change of methyl orange from yellow to red through a number of shades of orange. Many mixtures of two indicators have been recommended for improved color changes.

FEASIBILITY OF ACID-BASE TITRATIONS

We have previously mentioned that for a chemical reaction to be suitable for use in a titration, the reaction must be complete at the equivalence point. The degree of completeness of the reaction determines the size and sharpness of the vertical portion of the titration curve. The larger the equilibrium constant, the more complete the reaction, the larger the change in *p*H near the equivalence point, and the easier it is to locate the equivalence point with good precision. The completeness of the reaction is related to the practical feasibility of the titration. Theoretically, it may be possible to locate the equivalence point of a reaction which does not go well to completion, but practically, this may be a difficult problem.

The equilibrium constant for the titration of a strong acid with a strong base is quite large:

$$H_3O^+ + OH^- \rightleftharpoons 2H_2O \qquad K = \frac{1}{K_w} = 1.0 \times 10^{14}$$

We have noted the large ΔpH which occurs at the equivalence point, 5.20 units for $\Delta V = 0.10$ ml, and have pointed out that because of this large change, several indicators could be used to determine the equivalence volume with a precision of a few parts per thousand. Hence we say that the titration is *feasible.*

How large must the equilibrium constant be for a titration to be feasible? It is difficult to give an unequivocal answer to this question. The concentrations of the substance titrated and the titrant influence the magnitude of ΔpH, and under certain circumstances an analyst might be satisfied with less precision than we specified above. However, if we are given a specific set of conditions to be met, we can make a rather simple calculation to determine the magnitude of K. It is generally desired that essentially all of the substance titrated be converted into product at or near the equivalence point. In Chapter 5 (page 122) we calculated values of K for 99.9% and 99.99% conversion of the analyte into product at the equivalence point. It is also desirable that the *p*H change by 1 or 2 units for the addition of a few drops of titrant at the equivalence point if a visual indicator is to be employed. The following example illustrates a calculation of K_a for a weak acid and K for the titration reaction for a specific statement as to feasibility requirements.

Example. 50 ml of 0.10 M HA is titrated with 0.10 M strong base. (a) Calculate the minimum value of K so that, when 49.95 ml of titrant has been added, the reaction between HA and OH^- is essentially complete and the *p*H changes by 2.00 units on the addition of 2 more drops (0.10 ml) of titrant. (b) Repeat the calculation for $\Delta p\text{H} = 1.00$ unit.

(a) The *p*H 0.05 ml beyond the equivalence point can be calculated as follows:

$$[OH^-] = \frac{0.05 \times 0.10}{100.05} = 5 \times 10^{-5}\,M$$

$$p\text{OH} = 4.30$$

$$p\text{H} = 9.70$$

If $\Delta p\text{H}$ is to be 2.00 units, the $p\text{H}$ 0.05 ml before the equivalence point must be 7.70. At this point, if the reaction is complete, we have only 0.005 mmol of HA unreacted. Hence

$$pH = pK_a - \log\frac{[HA]}{[A^-]}$$

$$7.70 = pK_a - \log\frac{0.005}{4.995}$$

$$pK_a = 4.70$$

$$K_a = 2.0 \times 10^{-5}$$

$$K = \frac{K_a}{K_w} = \frac{2.0 \times 10^{-5}}{1.0 \times 10^{-14}} = 2.0 \times 10^{9}$$

(b) If $\Delta p\text{H} = 1.00$, then

$$8.70 = pK_a - \log\frac{0.005}{4.995}$$

$$pK_a = 5.70$$

$$K_a = 2.0 \times 10^{-6}$$

$$K = 2.0 \times 10^{8}$$

Effect of Concentration

The magnitude of $\Delta p\text{H}$ at the equivalence point also depends upon the concentrations of the analyte and the titrant. The effect of concentration on the change in $p\text{H}$ for the strong acid–strong base titration is shown in Fig. 6.3. The $\Delta p\text{H}$ decreases as the concentrations of analyte and titrant decrease.

For weak acids the effect of concentration as well as the magnitude of K_a on $\Delta p\text{H}$ is shown in Table 6.4. The following conclusions can be drawn from the table:

1. The smaller the value of K_a, the higher the $p\text{H}$ at the equivalence point and the smaller $\Delta p\text{H}$.
2. (a) Increasing the amount of HA titrated in the same initial volume decreases $\Delta p\text{H}$. However, this increases the volume of titrant required, rendering a given error in determining the end point a smaller relative error. (b) If the same amount of HA is titrated but the initial volume is decreased, $\Delta p\text{H}$ is increased. This is caused primarily by the fact that excess titrant is in a smaller volume. (See below.)
3. Increasing the concentration of titrant increases $\Delta p\text{H}$. This decreases the volume of titrant required, thus making a given error a larger relative error.

For the titration of a given amount of a certain weak acid, the procedure recommended to increase $\Delta p\text{H}$ is as described above in 2(b). Starting with a smaller volume, that is, an increased concentration of HA, will increase $\Delta p\text{H}$ at the equivalence point while using the same volume of titrant.

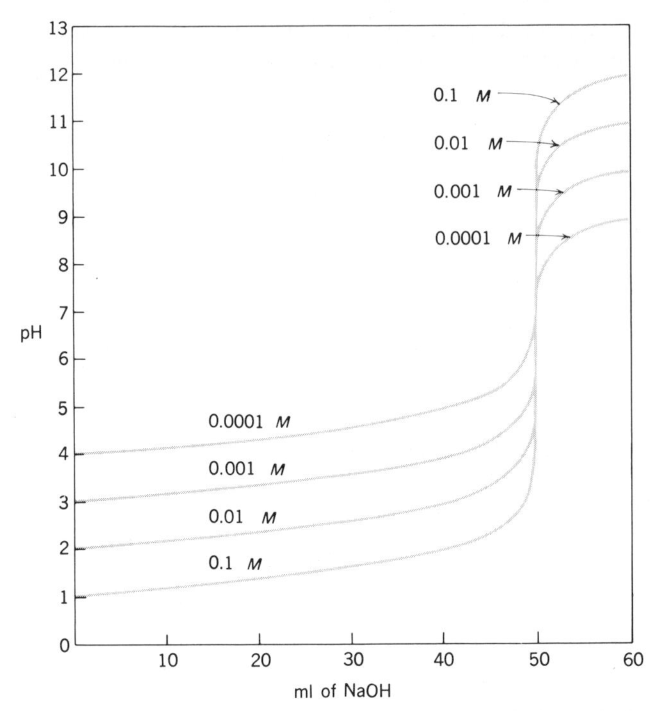

Figure 6.3 Effect of concentration on titration curves of strong acids with strong bases. 50 ml acid titrated with base of same molarity.

TABLE 6.4 ΔpH for Titration of Weak Acid, HA, with 0.1 M Strong Base

K_a of HA	mmol HA titrated	Initial volume, ml	pH, 0.05 ml before eq. pt.	pH at eq. pt.	pH, 0.05 ml after eq. pt.	ΔpH for 0.10 ml
1×10^{-5}	2.5	75	7.70	8.70	9.70	2.00
	2.5	50	7.70	8.76	9.82	2.12
	2.5	25	7.70	8.85	10.00	2.30
	5.0	75	8.00	8.80	9.60	1.60
	5.0	50	8.00	8.85	9.70	1.70
	5.0	25	8.00	8.91	9.82	1.82
5×10^{-6}	2.5	75	8.00	8.85	9.70	1.70
1×10^{-6}	2.5	75	8.70	9.20	9.70	1.00
5×10^{-7}	2.5	75	9.00	9.35	9.70	0.70
*1×10^{-7}	2.5	75	9.70	9.70	9.70	0.00

* Usual approximate calculations.

As a general rule it can be said that a precision of a few parts per thousand can be obtained in the titration of a 0.05 M solution of a weak acid or base of dissociation constant as low as 1×10^{-6}, using 0.10 M titrant. This corresponds to a value of K of about 1×10^8. Still weaker acids or bases may be titrated with sacrifice of precision in determining the end point.

It might also be noted that in aqueous solutions the most complete acid-base reaction is that between H_3O^+ and OH^- and the value of K is 1×10^{14}. The large change in pH which occurs in this case is roughly 6 units, from about pH 4 to 10. One can conclude that if the pH at the equivalence point of a titration falls below 4 or above 10, the magnitude of $\Delta p\text{H}/\Delta V$ will not be very large. It is doubtful that the titration will be considered feasible. One might also note that the closer the equivalence point pH is to 4 or 10, the smaller will be the value of $\Delta p\text{H}/\Delta V$.

BUFFER SOLUTIONS

A solution which resists large changes in pH when an acid or base is added or when the solution is diluted is called a *buffer solution*. A solution containing a conjugate acid-base pair is an example of a buffer. The acid reacts with any hydroxide ions added to the solution, and the conjugate base combines with hydrogen ions. Consider, for example, a solution of acetic acid and sodium acetate. Hydrogen and hydroxide ions are removed by the reactions:

$$\text{HOAc} + \text{OH}^- \rightleftharpoons \text{OAc}^- + \text{H}_2\text{O}$$

$$\text{OAc}^- + \text{H}_3\text{O}^+ \rightleftharpoons \text{HOAc} + \text{H}_2\text{O}$$

The pH is dependent upon the logarithm of the ratio of acid to salt (base),

$$p\text{H} = pK_a - \log \frac{[\text{HOAc}]}{[\text{OAc}^-]}$$

and it is necessary to change this ratio by a factor of 10 to change the pH by one unit. The flat portion of the titration curve (Fig. 6.2) for weak acids results from the buffering action just described. It should also be noted in the above equation for pH that dilution should have no effect on pH, at least theoretically, since the volume terms cancel out.

The effectiveness of a buffer solution in resisting change in pH per unit of strong acid or base added is greatest when the ratio of buffer acid to salt is unity. In the titration of a weak acid this point of maximum effectiveness is reached when the acid is half-neutralized, or the $p\text{H} = pK_a$. This can be seen from the following calculation:

Example 1. Calculate the slope of the titration curve of the weak acid, HA, titrated with OH^-, and find its minimum value. Let a = original mmol of HA and b = mmol of OH^- added. Then

$$[\text{HA}] = \frac{(a - b)}{v}$$

$$[\text{A}^-] = \frac{b}{v}$$

where v is the volume of solution.

$$pH = pK_a - \log\frac{a-b}{b} = pK_a - \log(a-b) + \log b$$

Differentiating, the slope is

$$\frac{dpH}{db} = \frac{0.43}{a-b} + \frac{0.43}{b} = \frac{0.43a}{b(a-b)}$$

To find the minimum value of the slope, differentiate the preceding expression and equate to zero,

$$\frac{d^2pH}{db^2} = -\frac{0.43a(a-2b)}{b^2(a-b)^2} = 0$$

$$b = \frac{a}{2}$$

That is, [HA] = [A^-], and at this point $pH = pK_a$.

It may be instructive to examine Table 6.5 which shows the change in *p*H produced during the titration of two different amounts of acetic acid at intervals of 1 mmol of added base. It is apparent that at the start of the titration the solution is not well buffered, and the *p*H rises rapidly as base is added. This explains the initial rapid rise in the titration curves of weak acids shown in Fig. 6.2. The rate of rise in *p*H decreases, passes through a minimum at $pH = pK_a$, and then slowly increases again. At the equivalence point, a large change occurs since the acid is exhausted and the solution is no longer buffered.

TABLE 6.5 Change in *p*H during Titration of Acetic Acid

mmol OH^- added to 10 mmol HOAc	*p*H*	Δ*p*H	mmol OH^- added to 20 mmol HOAc	*p*H*	Δ*p*H
0	2.87	—	0	2.72	—
1	3.79	0.92	1	3.46	0.74
2	4.14	0.35	2	3.79	0.33
3	4.37	0.23	3	3.99	0.20
4	4.56	0.19	9	4.65	—
5	4.74	0.18	10	4.74	0.09
6	4.92	0.18	11	4.83	0.09
7	5.11	0.19	12	4.92	0.09
8	5.34	0.23	18	5.69	—
9	5.69	0.35	19	6.02	0.33
10	8.87	3.18	20	9.02	3.00

* Calculated for 100 ml volume, assuming no change in volume as base added.

Concentrated solutions of strong acids and bases resist large changes in *p*H, and the titration curves are flat over a wide range of *p*H (Fig. 6.1). However, such solutions are sometimes not regarded as buffers in the strictest sense, since the *p*H is markedly changed by dilution.

The capacity of a buffer is a measure of its effectiveness in resisting changes in *p*H upon the addition of acid or base. The greater the concentrations of the acid and conjugate base, the greater is the capacity of the buffer. This is evident from Table 6.5 in that twice as much base is required to increase the *p*H of the more concentrated solution from 3.79 to 4.74 than is needed for the more dilute solution. Buffer capacity can be defined more quantitatively as the number of moles of strong base required to change the *p*H of 1 liter of solution by 1 *p*H unit. The term "range" of a buffer is ill-defined, but it is evident from Table 6.5 that little buffering action is obtained if the acid–salt ratio is greater than 9:1 or less than 1:9.

In preparing a buffer of a desired *p*H, the analyst should select an acid–salt (or base–salt) system in which the pK_a of the acid is as close as possible to the desired *p*H. By this selection, the ratio of acid to salt is near unity, and maximal effectiveness against increase or decrease in *p*H is obtained. The actual concentrations of acid and salt employed depend upon the desired resistance to change in *p*H. These points are illustrated in the following example.

Example 2. (a) It is desired to prepare 100 ml of a buffer of *p*H 5.00. Acetic, benzoic, and formic acids and their salts are available for use. Which acid should be used for maximum effectiveness against increase or decrease in *p*H? What acid–salt ratio should be used?

The pK_a values of these acids are: acetic, 4.74; benzoic, 4.18; and formic, 3.68. The pK_a of acetic is closest to the desired *p*H, and this acid and its salt should be used. Then

$$pH = pK_a - \log \frac{[HOAc]}{[OAc^-]}$$

$$5.00 = 4.74 - \log \frac{[HOAc]}{[OAc^-]}$$

Taking antilogs $[HOAc]/[OAc^-] = 1/1.8$, or the salt–acid ratio should be 1.8:1. This is the molar ratio, not the ratio of grams.

(b) If it is desired that the change in *p*H of the buffer be no more than 0.10 unit for the addition of 1 mmol of either acid or base, what minimum concentrations of the acid and salt should be used?

Since there is less acid present than salt, a greater change in *p*H will result if base is added. Hence if we calculate on the basis that base is added, the condition will be more than satisfied for the addition of acid. Thus, if

$$x = \text{mmol acid originally present}$$

$$1.8x = \text{mmol salt originally present}$$

If 1 mmol OH^- is added, then

$$x - 1 = \text{mmol acid remaining}$$

$$1.8x = \text{mmol salt}$$

Then

$$5.10 = 4.74 - \log \frac{x - 1}{1.8x + 1}$$

Solving gives $x = 6.6$ mmol, and $1.8x = 11.9$ mmol. The molar concentrations are then, [HOAc] = 0.066 mmol/ml, and $[OAc^-] = 0.119$ mmol/ml.

PHYSIOLOGICAL BUFFERS

It is of interest to point out that the principles of acid–base chemistry discussed in this chapter are of direct significance in such fields as biochemistry and physiology. The great physiologist Claude Bernard was the first to emphasize that the fluids of the body provide an "internal environment" in which the body cells live and perform their many functions protected from the inconstancy of the external environment. Living tissues are extremely sensitive to changes in the composition of the fluids that bathe them, and the regulatory mechanisms within the body which maintain the constancy of the internal environment comprise one of the most important phases in the study of the biological sciences.

A very important aspect of this regulation is the maintenance of a nearly constant *p*H in the blood and other fluids of the body. Substances that are acidic or alkaline in character are ingested in the diet and are formed continually by metabolic reactions, but the *p*H of the blood normally remains constant within about 0.1 *p*H unit (7.35 to 7.45).

The two principal routes for the elimination of acids from the body are the lungs and kidneys. It is estimated that in one day the normal human adult eliminates the equivalent of about 30 liters of 1 *M* acid by way of the lungs, and about 100 ml of 1 *M* acid through the kidneys.[1] To handle such large amounts of acid, the normal adult has enough buffers in his approximately 5 liters of blood to absorb about 150 ml of 1 *M* acid. The proton acceptors found in tissues, such as the muscles, can handle about five times as much acid as the blood buffers.

The principal buffers in the blood are proteins, bicarbonate, phosphates, hemoglobin (HHb), and oxyhemoglobin ($HHbO_2$). Carbon dioxide is formed metabolically in the tissues and is carried away by the blood primarily as bicarbonate ion. A typical reaction is

$$\underset{}{H_2O} + CO_2(aq) + \underset{\text{Base}}{Hb^-(aq)} \rightleftharpoons \underset{\text{Acid}}{HHb(aq)} + \underset{\llcorner\rightarrow \text{To lungs}}{HCO_3^-}$$

Note that H_2CO_3 is a stronger acid ($pK_{a_1} = 6.1$ under conditions in the blood) than is hemoglobin ($pK_a = 7.93$); hence the reaction above tends to go to the right. In the blood, at *p*H 7.4, the ratio of bicarbonate to free CO_2 can be calculated from the equation

$$7.4 = 6.1 - \log\frac{[CO_2]}{[HCO_3^-]}$$

The ratio $[HCO_3^-]/[CO_2]$ is about 20 : 1, showing that the predominant form in the blood is bicarbonate ion.

[1] W. R. Frisell, *Acid–Base Chemistry in Medicine*, Macmillan Publishing Co., Inc., New York, 1968.

In the lungs carbon dioxide is released by the reaction

$$HCO_3^-(aq) + HHbO_2(aq) \rightleftharpoons \underset{\text{To tissues}}{HbO_2^-(aq)} + H_2O + \underset{\text{Exhaled}}{CO_2(g)}$$

When blood is oxygenated in the lungs, hemoglobin is converted into oxyhemoglobin. Since oxyhemoglobin is a stronger acid ($pK_a = 6.68$) than hemoglobin, this facilitates the conversion of HCO_3^- to CO_2 by the reaction above.

The phosphate buffer system is found mostly in the red cells. Its reaction is

$$H_2PO_4^- + H_2O \rightleftharpoons HPO_4^{2-} + H_3O^+$$

The pK_a of $H_2PO_4^-$ is about 7.2; hence this system exhibits its maximal effectiveness very close to physiological pH.

Disturbances in the pH of the blood are seen clinically in certain diseases. For example, untreated diabetes sometimes give rise to an acidosis which may be fatal. Kidney failure, or chronic nephritis, leads to retention of $H_2PO_4^-$ and an increase in the amount of carbon dioxide in the blood:

$$H_2PO_4^- + HCO_3^- \rightleftharpoons HPO_4^{2-} + H_2O + CO_2$$

APPLICATIONS OF ACID–BASE TITRATIONS

Acid–base titrations are widely used for chemical analyses. In most applications water is used as the solvent, and we shall restrict our discussion at this point to aqueous solutions. In the next section we will discuss the use of nonaqueous solvents.

Acid–Base Reagents

In laboratory practice it is customary to prepare and standardize one solution of an acid and one of a base. These two solutions can then be used to analyze unknown samples of acids and bases. Since acid solutions are more easily preserved than basic solutions, an acid is normally chosen as a permanent reference standard in preference to a base.

In choosing an acid to use in a standard solution, the following factors should be considered. (1) The acid should be strong, that is, highly dissociated. (2) The acid should not be volatile. (3) A solution of the acid should be stable. (4) Salts of the acid should be soluble. (5) The acid should not be a sufficiently strong oxidizing agent to destroy organic compounds used as indicators.

Hydrochloric and sulfuric acids are most widely employed for standard solutions, although neither satisfies all the foregoing requirements. The chloride salts of silver, lead, and mercury(I) ion are insoluble, as are the sulfates of the alkaline earth metals and lead. This does not normally lead to trouble, however, in most applications of acid–base titrations. Hydrogen chloride is a gas, but is not appreciably volatile from solutions in the concentration range normally employed because it is so highly dissociated in aqueous solution. A solution as concentrated as 0.5 *N* can be boiled for some time without losing hydrogen chloride if the solution is not allowed to concentrate by evaporation. Nitric acid is seldom used, because it is a strong oxidizing agent, and its solutions decompose when heated or exposed to light. Perchloric is a strong acid, nonvolatile and stable toward reduction in dilute solutions. The potassium and ammonium salts may precipitate from concentrated solutions when formed during a titration. Perchloric acid is commonly preferred for nonaqueous titrations. It is inherently a stronger acid than hydrochloric and is more strongly dissociated in an acidic solvent, such as glacial acetic acid.

Sodium hydroxide is the most commonly used base. Potassium hydroxide offers no advantage over sodium hydroxide and is more expensive. Sodium hydroxide is always contaminated by small amounts of impurities, the most serious of which is sodium carbonate. When CO_2 is absorbed by a solution of NaOH the following reaction occurs:

$$CO_2 + 2OH^- \longrightarrow CO_3^{2-} + H_2O$$

Carbonate ion is a base, but it combines with hydrogen ion in two steps:

$$CO_3^{2-} + H_3O^+ \longrightarrow HCO_3^- + H_2O \quad (1)$$

$$HCO_3^- + H_3O^+ \longrightarrow H_2CO_3 + H_2O \quad (2)$$

If phenolphthalein is employed as the indicator, the color change occurs when reaction (1) is complete; that is, the carbonate ion has reacted with only one H_3O^+ ion. This results in an error since two OH^- ions were used in the formation of one CO_3^{2-}. If methyl orange is used as the indicator, the color change occurs when reaction (2) is complete and no error occurs since each CO_3^{2-} ion combines with two H_3O^+ ions. However, in the titration of weak acids phenolphthalein is the proper indicator to use and if CO_2 has been absorbed by the titrant an error will occur.

There are several ways to minimize the "carbonate error." Barium hydroxide can be used as the titrant. If CO_2 is absorbed by a solution of this base, a precipitate of barium carbonate is apparent:

$$Ba^{2+} + 2OH^- + CO_2 \longrightarrow BaCO_3(s) + H_2O$$

Since barium hydroxide is of limited solubility in water, solutions cannot be more concentrated than about 0.05 *N*.

The most common method used to avoid the carbonate error is to prepare carbonate-free sodium hydroxide and then protect the solution from the uptake of CO_2 from the air. Carbonate-free sodium hydroxide can be readily prepared from a concentrated solution of the base, one which is about 50% by weight NaOH. Sodium carbonate is insoluble in the concentrated NaOH solution and settles to the bottom of the container. The solution is decanted from the solid Na_2CO_3 and diluted to the desired concentration. It is then stored in a bottle equipped with a tube containing a solid material (soda-lime or Ascarite) which absorbs CO_2 from any air that enters.

The acid–base solutions used in the laboratory are usually in the concentration range of about 0.05 to 0.5 *N*, most often about 0.1 *N*. Solutions of such concentrations require reasonable volumes (30 to 50 ml) for titration of samples which are of convenient size to weigh on the analytical balance. For example, 0.6000 g of a pure substance of equivalent weight 200 will require 30 ml of a 0.1 *N* solution for titration.

Primary Standards

In laboratory practice it is customary to prepare solutions of an acid and a base of approximately the desired concentration and then to standardize the solutions against a primary standard. It is possible to prepare a standard solution of hydrochloric acid by direct weighing of a portion of constant-boiling HCl of known density, followed by dilution in a volumetric flask. More frequently, however, solutions of this acid are standardized in the customary manner against a primary standard.

The reaction between the substance selected as a primary standard and the acid or base should obviously fulfill the requirements for titrimetric analysis. In addition, the primary standard should have the following characteristics:

1. It should be readily available in a pure form or in a state of known purity. In general, the total amount of impurities should not exceed 0.01 to 0.02%, and it should be possible to test for impurities by qualitative tests of known sensitivity.
2. The substance should be easy to dry and should not be so hygroscopic that it takes up water during weighing. It should not lose weight on exposure to air. Salt hydrates are not normally employed as primary standards.
3. It is desirable that the primary standard have a high equivalent weight in order to minimize the consequences of errors in weighing.
4. It is preferable that the acid or base be strong, that is, highly dissociated. However, a weak acid or base may be employed as a primary standard with no great disadvantage, especially when the standard solution is to be used to analyze samples of weak acids or bases.

The compound potassium hydrogen phthalate (page 57), $KHC_8H_4O_4$ (abbreviated KHP), is an excellent primary standard for base solutions. It is readily available in purity of 99.95% or better from the Bureau of Standards and from chemical supply houses. It is stable on drying, nonhygroscopic, and has a high equivalent weight, 204.2 g/eq. It is a weak, monoprotic acid, but since base solutions are frequently used to determine weak acids, this is no disadvantage. Phenolphthalein indicator is employed in the titration, and the base solution should be carbonate-free.

Sulfamic acid, HSO_3NH_2, is a strong monoprotic acid, and either phenolphthalein or methyl red indicator can be employed in the titration with a strong base. It is readily available, inexpensive, and easily purified by recrystallization from water. It is a white crystalline solid, nonhygroscopic, and stable at temperatures up to 130°C. Its equivalent weight is 97.09, considerably smaller than that of KHP. However, the weight which would normally be employed to standardize solutions 0.1 *N* or greater is sufficiently large to keep weighing errors small. Sulfamic acid is readily soluble in water, and most of its salts are soluble.

The compound potassium hydrogen iodate, $KH(IO_3)_2$, a strong monoprotic acid, is also an excellent primary standard for base solutions. It is readily available in a form sufficiently pure for use as a primary standard. It is a white, crystalline, nonhygroscopic solid, and it has a high equivalent weight, 389.91. It is sufficiently stable to be dried at 110°C.

Sodium carbonate, Na_2CO_3, is widely used as a primary standard for solutions of strong acids. It is readily available in a very pure state, except for small amounts of sodium bicarbonate, $NaHCO_3$. The bicarbonate can be converted completely into carbonate by heating the substance to constant weight at 270 to 300°C. Sodium carbonate is somewhat hygroscopic but can be weighed without great difficulty. The carbonate can be titrated to sodium bicarbonate, using phenolphthalein indicator, and the equivalent weight is the molecular weight, 106.0. More commonly it is titrated to carbonic acid using methyl orange indicator. The equivalent weight in this case is one-half the molecular weight, 53.00.

The organic base tris(hydroxymethyl)aminomethane, $(CH_2OH)_3CNH_2$, also called TRIS or THAM, is an excellent primary standard for acid solutions. It is available commercially in purity of 99.95% and is readily dried and weighed. Its reaction with hydrochloric acid is

$$(CH_2OH)_3CNH_2 + H_3O^+ \longrightarrow (CH_2OH)_3CNH_3^+ + H_2O$$

and its equivalent weight is 121.06 g/eq.

Analyses Using Acid–Base Titrations

A wide variety of acid and basic substances, both inorganic and organic, can be determined by an acid-base titration. There are also many examples in which the analyte can be converted chemically into an acid or base and then determined by titration. We shall discuss briefly a few examples.

Nitrogen

The determination of nitrogen by titration of ammonia with a strong acid is an important application of acid–base titrations. The procedure depends upon the oxidation state of nitrogen in the compound to be analyzed. If nitrogen is present as the ammonium salt, oxidation state of −3, ammonia can be liberated by the addition of strong base:

$$NH_4^+ + OH^- \longrightarrow NH_3(g) + H_2O$$

The sample is heated in a distilling flask with excess base and the evolved ammonia is caught in excess standard sulfuric or hydrochloric acid. The excess acid is then titrated with standard base.

If, on the other hand, the nitrogen is attached to carbon in many organic compounds (proteins, etc.), ammonia is not readily evolved when the compound is heated with strong base. A more drastic treatment is required to break the carbon–nitrogen bond. Kjeldahl, in 1883, suggested a preliminary treatment of the nitrogen compound with hot concentrated sulfuric acid. The organic material is dehydrated, the carbon oxidized to CO_2, and the nitrogen converted to ammonium sulfate. Addition of strong alkali then liberates ammonia which can be absorbed and titrated as mentioned above.

The Kjeldahl method has been widely studied and various modifications have been proposed. Amines, amides, nitriles, cyanates, and isocyanates are particularly suited to the method. If a preliminary reduction step is employed, the procedure will also handle compounds containing nitro, nitroso, and azo groups. The method is the standard procedure for the determination of the protein content of certain grains, meat, and animal food.

Sulfur

This element can be determined in organic substances by burning the sample in a stream of oxygen, converting sulfur into SO_2 and SO_3. The gas formed by the reaction is passed through an aqueous solution of H_2O_2 to oxidize SO_2 to SO_3. The sulfuric acid is titrated with standard base:

$$H_2SO_4 + 2OH^- \longrightarrow SO_4^{2-} + 2H_2O$$

Boron

This element can be determined in organic compounds by combustion in a nickel bomb to convert boron to boric acid. Boric acid is too weak to titrate feasibly, but the addition of mannitol forms a strong acid which can be titrated with aqueous NaOH.

Carbonate Mixtures

We have previously mentioned that carbonate ion is titrated in two steps:

$$CO_3^{2-} + H_3O^+ \longrightarrow HCO_3^- + H_2O \qquad \text{(phenolphthalein)}$$

$$HCO_3^- + H_3O^+ \longrightarrow H_2CO_3 + H_2O \qquad \text{(methyl orange)}$$

Phenolphthalein serves as the indicator for the first step in the titration and methyl orange for the second. The titration of NaOH is complete at the phenolphthalein end point and only a drop or two of additional titrant is required to reach the methyl orange end point. (See Fig. 6.2.)

Sodium hydroxide is commonly contaminated with sodium carbonate; sodium carbonate and sodium bicarbonate often occur together. It is possible to analyze mixtures of these compounds by titration with standard acid, using the two indicators mentioned above. We shall discuss details of the calculations in Chapter 7.

Organic Functional Groups

A number of organic functional groups can be determined by acid-base titration. Carboxylic acids, R—COOH, generally have pK_a values of about 4 to 6 and are readily titrated. Sulfonic acids, $R—SO_3H$, are generally strong and readily soluble in water. They can be titrated with standard base.

Alcohols can be determined by the addition of excess acetic anhydride:

$$\underset{\text{Acetic anhydride}}{(CH_3CO)_2O} + \underset{\text{Alcohol}}{ROH} \longrightarrow \underset{\text{Ester}}{CH_3COOR} + \underset{\text{Acetic acid}}{CH_3COOH}$$

The excess anhydride is hydrolyzed to acetic acid,

$$(CH_3CO)_2O + H_2O \longrightarrow 2CH_3COOH$$

and the total acid produced by the two reactions is titrated with standard base.

Aliphatic amines, such as CH_3NH_2, generally have pK_b values of about 5 and can be titrated directly with standard acid. Aromatic amines, such as aniline, $C_6H_5NH_2$, have pK_b values of about 10 and are too weak to titrate in aqueous solution.

Esters can be determined by first hydrolyzing the compound with excess base:

$$R_1COOR_2 + OH^- \longrightarrow R_1COO^- + R_2OH$$

The excess base is then titrated with standard acid.

NONAQUEOUS TITRATIONS

Consider an acid, HB, which we wish to titrate with base, say NaOH. We have discussed the feasibility of this titration in terms of the "strength" of HB, using its dissociation constant, K_a, as a measure. But, as pointed out earlier, in terms of the Brønsted theory, K_a is really a measure of the tendency of HB to transfer a proton to the solvent, water:

$$HB + H_2O \rightleftharpoons H_3O^+ + B$$

That is, K_a is not a measure of an "intrinsic" acid strength of HB, because the basicity of water is also involved in this reaction. The same acid might dissociate to a much greater degree in a more basic solvent, say an organic amine:

$$HB + RNH_2 \rightleftharpoons RNH_3^+ + B$$

That is, there will be a greater concentration of solvated protons in the latter solvent. Thus it might appear that if HB were too weak an acid to be titrated feasibly in aqueous solution, we could enhance its "acidity" and hence its "titratibility" by choosing a solvent more basic than water.

Actually, in a practical sense, this is often the case, but the above discussion is misleading as it stands. In fact, dissociation is not at all necessary for successful acid-base titrations. Excellent titrations have been performed in nonpolar solvents like benzene or chloroform which do not promote dissociation to any appreciable extent. Indeed, it is *not* the greater basicity of the organic amine that makes it a better solvent than water for the titration of the very weak acid, HB. It is a better solvent for this titration because it is a *weaker acid* than water. In the aqueous system the titration reaction is

$$HB + OH^- \rightleftharpoons H_2O + B^- \qquad K = \frac{K_a}{K_w} \tag{1}$$

Water is a product of the titration reaction, and furthermore it is present in large excess. Thus, to the extent that water is acidic, it competes against the acid we wish to titrate and prevents the titration reaction from going to completion unless HB is itself sufficiently strong. This can be seen from the constant K; the constant is larger the larger K_a and the smaller the autoprotolysis constant of the solvent. In general terms, we wish the following reaction to go to completion:

$$HB + S^- \rightleftharpoons HS + B^- \qquad K = \frac{K_a}{K_{auto}} \tag{2}$$

Here HS is the solvent, S^- the conjugate base, and K_{auto} the autoprotolysis constant of the solvent. If HS is a weaker acid than water, K for reaction (2) will be larger than K for reaction (1). It often happens that the solvent is also more basic than water, but it is not correct to fixate upon this latter aspect.

In any case, we find that many titrations of weak acids and bases which are not feasible in water solution can be performed in other solvents. A variety of solvents have now been studied, and various methods of end-point detection are available. Much of the work is empirical because we do not have acidity scales in all these solvents as we have for water. But even on this basis, the field of nonaqueous titrations has become important in analytical chemistry.

Solvent Systems

Several classifications of solvents have been proposed. Laitinen[2] considers four types. *Amphiprotic* solvents possess both acidic and basic properties as does water. They undergo autoprotolysis, and, as we noted above, the degree to which the titration reaction goes to completion is a function of this reaction. Some, such as methanol and ethanol, have acid-base properties comparable to water and, along with water, are called *neutral* solvents. Others, called *acid* solvents, such as acetic acid, formic acid, and sulfuric acid, are much stronger acids and weaker bases than water. *Basic* solvents such as liquid ammonia and ethylenediamine have greater basicity and weaker acidity than water.

[2] H. A. Laitinen, *Chemical Analysis*, McGraw-Hill Book Company, New York, 1960, p. 60.

Aprotic, or inert, solvents are neither appreciably acidic nor basic and hence show little or no tendency to undergo autoprotolysis reactions. Examples are benzene, carbon tetrachloride, and chloroform.

Another group of solvents, called basic solvents, have a strong affinity for protons but are not appreciably acidic. Examples are ether, pyridine, and various ketones. Pyridine, for example, can accept a proton from an acid such as water:

$$C_5H_5N + H_2O \rightleftharpoons C_5H_5N^+H + OH^-$$

On the other hand, pyridine has no tendency to furnish a proton. Consequently, no autoprotolysis reaction can be written.

A fourth class of solvents would be those with acidic but no basic properties. No examples of such solvents are known.

Differentiating Ability of a Solvent

We have previously pointed out that water levels the mineral acids perchloric, hydrochloric, and nitric (page 108). That is, in aqueous solution these acids appear equally strong. However, in an acidic solvent such as acetic acid, the greater strength of perchloric acid over, say, hydrochloric, allows it to be titrated in a separate step from the latter acid. Of the two equilibria,

$$HClO_4 + HOAc \rightleftharpoons H_2OAc^+ + ClO_4^- \quad (1)$$

$$HCl + HOAc \rightleftharpoons H_2OAc^+ + Cl^- \quad (2)$$

the first goes much farther to the right than the second. Hence, in a titration of a mixture of the two acids in acetic acid solvent, two breaks in the titration curve are found, and the acids are said to be *differentiated.*

There are two properties of the solvent which determine its leveling or differentiating ability. One is the intrinsic acid-base character of the solvent, and the second is the autoprotolysis constant. For example, water is sufficiently strong a base to level HCl and $HClO_4$, but not HCl and HOAc. The latter two acids are differentiated in aqueous solution. Acetic acid is a weaker base than water and differentiates $HClO_4$ and HCl. Ammonia, however, is a stronger base than water and levels not only HCl and $HClO_4$ but also HCl and HOAc. An inert solvent, having no appreciable acidic or basic properties, exerts no leveling effect and hence is very suitable for differentiating mixtures of compounds of varying acidity.

We have previously pointed out that the neutralization of a strong acid with a strong base in aqueous solution is simply the reverse of the autoprotolysis reaction of the solvent:

$$H_3O^+ + OH^- \rightleftharpoons 2H_2O$$

The K for this reaction is $1/K_w = 1 \times 10^{14}$. The magnitude of this constant determines the size of the break at the equivalence point or the useful range over which breaks in titration curves can be detected. For the titration of 0.1 N reagents, the steep break is about 6 pH units, from pH 4 to 10. Below pH 4 and above pH 10 water levels acids and bases which are dissolved in it. These two pH extremes correspond to concentrations of H_3O^+ and OH^- of 10^{-4} M each.

Consider ethanol, C_2H_5OH ($pK_{auto} = 19.5$) as a solvent. The autoprotolysis constant is 3×10^{-20} and hence K for the reaction of strong acid with strong base,

$$C_2H_5OH_2^+ + OC_2H_5^- \rightleftharpoons 2C_2H_5OH$$

is $1/K_{auto}$ or 3×10^{19}. Reasoning as in the case of water above, the large break in the titration curve would occur between the limits of 10^{-4} M strong acid and strong base. That is, the useful pH range in ethanol is roughly $19.5 - 2 \times 4 = 11.5$ units, almost twice the useful range of water. In general, we can conclude that the useful pH range for a solvent in differentiating acids and bases increases as the autoprotolysis constant becomes smaller.

Dielectric Constant

Another property of a solvent which is of importance in nonaqueous titrations is the dielectric constant. In amphiprotic solvents the dissociation of a weak acid into separate ions is thought to occur as follows:

$$HB + HS \overset{1}{\rightleftharpoons} \underset{\text{Ion pair}}{\{H_2S^+B^-\}} \overset{2}{\rightleftharpoons} \underset{\text{Separate ions}}{H_2S^+ + B^-}$$

The first step is called ionization, and the product is called an *ion pair.* In the second step, complete separation of the ion occurs. Solvents with high dielectric constants encourage complete dissociation into ions by lessening the energy required for the process. In solvents of low dielectric constant, considerable ion pairing occurs.

The acidity of an ion such as NH_4^+ is not greatly affected by the dielectric constant of the solvent since no ion-pair production occurs:

$$NH_4^+ + H_2O \rightleftharpoons \{NH_3H_3O^+\} \rightleftharpoons NH_3 + H_3O^+$$

On the other hand, the autoprotolysis constant of the solvent is increased, the larger the dielectric constant:

$$HS + HS \rightleftharpoons \{H_2S^+S^-\} \rightleftharpoons H_2S^+ + S^-$$

since charge separation does occur.

Generally, a high dielectric constant is desirable for amphiprotic solvents. A factor of prime importance is solubility; a high dielectric constant generally favors the solubility of polar reagents and samples. Water is a unique solvent in having a very high dielectric constant and a relatively small autoprotolysis constant.

Completeness of the Titration Reaction

In general we represent the titration of a weak acid, HX, with the solvent anion (base), S^-, as follows:

$$HX + S^- \rightleftharpoons HS + X^-$$

In solvents of low dielectric constant where ion-pair formation may occur, we can represent the reaction as

$$H^+X^- + M^+S^- \rightleftharpoons HS + M^+X^-$$

From Le Châtelier's principle we can conclude that the reaction goes further to completion the more highly dissociated the ion pairs H^+X^- and M^+S^- and the more highly associated the ion pairs of the salt M^+X^-. In addition, the lower the autoprotolysis constant of the solvent, the larger K for the titration reaction. Concentration factors (page 143) also are to be considered in nonaqueous systems.

Titrants

Perchloric acid is by far the most widely used acid for the titration of weak bases, because it is a very strong acid which is readily available. It is normally obtained commercially as 72% $HClO_4$ by weight, the remainder being water; this is an azeotrope of $HClO_4$ and H_2O, and it represents approximately the composition $HClO_4 \cdot H_2O$ which some writers formulate as hydronium perchlorate $H_3O^+ClO_4^-$. Weak bases are titrated most often in glacial acetic acid solution. In such cases, the titrant is perchloric acid, say 0.1 M, in the same solvent. Because the presence of water may be deleterious (see above), the desired quantity of 72% $HClO_4$ is mixed with acetic acid, and then acetic anhydride is added in approximately the correct amount to react with the water estimated to be present. The product of this reaction is, of course, acetic acid.

A somewhat larger variety of strong bases are used, including alkali hydroxides, tetraalkylammonium hydroxides, and sodium or potassium methoxide or ethoxide. Common solvents for these bases are lower alcohols and mixtures of benzene with methanol or ethanol.

Normally the effect of temperature upon measured titrant volumes can be ignored with aqueous solutions under ordinary room temperature variations. Organic solvents such as acetic acid, benzene, or methanol, on the other hand, have fairly large coefficients of thermal expansion, and the volume changes may not be negligible if the titrant is at a different temperature from that at which it was standardized. Correction for the effect of a temperature change upon the volume of titrant may be made by means of an equation of the type

$$V_t = V_0(1 + \alpha t + \beta t^2 + \gamma t^3)$$

where V_0 is the volume at 0°C and V_t the volume at t°C. Values of α, β, and γ for various liquids may be found in handbooks. Practically, β and γ are usually small enough so that βt^2 and γt^3 may be ignored. Suppose the titrant were at 30°C when an unknown was titrated, whereas it had been standardized at 25°C. Neglecting the higher-order terms in the above equation and eliminating V_0 between the two temperatures involved gives

$$V_{25} = V_{30} \times \frac{1 + 25\alpha}{1 + 30\alpha}$$

Using a handbook value for α, the volume of titrant that would have been consumed had the titration been performed at 25°C can be readily calculated. For a mixed solvent such as benzene-methanol, a value for α may be used, weighted according to the volume fractions or the mole fractions of the two solvents in the mixture (if the mixture is nonideal, an exact, theoretically valid value of α cannot be calculated from the information that is normally available, but an adequate value can be obtained with weighted means).

End-Point Detection

A number of visual indicators are available, generally under trivial names such as cresol red, methyl red, azo violet, and crystal violet. The rationale of indicator selection is not on a good theoretical base, and the choice is often best made on the basis of experience, trial and error, or reference to analogous cases which may be found in the literature.

Potentiometric end-point methods (Chapter 12) are frequently employed, although, in general, electrode behavior in nonaqueous solvents is not well understood. Again, the safest approach is to see what other workers have used in similar situations. Other instrumental end points such as conductometric and photometric (Chapter 14) have been used successfully.

Applications

The number of compounds which have been titrated in nonaqueous media is much too large for listing here. Very weak acids, such as phenols, have been titrated in ethylenediamine. Carboxylic acids are sufficiently strong so that only moderately basic solvents such as methanol or ethanol can be employed. Nonaqueous titrations have become important in the pharmaceutical industry. For example, most of the well-known sulfa drug group can be determined by titration as acids (the acidity is conferred by the sulfonamide group, $—SO_2—NH—$) with alkali methoxide in benzene–methanol or dimethylformamide solution.

Weak bases, such as amines, amino acids, and anions of weak acids, have been titrated in glacial acetic acid solution using perchloric acid. Alkaloids have very weak basic properties and can be titrated in acidic or inert solvents.

The solvent methyl isobutyl ketone, a basic but not acidic solvent, has been used for the titration of a wide range of acids and bases. It has been used to differentiate a five-component acid mixture: perchloric, hydrochloric, salicyclic, acetic, and phenol. (See Fig. 6.4.) This mixture ranges from the strongest mineral acid, perchloric, to phenol, a very weak acid. The titrant for acids is a solution of tetrabutylammonium hydroxide in isopropanol; for the titration of bases, perchloric acid dissolved in dioxane is usually employed.

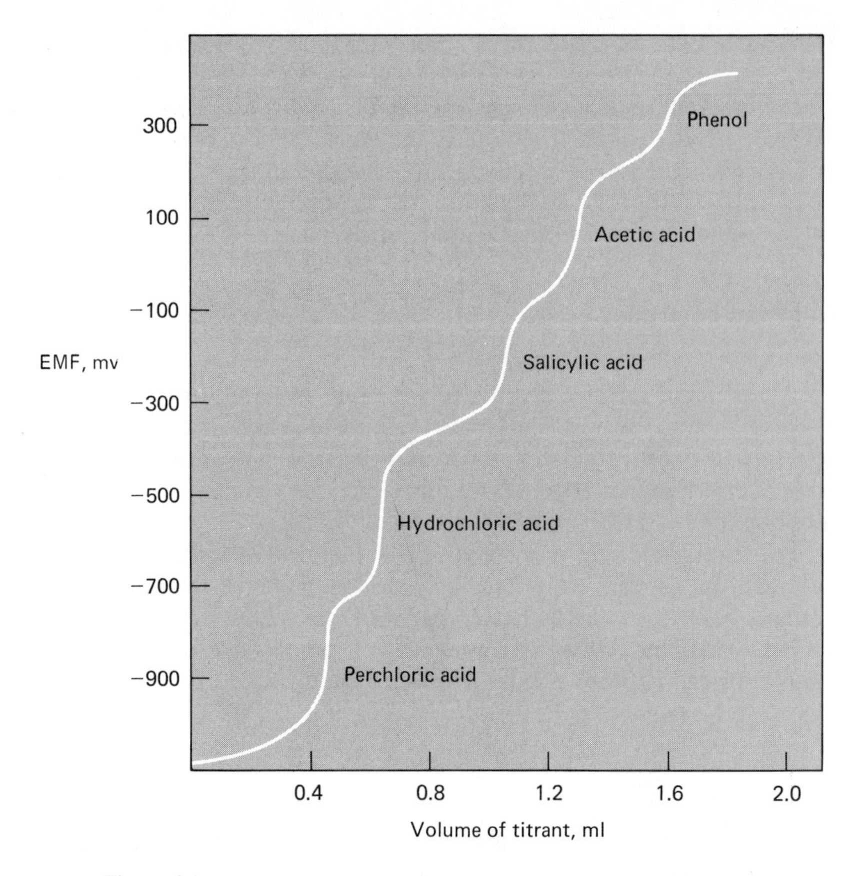

Figure 6.4 Titration of a five-component acid mixture in methyl isobutyl ketone using 0.2 *N* tetrabutylammonium hydroxide titrant, glass-platinum electrodes. (Courtesy of D. B. Bruss and G. E. A. Wyld, *Anal. Chem.*, *29*, 232 (1957).)

REFERENCES

J. N. BUTLER, *Ionic Equilibrium. A Mathematical Approach*, Addison-Wesley Publishing Co., Inc., Reading. Mass., 1964

WALTER HUBER, *Titrations in Nonaqueous Solvents*, Academic Press, Inc., New York, 1967.

H. A. LAITINEN and W. E. HARRIS, *Chemical Analysis*, 2nd ed., McGraw-Hill Book Company, New York, 1975.

L. G. SILLEN, "Graphic Presentation of Equilibrium Data." I. M. KOLTHOFF, "Concepts of Acids and Bases." S. BRUCKENSTEIN and I. M. KOLTHOFF, "Acid–Base Strength and Protolysis Curves in Water." and I. M. KOLTHOFF and S. BRUCKENSTEIN, "Acid–Base Equilibria in Nonaqueous Solution," Chaps. 8, 11, 12, and 13 of I. M. KOLTHOFF and P. J. ELVING, eds., *Treatise on Analytical Chemistry*, Part I, Vol. 1, Interscience Publishers, Inc., New York, 1959.

QUESTIONS

1. *Terms.* Explain what is meant by the following terms: (a) electroneutrality condition; (b) range of an indicator; (c) effectiveness of a buffer; (d) an indicator blank; (e) autoprotolysis constant; (f) aprotic solvent.

2. *Standard solutions.* Point out the factors to be considered in choosing an acid to use as a standard solution in the laboratory.

3. *Primary standards.* What characteristics should a substance have to be used as a primary standard for acid or base solutions? Name several commonly used primary standards.

4. *Carbonate error.* Explain why the "carbonate error" occurs and how it can be minimized in the laboratory.

5. *Physiological buffers.* (a) What are the principal buffers in the blood? (b) How are acids eliminated from the body? (c) How does oxygenating the blood in the lungs facilitate the liberation of CO_2?

6. *Feasibility.* Explain what is meant by a feasible acid–base titration. What factors determine the size of the ΔpH at the equivalence point?

Multiple-choice: In the following multiple-choice questions, select the *one best* answer.

7. Given the following acids, bases, and indicators, answer the questions below.

Acid	pK_a	Base	pK_b	Indicator
HA	Strong	MOH	Strong	I (acid), $pK_a = 9$
HB	8.70	NOH	4.60	II (base), $pK_b = 5$
HC	4.30	ROH	9.10	III (acid), $pK_a = 5$

Which of the following titrations is/are feasible (0.10 *M* solutions)? (a) HB with NaOH; (b) HC with NaOH; (c) ROH with HCl; (d) none of the above.

8. Which of the following titrations is/are feasible (0.10 *M* solutions)? (a) NCl with NaOH; (b) NaC with HCl; (c) NaB with HCl; (d) MA with NaOH; (e) more than one of the above.

9. HC is titrated with 0.10 *M* NaOH. Which is the proper indicator? (a) I; (b) II; (c) III; (d) I or II; (e) II or III.

10. Which of the following titrations (0.10 *M* solutions) will give the largest change in *p*H at the equivalence point? (a) ROH with HCl; (b) HC with NaOH; (c) NOH with HCl; (d) HB with NaOH.

11. A weak acid HX is titrated with 0.10 *M* NaOH. In which of the following titrations will the change in *p*H at the equivalence point be greatest? (a) 2.5 mmol of HX in 100 ml of water; (b) 5.0 mmol of HX in 100 ml of water; (c) 5.0 mmol of HX in 50 ml of water; (d) 2.5 mmol of HX in 50 ml of water.

12. Which of the following statements is/are true? (a) An amphiprotic solvent possesses both acidic and basic properties; (b) Dissociation into ions is not necessary for

successful acid-base titrations; (c) The titration reaction is more complete the smaller the autoprotolysis constant of the solvent; (d) all of the above.

PROBLEMS

1. *Titration curve.* 40.00 ml of 0.1100 *M* HCl is diluted to 100 ml with water and titrated with 0.1000 *M* NaOH. Calculate the *p*H after the addition of the following volumes (ml) of titrant: (a) 0.00; (b) 10.00; (c) 22.00; (d) 40.00; (e) 43.95; (f) 44.00; (g) 44.05; (h) 50.00. Plot the titration curve. Select an indicator from Table 6.3.

2. *Titration curve.* Repeat Problem 1 for the titration of 40.00 ml of 0.1100 *M* benzoic acid with 0.1000 *M* NaOH.

3. *Titration curve.* Repeat Problem 1 for the titration of 40.00 ml of 0.1100 *M* HCN with 0.1000 *M* NaOH.

4. *Titration curve.* 40.00 ml of 0.0900 *M* NaOH is diluted to 100 ml and titrated with 0.1000 *M* HCl. Calculate the *p*H after the addition of the following volumes of titrant (ml); (a) 0.00; (b) 10.00; (c) 18.00; (d) 30.00; (e) 35.95; (f) 36.00; (g) 36.05; (h) 40.00. Plot the titration curve. Select an indicator from Table 6.3.

5. *Titration curve.* Repeat Problem 4 for the titration of 40.00 ml of 0.0900 *M* NH_3 with 0.1000 *M* HCl.

6. *Fraction neutralized.* A solution of HF is titrated with NaOH. Calculate the *p*H when the following percentages of the acid have been neutralized: (a) 25%; (b) 33%; (c) 50%; (d) 75%; (e) 99%; (f) 99.9%; (g) 99.99%.

7. *Fraction neutralized.* A solution of ammonia is titrated with HCl. Calculate the percentage of base neutralized at the following *p*H values: (a) 9.74; (b) 9.56; (c) 9.26; (d) 8.66; (e) 6.26.

8. *Fraction neutralized.* 50 ml of 0.10 *M* HCl is titrated with 0.10 *M* NaOH. Calculate the fraction of the HCl neutralized at the following *p*H values: (a) 4.00; (b) 5.00; (c) 6.00; (d) 7.00. What volume of NaOH is required to change the *p*H from 4.00 to 7.00?

9. *Derivations.* Derive the following expressions for calculating the *p*H of a solution at the specified conditions: (a) Equivalence point in the titration of a weak acid HA with NaOH:

$$pH = \tfrac{1}{2}pK_w + \tfrac{1}{2}pK_a + \tfrac{1}{2}\log[A^-]$$

(b) Equivalence point in the titration of a weak base BOH with HCl:

$$pH = \tfrac{1}{2}pK_w - \tfrac{1}{2}pK_b - \tfrac{1}{2}\log[B^+]$$

(c) Solution of a weak acid HA:

$$pH = \tfrac{1}{2}pK_a - \tfrac{1}{2}\log[HA]$$

10. *pH calculations.* Calculate the *p*H of the following solutions: (a) 0.25 g of formic acid + 0.50 g of sodium formate in 500 ml of solution; (b) 0.50 g of formic acid dissolved in 100 ml of 0.120 *M* NaOH; (c) 0.25 g of formic acid + 0.50 g of sodium formate + 0.100 g of NaOH in 200 ml of solution.

11. *pH calculations.* Calculate the *p*H of the following solutions: (a) 2.8 g of hydrazine (N_2H_4) + 2.8 g of hydrazinium chloride (N_2H_5Cl) in 1.0 liter of solution; (b) 0.40 g of hydrazine dissolved in 100 ml of H_2O; (c) 0.30 g of hydrazine + 0.30 g of hydrazinium chloride + 0.050 g of HCl in 200 ml of solution.

12. *Weak acids and bases.* The K_b of the weak base BOH is 2.0×10^{-9}. Calculate the *p*H of a solution made by mixing 40 ml of a 0.15 *M* solution of the salt BCl with 60 ml of 0.10 *M* NaOH.

13. *Weak acids and bases.* The pK_a of the weak acid HA is 4.30. The *p*H of a solution of the salt NaA is 8.30. Calculate the molarity of the salt in this solution.

14. *Titration of a weak base.* A sample of the weak base COH is titrated with 0.100 *M* HCl. 40.00 ml is required to reach the equivalence point. At the point where 16.0 ml of titrant had been added the *p*H was 6.20. Calculate the pK_b of COH.

15. *Indicator.* Suppose that the *p*H range of an indicator HIn is 2.60 *p*H units and the ratio of [HIn] to [In^-] required to see the acid color is the same as the ratio of [In^-] to [HIn] to see the basic color. What percentage (minimum) of the indicator must be in the HIn form for the eye to detect only the HIn color?

16. *Indicator error.* A student is to titrate a sample containing the weak acid HA, pK_a = 5.00. (a) What will be the *p*H of his solution when 99.9% of HA has been converted to A^-? (b) Suppose that he uses the wrong indicator and stops the titration when the *p*H of the solution is 6.00. What fraction of HA is converted into A^- at this *p*H? (c) If the sample actually contains 26.40% HA, what value will he find if he makes the error described in part (b)?

17. *Carbonate error.* 100 ml of 0.300 *N* NaOH absorbs 2.00 mmol of CO_2 from the air. If this solution is now titrated with HCl using phenolphthalein indicator, what normality will be found?

18. *Buffer solution.* A student wishes to prepare 500 ml of a buffer of *p*H 4.50 using acetic acid and sodium acetate. (a) What should be the ratio of acid to salt for the buffer? (b) If he makes the solution 0.10 *M* in HOAc, how many grams of NaOAc should he add to get the desired *p*H?

19. *Buffer solution.* Suppose that the student in Problem 18 has available some glacial acetic acid (17 *M*) and a 0.080 *M* NaOH solution. How many milliliters of these two solutions should be mixed and diluted to 500 ml to give him his buffer of *p*H 4.50. Note that he wishes the final solution to be 0.10 *M* in HOAc.

20. *Buffer solution.* A chemist wishes to prepare 100 ml of a buffer of *p*H 10.00 using the base methyl amine, CH_3NH_2, and its salt, CH_3NH_3Cl. (a) What should be the ratio of base to salt to obtain a *p*H of 10.00? (b) If it is desired that the change in *p*H be no more than 0.10 unit for the addition of 1.0 mmol of either H_3O^+ or OH^-, what minimum concentrations of the acid and salt should be used?

21. *Buffer solution.* A buffer solution is prepared by dissolving 4.70 g of nitrous acid, HNO_2, and 13.8 g of sodium nitrite, $NaNO_2$, in 1.00 liter of solution. (a) Calculate the *p*H of the buffer. (b) Calculate the *p*H of the solution which results when the following are added to separate 100 ml portions of the buffer: (i) 5.00 mmol of HCl; (ii) 5.00 mmol of NaOH; (iii) 400 mg of NaOH; (iv) 730 mg of HCl; (v) 520 mg of NaOH. (c) Note the ΔpH in each case in part (b).

22. *Dilution of a buffer.* Given two solutions: A is a buffer of pH 4.00, made from the acid HX, $pK_a = 4.00$, and the salt NaX. It is 0.10 M in HX and 0.10 M in NaX. B is a 1.0×10^{-4} M solution of HCl. Calculate the pH values of the solutions resulting from the addition of 100 ml of water to 100 ml of A, and 100 ml of water to 100 ml of B.

23. *Buffer solution.* A buffer solution prepared from the weak base BOH, $pK_b = 4.30$, and its salt BCl has a pH of 10.00. When 60 mmol of HCl is added to 200 ml of the buffer, the pH changes to 9.00. Calculate the molarities of BOH and BCl in the original buffer.

24. *Buffer solution.* A buffer solution of pH 5.00 is prepared from a weak acid HX, $pK_a = 5.30$, and its salt NaX. To 100 ml of the buffer is added 10 mmol of HCl. What must be the original concentrations of HX and NaX in the buffer if the pH changes by 0.30 unit when the acid is added?

25. *Buffer solution.* The electrolytic reduction of an organic nitro compound was carried out in a solution buffered by acetic acid and sodium acetate. The reaction was

$$RNO_2 + 4H_3O^+ + 4e \longrightarrow RNHOH + 5H_2O$$

300 ml of a 0.0100 M solution of RNO_2 buffered initially at pH 5.00 was reduced, with the reaction above going to completion. The total acetate concentration, [HOAc] + [OAc^-], was 0.500 M. Calculate the pH of the solution after the reduction is complete.

26. *Feasibility of a titration.* 25.00 ml of a 0.1200 M solution of a hypothetical weak acid HX is diluted to 100 ml and titrated with 0.1000 M NaOH. Calculate the value of the equilibrium constant for the reaction

$$HX + OH^- \rightleftharpoons X^- + H_2O$$

so that 1 drop (0.05 ml) before the equivalence point the reaction is essentially complete and the pH changes by 1.60 units on the addition of 2 more drops (0.10 ml) of titrant.

27. *Feasibility of a titration.* 2.5 mmol of the weak acid HY, $pK_a = 7.00$, is dissolved in 75 ml of water and titrated with 0.10 M NaOH. Using the usual approximations, calculate the pH (a) 0.05 ml before the equivalence point; (b) at the equivalence point; (c) 0.05 ml after the equivalence point. Comment on the feasibility of the titration and the validity of the approximations.

28. *Effect of concentration.* 2.5 mmol of the weak acid HZ, $K_a = 1 \times 10^{-5}$, is titrated with 0.10 M NaOH. Calculate the pH 0.05 ml before the equivalence point, at the equivalence point, and 0.05 ml after the equivalence point for the cases where the acid was dissolved initially in a volume of (a) 75 ml; (b) 50 ml; (c) 25 ml of solution.

29. *Effect of concentration.* Repeat Problem 28 but starting with 5.0 mmol of the same weak acid.

30. *Approximations.* Calculate the concentrations of all the species in a 0.10 M solution of the weak acid HA, $K_a = 1 \times 10^{-3}$, using the usual approximations. Calculate the magnitude of the errors made using the procedure on page 132.

31. *Approximations.* Repeat Problem 30 for a 0.10 M solution of the salt NaA, where K_a of HA $= 1 \times 10^{-3}$. Follow the procedure on page 135.

32. *Nonfeasible titration.* 30 ml of a 0.10 M solution of NaOAc is diluted to 70 ml with water and titrated with 0.10 M HCl. (a) What is the value of the equilibrium constant for the titration reaction? (b) Calculate the pH at the equivalence point; (c) Calculate the pH 2 drops (0.10 ml) after the equivalence point. Comment on the feasibility of the titration.

33. *Ion trapping.* A model used to explain the absorption of a drug such as aspirin (a weak acid, designated HAsp), is as follows:

Blood plasma	Membrane	Stomach
pH = 7.4		pH = 1.0

$$H^+ + Asp^- \rightleftharpoons HAsp \rightleftharpoons HAsp \rightleftharpoons H^+ + Asp^-$$

It is assumed that ions such as H^+ and Asp^- do not penetrate the membrane, but that the undissociated form, HAsp, equilibrates freely across the membrane. At equilibrium the concentration of HAsp is the same on both sides of the membrane, but there is more *total drug* on the side where the degree of dissociation is greater. The mechanism is known as *ion trapping.*

Aspirin is a weak acid with a pK_a of 3.5. Calculate the ratio of total drug, [HAsp] + [Asp^-], in the blood plasma to total drug in the stomach, assuming that the model above is correct.

34. *Titration.* A 1.600-g sample containing a weak acid HX (MW 82.0) is dissolved in 60.0 ml of water and titrated with 0.250 M NaOH. When half of the acid is neutralized the pH is 5.00; at the equivalence point the pH is 9.00. Calculate the percentage of HX in the sample.

CHAPTER SEVEN

acid-base equilibria in complex systems

In the previous chapter we confined our attention to acids and bases which furnish or react with a single hydrogen ion. An acid which furnishes only one proton is called a *monoprotic* acid. Carbonic acid, H_2CO_3, is a *diprotic* acid, H_3PO_4 is triprotic, etc.; in general, acids which furnish two more protons are called *polyprotic*. Phosphoric acid, H_3PO_4, is one of the most important polyprotic acids, since phosphates are involved in buffers in the body fluids of living systems.

As might be expected, the equilibrium calculations involving polyprotic acids are more complex than those of monoprotic acids. Reasonable assumptions can be made, however, which enable the chemist to make good approximations of the pH values of solutions of such acids and their salts. The purpose of this chapter is to examine a few of the more important calculations which involve equilibria of polyprotic acids and their salts.

POLYPROTIC ACIDS

A solution of the diprotic acid H_2B contains two acids, H_2B and HB^-, dissociating as follows:

$$H_2B + H_2O \rightleftharpoons H_3O^+ + HB^- \qquad K_{a1} = \frac{[H_3O^+][HB^-]}{[H_2B]}$$

$$HB^- + H_2O \rightleftharpoons H_3O^+ + B^{2-} \qquad K_{a2} = \frac{[H_3O^+][B^{2-}]}{[HB^-]}$$

With a triprotic acid (e.g., phosphoric, H_3PO_4), there are three stages of dissociation; occasionally, even a tetraprotic acid is encountered, the best-known example being the chelon ethylenediaminetetraacetic acid (EDTA) which is discussed in detail in Chapter 8. A complete treatment of all the equilibria involved in solutions of polyprotic acids and their several salts is complicated and beyond the scope of this text. However, for many purposes, fairly valid approximations may be made which greatly simplify the treatment.

Frequently the successive K_a values for a polyprotic acid differ by several orders of magnitude, as may be seen in Table 1, Appendix I. As a result, the pH of a solution of an acid H_2B can be calculated accurately enough by considering only K_{a_1} and ignoring the further stages of dissociation. The problem thus reduces to one that we have already considered. Similarly, the K_b values of the conjugate bases often differ sufficiently to permit a fairly good calculation of the pH of a B^{2-} solution on the basis of K_{b_1} alone. (Recalling the relationship for a conjugate pair, $K_a \times K_b = K_w$, the student should note that K_{b_1} for the species B^{2-} is K_w/K_{a_2} and that K_{b_2} is K_w/K_{a_1}, where K_{a_1} and K_{a_2} are the successive constants for the acid H_2B.) The student may acquire a better "feeling" for this matter after reading the next section of this chapter on distribution of species as a function of pH.

Let us next consider the pH of a solution in which the intermediate species HB^- predominates. This might have been obtained simply by dissolving the salt NaHB in water, or by mixing equimolar quantities of H_2B and NaOH. Suppose we calculate the pH of a 0.10 M solution of NaHB for the case where K_{a_1} and K_{a_2} are 1.0×10^{-3} and 1.0×10^{-7}, respectively. There are three equilibria that may be considered, the dissociations of the two acids H_2B and HB^- as written above, and the dissociation of water. The equilibrium expressions for these processes may be combined in various ways, but it is most convenient to focus attention upon the fate of the principal species HB^-. This ion can react in three ways, all of which are of course acid-base reactions in the Brønsted sense.

Disproportionation:

$$HB^- + HB^- \rightleftharpoons H_2B + B^{2-} \qquad K = \frac{K_{a2}}{K_{a1}} = 1.0 \times 10^{-4}$$

Dissociation as a base:

$$HB^- + H_2O \rightleftharpoons H_2B + OH^- \qquad K_{b2} = \frac{K_w}{K_{a1}} = 1.0 \times 10^{-11}$$

Dissociation as an acid:

$$HB^- + H_2O \rightleftharpoons H_3O^+ + B^{2-} \qquad K_{a2} = 1.0 \times 10^{-7}$$

From the magnitudes of the constants of the three reactions, it is evident that the first reaction (disproportionation) proceeds farthest to the right. This suggests that we might neglect the amounts of H_2B and B^{2-} formed by the second and third reactions as compared with the quantities formed by the first one. Thus, at least approximately, we may write

$$[H_2B] \cong [B^{2-}]$$

Note that the product of the two dissociation constants contains these two terms:

$$K_{a_1} \times K_{a_2} = \frac{[H_3O^+]^2[B^{2-}]}{[H_2B]}$$

Thus

$$[H_3O^+]^2 = K_{a_1} \times K_{a_2}$$

$$[H_3O^+] = \sqrt{K_{a_1} \times K_{a_2}}$$

or

$$pH = \tfrac{1}{2}(pK_{a_1} + pK_{a_2})$$

In our example, this gives a pH value of 5.00 for the solution of NaHB.

Let us examine the magnitude of the error in the approximation we have made. There are six species in the solution: H_3O^+, OH^-, Na^+, H_2B, HB^-, and B^{2-}. The six equations are K_{a_1}, K_{a_2}, K_w, the charge balance equation,

$$[Na^+] + [H_3O^+] = [OH^-] + [HB^-] + 2[B^{2-}] \quad (1)$$

and the mass balance equations,

$$[H_2B] + [HB^-] + [B^{2-}] = 0.10 \quad (2)$$

$$[Na^+] = 0.10 \quad (3)$$

Adding equations (1) and (2) and noting that $[Na^+] = 0.10$ gives

$$[H_2B] + [H_3O^+] = [OH^-] + [B^{2-}] \quad (4)$$

Since the solution is acidic $[H_3O^+] > [OH^-]$, we may assume that

$$[H_2B] + [H_3O^+] \cong [B^{2-}] \quad (5)$$

Since HB^- is the principal species in a solution of the salt NaHB, we may assume that $[H_2B]$ and $[B^{2-}]$ can be dropped in equation [2], giving

$$[HB^-] \cong 0.10 \quad (6)$$

Substituting in the expression for K_{a_1} gives

$$[H_3O^+] = 0.01[H_2B] \quad (7)$$

Substitution of (7) into (5) gives

$$1.01[H_2B] = [B^{2-}]$$

In other words, the assumption that $[H_2B] \cong [B^{2-}]$ is in error by about 1%. The assumption in equation (2) can be checked by solving for $[H_2B]$ and $[B^{2-}]$ from K_{a_1} and K_{a_2}. Both are approximately equal to 0.001 M. These values can be substituted in equation (2), giving

$$0.001 + 0.10 + 0.001 = 0.10$$

The difference in the two sides of the equation is 2%.

A more exact equation for the hydrogen ion concentration in a solution of a salt such as NaHB can be derived. This equation is

$$[H_3O^+] = \sqrt{\frac{K_{a1}K_{a2}[HB^-] + K_{a1}K_w}{[HB^-] + K_{a1}}} \tag{8}$$

This equation is a good approximation if K_{a_1} and K_{a_2} are small and $[HB^-]$ is not too low. Under conditions where $[HB^-] \gg K_{a_1}$ and $K_{a_1}K_{a_2}[HB^-] \gg K_{a_1}K_w$, it reduces to the simpler equation

$$[H_3O^+] = \sqrt{K_{a1}K_{a2}}$$

Some examples illustrating the use of these two equations are given in Problem 1 at the end of the chapter.

Titration Curves of Polyprotic Acids

As previously mentioned, the *p*H of a solution of an acid, H_2B, can be calculated accurately enough by considering only K_{a_1}, provided that K_{a_1} is several orders of magnitude larger than K_{a_2}. Let us calculate the titration curve of an acid for which this is true. Then we shall consider an example where the two constants are closer in value.

Example. 50.0 ml of 0.100 *M* H_2B is titrated with 0.100 *M* NaOH. The dissociation constants are: $K_{a1} = 1.0 \times 10^{-3}$, $K_{a2} = 1.0 \times 10^{-7}$. Calculate the *p*H at various stages of the titration, using the usual approximation methods, and plot the titration curve.

(a) *Initial p*H. Consider only the first step in dissociation:

$$H_2B + H_2O \rightleftharpoons H_3O^+ + HB^-$$

$$\frac{[H_3O^+]^2}{0.10} = 1.0 \times 10^{-3}$$

$$[H_3O^+] = 1.0 \times 10^{-2}$$

and

$$pH = 2.00$$

Note that the value of K_{a1}, 1.0×10^{-3}, is considerably larger than that of the acid for which we made calculations in Chapter 6. The student might wish to calculate the error involved in assuming that $[H_2B] \cong 0.10$ *M*. If $[H_3O^+]$ is not dropped and if the complete quadratic is solved, the value of $[H_3O^+]$ is found to be 9.5×10^{-3} *M*.

(b) *p*H *after addition of 10.0 ml of base*

$$\text{mmol } H_2B = 50.0 \times 0.100 - 10.0 \times 0.100 = 4.00$$

$$\text{mmol } HB^- \text{ formed} = 10.0 \times 0.100 = 1.00$$

$$pH = pK_{a1} - \log\frac{[H_2B]}{[HB^-]}$$

$$pH = 3.00 - \log\frac{4.00}{1.00}$$

$$pH = 2.40$$

The pH at other points up to the first equivalence point is calculated in the same manner.

(c) pH *at first equivalence point.* 50.0 ml of base has been added and the species HB^- is the predominant one. The pH is readily approximated (see page 168) by the expression

$$pH = \tfrac{1}{2}(pK_{a_1} + pK_{a_2})$$

Hence

$$pH = \tfrac{1}{2}(3.00 + 7.00)$$

$$pH = 5.00$$

(d) pH *during titration of* HB^-*: 60.0 ml of base added.* The second acid, HB^-, is now being neutralized,

$$HB^- + OH^- \rightleftharpoons H_2O + B^{2-}$$

and the pH is calculated from the dissociation constant for HB^-,

$$HB^- + H_2O \rightleftharpoons H_3O^+ + B^{2-} \qquad K_{a_2} = 1.0 \times 10^{-7}$$

$$pH = pK_{a_2} - \log\frac{[HB^-]}{[B^{2-}]}$$

Here

$$\text{mmol } HB^- = 50.0 \times 0.100 - 10 \times 0.100 = 4.00$$

$$\text{mmol } B^{2-} = 10.0 \times 0.100 = 1.00$$

$$pH = 7.00 - \log\frac{4.00}{1.00}$$

$$pH = 6.40$$

The pH at other points up to the second equivalence point is calculated in the same manner.

(e) pH *at second equivalence point.* 100 ml of base has been added. The pH can be approximated by considering the first step in the hydrolysis of B^{2-},

$$B^{2-} + H_2O \rightleftharpoons HB^- + OH^-$$

$$[B^{2-}] = \frac{50.0 \times 0.100}{150} = 0.0333\ M$$

$$\frac{[HB^-][OH^-]}{[B^{2-}]} = \frac{K_w}{K_{a2}}$$

Since

$$[HB^-] \cong [OH^-]$$

$$\frac{[OH^-]^2}{0.0333} = \frac{1.0 \times 10^{-14}}{1.0 \times 10^{-7}}$$

$$[OH^-] = 5.8 \times 10^{-5}$$

$$pOH = 4.24$$

$$pH = 9.76$$

Values beyond the second equivalence point are calculated from the amount of excess base.

The titration curve is shown in Fig. 7.1, curve A, where the two titration steps are reasonably distinct. B in Fig. 7.1 is the curve for a diprotic acid in which the ratio of K_{a_1} to K_{a_2} is 10^2. In this case there is only a slight indication of two separate steps in the titration, a slightly more rapid rise in pH occurring at the first equivalence point. In curve C, for H_2SO_4, where both H_2SO_4 and HSO_4^- are strongly dissociated ($K_{a_2} = 0.012$), the shape is essentially the same as that for a strong monoprotic acid.

In general, one can conclude that in order for the steps in the titration of a polyprotic acid to be distinct, the successive constants must differ by a factor of at least 10^4, or the pK_a values must differ by 4 units. Maleic acid has pK_a values differing by about 4.3 units and hence titrates in two distinct steps. In the case of oxalic acid, the two pK_a values differ by only 3.0 units and the two steps are not sharply separated. Phosphoric is a triprotic acid with pK_a values of 2.12, 7.21, and 12.32. At the first equivalence point, about pH 4.62, the value of $\Delta p\text{H}/\Delta V$ is fairly large and methyl red is a suitable indicator for this step. At the second equivalence point, about pH 9.72, $\Delta p\text{H}/\Delta V$ is not as large because $H_2PO_4^-$ is a weaker acid than H_3PO_4. Phenolphthalein can be used to detect this equivalence

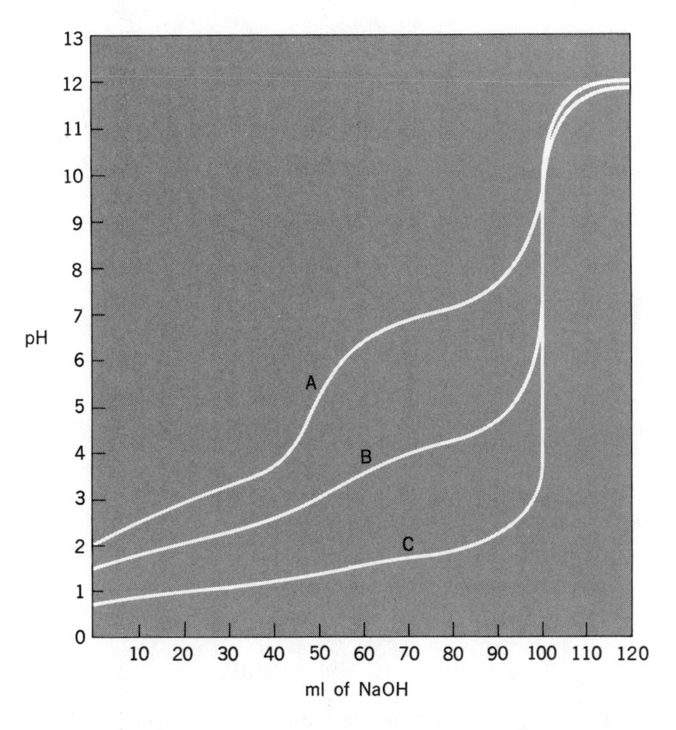

Figure 7.1 Titration curves for diprotic acids. *A*, $K_{a1}/K_{a2} = 10^4$; *B*, $K_{a1}/K_{a2} = 10^2$; *C*, H_2SO_4, K_{a_1} large, $K_{a2} = 0.012$.

point. The third acid, HPO_4^{2-}, is too weak for feasible titration (page 142). The value of K for the reaction

$$HPO_4^{2-} + OH^- \rightleftharpoons PO_4^{3-} + H_2O$$

is only $4.8 \times 10^{-13}/1.0 \times 10^{-14} = 48$.

Figure 7.2 shows the titration curve of phosphoric acid. The equations or reactions needed to calculate the pH at various stages in the titration are shown on the figure. Table 7.1 contains the approximate relations between the concentrations of the various species at the start of the titration of each species, halfway to the equivalence point, and at the equivalence point. The approximations are poorest at the two ends of the curve because H_3PO_4 has a rather large K_{a_1}, 7.5×10^{-3}, and PO_4^{3-} is a rather strong base, $K_{b_1} = 2.1 \times 10^{-2}$.

Titration of carbonates

It was pointed out in Chapter 6 that when CO_2 is absorbed by a standard solution of NaOH, the normality of the solution will be affected if phenolphthalein indicator is employed. It was also mentioned that mixtures of carbonate and

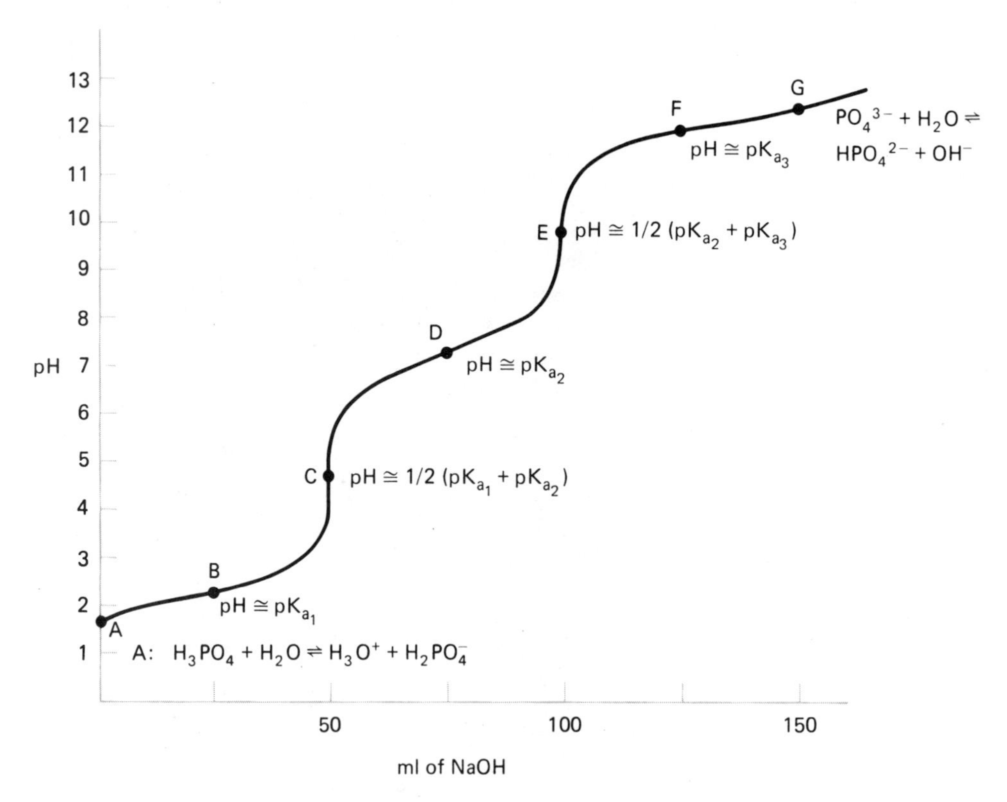

Figure 7.2 Titration curve of phosphoric acid: 50 ml of 0.10 M acid titrated with 0.10 M base. A, start of titration; B, halfway to first E.Pt.; C, first E.Pt.; D, half-way to second E.Pt.; E, second E.Pt.; F, half-way to third E.Pt.; G, third E.Pt.

TABLE 7.1 Approximate Relations between the Concentrations of Species in Phosphate Solutions

Place on titration curve*	Approximate relation of principal species
A	$[H_2PO_4^-] \cong [H_3O^+]$
B	$[H_3PO_4] \cong [H_2PO_4^-]$
C	$[H_3PO_4] \cong [HPO_4^{2-}]$
D	$[H_2PO_4^-] \doteqdot [HPO_4^{2-}]$
E	$[H_2PO_4^-] \cong [PO_4^{3-}]$
F	$[HPO_4^{2-}] \cong [PO_4^{3-}]$
G	$[HPO_4^{2-}] \cong [OH^-]$

* Letters correspond to position on titration curve in Fig. 7.2.

hydroxide, or carbonate and bicarbonate, can be determined by titration using phenolphthalein and methyl orange indicators. We would like to examine this topic in more detail now that we have discussed polyprotic acids.

The first pK_a of carbonic acid is 6.34 and the second 10.36, making the difference 4.02 units. We might expect a fair break between the two curves in this case, but K_{a_1} is so small that the break at the first equivalence point is poor. Usually, the carbonate ion is titrated as a base with a strong acid titrant, in which case two fair breaks are obtained, as shown in Fig. 7.3, corresponding to the reactions

$$CO_3^{2-} + H_3O^+ \rightleftharpoons HCO_3^- + H_2O$$

$$HCO_3^- + H_3O^+ \rightleftharpoons H_2CO_3 + H_2O$$

Phenolphthalein, pH range 8.0 to 9.6, is a suitable indicator for the first end point since the pH of a solution of $NaHCO_3$ is $\frac{1}{2}(pK_{a_1} + pK_{a_2})$ or 8.35. Methyl orange, pH range 3.1 to 4.4, is suitable for the second end point. A saturated solution of CO_2 has a pH of about 3.9. Neither end point is very sharp, but the second one can be greatly improved by removal of CO_2. Usually, samples containing only sodium carbonate (soda ash) are neutralized to the methyl orange point, and excess acid is added. Carbon dioxide is removed by boiling the solution, and the excess acid is titrated with standard base.

Mixtures of carbonate and bicarbonate, or of carbonate and hydroxide, can be titrated with standard HCl to the two end points mentioned above. As noted in Fig. 7.4, at the phenolphthalein end point NaOH is completely neutralized, Na_2CO_3 is half-neutralized, and HCO_3^- has not yet reacted. From the phenolphthalein to the methyl orange end point, bicarbonate is being neutralized. Only a few drops of titrant would be required by the NaOH to go from a pH of 8 to 4,

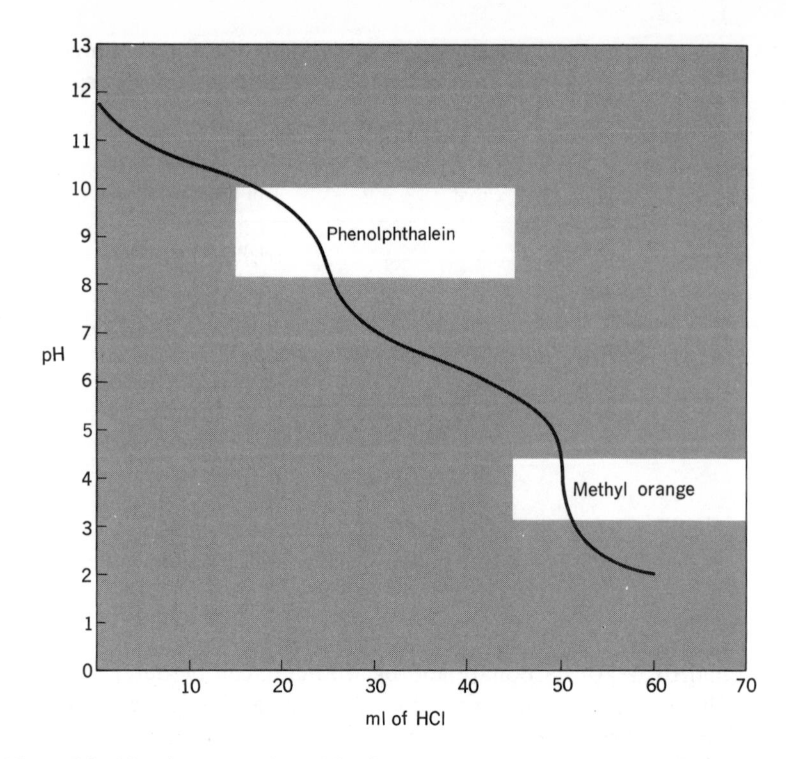

Figure 7.3 Titration curve of Na_2CO_3; 2.5 mmol Na_2CO_3 titrated with 0.10 *M* HCl.

and this can be corrected by an indicator blank. In Fig. 7.4 and Table 7.2, v_1 is the volume of acid in milliliters used from the start of the titration to the phenolphthalein end point, and v_2 is the volume from the phenolphthalein end point to the methyl orange.

In Table 7.2 are listed the relations between the volumes of acid used to the two end points for single components and mixtures. The molarity of the HCl is designated by *M.* The student should be able to verify these relationships, recalling that NaOH reacts completely in the first step, $NaHCO_3$ reacts only in the second step, and Na_2CO_3 reacts in both steps using equal volumes of titrant in the two steps. The mixture of NaOH and $NaHCO_3$ is not considered since these two compounds react:

$$HCO_3^- + OH^- \rightleftharpoons CO_3^{2-} + H_2O$$

The resulting product is either a mixture of CO_3^{2-} and OH^-, HCO_3^- and CO_3^{2-}, or CO_3^{2-} alone, depending upon the relative amounts of the two compounds in the sample.

The following examples illustrate the use of the two-indicator method and the effect of CO_2 absorption on the normality of a sodium hydroxide solution.

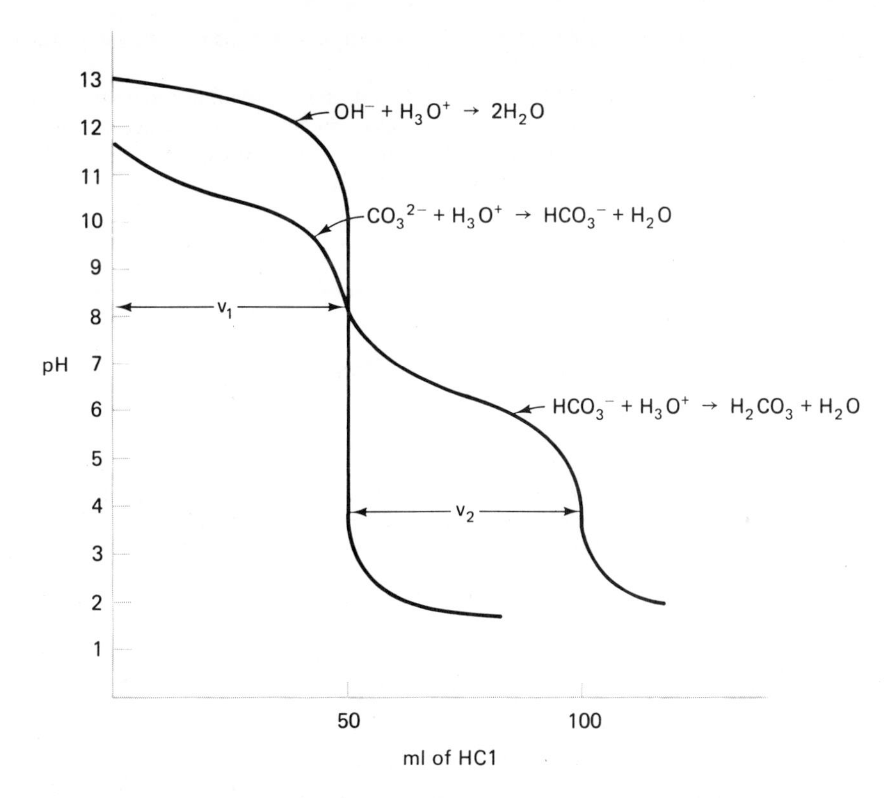

Figure 7.4 Titration curves of NaOH and Na_2CO_3: 50 ml of 0.10 *M* solution titrated with 0.10 *M* HCl.

TABLE 7.2 Volume Relations in Carbonate Titrations

Substance	Relation for qualitative identification	Millimoles of substance present
NaOH	$v_2 = 0$	$M \times v_1$
Na_2CO_3	$v_1 = v_2$	$M \times v_1$
$NaHCO_3$	$v_1 = 0$	$M \times v_2$
NaOH + Na_2CO_3	$v_1 > v_2$	NaOH: $M(v_1 - v_2)$
		Na_2CO_3: $M \times v_2$
$NaHCO_3$ + Na_2CO_3	$v_1 < v_2$	$NaHCO_3$: $M(v_2 - v_1)$
		Na_2CO_3: $M \times v_1$

Example 1. A 0.6234 g sample that might contain NaOH, Na_2CO_3, $NaHCO_3$, or a mixture of NaOH + Na_2CO_3 or Na_2CO_3 + $NaHCO_3$ is titrated with 0.1062 *M* HCl by the two-indicator method. It is found that 40.38 ml of the acid are required to reach the phenolphthalein end point. Methyl orange is then added to the solution and the titration continued using an additional 12.83 ml of the acid. (a) Identify the

base or mixture of bases in the sample. (b) Calculate the percentage of each in the sample.

(a) Since 40.38 ml > 12.83 ml, the sample must contain NaOH and Na_2CO_3.

(b) The volume of titrant used by Na_2CO_3 in the second step is 12.83 ml. An equal volume must have been used in the first step. Hence the volume used by the NaOH is 40.38 − 12.83 = 27.55 ml.

Then

$$\% \; Na_2CO_3 = \frac{12.83 \times 0.1062 \times 106.0}{623.4} \times 100 = 23.17$$

and

$$\% \; NaOH = \frac{27.55 \times 0.1062 \times 40.00}{623.4} \times 100 = 18.77$$

Example 2. A bottle which contains 200 ml of 0.100 *M* NaOH absorbs 1.00 mmol of CO_2 from the air. If the solution is now titrated with standard acid using phenolphthalein indicator, what normality will be found?

The solution contains

$$200 \text{ ml} \times 0.100 \text{ mmol/ml} = 20.0 \text{ mmol NaOH}$$

The 1.00 mmol of CO_2 reacts with 2.00 mmol of NaOH:

$$2NaOH + CO_2 \longrightarrow Na_2CO_3 + H_2O$$

The resulting solution contains 18.0 mmol of NaOH and 1.00 mmol of Na_2CO_3. On titration to the phenolphthalein end point the NaOH will use 18.0 mmol of H_3O^+ and the Na_2CO_3 will use 1.00 mmol. Hence the normality found will be

$$\frac{(18.0 + 1.00) \text{ meq}}{200 \text{ ml}} = 0.095 \; N$$

Note the "carbonate error." Had methyl orange indicator been used, the Na_2CO_3 would have used 2.00 mmol of acid and the normality would have been found to be 0.100.

TITRATION OF A MIXTURE OF TWO ACIDS

The conclusions we drew in the previous section concerning the titration of the acid H_2B in two steps apply in a similar manner to the titration of a mixture of two weak acids, HX and HY, provided that the initial concentrations of the two acids are the same. If HX, K_{a_1}, is the stronger acid, and HY, K_{a_2}, is the weaker, $pK_{a_1} - pK_{a_2}$ must be at least 4 units for the two titration steps to be reasonably distinct. If the difference in pK_a values is less than this, the two steps are not as distinct, as is indicated in Fig. 7.1. The *p*H at the first equivalence point is $\frac{1}{2}(pK_{a_1} + pK_{a_2})$ if the initial concentrations of HX and HY are the same. If these concentrations are not equal, the *p*H at this point is given by the expression

$$pH = \tfrac{1}{2}(pK_{a_1} + pK_{a_2}) - \tfrac{1}{2}\log\frac{[HY]}{[X^-]}$$

The principal application of this type of titration is in the case of titrating mixtures of a strong acid and a weak acid, such as hydrochloric and acetic. The HCl is titrated first, and if one calculates the *p*H during the titration it is logical to disregard the H_3O^+ contributed by the weak HOAc. This follows from LeChâtelier's Principle, excess H_3O^+ repressing the dissociation of the weak acid. This assumption becomes less valid as the first equivalence point is approached since the excess H_3O^+ is decreasing in concentration. At the first equivalence point the HCl is essentially used up and the *p*H is determined by the dissociation of HOAc. Beyond the first equivalence point the titration curve is that of acetic acid.

Figure 7.5 shows the curve for the titration of 50 ml of a solution which is 0.10 *M* in HCl and 0.10 *M* in HOAc. It can be seen that the first equivalence point is a poor one, the value of $\Delta p\text{H}/\Delta V$ not being very large. The *p*H of a 0.05 *M* solution of HOAc is approximately 3. Since this is below *p*H 4 it is not likely to be considered a feasible titration if a visual indicator is used. (See page 145.) The second step in the titration involves the reaction of HOAc with strong base and is a feasible titration.

Figure 7.5 also includes a curve for the titration of a mixture of HCl and an acid, HX, K_a of 1.0×10^{-8}. It can be seen that the value of $\Delta p\text{H}/\Delta V$ for the HCl is much larger in the presence of the weaker acid, and this titration step is definitely feasible. The $\Delta p\text{H}/\Delta V$ is not so large in the second step since HX is such a weak acid.

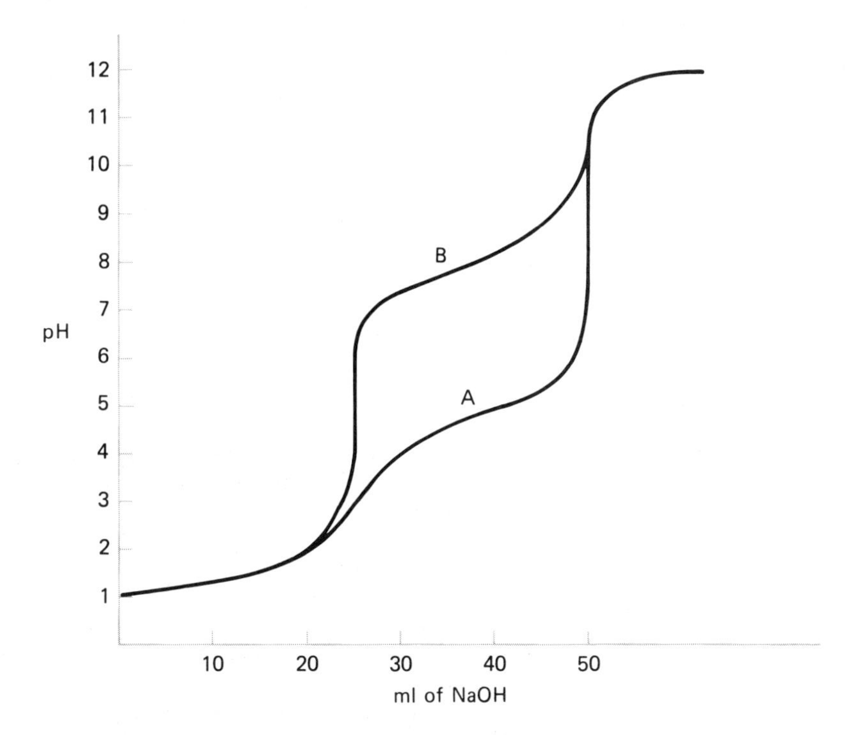

Figure 7.5 Titration curves: *A*, 50 ml of 0.10 *M* HCl and 0.10 *M* HOAc; *B*, 50 ml of 0.10 *M* HCl and 0.10 *M* HX, $K_a = 1 \times 10^{-8}$, titrated with 0.20 *M* NaOH.

DISTRIBUTION OF ACID–BASE SPECIES AS A FUNCTION OF pH

It is convenient for various purposes to be able to see at a glance the status of dissociation of common acid-base species as a function of pH. Graphs which show this enable us to determine which of several possible species predominate at a given pH, and they aid in selecting the regions of buffer effectiveness for mixtures of acids or bases and their salts. For example, the pH of blood plasma is held at about 7; it might be of interest to know whether plasma phosphate exists as H_3PO_4, $H_2PO_4^-$, HPO_4^{2-}, PO_4^{3-}, or as some mixture of these species at physiological pH. The type of graph we discuss below can provide answers to such questions almost instantly. It will also be useful in later chapters to be able to calculate the concentration of a particular species in the solution at a given pH. The following examples show the derivation of expressions for these fractions in the cases of a monoprotic and a diprotic acid.

Example 1. In a solution of acetic acid, calculate the fraction present as HOAc molecules and as OAc^- ions at various pH values. Draw an appropriate graph.

Let c_a represent the *analytical concentration* (page 51). This is the total concentration of all species arising from acetic acid and is simply a mass balance as used previously:

$$c_a = [\mathrm{HOAc}] + [\mathrm{OAc}^-]$$

From the dissociation constant expression for HOAc, we obtain

$$[\mathrm{OAc}^-] = \frac{[\mathrm{HOAc}]K_a}{[\mathrm{H_3O^+}]}$$

Substitution into the expression for c_a gives

$$c_a = [\mathrm{HOAc}] + \frac{[\mathrm{HOAc}]K_a}{[\mathrm{H_3O^+}]}$$

$$c_a = [\mathrm{HOAc}]\left\{1 + \frac{K_a}{[\mathrm{H_3O^+}]}\right\}$$

$$\frac{[\mathrm{HOAc}]}{c_a} = \frac{1}{1 + (K_a/[\mathrm{H_3O^+}])} = \frac{[\mathrm{H_3O^+}]}{[\mathrm{H_3O^+}] + K_a}$$

$[\mathrm{HOAc}]/c_a$ is the fraction of total acetate present in the undissociated form. By a similar approach, it may be shown that the fraction of the acetic acid in the dissociated form is given by

$$\frac{[\mathrm{OAc}^-]}{c_a} = \frac{K_a}{[\mathrm{H_3O^+}] + K_a}$$

Graphs of these fractions vs pH are shown in Fig. 7.6. Notice that at a pH roughly two units below pK_a, practically all of the acetate (about 99%) is in the undissociated form, HOAc, and that the acid is almost completely dissociated at a pH of $(pK_a + 2)$. At the intersection of the two curves, $[\mathrm{OAc}^-]/c_a = [\mathrm{HOAc}]/c_a = 0.5$ and $p\mathrm{H} = pK_a$ or $[\mathrm{H_3O^+}] = K_a$.

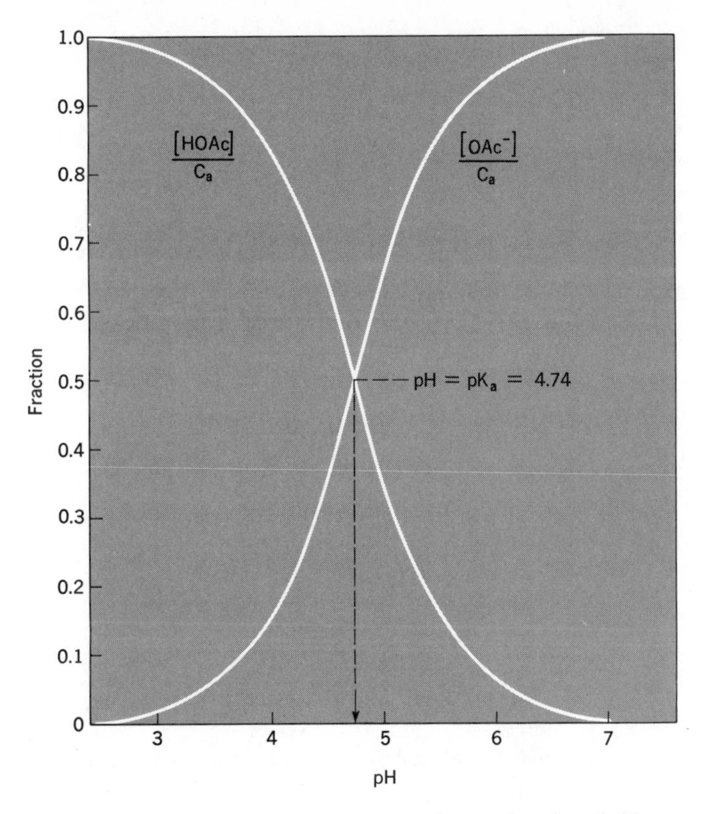

Figure 7.6 Distribution of acetate species as a function of pH.

Example 2. In a solution of the diprotic acid, oxalic, (H_2Ox), calculate the fractions present as H_2Ox molecules and as HOx^- and Ox^{2-} ions as a function of pH. Draw an appropriate graph.

Here the analytical concentration is given by

$$c_a = [H_2Ox] + [HOx^-] + [Ox^{2-}]$$

We also have the two dissociation expressions:

$$K_{a_1} = \frac{[H_3O^+][HOx^-]}{[H_2Ox]}$$

$$K_{a_2} = \frac{[H_3O^+][Ox^{2-}]}{[HOx^-]}$$

Rearrangement of the two K_a expressions gives

$$[HOx^-] = \frac{[H_2Ox]K_{a_1}}{[H_3O^+]}$$

$$[Ox^{2-}] = \frac{[HOx^-]K_{a_2}}{[H_3O^+]} = \frac{[H_2Ox]K_{a_1}K_{a_2}}{[H_3O^+]^2}$$

Substitution into the expression for the analytical concentration yields

$$c_a = [H_2Ox] + \frac{[H_2Ox]K_{a1}}{[H_3O^+]} + \frac{[H_2Ox]K_{a1}K_{a2}}{[H_3O^+]^2}$$

whence

$$c_a = [H_2Ox]\left\{1 + \frac{K_{a1}}{[H_3O^+]} + \frac{K_{a1}K_{a2}}{[H_3O^+]^2}\right\}$$

$$\frac{[H_2Ox]}{c_a} = \frac{1}{1 + \dfrac{K_{a1}}{[H_3O^+]} + \dfrac{K_{a1}K_{a2}}{[H_3O^+]^2}}$$

$$\frac{[H_2Ox]}{c_a} = \frac{[H_3O^+]^2}{[H_3O^+]^2 + [H_3O^+]K_{a1} + K_{a1}K_{a2}}$$

With no more difficulty, the expressions for the fractions present as HOx^- and Ox^{2-} can be derived.

$$\frac{[HOx^-]}{c_a} = \frac{[H_3O^+]K_{a1}}{[H_3O^+]^2 + [H_3O^+]K_{a1} + K_{a1}K_{a2}}$$

$$\frac{[Ox^{2-}]}{c_a} = \frac{K_{a1}K_{a2}}{[H_3O^+]^2 + [H_3O^+]K_{a1} + K_{a1}K_{a2}}$$

Fractions of total oxalate present as H_2Ox, HOx^-, and Ox^{2-} are shown as functions of *p*H in Fig. 7.7.

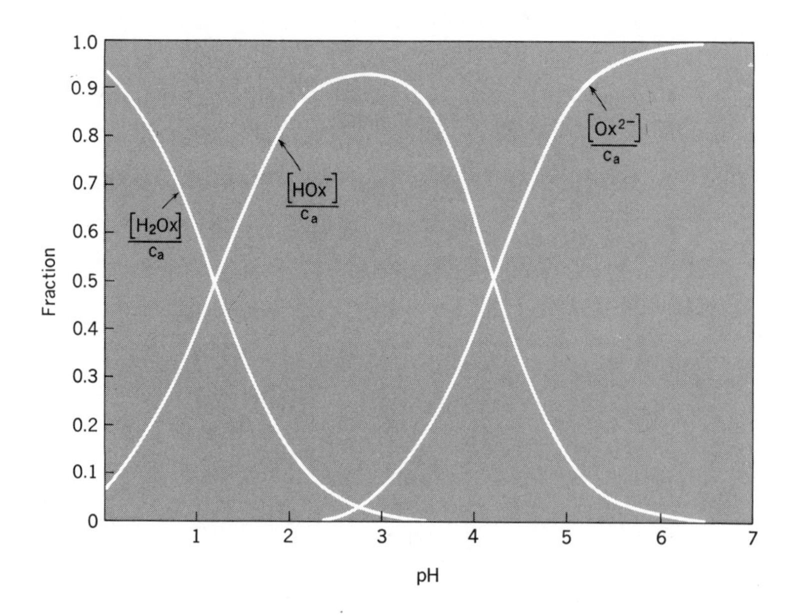

Figure 7.7 Distribution of oxalate species as a function of *p*H.

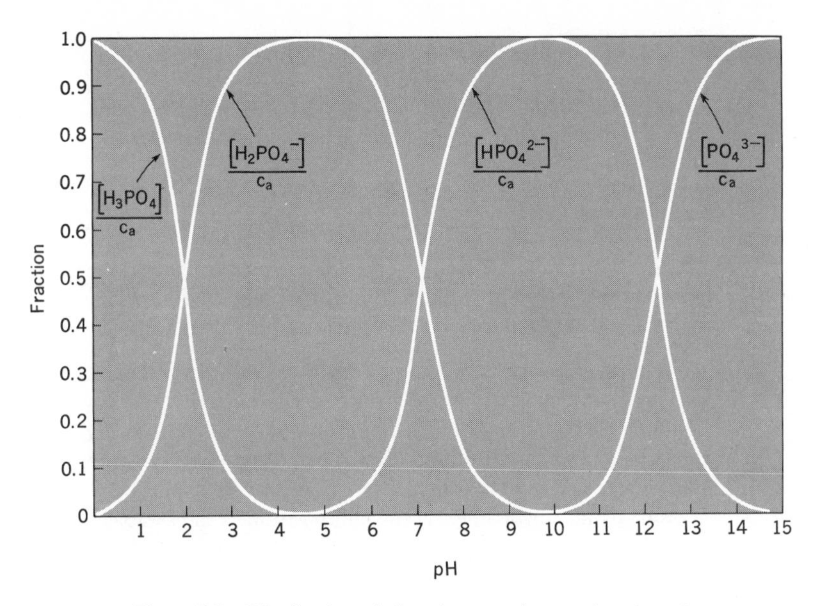

Figure 7.8 Distribution of phosphate species as a function of *p*H.

Derivation of similar equations for tri- or even tetraprotic acids, H_3B or H_4B, is more tedious but no more difficult than the above. Figure 7.8 shows the distribution of phosphoric acid species as a function of *p*H. It may be seen that below *p*H 5 or so, the only species present in significant concentration are H_3PO_4 and its first dissociation product, $H_2PO_4^-$. Thus the *p*H of an H_3PO_4 solution can be safely calculated on the basis of the first dissociation constant, as though the acid were monoprotic. As a matter of fact, at no *p*H are more than two species present in appreciable amount. In the case of oxalic acid, the two pK_a values are closer than are any pair of H_3PO_4 values; however, only in the *p*H range 2.5 to 3.0 are all three species discernible in Fig. 7.7 and even here one of the three is predominant.

REFERENCES

T. R. BLACKBURN, *Equilibrium, A Chemistry of Solutions,* Holt, Rinehart and Winston, Inc., New York, 1969.

J. N. BUTLER, *Ionic Equilibrium, A Mathematical Approach,* Addison-Wesley Publishing Co., Inc., Reading, Mass., 1964.

H. A. LAITINEN and W. E. HARRIS, *Chemical Analysis,* 2nd ed. McGraw-Hill Book Company, New York, 1975.

QUESTIONS

1. *Disproportionation.* The dissociation constants for H_3PO_4 are K_{a1}, K_{a2}, and K_{a3}. Derive an expression for the equilibrium constant of each of the following reactions in terms of these three constants.
(a) $H_2PO_4^- + H_2PO_4^- \rightleftharpoons H_3PO_4 + HPO_4^{2-}$
(b) $HPO_4^{2-} + HPO_4^{2-} \rightleftharpoons H_2PO_4^- + PO_4^{3-}$

2. *Species distribution.* Examine Fig. 7.8 and answer the following questions. Which phosphate species is in highest concentration at the following *p*H values? (a) 3.0; (b) 5.0; (c) 7.21; (d) 9.5; (e) 13. Between what *p*H values is HPO_4^{2-} the principal species?

3. *Carbonate mixtures.* If v_1 and v_2 are the milliliters of acid used in the titration of carbonate mixtures as indicated in Fig. 7.4, answer the following: (a) If the sample contains an equal number of mmol of $NaHCO_3$ and Na_2CO_3, what is the relation between v_1 and v_2? (b) If the relation between the two volumes is $2v_2 = v_1$, what can you conclude about the composition of the sample? (c) If $v_1 = 10$ ml and $v_2 = 30$ ml, what can you conclude about the composition of the sample?

Multiple-choice: In the following multiple-choice questions, select the *one best* answer.

4. In the titration of H_3PO_4 with NaOH at *p*H 4.67, (a) $[H_3PO_4] \cong [H_2PO_4^-]$; (b) $[H_2PO_4^-] \cong [HPO_4^{2-}]$; (c) $[HPO_4^{2-}] \cong [H_3PO_4]$; (d) $[H_3PO_4] \cong 2[H_2PO_4^-]$.

5. To 25 ml of 0.10 *M* NaH_2PO_4 is added 25 ml of 0.10 *M* NaOH. In the resulting solution, (a) $[H_3PO_4] \cong [HPO_4^{2-}]$; (b) $[H_2PO_4^-] \cong [PO_4^{3-}]$; (c) $[H_2PO_4^-] \cong [HPO_4^{2-}]$; (d) $[HPO_4^{2-}] \cong [PO_4^{3-}]$.

6. To 50 ml of 0.10 *M* Na_3PO_4 is added 50 ml of 0.15 *M* HCl. In the resulting solution, (a) $[H_2PO_4^-] \cong [HPO_4^{2-}]$; (b) $[H_2PO_4^-] \cong [H_3PO_4]$; (c) $[H_3PO_4] \cong [HPO_4^{2-}]$; (d) $[HPO_4^{2-}] \cong [PO_4^{3-}]$.

7. What is the value of the equilibrium constant for the reaction

$$HPO_4^{2-} + OH^- \rightleftharpoons PO_4^{3-} + H_2O$$

where K_{a1}, K_{a2}, and K_{a3} are the three dissociation constants of H_3PO_4? (a) $K_{a2} \times K_{a3}$; (b) K_w/K_{a3}; (c) K_{a3}/K_w; (d) K_{a2}/K_w.

8. What is the value of the equilibrium constant for the following reaction?

$$HPO_4^{2-} + H_2O \rightleftharpoons H_2PO_4^- + OH^-$$

(a) K_w/K_{a3}; (b) K_w/K_{a2}; (c) K_w/K_{a1}; (d) K_{a2}/K_w.

9. To 60 ml of 0.10 *M* Na_2CO_3 is added 40 ml of 0.15 *M* HCl. In the resulting solution (a) $[H_2CO_3] \cong [HCO_3^-]$; (b) $[H_2CO_3] \cong [CO_3^{2-}]$; (c) $[HCO_3^-] \cong [CO_3^{2-}]$; (d) $[HCO_3^-] \cong [H_3O^+]$.

10. For the acid H_2B, $pK_{a1} = 4.00$ and $pK_{a2} = 10.0$. Which of the following indicators is most suitable for the titration

$$H_2B + OH^- \longrightarrow HB^- + H_2O?$$

(a) Methyl orange; (b) bromcresol green; (c) *p*-nitrophenol; (d) phenolphthalein?

PROBLEMS

1. *Approximate equations.* Calculate the hydrogen ion concentration and pH of the following solutions using equation (8), page 169, and the simpler equation on page 168. (a) 0.10 M NaHB, $pK_{a_1} = 3.0$; $pK_{a_2} = 7.0$; (b) 0.0010 M NaHB; (c) 0.010 M NaH_2PO_4; (d) 0.010 M Na_2HPO_4; (e) 0.010 M $NaHCO_3$; (f) 0.010 M NaHS.

2. *Mixtures of solutions.* Calculate the pH of the following solutions, using the approximate equation given on page 168. (a) 40 ml of 0.15 M H_3PO_4 + 60 ml of 0.10 M NaOH; (b) 40 ml of 0.15 M H_3PO_4 + 90 ml of 0.10 M NaOH; (c) 40 ml of 0.15 M H_3PO_4 + 60 ml of 0.20 M NaOH; (d) 40 ml of 0.15 M H_3PO_4 + 60 ml of 0.25 M NaOH.

3. *Mixtures of solutions.* Calculate the pH of the following solutions, using the approximate equation given on page 168. (a) 50 ml of 0.10 M Na_3PO_4 + 50 ml of 0.30 M HCl; (b) 50 ml of 0.10 M Na_3PO_4 + 50 ml of 0.25 M HCl; (c) 50 ml of 0.10 M Na_3PO_4 + 50 ml of 0.20 M HCl; (d) 50 ml of 0.10 M Na_3PO_4 + 50 ml of 0.10 M HCl.

4. *Mixtures of solutions.* Calculate the pH of the following solutions using the approximate equation given on page 168. (a) 3.00 mmol of Na_2CO_3 + 60 ml of 0.10 M HCl + 40 ml of H_2O; (b) 40 ml of 0.020 M H_2CO_3 + 20 ml of 0.040 M NaOH; (c) 40 ml of 0.10 M Na_2HPO_4 + 60 ml of 0.10 M HCl; (d) 50 ml of 0.10 M H_3PO_4 + 50 ml of 0.050 M Na_2HPO_4.

5. *Titration curve.* 50.00 ml of 0.100 M H_3PO_4 is titrated with 0.100 M NaOH. (a) Calculate the pH after the addition of the following volumes of titrant: 0.00, 10.00, 25.00, 50.00, 65.00, 75.00, 100.00, and 110.00. Note that since K_{a_1} is relatively large, the complete quadratic should be solved for the first three volumes. (b) Plot the titration curve. (c) Select suitable indicators for the first and second equivalence points.

6. *Equilibrium constants.* Calculate the numerical values for the equilibrium constants of the following reactions:
(a) $PO_4^{3-} + H_2O \rightleftharpoons HPO_4^{2-} + OH^-$
(b) $H_2PO_4^- + OH^- \rightleftharpoons HPO_4^{2-} + H_2O$
(c) $H_3PO_4 + HPO_4^{2-} \rightleftharpoons 2H_2PO_4^-$
(d) $H_3PO_4 + PO_4^{3-} \rightleftharpoons H_2PO_4^- + HPO_4^{2-}$

7. *Phosphate buffer.* A phosphate buffer is prepared by dissolving 14.2 g of Na_2HPO_4 and 24.0 g of NaH_2PO_4 in 1.00 liter of solution. (a) Calculate the pH of the buffer. (b) 10.0 mmol of HCl is added to 100 ml of the buffer. Calculate the pH of the resulting solution.

8. *Phosphate solutions.* The pH of a phosphate solution is 7.81. The analytical concentration of phosphate is 0.10 M. What are the two principal species in the solution and what are their concentrations?

9. *Phosphate in the blood.* If the pH of blood is 7.40, what are the principal phosphate species at this pH? What is the ratio of the concentrations of the two principal species?

10. *Isoelectric point.* The amino acid glycine, NH_2CH_2COOH, dissociates in aqueous solution as follows:

$$NH_3CH_2COOH^+ \rightleftharpoons NH_3CH_2COO + H^+ \qquad pK_{a_1} = 2.35$$

$$NH_3CH_2COO \rightleftharpoons NH_2CH_2COO^- + H^+ \qquad pK_{a_2} = 9.78$$

Calculate the approximate pH at the isoelectric point, that is, the point at which the molarities of $NH_3CH_2COOH^+$ and $NH_2CH_2COO^-$ are equal.

11. *Species distribution.* Citric acid, $H_3C_6H_5O_7$, is a triprotic acid. Calculate the fraction of the acid in the molecular and various ionic forms at integral pH values of 1 to 8. Plot a curve similar to Fig. 7.8 and select the best buffering ranges.

12. *Carbonate mixtures.* The samples listed below contain the indicated number of millimoles of NaOH, Na_2CO_3, and $NaHCO_3$.

Sample	Na_2CO_3, mmol	NaOH, mmol	$NaHCO_3$, mmol
A	2.0	3.0	0
B	2.5	0	1.5
C	0	2.5	1.5
D	3.0	0	0
E	0	2.0	0
F	0	0	3.5

If these samples are titrated with 0.10 M HCl using the two-indicator method, calculate the values of v_1, the milliliters to the phenolphthalein end point, and v_2, the milliliters from the phenolphthalein to the methyl orange end point, for each sample.

13. *Carbonate mixtures.* A sample that may be a mixture of Na_2CO_3 and $NaHCO_3$ or NaOH and Na_2CO_3 is titrated using the two-indicator method. A 0.8432-g sample required 28.56 ml of 0.1206 M HCl to reach the phenolphthalein end point, and an additional 12.18 ml to reach the methyl orange end point. Identify the mixture and calculate the percentage of each component.

14. *Carbonate mixtures.* A 0.9205-g sample of a carbonate mixture required 18.73 ml of 0.1124 M HCl for titration to the phenolphthalein end point, and an additional 34.67 ml to reach the methyl orange end point. Identify the mixture and calculate the percentage of each component.

15. *Carbonate mixtures.* Suppose that a 0.3000-g sample consists of 50.00% by weight NaOH and 50.00% by weight Na_2CO_3. (a) How many milliliters of 0.1000 M HCl would be required for titration to the phenolphthalein end point? (b) How many milliliters would have been required had methyl orange indicator been used instead of phenolphthalein? (c) Repeat parts (a) and (b) for a sample which is 50–50 Na_2CO_3 and $NaHCO_3$.

16. *Carbonate mixtures.* Suppose that a 0.3000-g sample contains only NaOH and Na_2CO_3 and has an equal number of moles of the two compounds. (a) How many milliliters of 0.1000 M HCl would be required for titration to the phenolphthalein end point? (b) How many additional milliliters of HCl would be required to reach the methyl orange end point? (c) Repeat parts (a) and (b) for a sample containing an equal number of moles of Na_2CO_3 and $NaHCO_3$.

17. *Carbonate error.* A student prepared and standardized a solution of NaOH, finding the normality to be 0.1026. He left exactly 1.000 liter of this solution unprotected and it absorbed 0.1000 g of CO_2 from the air. Later the student titrated a 50.00-ml aliquot of the base with 0.1142 N HCl using phenolphthalein indicator. (a) How many milliliters of HCl was required? (b) how many milliliters would have been required had the solution not absorbed the CO_2?

18. *Carbonate mixtures.* A sample of pure $NaHCO_3$ weighing 0.5040 g is dissolved in water and 0.5600 g of pure NaOH is added to the solution. The solution is then diluted to 250.0 ml in a volumetric flask. A 50.00-ml aliquot is withdrawn and titrated with 0.1000 M HCl using phenolphthalein indicator. (a) How many milliliters of HCl is required? (b) How many milliliters will be required if methyl orange indicator is used instead of phenolphthalein?

19. *Carbonate mixtures.* A sample consists of only $NaHCO_3$ and Na_2CO_3. A portion weighing 0.3380 g requires 35.00 ml of 0.1500 M HCl for titration to the methyl orange end point. Calculate (a) the volume of HCl that would be required if phenolphthalein were used as the indicator; (b) the percentage by weight of $NaHCO_3$.

20. *Carbonate mixtures.* A 0.398-g sample consisting of only NaOH and Na_2CO_3 is dissolved in 50.0 ml of solution and titrated with standard HCl using phenolphthalein indicator. The concentration of the base is found to be 0.100 N. What normality would be found if methyl orange is used as the indicator?

21. *Titration of a mixture of two acids.* 50 ml of a solution which is 0.10 M in the acid HA, $pK_a = 4.00$, and 0.050 M in HB, $pK_a = 8.00$, is titrated with 0.20 M NaOH. Calculate the pH at (a) the first equivalence point; (b) at the second equivalence point.

22. *Titration of a mixture of two acids.* 50.0 ml of a solution which is 0.100 M in HCl and 0.100 M in the weak acid HX is titrated with 0.200 M NaOH. (a) Calculate the pH at which 99.9% of the HCl has been neutralized. (b) Calculate the percentage of HX which has reacted at the pH in part (a) if the pK_a of HX is (i) 4.00; (ii) 5.00; (iii) 6.00; (iv) 7.00; (v) 8.00.

CHAPTER EIGHT

complex formation titrations

One of the types of chemical reactions which may serve as the basis of a titrimetric determination involves the formation of a soluble but slightly dissociated *complex* or *complex ion*. An example is the reaction of silver ion with cyanide ion to form the very stable $Ag(CN)_2^-$ complex ion:

$$Ag^+ + 2CN^- \rightleftharpoons Ag(CN)_2^-$$

The complexes we wish to consider in this chapter are formed by the reaction of a metal ion, a cation, with an anion or neutral molecule. The metal ion in the complex is called the *central atom*, and the group attached to the central atom is called a *ligand*. The number of bonds formed by the central metal atom is called the *coordination number* of the metal. In the complex above, silver is the central metal atom with a coordination number of two, and cyanide is the ligand.

The reaction by which a complex is formed can be regarded as a Lewis acid-base reaction with the ligand acting as the base, donating a pair of electrons to the cation, which is the acid. The bond formed between the central metal atom and the ligand is often covalent, but in some cases the interaction may be one of coulombic attraction. Some complexes undergo substitution reactions very rapidly and the complex is said to be *labile*. An example is

$$\underset{\text{Light blue}}{Cu(H_2O)_4^{2+}} + 4NH_3 \rightleftharpoons \underset{\text{Dark blue}}{Cu(NH_3)_4^{2+}} + 4H_2O$$

The reaction goes readily to the right by the addition of ammonia to the aquo-complex; addition of a strong acid which neutralizes the ammonia shifts the equilibrium rapidly back to the aquo-complex. Some complexes undergo substitution reactions only very slowly and are said to be *nonlabile* or *inert.* Almost all complexes formed by cobalt and chromium in the +3 oxidation state are inert, whereas most of the other complexes of the first series of transition metals are labile. Some examples of typical complexes along with some properties are listed in Table 8.1.

TABLE 8.1 Some Typical Complexes

Metal	Ligand	Complex	Coordination number of metal	Geometry	Reactivity
Ag^+	NH_3	$Ag(NH_3)_2{}^+$	2	Linear	Labile
Hg^{2+}	Cl^-	$HgCl_2$	2	Linear	Labile
Cu^{2+}	NH_3	$Cu(NH_3)_4{}^{2+}$	4	Tetrahedral	Labile
Ni^{2+}	CN^-	$Ni(CN)_4{}^{2-}$	4	Square planar	Labile
Co^{2+}	H_2O	$Co(H_2O)_6{}^{2+}$	6	Octahedral	Labile
Co^{3+}	NH_3	$Co(NH_3)_6{}^{3+}$	6	Octahedral	Inert
Cr^{3+}	CN^-	$Cr(CN)_6{}^{3-}$	6	Octahedral	Inert
Fe^{3+}	CN^-	$Fe(CN)_6{}^{3-}$	6	Octahedral	Inert

Molecules or ions which act as ligands generally contain an electronegative atom, such as nitrogen, oxygen, or one of the halogens. Ligands which have only one unshared pair of electrons, for example $:NH_3$, are said to be *unidentate.* Ligands which have two groups capable of forming two bonds with the central atom are said to be *bidentate.* An example is ethylene diamine, $NH_2CH_2CH_2NH_2$, where both nitrogen atoms have unshared electron pairs. Copper(II) ion forms a complex with two molecules of ethylene diamine as follows:

$$Cu^{2+} + 2NH_2CH_2CH_2NH_2 \rightleftharpoons \left[\begin{array}{ccccc} CH_2 - NH_2 & & & & NH_2 - CH_2 \\ | & \searrow & Cu & \swarrow & | \\ & \nearrow & & \nwarrow & \\ CH_2 - NH_2 & & & & NH_2 - CH_2 \end{array}\right]^{2+}$$

Heterocyclic rings formed by the interaction of a metal ion with two or more functional groups in the same ligand are called *chelate rings*; the organic molecule is a *chelating* agent, and the complexes are called *chelates* or *chelate* compounds. Analytical applications based on the use of chelating agents as titrants for metal ions have shown remarkable growth in recent years.

STABILITY OF COMPLEXES

We saw in Chapter 5 how the equilibrium constant was formulated for complex formation reactions using the silver–ammonia complex ion as an example. Most of our discussion in this chapter will center on reactions of metal ions with chelating agents. These are generally 1 : 1 reactions in which a soluble complex is formed. We can represent such a reaction in a general manner as

$$M + L \rightleftharpoons ML$$

where M is the central metal cation, L the ligand, and ML the complex. The *stability constant* of the complex is

$$K = \frac{[ML]}{[M][L]}$$

As mentioned in Chapter 5, we shall always write the *stability* or *formation* constant of the complex rather than its reciprocal, the *instability* or *dissociation* constant.

We also noted in Chapter 5 (Table 5.6) that the form of the stability constant is the same as that for the titration of a weak acid with a strong base:

$$HOAc + OH^- \rightleftharpoons OAc^- + H_2O \qquad K = \frac{[OAc^-]}{[HOAc][OH^-]}$$

We have now seen (page 143) that such a reaction with a K of about 10^8 is sufficiently complete at the equivalence point for a feasible titration. We can predict that a reaction yielding a complex of the form ML with a stability constant of the same order of magnitude should give a feasible titration under comparable conditions of concentrations.

Stepwise Formation Constants

As noted in Chapter 5, the reaction of cations with ligands such as ammonia usually proceeds stepwise. For example, the formation of the complex $Cu(NH_3)_4^{2+}$ proceeds in four steps:

$$Cu^{2+} + NH_3 \rightleftharpoons CuNH_3^{2+} \qquad K_1 = 1.9 \times 10^4$$

$$CuNH_3^{2+} + NH_3 \rightleftharpoons Cu(NH_3)_2^{2+} \qquad K_2 = 3.6 \times 10^3$$

$$Cu(NH_3)_2^{2+} + NH_3 \rightleftharpoons Cu(NH_3)_3^{2+} \qquad K_3 = 7.9 \times 10^2$$

$$Cu(NH_3)_3^{2+} + NH_3 \rightleftharpoons Cu(NH_3)_4^{2+} \qquad K_4 = 1.5 \times 10^2$$

Considering the overall reaction,

$$Cu^{2+} + 4NH_3 \rightleftharpoons Cu(NH_3)_4^{2+}$$

$$K = \frac{[Cu(NH_3)_4^{2+}]}{[Cu^{2+}][NH_3]^4} = K_1K_2K_3K_4 = 8.1 \times 10^{12}$$

the equilibrium constant seems large enough for a feasible titration. The titration of a strong acid with ammonia, $H_3O^+ + NH_3 \rightleftharpoons NH_4^+ + H_2O$, where the K is 1.8×10^9, is feasible. However, as shown in Fig. 8.1, the titration of strong acid with ammonia gives a large increase in pH at the equivalence point, whereas the titration of Cu^{2+} with ammonia does not.

It may be seen in Fig. 8.1 that the $\Delta p\text{Cu}/\Delta V$ in the copper titration would be better if the pCu ($p\text{Cu} = -\log[Cu^{2+}]$) remained lower in the early stages of the

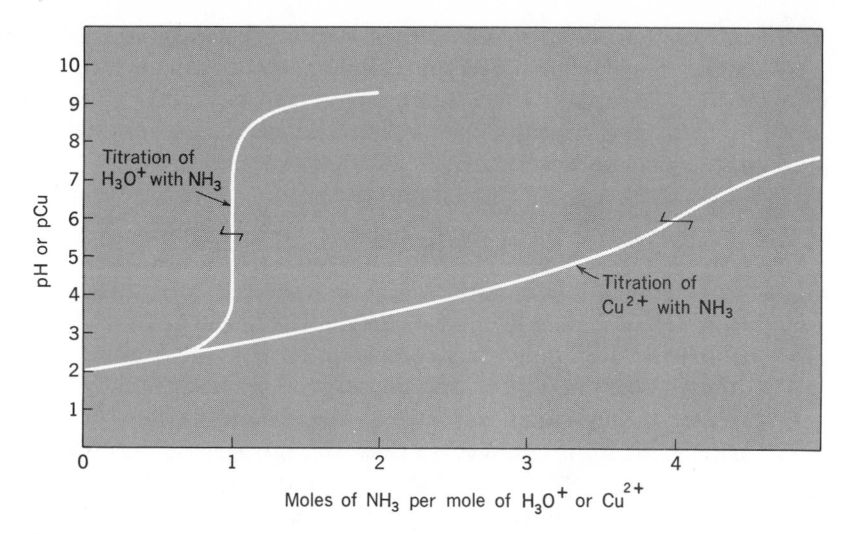

Figure 8.1 Titration of strong acid and of copper(II) ion with ammonia calculated for 10^{-2} M H_3O^+ and Cu^{2+}, assuming no volume change.

titration, as does the *p*H in the H_3O^+ titration. For example, 75% of the way to the equivalence point, if the ratio of $[Cu^{2+}]$ to $[Cu(NH_3)_4{}^{2+}]$ were close to 1 : 3, the *p*Cu would be about 2.6, whereas in fact it is about 4.5 at this point. The reason that the *p*Cu is too high lies in the fact that all of the added ammonia has not been used to form the complex $Cu(NH_3)_4{}^{2+}$. Rather, lower complex species such as $Cu(NH_3)^{2+}$ have formed, lowering the free $[Cu^{2+}]$ below a desirable value for a feasible titration. Such behavior is predictable from the formation constants of the individual steps given above. It is seen, for example, that there is less tendency for $Cu(NH_3)^{2+}$ to add a second ammonia than for free Cu^{2+} to bind the first one.

The difficulty arising from lower complexes can be averted by the use of chelating agents as titrants. Consider, for example, the compound triethylenetetramine, a quadridentate ligand, often abbreviated "trien." Here, four nitrogen atoms are linked by ethylene bridges in a single molecule which can satisfy copper's normal coordination number of four in one step:

$$\left[\begin{array}{ccc} & CH_2CH_2 & \\ H_2N & & NH-CH_2 \\ & Cu & \\ H_2N & & NH-CH_2 \\ & CH_2CH_2 & \end{array}\right]^{2+}$$

It may be supposed that the formation of the first nitrogen-copper bond brings the other nitrogens of the trien molecule into such proximity that the formation of additional bonds involving these nitrogens is much more probable than the formation of bonds between the copper and other trien molecules. Similarly, it is unlikely that one trien molecule will coordinate with more than one copper. Thus, under ordinary conditions, the stoichiometry of complex formation in this system is 1 Cu^{2+} : 1 trien. The resulting five-membered rings shown in the structural formula are relatively free of strain. The complex is very stable, as shown by its formation constant:

$$Cu^{2+} + \text{trien} \rightleftharpoons Cu(\text{trien})^{2+} \qquad K = \frac{[Cu(\text{trien})^{2+}]}{[Cu^{2+}][\text{trien}]} = 2.5 \times 10^{20}$$

Thus trien is a good titrant for copper: The ligand and the complex ion are both soluble in water, only a 1 : 1 complex is formed, the equilibrium constant for the titration reaction is large, and the reaction proceeds rapidly.

Only a few metal ions such as copper, cobalt, nickel, zinc, cadmium, and mercury(II) form stable complexes with nitrogen ligands such as ammonia and trien. Certain other metal ions (e.g., aluminum, lead, and bismuth) are better complexed with ligands containing oxygen atoms as electron donors. Certain chelating agents which contain both oxygen and nitrogen are particularly effective in forming stable complexes with a wide variety of metals. Of these, the best known is *e*thylene*d*iamine*t*etra*a*cetic acid, sometimes designated (ethylenedinitrilo)tetraacetic acid, and often abbreviated EDTA:

```
HOOCCH2               CH2COOH
       \             /
        NCH2CH2N
       /             \
HOOCCH2               CH2COOH
```

The term *chelon* (pronounced "key-loan") has been proposed as a generic name for the entire class of reagents, including polyamines such as trien, polyaminocarboxylic acids such as EDTA, and related compounds that form stable, water-soluble, 1 : 1 complexes with metal ions and which hence may be employed as titrants for metals. The complexes, a special class of chelate compounds, are called *metal chelonates*, and the titrations are termed *chelometric titrations*. Chelons have practically revolutionized the analytical chemistry of many of the metallic elements, and they are of great importance in many fields.

CHELOMETRIC TITRATIONS

The suitability of chelons such as EDTA as titrants for metal ions has been mentioned above. We wish here to examine some of the equilibria involved in these titrations, consider end-point techniques, and show some representative applications. Our discussion will be limited largely to EDTA.

EDTA is potentially a sexidentate ligand which may coordinate with a metal ion through its two nitrogens and four carboxyl groups. It is known from infrared spectra and other measurements that this is the case, for example, with the cobalt(II) ion, which forms an octahedral EDTA complex whose structure is somewhat as shown below:

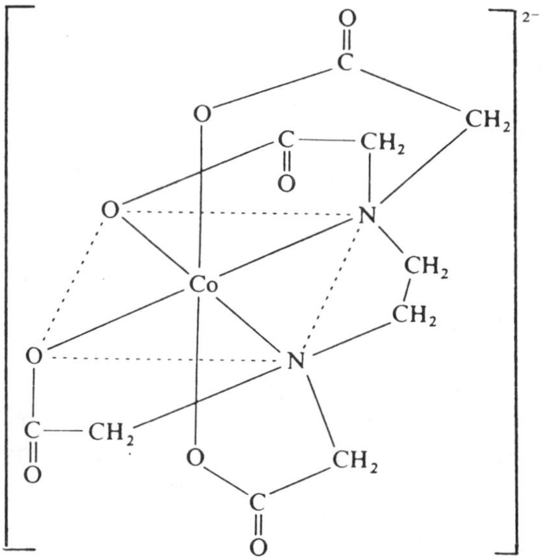

In other cases, EDTA may behave as a quinquedentate or quadridentate ligand having one or two of its carboxyl groups free of strong interaction with the metal.

For convenience, the free acid form of EDTA is often abbreviated H_4Y. The cobalt complex shown above is then written CoY^{2-}, and other complexes become CuY^{2-}, FeY^{-}, CaY^{2-}, etc. In solutions which are fairly acidic, partial protonation of EDTA without complete rupture of the metal complex may occur, leading to species such as $CuHY^{-}$; but under the usual conditions all four hydrogens are lost when the ligand is coordinated with a metal ion. At very high *p*H values, hydroxide ion may penetrate the coordination sphere of the metal and complexes such as $Cu(OH)Y^{3-}$ may exist.

Equilibria Involved in EDTA Titrations

We may consider a metal ion such as Cu^{2+}, which is seeking electrons in its reactions, to be analogous to an acid like H_3O^+, and the EDTA anion Y^{4-}, which is an electron donor, to be a base. Then the reaction $Cu^{2+} + Y^{4-} \rightleftharpoons CuY^{2-}$ is analogous to an ordinary neutralization reaction, and it should be a simple matter to calculate *p*Cu values under various conditions, calculate titration curves, discuss feasibility, etc. As a matter of fact, however, the situation is more complicated than this because of the intrusion of other equilibria into the titration situation. We shall discuss some of these in the sections below.

The Absolute Stability or Formation Constant

It is customary to tabulate for various metal ions and various chelons such as EDTA, values of the equilibrium constants for reactions formulated as follows:

$$M^{n+} + Y^{4-} \rightleftharpoons MY^{-(4-n)} \qquad K_{abs} = \frac{[MY^{-(4-n)}]}{[M^{n+}][Y^{4-}]}$$

K_{abs} is called the *absolute stability constant* or the *absolute formation constant.* Values of some of these constants may be found in Table 2, Appendix I.

The pH Effect

The four dissociation constants of the acid H_4Y are as follows:

$$H_4Y + H_2O \rightleftharpoons H_3O^+ + H_3Y^- \qquad K_{a_1} = 1.02 \times 10^{-2}$$

$$H_3Y^- + H_2O \rightleftharpoons H_3O^+ + H_2Y^{2-} \qquad K_{a_2} = 2.14 \times 10^{-3}$$

$$H_2Y^{2-} + H_2O \rightleftharpoons H_3O^+ + HY^{3-} \qquad K_{a_3} = 6.92 \times 10^{-7}$$

$$HY^{3-} + H_2O \rightleftharpoons H_3O^+ + Y^{4-} \qquad K_{a_4} = 5.50 \times 10^{-11}$$

The distributions of the five EDTA species as functions of *p*H are shown in Fig. 8.2. It may be seen that only at *p*H values greater than about 12 does most of the EDTA exist as the tetraanion Y^{4-}. At lower *p*H values, the protonated species HY^{3-}, etc., predominate. We may consider that H_3O^+, then, competes with a metal ion for EDTA, and it is clear that the real tendency to form the metal chelonate at any particular *p*H value is not discernible directly from K_{abs}. For example, at *p*H 4 the predominant EDTA species is H_2Y^{2-}, and the reaction with a metal such as copper may be written

$$Cu^{2+} + H_2Y^{2-} \rightleftharpoons CuY^{2-} + 2H^+$$

Obviously, as the *p*H goes down, the equilibrium is shifted away from the

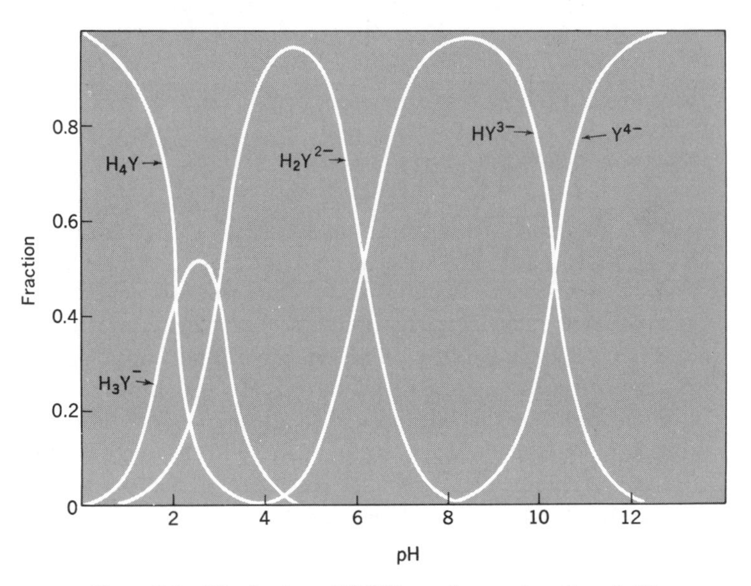

Figure 8.2 Distribution of EDTA species as a function of *p*H.

formation of the chelonate CuY^{2-}, and we may expect that there will be a pH value below which the titration of copper with EDTA will not be feasible. We wish to be able to estimate what this value is. Clearly a calculation will involve K_{abs} and the appropriate K_a values of EDTA. Actually, as shown below, it is possible to estimate very easily the minimal pH for a feasible metal ion titration from the K_{abs} value and a simple graph.

The expression for the fraction of EDTA in the Y^{4-} form can be obtained in the same manner as was done for oxalic acid (page 179). Let c_Y represent the total concentration of the uncomplexed EDTA:

$$c_Y = [Y^{4-}] + [HY^{3-}] + [H_2Y^{2-}] + [H_3Y^-] + [H_4Y]$$

Substituting for the concentrations of the various species in terms of the dissociation constants and solving for the fraction in the Y^{4-} form gives

$$\frac{[Y^{4-}]}{c_Y} = \frac{K_{a1}K_{a2}K_{a3}K_{a4}}{[H_3O^+]^4 + [H_3O^+]^3K_{a1} + [H_3O^+]^2K_{a1}K_{a2} + [H_3O^+]K_{a1}K_{a2}K_{a3} + K_{a1}K_{a2}K_{a3}K_{a4}}$$

Giving the fraction of EDTA in the Y^{4-} form the symbol α_4, we may write

$$\frac{[Y^{4-}]}{c_Y} = \alpha_4$$

or

$$[Y^{4-}] = \alpha_4 c_Y$$

The value of α_4 may obviously be calculated at any desired pH for any chelon whose dissociation constants are known. Shortcuts may be taken in the calculation; for example, it is obvious that at very high pH values, the term containing $[H_3O^+]^4$ will be negligible. In any case, the work has already been done, and graphs or tables showing α values as functions of pH for a number of chelons may be found in the literature.[1] Because the values extend over a wide range of magnitudes, $-\log \alpha_4$ is usually plotted vs. pH. Such a graph for EDTA is shown in Fig. 8.3. Some values are also given in Table 8.2.

Substitution of $\alpha_4 c_Y$ in the absolute stability constant expression given above yields

$$K_{abs} = \frac{[MY^{-(4-n)}]}{[M^{n+}]\alpha_4 c_Y}$$

or

$$K_{abs}\alpha_4 = \frac{[MY^{-(4-n)}]}{[M^{n+}]c_Y} = K_{eff}$$

K_{eff} is called the *effective* or *conditional stability constant.* Unlike K_{abs}, K_{eff} varies with pH because of the pH dependence of α_4. In certain regards K_{eff} is more immediately useful than K_{abs}, because it shows the actual tendency to form the

[1] For example, see C. N. Reilley, R. W. Schmid, and F. S. Sadek, *J. Chem. Educ.*, **36**, 555 (1959).

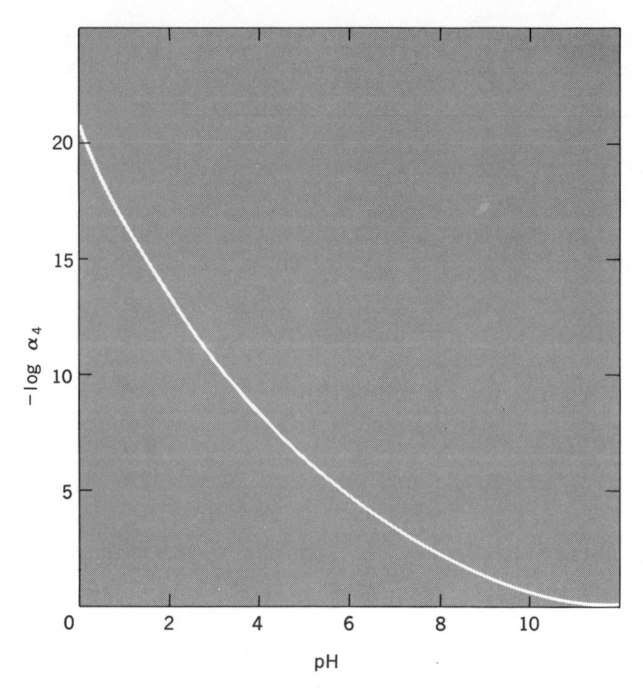

Figure 8.3 Variation of $-\log \alpha_4$ with *p*H for EDTA.

TABLE 8.2 Values of α_4 for EDTA

*p*H	α_4	$-\log \alpha_4$
2.0	3.7×10^{-14}	13.44
2.5	1.4×10^{-12}	11.86
3.0	2.5×10^{-11}	10.60
4.0	3.3×10^{-9}	8.48
5.0	3.5×10^{-7}	6.45
6.0	2.2×10^{-5}	4.66
7.0	4.8×10^{-4}	3.33
8.0	5.1×10^{-3}	2.29
9.0	5.1×10^{-2}	1.29
10.0	0.35	0.46
11.0	0.85	0.07
12.0	0.98	0.00

metal chelonate at the *p*H value in question. Although K_{eff} values are not customarily tabulated, it is apparent that they may be estimated readily from values of K_{abs}, which are found in tables of constants, and α_4 values obtained from tables such as Table 8.2.

It may be noted that, as the *p*H goes down, α_4 becomes smaller, and hence K_{eff} becomes smaller. Remember that α_4 is the fraction of EDTA in the Y^{4-} form.

Thus at pH values above 12 or so, where EDTA is essentially completely dissociated, α_4 approaches unity ($-\log \alpha_4$ approaches zero), and K_{eff} approaches K_{abs}.

Normally, the solutions of metal ions to be titrated with EDTA are buffered so that the pH will remain constant despite the release of H_3O^+ as the complexes are formed. Thus there is usually a definite basis for estimating K_{eff}, and with this value at hand, it is easy to calculate the titration curve, from which a judgment of feasibility may be made just as with acid-base titrations. The pH is often adjusted to as low a value as is consistent with feasibility. At high pH many metal ions tend to hydrolyze and even precipitate as hydroxides. In most titrations the concentration of the cation is kept as low as 0.010 to 0.0010 M to decrease the chances of precipitation.

The following example shows the calculations involved in deriving a titration curve for Ca^{2+} titrated with EDTA.

Example 1. 50.0 ml of a solution which is 0.0100 M in Ca^{2+} and buffered at pH 10.0 is titrated with a 0.0100 M EDTA solution. Calculate values of pCa at various stages of the titration and plot the titration curve.

K_{abs} for CaY^{2-} is 5.0×10^{10}. Referring to Table 8.2 α_4 at pH 10.0 is 0.35. Hence K_{eff} is $5.0 \times 10^{10} \times 0.35 = 1.8 \times 10^{10}$.

(a) *Start of titration*

$$[Ca^{2+}] = 0.0100\,M$$

$$pCa = -\log[Ca^{2+}] = 2.00$$

(b) *After addition of 10.0 ml of titrant.* There is a considerable excess of Ca^{2+} at this point, and with a K value of the order of 10^{10}, we may assume that the reaction goes to completion. Thus

$$[Ca^{2+}] = \frac{(0.50 - 0.10)\text{ mmol}}{60.0\text{ ml}} = 0.0067\,M$$

$$pCa = 2.17$$

Similar calculations can be made at various intervals before the equivalence point. In the vicinity of the equivalence point, more accurate calculations can be made by not assuming complete reaction, that is, by taking into account calcium ions produced by dissociation of CaY^{2-}, and solving the usual quadratic equation. The data in Table 8.3 were calculated by the approximate method.

(c) *Equivalence point*

$$[Ca^{2+}] = c_Y$$

$$[CaY^{2-}] \cong \frac{0.500\text{ mmol}}{100\text{ ml}} \cong 5.0 \times 10^{-3}\,M$$

$$K_{eff} = \frac{5.0 \times 10^{-3}}{[Ca^{2+}]^2} = 1.8 \times 10^{10}$$

$$[Ca^{2+}] = 5.2 \times 10^{-7}$$

$$pCa = 6.28$$

(d) *After addition of 60.0 ml titrant.* Excess EDTA = 0.100 mmol

$$c_Y = \frac{0.100 \text{ mmol}}{110 \text{ ml}} = 9.1 \times 10^{-4} M$$

$$[CaY^{2-}] = \frac{0.500 \text{ mmol}}{110 \text{ ml}} = 4.55 \times 10^{-3} M$$

$$\frac{4.55 \times 10^{-3}}{[Ca^{2+}]9.1 \times 10^{-4}} = 1.8 \times 10^{10}$$

$$[Ca^{2+}] = 2.8 \times 10^{-10}$$

$$pCa = 9.55$$

The data for this titration are given in Table 8.3 and the titration curve is plotted in Fig. 8.4. The titration curve is of familiar shape, with a sharp increase in the value of *p*Ca at the equivalence point. Also shown in the figure are curves for the titration done in solutions of *p*H 8 and *p*H 12. In these solutions the values of K_{eff} (same as K for the titration) are 2.6×10^8 and 4.9×10^{10}, respectively. Note that the curves are the same up to the equivalence point. The larger increase in *p*Ca is obtained at the higher *p*H since K_{eff} is larger in solutions of low hydrogen ion concentration. At low *p*H, K_{eff} becomes so small that the titration is not feasible.

The magnitude of K_{eff} or K required for a feasible titration can be calculated as was done for an acid–base titration (page 142). The following example is an illustration.

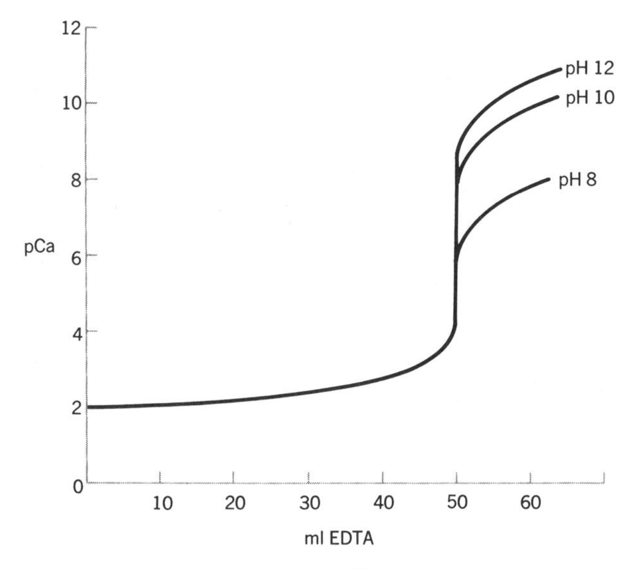

Figure 8.4 Titration curves: 50 ml 0.0100 M Ca^{2+} titrated with 0.0100 M EDTA at *p*H 8, 10, and 12.

TABLE 8.3 Titration of 50.0 ml of 0.0100 *M* Ca^{2+} with 0.0100 *M* EDTA

EDTA, ml	$[Ca^{2+}]$	*p*Ca	% Ca^{2+} reacted
0.00	0.0100	2.00	0.0
10.0	0.0067	2.17	20.0
20.0	0.0043	2.37	40.0
30.0	0.0025	2.60	60.0
40.0	0.0011	2.96	80.0
49.0	1.0×10^{-4}	4.00	98.0
49.9	1.0×10^{-5}	5.00	99.8
50.0	5.2×10^{-7}	6.28	100.0
50.1	2.8×10^{-8}	7.55	100.0
60.0	2.8×10^{-10}	9.55	100.0

Example 2. 50 ml of 0.010 *M* M^{2+} is titrated with 0.010 *M* EDTA. Calculate the value of K_{eff} so that when 49.95 ml of titrant has been added, the reaction is essentially complete, and the *p*M changes by 2.00 units on the addition of 2 more drops (0.10 ml) of titrant.

One drop before the equivalence point, 0.4995 mmol of EDTA has been added. We started with $50 \times 0.010 = 0.50$ mmol of M^{2+}. There must remain 0.00050 mmol. Hence

$$[M^{2+}] = \frac{0.00050 \text{ mmol}}{99.95 \text{ ml}} = 5 \times 10^{-6}\,M$$

$$pM = 5.30$$

If $\Delta pM = 2.00$ units, then $pM = 7.30$ and $[M^{2+}] = 5 \times 10^{-8}\,M$ when 50.05 ml of titrant is added. At this point

$$c_Y = \frac{0.05 \times 0.010}{100.05} \cong 5 \times 10^{-6}$$

$$[MY^{2-}] \cong \frac{0.5}{100} \cong 5 \times 10^{-3}$$

Hence

$$K_{eff} = \frac{5 \times 10^{-3}}{(5 \times 10^{-8})(5 \times 10^{-6})}$$

$$K_{eff} = K = 2 \times 10^{10}$$

The student should confirm that, for $\Delta pM = 1.00$, K_{eff} should be 2×10^9.

Once a value of *K* is selected for feasibility, it is easy to determine the lowest *p*H at which the titration can be carried out. For example, suppose one decides that in a titration of Zn^{2+} with EDTA he wishes $\log K_{eff}$ to be at least 8.00. From Table 2, Appendix I, we find that $\log K_{abs}$ for ZnY^{2-} is 16.26. Since

$$\log K_{eff} = \log K_{abs} + \log \alpha_4$$

we can calculate values of log K_{eff} at different pH values using the data of Table 8.2. We find that log K_{eff} = 7.78 at pH 4 and 9.81 at pH 5. Hence the titration can be carried out at a pH slightly above 4 with the desired feasibility.

The Complex Effect

Substances other than the chelon titrant which may be present in the metal ion solution may form complexes with the metal and thus compete against the desired titration reaction. Actually, such complexing is sometimes used deliberately to overcome interferences, in which case the effect of the complexer is called *masking.* For example, nickel forms a very stable complex ion with cyanide, $Ni(CN)_4^{2-}$, whereas lead does not. Thus, in the presence of cyanide, lead can be titrated with EDTA without interference from nickel, despite the fact that the stability constants for NiY^{2-} and PbY^{2-} are nearly the same (log K_{abs} values are 18.6 and 18.3, respectively).

With certain metal ions that hydrolyze readily, it may be necessary to add complexing ligands in order to prevent precipitation of the metal hydroxide. As mentioned above, the solutions are frequently buffered, and buffer anions or neutral molecules such as acetate or ammonia may form complex ions with the metal. Just as the interaction of hydrogen ions with Y^{4-} lowers K_{eff}, so it is lowered by ligands which complex the metal ion. If the stability constants for all the complexes are known, then the effect of the complexers upon the EDTA titration reaction can be calculated. For example, Zn^{2+} forms four complexes with ammonia:

$$Zn^{2+} + NH_3 \rightleftharpoons Zn(NH_3)^{2+} \qquad K_1 = 190$$

$$Zn(NH_3)^{2+} + NH_3 \rightleftharpoons Zn(NH_3)_2^{2+} \qquad K_2 = 210$$

$$Zn(NH_3)_2^{2+} + NH_3 \rightleftharpoons Zn(NH_3)_3^{2+} \qquad K_3 = 250$$

$$Zn(NH_3)_3^{2+} + NH_3 \rightleftharpoons Zn(NH_3)_4^{2+} \qquad K_4 = 110$$

These constants are for an ionic strength of zero. If we designate the total or analytical concentration of all species containing zinc as c_{Zn}, then

$$c_{Zn} = [Zn^{2+}] + [Zn(NH_3)^{2+}] + [Zn(NH_3)_2^{2+}] + [Zn(NH_3)_3^{2+}] + [Zn(NH_3)_4^{2+}]$$

$$c_{Zn} = [Zn^{2+}]\{1 + K_1[NH_3] + K_1K_2[NH_3]^2 + K_1K_2K_3[NH_3]^3 + K_1K_2K_3K_4[NH_3]^4\}$$

Let us designate the fraction of zinc in the uncomplexed form as β_4:

$$\frac{[Zn^{2+}]}{c_{Zn}} = \beta_4$$

or

$$[Zn^{2+}] = \beta_4 c_{Zn}$$

The term β_4 is simply the reciprocal of the terms in the bracket of the equation for the total concentration of all species containing zinc. It can be evaluated from the various equilibrium constants and the concentration of NH_3.

For the reaction of Zn^{2+} with EDTA in the presence of ammonia,

$$Zn^{2+} + Y^{4-} \rightleftharpoons ZnY^{2-}$$

$$K_{abs} = \frac{[ZnY^{2-}]}{[Zn^{2+}][Y^{4-}]} = \frac{[ZnY^{2-}]}{\beta_4 c_{Zn} \alpha_4 c_Y}$$

$$K_{abs}\alpha_4\beta_4 = K_{eff} = \frac{[ZnY^{2-}]}{c_{Zn}c_Y}$$

The following is an example illustrating the calculation of K_{eff} in a solution which contains ammonia.

Example. Given the four constants, K_1, K_2, K_3, and K_4, for the reaction of Zn^{2+} with NH_3 (above) and that K_{abs} for the reaction of Zn^{2+} with EDTA is 1.8×10^{16}. Calculate the value of K_{eff} for the reaction of Zn^{2+} with EDTA in a buffer of *p*H 9.0. Assume that the concentration of free NH_3 in the buffer is 0.10 *M*.

The value of β_4 is given by

$$\beta_4 = \frac{1}{1 + 190 \times 0.10 + 190 \times 210 \times (0.10)^2 + 190 \times 210 \times 250 \times (0.10)^3 + 190 \times 210 \times 250 \times 110 \times (0.10)^4}$$

$$\beta_4 = 8.3 \times 10^{-6}$$

At *p*H 9.0, α_4 is 5.1×10^{-2}. Hence

$$K_{eff} = K_{abs} \times \alpha_4 \times \beta_4$$

$$K_{eff} = 1.8 \times 10^{16} \times 0.051 \times 8.3 \times 10^{-6}$$

$$K_{eff} = 7.6 \times 10^{9}$$

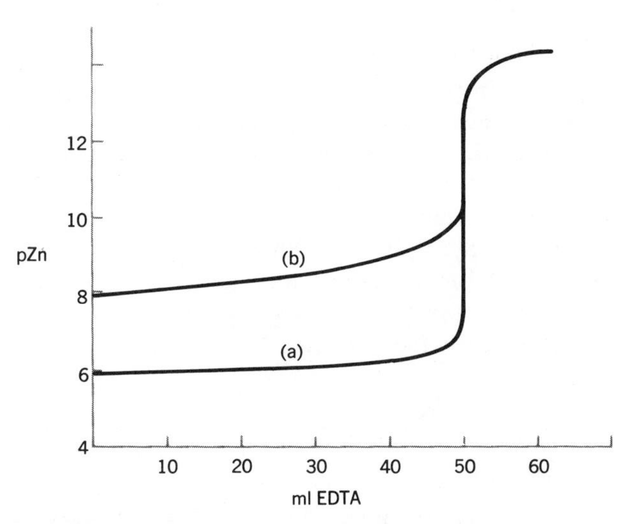

Figure 8.5 Titration of 0.0010 *M* Zn^{2+} with 0.0010 *M* EDTA at *p*H 9; (a) 0.010 *M* in NH_3, (b) 0.10 *M* in NH_3.

The effect of the concentration of ammonia on the titration curve on Zn^{2+} with EDTA at *p*H 9.0 is shown in Fig. 8.5. It can be seen that the break at the equivalence point is smaller the higher the concentration of ammonia. It may be noted that the addition of too much buffer is a common error in EDTA titrations, the resulting complexing action often worsening the end point unnecessarily.

Hydrolysis Effect

Hydrolysis of metal ions may compete with the chelometric titration process. Raising the *p*H makes this effect worse by shifting toward the right equilibria of the type

$$M^{2+} + H_2O \rightleftharpoons M(OH)^+ + H^+$$

Extensive hydrolysis may lead to the precipitation of hydroxides which react only slowly with EDTA even when equilibrium considerations favor the formation of the metal chelonate. Frequently, the appropriate hydrolysis constants for metal ions are not at hand, and hence these effects often cannot be calculated accurately. But of course there is much empirical information which serves experienced persons in deciding how high the *p*H may be for EDTA titrations of various metal ions. Solubility product constants may sometimes be used to predict where precipitation may occur, although often these constants are quite inaccurate in the case of metal hydroxides.

Sometimes precipitation is actually utilized as a sort of masking in order to circumvent a particular interference. For example, at *p*H 10, both calcium and magnesium are titrated together with EDTA, only the sum of the two being obtainable. But, if strong base is added to raise the *p*H above 12 or so, $Mg(OH)_2$ precipitates and calcium alone can be titrated.

Chelons Other than EDTA

Many other chelons have been synthesized. A few of these offer advantages over EDTA in particular situations, although none is so frequently used. The all-nitrogen chelons such as triethylenetetramine, mentioned in the introduction of this chapter, are more selective than EDTA. For example, copper can be titrated with trien in the presence of nickel, zinc, and cadmium, whereas with EDTA these metals interfere.

Ethylene glycol-bis-(β-aminoethyl ether)-*N,N'*-tetraacetic acid (EGTA) (below) forms a much more stable chelonate with calcium than with magnesium ($\log K_{abs}$ = 11.0 vs. 5.4), whereas with EDTA, as noted above, the stabilities are much more nearly the same ($\log K_{abs}$ = 10.7 vs. 8.7). Thus

$$\begin{array}{lcr} HOOCCH_2 & & CH_2COOH \\ \quad\diagdown & & \diagup\quad \\ & NCH_2CH_2{-}O{-}CH_2CH_2{-}O{-}CH_2CH_2N & \\ \quad\diagup & & \diagdown\quad \\ HOOCCH_2 & & CH_2COOH \end{array}$$

Ethylene glycol-bis-(β-aminoethyl ether)-*N,N'*-tetraacetic acid (EGTA)

calcium can be titrated selectively with EGTA in the presence of magnesium, whereas only the sum of the two can be obtained with EDTA unless the magnesium is precipitated as noted before.

Indicators for Chelometric Titrations

When EDTA was first introduced as a titrant, there was a dearth of good visual indicators, and various instrumental end-point techniques were frequently employed. The latter are still valuable in certain situations, but a wide variety of good visual indicators is now available, and usually the visual titrations are the more convenient. We have seen above, using calcium as an example, that there is a large and abrupt break in *p*M in the vicinity of the equivalence point in a feasible chelometric titration. We wish to convert this into a color change just as acid–base indicators respond to *p*H changes by changing color. A variety of chemical substances, often called *metallochromic indicators*, are now available for this purpose. Whereas all *p*H indicators need respond only to hydrogen ion, for chelometric titrations we need a series of substances responsive to *p*Mg, *p*Ca, *p*Cu, etc., although often one indicator may be useful with more than one metal ion.

Basically, the metallochromic indicators are colored organic compounds which themselves form chelates with metal ions. The chelate must have a different color from the free indicator, of course, and if large indicator blanks are to be avoided and sharp end points obtained, the indicator must release the metal ion to the EDTA titrant at a *p*M value very close to that of the equivalence point. This may be considered as analogous to the action of an indicator acid in releasing hydrogen ion to hydroxide ion in the titration of an acid. A complete treatment of the equilibria involved is somewhat more complicated than the analogous discussion of acid-base indicators, however, because of the circumstance that the common metallochromic indicators also have acid-base properties and respond as *p*H indicators as well as indicators for *p*M. Thus, in order to specify the color that a metallochromic indicator will assume in a certain solution, we generally must know both the *p*H value and the *p*M value for the particular metal ion which is present. A thorough discussion of the equilibria involved in the action of metallochromic indicators has been given by Reilley and Schmid.[2] We shall present here a somewhat simplified discussion of one indicator, Eriochrome Black T, and then simply note some of the others which are available.

The structure of Eriochrome Black T is as follows:

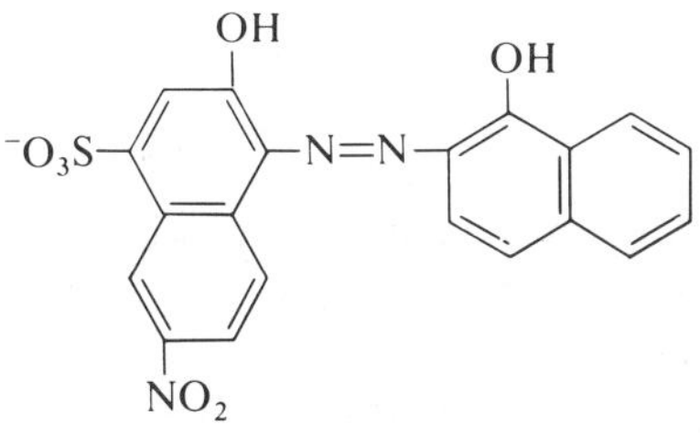

OH

OH

$^{-}O_3S$— —N=N—

NO_2

[2] C. N. Reilley and R. W. Schmid, *Anal. Chem.*, **31**, 887 (1959).

Metal chelates are formed with this molecule by loss of hydrogen ions from the phenolic —OH groups and the formation of bonds between the metal ions and the oxygen atoms as well as the azo group. The molecule is usually represented in abbreviated form as a tribasic acid, H_3In. The sulfonic acid group is shown in the figure as ionized; this is a strong acid group which is dissociated in aqueous solution regardless of *p*H, and thus the structure shown is that of the ion H_2In^-. This form of the indicator is red. The pK_a value for the dissociation of H_2In^- is 6.3. The latter species is blue. The pK_a value for the ionization of HIn^{2-} to form In^{3-} is 11.6; the latter ion is a yellowish-orange color. The indicator forms stable 1:1 complexes, which are wine-red in color, with a number of cations, such as Mg^{2+}, Ca^{2+}, Zn^{2+}, and Ni^{2+}. Many EDTA titrations are performed in buffer of *p*H 8 to 10, the range in which the predominant form of Eriochrome Black T is the blue HIn^{2-} form.

The reaction which results in the color change can be written

$$\underset{\text{Red}}{MIn^-} + HY^{3-} \rightleftharpoons MY^{2-} + \underset{\text{Blue}}{HIn^{2-}}$$

In order for the color change to occur at the proper value of *pM*, the stability of the MIn^- complex must be less than that of MY^{2-} so that the metal releases the indicator when only a slight excess of EDTA is added.

Eriochrome Black T is a suitable indicator for the titration of Zn^{2+} with EDTA in ammonia buffers of *p*H 9. The equilibrium constant for the reaction

$$Zn^{2+} + HIn^{2-} \rightleftharpoons ZnIn^- + H^+$$

is about 22. That is,

$$\frac{[ZnIn^-][H^+]}{[Zn^{2+}][HIn^{2-}]} = 22$$

or

$$[Zn^{2+}] = \frac{[ZnIn^-][H^+]}{22[HIn^{2-}]}$$

At the end point of the titration, the ratio of $[ZnIn^-]$ to $[HIn^{2-}]$ is about 1 to 10. Hence the zinc ion concentration at *p*H 9 is

$$[Zn^{2+}] = \frac{1 \times 10^{-9}}{22 \times 10}$$

$$[Zn^{2+}] = 5 \times 10^{-12}$$

$$p\text{Zn} = 11.3$$

Note in Fig. 8.5 that this value of *p*Zn occurs on the steep portion of the titration curve. Hence the color change occurs very close to the equivalence point of the titration.

The complex formed between calcium ion and Eriochrome Black T is not very stable and the color change occurs prematurely in the titration of Ca^{2+} with EDTA. The titration error can be reduced by standardizing the EDTA solution against a standard calcium ion solution using conditions identical to those used for the unknown.

Eriochrome Black T is, unfortunately, unstable in solution and solutions must be freshly prepared in order to obtain the proper color change. Another indicator, called Calmagite, has been developed. Its structure is as follows:

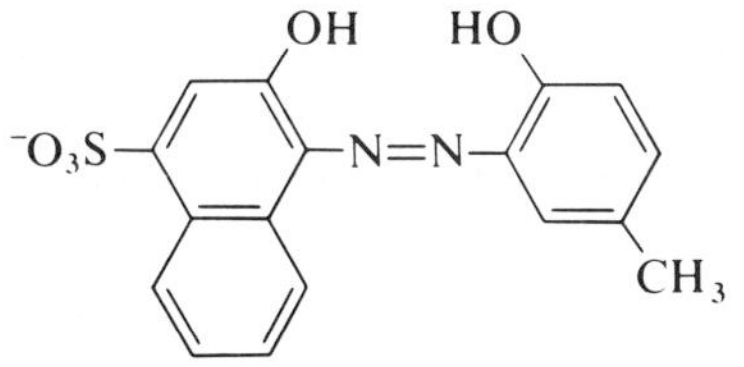

Calmagite is stable in aqueous solution and may be substituted for Eriochrome Black T in procedures which call for the latter indicator.

A number of other indicators are known which can be used for various cations. These are discussed in the article of Reilley and Schmid to which we referred in footnote 2.

Applications of Chelometric Titrations

Chelometric titrations have been carried out successfully on nearly all common cations. These titrations have virtually replaced the former tedious gravimetric analyses for many metals in a variety of samples. There are several procedures which are employed.

Direct titrations with EDTA can be carried out on at least 25 cations using metallochromic indicators.[3] Complexing agents, such as citrate and tartrate, are often added to prevent precipitation of metal hydroxides. An NH_3—NH_4Cl buffer of *p*H 9 to 10 is often used for metals which form complexes with ammonia.

The total hardness of water, calcium plus magnesium, can be determined by direct titration with EDTA using Eriochrome Black T or Calmagite indicator. As mentioned earlier, the complex between Ca^{2+} and the indicator is too weak for a proper color change to occur. However, magnesium forms a stronger complex with the indicator than does calcium and a proper end point is obtained is an ammonia buffer of *p*H 10. If the sample titrated does not contain magnesium, some magnesium salt can be added to the EDTA before this solution is standardized. The titrant then (*p*H 10) is a mixture of MgY^{2-} and Y^{4-}. As this is added to the solution containing Ca^{2+}, the more stable CaY^{2-} is formed, liberating Mg^{2+} to react with the indicator and form the red $MgIn^{-}$. After the calcium has been used up, additional titrant converts $MgIn^{-}$ to MgY^{2-} and the indicator reverts to the blue HIn^{2-} form.

Back-titrations are used when the reaction between the cation and EDTA is slow or when a suitable indicator is not available. Excess EDTA is added and

[3] F. J. Welcher, *The Analytical Use of Ethylenediaminetetraacetic Acid*, D. Van Nostrand Co., Inc., New York, 1958, Chapter 3. See also G. Schwarzenbach and H. Flaschka, *Complexometric Titrations*, 5th ed., trans. by H. M. N. H. Irving, Methuen & Co. Ltd., London, 1969.

the excess titrated with a standard solution of magnesium using Calmagite as the indicator. The magnesium–EDTA complex is of relatively low stability and the cation being determined is not displaced by magnesium. This method can also be used to determine metals in precipitates, such as lead in lead sulfate and calcium in calcium oxalate.

Replacement titrations are useful when no suitable indicator is available for the metal ion being determined. An excess of a solution containing magnesium–EDTA complex is added and the metal ion, say M^{2+}, displaces the magnesium from the relatively weak EDTA complex:

$$M^{2+} + MgY^{2-} \rightleftharpoons MY^{2-} + Mg^{2+}$$

The displaced Mg^{2+} is then titrated with a standard EDTA solution using Calmagite as the indicator.

Indirect determinations of several types have been reported. Sulfate has been determined by adding excess barium ion to precipitate $BaSO_4$. The excess Ba^{2+} is then titrated with EDTA. Phosphate has been determined by titration of the Mg^{2+} in equilibrium with the moderately soluble $MgNH_4PO_4$.

Since metal ions differ in the stabilities of their EDTA complexes, it is occasionally possible to obtain consecutive end points for more than one metal in a single titration. This situation is somewhat analogous to the titration with base of acids with different dissociation constants. It is necessary, of course, that the stability constants of the metal complexes differ sufficiently, and also the indicator problem is exceptionally critical in such cases. Iron(III) and copper(II) have been determined in a single EDTA titration, using photometric detection of the end points,[4] as have lead(II) and bismuth(III).[5]

Sometimes by adjusting the *p*H of a solution it is possible to obtain some degree of selectivity in titrations with EDTA. It is possible, for example, to titrate in solutions of low *p*H ions which form very stable complexes. At such low *p*H values ions which form less stable complexes will not interfere. The log K_{eff} of FeY^- at *p*H 2 is 11.7, quite large enough for a feasible titration. Iron(III) can be titrated at this *p*H in the presence of such divalent cations as calcium, barium, and magnesium. The chelates of the latter cations are not formed to any appreciable extent since the effective stability constants are so low. Nickel(II) can also be titrated in the presence of the alkaline earth cations at *p*H 3.5, whereas interference will occur at *p*H 10 in an ammonia buffer. Nickel forms stable complexes with ammonia while the alkaline earths do not. For this reason the effective stability constant of NiY^{2-} is about the same size as the constants of the alkaline earth complexes in the ammonia buffer and interference occurs.

Potentiometric EDTA titrations with a mercury indicator electrode are explained in Chapter 12, and photometric titrations of metal ions with EDTA are discussed briefly in Chapter 14.

[4] A. L. Underwood, *Anal. Chem.*, **25,** 1910 (1953).
[5] R. N. Wilhite and A. L. Underwood, *Anal. Chem.*, **27**, 1334 (1955).

METAL ION BUFFERS

We saw in Chapter 6 that certain systems containing Brønsted conjugate acid-base pairs resisted pH changes upon the addition of strong acids or bases. Such systems are said to be buffered. An analogous buffering action with respect to changes in pM is established in solutions containing a metal complex and excess complexing agent.

Consider the equilibrium involving a metal ion M, a ligand L, and a complex ML, where the charges are omitted for convenience:

$$M + L \rightleftharpoons ML$$

$$K_{eff} = \frac{[ML]}{[M]c_L}$$

Here c_L represents the analytical concentration of the ligand. Solving for [M] and then taking logs, we obtain

$$[M] = \frac{1}{K_{eff}} \times \frac{[ML]}{c_L}$$

$$\log [M] = \log \frac{1}{K_{eff}} + \log \frac{[ML]}{c_L}$$

$$pM = \log K_{eff} - \log \frac{[ML]}{c_L}$$

Compare this with the Henderson-Hasselbalch equation (page 114).

It is seen that the pM of such a solution is fixed by the value of K_{eff} and the molar ratio of metal complex to ligand. Introduction of additional metal ion to the solution will lead to formation of more ML; in other words, the solution resists the lowering of pM that would otherwise occur if L were absent. Similarly, removal of metal ion will be resisted by the dissociation of ML, and metal ion can be drained by some other reaction without a large rise in pM so long as the capacity of ML to furnish metal ion is not exhausted.

Metal ion buffers have found application in biology and biochemistry in studies of enzyme systems whose catalytic activity exhibits a metal ion dependence. Just as pK_a must be considered in the analogous case of pH buffers, a metal ion buffer will be most efficient if $\log K_{eff}$ is nearly the same as the desired pM value.

TITRATIONS INVOLVING UNIDENTATE LIGANDS

Because of the stepwise formation of successive complexes as noted above, unidentate ligands are only rarely suitable for the titration of metal ions. However, there are a few examples of important titrations based upon such ligands, and we shall consider briefly the two best-known cases.

Titration of Chloride with Mercury(II)

The mercury(II) ion–chloride system is unusual in that the last two of the successive complexes in the formation of $HgCl_4^{2-}$ are of much lesser stability than the first two, as shown by the following successive formation constants:[6]

$$Hg^{2+} + Cl^- \rightleftharpoons HgCl^+ \qquad K_1 = \frac{[HgCl^+]}{[Hg^{2+}][Cl^-]} = 5.5 \times 10^6$$

$$HgCl^+ + Cl^- \rightleftharpoons HgCl_2 \qquad K_2 = \frac{[HgCl_2]}{[HgCl^+][Cl^-]} = 3.0 \times 10^6$$

$$HgCl_2 + Cl^- \rightleftharpoons HgCl_3^- \qquad K_3 = \frac{[HgCl_3^-]}{[HgCl_2][Cl^-]} = 7.1$$

$$HgCl_3^- + Cl^- \rightleftharpoons HgCl_4^{2-} \qquad K_4 = \frac{[HgCl_4^{2-}]}{[HgCl_3^-][Cl^-]} = 10$$

Thus in the titration of a chloride solution with an ionized mercury salt such as mercury(II) nitrate or perchlorate, there is a sudden drop in pHg ($p\mathrm{Hg} = -\log[Hg^{2+}]$) when the formation of $HgCl_2$ is essentially complete.

One of the common indicators for this titration is sodium nitroprusside, $Na_2Fe(CN)_5NO$. This compound forms a white precipitate of mercury(II) nitroprusside, and the end point is taken as the appearance of a white turbidity in the formerly homogeneous solution. The pHg at the equivalence point of the titration is not so low as might otherwise be expected because of the consumption of mercury(II) ion in the following reaction:

$$Hg^{2+} + HgCl_2 \rightleftharpoons 2HgCl^+ \qquad K = \frac{K_1}{K_2} \cong 1.8$$

Actually, the mercury(II) nitroprusside precipitate is first seen somewhat after the equivalence point, and a correction must be applied in order to obtain the best results. The correction is not really the same as an indicator blank run with distilled water, because the reaction shown above does not then take place appreciably. The correction depends upon the final concentration of mercury(II) chloride, and hence varies with the quantity of sample and the final volume. The acidity of the solution also affects the correction, and there is further variation which depends upon the rate at which the titration is performed and the manner in which the individual analyst views the turbid solution. Typical correction values are given by Kolthoff and Stenger;[7] for example, where the final solution is 100 ml of 0.025 M $HgCl_2$, the correction is roughly 0.2 ml. An advantage of this particular method lies in the fact that the titration may be performed in solutions which are quite acidic, and it works well even in fairly dilute solution, for example, at levels of chloride (e.g., 10 mg/liter) which frequently occur in natural waters.

[6] A. Johnson, I. Quarfort, and L. G. Sillen, *Acta Chem. Scand.*, **1**, 461, 473 (1947).

[7] I. M. Kolthoff and V. A. Stenger, *Volumetric Analysis*, 2nd ed., Vol. II, Interscience Publishers, Inc., New York, 1947, p. 332.

Certain organic compounds which form colored complexes with mercury(II) ion have also been employed as indicators for the mercurimetric titration of chloride. The best known are diphenylcarbazide (colorless) and diphenylcarbazone (orange), which form intense violet mercury(II) complexes. With these indicators, it has been found important to control the *p*H of the solution being titrated. According to Roberts,[8] diphenylcarbazide performs best at *p*H 1.5 to 2.0, while Clark[9] found that diphenylcarbazone is best employed at *p*H 3.2 to 3.3.

It should be pointed out that bromide, thiocyanate, and cyanide may be determined by mercurimetric titration, although there is no advantage over the usual titrations of these ions with silver nitrate. Nitroprusside cannot be used as the indicator in the thiocyanate titration because the appearance of the mercury(II) nitroprusside precipitate is obscured by the slightly soluble thiocyanate. In this case, the usual indicator is iron(III) ion, which acts by the formation of red complexes with thiocyanate such as $FeSCN^{2+}$. The titration of iodide with mercury is largely unsatisfactory. The complex HgI_4^{2-} forms during the titration; later, a red precipitate of HgI_2 appears through the reaction

$$Hg^{2+} + HgI_4^{2-} \rightleftharpoons 2HgI_2$$

The appearance of this precipitate has been used as an end point, but actually it occurs much too early.

Titration of Cyanide with Silver Ion

Another titration of some practical importance involving a unidentate ligand and a metal ion is the so-called Liebig titration of cyanide with silver nitrate. The basis of the method is the formation of the very stable complex ion, $Ag(CN)_2^-$:

$$2CN^- + Ag^+ \rightleftharpoons Ag(CN)_2^-$$

The equilibrium constant for this reaction as written is about 10^{21}, and this is the only silver–cyanide complex of appreciable stability. Originally, the end point was based upon the appearance of turbidity due to the precipitation of silver cyanide, which may be written as

$$Ag^+ + Ag(CN)_2^- \rightleftharpoons 2AgCN$$

or

$$Ag^2 + Ag(CN)_2^- \rightleftharpoons Ag[Ag(CN)_2]$$

This precipitation occurs after $[CN^-]$ has dropped to a low value, although a calculation based upon the appropriate equilibria shows that it actually comes a little too early, corresponding to an end-point error of the order of 0.2 part per thousand. This error is small enough to be accepted, but there is an additional problem: Silver cyanide precipitated locally is slow to redissolve as the solution is stirred, and it is time consuming to perform the titration carefully. Also, there is some difficulty in seeing the silver cyanide precipitate.

[8] I. Roberts, *Ind. Eng. Chem., Anal. Ed.*, **8**, 365 (1936).
[9] F. E. Clark, *Anal. Chem.*, **22**, 553 (1950).

In the Deniges modification of Liebig's method, iodide ion is added as the indicator. Precipitated silver iodide is bulky and easy to see, and it is less soluble than silver cyanide and hence precipitates in place of the latter at the end point. This end point occurs, however, too early in the titration. For this reason, ammonia is added which, by forming the soluble species $Ag(NH_3)_2^+$, retards the precipitation of silver iodide until a more propitious time; ammonia does not prevent the formation of the much more stable $Ag(CN)_2^-$, and hence does not interfere with the titration reaction.

REFERENCES

H. A. LAITINEN and W. E. HARRIS, *Chemical Analysis*, 2nd ed., McGraw-Hill Book Company, New York, 1975.

A. RINGBOM, *Complexation in Analytical Chemistry*, Wiley–Interscience, New York, 1963.

G. SCHWARZENBACH and H. FLASCHKA, *Complexometric Titrations*, 5th ed., trans. by H. M. N. H. Irving, Methuen & Co. Ltd., London, 1969.

QUESTIONS

1. *Terms.* Explain the meaning of the following terms: (a) complex; (b) chelon; (c) quadridentate ligand; (d) chelating agent; (e) labile complex; (f) masking; (g) replacement titration; (h) metallochromic indicator.

2. *Stability constant.* Explain why the concept of an effective stability constant is useful in chelometric titrations.

3. *Indicators.* Explain why the pH of the solution is an important factor in the selection of an indicator for a chelometric titration.

4. *Titrations with mercury(II) ion.* (a) Why is the titration of I^- with Hg^{2+} not satisfactory? (b) Why cannot sodium nitroprusside be used as the indicator for the titration of SCN^- with Hg^{2+}?

5. *Stability constant.* Given $M + L \rightleftharpoons ML$; K_{abs} is the stability constant. Show that the following relation holds:

$$pM + pL - pML = \log K_{abs}$$

Multiple-choice: In the following multiple-choice questions, select the *one best* answer.

6. Which of the following statements concerning α_4, the fraction of EDTA in the Y^{4-} form, is correct? (a) α_4 decreases as the pH decreases; (b) $-\log \alpha_4$ increases as the $[H_3O^+]$ increases; (c) $-\log \alpha_4$ approaches zero at high pH; (d) all of the above; (e) none of the above.

7. 50 ml of 0.10 M Cu^{2+} is titrated with 0.10 M EDTA at pH 9.0. When 100 ml of titrant is added (a) $p\text{Cu} = \log K_{eff}$; (b) $p\text{Cu} = -pK_{eff}$; (c) $\log [Cu^{2+}] = pK_{eff}$; (d) all of the above; (e) none of the above.

8. Two solutions of 0.001 *M* Zn^{2+} at *p*H 9 are titrated with 0.001 *M* EDTA. The first solution is 0.010 *M* in NH_3; the second is 0.10 *M* in NH_3. The values of *p*Zn are about equal in the two titrations (a) at the start of the titration; (b) halfway to the equivalence point; (c) at the equivalence point; (d) 10 ml beyond the equivalence point.

9. In the titration of chloride ion with Hg^{2+} the principal species in the solution at the equivalence point is (a) $HgCl_4^{2-}$; (b) $HgCl_2$; (c) $HgCl^+$; (d) Hg_2Cl_2.

10. Consider the titration of Zn^{2+} with EDTA. Which statement below is true? (a) The reaction is more complete the lower the hydrogen ion concentration; (b) The titration is less feasible at low *p*H than at high *p*H; (c) The titration is more feasible in 0.010 *M* NH_3 than in 0.10 *M* NH_3; (d) all of the above; (e) none of the above.

PROBLEMS

1. *Standardization.* A sample of pure $CaCO_3$ weighing 0.2428 g is dissolved in hydrochloric acid and the solution diluted to 250.0 ml in a volumetric flask. A 50.00-ml aliquot requires 42.74 ml of an EDTA solution for titration. Calculate (a) the molarity of the EDTA solution; (b) the number of grams of $Na_2H_2Y \cdot 2H_2O$ (MW 372.2) required to prepare 1 liter of the solution.

2. *Hardness of water.* A 200-ml sample of water containing Ca^{2+} ions is titrated with 16.38 ml of the EDTA solution in Problem 1. Calculate the degree of hardness of water in ppm of $CaCO_3$. Recall that 1 ppm is 1 mg/liter.

3. *Hardness of water.* A 100-ml sample of water containing Ca^{2+} and Mg^{2+} ions is titrated with 15.28 ml of 0.01016 *M* EDTA in an ammonia buffer at *p*H 10.0 Another sample of 100 ml is treated with NaOH to precipitate $Mg(OH)_2$, and then titrated at *p*H 13 with 10.43 ml of the same EDTA solution. Calculate the parts per million of $CaCO_3$ and of $MgCO_3$ in the sample.

4. *Masking.* 50.00 ml of a solution containing both Ni^{2+} and Pb^{2+} ions requires 46.32 ml of a 0.02041 *M* EDTA solution for titration of both metals. A second 50.00-ml sample is treated with KCN to mask the nickel and then titrated with 30.28 ml of the same EDTA solution. Calculate the molarities of the Ni^{2+} and Pb^{2+} ions.

5. *Analysis of an alloy.* A 0.5745-g sample of an alloy containing principally bismuth and lead is dissolved in nitric acid and diluted to 250.0 ml in a volumetric flask. A 50.00-ml aliquot is withdrawn, the *p*H adjusted to 1.5, and the bismuth titrated with 30.26 ml of 0.01024 *M* EDTA. The *p*H of the solution is then increased to 5.0 and the lead titrated with 20.42 ml of the same EDTA solution. Calculate the percentages of lead and bismuth in the alloy.

6. *Liebig method.* A 0.7562-g sample containing NaCN is dissolved in water and then concentrated ammonia and some KI solution are added. The solution requires 23.58 ml of 0.0988 *M* $AgNO_3$ for titration. Calculate the percentage of NaCN in the sample.

7. *Liebig method.* A 0.4574-g sample containing only NaCN and KCN requires 41.65 ml of 0.1056 *M* $AgNO_3$ for titration. Calculate the percentage of NaCN in the sample.

8. *Value of α.* Verify the values of α_4 for EDTA given in Table 8.2 at *p*H 2, 5, 9, and 12.

9. *Equilibrium constants.* Calculate the equilibrium constants of the following reactions:
(a) $Hg^{2+} + 2Cl^- \rightleftharpoons HgCl_2$
(b) $HgCl^+ + 2Cl^- \rightleftharpoons HgCl_3^-$

10. *Effective stability constant.* Calculate the value of K_{eff} for the reaction of Zn^{2+} with EDTA in a buffer of pH 9.0. Assume the concentration of free NH_3 in the buffer is 0.010 *M*. Use the constants given on page 198 for the zinc–ammonia complexes.

11. *Mixtures of solutions.* Calculate the value of pCd in the following solutions, all at pH 9.0: (a) 50 ml of 0.010 *M* Cd^{2+} + 10 ml of 0.020 *M* EDTA; (b) 50 ml of 0.010 *M* Cd^{2+} + 25 ml of 0.020 *M* EDTA; (c) 50 ml of 0.010 *M* Cd^{2+} + 50 ml of 0.020 *M* EDTA.

12. *Derivations.* For the reaction $M + L \rightleftharpoons ML$ the stability constant is K_{abs}. Show that the following expressions hold at the equivalence point in the titration of M with L:
(a) $pM = \frac{1}{2}(pML - pK_{abs})$
(b) $\log [ML] = \log K_{abs} + 2 \log [M]$

13. *Equivalence point.* Calculate the value of pM at the equivalence point for the titration of 1.00 mmol of each of the following metals with EDTA at pH 5.0. The volume in each case is 100 ml. (a) Pb^{2+}; (b) Fe^{2+}; (c) Mg^{2+}.

14. *Titration curve.* 50.0 ml of a solution which is 0.0100 *M* in a metal ion M^{2+} and buffered at pH 10.0 is titrated with 0.0100 *M* EDTA. The value of K_{abs} for MY^{2-} is 4.0×10^{14}. Calculate the values of pM when the following volumes of titrant are added: (a) 0.00; (b) 25.0; (c) 49.9; (d) 50.0; (e) 50.1; (f) 55.0 ml. Plot the titration curve.

15. *Titration curve.* Repeat Problem 14 at pH 8 and 12. Compare the values of $\Delta pM/\Delta V$ at the equivalence point for the three different pH values.

16. *Feasibility of titration.* (a) 50.0 ml of a 0.0500 *M* solution of a cation N^{2+} is titrated with 0.0500 *M* EDTA. Calculate the value of K_{eff} for the formation of NY^{2-} so that when 49.95 ml of titrant is added the reaction is complete and that pN changes by 1.00 unit on the addition of 0.10 ml of additional titrant. (b) Repeat the calculation for $\Delta pN = 2.0$ units for 0.10 ml of titrant.

17. *Feasibility of titration.* Repeat Problem 16 for the titration of 0.0100 *M* N^{2+} with 0.0100 *M* EDTA.

18. *pH—feasibility.* Suppose a chemist decides that he wishes the K_{eff} for the reaction

$$M^{2+} + Y^{4-} \rightleftharpoons MY^{2-}$$

to be 5.0×10^{10} for the titration of several cations. Calculate the pH values at which titrations should be carried out for the following cations; (a) Fe^{3+}; (b) Cu^{2+}; (c) Fe^{2+}; (d) Ca^{2+}; (e) Ba^{2+}.

19. *pH—fraction complexed.* For the reaction of Ca^{2+} with EDTA

$$Ca^{2+} + Y^{4-} \rightleftharpoons CaY^{2-}$$

calculate the fraction of calcium in the CaY^{2-} form at the following pH values: (a) 4.0; (b) 5.0; (c) 6.0; (d) 8.0. Assume that 5.00 mmol of EDTA has been added to 5.00 mmol of Ca^{2+} and the volume of solution is 100 ml.

20. *Metal ion buffer.* A metal ion buffer is prepared which contains 4.00 mmol of Ca^{2+} and 6.00 mmol of EDTA per liter of solution at pH 10.0 (a) Calculate the pCa of the buffer. (b) To 100 ml of the buffer is added 0.0200 mmol of Ca^{2+}. Calculate the pCa of the resulting solution and the value of ΔpCa.

21. *Titrations.* Given the following data:

METAL		CHELON, L	
Complex	$\log K_{abs}$	pH	$-\log \alpha$
ML	24.60	2.0	14.40
NL	19.30	4.0	9.30
QL	14.70	6.0	5.10
RL	8.30	8.0	2.50
SL	2.70	10.0	0.70

(a) For the titration of 0.0100 M solutions of the metal ions with 0.0100 M L, which metals should give feasible titrations at pH (i) 2.0; (ii) 4.0; (iii) 8.0? (b) Calculate the values of K_{eff} for (i) ML at pH 2.0; (ii) RL at pH 4.0. (c) 50 ml of 0.020 M R is mixed with 50 ml of 0.010 M L. Calculate the value of pR if the pH of the solution is (i) 8.0; (ii) 10.0. (d) 50 ml of 0.0100 M Q is mixed with 50 ml of 0.0100 M L. Calculate the value of pQ if the pH is (i) 6.0; (ii) 8.0. (e) 50 ml of 0.020 M M is mixed with 50 ml of 0.020 M L at pH 4.0. (i) Calculate the value of pM. (ii) The value of pM in (i) is measured experimentally and found to be 7.15, rather than the calculated value. If this difference is caused by M being complexed by an anion in the buffer, calculate the value of $-\log \beta$.

CHAPTER NINE

solubility equilibria

Precipitation reactions have been widely used in analytical chemistry in titrations, in gravimetric determinations, and in separating a sample into its component parts. Gravimetric methods do not occupy as prominent a place in analytical chemistry as they once did, and other methods of separation have been developed in recent years. Nevertheless, precipitation is still an important technique in many analytical procedures.

In this chapter we shall first discuss precipitation titrations and then the use of precipitation as a separation technique.

PRECIPITATION TITRATIONS

Titrations involving precipitation reactions are not nearly so numerous in titrimetric analysis as those involving redox or acid-base reactions. In fact, in a beginning course examples of such titrations are usually limited to those involving precipitation of silver ion with anions such as the halogens and thiocyanate. One of the reasons for the limited use of such reactions is the lack of suitable indicators. In some cases, particularly in the titration of dilute solutions, the rate of reaction is too slow for convenience of titration. As the equivalence point is approached and the titrant is added slowly, a high degree of supersaturation does not exist and the precipitation may be very slow. Another difficulty is that the composition of the precipitate is frequently not known because of coprecipitation effects. Although the latter can be minimized or partially corrected for by processes such as aging the precipitate, this is not possible in a direct titration involving the formation of a precipitate.

We shall limit our discussion here to precipitation titrations involving silver salts with particular emphasis on the indicators which have been successfully employed in such titrations.

Titration Curves

Titration curves for precipitation reactions can be constructed and are entirely analogous to those for acid-base and complex formation titrations. Equilibrium calculations are based on the solubility product constant, and this topic was reviewed in Chapter 5. The following example illustrates the calculations involved in the titration of chloride ion with silver ion.

Example. 50.0 ml of 0.100 *M* NaCl solution is titrated with 0.100 *M* $AgNO_3$. Calculate the chloride ion concentration at intervals during the titration and plot pCl vs. ml of $AgNO_3$. $p\text{Cl} = -\log[\text{Cl}^-]$, and K_{sp} of AgCl $= 1 \times 10^{-10}$.

(a) *Start of titration.* Since

$$[\text{Cl}^-] = 0.100 \text{ mmol/ml}$$

$$p\text{Cl} = 1.00$$

(b) *After addition of 10.0 ml $AgNO_3$.* Since the reaction goes well to completion,

$$[\text{Cl}^-] = \frac{[(50.0 \times 0.100) - (10.0 \times 0.100)] \text{ mmol}}{(50.0 + 10.0) \text{ ml}}$$

$$[\text{Cl}^-] = 0.067 \text{ mmol/ml}$$

$$p\text{Cl} = 1.17$$

(c) *After addition of 49.9 ml Ag NO_3*

$$[\text{Cl}^-] = \frac{[(50.0 \times 0.100) - (49.9 \times 0.100)] \text{ mmol}}{(50.0 + 49.9) \text{ ml}}$$

$$[\text{Cl}^-] = 1.00 \times 10^{-4} \text{ mmol/ml}$$

$$p\text{Cl} = 4.00$$

In these calculations we have disregarded the contribution of chloride ions to the solution by the solubility of the precipitate. This approximation is valid except within 1 or 2 drops of the equivalence point.

(d) *Equivalence point.* This point is reached when 50.0 ml of $AgNO_3$ has been added. There is neither excess chloride nor silver ion, and the concentration of each is given by the square root of K_{sp}.

$$[\text{Ag}^+] = [\text{Cl}^-]$$

$$[\text{Cl}^-]^2 = 1.0 \times 10^{-10}$$

$$[\text{Cl}^-] = 1.0 \times 10^{-5}$$

$$p\text{Cl} = 5.00$$

(e) *After addition of 60.0 ml of $AgNO_3$.* The concentration of excess silver ion is

$$[Ag^+] = \frac{[(60.0 \times 0.100) - (50.0 \times 0.100)]\ \text{mmol}}{(50.0 + 60.0)\ \text{ml}}$$

$$[Ag^+] = 9.1 \times 10^{-3}$$

$$pAg = 2.04$$

Since

$$pCl + pAg = 10.00$$

$$pCl = 7.96$$

TABLE 9.1 Titration of 50 ml of 0.10 *M* NaCl with 0.10 *M* $AgNO_3$

$AgNO_3$, ml	$[Cl^-]$	% Cl^- pptd.	*p*Cl
0.0	0.10	0.0	1.00
10.0	0.067	20.0	1.17
20.0	0.043	40.0	1.37
30.0	0.025	60.0	1.60
40.0	0.011	80.0	1.96
49.0	0.0010	98.0	3.00
49.9	1.0×10^{-4}	99.8	4.00
50.0	1.0×10^{-5}	100	5.00
50.1	1.0×10^{-6}	100	6.00
51.0	1.0×10^{-7}	100	7.00
60.0	1.1×10^{-8}	100	7.96

The data for this titration are given in Table 9.1 and the titration curve is plotted in Fig. 9.1. The curves for the titration of iodide and bromide ions with silver are also plotted in this figure. Note that the increase in *p*X (X = Cl, Br, or I) at the equivalence point is greatest for the titration of iodide, since silver iodide is the least soluble of the three salts. Note also that the value of *K* for the titration reaction

$$Ag^+ + X^- \rightleftharpoons AgX(s)$$

is

$$K = \frac{1}{[Ag^+][X^-]} = \frac{1}{K_{sp}}$$

Hence the smaller K_{sp} is, the larger is the *K* for the titration reaction. For the three salts shown in Fig. 9.1 the values of *K* are: AgCl, 1×10^{10}; AgBr, 2×10^{12}; and AgI, 1×10^{16}.

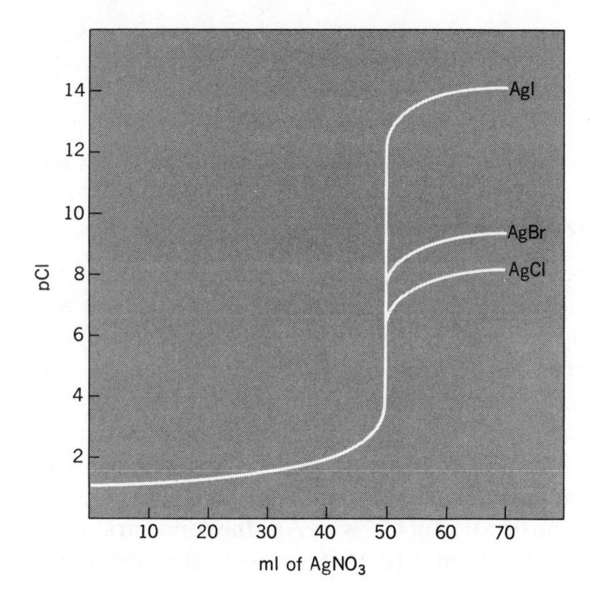

Figure 9.1 Titration curves of NaCl, NaBr, and NaI. 50 ml of 0.1 *M* salt titrated with 0.1 *M* $AgNO_3$.

Feasibility of Precipitation Titrations

The magnitude of K required for a feasible precipitation titration can be calculated as was done previously for acid-base (page 142) and complex formation titrations (page 197). The following example illustrates this calculation for a salt of the type MX.

Example. 50 ml of 0.10 *M* NaX is titrated with 50 ml of 0.10 *M* $AgNO_3$. Calculate the value of K and that of K_{sp} for AgX so that when 49.95 ml of titrant has been added the reaction is complete, and the pX changes by 2.00 units on the addition of 2 more drops (0.10 ml) of titrant. NaX is a completely dissociated salt, and the titration reaction is

$$Ag^+ + X^- \rightleftharpoons AgX(s) \qquad K = 1/K_{sp}$$

One drop before the equivalence point, 4.995 mmol of Ag^+ has been added. We started with $50 \times 0.10 = 5.0$ mmol of X^-. Hence 0.0050 mmol remains and

$$[X^-] = \frac{0.0050 \text{ mmol}}{99.95 \text{ ml}} \cong 5 \times 10^{-5}\, M$$

$$pX = 4.30$$

If $\Delta p\text{X} = 2.00$, then $p\text{X} = 6.30$ and $[X^-] = 5 \times 10^{-7}\, M$ when the volume of titrant is 50.05 ml. Since

$$[Ag^+] = \frac{0.05 \times 0.10}{100.05} \cong 5 \times 10^{-5}$$

$$K = \frac{1}{(5 \times 10^{-5})(5 \times 10^{-7})}$$

$$K = 4 \times 10^{10} \qquad \text{and} \qquad K_{sp} = 2.5 \times 10^{-11}$$

The student should confirm that for $\Delta pX = 1.00$, K should be 4×10^9.

Effect of Concentration

The magnitude of ΔpX at the equivalence point in the titration of X^- with Ag^+ depends upon the concentrations of the analyte and the titrant. The effect is exactly the same as that for the titration of a strong acid with a strong base (Fig. 6.3, page 144). The lower the concentration of X^-, the higher are the values of pX before the equivalence point and the smaller is the value of ΔpX at the equivalence point. As the concentration of titrant is lowered, the branch of the curve after the equivalence point is lowered, again resulting in a lower value of ΔpX at the equivalence point.

In the titration of Cl^- with Ag^+ the concentrations of both reactants should not be much less than 0.10 M if a reasonably good end point is to be obtained.

INDICATORS FOR PRECIPITATION TITRATIONS INVOLVING SILVER

It has been pointed out that one of the problems associated with precipitation titrations is that of finding a suitable indicator. In titrations involving silver salts there are three indicators which have been successfully employed for many years. The Mohr method uses chromate ion, CrO_4^{2-}, to precipitate brown Ag_2CrO_4. The Volhard method uses Fe^{3+} ion to form a colored complex with thiocyanate ion, SCN^-. And Fajans' method utilizes "adsorption indicators". We shall discuss these three methods briefly.

Formation of a Colored Precipitate: The Mohr Method

Just as an acid-base system can be used as an indicator for an acid-base titration, the formation of another precipitate can be used to indicate the completion of a precipitation titration. The best-known example of such a case is the so-called *Mohr* titration of chloride with silver ion, in which chromate ion is used as the indicator. The first permanent appearance of the reddish silver chromate precipitate is taken as the end point of the titration.

It is necessary, of course, that the precipitation of the indicator occur at or near the equivalence point of the titration. Silver chromate is more soluble (about 8.4×10^{-5} mol/liter) than silver chloride (about 1×10^{-5} mol/liter). If silver ions are added to a solution containing a large concentration of chloride ions and a small concentration of chromate ions, silver chloride will first precipitate; silver chromate will not form until the silver ion concentration increases to a large enough value to exceed the K_{sp} of silver chromate. One can readily calculate the concentration of chromate that will lead to precipitation of silver chromate at the equivalence point where $pAg = pCl = 5.00$. Since the K_{sp} of Ag_2CrO_4 is 2×10^{-12}, and $[Ag^+] = 1 \times 10^{-5}$ at the equivalence point, then

$$[Ag^+]^2[CrO_4^{2-}] = 2 \times 10^{-12}$$

$$[CrO_4^{2-}] = \frac{2 \times 10^{-12}}{(1 \times 10^{-5})^2} = 0.02\ M$$

Such a high concentration cannot be used in practice, however, since the yellow color of chromate ion makes it difficult to observe the formation of the colored precipitate. Normally, a concentration of 0.005 to 0.01 M chromate is employed. The error caused by using such a concentration is quite small. It can be corrected by running an indicator blank or by standardizing the silver nitrate against a pure chloride salt under conditions identical to those used in the analysis.

The Mohr titration is limited to solutions with pH values from about 6 to 10. In more alkaline solutions silver oxide precipitates. In acid solutions the chromate concentration is greatly decreased, since $HCrO_4^-$ is only slightly ionized. Furthermore, hydrogen chromate is in equilibrium with dichromate:

$$2H^+ + 2CrO_4^{2-} \rightleftharpoons 2HCrO_4^- \rightleftharpoons Cr_2O_7^{2-} + H_2O$$

A decrease in chromate ion concentration makes it necessary to add a large excess of silver ions to bring about precipitation of silver chromate, and thus leads to large errors. Dichromates are, in general, fairly soluble.

The Mohr method can also be applied to the titration of bromide ion with silver, and also cyanide ion in slightly alkaline solutions. Adsorption effects make the titration of iodide and thiocyanate ions not feasible. Silver cannot be titrated directly with chloride, using chromate indicator. The silver chromate precipitate, present initially, redissolves only slowly near the equivalence point. However, one can add excess standard chloride solution, and then back-titrate, using the chromate indicator.

Formation of a Colored Complex: The Volhard Method

The Volhard method is based on the precipitation of silver thiocyanate in nitric acid solution, with iron(III) ion employed to detect excess thiocyanate ion:

$$Ag^+ + SCN^- \rightleftharpoons AgSCN\ (s)$$

$$Fe^{3+} + SCN^- \rightleftharpoons FeSCN^{2+} \quad \text{(red)}$$

The method can be used for the direct titration of silver with standard thiocyanate solution, or for the indirect titration of chloride ion. In the latter case, an excess of standard silver nitrate is added, and the excess is titrated with standard thiocyanate. Other anions, such as bromide and iodide, can be determined by the same procedure. Anions of weak acids, such as oxalate, carbonate, and arsenate, the silver salts of which are soluble in acid, can be determined by precipitation at higher pH and filtration of silver salt. The precipitate is then dissolved in nitric acid and the silver titrated directly with thiocyanate.

The Volhard method is widely used for silver and chloride because of the fact that the titration can be done in acid solution. In fact, it is desirable to employ an acid medium to prevent hydrolysis of the iron(III)–ion indicator. Other common methods for silver and chloride require a nearly neutral solution for successful titration. Many cations precipitate under such conditions, and hence interfere in these methods. Mercury is the only common cation that interferes with the Volhard method. In fact, mercury can be determined by titration with thiocyanate, since mercury(II) thiocyanate is a very slightly dissociated compound. High concentrations of colored cations, such as cobalt(II), nickel(II), and copper(II), cause difficulty in observation of the end point. Nitrous acid interferes in the titration, since it reacts with thiocyanate to produce a transitory red color.

In the direct titration of silver with thiocyanate there are two sources of error, both of which are minor. In the first place, the silver thiocyanate precipitate adsorbs silver ions on its surface, thereby causing the end point to occur prematurely. This difficulty can be largely overcome by vigorous stirring of the mixture near the end point. Second, the color change which marks the end point occurs at a concentration of thiocyanate slightly in excess of the concentration of the equivalence point. The magnitude of this error is of the order of a few hundredths of a percent.

In the indirect method a more serious error is encountered if the silver salt of the anion being determined is more soluble than silver thiocyanate. Silver chloride, for example, is more soluble than silver thiocyanate, and the chloride tends to redissolve according to the reaction

$$AgCl(s) + SCN^- \rightleftharpoons AgSCN(s) + Cl^-$$

The equibrium constant of this reaction is given by the ratio of the solubility product constant of silver chloride to that of silver thiocyanate. Since the former constant is larger than the latter, the foregoing reaction has a strong tendency to proceed from left to right. Thus thiocyanate can be consumed not only by excess silver ions, but also by the silver chloride precipitate itself. If this occurs, low results will be obtained in the chloride analysis. This reaction can be prevented, however, by filtering off the silver chloride or adding nitrobenzene before titration with thiocyanate.[1] The nitrobenzene apparently forms an oily coating on the silver chloride surface, preventing the reaction with thiocyanate. Another method of decreasing this error is to use a sufficiently high concentration of iron(III) ion (about 0.2 *M*) so that the end-point color is reached at a lower concentration of thiocyanate.[2] A smaller amount of silver chloride is then redissolved and there is still a sufficiently high concentration of the red $FeSCN^{2+}$ complex to be visible. Swift et al.[2] found that the end-point error was reduced to 0.1% by this procedure.

In the determination of bromide and iodide by the indirect Volhard method, the reaction with thiocyanate does not cause any trouble, because silver bromide

[1] J. R. Caldwell and H. V. Moyer, *Ind. Eng. Chem., Anal. Ed.*, **7**, 38 (1935).

[2] E. H. Swift, G. M. Arcand, R. Lutwack, and D. J. Meier, *Anal. Chem.*, **22**, 306 (1950).

has about the same solubility as silver thiocyanate and silver iodide is considerably less soluble.

The following example illustrates the calculation involved in the indirect titration of chloride ion.

Example. A 0.8165-g sample containing chloride ion is analyzed by the Volhard method. The sample is dissolved in water and 50.00 ml of 0.1214 *M* $AgNO_3$ is added to precipitate the chloride ion. The excess $AgNO_3$ is titrated with 11.76 ml of 0.1019 *M* KSCN. Calculate the percentage of chloride in the sample.

$$\text{mmol AgNO}_3 = \text{mmol Cl}^- + \text{mmol KSCN}$$

$$50.00 \times 0.1214 = \text{mmol Cl}^- + 11.76 \times 0.1019$$

$$\text{mmol Cl}^- = 4.872$$

$$\% \text{ Cl}^- = \frac{4.872 \text{ mmol} \times 35.453 \text{ mg/mmol}}{816.5 \text{ mg}} \times 100$$

$$\% \text{ Cl} = 21.15$$

Adsorption Indicators

When a colored organic compound is adsorbed on the surface of a precipitate, modification of the organic structure may occur, and the color may be greatly changed and may become more intense. This phenomenon can be used to detect the end point of precipitation titrations of silver salts. The organic compounds thus employed are referred to as "adsorption indicators."

The mechanism by which such indicators work is different from any we have discussed so far. Fajans,[3] who discovered the fact that fluorescein and some substituted fluoresceins could serve as indicators for silver titrations, explained the process as follows. When silver nitrate is added to a solution of sodium chloride, the finely divided particles of silver chloride tend to hold to their surface (adsorb) some of the excess chloride ions in the solution. These chloride ions are said to form the primary adsorbed layer and thus cause the colloidal particles of silver chloride to be negatively charged. These negative particles then tend to attract positive ions from the solution to form a more loosely held secondary adsorption layer:

$(AgCl) \cdot Cl^-$	M^+	
Primary layer	Secondary layer	Excess chloride

If one continues to add silver nitrate until silver ions are in excess, these ions will displace chloride ions in the primary layer.[4] The particles then become

[3] K. Fajans and O. Hassel, *Z. Elektrochem.*, **29**, 495 (1923); see also I. M. Kolthoff, *Chem. Rev.*, **16**, 87 (1935), and K. Fajans, Chapter 7 of W. Bottger, ed., *Newer Methods of Volumetric Analysis*, D. Van Nostrand and Co., New York, 1938.

[4] A precipitate tends to adsorb most readily those ions that form an insoluble compound with one of the ions in the lattice. Thus silver or chloride ions will be more readily adsorbed by a silver chloride precipitate than will, say, sodium or nitrate ions.

positively charged, and anions in the solution are attracted to form the secondary layer:

$(AgCl) \cdot Ag^+$	X^-	
Primary layer	Secondary layer	Excess silver

Fluorescein is a weak organic acid, which we may represent as HFl. When fluorescein is added to the titration flask, the anion, Fl^-, is not adsorbed by colloidal silver chloride as long as chloride ions are in excess. However, when silver ions are in excess the Fl^- ions can be attracted to the surface of the positively charged particles, as

$$(AgCl) \cdot Ag^+ \vdots Fl^-$$

The resulting aggregate is pink, and the color is sufficiently intense to serve as a visual indicator.

A number of factors must be considered in choosing a proper adsorption indicator for a precipitation titration. These are summarized below.

1. Since the surface of the precipitate is the "active agent" in the operation of the indicator, the precipitate should not be allowed to coagulate into large particles and settle to the bottom of the titration flask. Coagulation of a silver chloride precipitate will occur at the equivalence point, where neither chloride nor silver ions are in excess, unless a substance such as dextrin is present. Dextrin acts as a "protective colloid," keeping the precipitate highly dispersed. In the presence of dextrin the color change is reversible, and if the end point is overrun, one can back-titrate with a standard chloride solution.
2. The degree to which different indicator ions are adsorbed varies considerably, and an indicator must be chosen that is not too strongly or too weakly adsorbed. Ideally, adsorption should start just before the equivalence point is reached and increase rapidly at the equivalence point. Some indicators are so strongly adsorbed that they will actually displace the primarily adsorbed ion well before the equivalence point is reached. Eosin, for example, cannot be used for titration of chloride with silver because of this effect. On the other hand, eosin can be used for the titration of iodide or bromide with silver, since these two anions are so strongly adsorbed that eosin does not displace them. If the indicator is too weakly adsorbed, the end point will occur, of course, after the equivalence point is passed.
3. Adsorption indicators are weak acids or bases, and thus the *p*H of the titration medium is of importance. The dissociation constant of fluorescein, for example, is about 10^{-7}. In solutions more acidic than *p*H 7, the concentration of the Fl^- anion is so small that no color change is observed. Fluorescein can be used only in the *p*H range of about 7 to 10. On the other hand, derivatives of fluorescein that are stronger acids can be used in solutions of lower *p*H. For example, dichlorofluorescein has a dissociation constant of about 10^{-4} and can be used in the *p*H range 4 to 10. The anion of dichlorofluorescein also is more strongly adsorbed than the anion of fluorescein. Eosin (tetra-

bromofluorescein) is a still stronger acid and can be used in bromide or iodide titrations even at a *p*H of 2.

4. It is preferable that the indicator ion be of opposite charge to the ion added as the titrant. Adsorption of the indicator will then not occur until excess titrant is present. For the titration of silver with chloride, methyl violet, the chloride salt of an organic base, can be employed. The cation is not adsorbed until excess chloride ions are present and the colloid is negatively charged. It is possible to use dichlorofluorescein in this case, but the indicator should not be added until just before the equivalence point.

A list of some adsorption indicators is given in Table 9.2.

TABLE 9.2 Some Adsorption Indicators

Indicator	Ion titrated	Titrant	Conditions
Dichlorofluorescein	Cl^-	Ag^+	*p*H 4
Fluorescein	Cl^-	Ag^+	*p*H 7–8
Eosin	Br^-, I^-, SCN^-	Ag^+	*p*H 2
Thorin	SO_4^{2-}	Ba^{2+}	*p*H 1.5–3.5
Bromcresol green	SCN^-	Ag^+	*p*H 4–5
Methyl violet	Ag^+	Cl^-	Acid solution
Rhodamine 6 G	Ag^+	Br^-	Sharp in presence of HNO_3 up to 0.3 *M*
Orthochrome T	Pb^{2+}	CrO_4^{2-}	Neutral 0.02 *M* solution
Bromphenol blue	Hg_2^{2+}	Cl^-	0.1 *M* solution

SEPARATIONS BY PRECIPITATION

Precipitation is a very valuable method for separating a sample into its component parts, and until recent years it was the analyst's most widely used separation technique. The process involved is one in which the substance being separated is used to construct a new phase—the solid precipitate. The discussion of this section will be devoted to a consideration of the equilibrium between the solid and its saturated solution. We shall consider the factors which affect solubility and hence the completeness of the separation that can be effected.

FACTORS AFFECTING SOLUBILITY

The important factors that affect the solubility of crystalline solids are temperature, nature of the solvent, and the presence of other ions in the solution. In the latter category are included ions that may be common or not common to ions in the solid, and ions that form slightly dissociated molecules or complex ions with ions of the solid.

Temperature

Most of the inorganic salts in which we are interested increase in solubility as the temperature is increased. It is usually advantageous to carry out the operations of precipitation, filtration, and washing with hot solutions. Particles of large size may result, filtration is faster, and impurities are dissolved more readily. Therefore, directions frequently call for employing hot solutions in those cases where the solubility of the precipitate is still negligible at the higher temperature. However, in the case of a fairly soluble compound, such as magnesium ammonium phosphate, the solution must be cooled in ice water before filtration. Quite an appreciable amount of this compound would be lost if the solution were filtered while hot.

The student may recall that lead chloride is separated from silver and mercury(I) chlorides in the qualitative analysis scheme by treatment with hot water. The lead salt dissolves at elevated temperature leaving the other two salts in the precipitate.

Solvent

Most inorganic salts are more soluble in water than in organic solvents. Water has a large dipole moment and is attracted to both cations and anions to form hydrated ions. We have already noted, for example, that the hydrogen ion in water is completely hydrated, forming the H_3O^+ ion. All ions are undoubtedly hydrated to some extent in water solutions, and the energy released by interaction of the ions and the solvent helps overcome the attractive forces tending to hold the ions in the solid lattice. The ions in a crystal do not have so large an attraction for organic solvents, and hence the solubilities are usually smaller than in water. The analyst can frequently utilize the decreased solubility in organic solvents to separate two substances which are quite soluble in water. For example, a dried mixture of calcium and strontium nitrates can be separated by treatment with a mixture of alcohol and ether. Calcium nitrate dissolves, leaving strontium nitrate. Potassium can be separated from sodium by precipitating K_2PtCl_6 from an alcohol–water mixed solvent.

Common-Ion Effect

A precipitate is generally more soluble in pure water than in a solution which contains one of the ions of the precipitate. In a solution of silver chloride, for example, the product of the concentrations of silver and chloride ions cannot exceed the value of the solubility product constant, 1×10^{-10}. In pure water, each ion has a concentration of 1×10^{-5} *M*, but if sufficient silver nitrate is added to make the silver ion concentration 1×10^{-4} *M*, the chloride ion concentration must decrease to a value of 1×10^{-6} *M*. The reaction

$$Ag^+ + Cl^- \rightleftharpoons AgCl(s)$$

is forced to the right by excess silver ion, resulting in the precipitation of additional salt, and decreasing the quantity of chloride remaining in the solution.

The importance of the common-ion effect in bringing about complete precipitation in quantitative analyses is readily apparent. In carrying out precipitations, the analyst always adds some excess of the precipitating agent to insure complete precipitation. In washing a precipitate where solubility losses may be appreciable, a common ion may be used in the wash liquid to diminish solubility. The ion should be that of the precipitating agent, of course, not the ion sought. Likewise, the salt used in the wash water should be such that any excess is removed by volatilization when the precipitate is finally heated to constant weight.

In the presence of a large excess of common ion, the solubility of a precipitate may be considerably greater than the value predicted by the solubility product constant. This effect will be discussed later. In general, directions call for adding about 10% excess precipitating agent.

The effect of a common ion on the solubility of a precipitate is illustrated in the following calculations.

Example. Calculate the molar solubility of CaF_2 in (a) water, (b) 0.010 M $CaCl_2$, (c) 0.010 M NaF solution, given the K_{sp} as 4×10^{-11} and neglecting hydrolysis of the fluoride ion.

(a) The equilibrium is

$$CaF_2(s) \rightleftharpoons Ca^{2+} + 2F^-$$

Let s = molar solubility of CaF_2. The mass balances are

$$[Ca^{2+}] = s$$

$$[F^-] = 2s$$

Since

$$[Ca^{2+}][F^-]^2 = K_{sp}$$

$$(s)(2s)^2 = 4 \times 10^{-11}$$

and

$$s = 2.1 \times 10^{-4} \text{ mol/liter}$$

(b) In 0.010 M $CaCl_2$ the mass balances are

$$[Ca^{2+}] = 0.010 + s$$

$$[F^-] = 2s$$

Hence

$$(0.01 + s)(2s)^2 = 4 \times 10^{-11}$$

Since $s \ll 0.01$, this becomes

$$4s^2 = 4 \times 10^{-9}$$

$$s = 3.2 \times 10^{-5} \text{ mol/liter}$$

(c) The mass balances are

$$[Ca^{2+}] = s$$

$$[F^-] = 0.01 + 2s$$

Hence

$$(s)(0.01 + 2s)^2 = 4 \times 10^{-11}$$

Since $2s \ll 0.01$, this becomes

$$s = 4 \times 10^{-7} \text{ mol/liter}$$

Note the extensive reduction in solubility brought about by the common ion. It should also be noted that excess F^- has a greater effect than excess Ca^{2+}.

Diverse-Ion Effect

It has been found that many precipitates show an increased solubility when salts that contain no ions in common with the precipitate are present in the solution. The effect is referred to by various names, such as *diverse-ion, neutral salt,* or *activity* effect. The data in Table 9.3 illustrate the magnitude of this increased solubility for silver chloride and barium sulfate in potassium nitrate solutions. It is seen that in 0.010 *M* KNO_3 the solubility of AgCl is increased from the value in water by about 12% and that of $BaSO_4$ by about 70%.

TABLE 9.3 Solubility of AgCl and $BaSO_4$ in KNO_3 Solutions*

Molarity KNO_3	Molarity $AgCl \times 10^5$	Molarity $BaSO_4 \times 10^5$
0.000	1.00	1.00
0.001	1.04	1.21
0.005	1.08	1.48
0.010	1.12	1.70

* From data of S. Popoff and E. W. Neuman, *J. Phys. Chem.*, **34**, 1853 (1930); and E. W. Neuman, *J. Am. Chem. Soc.*, **44**, 879 (1933). The K_{sp} of each salt was taken as 1×10^{-10}.

It was pointed out earlier (page 106) that we were justified in substituting molarity for activity only in very dilute solutions, where activity coefficients are approximately unity. In more concentrated solutions of electrolytes, activity coefficients decrease rapidly because of greater attraction between oppositely charged ions. The effectiveness of the ions in maintaining equilibrium conditions is thus decreased and additional precipitate must dissolve to restore this activity. The solubility product expression for AgCl is

$$a_{Ag^+} \times a_{Cl^-} = K^\circ_{sp}$$

where K°_{sp} is the equilibrium constant in terms of activities. In terms of concentrations this becomes

$$\gamma_{Ag^+}[Ag^+] \times \gamma_{Cl^-}[Cl^-] = K^\circ_{sp}$$

or

$$[Ag^+][Cl^-] = \frac{K^\circ_{sp}}{\gamma_{Ag^+}\gamma_{Cl^-}} = K_{sp}$$

It is apparent that the smaller the activity coefficients of the two ions, the larger is the product of the molar concentrations of the ions (K_{sp}). This increase in solubility is greater for $BaSO_4$ than for AgCl since activity coefficients of bivalent ions decrease more rapidly than those of univalent ions, as the electrolyte concentration increases. The decrease for tervalent ions is greater than that for bivalent ions. For example, the activity coefficients in solutions of ionic strength 0.001 and 0.1 are as follows: Ag^+, 0.924 and 0.75; Ba^{2+}, 0.744 and 0.38; and Fe^{3+}, 0.54 and 0.18. In very dilute solutions the activity coefficients approach unity, and K_{sp} is approximately the same as K°_{sp}.

The following example illustrates a calculation that the chemist can make in estimating activity effects on solubilities.

Example. Calculate the K_{sp} of $BaSO_4$ in 0.010 M KNO_3 solution and compare with the experimental value from Table 9.3. Use the Debye-Hückel limiting law to estimate activity coefficients.[5]

The Debye-Hückel expression for the activity coefficient of an ion in water at 25°C is

$$-\log \gamma_i = 0.5Z_i^2\sqrt{\mu}$$

where μ, the ionic strength, is given by

$$\mu = \tfrac{1}{2}\sum C_iZ_i^2$$

Z_i is the charge on the ion and C_i is the concentration. The ionic strength in this case can be calculated on the basis of the KNO_3 alone, since the concentrations of Ba^{2+} and $SO_4{}^{2-}$ are so small. Hence

$$\mu = \tfrac{1}{2}[0.01(1)^2 + 0.01(-1)^2] = 0.01$$

The K_{sp} expression above in logarithmic form is

$$\log K_{sp} = \log K^\circ_{sp} - \log \gamma_{Ba^{2+}} - \log \gamma_{SO_4{}^{2-}}$$

or

$$\log K_{sp} = \log K^\circ_{sp} + 0.5(2)^2\sqrt{0.01} + 0.5(-2)^2\sqrt{0.01}$$

$$\log K_{sp} = -10.0 + 4\sqrt{0.01} = -10.0 + 0.4 = -9.6$$

and

$$K_{sp} = 2.5 \times 10^{-10}$$

From Table 9.3, $K_{sp} = 2.9 \times 10^{-10}$.

The diverse-ion effect does not cause serious problems for the analyst since conditions are normally chosen so as to make the loss from solubility negligibly small. It is rarely necessary to make a precipitation from a salt solution of very

[5] P. Debye and E. Hückel, *Physik. Z.*, **24**, 185 (1923).

high concentration, and in such a case an estimate of the increased solubility can be made as illustrated above. Errors from other sources are normally more important.

Effect of *p*H

The solubility of the salt of a weak acid depends upon the *p*H of the solution. Some of the more important examples of such salts in analytical chemistry are oxalates, sulfides, hydroxides, carbonates, and phosphates. Hydrogen ion combines with the anion of the salt to form the weak acid, thereby enhancing the solubility of the salt. We shall limit our discussion in this section to solutions which are fairly acidic, so that the hydrogen ion concentration is not changed appreciably as the salt dissolved.

Let us consider first the simplest case, that of a salt MA of the weak acid HA. The equilibria to be considered are

$$\mathrm{MA(s)} \rightleftharpoons \mathrm{M^+ + A^-}$$

$$\mathrm{HA + H_2O} \rightleftharpoons \mathrm{H_3O^+ + A^-}$$

As we did previously (page 178), let us designate c_a as the total (analytical) concentration of all species related to the acid HA.

$$c_a = [\mathrm{A^-}] + [\mathrm{HA}]$$

$$c_a = [\mathrm{A^-}]\left\{\frac{[\mathrm{H_3O^+}] + K_a}{K_a}\right\}$$

The fraction in the A^- form is

$$\frac{[\mathrm{A^-}]}{c_a} = \frac{K_a}{[\mathrm{H_3O^+}] + K_a} = \alpha_1$$

Hence

$$[\mathrm{A^-}] = \alpha_1 c_a$$

The latter expression can be substituted in the K_{sp} giving

$$K_{sp} = [\mathrm{M^+}][\mathrm{A^-}] = [\mathrm{M^+}]\alpha_1 c_a$$

or

$$\frac{K_{sp}}{\alpha_1} = K_{\mathrm{eff}} = [\mathrm{M^+}]c_a$$

We have designated K_{eff} as the effective solubility product constant, in agreement with the terminology used on page 193 for the effective stability constant of complexes. The value of K_{eff} varies with *p*H because of the *p*H dependence of α_1.

The student should be able to show that, for a salt MA_2, the relation is

$$K_{\mathrm{eff}} = \frac{K_{sp}}{\alpha_1^2} = [\mathrm{M^{2+}}]c_a^2$$

and that for a diprotic acid, H_2A, the concentration of A^{2-} is given by $\alpha_2 c_a$, where

$$\alpha_2 = \frac{K_{a_1}K_{a2}}{[H_3O^+]^2 + [H_3O^+]K_{a_1} + K_{a_1}K_{a2}}$$

$$K_{\text{eff}} = \frac{K_{sp}}{\alpha_2} = [M^{2+}]c_a$$

The following examples illustrate some calculations based on the relations just described.

Example 1. Calculate the molar solubility of CaF_2 in an HCl solution, $pH = 3.00$, given that K_{sp} of $CaF_2 = 4 \times 10^{-11}$, and K_a of $HF = 6 \times 10^{-4}$.

First evaluate α_1:

$$\alpha_1 = \frac{6 \times 10^{-4}}{6 \times 10^{-4} + 1 \times 10^{-3}} = 0.38$$

$$\alpha_1^2 = 0.14$$

Hence

$$K_{\text{eff}} = \frac{4 \times 10^{-11}}{0.14} = 2.9 \times 10^{-10}$$

Let s = molar solubility of CaF_2. The mass balances are

$$[Ca^{2+}] = s$$

$$c_F = [HF] + [F^-] = 2s$$

and

$$(s)(2s)^2 = 2.9 \times 10^{-10}$$

$$s = 4.2 \times 10^{-4} \text{ mol/liter}$$

Example 2. Calculate the solubility of CaC_2O_4 in an HCl solution of pH 3.00, given $K_{sp} = 2 \times 10^{-9}$, $K_{a_1} = 6.5 \times 10^{-2}$, $K_{a_2} = 6.1 \times 10^{-5}$.

$$\alpha_2 = \frac{6.5 \times 10^{-2} \times 6.1 \times 10^{-5}}{6.5 \times 10^{-2} \times 6.1 \times 10^{-5} + 6.5 \times 10^{-2} \times 10^{-3} + (10^{-3})^2}$$

$$\alpha_2 = 0.057$$

Hence

$$K_{\text{eff}} = \frac{2 \times 10^{-9}}{0.057} = 3.5 \times 10^{-8}$$

The mass balances are

$$[Ca^{2+}] = s$$

$$c_{O_x} = s$$

Then

$$s^2 = 3.5 \times 10^{-8}$$

$$s = 1.9 \times 10^{-4} \text{ mol/liter}$$

The separation of metal sulfides, based upon the control of pH, has been used for many years in the qualitative analysis scheme. The metals which form the less soluble sulfides (Group II) are precipitated by H_2S in about 0.10 M HCl. Then the pH is raised to precipitate the metals of Group III. Hydrogen sulfide is a diprotic acid, and the expression for α_2 (above) is applicable. However, since the two acid constants are so small ($K_{a_1} = 1 \times 10^{-7}$, $K_{a_2} = 1 \times 10^{-15}$), the two terms in the denominator containing the acid constants are negligible compared to the square of the hydrogen ion concentration. The expression becomes (approximately)

$$\alpha_2 \cong \frac{K_{a1}K_{a2}}{[H_3O^+]^2}$$

Also, in strongly acidic solution, the analytical concentration of hydrogen sulfide is approximately

$$c_S = [H_2S] + [HS^-] + [S^{2-}] \cong [H_2S]$$

Hence the sulfide ion concentration, $\alpha_2 c_S$, becomes

$$[S^{2-}] = \frac{[H_2S]K_{a1}K_{a2}}{[H_3O^+]^2}$$

Since a saturated solution of H_2S is about 0.10 M, this gives

$$[S^{2-}] = \frac{1 \times 10^{-23}}{[H_3O^+]^2}$$

This is the usual expression employed to show how the sulfide ion concentration can be varied by changing the hydrogen ion concentration. The following example illustrates the separation of two metals by employing this principle.

Example 3. 100 ml of a solution that is 0.10 M in both Cu^{2+} and Mn^{2+} and 0.20 M in H_3O^+ is saturated with H_2S.

(a) Show which metal sulfide precipitates. K_{sp} of CuS is 4×10^{-38}, of MnS 1×10^{-16}.

The sulfide concentration is given by

$$[S^{2-}] = \frac{1 \times 10^{-23}}{(0.20)^2} = 2.5 \times 10^{-22}$$

The K_{sp} of CuS is greatly exceeded but that of MnS is not:

$$(0.10)(2.5 \times 10^{-32}) = 2.5 \times 10^{-23} \gg 4 \times 10^{-38}$$

$$= 2.5 \times 10^{-23} \ll 1 \times 10^{-16}$$

Hence CuS precipitates, but MnS does not.

(b) What must be the hydrogen ion concentration for MnS to start to precipitate?

The sulfide ion concentration needed in order for $[Mn^{2+}][S^{2-}]$ to equal the K_{sp} of MnS is

$$(0.10)[S^{2-}] = 1 \times 10^{-16}$$

$$[S^{2-}] = 1 \times 10^{-15}$$

Hence

$$1 \times 10^{-15} = \frac{1 \times 10^{-23}}{[H_3O^+]^2}$$

$$[H_3O^+] = 1 \times 10^{-4}\,M$$

The following example illustrates the separation of two metal hydroxides by control of pH.

Example 4. Calculate the pH at which the following hydroxides begin to precipitate if the solution is $0.1\,M$ in each cation: $Fe(OH)_3$, $K_{sp} = 1 \times 10^{-36}$; and $Mg(OH)_2$, $K_{sp} = 1 \times 10^{-11}$.

Iron(III) hydroxide:

$$[Fe^{3+}][OH^-]^3 = 1 \times 10^{-36}$$

$$(0.1)[OH^-]^3 = 1 \times 10^{-36}$$

$$[OH^-]^3 = 1 \times 10^{-35}$$

$$3pOH = 35$$

$$pOH = 11.7$$

$$pH = 2.3$$

Magnesium hydroxide:

$$[Mg^{2+}][OH^-]^2 = 1 \times 10^{-11}$$

$$(0.1)[OH^-]^2 = 1 \times 10^{-11}$$

$$[OH^-]^2 = 1 \times 10^{-10}$$

$$2pOH = 10.0$$

$$pOH = 5.0$$

$$pH = 9.0$$

Thus, if an acidic solution containing these two ions is slowly neutralized with base, iron(III) hydroxide will precipitate first. This precipitate can be separated by filtration before the pH is sufficiently high to precipitate magnesium hydroxide. In actual practice, however, the iron(III) hydroxide precipitate is likely to be contaminated by magnesium hydroxide. This arises from the fact that in the region where the two solutions mix, the solubility product constant of magnesium hydroxide may be temporarily exceeded. The magnesium hydroxide may not redissolve as the solution is stirred, and the separation is then not a clean one. Usually, a buffer solution of intermediate pH is employed to diminish the local increase in hydroxide ion concentration. Better still, the pH can be gradually increased by the hydrolysis of a substance such as urea.

Effect of Hydrolysis

In the previous section we limited our discussion to solutions of fairly high acidity, such that the anion of the weak acid did not change the pH appreciably.

Let us now consider the case in which the salt of a weak acid is dissolved, not in strong acid, but in water. The problem is more complex than the previous one since the change in hydrogen ion concentration may be of considerable magnitude.

For simplification let us consider that whatever the amount of salt MA which dissolves, the anion is completely hydrolyzed:

$$A^- + H_2O \rightleftharpoons HA + OH^-$$

This is a good approximation if HA is very weak and if MA is not very soluble, (i.e., if both K_a and K_{sp} are small). It should be noted that the lower the concentration of A^-, the more complete the hydrolysis reaction.

Let us further consider two extremes, depending upon the magnitude of the K_{sp}:

1. The solubility is so low that the *p*H of water is not changed appreciably by the hydrolysis.
2. The solubility is sufficiently large so that hydroxide ion contribution of water can be neglected.

These cases are illustrated in the following example.

Example. Calculate the molar solubilities in water of (a) CuS, $K_{sp} = 4 \times 10^{-38}$, and (b) MnS, $K_{sp} = 1 \times 10^{-16}$. Consider the hydrolysis reaction

$$S^{2-} + H_2O \rightleftharpoons HS^- + OH^-$$

(a) Since the solubility of CuS is so low, we shall neglect the OH^- produced by hydrolysis, taking $[OH^-] = 1 \times 10^{-7}$. Hence

$$\alpha_2 = \frac{1 \times 10^{-22}}{(1 \times 10^{-7})^2 + (1 \times 10^{-7})(1 \times 10^{-7}) + 1 \times 10^{-22}}$$

$$\alpha_2 = 5 \times 10^{-9}$$

$$K_{\text{eff}} = \frac{4 \times 10^{-38}}{5 \times 10^{-9}} = 8 \times 10^{-30}$$

Letting s = solubility, the mass balances are

$$[Cu^{2+}] = s$$

$$c_S = s$$

Hence

$$s^2 = 8 \times 10^{-30}$$

$$s = 3 \times 10^{-15}$$

(b) Since the hydrolysis is complete, we can write the reaction as

$$MnS(s) + H_2O \rightleftharpoons Mn^{2+} + HS^- + OH^-$$

the equilibrium constant for which is given by

$$K = \frac{K_{sp}K_w}{K_{a2}} = \frac{1 \times 10^{-16} \times 1 \times 10^{-14}}{1 \times 10^{-15}}$$

$$K = 1 \times 10^{-15}$$

Letting s = solubility, then

$$[Mn^{2+}] = s$$

$$[HS^-] = s$$

$$[OH^-] = s$$

Hence

$$s^3 = 1 \times 10^{-15}$$

$$s = 1 \times 10^{-5}$$

The cation of a salt can undergo hydrolysis just as can the anion, and this will also increase the solubility. Typical hydrolytic reactions of iron(III) ion are

$$Fe^{3+} + HOH \rightleftharpoons FeOH^{2+} + H^+$$

$$FeOH^{2+} + HOH \rightleftharpoons Fe(OH)_2{}^+ + H^+$$

Many metals have been found to form ionic species containing more than one metal atom, as, for example,

$$2Fe^{3+} + 2H_2O \rightleftharpoons Fe_2(OH)_2{}^{4+} + 2H^+$$

In the case of aluminum, species such as $Al_6(OH)_{15}{}^{3+}$ have been postulated to explain certain experimental data.

Because of the complexity of these processes, we shall not consider the topic further here.

Effect of Complexes

The solubility of a slightly soluble salt is also dependent upon the concentration of substances which form complexes with the cation of the salt. The effect of hydrolysis, mentioned above, is an example in which the complexing agent is hydroxide ion. The complexing agents normally considered under a heading such as this are neutral molecules and anions, both foreign and common to the precipitate.

One of the best-known examples in analytical chemistry is the effect of ammonia on the solubility of the silver halides, especially silver chloride. Silver chloride can be dissolved in ammonia, and this fact is utilized in separating silver from mercury in the first group of the traditional qualitative analysis scheme. Silver ion forms two complexes with ammonia,

$$Ag^+ + NH_3 \rightleftharpoons Ag(NH_3)^+ \qquad K_1 = 2.3 \times 10^3$$

$$Ag(NH_3)^+ + NH_3 \rightleftharpoons Ag(NH_3)_2{}^+ \qquad K_2 = 6.0 \times 10^3$$

Designating β_2 as the fraction of silver in the uncomplexed form as we did for zinc (page 198):

$$\beta_2 = \frac{1}{1 + K_1[NH_3] + K_1K_2[NH_3]^2} = \frac{[Ag^+]}{c_{Ag}}$$

where c_{Ag} is the analytical concentration of silver. Since

$$K_{sp} = [Ag^+][Cl^-]$$

$$K_{sp} = \beta_2 c_{Ag}[Cl^-]$$

or

$$\frac{K_{sp}}{\beta_2} = K_{eff} = c_{Ag}[Cl^-]$$

The following example illustrates a calculation of the effect of complexes on the solubility of AgCl.

Example. Calculate the molar solubility of AgCl in 0.010 M NH_3. (This is the final concentration of free NH_3 molecules in the solution.) Given K_{sp} of AgCl = 1.0×10^{-10} and stability constants: $K_1 = 2.3 \times 10^3$, $K_2 = 6.0 \times 10^3$.

Evaluating β_2:

$$\beta_2 = \frac{1}{1 + 2.3 \times 10^3(10^{-2}) + 1.4 \times 10^7(10^{-2})^2}$$

$$\beta_2 = 7.1 \times 10^{-4}$$

$$K_{eff} = \frac{1.0 \times 10^{-10}}{7.1 \times 10^{-4}} = 1.4 \times 10^{-7}$$

Letting s = molar solubility

$$s = c_{Ag} = [Cl^-]$$

Hence

$$s^2 = 1.4 \times 10^{-7}$$

$$s = 3.7 \times 10^{-4} \text{ mol/liter}$$

Many precipitates form soluble complexes with the ion of the precipitating agent itself. In such a case, the solubility first decreases because of the common-ion effect, passes through a minimum, and then increases as complex formation becomes appreciable. Silver chloride forms complexes with both silver and chloride ions, such as

$$AgCl + Cl^- \rightleftharpoons AgCl_2^-$$

$$AgCl_2^- + Cl^- \rightleftharpoons AgCl_3^{2-}$$

and

$$AgCl + Ag^+ \rightleftharpoons Ag_2Cl^+$$

In addition, there is a certain amount of undissociated AgCl molecules in solution. Figure 9.2 shows the solubility of AgCl in NaCl and $AgNO_3$ solutions. It is

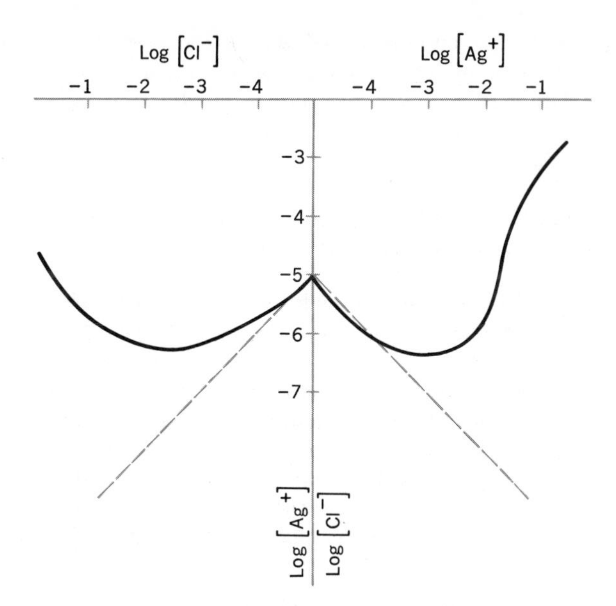

Figure 9.2 Solubility of AgCl in solutions of NaCl and $AgNO_3$. (From H. F. Walton, *Principles and Methods of Chemical Analysis*, 2nd ed., Prentice-Hall, Inc., Englewood Cliffs, N.J. Used by permission of the author and publisher.)

interesting to note that AgCl is actually more soluble in 0.1 *M* $AgNO_3$ and in 1 *M* NaCl than it is in water. It is because of such effects that only a reasonable excess (usually about 10%) of precipitating agent is used in quantitative precipitations.

REFERENCES

L. GORDON, M. L. SALUTSKY, and H. H. WILLARD, *Precipitation from Homogeneous Solution*, John Wiley & Sons, Inc., New York, 1959.

H. A. LAITINEN and W. E. HARRIS, *Chemical Analysis*, 2nd ed., McGraw-Hill Book Company, New York, 1975.

D. L. LEUSSING, "Solubility," and J. F. COETZEE, "Equilibria in Precipitation Reactions and Precipitation Lines," Chaps. 17 and 19 of I. M. Kolthoff, P. J. Elving, and E. B. Sandell, eds., *Treatise on Analytical Chemistry*, Part I, Vol. 1, Interscience Publishers, Inc., New York, 1959.

M. L. SALUTSKY, "Precipitates: Their Formation, Properties, and Purity," Chap. 18 of I. M. Kolthoff, P. J. Elving, and E. B. Sandell, eds., *Treatise on Analytical Chemistry*, Part I, Vol. 1, Interscience Publishers, Inc., New York, 1959.

QUESTIONS

1. *Adsorption indicators.* Explain clearly what is meant by an adsorption indicator and the mechanism by which such an indicator works. What is the function of dextrin? Why must the *p*H be controlled?

2. *Errors*. Explain how the following errors affect the indicated determination; that is, will the error make the result high, low, or have no effect? (a) Chloride (Mohr), solution of *p*H 2; (b) chloride (Volhard), failure to add nitrobenzene; (c) chloride (Fajans), eosin used as the indicator; (d) bromide (Mohr), $AgNO_3$ standardized against NaCl, no indicator blank.

3. *Solubility*. Point out the factors that affect the solubility of inorganic crystals and explain how each factor operates.

4. *Solubility*. Explain why the following statements are true:
   (a) Silver chloride is more soluble in 1 *M* KNO_3 than in water.
   (b) Silver chloride is more soluble in 1 *M* HCl than in water.
   (c) Silver chloride is more soluble in 1 *M* NH_3 than in water.
   (d) Silver chloride is less soluble in 0.001 *M* HCl than in water.
   (e) Iron(II) hydroxide is less soluble in 0.1 *M* NH_3 than in water.
   (f) Zinc hydroxide is more soluble in 0.1 *M* NH_3 than in water.
   (g) Calcium fluoride is more soluble at *p*H 3 than at *p*H 4.
   (h) Silver chromate is less soluble in 0.001 *M* $AgNO_3$ than in 0.001 *M* K_2CrO_4.

Multiple-choice: In the following multiple-choice questions, select the *one best* answer.

5. The titration of chloride ion with silver ion using dichlorofluorescein indicator is carried out at *p*H 2 by mistake. The use of this low *p*H means that (a) the results for chloride will be low; (b) no color change will be observed; (c) the indicator will be too strongly adsorbed; (d) more than one of the above.

6. An example of a back-titration is (a) the Mohr determination of Cl^-; (b) the Volhard determination of Cl^-; (c) the Volhard determination of Ag^+; (d) the determination of Cl^- using an adsorption indicator.

7. 50 ml of 0.10 *M* NaCl is titrated with 0.10 *M* $AgNO_3$. In a second titration 50 ml of 0.10 *M* NaBr is titrated with 0.10 *M* $AgNO_3$. Which of the following statements is false? (a) At the start of the titrations, $p\text{Cl} = p\text{Br}$; (b) When 25 ml of titrant is added in each titration, $p\text{Cl} = p\text{Br}$; (c) At the equivalence points, $p\text{Br} > p\text{Cl}$; (d) when 60 ml of titrant is added in each case, $p\text{Cl} = p\text{Br}$.

8. In a certain titration $[Cl^-] = 0.10\ M$, $[Br^-] = 0.00050\ M$, and $[CrO_4^{2-}] = 0.020\ M$. One drop (0.05 ml) of 0.010 *M* $AgNO_3$ is added to 100 ml of this solution, forming a precipitate. The precipitate is (a) AgCl; (b) AgBr; (c) Ag_2CrO_4; (d) AgCl + AgBr; (e) AgCl + AgBr + Ag_2CrO_4.

9. Which of the following statements is *false*? (a) The Mohr method can be used for the direct titration of Br^-; (b) The Volhard method can be used for the direct titration of Cl^-; (c) Mercury interferes with the Volhard method for silver; (d) Eosin is a suitable indicator for the titration of I^- with Ag^+.

10. The K_{sp} of Ag_2S is 4×10^{-48}; the K_{sp} of ZnS is 1×10^{-24}. If 1 mmol of Zn^{2+} is added to 100 ml of a solution containing solid Ag_2S in equilibrium with Ag^+ and S^{2-} ions, what will happen? (a) More Ag_2S will precipitate; (b) ZnS will precipitate; (c) More Ag_2S will dissolve; (d) both (b) and (c); (e) both (a) and (b).

PROBLEMS

1. *Solubility product constant.* Calculate the solubility product constants from the given solubilities: (a) AgBr, 0.118 mg/liter; (b) $Cu(OH)_2$, 3.59×10^{-6} g/100 ml; (c) Ag_2CrO_4, 2.63 mg/100 ml.

2. *Solubility.* From the solubility product constants listed in Table 3, Appendix I, calculate the following solubilities in water, neglecting such effects as hydrolysis: (a) SrC_2O_4, in mg/ml; (b) $Ag_2C_2O_4$, in g/100 ml; (c) $Pb(IO_3)_2$, in mg/100 ml.

3. *Common-ion effect.* Calculate the molar solubilities of the following neglecting such effects as hydrolysis: (a) $BaSO_4$ in 0.020 *M* K_2SO_4; (b) MgF_2 in 0.10 *M* NaF; (c) Ag_2CrO_4 in 0.010 *M* K_2CrO_4.

4. *Effect of pH.* Calculate the molar solubilities of the following: (a) $Fe(OH)_2$ at *p*H 11.70; (b) CuS at *p*H 2.30, solution saturated with H_2S; (c) HgS at *p*H 0.60, solution saturated with H_2S.

5. *Effect of pH.* Calculate the molar solubilities of the following: (a) CaF_2 in HCl, *p*H = 2.30; (b) $Ag_2C_2O_4$ in HNO_3, *p*H = 2.70; (c) $CdCO_3$ in HCl, *p*H = 4.00.

6. *Effect of complexes.* Calculate the solubilities in grams per liter of the following: (a) AgCl in 0.10 *M* NH_3; (b) AgBr in 3.0 *M* NH_3; (c) AgI in 10 *M* NH_3. The NH_3 concentrations are of the uncomplexed NH_3 at equilibrium.

7. *Solubility.* 50 ml of 0.060 *M* K_2CrO_4 is mixed with 50 ml of 0.14 *M* $AgNO_3$. Calculate the following: (a) the molar solubility of Ag_2CrO_4 in the solution; (b) the molarities of Ag^+, CrO_4^{2-}, K^+, and NO_3^-.

8. *Hydrolysis.* Calculate the molar solubilities in water of the following, taking into account hydrolysis of the anion: (a) $CaCO_3$; (b) $BaCrO_4$; (c) Ag_2CrO_4; (d) ZnS; (e) Ag_2S.

9. *Calculation of pX.* Calculate the following: (a) *p*Cl of 0.010 *M* $CaCl_2$; (b) *p*Br of 0.0050 *M* NaBr; (c) *p*Cl of 0.0020 *M* $CrCl_3$.

10. *Mixtures.* Calculate the values of *p*Cl and *p*Ag of the solutions made by mixing (a) 50 ml of 0.10 *M* NaCl + 50 ml of 0.20 *M* $AgNO_3$; (b) 60 ml of 0.10 *M* NaCl + 40 ml of 0.15 *M* $AgNO_3$; (c) 60 ml of 0.20 *M* NaCl + 40 ml of 0.10 *M* $AgNO_3$.

11. *Mixtures.* 50 ml of 0.080 *M* $CaCl_2$ is mixed with 50 ml of 0.10 *M* $AgNO_3$. Calculate the molarities of Ca^{2+}, Cl^-, Ag^+, and NO_3^- in the resulting solution.

12. *Mixtures.* (a) Calculate the value of *p*Cl of a solution made by mixing equal volumes of HCl, one with a *p*Cl of 2.00, the other with a *p*Cl of 3.00. (b) Calculate the value of *p*Ag of a solution made by mixing equal volumes of a solution of $AgNO_3$, *p*Ag = 2.00, and of $CaCl_2$, *p*Cl = 1.70.

13. *Titration curve.* 50.0 ml of 0.100 *M* NaBr is titrated with 0.100 *M* $AgNO_3$. Calculate the value of *p*Br when the following volumes of titrant have been added: (a) 0.00; (b) 25.0; (c) 49.9; (d) 50.0; (e) 50.1; (f) 60.0. Plot the titration curve.

14. *Titration curve.* 50.0 ml of 0.100 *M* K_2CrO_4 is titrated with 0.200 *M* $AgNO_3$. Calculate the value of *p*CrO_4 when the following volumes of titrant have been added: (a) 0.00; (b) 25.0; (c) 49.9; (d) 50.0; (e) 50.1; (f) 60.0. Plot the titration curve.

15. *Derivations.* For the titration of Ag^+ with CrO_4^{2-} show that the following relations are true at the equivalence point: (a) $3pAg = pK_{sp} - \log 2$; (b) $3pCrO_4 = pK_{sp} + 2 \log 2$.

16. *Approximations.* A 50-ml sample of a 0.10 *M* solution of the salt NaX is titrated with 0.10 *M* $AgNO_3$, forming the precipitate AgX. Calculate the value of *p*X after the addition of 49.9 ml of titrant two ways: (a) first, neglecting the solubility of AgX, and (b) not neglecting the solubility. Use the following values of K_{sp} of AgX; (a) 1×10^{-6}; (b) 1×10^{-8}; (c) 1×10^{-10}.

17. *Feasibility of titration.* 50 ml of 0.10 *M* NaX is titrated with 0.10 *M* $AgNO_3$. Calculate the value of the equilibrium constant for the reaction

$$Ag^+ + X^- \rightleftharpoons AgX(s)$$

so that when 49.9 ml of titrant is added, the reaction is complete, and the *p*X changes by 2.00 units for the addition of 0.20 ml more titrant. Also calculate the K_{sp} of AgX.

18. *Feasibility of titration.* 50 ml of 0.10 *M* Na_2Y is titrated with 0.20 *M* $AgNO_3$. (a) Calculate the value of the equilibrium constant for the reaction

$$2Ag^+ + Y^{2-} \rightleftharpoons Ag_2Y(s)$$

so that when 49.9 ml of titrant has been added the reaction is complete, and the *p*Y changes by 2.00 units for the addition of 0.20 ml more titrant. (b) Calculate the K_{sp} of Ag_2Y.

19. *Volhard method.* Calculate the equilibrium constants for the following reactions:
(a) $AgCl(s) + SCN^-(aq) \rightleftharpoons AgSCN(s) + Cl^-(aq)$
(b) $AgBr(s) + SCN^-(aq) \rightleftharpoons AgSCN(s) + Br^-(aq)$
(c) $AgI(s) + SCN^-(aq) \rightleftharpoons AgSCN(s) + I^-(aq)$
Why does this reaction cause trouble in the indirect Volhard method for chloride but not for bromide and iodide?

20. *Fraction precipitated.* (a) A base is added to a 0.10 *M* solution of $MgCl_2$, raising the *p*H gradually. Calculate the *p*H values when 50, 90, 99.9, and 99.99% of the magnesium is precipitated. (b) Repeat the calculation for a 0.10 *M* solution of Fe^{3+}.

21. *Solubility.* A 0.843-g sample containing 16.7% Na_2SO_4 is dissolved in water and treated with 40.0 ml of 0.0500 *M* $BaCl_2$ to precipitate $BaSO_4$. (a) Calculate the number of milligrams of Na_2SO_4 not precipitated if the volume of the solution is 100 ml. (b) The precipitate is filtered and washed with 100 ml of water at room temperature. Assuming that solubility equilibrium is reached, how many milligrams of $BaSO_4$ dissolve in the wash water?

22. *Sulfide precipitation.* (a) Calculate the *p*H required just to prevent the precipitation of MnS from a solution which is 0.10 *M* in Mn^{2+} and is saturated with H_2S ($[H_2S] = 0.10\ M$). (b) If it is desired to lower the Mn^{2+} concentration to 10^{-6} *M* by precipitation of MnS, what should the *p*H be?

23. *Sulfide precipitation.* 100 ml of a solution which is 0.10 *M* in Cd^{2+} and 0.20 *M* in H^+ is saturated with H_2S ($[H_2S] = 0.10\ M$). Calculate the milligrams of Cd^{2+} left in solution, taking into account the fact that the precipitation of CdS produces hydrogen ions.

24. *Separations.* It is desired to separate Cd^{2+} and Fe^{2+} by precipitation with H_2S while controlling the *p*H. Starting with the strongly acidic solution that is 0.10 *M* in each metal ion, the *p*H is gradually raised by the addition of base. (Assume that $[H_2S]$ remains 0.10 *M* throughout.) Make the following calculations: (a) At what *p*H does CdS begin to precipitate? (b) At what *p*H does $[Cd^{2+}] = 1 \times 10^{-6}$ *M*? (c) At what *p*H does FeS begin to precipitate? (d) At what *p*H does $[Fe^{2+}]$ reach 1×10^{-6} *M*? If the separation is considered successful if the concentration of one ion can be reduced to 1×10^{-6} *M* before the other starts to precipitate, is this separation feasible?

25. *Separations.* Repeat Problem 24 for 0.10 *M* solutions of (a) Cd^{2+} and Zn^{2+}; (b) Zn^{2+} and Fe^{2+}.

26. *Separations.* (a) A solution which is 0.10 *M* in two hypothetical cations, M^{2+} and N^{2+}, and 0.10 *M* in H^+ is saturated with H_2S (0.10 *M*). What is the minimum ratio of the K_{sp} of MS to the K_{sp} of NS so that the concentration of N^{2+} can be reduced to 1×10^{-6} *M* without precipitating M^{2+}? (b) Repeat part (a) where the sulfides are MS and N_2S.

27. *Separations.* A solution of *p*H 1.0 is 0.10 *M* in each of the following ions: Al^{3+}, Cu^{2+}, and Mg^{2+}. The *p*H is gradually raised by the addition of base. Calculate the *p*H values at which each metal begins to precipitate as the metal hydroxide, and the *p*H at which the metal ion concentration reaches 1×10^{-6} *M*. Is the separation of these three cations by hydroxide precipitation theoretically feasible? (See Problem 24.)

28. *Precipitation of carbonates.* A solution which is 0.10 *M* in Sr^{2+} and 0.05 *M* in Ca^{2+} is treated with solid Na_2CO_3, first precipitating $SrCO_3$. What percentage of the Sr^{2+} is precipitated when $CaCO_3$ begins to precipitate?

29. *Precipitation of lead.* 50 ml of a solution which is 0.10 *M* in Pb^{2+} is mixed with 50 ml of 0.60 *M* HCl. Calculate (a) the $[Pb^{2+}]$ in the resulting solution, assuming that the precipitation of $PbCl_2$ is complete; (b) the $[H^+]$ in the solution. The solution is now saturated with H_2S (0.10 *M*). Show whether or not PbS will precipitate.

30. *Complexes.* The successive stepwise formation constants for $Cd(NH_3)_4{}^{2+}$ are as follows: $k_1 = 550$, $k_2 = 162$, $k_3 = 23.5$, $k_4 = 13.5$. Calculate the molar solubility of CdS in a 0.20 *M* solution of ammonia, calculating first α_2 for the anion, β_4 for the cation, and then the solubility.

31. *Complexes.* 1 mmol of AgCl is dissolved in 500 ml of ammonia, the final concentration of uncomplexed NH_3 being 0.50 *M*. Calculate the concentration of uncomplexed Ag^+ in the solution.

32. *Buffer solution.* Manganese hydroxide, $Mn(OH)_2$, is precipitated from an NH_3–NH_4Cl buffer. What must be the concentration of NH_3 in the buffer if all but 0.1 mg of manganese is precipitated from 100 ml of a solution which is 0.15 *M* in NH_4Cl?

33. *Buffer solution.* A certain metal M^{2+} forms an insoluble sulfide, MS, $K_{sp} = 1 \times 10^{-20}$. A chemist wishes to precipitate MS and reduce the concentration of M^{2+} to 1×10^{-6} *M*. Knowing that hydrogen ions are generated by the reaction

$$M^{2+} + H_2S \longrightarrow MS(s) + 2H^+$$

he decides to buffer his solution with HOAc and NaOAc. If 100 ml of 0.010 *M* M^{2+} containing HOAc and NaOAc is saturated with H_2S (0.10 *M*), calculate the following: (a) the *p*H when $[M^{2+}] = 1 \times 10^{-6}$ *M*; (b) the $[OAc^-]$ so that the *p*H will not drop below the value calculated in part (a) (the original $[HOAc] = 0.10$ *M*); (c) the *p*H of the buffer before MS is precipitated.

CHAPTER TEN

oxidation-reduction equilibria

It will be recalled that oxidation is the loss of one or more electrons by an atom, ion, or molecule, while reduction is electron gain. There are no free electrons in ordinary chemical systems, and electron loss by one chemical species is always accompanied by electron gain on the part of another. Hence we speak of oxidation-reduction reactions (redox, for short); the term "electron transfer reaction" could be used.

It is interesting to compare redox with acid-base reactions—electron transfer, on the one hand, and proton transfer, on the other. We could think of Fe^{2+} and Fe^{3+}, for example, as a conjugate pair by analogy to conjugate Brønsted acids and bases. On the other hand, there are important differences. For instance, electrons can travel through wires while protons cannot. Thus to effect a proton transfer, the donor and the acceptor must encounter each other directly, whereas in redox reactions, as we shall see, the electron donor and the acceptor can be in separate solutions if we wish. (The direct reaction is also possible, of course.)

Second, there is often a contrast in regard to reaction rate. Acid-base reactions are so fast that they used to be called "instantaneous"; only in recent years have techniques been developed for measuring the rates of such reactions. Redox reactions, on the other hand, are sometimes slow, and titrimetric procedures based upon these reactions may require an elevated temperature, the addition of a catalyst, or perhaps an excess of reagent followed by a back-titration. Slowness reflects the greater complexity of many redox reactions as compared with acid-base; frequently, electron transfer is only one part of a multistep sequence which may involve forming or breaking covalent bonds, protonation, and various sorts of rearrangements. Compare, for example, the simple protonation of a base, $B + H^+ = BH^+$, with the following reactions:

$$MnO_4^- + 8H^+ + 5e \longrightarrow Mn^{2+} + 4H_2O$$

$$C_2O_4^{2-} \longrightarrow 2CO_2 + 2e$$

Third (but not unrelated to the above), we may note the difference between the acid-base and the redox properties of our common solvent, water. Protons are rapidly transferred to H_2O to form H_3O^+, and they are just as easily passed along to some other base to regenerate H_2O. Likewise, removal of a proton from water to form OH^- is readily reversed if protons are supplied by another acid. Further, all the species H_3O^+, H_2O, and OH^- are highly soluble. By contrast, the addition of electrons to water results in the formation of hydrogen, a slightly soluble gas:

$$2H_2O + 2e \longrightarrow 2OH^- + H_2$$

This product tends to escape from the solution, and moreover, even if it remained, many of its reactions are so slow under reasonable conditions that it would seldom be a suitable titrant. Similar considerations arise if water is oxidized:

$$2H_2O \longrightarrow O_2 + 4H^+ + 4e$$

Thus, while strong acids and bases that undergo proton exchanges with water are good titrants, reagents that oxidize or reduce water are usually avoided; that is, the strongest oxidants and reductants are impractical titrants. We do not encounter, then, a leveling effect such as we saw in acid-base chemistry. To be sure, reducing agents stronger than H_2 would be leveled to the power of that reagent, and oxidants stronger than O_2 would generate the latter, but we avoid such reagents in most cases. Thus typically each reagent exerts its own characteristic reactivity without mediation by the solvent. It should be noted that some reagents which are strong enough to oxidize or reduce water actually do so only very slowly; hence their aqueous solutions are sometimes stable enough to serve as titrants.

The material in this chapter may be important to the student for several reasons. Redox reactions are widely used for titrimetric determinations of both inorganic and organic substances. In addition, preliminary redox steps to establish desired oxidation states precede the measurements in many analyses. It will be found that electroanalytical techniques discussed in later chapters (e.g., potentiometry, electrolysis, and polarography) require an understanding of the principles explained here. Looking further ahead, we note that the theoretical material in this chapter is consistent with the discussion of similar topics to be enountered in physical chemistry courses. Finally, students whose futures lie in the biomedical area will see a direct relation between the substance of this chapter and the important topic of biological oxidations.

GALVANIC CELLS

Redox equilibria are conveniently treated in terms of the electromotive forces (emf values) of galvanic cells. We shall see, for instance, how to evaluate the equilibrium constant (and hence the degree of completeness) of a redox reaction, and how to select an indicator for a titration, using emf data.

Single Electrode Potentials

Suppose that we place a strip of metal, say, zinc, in contact with a solution containing zinc ions, as suggested in Fig. 10.1. Now, we may think of the metal itself as comprising zinc ions and electrons. In general, the activities of zinc ion in the metal and in the solution phases will be different, providing a driving force for the loss or gain of Zn^{2+} by the metal strip. In the figure we have shown the metal losing Zn^{2+} to the solution, leaving an excess of electrons behind while the solution acquires a positive charge, not unreasonable for an active metal such as zinc in a solution of moderate concentration. Perhaps most of the zinc ions which enter the solution, although acquiring some mobility, tend to remain near the negatively charged metal surface, giving rise to an electrical double layer at the interface, as suggested in the figure. As this charge separation develops, it will restrain the process of zinc ion transfer, and we may suppose that an equilibrium is shortly established in which the metal electrode has adopted a certain potential and the adjacent solution a different one. (We use the term "potential" here as defined in physics, in terms of the work required to move a charge from an infinite distance or from some other reference point to the site in question.) The process we have described may be written

$$Zn^{2+} + 2e \rightleftharpoons Zn$$

and in the discussion above we imagined that it proceeded somewhat from right to left. If the metal had been a less active one than zinc, say, copper, the process

$$Cu^{2+} + 2e \rightleftharpoons Cu$$

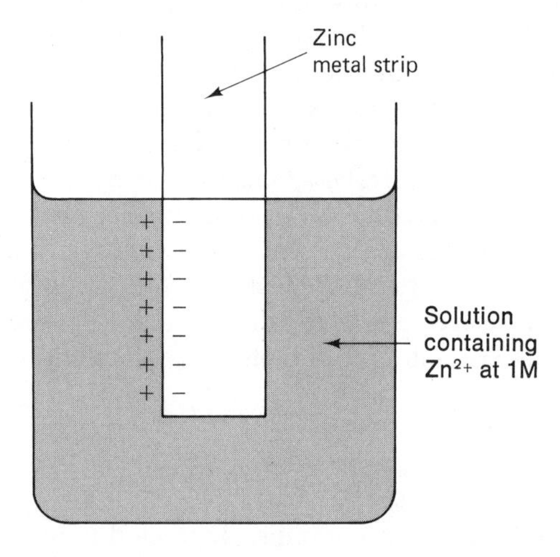

Figure 10.1 Schematic depiction of a single electrode potential.

would presumably have proceeded toward the right or at least not so far toward the left as in the zinc case, and the potential difference between metal and solution would not have been the same.

The student will have noted in the foregoing discussion our use of the words "perhaps," "suppose," "imagine," and "presumably." This was intentional because, in fact, we cannot measure potential differences of the sort described. Normally, to determine the potential difference between two points, we employ some kind of voltmeter, and this requires wires connecting the two points with the meter. But if we attempt to measure the potential difference between the strip of zinc and the solution by attaching one of the meter leads to the metal and inserting the other one into the solution, we introduce by that very action a second electrode. Some potential difference will presumably be established between the meter lead and the solution, and we shall be unable to correct for its contribution to the measured voltage. The *single electrode potential* between the zinc metal and the solution, then, cannot be measured.[1]

Cells

To obtain a useful system upon which meaningful measurements may be performed, we must combine two single electrodes to form a cell. An example is shown in Fig. 10.2, where the two half-cells mentioned in the preceding section have been paired. Now there is more than a hypothetical tendency for electron transfer to occur. The excess electrons remaining on the zinc electrode when Zn^{2+}

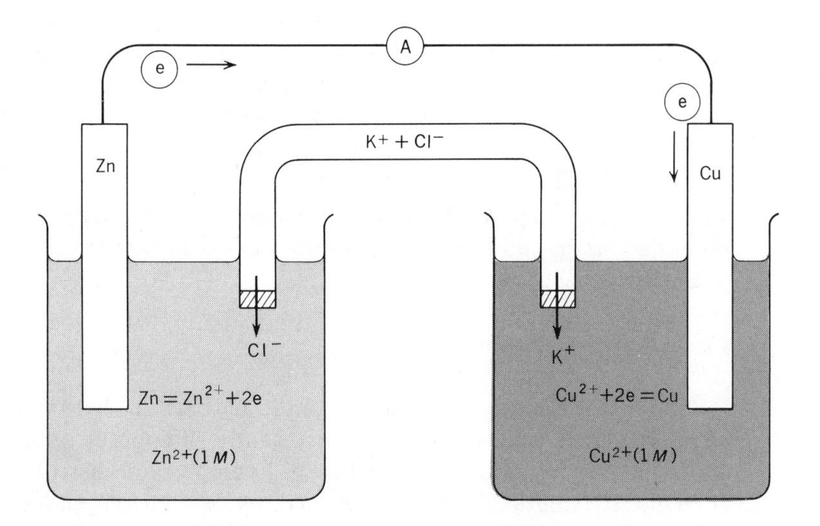

Figure 10.2 A galvanic cell.

[1] It used to be stated that single electrode potentials were unmeasurable in principle, but a hypothetical experiment has been proposed which theoretically would enable one to calculate a single electrode potential from measurements of the radiation emitted by the oscillating charges if the electrode were caused to vibrate. See I. Oppenheim, *J. Phys. Chem.*, **68**, 2959 (1964).

enters the solution find a place to go: they can travel through the external wiring to the copper electrode, where they can be consumed in the reduction of Cu^{2+}. The ammeter A in Fig. 10.2 would show an electrical current. This two-electrode system is an electrochemical cell. Instead of doing the experiment as a stunt to show a current in the ammeter, we might have used the cell to light a flashlight bulb, to run a small motor, or to do some other useful thing. A cell like this, where some of the energy released in a spontaneous chemical reaction appears as electrical energy which is available for performing work, is called a *galvanic cell.* (The contrasting type, the *electrolytic cell,* is encountered in a later chapter.)

Normally in analytical chemistry we are not interested in actually doing electrical work using galvanic cells. Rather, we are interested in the maximum voltage, which can be observed only if we do not allow the cell to run down while we are measuring it. Thus, in practice, the ammeter in Fig. 10.2 will be replaced with a voltage measuring device of very high electrical resistance so that the current drain, and the extent to which the redox reaction proceeds in the cell, are negligible.

Schematic Representation of Galvanic Cells

Instead of pictures such as Fig. 10.2, schematic representations of galvanic cells are normally employed. For example, the cell of Fig. 10.2 is written as follows:

$$\mathrm{Zn} \underset{E_l}{|} \mathrm{Zn}^{2+}(1M) \underset{E_{j1}}{|} \underset{E_{j2}}{|} \mathrm{Cu}^{2+}(1M) \underset{E_r}{|} \mathrm{Cu}$$

A vertical line (a "slant" when made with a typewriter) indicates a phase boundary across which we suppose that a potential exists. The emf of this cell may be regarded as the algebraic sum of four potential differences, one across each metal–solution interface and one at each end of the KCl salt bridge.

$$E_{\text{cell}} = E_l + E_{j1} + E_{j2} + E_r$$

E_l and E_r, the two potentials at the metal–solution interfaces, are single electrode potentials, as discussed above. The potentials designated by E_j are *liquid junction potentials* (see below).

Liquid Junction Potentials

Suppose we could prepare a quiet interface between pure water and a solution of hydrochloric acid, perhaps by very gently sliding out an ultrathin partition separating the two liquids. Immediately, hydrogen and chloride ions would move from the HCl solution into the water, and if we waited long enough, a uniform HCl concentration would prevail throughout. Stirring, of course, would speed the mixing, but we are not going to do that. Now, although both cations and anions migrate across the boundary, a potential will develop at the interface because hydrogen ions move much more rapidly than do chloride ions. Thus there is a slight tendency for a charge separation, as depicted in Fig. 10.3, with the water side of the interface positive and an excess of negative charge on the HCl side. This

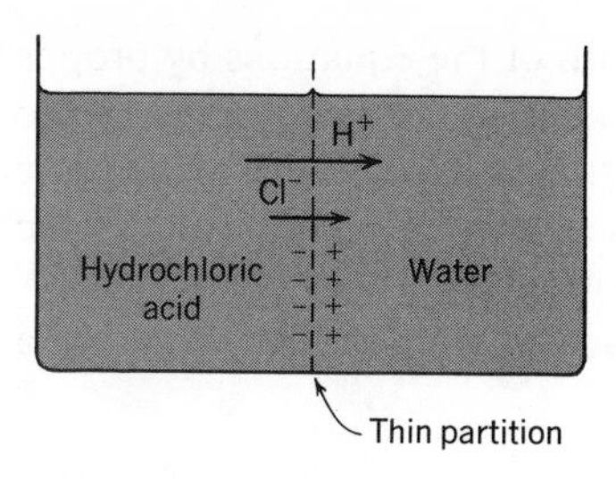

Figure 10.3 Development of a liquid junction potential.

potential, once established, will itself act upon the migrating ions, decelerating the faster ones and speeding up the slower ones. Thus we may visualize a "steady-state" situation, at least for a certain length of time, involving a constant charge separation.

A liquid junction potential may be expected to develop at any interface between two ionic solutions of different compositions as a result of the tendencies of different ions to diffuse at different rates across the boundary. It is possible to calculate the liquid junction potential for a simple case, but in complex electrolyte mixtures the value is generally unknown. Junction potentials may vary over a considerable range and can be quite appreciable in some cases, but typically they are of the order of a few millivolts (thousandths of a volt). The junction potential is minimized by using a salt bridge in which the electrolyte is quite concentrated and the cation and the anion have comparable mobilities. Potassium and chloride ions represent such a case, and salt bridges of saturated aqueous potassium chloride are widely used.

It is necessary to prevent the KCl solution from running out of the salt bridge, and for this purpose the ends are plugged with a porous material that soaks up solution and permits migration of ions but acts as a barrier to the heavy flow of a liquid stream. Small wads of cotton were used at one time; currently, thin disks of porous glass or small plugs of agar gel are often employed.

In our emf calculations, we shall neglect liquid junction potentials. This is done primarily because we usually do not know what they are, but in fact the errors we incur in this manner are negligible for most purposes.

Conventions

To represent the reaction which occurs in a galvanic cell if the cell is allowed to discharge spontaneously, proceed as follows:

1. Write the half-reaction for the right-hand electrode, with electrons on the left. For the cell in Fig. 10.2 this is

$$Cu^{2+} + 2e \rightleftharpoons Cu \qquad E_r^\circ = +0.34 \text{ V}$$

   Also record the standard potential.
2. Write the half-reaction and standard potential for the left-hand electrode in the same manner:

$$Zn^{2+} + 2e \rightleftharpoons Zn \qquad E_l^\circ = -0.76 \text{ V}$$

3. If necessary, multiply one or both of the equations by proper integers so that the number of electrons is the same in both equations. Since $n = 2$ for both the copper and the zinc couples, this is unnecessary in the present example. Do not multiply the potentials; these are experimental values which do not depend upon how the equations are written.
4. Subtract the left-hand half-reaction from the right-hand half-reaction, and also subtract the potentials:

$$Cu^{2+} + 2e \rightleftharpoons Cu \qquad E_r^\circ = +0.34\ V$$

$$Zn^{2+} + 2e \rightleftharpoons Zn \qquad E_l^\circ = -0.76\ V$$

$$Cu^{2+} + Zn \rightleftharpoons Cu + Zn^{2+} \qquad E_{cell}^\circ = +1.10\ V$$

5. The sign of the emf, E_{cell}°, gives the polarity of the right-hand electrode. In the present example, the copper electrode is positive, the zinc electrode negative.
6. The sign of the E_{cell}° indicates the direction of spontaneous reaction. If the sign is positive, the reaction goes from left to right as written. In the cell of Fig. 10.2, Cu^{2+} ion is reduced to copper metal and zinc metal is oxidized to Zn^{2+} ion when the cell discharges. A negative sign indicates, of course, that the reaction proceeds from right to left.

Students sometimes ask what would happen if they had written the cell in the reverse manner, that is, with zinc as the right-hand electrode and copper the left. If the procedure just outlined is followed, the cell potential will be -1.10 V, but the cell reaction will be reversed. Hence the conclusions regarding the direction of spontaneous reaction and polarity of the electrodes are the same as before. It is true that, for a cell such as ours, the direction of spontaneous reaction would be obvious ahead of time to most chemists. In cases where this is so, many people would probably formulate the cell in such a way that E_{cell}° came out positive. The rationale for this is that we read English from left to right, and it is convenient if the direction of spontaneous change comes out the same way we read. But there is no reason to endorse this; indeed, when we consider the effect of concentrations upon the cell emf values, we shall see that the sign of E_{cell}° may not be obvious ahead of time.

The Standard Hydrogen Electrode

The student will have noticed that, after carefully explaining why single electrode potentials could not be measured, we inserted actual numbers for E_l and E_r in the above section. We must now explain what we mean by these.

If we measure the potential of the cell in Fig. 10.2, we find that the value is 1.10 V. We can also determine experimentally that the zinc electrode is negative with respect to copper, meaning that electrons will flow from zinc to copper if we short-circuit the cell as discussed above.

The quantity we have measured is the *difference* in potential between the zinc and copper redox systems. We do not know the absolute value of either single electrode potential; in order to measure the tendency of one redox couple to lose or gain electrons, we must introduce another couple for the purpose of comparison. Since definite potential values are convenient to use, the chemist arbitrarily assigns a value to a certain single electrode. All other single electrodes are then referred to this standard by measuring the emf of galvanic cells in which one electrode is the standard and the other the electrode in question.

The redox couple selected as the reference is H_2–H^+:

$$2H^+ + 2e \rightleftharpoons H_2$$

This couple is assigned a potential of exactly zero volts at all temperatures when the pressure of hydrogen gas is 1 atmosphere and the hydrogen ion activity is unity. Physically, the electrode is somewhat as suggested schematically in Fig. 10.4. A platinum surface, rough so as to have a large area, provides an electrical connection to the external circuit and serves as a catalyst for the combination of H atoms formed in the electron transfer step.

If we set up a cell in which one electrode is Zn–Zn^{2+} (1 *M*) and the other is the standard hydrogen electrode, we find that the potential at 25°C is 0.76 V. We also can determine that the polarity of the zinc electrode is negative, hydrogen positive. If we set up another cell consisting of the Cu–Cu^{2+} (1 *M*) electrode and the standard hydrogen, we find that the potential is 0.34 V, with copper positive and hydrogen negative. We say that the standard potential of the Zn–Zn^{2+} couple is −0.76 V and that of the Cu–Cu^{2+} couple is 0.34 V referred to the hydrogen electrode. By international agreement the electrode reactions are written as reduction (left to right):

$$Zn^{2+} + 2e \rightleftharpoons Zn \qquad E^\circ = -0.76 \text{ V}$$

$$2H^+ + 2e \rightleftharpoons H_2 \qquad E^\circ = 0.00 \text{ V}$$

$$Cu^{2+} + 2e \rightleftharpoons Cu \qquad E^\circ = +0.34 \text{ V}$$

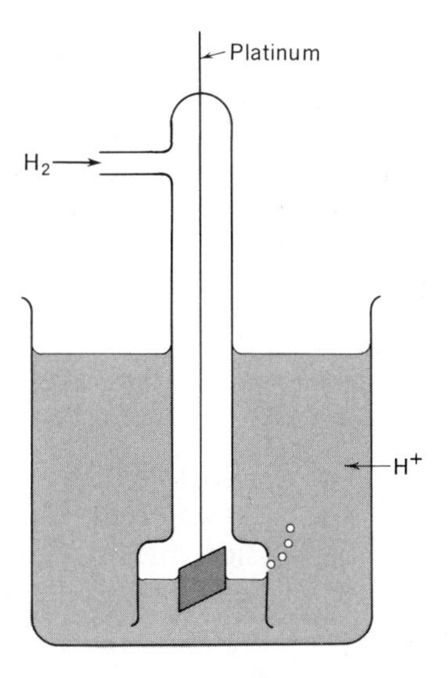

Figure 10.4 Hydrogen electrode.

The voltages are given positive or negative signs in accordance with experimentally determined polarities with respect to hydrogen. Zinc is more negative than hydrogen; this means that the reaction

$$Zn + 2H^+ \longrightarrow Zn^{2+} + H_2$$

will occur spontaneously rightward if both H^+ and Zn^{2+} are at unit activity; zinc loses electrons to hydrogen in this reaction. On the other hand, hydrogen is more negative than copper, the reaction

$$H_2 + Cu^{2+} \longrightarrow 2H^+ + Cu$$

occurring spontaneously rightward if both ions are at unit activity and the pressure of H_2 is 1 atm.

Voltage Measurements

As noted on page 242, we usually desire, in analytical chemistry, to measure the voltage of a galvanic cell under conditions where the cell actually does no electrical work. That is, we wish to measure the *tendency* of the cell reaction to proceed without allowing it in fact to proceed, at least appreciably. In other words, we do not want the measurement process to lower the voltage we are trying to measure. This generally precludes the use of simple voltmeters where the galvanic cell itself provides the power to move the galvanometer coil and pointer. A simple device called a potentiometer used to be employed to measure voltages of galvanic cells. Essentially, the cell voltage was opposed by an external, variable voltage provided by a battery, and a balance point was determined in a manner that drew very little current from the test cell. Potentiometers worked very well in many cases but were unsuitable for measurements of cells involving very high internal resistances such as are encountered in *p*H measurements with glass electrodes (Chapter 12). Electronic voltmeters with very high input resistances were developed after the science of electronics came of age. Not only did these work with cells of high resistance; they also became so inexpensive and so accurate that they have replaced the potentiometer even for applications where the latter served well. The circuitry of these meters is beyond the scope of this textbook; in principle, the voltage to be measured *controls*, but does not *power*, the read-out device through a circuit that draws currents of the order of 10^{-12} amperes or even less from the test cell. Electronic voltmeters with solid-state components that provide direct digital read-out are now very common.

THE NERNST EQUATION

The potential of a galvanic cell depends upon the activities of the various species which undergo reaction in the cell. The equation which expresses this relationship is called the Nernst equation, after the physical chemist, Nernst, who in 1889 first used the equation to express the relation between the potential of a metal-metal ion electrode and the concentration of the ion in solution.

In a chemical reaction such as

$$aA + bB \rightleftharpoons cC + dD$$

the change in free energy is given by the equation

$$\Delta G = \Delta G^\circ + 2.3RT \log \frac{a_C^c \times a_D^d}{a_A^a \times a_B^b}$$

where ΔG° is the free energy change when all the reactants and products are in their standard states (unit activity). R is the gas constant, 8.314 joules/deg-mol, and T is the absolute temperature.

The free energy change, or work done, by driving Avogadro's number of electrons through a voltage E is $(Ne)E$, where N is Avogadro's number and e is the charge on the electron. The product Ne is 96,500 coulombs, called 1 faraday, or F. Hence

$$\Delta G = -nFE$$

where n is the number of moles of electrons involved in the reaction. If all reactants and products are in their standard states, this becomes

$$\Delta G^\circ = -nFE^\circ$$

Hence

$$-nFE = -nFE^\circ + 2.3RT \log \frac{[C]^c[D]^d}{[A]^a[B]^b}$$

where concentrations are substituted for activities. This can be written as

$$E = E^\circ - \frac{2.3RT}{nF} \log \frac{[C]^c[D]^d}{[A]^a[B]^b}$$

at 298° K the equation becomes

$$E = E^\circ - \frac{0.059}{n} \log \frac{[C]^c[D]^d}{[A]^a[B]^b}$$

This is the form in which we commonly use the Nernst equation. Note that at equilibrium, $E = 0$, $\Delta G = 0$, and the logarithmic term is the equilibrium constant. Hence

$$\Delta G^\circ = -2.3RT \log K$$

or

$$E^\circ = \frac{0.059}{n} \log K$$

If we know the standard potentials of two redox couples, we can calculate the equilibrium constant for the reaction between the couples using the preceding equation. We can then judge whether the reaction goes sufficiently to completion to be useful in a titrimetric procedure. Before considering titrations, we will illustrate the application of the Nernst equation, as well as other points covered thus far.

Example 1. A cell is set up as follows:

$$\text{Fe}|\text{Fe}^{2+}(a = 0.1)||\text{Cd}^{2+}(a = 0.001)|\text{Cd}$$

(a) Write the cell reaction. (b) Calculate the voltage of the cell, the polarity of the electrodes, and the direction of spontaneous reaction. (c) Calculate the equilibrium constant of the cell reaction.

(a) The electrode reactions and standard potentials are

$$\begin{array}{llr} & \text{Cd}^{2+} + 2e \rightleftharpoons \text{Cd} & E_r^\circ = -0.40\ \text{V} \\ & \text{Fe}^{2+} + 2e \rightleftharpoons \text{Fe} & E_l^\circ = -0.44\ \text{V} \\ \hline \text{Subtracting,} & \text{Fe} + \text{Cd}^{2+} \rightleftharpoons \text{Fe}^{2+} + \text{Cd} & E_{\text{cell}}^\circ = +0.04\ \text{V} \end{array}$$

(b) The cell potential can be calculated from the single electrode potentials:

$$E_r = -0.40 - \frac{0.059}{2} \log \frac{1}{0.001} = -0.49\ \text{V}$$

$$E_l = -0.44 - \frac{0.059}{2} \log \frac{1}{0.1} = -0.47\ \text{V}$$

Thus

$$E_r - E_l = -0.02\ \text{V}$$

Alternatively, the cell potential can be evaluated from the expression

$$E_{\text{cell}} = E_{\text{cell}}^\circ - \frac{0.059}{2} \log \frac{a_{\text{Fe}^{2+}}}{a_{\text{Cd}^{2+}}}$$

$$E_{\text{cell}} = +0.04 - \frac{0.059}{2} \log \frac{0.1}{0.001}$$

$$E_{\text{cell}} = +0.04 - 0.06 = -0.02\ \text{V}$$

Therefore, the cell reaction, as written above, tends to occur spontaneously from right to left at the given activities. The cadmium electrode is negative, the iron positive. Note that if both ions are at unit activity, $E_{\text{cell}}^\circ = +0.04$ V, and the direction is from left to right. The polarities of the electrodes are reversed also.

(c) The equilibrium constant is given by

$$E_{\text{cell}}^\circ = \frac{0.059}{n} \log K$$

$$+0.04 = \frac{0.059}{2} \log K$$

$$\log K = 1.36$$

$$K = 23$$

As a further illustration, consider the following example.

Example 2. Calculate the potential of the following cell, giving the polarities of the electrodes and the direction of spontaneous reaction. Calculate the equilibrium constant of the cell reaction.

$$\text{Pt, } H_2(0.9 \text{ atm})|H^+(0.1\ M)|\,|KCl(0.1\ M), \text{AgCl}|\text{Ag}$$

The electrode reactions are

$$\begin{array}{ll} 2AgCl + 2e \rightleftharpoons 2Ag + 2Cl^- & E_r^\circ = 0.22 \text{ V} \\ 2H^+ + 2e \rightleftharpoons H_2 & E_l^\circ = 0.00 \text{ V} \\ \hline 2AgCl + H_2 \rightleftharpoons 2H^+ + 2Ag + 2Cl^- & E_{cell}^\circ = +0.22 \text{ V} \end{array}$$

$$E_{cell} = +0.22 - \frac{0.059}{2} \log \frac{(a_{H^+})^2 (a_{Ag})^2 (a_{Cl^-})^2}{(a_{AgCl})^2 (a_{H_2})}$$

In accordance with our conventions regarding activities:

a_{AgCl} = 1, since AgCl is a pure solid
a_{H_2} = 0.9, since this is the partial pressure of the gas in atmospheres
a_{H^+} = 0.1, since this is a soluble electrolyte
a_{Ag} = 1, since silver is a pure solid
a_{Cl^-} = 0.1, since this is a soluble electrolyte

Substituting these values and solving gives

$$E_{cell} = +0.34 \text{ V}$$

Hence the silver–silver chloride electrode is positive and the reaction is spontaneous left to right.

The equilibrium constant is given by

$$0.22 = \frac{0.059}{2} \log K$$

$$\log K = 7.46$$

$$K = 2.9 \times 10^7$$

TYPES OF ELECTRODES

A metal electrode whose potential responds to the activity of its metal ion, such as the zinc or copper electrodes discussed earlier, is sometimes called an "electrode of the first kind." This is not, however, the only type. In fact, not all electrodes involve electroactive metals. Several other types are described in Chapter 12. At this point it is useful to discuss electrodes of the second kind because they provide the commonest laboratory reference electrodes.

Laboratory Reference Electrodes

The hydrogen electrode is inconvenient for routine, practical measurements in the laboratory. It requires a tank of compressed gas, which is heavy and awkward, explosive mixtures of hydrogen and air may be formed, and the catalytic platinum surface is easily poisoned, that is, contaminated with adsorbed substances that inhibit catalytic activity. Thus, in the laboratory, more convenient reference electrodes are commonly employed for measuring the potentials of other half-cells. The potentials of these reference electrodes have themselves been measured against the standard hydrogen electrode. Regardless of what reference electrode is actually employed, it is customary to report any potential as though it had been measured against the standard hydrogen electrode. The commonest reference electrodes are the calomel and the silver–silver chloride electrodes.

The Calomel Electrode

One form of the calomel electrode is shown in Fig. 10.5. External contact is made through a wire from a pool of mercury. The mercury is in contact with a moist paste prepared by intimately mixing mercury, mercury(I) chloride (calomel), and aqueous KCl solution. As with an electrode of the first kind, the potential is established by the simple redox couple

$$Hg_2^{2+} + 2e \rightleftharpoons 2Hg$$

[It may be recalled that the mercury(I) ion is a dimeric species; i.e., we are dealing with Hg(I), but the ion in solution is Hg_2^{2+}.] According to the Nernst equation, the potential is given by

$$E = E^\circ - \frac{0.059}{2} \log \frac{1}{[Hg_2^{2+}]}$$

But there is more to the story, because Hg_2^{2+} is involved in another equilibrium: Hg_2Cl_2 is a slightly soluble salt, for which we may write a solubility product constant, K_{sp} (Chapter 9).

$$K_{sp} = [Hg_2^{2+}][Cl^-]^2$$

Rearranging, we obtain

$$[Hg_2^{2+}] = \frac{K_{sp}}{[Cl^-]^2}$$

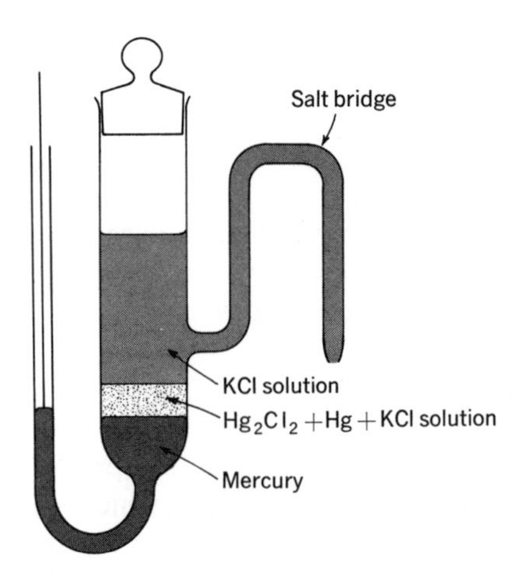

Figure 10.5 Calomel electrode.

and substituting into the Nernst equation above yields

$$E = E° - \frac{0.059}{2} \log \frac{[Cl^-]^2}{K_{sp}}$$

Since $E°$ and K_{sp} are constants, it is seen that the potential reflects the concentration of Cl^-.

Most commonly, the solution is saturated with KCl; the electrode is then called the saturated calomel electrode or SCE, and its single electrode potential is +0.2458 V at 25°C, as determined against a standard hydrogen electrode. If the KCl solution is exactly 1 *M*, the potential is +0.2847 V; this is sometimes called a standard or normal calomel electrode.

The student should note that we may write the Nernst equation as we did above, in which case $E°$ is the value for the $Hg\text{–}Hg_2^{2+}$ couple, +0.79 V. Alternatively, we can formulate the half-reaction as

$$Hg_2Cl_2 + 2e \rightleftharpoons 2Hg + 2Cl^-$$

whereupon the $E°$ value becomes +0.2847 V. Changing this value to +0.2458 V (paragraph above) allows for the chloride concentration in a saturated KCl solution.

Example. Using the potentials given in the preceding discussion, calculate the solubility product constant for mercury(I) chloride and the molar solubility of KCl.

$$E = E° - \frac{0.059}{2} \log \frac{1}{[Hg_2^{2+}]}$$

If we choose $E = +0.28$ V, then $[Cl^-] = 1$ *M*. The mercury(I) ion concentration which obtains in this solution is found as follows:

$$+0.28 = +0.79 - \frac{0.059}{2} \log \frac{1}{[Hg_2^{2+}]}$$

$$[Hg_2^{2+}] = 5.01 \times 10^{-18}$$

$$K_{sp} = [Hg_2^{2+}][Cl^-]^2 = 5.01 \times 10^{-18} \times 1^2 = 5.01 \times 10^{-18}$$

For the SCE:

$$+0.25 = +0.79 - \frac{0.059}{2} \log \frac{[Cl^-]^2}{5.01 \times 10^{-18}}$$

$$[Cl^-] = 3.2\ M = \text{solubility}$$

Calomel electrodes may be fabricated in a variety of sizes and shapes. Some, of the sort shown in Fig. 10.5, are quite large, while some are small enough to be inserted into blood vessels of experimental animals through syringe needles. It should be emphasized that the potential of an electrode does not depend upon its size. Likewise, the liquid junction between the test solution and the KCl solution in the electrode may be physically established in various ways. Sometimes it involves an agar plug or a sintered glass frit, sometimes a pinhole, sometimes a small, wet fiber; the point is to permit the migration of ions without allowing excessive solution to pour across the interface. Although contamination from calomel electrodes is inconsequential for most purposes, in critical cases it must be remembered that a small amount of KCl may leak into the test solution. Similarly, a test solution with a large hydrostatic head may contaminate the contents of a calomel electrode.

The Silver–Silver Chloride Electrode

This electrode, also widely used, is similar in principle to the calomel electrode. Usually, a silver wire is coated with a retentive layer of silver chloride by anodizing the silver in a chloride solution. In the finished electrode, the coated wire dips into a potassium chloride solution. As in the calomel case, the potential is basically established by a metal–metal ion electron transfer reaction

$$Ag^+ + e \rightleftharpoons Ag$$

but the concentration of silver ion is, in turn, governed by the chloride concentration via the K_{sp} for AgCl. The overall half-reaction may thus be written

$$AgCl + e \rightleftharpoons Ag + Cl^-$$

For chloride ion at unit activity, the standard potential is +0.2221 V at 25°C. The following example illustrates the use of a reference electrode.

Example. (a) The potential of a cell made up of an electrode of unknown potential and a standard calomel electrode is 1.04 V, with the calomel electrode positive. Calculate the potential of the unknown electrode referred to the hydrogen electrode.

If we consider our cell as having the unknown electrode on the left, calomel on the right, we must give a positive sign to the potential. That is,

$$E_{cell} = E_r - E_l$$

$$+1.04 = +0.28 - E_l$$

$$E_l = -0.76 \text{ V}$$

(b) In a second measurement with another unknown electrode, the potential of the cell is found to be 0.06 V, with the calomel negative. Calculate the potential of the unknown electrode referred to hydrogen.

In this case, if the calomel is the right-hand electrode as before, the potential of the cell must be written −0.06 V. Hence

$$-0.06 = 0.28 - E_l$$

$$E_l = +0.34 \text{ V}$$

The relation between these potentials is shown schematically in Fig. 10.6.

Inert Electrodes

So far most of our examples of redox systems have consisted of metals in equilibrium with their ions. There are many important examples of redox systems

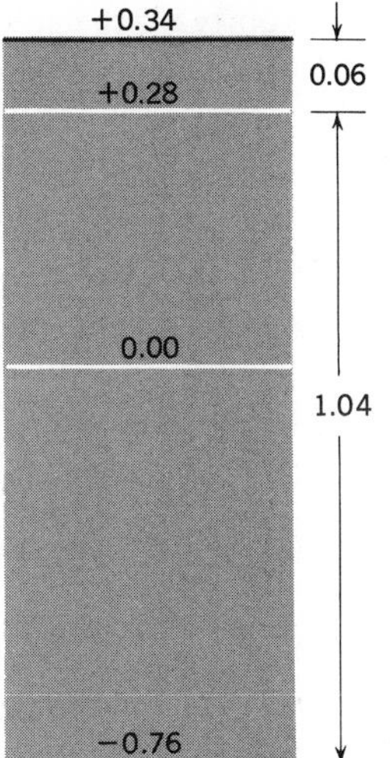

Figure 10.6 Potential relations.

involving only different oxidation states of ions in solution. A familiar example is the $Fe^{3+} - Fe^{2+}$ system,

$$Fe^{3+} + e \rightleftharpoons Fe^{2+}$$

A solution of Fe^{3+} and Fe^{2+} ions is a possible source of electrons, but some metal must be inserted into the solution to act as a conductor. A metal such as platinum is normally used because it is not easily attacked by most solutions. The platinum is said to be an *inert* electrode, since it does not enter into the reaction.

A cell made up of the Fe^{3+}–Fe^{2+} and Ce^{4+}–Ce^{3+} systems is written as

$$\mathrm{Pt|Fe^{3+}}(xM) + \mathrm{Fe^{2+}}(yM)||\mathrm{Ce^{4+}}(aM) + \mathrm{Ce^{3+}}(bM)|\mathrm{Pt}$$

The student should confirm that the standard potential of this cell as written is +0.84 V, and that Ce^{4+} ion oxidizes Fe^{2+} ion spontaneously if the concentration of each reactant is 1 *M*.

EQUILIBRIUM CONSTANTS FROM STANDARD POTENTIALS

In the discussion of the Nernst equation (page 247), we pointed out the relationship between $E°$ for a cell and the equilibrium constant for the cell reaction. In the treatment of calomel electrodes, we used potentials to calculate K_{sp} for Hg_2Cl_2, one type of equilibrium constant. Let us formulate what we did on page 251 a little differently (although it has to amount to the same thing):

$$\begin{array}{ll} Hg_2Cl_2 + 2e \rightleftharpoons 2Hg + 2Cl^- & E° = +0.28\ V \\ Hg_2^{2+} + 2e \rightleftharpoons 2Hg & E° = +0.79\ V \\ \hline Hg_2Cl_2 \rightleftharpoons Hg_2^{2+} + 2Cl^- & E°_{cell} = -0.51\ V \end{array}$$

$$E = E° - \frac{0.059}{2}\log\,[Hg_2^{2+}][Cl^-]^2$$

(In the Nernst equation above, we take unity for the activity of solid Hg_2Cl_2.) Now at equilibrium, $E = 0$ and $[Hg_2^{2+}][Cl^-]^2 = K_{sp}$. Thus

$$0 = -0.51 - \frac{0.059}{2}\log K_{sp}$$

$$\log K_{sp} = -17.3$$

$$K_{sp} = 5 \times 10^{-18}$$

If we had not just seen these $E°$ values in the previous section and had been asked to start from "scratch" to calculate K_{sp}, what would we have done? It is clear that we would have searched through the table of half-reactions and $E°$ values in Appendix I to find two half-reactions which, when subtracted, yielded the reaction

$$Hg_2Cl_2 \rightleftharpoons Hg_2^{2+} + 2Cl^-$$

This may be considered "trial-and-error," although one may become more adept at it with practice. Let us consider another case.

Example. Calculate K_w, the ion-product constant of water, from data in Appendix I, Table 4.

We seek two half reactions which, when subtracted, yield

$$H_2O \rightleftharpoons H^+ + OH^-$$

After searching the table (and perhaps making a false start or two if inexperienced), we come up with:

$$\begin{array}{ll} H_2O + e \rightleftharpoons \frac{1}{2}H_2 + OH^- & E° = -0.83\text{ V} \\ H^+ + e \rightleftharpoons \frac{1}{2}H_2 & E° = 0.00 \\ \hline H_2O \rightleftharpoons H^+ + OH^- & E°_{cell} = -0.83\text{ V} \end{array}$$

$$E = -0.83 - 0.059\log[H^+][OH^-]$$

At equilibrium, $E = 0$ and $[H^+][OH^-] = K_w$. Thus

$$\log K_w = \frac{-0.83}{0.059} = -14.0$$

$$K_w = 1 \times 10^{-14}$$

Any number of examples could be given, of course, but the principle is already clear: having a table of standard potentials is tantamount to the possession of a great many equilibrium constants.

FORMAL POTENTIALS

Consider an oxidant and a reductant that constitute a redox couple, say, Fe^{3+} and Fe^{2+}:

$$Fe^{3+} + e \rightleftharpoons Fe^{2+}$$

For the single electrode potential we have

$$E = E^\circ - 0.059 \log \frac{a_{Fe^{2+}}}{a_{Fe^{3+}}}$$

When $a_{Fe^{2+}} = a_{Fe^{3+}}$, then $E = E^\circ$; that is, the potential is the standard potential, and presumably we could measure this against a reference electrode. But, although this appears to be simple enough, in fact it is not easy experimentally to prepare a solution in which we know that the two activities are equal, and determining a true standard potential is often difficult.

In our example here, hydroxides of the metal ions are very insoluble, and to prevent the hydrolysis of Fe^{3+} and Fe^{2+} it is necessary to acidify the solution. But we then introduce the possibility of complex formation between the metal ions and the anion of the acid. If, for example, we use HCl, complexes such as $FeCl^{2+}$, $FeCl_2{}^+$, etc. (up to $FeCl_6{}^{3-}$) may form in varying degrees depending upon the chloride concentration. These iron(III) complexes are stronger than the corresponding iron(II) complexes, $FeCl^+$, etc. Thus complexing with Cl^- tends to stabilize the Fe^{3+} state, in effect making Fe^{3+} a weaker oxidant or Fe^{2+} a stronger reductant than would be the case in the absence of the complexing agent.

In the kinds of solutions encountered in analytical chemistry, there is often complexing such as that described above, and electrolyte concentrations are often such that activity coefficients can scarcely be guessed at. Fortunately, however, what we really want to know about redox reagents is how potent they are in these very solutions, not what they might be in some hypothetical state, perhaps at infinite dilution. Thus it is convenient to write our Nernst equation as follows:

$$E = E^{\circ\prime} - 0.059 \log \frac{C_{Fe^{2+}}}{C_{Fe^{3+}}}$$

where the C terms are analytical concentrations and $E^{\circ\prime}$ is called a *formal potential.* For example, if we started with an iron(III) solution in 0.1 M HCl, and reduced exactly half of the iron to Fe^{2+}, then the potential E would be equal to the formal potential, $E^{\circ\prime}$, for this redox couple in this particular solution. Changing the HCl concentration, or switching from HCl to, say, sulfuric acid, would change $E^{\circ\prime}$.

$E^{\circ\prime}$ values are used like standard potentials in calculations involving the Nernst equation. Calculated and experimental E values will agree better if $E^{\circ\prime}$ is used instead of E°, provided that the formal potential is the correct one for the particular conditions that obtain.

Formal potentials are often employed in connection with organic compounds, especially by biochemists. Frequently, hydrogen ion participates in the overall redox reaction:

$$\text{Ox} + m\text{H}^+ + ne \rightleftharpoons \text{Red}$$

In such a case, H^+ appears in the Nernst equation:

$$E = E^\circ - \frac{0.059}{n} \log \frac{a_{\text{Red}}}{a_{\text{Ox}} \times a_{\text{H}^+}^m}$$

Now, in order that $E = E°$, not only must the activities of the oxidant and the reductant be the same but the hydrogen ion activity must be unity. This represents a very acidic solution, of the order of 1 M in strong acid. Because many compounds of biological importance decompose under such conditions, it is often impossible to measure standard potentials directly. Moreover, the potential at a physiological pH is of far more interest anyway. Thus biochemistry books often list $E^{\circ\prime}$ values for redox couples at pH 7, and although the term "formal potentials" may not be used, this is what they are.

Because it is often difficult to measure standard potentials accurately, many of the values in the literature which are called standard potentials are, in fact, formal potentials. A critical evaluation of the original data is often required in order to determine exactly what the investigators actually measured, regardless of what they chose to call it. Table 4, Appendix I, doubtless contains both standard and formal potentials. For many purposes in analytical chemistry, calculations based upon these values are sufficiently accurate; for example, in this chapter, uncertainty regarding the exact significance of the potentials employed would not affect conclusions about the feasibility of titrations except for cases which were borderline in any event. A short list of formal potentials is given in Table 5, Appendix I.

OTHER APPROXIMATIONS

The mechanism of many redox reactions is complex, and we do not always know the exact nature of the reaction which determines the potential at an electrode surface. For example, the overall reaction for the reduction of dichromate ion to chromium(III) ion is

$$Cr_2O_7^{2-} + 14H^+ + 6e \rightleftharpoons 2Cr^{3+} + 7H_2O$$

The standard potential, +1.33 V, was obtained indirectly, not from galvanic cell measurements. The reaction above simply represents the correct stoichiometry. It is thought that the reaction proceeds in steps via an unstable intermediate which is converted into the products. The Nernst expression

$$E = 1.33 - \frac{0.059}{6} \log \frac{[Cr^{3+}]^2}{[Cr_2O_7^{2-}][H^+]^{14}}$$

is not followed. Actually, the potential is practically independent of the concentration of chromium(III) ions.

Similar behavior is found with other complex redox systems, the permanganate–manganese(II) system being one important example. With reactions which involve a large number of hydrogen ions, it is normally found that the potential is strongly dependent upon the hydrogen ion concentration but that the dependence cannot be predicted from the coefficients in the balanced equations. Hence calculations of the potentials of such systems on the basis of these equations give incorrect results. In many cases, however, the error is small enough that conclusions regarding the feasibility of titrations may still be valid.

TITRATION CURVES

Titration of Iron(II) with Cerium(IV) Ion

Let us now consider the following example.

Example. 5.0 mmol of an iron(II) salt is dissolved in 100 ml of sulfuric acid solution and titrated with 0.10M cerium(IV) sulfate. Calculate the potential of an inert electrode in the solution at various intervals in the titration and plot a titration curve. Use 0.68 V as the formal potential of the Fe^{2+}–Fe^{3+} system in sulfuric acid and 1.44 V for the Ce^{3+}–Ce^{4+} system.

(a) *Start of titration.* The potential is determined by the Fe^{2+}–Fe^{3+} ratio, that is,

$$E = 0.68 - 0.059 \log \frac{[Fe^{2+}]}{[Fe^{3+}]}$$

However, we do not know the iron(III) ion concentration, this being dependent upon how the iron(II) salt was prepared, how much has been oxidized by air, etc. Let us assume that no more than 0.1% of the iron remains in the +3 state, that is, the iron(II) to iron(III) ion ratio is 1000 : 1. For such a condition the potential can be calculated:

$$E = 0.68 - 0.059 \log 1000$$

$$E = 0.50 \text{ V}$$

[If the iron(III) ion concentration were actually zero, what would be the value of the potential?]

(b) *10-ml cerium(IV) solution added.* We now have mixed our two redox systems and allowed them to react and reach equilibrium,

$$Fe^{2+} + Ce^{4+} \rightleftharpoons Fe^{3+} + Ce^{3+}$$

We can calculate the potentials from the expression for either redox system, that is,

$$E = 0.68 - 0.059 \log \frac{[Fe^{2+}]}{[Fe^{3+}]}$$

$$E = 1.44 - 0.059 \log \frac{[Ce^{3+}]}{[Ce^{4+}]}$$

The system is at equilibrium; that is, each redox couple has the same potential.[2] It is simpler at this stage of the titration to use the expression for the iron(II)–iron(III) system since we can estimate the concentrations of these two ions more readily than

[2] Some students ask why this is so. They have become accustomed to calculating the potential of a galvanic cell and think that the potentials of the two half-cells are different. They usually are, but at equilibrium the galvanic cell potential is zero, meaning that the two single electrodes have the same potential, here designated by E. In the titration we are placing the two reagents in the same container and allowing them to come to equilibrium. The solution has only one potential and it can be calculated from either redox system. This is the potential you would observe if you made your titration solution one half of the galvanic cell, and the other electrode was the standard hydrogen electrode.

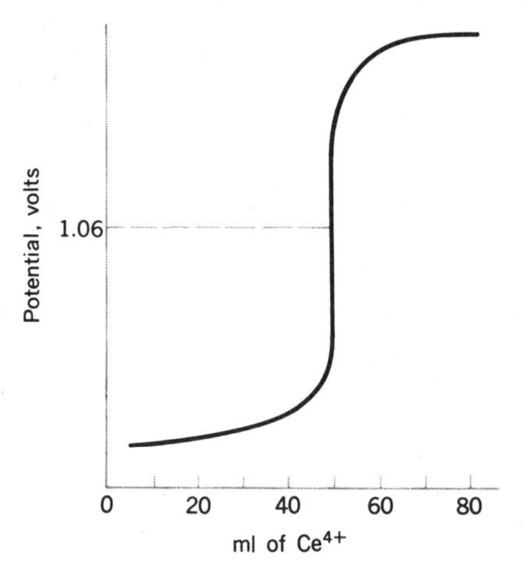

Figure 10.7 Titration of iron(II) with cerium(IV) ion.

that of the ceric ion as seen below:

$$[Fe^{2+}] = \left\{\frac{4.0}{110} + x\right\} \text{mmol/ml}$$

$$[Fe^{3+}] = \left\{\frac{1.0}{110} - x\right\} \text{mmol/ml}$$

$$[Ce^{3+}] = \left\{\frac{1.0}{110} - x\right\} \text{mmol/ml}$$

$$[Ce^{4+}] = x \text{ mmol/ml}$$

The value of x can be calculated, of course, from the equilibrium constant of the reaction. Assuming that the reaction goes well to completion, x is small and may be disregarded in estimating the iron(III) and iron(II) ion concentrations. Hence

$$E = 0.68 - 0.059 \log \frac{4.0/110}{1.0/110}$$

or

$$E = 0.64 \text{ V}$$

(Note that the volume term cancels; that is, the potential is independent of volume.)

Values of the potential of all other points before the equivalence point are calculated in the same manner. In Table 10.1 is a list of such values, and these are plotted in Fig. 10.7. Note that the potential rises slowly in the earlier stages of the titration and begins to increase more rapidly as the equivalence point is approached.

(c) *Equivalence point.* This point is reached when 50 ml of cerium(IV) solution are added. The concentrations of the reactants and products are then

$$[Fe^{2+}] = [Ce^{4+}] = x$$

$$[Fe^{3+}] = [Ce^{3+}] = \frac{5.0}{150} - x$$

TABLE 10.1 Redox Potential during Titration of 5.0 mmol of Fe^{2+} with 0.10 *M* Ce^{4+}

Ce^{4+}, ml	Fe^{2+} unoxidized, mmol	Fe^{2+} oxidized, %	E, V
0.00	5.0	0	—
10.0	4.0	20	+0.64
20.0	3.0	40	0.67
30.0	2.0	60	0.69
40.0	1.0	80	0.72
45.0	0.50	90	0.74
49.50	0.05	99	0.80
49.95	0.005	99.9	0.86
50.0	—	100	1.06
	Ce^{4+} excess, mmol		
50.05	0.005		+1.26
50.50	0.05		1.32
51.0	0.10		1.34
55.0	0.50		1.38
60.0	1.0		1.40

If either of the following expressions is employed, it is necessary to evaluate x, using the equilibrium constant, in order to calculate the potential:

$$E = 0.68 - 0.059 \log \frac{[Fe^{2+}]}{[Fe^{3+}]}$$

$$E = 1.44 - 0.059 \log \frac{[Ce^{3+}]}{[Ce^{4+}]}$$

Notice, however, that if the two equations are added, giving

$$2E = 2.12 - 0.059 \log \frac{[Fe^{2+}][Ce^{3+}]}{[Fe^{3+}][Ce^{4+}]}$$

the logarithmic term is zero, since at the equivalence point

$$[Fe^{2+}] = [Ce^{4+}] \qquad \text{and} \qquad [Fe^{3+}] = [Ce^{3+}]$$

Hence

$$2E = 2.12 \qquad \text{or} \qquad E = 1.06 \text{ V}$$

For any reaction in which the number of electrons lost by the reductant is the same as the number gained by the oxidant, the potential at the equivalence point is simply the arithmetic mean of the two standard potentials:

$$E_{\text{eq pt}} = \frac{E_1^\circ + E_2^\circ}{2}$$

(d) *60-ml cerium(IV).* After the addition of 60 ml of titrant, the concentrations are

$$[Fe^{2+}] = x$$

$$[Fe^{3+}] = \left\{\frac{5.0}{160} - x\right\} \text{mmol/ml}$$

$$[Ce^{3+}] = \left\{\frac{5.0}{160} - x\right\} \text{mmol/ml}$$

$$[Ce^{4+}] = \left\{\frac{1.0}{160} + x\right\} \text{mmol/ml}$$

It is now the iron(II) ion concentration that must be evaluated from the equilibrium constant. Hence it is more convenient to employ the expression for the cerium(IV)–cerium(III) system:

$$E = 1.44 - 0.059 \log \frac{[Ce^{3+}]}{[Ce^{4+}]}$$

Noting that x is small, we write

$$E = 1.44 - 0.059 \log \frac{5.0/160}{1.0/160} = 1.40 \text{ V}$$

Other values beyond the equivalence point are calculated in the same manner. The curve is plotted in Fig. 10.7, where its similarity to a strong-base titration curve may be seen. A large change in potential occurs in the vicinity of the equivalence point of the titration. The data are given in Table 10.1.

Titration of Tin(II) with Cerium(IV) Ion

Let us consider now an example where the reductant loses two electrons while the oxidant gains one.

Example. Calculate the potential at the equivalence point in the titration of tin(II) ion with cerium(IV) ion:

$$Sn^{2+} + 2Ce^{4+} \rightleftharpoons Sn^{4+} + 2Ce^{3+}$$

The potential is given by either of the following expressions:

$$E = 0.15 - \frac{0.059}{2} \log \frac{[Sn^{2+}]}{[Sn^{4+}]}$$

or

$$E = 1.44 - 0.059 \log \frac{[Ce^{3+}]}{[Ce^{4+}]}$$

Multiplying the first equation by 2 and adding it to the second gives

$$3E = 1.74 - 0.059 \log \frac{[Sn^{2+}][Ce^{3+}]}{[Sn^{4+}][Ce^{4+}]}$$

The logarithmic term is zero since at the equivalence point[3]

$$[Ce^{4+}] = 2[Sn^{2+}] \qquad \text{and} \qquad [Ce^{3+}] = 2[Sn^{4+}]$$

Hence

$$E = \frac{1.74}{3} = 0.58 \text{ V}$$

For the case in which one redox system gains or loses one electron and the other loses or gains two, the potential of the equivalence point is

$$E = \frac{E_1^\circ + 2E_2^\circ}{3}$$

where E_1° is the standard potential of the first system and E_2° that of the second. It should be noted that the titration curve is not symmetrical about the equivalence point in this case as it is in the titration of iron(II) with cerium(IV) ion. This is always true when the two redox systems exchange a different number of electrons per molecule.

A completely general expression for the potential at the equivalence point can be derived in the same manner that we have employed here (see Problem 8).

Titration of Other Redox Couples

Figure 10.8 shows the variation of potential with fraction of reagent in the oxidized form for several redox couples. The curves for couples with oxidation potentials greater than 1 V have been plotted on a scale to the right of those couples with potentials less than 1 V. This is done to emphasize the fact that a single titration curve is a combination of two of the branches plotted here.

In comparing Fig. 10.8 with the titration curves for acid-base reactions, one should keep in mind the fact that the acid-base properties of our solvent, water, are quite different from the redox properties. As noted earlier, we do not ordinarily encounter a leveling effect in redox titrations.

Several points about Fig. 10.8 should be noted.

1. The change in potential at the equivalence point in the titration of, say, iron(II) ion depends upon the oxidant used. The oxidants are not leveled by water as pointed out before.
2. The shape of a curve depends upon the value of n, the number of electrons gained or lost by the oxidant or reductant. Note that the iron(II)–iron (III) curve ($n = 1$) is steeper than the tin(II)–tin(IV) curve ($n = 2$). Also note the flatness of the permanganate and dichromate curves, where n is 5 and 6, respectively. The shape of a titration curve will obviously be determined by that of the two halves which make up the curve.

[3] Remember that the concentrations are molarities. Obviously, the number of equivalents of cerium(IV) ion is the same as that of tin(II) ion. But there are twice as many moles of cerium(IV) as of tin(II) ion.

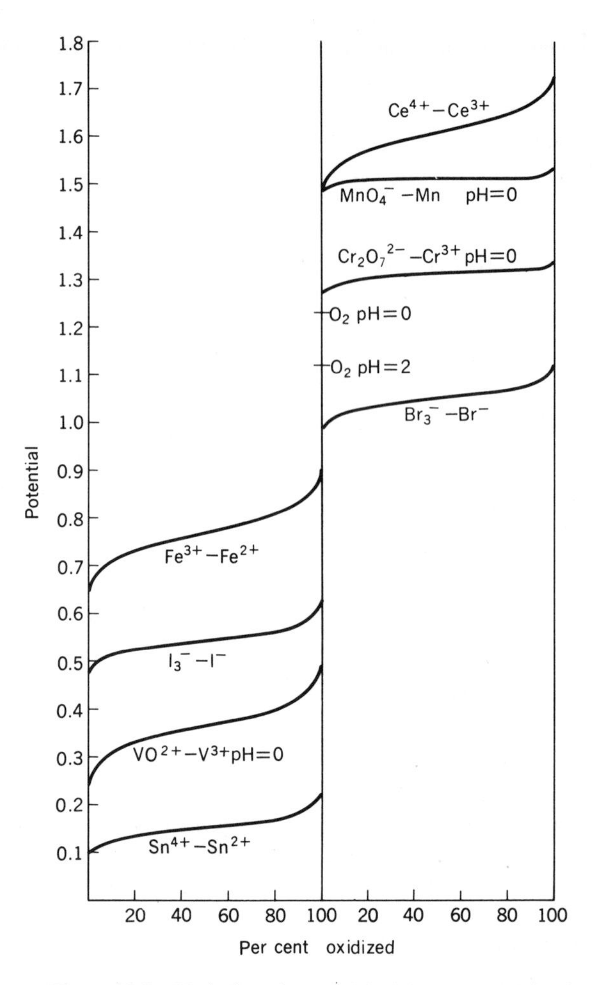

Figure 10.8 Variation of potential with percent oxidized.

3. The curves are asymptotic to the vertical axis at zero and 100% oxidation. The curves are flattest near the midpoints (50% oxidation), which corresponds to the standard potential for a couple such as Fe^{3+}–Fe^{2+}. The stabilization of the potential in this region is analogous to buffering action of an acid-base pair about the *p*H region which corresponds to the pK_a. A redox couple is said to be *poised* in this region where the potential is stabilized.
4. In Fig. 10.8 the potential of the reaction

$$O_2 + 4H^+ + 4e \rightleftharpoons 2H_2O$$

is indicated at *p*H 0 and 2. Redox couples with standard potentials more positive than these values should not be stable in water at *p*H 0 and 2. Solutions of these couples are often stable, however, since the reactions with water to liberate oxygen are generally slow.

FEASIBILITY OF REDOX TITRATIONS

The magnitude of the equilibrium constant required for a feasible redox titration can be calculated as was done previously for acid-base (page 142), complex (page 197), and precipitation titrations (page 215). The following example illustrates this.

Example. For the redox reaction

$$Ox_1 + Red_2 \rightleftharpoons Red_1 + Ox_2$$

where

$$Ox_1 + e \rightleftharpoons Red_1 \quad E_1^\circ$$

$$Ox_2 + e \rightleftharpoons Red_2 \quad E_2^\circ$$

(a) Calculate the value of the equilibrium constant for the following conditions: 50 ml of 0.10 M Red_2 is titrated with 0.10 M Ox_1. When 49.95 ml of titrant is added, the reaction is complete. On the addition of 2 more drops (0.10 ml) of titrant, the value of $p\mathrm{Red}_2$ changes by 2.00 units.

(b) What is the difference in standard potentials of the two redox couples for this value of K?

(a) We start with $50 \times 0.10 = 5.0$ mmol of Red_2, and at 49.95 ml of titrant 0.0050 mmol remains unreacted. Hence

$$[\mathrm{Red}_2] = \frac{0.0050\ \text{mmol}}{99.95\ \text{ml}} \cong 5.0 \times 10^{-5}\ M$$

$$p\mathrm{Red}_2 = 4.30$$

For a change of 2.00 units, $p\mathrm{Red}_2 = 6.30$ and $[\mathrm{Red}_2] = 5 \times 10^{-7}\ M$ when the volume of titrant is 50.05 ml. The other concentrations are

$$[\mathrm{Ox}_1] = \frac{0.05 \times 0.10}{100.05} \cong 5.0 \times 10^{-5}$$

$$[\mathrm{Red}_1] = [\mathrm{Ox}_2] \cong \frac{5.0\ \text{mmol}}{100.05\ \text{ml}} \cong 5.0 \times 10^{-2}\ M$$

Hence

$$K = \frac{(5.0 \times 10^{-2})(5.0 \times 10^{-2})}{(5.0 \times 10^{-5})(5.0 \times 10^{-7})}$$

$$K = 1.0 \times 10^{8}$$

(b) Since

$$E_1^\circ - E_2^\circ = \frac{0.059}{1} \log K$$

$$E_1^\circ - E_2^\circ = 0.059 \log 1 \times 10^{8}$$

$$E_1^\circ - E_2^\circ = 0.47\ \text{V}$$

Note that in this example the number of electrons, n, is 1 for both redox couples. See Problems 20 and 21 for calculations involving other values of n.

REDOX INDICATORS

There are several types of indicators which may be used in redox titrations:

1. A colored substance may act as its own indicator. For example, potassium permanganate solutions are so deeply colored that a slight excess of this reagent in a titration can be easily detected.
2. A *specific* indicator is a substance which reacts in a specific manner with one of the reagents in a titration to produce a color. Examples are starch, which forms a deep-blue color with iodine, and thiocyanate ion, which forms a red color with iron(III) ion.
3. External, or *spot test*, indicators were once employed when no internal indicator was available. The ferricyanide ion was used to detect iron(II) ion by formation of iron(II) ferricyanide (Turnbull's blue) on a spot plate outside the titration vessel.
4. The redox potential can be followed during a titration and the equivalence point detected from the large change in potential in the titration curve. Such a procedure is called potentiometric titration (see Chapter 12), and the titration curve may be plotted manually or automatically recorded.
5. Finally, an indicator which itself undergoes oxidation-reduction may be employed. We shall refer to such a substance as a true redox indicator, and it is with such a reagent that the rest of our discussion is concerned.

For simplicity, let us designate the redox couple as follows:

$$\underset{\text{Color A}}{\text{In}^+} + e \rightleftharpoons \underset{\text{Color B}}{\text{In}}$$

where one electron is gained by the oxidant and no hydrogen ions are involved in the reaction. Let us also say that the colors of the oxidized and reduced forms are different, as indicated above. The equation for the potential of this system is

$$E = E_i^\circ - 0.059 \log \frac{[\text{In}]}{[\text{In}^+]}$$

where E_i° is the standard potential of the indicator couple.

Let us now assume that, if the ratio $[\text{In}]/[\text{In}^+]$ is 10 to 1 or greater, only color B can be seen by the eye. (See page 139 for similar treatment of acid-base indicators.) Also, if the ratio is 1:10 or smaller, only color A is observed. That is,

$$\begin{aligned}
\text{Color B:}\quad E &= E_i^\circ - 0.059 \log 10/1 = E_i^\circ - 0.059 \\
\text{Color A:}\quad E &= E_i^\circ - 0.059 \log 1/10 = E_i^\circ + 0.059 \\
\hline
\text{Subtracting,}\quad \Delta E &= \pm 2 \times 0.059 = \pm 0.12 \text{ V}
\end{aligned}$$

Thus a change in potential of about 0.12 V is required to bring about a change in color of the indicator, if our assumptions are reasonable.

Table 10.2 lists some true redox indicators, with the observed colors and "transition potentials" of the redox couples. The treatment in the above paragraph supposes that the two colored forms of the indicator are equally intense

TABLE 10.2 Transition Potentials of Some Redox Indicators

Indicator	Color reductant	Color of oxidant	Transition potential, volts	Conditions
Phenosafranine	Colorless	Red	+0.28	1 *M* acid
Indigo tetrasulfonate	Colorless	Blue	0.36	1 *M* acid
Methylene blue	Colorless	Blue	0.53	1 *M* acid
Diphenylamine	Colorless	Violet	0.76	1 *M* H_2SO_4
Diphenylbenzidine	Colorless	Violet	0.76	1 *M* H_2SO_4
Diphenylaminesulfonic acid	Colorless	Red-violet	0.85	Dilute acid
5,6-Dimethylferroin			0.97	1 *M* H_2SO_4
Erioglaucin A	Yellow-green	Bluish-red	0.98	0.5 *M* H_2SO_4
5-Methylferroin			1.02	1 *M* H_2SO_4
Ferroin	Red	Faint blue	1.11	1 *M* H_2SO_4
Nitroferroin	Red	Faint blue	1.25	1 *M* H_2SO_4

to the eye. This is not always the case, and hence the transition potentials may not represent exactly 50% conversion of one indicator form to the other; i.e., they may not be the same as the formal potentials. With ferroin, for example, the formal potential in 1 *M* acid is about +1.06 V but, because the color changes from deep red to pale blue, the analyst will not perceive a visual end point until a potential of about +1.11 V is reached.

Selection of Indicator

Obviously, an indicator should change color at or near the equivalence potential. If the titration is feasible, there will be a large change in potential at the equivalence point, and this should be sufficient to bring about the change in color of the indicator. The following example illustrates more precisely the procedure that may be followed to select the proper indicator.

Example. (a) In the titration of iron(II) iron with cerium(IV) sulfate in 1 *M* sulfuric acid, what indicator should be used? We have calculated the potential at the equivalence point to be 1.06 V (page 259). Referring to Table 10.2, it is seen that ferroin, with a transition potential of 1.11 V, is a suitable indicator. The standard potential of ferroin is 1.06 V, but the color change occurs at 1.11 V since it is necessary to have more of the indicator in the oxidized form (light blue) than the reduced form (dark red).

(b) Iron(II) ion is titrated with an oxidizing agent in a sulfuric-phosphoric acid medium. What should be the transition potential of an indicator which changes color when all but 0.1% of iron(II) ion is oxidized to iron(III)?

The formal potential of the $Fe^{3+} - Fe^{2+}$ couple in 1 *F* H_2SO_4 and 0.5 *F* H_3PO_4 is 0.61 V. Hence

$$E = 0.61 - 0.059 \log \frac{[Fe^{2+}]}{[Fe^{3+}]}$$

$$E = 0.61 - 0.059 \log \frac{1}{1000}$$

$$E = 0.61 + 0.18 = 0.79 \text{ V}$$

The indicator diphenylaminesulfonic acid is frequently used when iron is titrated with potassium dichromate in sulfuric-phosphoric acid media. Note (Table 10.2) that its transition potential is 0.85 V, and hence it changes color when even less than 0.1% of Fe^{2+} remains unoxidized.

STRUCTURAL CHEMISTRY OF REDOX INDICATORS

The redox indicators to which we have referred in this chapter are organic molecules that undergo structural changes upon being oxidized or reduced. There are fewer such indicators than there are acid-base indicators, and their chemistry has not been as widely studied. Nevertheless, the structural changes which account for the different colors are known for a number of substances. We shall consider only two examples here, sodium diphenylaminesulfonate and iron(II) orthophenanthroline (ferroin).

Diphenylamine was one of the first redox indicators to be widely used in titrimetric analysis. Since this compound is difficultly soluble in water, and since tungstate ion and mercury(II) chloride interfere with its action, the barium or sodium salt of diphenylaminesulfonic acid is more commonly used. The reduced form of this indicator is colorless, the oxidized form a deep violet. The mechanism of the color change has been shown to be as follows, using diphenylamine as the example:[4]

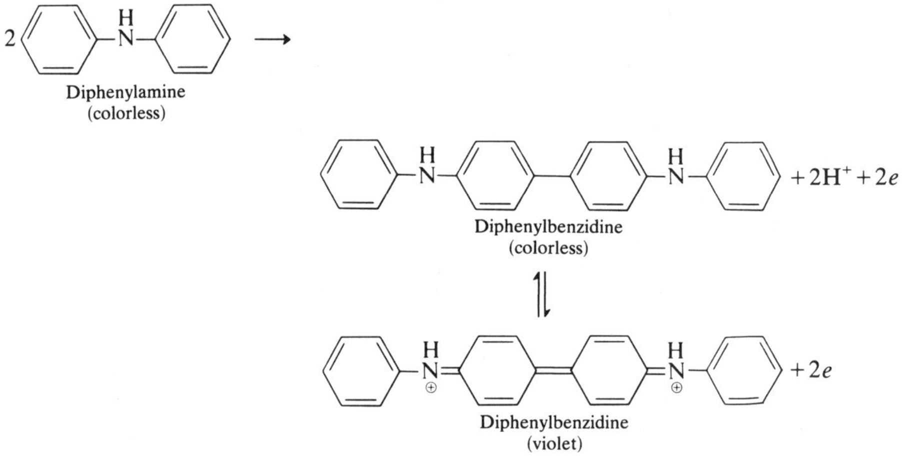

2 C_6H_5–NH–C_6H_5 ⟶

Diphenylamine
(colorless)

C_6H_5–NH–C_6H_4–C_6H_4–NH–C_6H_5 $+ 2H^+ + 2e$

Diphenylbenzidine
(colorless)

⇅

C_6H_5–$\overset{\oplus}{N}$H=C_6H_4=C_6H_4=$\overset{\oplus}{N}$H–C_6H_5 $+ 2e$

Diphenylbenzidine
(violet)

The presence of a long conjugated system, such as that in the diphenylbenzidine ion, leads to absorption of light in the visible region, and hence the ion is colored.

[4] A. M. Kolthoff and L. A. Sarver, *J. Am. Chem. Soc.*, **52**, 4179 (1930).

The indicator ferroin is the iron(II) complex of the organic compound 1,10-phenanthroline,

$$3\,(\text{C}_{12}\text{H}_8\text{N}_2) + Fe^{2+} \rightleftharpoons \{(\text{C}_{12}\text{H}_8\text{N}_2)_3\}Fe^{2+}$$

1,10-Phenanthroline   Iron(II) 1,10-Phenanthroline

Each of two nitrogen atoms in 1,10-phenanthroline has an unshared pair of electrons that can be shared with iron(II) ion. Three such molecules of the organic compound attach themselves to the metallic ion to form a blood-red complex ion. The iron(II) ion can be oxidized to iron(III), and the latter ion also forms a complex with three molecules of 1,10-phenanthroline. The color of the iron(III) complex is light blue, and hence a sharp color change occurs when iron(II) is oxidized to iron(III) in the presence of 1,10-phenanthroline:

$$\underset{\text{Light blue}}{Ph_3Fe^{3+}} + e \rightleftharpoons \underset{\text{Dark red}}{Ph_3Fe^{2+}} \qquad E° = 1.06\ V$$

The indicator is prepared by mixing equivalent quantities of iron(II) sulfate and 1,10-phenanthroline. The complex salt is called *ferroin*; the complex salt of iron(III) ions is called *ferriin*. As previously mentioned, the color change occurs at about 1.11 V, since the color of ferroin is so much more intense than that of ferriin.

Substituted 1,10-phenanthrolines also form complexes with iron(II) and iron(III) ions and act as redox indicators. The redox potentials are different from that of the ferroin-ferriin system. A few examples are included in Table 10.2, where a partial list of redox indicators is given.

REFERENCES

R. G. BATES, "Electrode Potentials," and F. R. DUKE, "Oxidation-Reduction Equilibria and Titration Curves," Chaps. 9 and 16, of I. M. Kolthoff, P. J. Elving, and E. B. Sandell, eds., *Treatise on Analytical Chemistry*, Part I, Vol. 1, Interscience Publishers, Inc., New York, 1959.

H. A. LAITINEN and W. E. HARRIS, *Chemical Analysis*, 2nd ed., McGraw-Hill Book Company, New York, 1975.

J. J. LINGANE, *Electroanalytical Chemistry*, 2nd ed., Interscience Publishers, Inc., New York, 1958.

H. F. WALTON, *Principles and Methods of Chemical Analysis*, 2nd ed., Prentice-Hall, Inc., Englewood Cliffs, N.J., 1964.

QUESTIONS

1. *Standard potentials.* Consult the table of standard potentials (Table 4, Appendix I). At standard conditions, which is the strongest oxidant in the table? The weakest? Which is the strongest reductant? The weakest?

2. *Effect of pH.* Which is the better oxidant at unit activity, I_2 or H_3AsO_4? Explain how the titration of $HAsO_2$ with I_2 is carried out feasibly.

3. *Leveling effect.* Explain why the leveling effect is not observed with redox reagents as it is with acids and bases.

4. *Cell conventions.* Suppose that we chose to write redox reactions as oxidations rather than reductions, that is,

$$Zn \rightleftharpoons Zn^{2+} + 2e$$

with $E° = -0.76$ V. If we followed the same conventions for obtaining the cell reaction as on page 243, what changes in the conclusions drawn from the sign of $E°$ are necessary?

5. *Reference electrode.* Suppose that the saturated calomel electrode ($E° = +0.25$ V) were adopted as the primary reference electrode and assigned a value of 0.00 V. What would be the potentials of the following redox couples?
(a) $2H^+ + 2e \rightleftharpoons H_2(g)$
(b) $Cu^{2+} + 2e \rightleftharpoons Cu$
(c) $Zn^{2+} + 2e \rightleftharpoons Zn$

Multiple-choice: In the following multiple-choice questions, select the *one best* answer.

6. Given the redox couple

$$TiO^{2+} + 2H^+ + e \rightleftharpoons Ti^{3+} + H_2O \qquad E° = +0.10 \text{ V}$$

If the pH is increased by 1 unit, the potential of this couple will (a) become more negative by 2×0.059 V; (b) become more positive by 2×0.059 V; (c) become more negative by 0.059 V; (d) become more positive by 0.059/2 V.

7. Given the following redox couples:

$$A^{3+} + e \rightleftharpoons A^{2+} \qquad E° = +1.70 \text{ V}$$

$$B^{4+} + 2e \rightleftharpoons B^{2+} \qquad E° = +0.50 \text{ V}$$

$$C^{3+} + e \rightleftharpoons C^{2+} \qquad E° = +0.80 \text{ V}$$

Which of the following titrations is feasible? (a) A^{2+} with B^{4+}; (b) A^{3+} with C^{2+}; (c) A^{2+} with C^{3+}; (d) C^{2+} with B^{4+}.

8. In the titration of Fe^{2+} with Ce^{4+} the equilibrium potentials developed by the Fe^{3+}–Fe^{2+} and the Ce^{4+}–Ce^{3+} couples will be equal (a) only at the equivalence point; (b) halfway to the equivalence point; (c) throughout the titration; (d) only before any Ce^{4+} is added.

9. The cell $Zn|Zn^{2+}(1\ M)|\ |Cu^{2+}(1\ M)|Cu$ has a potential of +1.10 V. This means that (a) The Zn electrode is positive with respect to Cu; (b) The spontaneous cell reaction is

$Zn + Cu^{2+} \rightleftharpoons Zn^{2+} + Cu$; (c) At equilibrium $E° = 0$; (d) At equilibrium $E = E°$; (e) more than one of the above.

10. Which of the following expressions is correct? (a) $RT/nF = 0.059/n$ at 0°C; (b) $-\log K = 17nE°$; (c) $\Delta G° = nFE°$; (d) $nE° = 0.059 \log K$.

PROBLEMS

1. *Cell potentials.* Calculate the potentials of the following cells. Write the cell reaction and indicate the polarities of the electrodes.
(a) $Fe|Fe^{2+}(0.004\ M)||Cd^{2+}(0.001\ M)|Cd$
(b) $Fe|Fe^{2+}(0.2\ M)||Cd^{2+}(0.001\ M)|Cd$
(c) $Ag|AgCl, HCl(0.10\ M)|H_2(1\ atm), Pt$
(d) $Pt|Fe^{2+}(0.04\ M) + Fe^{3+}(0.10\ M)||Ce^{3+}(0.08\ M) + Ce^{4+}(0.01\ M)|Pt$

2. *Cell potentials.* Calculate the potentials of the following cells. Write the cell reaction and indicate the polarities of the electrodes.
(a) $Zn|Zn^{2+}(0.002\ M)||Cr^{3+}(0.1\ M)|Cr$
(b) $Zn|Zn^{2+}(0.2\ M)||Cr^{3+}(0.001\ M)|Cr$
(c) $Ag|AgCl, HCl(0.1\ M)||KCl(1.0\ M), Hg_2Cl_2|Hg$
(d) $Pt, H_2(0.50\ atm)|HCl(0.10\ M)|Cl_2(1\ atm)|Pt$

3. *Single electrode potentials.* Calculate the following single electrode potentials at the indicated concentrations.
(a) $H^+(10^{-7}\ M) + e \rightleftharpoons \frac{1}{2}H_2(1\ atm)$
(b) $2H_2O + 2e \rightleftharpoons H_2(1\ atm) + 2OH^-(1\ M)$
(c) $O_2(1\ atm) + 4H^+(10^{-7}\ M) + 4e \rightleftharpoons 2H_2O$
(d) $H_3AsO_4(0.1\ M) + 2H^+(10^{-9}\ M) + 2e \rightleftharpoons HAsO_2(0.1\ M) + 2H_2O$

4. *Single electrode potentials.* Suppose that one could lower the Fe^{3+} concentration in a solution to only one ion per liter. (a) If the Fe^{2+} concentration is 0.10 *M*, what is the potential of the Fe^{3+}–Fe^{2+} couple? (b) Calculate the $[Fe^{3+}]$ needed to lower the potential of the couple to −2.50 V, the $[Fe^{2+}]$ being 0.10 *M*.

5. *Equilibrium constant.* (a) Calculate the equilibrium constant for the reaction in the zinc–copper galvanic cell:

$$Zn(s) + Cu^{2+} \rightleftharpoons Cu(s) + Zn^{2+}$$

(b) Suppose that a galvanic cell is prepared with the initial concentrations of Cu^{2+} and Zn^{2+} ions being 0.10 *M* and zero, respectively. The cell is allowed to discharge until equilibrium is reached. Calculate the equilibrium concentrations of Cu^{2+} and Zn^{2+}. (c) Calculate the single electrode potentials of the copper and zinc couples at equilibrium.

6. *Concentration cell.* Calculate the potential of the following galvanic cell:

$$Ag|AgCl, HCl(0.10\ M)||HCl(0.0010\ M), AgCl|Ag$$

Indicate the polarities of the electrodes.

7. *Free energy change.* (a) Calculate the equilibrium constant of the following reaction:

$$Ag(s) + Fe^{3+} \rightleftharpoons Ag^+ + Fe^{2+}$$

(b) If each reactant and product is at unit activity, in which direction is the reaction spontaneous? (c) Calculate the value of $\Delta G°$. (d) Calculate the value of ΔG for the following concentrations: $[Ag^+] = 0.01\ M$, $[Fe^{2+}] = 0.002\ M$, $[Fe^{3+}] = 0.10\ M$. In which direction is the reaction spontaneous?

8. *Equivalence point.* Show that the potential at the equivalence point in the titration of Red_1 with Ox_2 is

$$E = \frac{aE_1^\circ + bE_2^\circ}{a + b}$$

where

$$Ox_1 + ae \rightleftharpoons Red_1 \qquad E_1^\circ$$

and

$$Ox_2 + be \rightleftharpoons Red_2 \qquad E_2^\circ$$

9. *Equivalence point.* Show that the potential at the equivalence point in the titration of Fe^{2+} with MnO_4^- is

$$E = \frac{E_1^\circ + 5E_2^\circ}{6} - 0.08\ pH$$

where E_1° is the standard potential of the Fe^{3+}–Fe^{2+} couple and E_2° is that of the MnO_4^-–Mn^{2+} couple.

10. *Mixtures of solutions.* In the following problem the solutions are mixed and the reaction reaches equilibrium. A platinum electrode is inserted into the solution to make a half-cell. This half-cell is connected to a standard hydrogen electrode and the potential measured. In each case calculate what the potential should be.
(a) 50 ml of 0.050 M Sn^{2+} + 10 ml of 0.10 M Ce^{4+}
(b) 50 ml of 0.050 M Sn^{2+} + 25 ml of 0.10 M Ce^{4+}
(c) 50 ml of 0.050 M Sn^{2+} + 50 ml of 0.10 M Ce^{4+}
(d) 50 ml of 0.050 M Sn^{2+} + 100 ml of 0.10 M Ce^{4+}

11. *Mixtures of solutions.* Repeat Problem 10 for the following:
(a) 60 ml of 0.10 M Fe^{2+} + 10 ml of 0.02 M $KMnO_4$, $[H^+] = 1.0\ M$;
(b) 60 ml of 0.10 M Fe^{2+} + 40 ml of 0.03 M $KMnO_4$, $[H^+] = 1.0\ M$;
(c) 50 ml of 0.10 M Sn^{2+} + 30 ml of 0.06 M Fe^{3+};
(d) 30 ml of 0.10 M Sn^{2+} + 40 ml of 0.20 M Fe^{3+}

12. *Fraction oxidized.* A solution of Fe^{2+} is titrated with an oxidizing agent. Calculate the potential of the Fe^{3+}–Fe^{2+} couple when the following percentages of Fe^{2+} have been oxidized: (a) 10%; (b) 25%; (c) 33%; (d) 50%; (e) 99%; (f) 99.9%.

13. *Fraction oxidized.* Repeat Problem 12 for the oxidation of Sn^{2+} to Sn^{4+}.

14. *Fraction oxidized.* A galvanic cell is assembled with two platinum electrodes dipping into separate beakers which are connected by a salt bridge. Each beaker contains 25.00 ml of a mixture of Fe^{2+} and Fe^{3+} ions. Each ion in each beaker is 0.10 M. Now 5.00 ml of a reducing agent is added to one beaker, reducing some Fe^{3+} to Fe^{2+}. The voltage reading between the two electrodes changes from 0.00 to 0.0295 V. Calculate the normality of the reducing agent.

15. *Titration curve.* 4.00 mmol of Fe^{3+} is dissolved in 100 ml of an acid solution and titrated with 0.0500 M Sn^{2+}. Calculate the potential of the solution after the addition

of the following volumes (ml) of titrant: (a) 10.0; (b) 20.0; (c) 39.95; (d) 40.0; (e) 40.05; (f) 50.0. Plot the titration curve.

16. *Titration.* 8.00 mmol of Fe^{2+} is dissolved in 100 ml of solution and titrated with 0.10 M B^{4+}, where $B^{4+} + 2e \rightleftharpoons B^{2+}$, $E° = +1.28$ V. Calculate the potential of the solution referred to hydrogen (a) after the addition of 10.0 ml of titrant; (b) at the equivalence point; (c) after the addition of 60.0 ml of titrant. (d) Calculate the ratio of Fe^{3+} to Fe^{2+} concentrations at the equivalence point. (e) What percentage of the Fe^{2+} remains unoxidized at the equivalence point?

17. *Titration.* 6.00 mmol of Fe^{2+} is titrated with 0.025 M $K_2Cr_2O_7$, the final volume being 100 ml and the $[H^+] = 1.0$ M. Calculate the number of milligrams of Fe^{2+} remaining in the solution when 0.10 ml of titrant is in excess.

18. *Indicator.* The indicator ferroin shows a color change at +1.11 V. The standard potential is +1.06 V. Calculate the percentage of the indicator in the oxidized form at the potential of the color change. A one-electron change is involved.

19. *Reference electrode.* The potential of a cell made up of an electrode of unknown potential and a saturated calomel electrode ($E = 0.25$ V) is 0.84 V. Calculate the potential of the unknown electrode referred to the standard hydrogen electrode if the polarity of the calomel electrode is (a) positive; (b) negative.

20. *Feasibility of titration.* For the redox titration

$$A^{2+} + 2M^{3+} \rightleftharpoons A^{4+} + 2M^{2+}$$

(a) Calculate the value of the equilibrium constant for the following conditions: 50 ml of 0.10 M A^{2+} is titrated with 0.20 M M^{3+}. When 49.95 ml of titrant is added the reaction between A^{2+} and M^{3+} is complete. On addition of 2 more drops (0.10 ml) of titrant, the value of pA changes by 2.00 units. (b) What is the difference in standard potentials of the two redox couples for this value of K?

21. *Feasibility of titration.* Repeat problem 20 for the titration

$$A^{2+} + N^{4+} \rightleftharpoons A^{4+} + N^{2+}$$

Use 0.10 M A^{2+} and 0.10 M N^{4+}.

22. *Feasibility of titration.* (a) Calculate the equilibrium constant for the reaction

$$Fe^{2+} + B^{3+} \rightleftharpoons Fe^{3+} + B^{2+}$$

where the $E°$ of the B^{3+}–B^{2+} system is +1.07 V. Do you expect the titration to be feasible? (b) Calculate the number of milligrams of Fe^{2+} which remain unoxidized when 5.0 mmol of Fe^{2+} is titrated with 0.10 M B^{3+}, 1 drop (0.05 ml) of titrant in excess. The final volume is 100 ml.

23. *pH—cell potential.* Given the cell

$$\text{Pt, } H_2(1 \text{ atm})|H^+(x\,M)|\,|KCl(\text{sat.}), Hg_2Cl_2|Hg$$

(a) Derive an equation relating the potential of this cell to the pH of the solution in the left-hand electrode. (b) If the solution in the left-hand electrode is 0.10 M HCl, what is the potential of the cell? (c) If the solution in the left-hand electrode is 0.10 M NaOH, what is the potential?

24. *pH—cell potential.* Given the same cell as in Problem 23, calculate the cell potential if the solution in the left-hand electrode is (a) 0.10 M HF; (b) 0.10 M NaF; (c) 0.10 M NH_3.

25. *Buffer solution.* Given the same cell as in Problem 23, with the solution in the left-hand electrode consisting of 50 ml of a buffer which is 0.10 M in HOAc and 0.20 M in NaOAc. (a) Calculate the potential of the cell. (b) 50 ml of 0.10 M HCl is added to the buffer. What is the potential of the cell after the reaction has occurred?

26. *Dissociation constant.* The following cell has a potential of 0.53 V when the solution in the left-hand electrode is 0.20 M in the salt BCl. BOH is a weak base.

$$\text{Pt, } H_2(1 \text{ atm})|H^+(x\,M)||HCl(1\,M), AgCl|Ag$$

The Ag electrode is positive. Calculate the dissociation constant of BOH.

27. *Solubility product constant.* Given the following cell:

$$Pb|Pb^{2+}(x\,M)||HCl(1\,M), AgCl|Ag$$

Sodium hydroxide is added to the solution in the left-hand electrode, precipitating $Pb(OH)_2$. The final pH of the solution is 10.00. The voltage of the cell is now 0.58 V with the silver electrode positive. Calculate the K_{sp} of $Pb(OH)_2$.

28. *Solubility product constant.* Given the single electrode potentials

$$Mn^{2+} + 2e \rightleftharpoons Mn \qquad E° = -1.18 \text{ V}$$

and

$$Mn(OH)_2(s) + 2e \rightleftharpoons Mn + 2OH^- \qquad E° = -1.59 \text{ V}$$

calculate the K_{sp} of $Mn(OH)_2$.

29. *Solubility product constant.* Given the single electrode potential

$$CuCl(s) + e \rightleftharpoons Cu(s) + Cl^- \qquad E° = +0.14 \text{ V}$$

and the K_{sp} of CuCl $= 3 \times 10^{-7}$, calculate the $E°$ of the half-reaction

$$Cu^+ + e \rightleftharpoons Cu$$

30. *Complex formation.* (a) Given the following cell:

$$Ni|Ni^{2+}(0.20\,M)||Pb^{2+}(0.050\,M)|Pb$$

Calculate the potential, the direction of the cell reaction, and indicate the polarities of the electrodes. (b) Suppose that the solution in the right-hand electrode contains 5.00 mmol of Pb^{2+} and the pH $= 5.00$. We add 6.00 mmol of EDTA to the solution to complex the Pb^{2+}. The volume is 100 ml. Repeat the calculations of part (a) under the new conditions.

31. *Complex formation.* (a) Calculate the potential of the following cell:

$$Ag|Ag^+(0.20\,M)||H^+(1.0\,M)|H_2(1 \text{ atm}), \text{Pt}$$

(b) Ammonia is added to the electrode containing Ag^+ until the final concentration of the uncomplexed NH_3 is 1.0 M. Calculate the potential of the cell under these conditions.

32. *Stability constant.* A solution is prepared by mixing 30 ml of 0.10 M Hg^{2+} with 70 ml of 0.10 M EDTA. The pH of the solution is 4.00. Into the solution are inserted a mercury and a saturated calomel electrode ($E = 0.25$ V), giving the cell

$$Hg|Hg^{2+} + EDTA||KCl(sat.), Hg_2Cl_2|Hg$$

The potential of the cell is 0.20 V and the calomel electrode is negative. Calculate the value of K_{abs} for the HgY^{2-} complex.

33. *Disproportionation.* (a) From the table of standard potentials, determine whether the following disproportionation occurs spontaneously or not when all reactants are at unit activity:

$$3Br_2(aq) + 3H_2O \rightleftharpoons BrO_3^- + 5Br^- + 6H^+$$

(b) Calculate the equilibrium constant of the reaction.

34. *Disproportionation.* In basic solution the reaction in Problem 33 is

$$3Br_2(aq) + 6OH^- \rightleftharpoons BrO_3^- + 5Br^- + 3H_2O$$

(a) Calculate the equilibrium constant for this reaction. (b) Calculate the pH at which the reaction is spontaneous. The activities of Br_2, BrO_3^-, and Br^- are each unity.

35. *Disproportionation.* (a) Refer to the table of standard potentials (Table 4, Appendix I) and calculate whether or not copper(I) ion disproportionates (unit activities):

$$2Cu^+ \rightleftharpoons Cu^{2+} + Cu$$

(b) Calculate the equilibrium constant for this reaction. (c) How large a concentration of Cu^+ ions can exist in equilibrium with Cu^{2+} ions at 1 M concentration?

CHAPTER ELEVEN

applications of oxidation-reduction titrations

We have previously mentioned that chemical reactions involving oxidation–reduction are widely used in titrimetric analyses. The ions of many elements can exist in different oxidation states, resulting in the possibility of a very large number of redox reactions. Many of these satisfy the requirements for use in titrimetric analyses and applications are quite numerous.

In this chapter we shall discuss some of the more widely used redox reagents, their properties, methods used to standardize solutions, and applications to analyses.

REAGENTS USED FOR PRELIMINARY REDOX REACTIONS

In many analytical procedures the analyte is in more than one oxidation state and must be converted to a single oxidation state before titration. A common example occurs in the determination of iron in an ore. Once the ore is dissolved, iron is present in both the $+2$ and $+3$ oxidation states. It must be reduced completely to the $+2$ state before titration with a standard solution of an oxidizing agent. The redox reagent used in this preliminary step must be able to convert the analyte completely and rapidly into the desired oxidation state. Excess of the reagent is normally added and one must be able to remove the excess conveniently so that it will not react with the titrant in the subsequent titration.

The following are a few of the more common reagents which are used in preliminary steps.

Oxidizing Agents

Sodium and Hydrogen Peroxide

Hydrogen peroxide is a good oxidizing agent with a large positive standard potential:

$$H_2O_2 + 2H^+ + 2e \rightleftharpoons 2H_2O \qquad E^\circ = +1.77\text{ V}$$

In acidic solution it will oxidize Fe(II) to Fe(III). In alkaline solution it will oxidize Cr(III) to CrO_4^{2-} and Mn(II) to MnO_2. The excess reagent is easily removed by boiling the solution for a few minutes:

$$2H_2O_2 \rightleftharpoons 2H_2O + O_2(g)$$

Potassium and Ammonium Peroxodisulfate

The peroxodisulfate ion is a powerful oxidizing agent in acidic solutions:

$$S_2O_8^{2-} + 2e \rightleftharpoons 2SO_4^{2-} \qquad E^\circ = +2.01\text{ V}$$

It will oxidize Cr(III) to $Cr_2O_7^{2-}$, Ce(III) to Ce(IV), and Mn(II) to MnO_4^-. The reaction is usually catalyzed by a trace of silver(I) ion. After the oxidation is complete the excess reagent can be removed by boiling the solution:

$$2S_2O_8^{2-} + 2H_2O \longrightarrow 4SO_4^{2-} + O_2(g) + 4H^+$$

Sodium Bismuthate

The formula of this compound is not known with certainty but is usually written $NaBiO_3$. It is a powerful oxidizing agent, oxidizing Mn(II) to MnO_4^-, Cr(III) to $Cr_2O_7^{2-}$, and Ce(III) to Ce(IV). The bismuth is reduced to Bi(III). The compound is sparingly soluble and the solution of the substance to be oxidized is heated with excess solid. After the reaction is complete the excess bismuthate is removed by filtration.

Reducing Agents

Sulfur Dioxide and Hydrogen Sulfide

Both of these gases are relatively mild reducing agents:

$$SO_4^{2-} + 4H^+ + 2e \rightleftharpoons H_2SO_3 + H_2O \qquad E^\circ = +0.17\text{ V}$$

$$S + 2H^+ + 2e \rightleftharpoons H_2S \qquad E^\circ = +0.14\text{ V}$$

They are readily soluble in water and the excess reagent can be easily removed by boiling the solution. They will reduce Fe(III) to Fe(II), V(V) to V(IV), and Ce(IV) to Ce(III). Both gases are toxic and have an unpleasant odor. They are rarely employed in the elementary laboratory.

Tin(II) chloride

This reagent is used almost exclusively for the reduction of Fe(III) to Fe(II) in samples which have been dissolved in hydrochloric acid. We shall discuss the reagent in detail later.

Metals and Amalgams

A number of metals can be used as reducing agents. Several metals, in particular silver, zinc, cadmium, aluminum, nickel, copper, and mercury, have been widely used in analytical procedures. Sometimes the metal can be used in the form of a rod or a coil of wire and is inserted directly in the solution of the analyte. When reduction is complete the unused metal is removed from the solution and washed thoroughly. An alternative procedure which ensures more thorough contact between the solution and the metal is to prepare a *reductor*, a glass column containing granules of the metal used as the reductant. The solution to be reduced is poured through the column and caught in the titration flask. A typical setup is shown in Fig. 11.1.

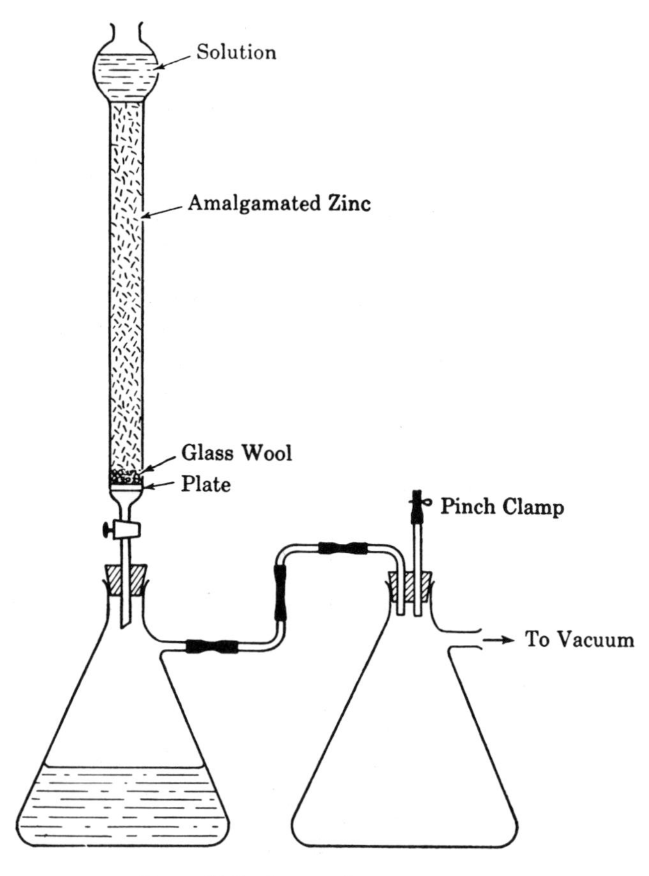

Figure 11.1 Jones reductor.

Very active metals, such as zinc, cadmium, and aluminum, not only reduce the analyte but also dissolve in acidic solutions with the evolution of hydrogen. This side reaction is not desirable, since it can consume large amounts of the metal and introduce a considerable quantity of metallic ion into the sample solution. The reaction can be largely prevented by amalgamating the metal with mercury. Amalgamated zinc is used in the *Jones reductor.* Granulated zinc is treated with a dilute solution of mercury(II) chloride and mercury is displaced, forming a coating of amalgam on the surface:

$$Zn + Hg^{2+} \rightleftharpoons Zn^{2+} + Hg$$

Hydrogen is not as easily displaced by zinc on the amalgamated surface because of the high overvoltage of hydrogen on mercury (Chapter 13). Amalgamated zinc can be used in very acidic solutions and makes an ideal packing for a reductor.

Silver metal in the presence of hydrochloric acid is widely used as the packing in a metal reductor. Silver is a poor reducing agent, but in the presence of hydrochloric acid its reducing ability is increased:

$$Ag(s) + Cl^- \rightleftharpoons AgCl(s) + e$$

Although it is not as strong a reductant as amalgamated zinc, silver is somewhat more selective than the latter metal.

The Jones and the silver reductors are the two most widely used in analytical procedures. Table 11.1 shows some of the applications of the two reductors.

TABLE 11.1 Comparison of Reduction Products in the Jones and Silver Reductors*

	Reduction Product	
Metal ion	Silver reductor†	Jones reductor‡
Titanium(IV)	Not reduced	Titanium(III)
Vanadium(V)	Vanadium(IV)	Vanadium(II)
Chromium(III)	Not reduced	Chromium(II)
Molybdenum(VI)	Molybdenum(V)	Molybdenum(III)
Iron(III)	Iron(II)	Iron(II)
Copper(II)	Copper(I)	Copper(0)
Uranium(VI)	Uranium(IV)	Uranium(IV and III)

* I. M. Kolthoff and R. Belcher, *Volumetric Analysis*, Vol. 3, Interscience Publishers, Inc., 1957, New York, p. 12.

† $Ag(s) + Cl^- \rightleftharpoons AgCl(s) + e$.

‡ $Zn(Hg) \rightleftharpoons Zn^{2+} + 2e + Hg$.

POTASSIUM PERMANGANATE

Properties

Potassium permanganate has been widely used as an oxidizing agent for over a hundred years. It is a reagent that is readily available, inexpensive, and requires no indicator unless very dilute solutions are used. One drop of 0.1 N permanganate imparts a perceptible pink color to the volume of solution usually used in a titration. This color is used to indicate excess of the reagent. Permanganate undergoes a variety of chemical reactions since manganese can exist in oxidation states of +2, +3, +4, +6, and +7. These reactions are summarized below:

$$MnO_4^- + 8H^+ + 5e \rightleftharpoons Mn^{2+} + 4H_2O \qquad E° = +1.51\ V \tag{1}$$

This is the reaction that takes place in very acidic solutions (0.1 N or greater).

$$MnO_4^- + 4H^+ + 3e \rightleftharpoons MnO_2(s) + 2H_2O \qquad E° = +1.70\ V \tag{2}$$

This reaction takes place in solutions of lower acidity. It is predominant in the pH range of about 2 to 12.

$$MnO_4^- + 3H_2P_2O_7^{2-} + 8H^+ + 4e \rightleftharpoons Mn(H_2P_2O_7)_3^{3-} + 4H_2O \qquad E° = +1.50\ V \tag{3}$$

The +3 oxidation state of manganese is not stable, but complexing anions, such as pyrophosphate or fluoride, will stabilize the ion.

$$MnO_4^- + e \rightleftharpoons MnO_4^{2-} \qquad E° = +0.54\ V \tag{4}$$

This reaction takes place only in very alkaline solution, 1 M or so in OH^- ions. In solutions of lower pH, reaction (2) will occur. Barium chloride is usually added to precipitate $BaMnO_4$, thus removing the green color of the MnO_4^{2-} ion and also preventing further reduction from occurring.

The most common reaction encountered in the introductory laboratory is the first, that in very acidic solution. Permanganate reacts rapidly with many reducing agents according to reaction (1), but some substances require heating or the use of a catalyst to speed up the reaction. Were it not for the fact that many reactions of permanganate are slow, more difficulties would be encountered in the use of this reagent. For example, permanganate is a sufficiently strong oxidizing agent to oxidize Mn(II) to MnO_2 according to the equation

$$3Mn^{2+} + 2MnO_4^- + 2H_2O \longrightarrow 5MnO_2(s) + 4H^+$$

The slight excess of permanganate present at the end point of a titration is sufficient to bring about the precipitation of some MnO_2. However, since the reaction is slow, MnO_2 is not normally precipitated at the end point of permanganate titrations.

Special precautions must be taken in the preparation of permanganate solutions. Manganese dioxide catalyzes the decomposition of permanganate solutions. Traces of MnO_2 initially present in the permanganate, or formed by the reaction of permanganate with traces of reducing agents in the water, lead to decomposition. Directions usually call for dissolving the crystals, heating to destroy reducible substances, and filtering through asbestos or sintered glass (nonreducing filters) to remove MnO_2. The solution is then standardized, and if kept in the dark and not acidified, its concentration will not change appreciably over a period of several months.

Acidic solutions of permanganate are not stable because permanganic acid is decomposed according to the equation

$$4MnO_4^- + 4H^+ \longrightarrow 4MnO_2(s) + 3O_2(g) + 2H_2O$$

This is a slow reaction in dilute solutions at room temperature. However, one should never add excess permanganate to a reducing agent and then raise the temperature to hasten oxidation, because the foregoing reaction will then occur at an appreciable rate.

Standardization

Arsenic(III) Oxide

The compound As_2O_3 is an excellent primary standard for permanganate solutions. It is stable, nonhygroscopic, and is readily available in a high degree of purity. The oxide is dissolved in sodium hydroxide, the solution acidified with hydrochloric acid, and titrated with permanganate:

$$5HAsO_2 + 2MnO_4^- + 6H^+ + 2H_2O \longrightarrow 2Mn^{2+} + 5H_3AsO_4$$

(The acid produced by dissolving As_2O_3 behaves as a monoprotic weak acid. We shall write its formula $HAsO_2$ rather than H_3AsO_3.) The reaction is slow at room temperature unless a catalyst is added. Potassium iodide, KI, potassium iodate, KIO_3, and iodine monochloride, ICl, have been used as catalysts.

Sodium Oxalate

This compound, $Na_2C_2O_4$, is also a good primary standard for permanganate in acid solution. It can be obtained in a high degree of purity, is stable on drying, and is nonhygroscopic. The reaction with permanganate is somewhat complex, and even though many investigations have been made, the exact mechanism is not clear. The reaction is slow at room temperature and hence the solution is normally heated to about 60°C. Even at elevated temperature the reaction starts slowly, but the rate increases as manganese(II) ion is formed. Manganese(II) acts as a catalyst and the reaction is termed "autocatalytic" since the catalyst is produced in the reaction itself. The ion may exert its catalytic effect by rapidly reacting with permanganate to form manganese of intermediate oxidation states (+ 3 or + 4), which in turn rapidly oxidize oxalate ion, returning to the divalent state.

The equation for the reaction between oxalate and permanganate is

$$5C_2O_4^{2-} + 2MnO_4^- + 16H^+ \longrightarrow 2Mn^{2+} + 10CO_2 + 8H_2O$$

For a number of years analysts employed the procedure recommended by McBride,[1] which called for the entire titration to be carried out slowly at elevated temperature with vigorous stirring. Later, Fowler and Bright[2] made a very

[1] R. S. McBride, *J. Am. Chem. Soc.*, **34**, 393 (1912).

[2] R. M. Fowler and H. A. Bright, *J. Res. Nat. Bur. Standards*, **15**, 493 (1935).

thorough investigation of possible errors in the titration. They found some evidence for peroxide formation,

$$O_2 + H_2C_2O_4 \longrightarrow H_2O_2 + 2CO_2$$

and that if the peroxide decomposes before reacting with the permanganate, too little of the latter solution is used and the normality found is high. Fowler and Bright recommended that almost all the permanganate be added rapidly to the acidified solution (about 1.5 N) at room temperature. After the reaction is complete, the solution is heated to 60°C and the titration completed at this temperature. This procedure eliminates the foregoing error. Other possible errors in the procedure were found to be negligible.

Iron

Iron wire of a high degree of purity is available as a primary standard. It is dissolved in dilute hydrochloric acid, and any iron(III) produced during the dissolving process is reduced to iron(II). If the solution is then titrated with permanganate, an appreciable amount of chloride ion is oxidized in addition to the iron(II). The oxidation of chloride ion by permanganate is slow at room temperature. However, in the presence of iron, the oxidation occurs more rapidly. Although iron(II) is a stronger reducing agent than chloride ion, the latter ion is oxidized simultaneously with the iron. It is frequently said that iron "induces" the oxidation of chloride ion. No such difficulty is encountered in the oxidation of As_2O_3 or $Na_2C_2O_4$ in hydrochloric acid solution.

A solution of manganese(II) sulfate, sulfuric acid, and phosphoric acid, called "preventive," or Zimmermann–Reinhardt solution, can be added to the hydrochloric acid solution of iron before titration with permanganate. Phosphoric acid lowers the concentration of iron(III) ion by formation of a complex, helping force the reaction to completion, and also removes the yellow color which iron(III) shows in chloride media. The phosphate complex is colorless and the end point is made somewhat clearer.

Determinations with Permanganate

Iron

The determination of iron in iron ores is one of the most important applications of permanganate titrations. The principal ores of iron are the oxides or hydrated oxides: hematite, Fe_2O_3; magnetite, Fe_3O_4; goethite, $Fe_2O_3 \cdot H_2O$; and limonite, $2Fe_2O_3 \cdot 3H_2O$. The carbonate, $FeCO_3$ (siderite), and the sulfide, FeS_2 (pyrite), are of lesser importance. The best acid for dissolving these ores is hydrochloric. The hydrated oxides dissolve readily, whereas magnetite and hematite dissolve rather slowly. The addition of tin(II) chloride aids in the dissolving of the latter oxides. The residue of silica, remaining after the sample has been heated with acid, may retain some iron. The silica can be fused with sodium carbonate and then treated with hydrochloric acid to recover the iron.

Before titrating with permanganate any iron(III) must be reduced to iron(II). This reduction can be done with the Jones reductor (page 277) or with tin(II) chloride. The Jones reductor is preferred if the acid present is sulfuric since no chloride ion is introduced.

If the solution contains hydrochloric acid, as is often the case, reduction with tin(II) chloride is convenient. The chloride is added to a hot solution of the sample and the progress of the reduction is followed by noting the disappearance of the yellow color of the iron(III) ion:

$$Sn^{2+} + 2Fe^{3+} \longrightarrow Sn^{4+} + 2Fe^{2+}$$

A slight excess of tin(II) chloride is added to ensure completeness of reduction. This excess must be removed or it will react with the permanganate upon titration. For this purpose, the solution is cooled and mercury(II) chloride is added rapidly to oxidize excess tin(II) ion:

$$2HgCl_2 + Sn^{2+} \longrightarrow Hg_2Cl_2(s) + Sn^{4+} + 2Cl^-$$

Iron(II) is not oxidized by the mercury(II) chloride. The precipitate of mercury(I) chloride, if small, does not interfere in the subsequent titration. However, if too large an excess of tin(II) chloride is added, mercury(I) chloride may be further reduced to free mercury:

$$Hg_2Cl_2(s) + Sn^{2+} \longrightarrow 2Hg(l) + 2Cl^- + Sn^{4+}$$

Mercury, which is produced in a finely divided state under these conditions, causes the precipitate to appear gray to black. If the precipitate is dark, the sample should be discarded since mercury, in the finely divided state, will be oxidized during the titration. The tendency toward further reduction of Hg_2Cl_2 is diminished if the solution is cool and the $HgCl_2$ is added rapidly. Of course, if insufficient $SnCl_2$ is added, no precipitate of Hg_2Cl_2 will be obtained. In such a case, the sample must be discarded.

Tin(II) chloride is normally used for reduction of iron in samples which have been dissolved in hydrochloric acid. Zimmermann-Reinhardt preventive solution is then added if the titration is to be done with permanganate.

Other Reducing Agents

Many reducing agents in addition to iron(II) can be determined by titration with permanganate in acid solution. As mentioned in the discussion of standardization processes, arsenic can be titrated in hydrochloric acid. Antimony can also be determined, as well as nitrites (oxidized to nitrates) and hydrogen peroxide. The peroxide acts as a reducing agent in the reaction

$$2MnO_4^- + 5H_2O_2 + 6H^+ \longrightarrow 2Mn^{2+} + 5O_2(g) + 8H_2O$$

We have also seen that oxalates can be titrated with permanganate. The titrimetric determination of calcium in limestone is frequently used as a student exercise. Calcium is precipitated as the oxalate, CaC_2O_4. After filtration and washing, the precipitate is dissolved in sulfuric acid and the oxalate titrated with permanganate. The procedure is more rapid than the gravimetric one in which CaC_2O_4 is ignited to CaO and weighed. However, care must be taken in the precipitation to avoid contamination of the precipitate with other oxalates or oxalic acid. Such contamination, of course, leads to high results. Other cations which form insoluble oxalates can be determined in the same manner. These include manganese(II), zinc(II), cobalt(II), barium(II), strontium(II), and lead(II).

Table 11.2 summarizes some of the more common determinations that can be made by direct titration with permanganate in acid solution.

TABLE 11.2 Some Applications of Direct Titrations with Permanganate

Analyte	Half-reaction of substance oxidized
Antimony(III)	$HSbO_2 + 2H_2O \longrightarrow H_3SbO_4 + 2H^+ + 2e$
Arsenic	$HAsO_2 + 2H_2O \longrightarrow H_3AsO_4 + 2H^+ + 2e$
Bromine	$2Br^- \longrightarrow Br_2 + 2e$
Hydrogen peroxide	$H_2O_2 \longrightarrow O_2 + 2H^+ + 2e$
Iron(II)	$Fe^{2+} \longrightarrow Fe^{3+} + e$
Tin(II)	$Sn^{2+} \longrightarrow Sn^{4+} + 2e$
Nitrite	$HNO_2 + H_2O \longrightarrow NO_3^- + 3H^+ + 2e$
Oxalate	$H_2C_2O_4 \longrightarrow 2CO_2 + 2H^+ + 2e$

Oxidizing Agents

A standard solution of $KMnO_4$ can also be employed indirectly in the determination of oxidizing agents, particularly the higher oxides of such metals as lead and manganese. Such oxides are difficult to dissolve in acids or bases without reduction of the metal to a lower oxidation state. It is impractical to titrate these substances directly, because the reaction of the solid with a reducing agent is slow. Hence the sample is treated with a known excess of some reducing agent and heated to complete the reaction. Then the excess reducing agent is titrated with standard permanganate. Various reducing agents may be employed, such as As_2O_3 and $Na_2C_2O_4$. The analysis of pyrolusite, an ore containing MnO_2, is a common student exercise. The reaction of MnO_2 with $HAsO_2$ is

$$MnO_2(s) + HAsO_2 + 2H^+ \longrightarrow Mn^{2+} + H_3AsO_4$$

COMPOUNDS OF CERIUM

Properties

The element cerium (atomic number 58) can exist in only two oxidation states, +4 and +3. In the quadrivalent state it is a powerful oxidizing agent, undergoing a single reaction,

$$Ce^{4+} + e \rightleftharpoons Ce^{3+}$$

The Ce(IV) ion is used in solutions of high acidity because hydrolysis leads to precipitation in solutions of low hydrogen ion concentration. The redox potential of the Ce(IV)/Ce(III) couple is dependent upon the nature and concentration of the acid present (page 255). The formal potentials in 1 *M* solutions of the common acids are: $HClO_4$, +1.70 V; HNO_3, +1.61 V; H_2SO_4, +1.44 V; HCl, +1.28 V.

It is known that both cerium(IV) and cerium(III) ions form stable complexes with various anions. Some chemists name the acids and salts of cerium to indicate that the element is present as a complex anion, rather than as a cation. For example, the salt $(NH_4)_2Ce(NO_2)_6$ is called *ammonium hexanitratocerate.* For simplicity, we shall call such a compound cerium(IV) ammonium nitrate and write the formula $Ce(NO_3)_4 \cdot 2NH_4NO_3$.

Although cerium is a rare earth element, its compounds are readily available for analytical use at a reasonable cost. Since 1928, beginning with the work of N.H. Furman at Princeton and H. H. Willard at Michigan, the reagent has found an ever increasing use as an oxidizing agent in analytical chemistry.[3] It is usually necessary to employ a redox indicator, and the compound *ferroin* (page 267) has been developed for this purpose.

The Ce(IV) ion can be employed in most titrations where permanganate is used, and it possesses properties which often make it a better choice as an oxidizing agent than permanganate. The principal advantages can be summarized as follows:

1. There is only one oxidation state, Ce(III), to which the Ce(IV) ion is reduced.
2. It is a very strong oxidizing agent, and, as pointed out above, one can vary the intensity of its oxidizing power by choice of the acid employed.
3. Sulfuric acid solutions of Ce(IV) ion are extremely stable. Solutions can be kept indefinitely without change in concentration. Solutions in nitric and perchloric acids decompose, but only slowly.
4. The chloride ion, in moderate concentration, is not readily oxidized, even in the presence of iron. The reagent can thus be used for iron titrations in hydrochloric acid solution without need of Zimmermann–Reinhardt preventive solution. Cerium(IV) solutions can be employed, even in the presence of chloride ion, for oxidations which must be carried out by use of excess reagent at elevated temperature. However, chloride ion is oxidized if the solution is boiled. Also, Ce(IV) solutions in hydrochloric acid are unstable if the concentration of acid is more than about 1 *M.*
5. A salt, cerium(IV) ammonium nitrate, sufficiently pure to be weighed directly for preparing standard solutions, is available.
6. Although the Ce(IV) ion is yellow, the color does not cause difficulty in reading a buret, unless the concentration is greater than about 0.1 *M.* The Ce(III) ion is colorless.

[3] See the booklet entitled *Cerate Oxidimetry*, published by the G. Frederick Smith Chemical Co., Columbus, Ohio, 1942, and similar booklets; see also I. M. Kolthoff and R. Belcher, *Volumetric Analysis*, Vol. 3, Interscience Publishers, Inc., New York, 1957, p. 121.

Standardization of Solutions

Solutions of Ce(IV) are usually prepared from cerium(IV) hydrogen sulfate, $Ce(HSO_4)_4$, cerium(IV) ammonium sulfate, $Ce(SO_4)_2 \cdot 2(NH_4)_2SO_4 \cdot 2H_2O$, or cerium(IV) hydroxide, $Ce(OH)_4$. The compounds are dissolved in 0.2 to 0.5 *M* strong acid to prevent hydrolysis and the formation of slightly soluble basic salts. The solution is then standardized against one of the primary standards listed below.

As mentioned above, the compound cerium(IV) ammonium nitrate is available as a primary standard and standard solutions may be made by direct weighing, followed by dilution in a volumetric flask. This salt can also be obtained as ordinary reagent-grade material, in which case the solution must be standardized.

The same primary standards used for potassium permanganate can be used to standardize Ce(IV) solutions. These are:

1. *Arsenic(III) oxide.* The reaction is slow unless osmium tetroxide, OsO_4, or iodine monochloride, ICl, is used as a catalyst. With OsO_4 catalyst the titration can be made at room temperature. Ferroin is a suitable indicator. With ICl catalyst the solution is heated to 50°C. Swift[4] has found that the titration can be performed at room temperature using ICl as the catalyst if the solution is 4*M* in hydrochloric acid. The ICl also acts as an indicator.
2. *Sodium oxalate.* The reaction of oxalate and Ce(IV) ion is slow in sulfuric acid at room temperature, and again either OsO_4 or ICl is a suitable catalyst. However, if the solution is 2 *M* in perchloric acid, the reaction proceeds rapidly at room temperature.
3. *Iron.* Pure iron wire can be employed as a primary standard, the reaction being rapid at room temperature. Ferroin is a suitable indicator.

Determinations with Cerium(IV) Solutions

As previously mentioned, Ce(IV) solutions can be employed in most titrations where permanganate is used. The more important applications include the determination of iron, arsenic, antimony, oxalates, ferrocyanide, titanium, chromium, vanadium, molybdenum, uranium, and the oxides of lead and manganese.

POTASSIUM DICHROMATE

Potassium dichromate is a fairly strong oxidizing agent, the standard potential of the reaction

$$Cr_2O_7^{2-} + 14H^+ + 6e \rightleftharpoons 2Cr^{3+} + 7H_2O$$

[4] E. H. Swift, *Introductory Quantitative Analysis*, Prentice-Hall, Inc., Englewood Cliffs, N.J., 1950, p. 161.

being +1.33 V. It is not, however, as strong as potassium permanganate or the cerium(IV) ion. It has the advantages of being inexpensive, very stable in solution, and available in sufficiently pure form for preparing standard solutions by direct weighing. It is frequently used as a primary standard for sodium thiosulfate solutions (see the next section).

Dichromate solutions have not been so widely used as permanganate or cerium(IV) solutions in analytical procedures because they are not so strong an oxidizing agent and because of the slowness of some of its reactions. The principal use has been in the titration of iron in hydrochloric acid solution, since no difficulty is encountered in the oxidation of chloride ion if the hydrochloric acid concentration is less than 2 *M*. The compound diphenylaminesulfonic acid is a suitable indicator when iron is titrated in a sulfuric–phosphoric acid medium (page 265). This indicator has a transition potential of +0.85 V and is oxidized to a deep purple color by excess dichromate. This color is sufficiently intense to be readily detected even in the presence of green chromium(III) ion produced by the reduction of dichromate during the titration. Sodium diphenylbenzidine sulfonate ($E° = +0.87$ V) is also a suitable indicator.

As mentioned above, the principal use of dichromate solutions is in the titration of iron in hydrochloric acid solutions. An indirect method for determining oxidizing agents involves treating the sample with a known excess of iron(II), then titrating the excess with standard dichromate. Such oxidizing agents as nitrate, NO_3^-, chlorate, ClO_3^-, and hydrogen peroxide, H_2O_2, have been determined in this manner. Copper(I) has been determined by reaction with standard Fe(III) solution, followed by titration with dichromate of the resulting Fe(II).

IODINE

The iodine (triiodide)–iodide redox system,[5]

$$I_3^- + 2e \rightleftharpoons 3I^-$$

has a standard potential of +0.54 V. Iodine, therefore, is a much weaker oxidizing agent than potassium permanganate, cerium(IV) compounds, and potassium dichromate. On the other hand, iodide ion is a reasonably strong reducing agent, stronger, for example, then Fe(II) ion. In analytical processes, iodine is used as an oxidizing agent (*iodimetry*) and iodide ion is used as a reducing agent (*iodometry*). Relatively few substances are sufficiently strong reducing agents to be titrated directly with iodine. Hence the number of iodimetric determinations is small. However, many oxidizing agents are sufficiently strong to react completely with iodide ion, and there are many applications of iodometric processes. An excess of iodide ion is added to the oxidizing agent being determined, liberating iodine, which is then titrated with sodium thiosulfate solution. The reaction between

[5] The principal species in a solution of iodine and potassium iodide is the triiodide ion, I_3^-, and many chemists refer to *triiodide* solutions rather than *iodine* solutions. For simplicity, we shall continue to use the term *iodine* solutions and write equations using I_2 rather than I_3^-.

iodine and thiosulfate goes well to completion. This reaction is discussed below. It should be pointed out that some chemists prefer to avoid the term *iodimetry*, and instead, speak of direct and indirect *iodometric* processes.

Direct, or Iodimetric Processes

The more important substances that are sufficiently strong reducing agents to be titrated directly with iodine are thiosulfate, arsenic(III), antimony(III), sulfide, sulfite, tin(II), and ferrocyanide. The reducing power of several of these substances depends upon the hydrogen ion concentration, and only by proper adjustment of pH can the reaction with iodine be made quantitative.

Preparation of Iodine Solution

Iodine is only slightly soluble in water (0.00134 mol/liter at 25°C), but is quite soluble in solutions containing iodide ion. Iodine forms the triiodide complex with iodide,

$$I_2 + I^- \rightleftharpoons I_3^-$$

the equilibrium constant being about 710 at 25°C. An excess of potassium iodide is added to increase the solubility and to decrease the volatility of iodine. Usually about 3 to 4% by weight of KI is added to a 0.1 N solution, and the bottle containing the solution is well stoppered.

Iodine tends to hydrolyze in water, forming hydroiodic and hypoiodous acids,

$$I_2 + H_2O \longrightarrow HIO + H^+ + I^-$$

Conditions which increase the degree of hydrolysis should be avoided. Titrations cannot be made in very basic solutions, and standard solutions of iodine must be kept in dark bottles to prevent decomposition of HIO by sunlight,

$$2HIO \longrightarrow 2H^+ + 2I^- + O_2(g)$$

Hypoiodous acid may also be converted into iodate in basic solution,

$$3HIO + 3OH^- \longrightarrow 2I^- + IO_3^- + 3H_2O$$

Standardization

Standard iodine solutions can be prepared by direct weighing of pure iodine and dilution in a volumetric flask. The iodine is purified by sublimation and is added to a concentrated solution of KI, which is accurately weighed before and after the addition of iodine. Usually, however, the solution is standardized against a primary standard, As_2O_3 being most commonly used. The reducing power of $HAsO_2$ depends upon the pH, as shown by the following equation:

$$HAsO_2 + I_2 + 2H_2O \rightleftharpoons H_3AsO_4 + 2H^+ + 2I^-$$

The value of the equilibrium constant for this reaction is 0.17; hence the reaction does not go to completion at the equivalence point. However, by lowering the hydrogen ion concentration, the reaction is forced to the right and can be made sufficiently complete to be suitable for a titration.

From equilibrium considerations, pH values between about 5 and 11 are permissible for the titration of $HAsO_2$ with I_2.[6] However, a pH of about 8 is usually employed for the following reasons. The rate of reaction between I_2 and $HAsO_2$ is slow if the pH is less than about 7. If the pH is much greater than 9, hydrolysis and subsequent formation of IO_3^- may occur locally, where drops of titrant strike the solution. Sodium bicarbonate is usually added to the $HAsO_2$ solution before titration with I_2. The solution is then buffered at a pH slightly above 8 and the titration gives excellent results.

Starch Indicator

The color of a 0.1 N solution of iodine is sufficiently intense that iodine can act as its own indicator. Iodine also imparts an intense purple or violet color to such solvents as carbon tetrachloride or chloroform, and sometimes this is utilized in detecting the end point of titrations. More commonly, however, a solution (colloidal dispersion) of starch is employed, since the deep blue color of the starch–iodine complex serves as a very sensitive test for iodine. The sensitivity is greater in slightly acid than in neutral solutions and is greater in the presence of iodide ions.

The exact mechanism of the formation of the colored complex is not known. However, it is thought that molecules of iodine are held on the surface of β-amylose, a constituent of starch.[7] Another constituent of starch, α-amylose, or amylopectin, forms reddish complexes with iodine which are not easily decolorized. Therefore, starches that contain much amylopectin should not be used. The commercial product "soluble starch" is principally β-amylose.

Starch solutions are easily decomposed by bacteria, a process which may be retarded by sterilization or addition of a preservative. The decomposition products consume iodine and turn reddish. Mercury(II) iodide, boric acid, or furoic acid can be used as preservatives. Conditions that lead to hydrolysis or coagulation of starch should be avoided. The indicator sensitivity is decreased by increasing temperature, and by some organic materials, such as methyl and ethyl alcohols.

Determinations with Iodine

Some of the determinations that can be done by direct titration with a standard iodine solution are listed in Table 11.3. The determination of antimony is similar to that of arsenic except that tartrate ions, $C_4H_4O_6^{2-}$, are added to complex antimony and avoid precipitation of salts such as SbOCl, when the solution is neutralized. The titration is carried out in a bicarbonate buffer of pH about 8. In the determination of tin and sulfites the solution being titrated must be protected from oxidation by air. The hydrogen sulfide titration is frequently used to determine sulfur in iron or steel.

[6] E. W. Washburn, *J. Am. Chem. Soc.*, **30**, 31 (1908); R. K. McAlpine, *J. Chem. Ed.*, **26**, 362 (1949).

[7] R. E. Rundle, J. F. Foster, and R. R. Baldwin, *J. Am. Chem. Soc.*, **66**, 2116 (1944).

TABLE 11.3 Determinations by Direct Iodine Titrations

Substance determined	Reaction
Antimony	$HSbOC_4H_4O_6 + I_2 + H_2O \longrightarrow HSbO_2C_4H_4O_6 + 2H^+ + 2I^-$
Arsenic	$HAsO_2 + I_2 + 2H_2O \longrightarrow H_3AsO_4 + 2H^+ + 2I^-$
Sulfur (sulfide)	$H_2S + I_2 \longrightarrow 2H^+ + 2I^- + S$
Sulfur (sulfite)	$H_2SO_3 + I_2 + H_2O \longrightarrow H_2SO_4 + 2H^+ + 2I^-$
Tin	$Sn^{2+} + I_2 \longrightarrow Sn^{4+} + 2I^-$
Ferrocyanide	$2Fe(CN)_6^{4-} + I_2 \longrightarrow 2Fe(CN)_6^{3-} + 2I^-$

Indirect, or Iodometric Processes

Many strong oxidizing agents can be analyzed by adding potassium iodide in excess and titrating the liberated iodine. Since many oxidizing agents require acidic solutions for reactions with iodide, sodium thiosulfate is commonly used as the titrant. As discussed above, the titration with arsenic(III) requires a slightly alkaline solution.

Some precautions must be taken in handling solutions of potassium iodide to avoid errors. For example, iodide ion is oxidized by oxygen of the air:

$$4H^+ + 4I^- + O_2 \longrightarrow 2I_2 + 2H_2O$$

This reaction is slow in neutral solution but is faster in acid and is accelerated by sunlight. After addition of potassium iodide to an acid solution of an oxidizing agent, the solution should not be allowed to stand for a long time in contact with air, since additional iodine will be formed by the foregoing reaction. Nitrites should not be present because they are reduced by iodide ion to nitric oxide, which in turn is oxidized back to nitrite by oxygen of the air:

$$2HNO_2 + 2H^+ + 2I^- \longrightarrow 2NO + I_2 + 2H_2O$$

$$4NO + O_2 + 2H_2O \longrightarrow 4HNO_2$$

The potassium iodide should be free of iodates since these two substances react in acid solution to liberate iodine:

$$IO_3^- + 5I^- + 6H^+ \longrightarrow 3I_2 + 3H_2O$$

Sodium Thiosulfate

The standard solution employed in most iodometric processes is sodium thiosulfate. This salt is commonly purchased as the pentahydrate, $Na_2S_2O_3 \cdot 5H_2O$. Solutions should not be standardized by direct weighing, but should be standardized against a primary standard.

Sodium thiosulfate solutions are not stable over long periods of time. Sulfur-consuming bacteria find their way into the solution, and their metabolic processes lead to the formation of SO_3^{2-}, SO_4^{2-}, and colloidal sulfur. The presence of the latter causes turbidity, the appearance of which justifies discarding the solution. Normally, the water used to prepare thiosulfate solutions is boiled to render it sterile, and frequently borax or sodium carbonate is added as a preservative. Air oxidation of thiosulfate is slow. However, a trace of copper, which is sometimes present in distilled water, will catalyze oxidation by air.

Thiosulfate is decomposed in acidic solutions, forming sulfur as a milky precipitate:

$$S_2O_3^{2-} + 2H^+ \longrightarrow H_2S_2O_3 \longrightarrow H_2SO_3 + S(s)$$

The reaction is slow, however, and does not occur when thiosulfate is titrated into acidic solutions of iodine if the solutions are well stirred. The reaction between iodine and thiosulfate is much more rapid than the decomposition reaction.

Iodine oxidizes thiosulfate to the tetrathionate ion:

$$I_2 + 2S_2O_3^{2-} \longrightarrow 2I^- + S_4O_6^{2-}$$

The reaction is rapid, goes well to completion, and there are no side reactions. The equivalent weight of $Na_2S_2O_3 \cdot 5H_2O$ is the molecular weight, 248.17, since one electron per molecule is lost. If the pH of the solution is above 9, thiosulfate is oxidized partially to sulfate:

$$4I_2 + S_2O_3^{2-} + 5H_2O \longrightarrow 8I^- + 2SO_4^{2-} + 10H^+$$

In neutral, or slightly alkaline solution, oxidation to sulfate does not occur, especially if iodine is used as the titrant. Many strong oxidizing agents, such as permanganate, dichromate, and cerium(IV) salts, oxidize thiosulfate to sulfate, but the reaction is not quantitative.

Standardization of Thiosulfate Solutions

A number of substances can be used as primary standards for thiosulfate solutions. Pure iodine is the most obvious standard, but is seldom used because of difficulty in handling and weighing. More often, use is made of a strong oxidizing agent which will liberate iodine from iodide, an iodometric process.

Potassium dichromate. This compound can be obtained in a high degree of purity. It has a fairly high equivalent weight, it is nonhygroscopic, and the solid and its solutions are very stable. The reaction with iodide is carried out in about 0.2 to 0.4 M acid, and is complete in 5 to 10 minutes:

$$Cr_2O_7^{2-} + 6I^- + 14H^+ \longrightarrow 2Cr^{3+} + 3I_2 + 7H_2O$$

The equivalent weight of potassium dichromate is one-sixth of the molecular weight, or 49.03 g/eq. At acid concentrations greater than 0.4 M, air oxidation of potassium iodide becomes appreciable. For best results, a small portion of sodium bicarbonate or dry ice is added to the titration flask. The carbon dioxide produced displaces the air, after which the mixture is allowed to stand until the reaction is complete.

Potassium iodate and potassium bromate. Both of these salts oxidize iodide quantitatively to iodine in acid solution:

$$IO_3^- + 5I^- + 6H^+ \longrightarrow 3I_2 + 3H_2O$$

$$BrO_3^- + 6I^- + 6H^+ \longrightarrow 3I_2 + Br^- + 3H_2O$$

The iodate reaction is quite rapid; it also requires only a slight excess of hydrogen ions for complete reaction. The bromate reaction is rather slow, but the speed can be increased by increasing the hydrogen ion concentration. Usually, a small amount of ammonium molybdate is added as a catalyst.

The principal disadvantage of these two salts as primary standards is that the equivalent weights are small. In each case the equivalent weight is one-sixth of the molecular weight,[8] that of KIO_3 being 35.67, and that of $KBrO_3$ being 27.84. In order to avoid a large error in weighing, directions usually call for weighing a large sample, diluting in a volumetric flask, and withdrawing aliquots. The salt potassium acid iodate, $KIO_3 \cdot HIO_3$, can also be used as a primary standard, but its equivalent weight is also small, one-twelfth the molecular weight, or 32.49.

Copper. Pure copper can be used as a primary standard for sodium thiosulfate, and is recommended when the thiosulfate is to be used for the determination of copper. The standard potential of the Cu(II)–Cu(I) couple,

$$Cu^{2+} + e \rightleftharpoons Cu^{+}$$

is +0.15 V, and thus iodine, $E° = +0.53$ V, is a better oxidizing agent than Cu(II) ion. However, when iodide ions are added to a solution of Cu(II), a precipitate of CuI is formed,

$$2Cu^{2+} + 4I^{-} \longrightarrow 2CuI(s) + I_2$$

The reaction is forced to the right by the formation of the precipitate and also by the addition of excess iodide ion.

The pH of the solution must be maintained by a buffer system, preferably between 3 and 4. At higher pH values partial hydrolysis of Cu(II) ion takes place, and the reaction with iodide ion is slow. In very acidic solution the copper-catalyzed oxidation of iodide ion occurs at an appreciable rate.

If the anion (such as acetate) used in the buffer forms a fairly stable complex with Cu(II) ion, the reaction between Cu(II) and iodide ions may be prevented from going to completion. As iodine is removed by titration with thiosulfate, the Cu(II) complex dissociates to form more Cu(II) ion, which in turn reacts with iodide to liberate more iodine. This results in a recurring end point.

It has been found that iodine is held by adsorption of the surface of the copper(I) iodide precipitate, rendering it grayish rather than white. Unless the iodine is displaced, the end point is reached too soon and may recur if iodine is slowly desorbed from the surface. Foote and Vance[9] found that addition of potassium thiocyanate, just before the end point is reached, gives a sharper color change. There are two reasons for this. First, copper(I) thiocyanate is less soluble than copper(I) iodide, and thus thiocyanate aids in forcing the reaction to

[8] It should be noted that the iodate ion gains five electrons in the reaction with iodide ions, and therefore its equivalent weight *in this reaction* is one-fifth of the molecular weight. However, the reaction involved in the titration is that between iodine and thiosulfate. Since 1 mmol of iodate produces 3 mmol or 6 meq of iodine, the equivalent weight of iodate for the complete process is one-sixth of the molecular weight.

[9] H. W. Foote, and J. E. Vance, *J. Am. Chem. Soc.*, **57**, 845 (1935); *Ind. Eng. Chem., Anal. Ed.*, **8**, 119 (1936).

completion, even in the presence of complexing anions,

$$2Cu^{2+} + 2I^- + 2SCN^- \longrightarrow 2CuSCN(s) + I_2$$

Second, copper(I) thiocyanate may be formed on the surface of copper(I) iodide particles already precipitated:

$$CuI(s) + SCN^- \longrightarrow CuSCN(s) + I^-$$

Thiocyanate ion is more strongly adsorbed than is I_3^- ion on the CuSCN surface. Adsorbed iodine is displaced from the surface and is able to react rapidly with thiosulfate. The thiocyanate should not be added until most of the iodine has been titrated, since iodine may oxidize thiocyanate.

TABLE 11.4 Determinations by Indirect Iodine Titrations

Substance determined	Reaction
Copper	$2Cu^{2+} + 4I^- \longrightarrow 2CuI(s) + I_2$
Iron	$2Fe^{3+} + 2I^- \longrightarrow 2Fe^{2+} + I_2$
Chromium	$Cr_2O_7^{2-} + 6I^- + 14H^+ \longrightarrow 2Cr^{3+} + 3I_2 + 7H_2O$
Arsenic	$AsO_4^{3-} + 2I^- + 2H^+ \longrightarrow AsO_3^{3-} + I_2 + H_2O$
Chlorine	$Cl_2 + 2I^- \longrightarrow 2Cl^- + I_2$
Bromine	$Br_2 + 2I^- \longrightarrow 2Br^- + I_2$
Hydrogen peroxide	$H_2O_2 + 2I^- + 2H^+ \longrightarrow I_2 + 2H_2O$
Chlorate	$ClO_3^- + 6I^- + 6H^+ \longrightarrow Cl^- + 3I_2 + 3H_2O$
Bromate	$BrO_3^- + 6I^- + 6H^+ \longrightarrow Br^- + 3I_2 + 3H_2O$
Iodate	$IO_3^- + 5I^- + 6H^+ \longrightarrow 3I_2 + 3H_2O$
Nitrite	$2HNO_2 + 2I^- + 2H^+ \longrightarrow 2NO + I_2 + 2H_2O$

Iodometric Determinations

There are many applications of iodometric processes in analytical chemistry. Some of these are listed in Table 11.4. The iodometric determination of copper is widely used for both ores and alloys. The method gives excellent results and is more rapid than the electrolytic determination of copper. Copper ores commonly contain iron, arsenic, and antimony. These elements in their higher oxidation states (usually so from the dissolving process) will oxidize iodide and thus interfere. The interference of iron can be prevented by addition of ammonium bifluoride, NH_4HF_2, which converts iron(III) ion into the stable complex, FeF_6^{3-}. As previously mentioned, antimony and arsenic will not oxidize iodide ions except in solutions of high acidity. By adjusting the pH to about 3.5 with a buffer, interference from these two elements is eliminated. Park[10] suggested using a phthalate buffer for this purpose. However, later investigations[11] showed that a solution of bifluoride ion, HF_2^-, added to complex iron, gives a buffer of approximately the desired pH, so that no additional buffer is needed.

[10] B. Park, *Ind. Eng. Chem., Anal. Ed.*, **3**, 77 (1931).

[11] W. R. Crowell, T. E. Hillis, S. P. Rittenberg, and R. F. Svenson, *Ind. Eng. Chem., Anal. Ed.*, **8**, 9 (1936).

The classical method of Winkler[12] is a sensitive method for determining oxygen dissolved in water. To the water sample is added an excess of a manganese(II) salt, sodium iodide, and sodium hydroxide. White $Mn(OH)_2$ is precipitated and is quickly oxidized to brown $Mn(OH)_3$:

$$4Mn(OH)_2(s) + O_2 + 2H_2O \longrightarrow 4Mn(OH)_3(s)$$

The solution is acidified and the $Mn(OH)_3$ oxidizes iodide to iodine.

$$2Mn(OH)_3(s) + 2I^- + 6H^+ \longrightarrow 2Mn^{2+} + I_2 + 6H_2O$$

The liberated iodine is titrated with a standard solution of sodium thiosulfate.

PERIODIC ACID

The compound paraperiodic acid, H_5IO_6, is a powerful oxidizing agent which is extremely useful in performing selective oxidations of organic compounds with certain functional groups. The standard potential of the reaction

$$H_5IO_6 + H^+ + 2e \rightleftharpoons IO_3^- + 3H_2O$$

is about + 1.6 to 1.7 V.

Preparation and Standardization of Solutions

Three compounds are available commercially for the preparation of periodate solutions: H_5IO_6, paraperiodic acid; $NaIO_4$, sodium metaperiodate; and KIO_4, potassium metaperiodate. Of these compounds, $NaIO_4$ is generally preferred because of its relatively high solubility and ease of purification. Solutions as concentrated as 0.06 *M* can be prepared. Periodic acid solutions slowly oxidize water to oxygen and ozone. Solutions containing an excess of sulfuric acid are the most stable.

Periodate solutions are standardized by an iodometric procedure:

$$IO_4^- + 2I^- + H_2O \longrightarrow IO_3^- + I_2 + 2OH^-$$

Excess potassium iodide is added to an aliquot of the periodate and the liberated iodine is titrated with a standard solution of arsenic(III):

$$I_2 + AsO_2^- + 2H_2O \longrightarrow HAsO_4^{2-} + 2I^- + 3H^+$$

The solution is buffered at a *p*H of 8 to 9 with borax or sodium bicarbonate.

The Malaprade Reaction

In 1928 Malaprade[13] reported that periodic acid could be used for the selective oxidation of organic compounds with hydroxyl groups on *adjacent*

[12] L. W. Winkler, *Ber.*, **21**, 2843 (1888); *Standard Methods for the Examination of Water and Sewage*, 9th ed., American Public Health Association, New York, 1946, p. 124.

[13] L. Malaprade, *Compt. Rend.*, **186**, 382 (1928); *Bull. Soc. Chim. France*, (4)**43**, 683 (1928).

carbon atoms. The reaction with ethylene glycol is

$$\begin{matrix} H_2C{-}OH \\ | \\ H_2C{-}OH \end{matrix} + H_4IO_6^- \longrightarrow 2H_2C{=}O + IO_3^- + 3H_2O$$

Ethylene glycol — Formaldehyde

The carbon–carbon bond in the glycol is broken and two molecules of formaldehyde are produced. Excess periodate is used and after the reaction is complete, the excess can be determined by the iodometric procedure used for standardization. The reaction is carried out at room temperature, usually requiring from 30 minutes to 1 hour. At higher temperatures undesired side reactions may occur and the selectivity of the periodate oxidation is not attained. Aqueous solutions which are neutral, slightly acid, or slightly basic can often be employed, although an organic solvent may be required if the compound is not soluble in water.

Organic compounds which contain carbonyl groups ($>C{=}O$) on adjacent carbon atoms are also oxidized by periodate. The reaction with glyoxal is

$$\begin{matrix} H{-}C{=}O \\ | \\ H{-}C{=}O \end{matrix} + H_4IO_6^- \longrightarrow 2H{-}\underset{\displaystyle OH}{\underset{|}{C}}{=}O + IO_3^- + H_2O$$

Glyoxal — Formic acid

The carbon–carbon bond is broken and two molecules of formic acid are produced. A compound which contains a hydroxyl and a carbonyl group on adjacent carbon atoms is oxidized to an acid and an aldehyde:

$$\begin{matrix} H \\ | \\ CH_3{-}C{-}OH \\ | \\ CH_3{-}C{=}O \end{matrix} + H_4IO_6^- \longrightarrow CH_3{-}\overset{\displaystyle H}{\overset{|}{C}}{=}O + CH_3{-}\underset{\displaystyle OH}{\underset{|}{C}}{=}O + IO_3^- + 2H_2O$$

Acetoin — Acetaldehyde — Acetic acid

The compound glycerol is oxidized to 2 mol of formaldehyde and 1 mol of formic acid. The products can be rationalized by picturing the reaction occurring in two steps. The first step produces two aldehydes,

$$\begin{matrix} H \\ | \\ H{-}C{-}OH \\ | \\ H{-}C{-}OH \\ | \\ H{-}C{-}OH \\ | \\ H \end{matrix} \longrightarrow H{-}\overset{\displaystyle H}{\overset{|}{C}}{=}O + \begin{matrix} H{-}C{=}O \\ | \\ H{-}C{-}OH \\ | \\ H \end{matrix} \longrightarrow H{-}\overset{\displaystyle H}{\overset{|}{C}}{=}O + H{-}\underset{\displaystyle OH}{\underset{|}{C}}{=}O$$

Glycerol — Formaldehyde — Glycolic aldehyde — Formaldehyde — Formic acid

One of these aldehydes, glycolic aldehyde, contains a carbonyl and a hydroxyl group on adjacent carbon atoms and is therefore oxidized to an aldehyde and an acid.

The Malaprade reaction has found many applications in the determination of organic compounds. Glycerol and ethylene glycol can be easily determined by the reactions described above. Mixtures of glycerol, ethylene glycol and 1,2-propyleneglycol have been determined in fats and waxes used in cosmetic materials. In addition, compounds containing a hydroxyl group and an amino group ($—NH_2$) on adjacent carbon atoms are readily oxidized to an aldehyde and ammonia. The ammonia can be distilled from the alkaline reaction solution and titrated with standard acid. This procedure has been used in the analysis of α-hydroxyamino compounds in various proteins.

POTASSIUM BROMATE

Potassium bromate, $KBrO_3$, is a strong oxidizing agent, the standard potential of the reaction

$$BrO_3^- + 6H^+ + 6e \longrightarrow Br^- + 3H_2O$$

being +1.44 V. The reagent can be employed in two ways, as a direct oxidant for certain reducing agents, and for the generation of known quantities of bromine.

Direct Titrations

A number of reducing agents, such as arsenic(III), antimony(III), iron(II), and certain organic sulfides and disulfides can be titrated directly with a solution of potassium bromate. The reaction with arsenic(III) is

$$BrO_3^- + 3HAsO_2 \longrightarrow Br^- + 3HAsO_3$$

The solution is usually about 1 *M* in hydrochloric acid. The end point of the titration is marked by the appearance of bromine, according to the reaction

$$BrO_3^- + 5Br^- + 6H^+ \longrightarrow 3Br_2 + 3H_2O$$

The appearance of bromine is sometimes suitable for determining the end point of the titration. Several organic indicators which react with bromine to give a color change have been studied.[14] The color change is usually not reversible and considerable care must be taken to obtain good results. There are three indicators which have been found to behave reversibly: α-naphthoflavone, quinoline yellow, and *p*-ethoxychrysoidine. These are commercially available.

Bromination of Organic Compounds

A standard solution of potassium bromate can be employed for the generation of known quantities of bromine. The bromine can then be used to quantitatively brominate various organic compounds. Excess bromide (with respect to bromate) is present in such cases, so that the quantity of bromine generated can be

[14] G. F. Smith and H. H. Bliss, *J. Am. Chem. Soc.*, **53**, 209 (1931).

calculated from the quantity of $KBrO_3$ taken. Usually, bromine is generated in excess of the quantity required to brominate the organic compound in order to help force this reaction to completion.

The reaction of bromine with the organic compound is either one of substitution or addition. The reaction with 8-hydroxyquinoline is a substitution reaction:

$$\text{OH, N (8-hydroxyquinoline)} + 2Br_2 \longrightarrow \text{Br, OH, N, Br (dibromo product)} + 2HBr$$

The reaction with ethylene is an addition reaction:

$$H_2C{=}CH_2 + Br_2 \longrightarrow H_2CBr{-}CBrH_2$$

In the analysis of an organic compound, a measured excess of a $KBr{-}KBrO_3$ mixture is added and the mixture acidified, liberating Br_2. After the bromination reaction is complete, the excess bromine is determined by the addition of potassium iodide, followed by the titration of the liberated iodine with standard sodium thiosulfate:

$$Br_2 + 2I^- \longrightarrow I_2 + 2Br^-$$

$$I_2 + 2S_2O_3^{2-} \longrightarrow 2I^- + S_4O_6^{2-}$$

A common application is the determination of metals with 8-hydroxyquinoline (page 87). A metal such as aluminum is precipitated with the organic reagent, and the precipitate is filtered, washed, and dissolved in hydrochloric acid. Potassium bromide and standard potassium bromate are then added. The reactions with aluminum (8-hydroxyquinoline abbreviated HQ) are as follows:

$$Al^{3+} + 3HQ \longrightarrow AlQ_3(s) + 3H^+ \qquad \text{(precipitation)}$$

$$AlQ_3(s) + 3H^+ \longrightarrow Al^{3+} + 3HQ \qquad \text{(redissolving)}$$

$$3HQ + 6Br_2 \longrightarrow 3HQBr_2 + 6HBr \qquad \text{(bromination)}$$

The number of equivalents of bromate is the same as the equivalents of aluminum. Here the equivalent weight of aluminum is one-twelfth its atomic weight, since $1Al^{3+} = 3HQ = 6Br_2 = 12$ electrons.

Addition reactions of bromine are used primarily in the determination of unsaturation in petroleum products and fats and oils. Many examples are found in the literature.

REDUCING AGENTS

Standard solutions of reducing agents are not as widely used as are those of oxidizing agents, because most reducing agents are slowly oxidized by oxygen of the air. Sodium thiosulfate is the only common reducing agent that can be kept for long periods of time without undergoing air oxidation. This reagent is used exclusively for iodine titrations, and its properties have already been discussed. The following are other reducing agents that are sometimes employed in the laboratory.

Iron(II)

Solutions of iron(II) ions in 0.5 to 1 *N* sulfuric acid are only slowly oxidized by air, and can be employed as standard solutions. The normality should be checked at least daily. Permanganate, cerium(IV), or dichromate solutions are suitable for titration of the iron(II) solution.

Chromium(II)

The chromium(II) ion is a powerful reducing agent, the standard potential of the reaction

$$Cr^{3+} + e \rightleftharpoons Cr^{2+}$$

being −0.41 V. Solutions are oxidized rapidly by air and extreme care must be employed in their use. Many substances have been determined by titration with either chromium(II) chloride or sulfate, including iron, copper, silver, gold, bismuth, uranium, and tungsten.

Titanium(III)

Salts of titanium(III) are also strong reducing agents, the standard potential of the reaction

$$TiO^{2+} + 2H^+ + e \longrightarrow Ti^{3+} + H_2O$$

being +0.04 V. Solutions of these salts are readily oxidized by air, but are easier to handle than solutions of chromium(II) salts. The principal use of titanium(III) solutions is in titrating solutions of iron(III). Other substances which can be determined include copper, tin, chromium, and vanadium.

Oxalate and Arsenic(III)

The reactions of sodium oxalate and arsenic(III) acid have already been discussed. Standard solutions of oxalic acid are fairly stable; those of sodium oxalate are much less stable. Neutral or weakly acid solutions of $HAsO_2$ are fairly stable, but alkaline solutions are slowly oxidized by air.

QUESTIONS

1. *Potassium permanganate.* List some advantages and disadvantages in using standard solutions of potassium permanganate as an oxidizing agent. What reactions would interfere with its use were it not for the fact that the rates are slow?

2. *Oxidizing agents.* Point out the advantages and disadvantages of solutions of cerium(IV) salts, potassium dichromate, and iodine as standard solutions of oxidizing agents.

3. *Equivalent weights.* Calculate the equivalent weights of the following substances when determined by the reactions given in Table 11.4: (a) copper; (b) chromium; (c) bromine; (d) $NaClO_3$; (e) KIO_3.

4. *Errors.* Point out the errors, if any, in the following procedures. Explain the effect of the error on the determination being carried out.
(a) Copper was being determined iodometrically; the solution was strongly basic when the titration of iodine with thiosulfate was carried out.
(b) After an iodine solution was standardized, it was not placed in a brown bottle and was exposed to sunlight for several days before it was used to analyze an arsenic unknown.
(c) A potassium iodide solution was prepared for use in the determination of iron(III), and then left in an open beaker on a desktop for several hours before use.
(d) An iron ore sample was dissolved and reduced with tin(II) chloride. Mercury(II) chloride was then added slowly while the solution was still hot.
(e) In the iodometric determination of copper, no potassium thiocyanate was added.
(f) An iron ore sample was dissolved in hydrochloric acid and titrated with permanganate without addition of preventive solution.

5. *Procedures.* Explain the reasons for the following procedures:
(a) Heating and filtering a permanganate solution before standardization.
(b) Maintaining the pH between 3 and 4 in the iodometric determination of copper.
(c) Addition of NH_4HF in the iodometric determination of copper.
(d) Amalgamating the zinc for use in a Jones reductor.
(e) Allowing a cerium(IV) sulfate solution to stand for several days before standardization.
(f) Addition of both sulfuric and phosphoric acids in the titration of iron with dichromate.

Multiple-choice: In the following multiple-choice questions, select the *one best* answer.

6. Which of the following reactions does not require a catalyst to obtain a convenient rate? (a) Titration of arsenic(III) with $KMnO_4$; (b) Titration of arsenic(III) with cerium(IV); (c) Titration of oxalate with cerium(IV); (d) Titration of arsenic(III) with iodine at pH 8.5.

7. Which of the following statements is false? (a) Acidic solutions of permanganate are not stable; (b) In the Fowler-Bright method the titration is carried out slowly at elevated temperature; (c) MnO_2 catalyzes the decomposition of a $KMnO_4$ solution; (d) The oxidation of Cl^- by permanganate is fast in the presence of iron(II).

8. Which of the following statements is true? (a) In an iodometric process iodine is used as an oxidizing agent; (b) $K_2Cr_2O_7$ is a stronger oxidizing agent than $KMnO_4$ in 1 M acid; (c) the Ce^{4+} ion hydrolyzes readily in solutions of low pH; (d) Soluble starch is principally β-amylose.

9. Iron(III) interferes in the iodometric determination of copper. The interference can be prevented by the addition of (a) NH_4Cl; (b) NH_4HF_2; (c) $Na_2S_2O_3$; (d) KI.

10. When 1 mol of glycerol is oxidized by periodic acid the organic products are (a) 2 mol of formaldehyde; (b) 2 mol of formic acid; (c) 1 mol of formaldehyde + 2 mol of formic acid; (d) 2 mol of formaldehyde +1 mol of formic acid.

PROBLEMS

1. *Preliminary redox reactions.* Write a balanced equation for the following redox reactions which may be involved in a preliminary step of an analysis. Add H_2O, H^+, or OH^- as needed. (See Appendix II.)
   (a) Oxidation of Fe^{2+} to Fe^{3+} by H_2O_2 in acid medium.
   (b) Oxidation of Mn^{2+} to MnO_2 by H_2O_2 in basic medium.
   (c) Oxidation of Mn^{2+} to MnO_4^- by $S_2O_8^{2-}$ in acid medium.
   (d) Oxidation of Cr^{3+} to $Cr_2O_7^{2-}$ by BiO_3^- in acid medium.
   (e) Reduction of Fe^{3+} to Fe^{2+} by SO_2 in acid medium.

2. *Malaprade reaction.* Write the balanced half-reaction (Appendix II) for the following reactions:
   (a) Oxidation of glyoxal, $H_2C_2O_2$, to formic acid, H_2CO_2, in acid medium.
   (b) Oxidation of glycerol, $C_3H_8O_3$, to 2 mol of formaldehyde, H_2CO, and 1 mol of formic acid, H_2CO_2, in acid medium.
   Calculate the equivalent weights of glyoxal and glycerol in these reactions.

3. *Oxidation of ethylene glycol.* When ethylene glycol, $C_2H_6O_2$, is oxidized by acidic permanganate the oxidation products are CO_2 and H_2O. When $Ce(ClO_4)_4$ is used the product is formic acid, H_2CO_2. When periodic acid is used the glycol is oxidized to formaldehyde, H_2CO, and the periodic acid is reduced to IO_3^-. Write complete balanced equations for these three reactions and calculate the equivalent weight of ethylene glycol (MW 62.068) in each case.

4. *Titer.* A solution of $KMnO_4$ is 0.0230 *M*. Calculate the titer of this solution for reduction to Mn^{2+} in terms of the following substances: (a) $Na_2C_2O_4$ (oxidized to CO_2); (b) As_2O_3 [oxidized to As(V)]; (c) SO_2 (oxidized to H_2SO_4); (d) Fe (oxidized from Fe^{2+} to Fe^{3+}).

5. *Titer.* The copper titer of a solution of sodium thiosulfate is 13.24 mg/ml. (a) How many milligrams of $Na_2S_2O_3 \cdot 5H_2O$ does each milliliter contain? (b) Calculate the titer of the thiosulfate solution in terms of $K_2Cr_2O_7$, KIO_3, and H_2O_2.

6. *Potassium permanganate.* A solution of $KMnO_4$ has a normality of 0.0600 when used in a solution of *p*H 10.0. (a) What is the half-reaction of MnO_4^- at this *p*H? (b) What is the normality of solution if the permanganate is allowed to react in a 0.20 *N* solution of H_2SO_4? (c) What is the normality if the reaction is carried out in 1 *M* NaOH? (d) What is the molarity of the solution?

7. *Reduction of iron.* Calculate the volume of 0.10 *M* $SnCl_2$ required to reduce the iron in a 0.50-g sample that contains 20% goethite, $Fe_2O_3 \cdot H_2O$. The iron is reduced from Fe^{3+} to Fe^{2+} and the tin is oxidized from Sn^{2+} to Sn^{4+}.

8. *Hydrogen peroxide.* What volume of 0.0300 *M* $KMnO_4$ is required to react in acid solution with 5.00 ml of H_2O_2 that has a density of 1.01 g/ml and contains 3.00% by weight H_2O_2? The permanganate is reduced to Mn^{2+} and the H_2O_2 oxidized to O_2.

9. *Oxidation of sulfur dioxide.* The reaction between potassium permanganate and sulfur dioxide in acid solution is as follows:

$$MnO_4^- + SO_2 + H_2O \longrightarrow Mn^{2+} + H^+ + SO_4^{2-}$$

(a) Balance the equation. (b) If 45.00 ml of 0.04000 *M* $KMnO_4$ is treated with SO_2, how many milliliters of 0.100 *N* NaOH is required to neutralize the acid formed? The excess SO_2 is removed by boiling.

10. *Oxalates.* How many grams of solid $Na_2C_2O_4$ should be added to 250 ml of a 0.200 *M* solution of the salt $KHC_2O_4 \cdot H_2C_2O_4$ in order that the normality of the solution as a reducing agent be four times that as an acid? The oxalate is oxidized to CO_2.

11. *Calcium in limestone.* A student's limestone sample contained 30.00% CaO. He precipitated the calcium as CaC_2O_4, dissolved the precipitate in H_2SO_4, and titrated the solution with standard $KMnO_4$. He obtained a value of 32.50% CaO in the sample. If his error was caused by the precipitate being contaminated with MgC_2O_4, what percentage of the weight of the precipitate he obtained was MgC_2O_4?

12. *Determination of copper.* A sample of copper oxide weighing 2.013 g is dissolved in acid, the *p*H adjusted, and excess KI added, liberating I_2. The I_2 is titrated with 29.68 ml of 0.1058 *N* $Na_2S_2O_3$. Calculate the percentage of CuO in the sample.

13. *Determination of vitamin C.* Ascorbic acid (vitamin C, MW 176.126) is a reducing agent, reacting as follows:

$$C_6H_8O_6 \longrightarrow C_6H_6O_6 + 2H^+ + 2e$$

It can be determined by oxidation with a standard solution of iodine. A 200.0-ml sample of a citrus fruit drink was acidified with sulfuric acid, and 10.00 ml of 0.0250 *M* I_2 added. After the reaction was complete, the excess I_2 was titrated with 4.60 ml of 0.0100 *M* $Na_2S_2O_3$. Calculate the number of milligrams of ascorbic acid per milliliter of the drink.

14. *Bleaching powder.* Bleaching powder, Ca(OCl)Cl, reacts with iodide ion in acid medium liberating iodine:

$$OCl^- + I^- + H^+ \longrightarrow I_2 + Cl^- + H_2O$$

(a) Balance the equation. (b) If 35.24 ml of 0.1084 *N* $Na_2S_2O_3$ is required to titrate the iodine liberated from a 0.6000-g sample of bleaching powder, calculate the percentage of Cl in the sample.

15. *Cerium(IV) solutions.* 32.0 g of the salt $Ce(SO_4)_2 \cdot 2(NH_4)_2SO_4 \cdot 2H_2O$ is dissolved in 500 ml of solution. Calculate (a) the normality of the solution; (b) the $Na_2C_2O_4$ titer.

16. *Cerium(IV) solutions.* 41.1 g of the salt $Ce(NO_3)_4 \cdot 2(NH_4)_2NO_3$ is dissolved in 750 ml of 0.2 *M* H_2SO_4. Calculate the number of grams of As_2O_3 which would be required to react with 50.0 ml of this solution.

17. *Factor weight solution.* How many grams of $K_2Cr_2O_7$ should be dissolved in 300.0 ml of solution in order that the volume used in a titration equal the percentage of Fe_2O_3 in a 2.000-g sample?

18. *Mixture of iron and vanadium.* A 50.00-ml sample of a solution containing iron(III) and vanadium(V) was passed through a silver reductor and then titrated with 30.42 ml of 0.0982 *N* Ce^{4+} solution. A second 50.00-ml sample was passed through a Jones reductor and then titrated with 42.34 ml of the same Ce^{4+} solution. Calculate the molarities of iron(III) and vanadium(V) in the original solution. See Table 11.1 for reactions in the reductors. Assume that vanadium is reoxidized to vanadium(V) by the Ce^{4+}.

19. *Mixture of chromium and iron.* A 25.00-ml aliquot of a solution containing iron(III) and chromium(III) is passed through a Jones reductor and then titrated with 35.83 ml of 0.1016 *N* $KMnO_4$ in strong acid solution. A second 25.00-ml sample of the same solution required 14.65 ml of the permanganate solution for titration after passing through a silver reductor. Calculate the molarities of the iron(III) and chromium(III) in the original solution. See Table 11.1 for reactions in the reductors.

20. *Malaprade reaction.* 50.00 ml of a solution containing ethylene glycol was treated with 50.00 ml of a 0.03010 *M* solution of paraperiodic acid and the solution allowed to stand for 1 hr. The solution was then buffered at *p*H 8 with $NaHCO_3$ and excess KI added. The liberated I_2 was titrated with 25.38 ml of 0.0602 *N* arsenic(III) solution. Calculate the number of milligrams of ethylene glycol in the original solution.

21. *Determination of sodium formate.* Sodium formate, $NaCHO_2$, reacts with permanganate in neutral solution according to the equation (unbalanced)

$$CHO_2^- + MnO_4^- + H_2O \longrightarrow MnO_2(s) + CO_2 + OH^-$$

(a) Balance the equation. (b) A 0.500-g sample containing sodium formate is treated with 50.00 ml of 0.0600 *M* MnO_4^- (an excess) in neutral solution, and the reaction given above is allowed to proceed to completion. The solution is filtered to remove the MnO_2, acidified with H_2SO_4, and titrated with 30.00 ml of 0.1000 *M* $H_2C_2O_4$. The MnO_4^- is reduced to Mn^{2+} and the oxalate oxidized to CO_2. Calculate the percentage of $NaCHO_2$ in the example.

22. *Potassium bromate.* How many milliliters of a 0.050 *M* $KBrO_3$ solution is needed to furnish sufficient Br_2 to react with the 8-hydroxyquinoline precipitate obtained from 10.0 mg of aluminum?

23. *Potassium iodate.* 35.00 ml of a solution of KIO_3 reacted with exactly 35.00 ml of a 0.1000 *M* solution of KI in acid medium according to the equation (unbalanced)

$$IO_3^- + I^- + H^+ \longrightarrow I_2 + H_2O$$

(a) Balance the equation and calculate the molarity of the KIO_3 solution. (b) 50.00 ml of this KIO_3 is treated with excess pure KI in acid solution and the liberated I_2 is titrated with 0.1000 *N* $Na_2S_2O_3$. What volume of titrant is required?

24. *Determination of potassium iodide.* A sample of impure KI weighing 0.600 g is dissolved in water, the solution acidified, and 25.00 ml of 0.0400 *M* KIO_3 (an excess) is added. The iodate is reduced to I_2 and the iodide is oxidized to I_2. The I_2 is boiled off, the solution cooled, and an excess of pure KI is added to react with the unused KIO_3. The I_2 produced is titrated with 40.38 ml of 0.1053 *N* thiosulfate. Write the equations for the reactions which occurred and calculate the percentage of KI in the sample.

25. 8-*Hydroxyquinoline.* A 0.3000-g sample containing MgO is dissolved and the magnesium precipitated with 8-hydroxyquinoline. The precipitate is filtered, dissolved in acid, and 50.00 ml of 0.0400 *M* $KBrO_3$ and 3 g of KBr are added. After the reaction between bromine and 8-hydroxyquinoline is complete, 3 g of KI is added and the liberated I_2 is titrated with 40.00 ml of 0.0500 M $Na_2S_2O_3$. Calculate the percentage of MgO in the sample.

26. *Permanganate end point.* It was pointed out on page 278 that the slight excess of permanganate normally present at the end point of a titration is sufficient to cause the reaction

$$3Mn^{2+} + 2MnO_4^- + 2H_2O \longrightarrow 5MnO_2(s) + 4H^+$$

to occur. (a) Write the two half-reactions and calculate the equilibrium constant for the reaction. (b) Show that the reaction does tend to go to the right, assuming the following concentrations at the end point: $[Mn^{2+}] = 0.01\ M$, $[MnO_4^-] = 1 \times 10^{-5}\ M$, and $[H^+] = 1\ M$.

27. *Arsenic(III)–I_2 reaction.* (a) Calculate the equilibrium constant for the reaction

$$HAsO_2 + I_3^- + 2H_2O \rightleftharpoons 3I^- + H_3AsO_4 + 2H^+$$

and compare it with the experimental value (page 286). (b) If 2.0 mmol of $HAsO_2$ is titrated with I_2 at pH 5.0, calculate (using the calculated K) the number of milligrams of the acid that remain unoxidized at the end point. Assume that at the end point: volume = 100 ml, $[I^-] = 0.10\ M$, $[I_3^-] = 2 \times 10^{-5}\ M$. (c) Why is the titration not carried out at pH 5.0?

CHAPTER TWELVE

potentiometric methods of analysis

Since the potentials of galvanic cells depend upon the activities of certain ionic species in the cell solutions measurements of cell potentials are of considerable importance in analytical chemistry. In many cases a cell can be devised whose potential depends upon the activity of a single ionic species in the solution. One of the electrodes of the cell must be such that its potential depends upon the activity of the ion being determined; it is called an *indicator electrode*. The other electrode is a reference, such as calomel, whose potential is known and remains constant during the measurement.

There are two methods used in making experimental measurements. First, a single measurement of the potential of the cell is made; this is sufficient to determine the activity of the ion of interest. Second, the ion may be titrated and the potential measured as a function of the volume of titrant. The first method is called *direct potentiometry* and has been used principally in the measurement of *p*H of aqueous solutions. Today it is also widely applied to the determination of other ions through the use of *ion-selective electrodes*. The second method, called *potentiometric titration*, employs the measurement of potential to detect the equivalence point of a titration. It can be applied to all types of reactions we have found suitable for titrimetric analysis.

Direct potentiometric measurements are very useful for determining the activity of a species in an equilibrium mixture, since the equilibrium is not disturbed by the measurement. For example, the *p*H of a 0.10 F solution of acetic acid might be measured and the hydrogen ion concentration (estimated from the activity) found to be 0.0013 M. On the other hand, if the solution is titrated, we will find that the concentration is 0.10 M. The titration yields stoichiometric information on the total number of available protons, whereas the direct measurement gives the equilibrium activity of protons in the solution at any instant.

Because of the logarithmic relation of potential to activity, determinations by direct potentiometry are usually not highly accurate unless special measures are taken. Playing with numbers in the Nernst equation easily shows that errors of the order of a few percent in an ion activity may result from a 1-mV error in measuring the potential.[1]

There is another problem with direct potentiometry. For certain purposes (e.g., determining thermodynamic equilibrium constants) the *activity* of an electroactive species may be desired, but in analytical work we usually want to know the *concentration.* Unless the composition of the solution with respect to *all* ions is specified, converting an activity into a concentration is a risky game, and even then one may well be unable to find a suitable activity coefficient in the literature for the stated conditions. This problem can be countered by calibration if all the unknown samples have the same gross composition. A series of standards is prepared in which the ion to be determined is varied but which are as similar as possible to the unknowns in every other regard. Potential readings are then converted to concentrations using a graph of E vs. log C plotted from measurements on the standards. (An example might be determining a minor metal ion in seawater; the standards would be synthetic seawater as close as possible to the real thing in regard to salt and other major solutes, spiked with known quantities of the metal ion.) This type of problem—the effect of the overall composition of a sample, particularly in regard to major components, upon the response to the element being determined—is sometimes called a *matrix effect.* We shall see examples of matrix effects with other techniques in later chapters; they are an important concern in many practical analytical situations. Sometimes, if a series of samples are too variable in gross composition, the analyst in effect creates his own matrix by adding a large quantity of some solute, thereby swamping out the smaller differences between samples.

Despite such problems, direct potentiometry is attractive because it is fast, not terribly expensive, easily automated, and nondestructive of the sample. Where matrix effects can be compensated for, and when the highest accuracy is not required, these methods find wide acceptance.

In this chapter we shall discuss the types of indicator electrodes which are available for various ions. The measurement of *p*H by direct potentiometry will be examined in some detail. Then we shall discuss the application of potentiometric methods to the four types of reactions used in titrimetric analyses.

INDICATOR ELECTRODES

Metallic Electrodes

Some metals, such as silver, mercury, copper, and lead, can act as indicator electrodes when they are in contact with a solution of their ions. For example, the

[1] For an example of a special technique for lowering the errors in direct potentiometry, see H. V. Malmstedt and J. D. Winefordner, *Anal. Chim. Acta*, **20**, 283 (1959).

potential developed at a piece of silver wire dipping in a solution of silver nitrate varies with the activity of silver ion in accordance with the prediction of the Nernst equation. Apparently, reversible electron transfer takes place between the metal surface and the ions in solution:

$$Ag^+ + e \rightleftharpoons Ag \qquad E^\circ = +0.80 \text{ V}$$

and the potential is given by the equation

$$E = 0.80 - 0.059 \log \frac{1}{a_{Ag+}}$$

Electrodes of this type, in which the ion exchanges electrons directly with the metal, are called "electrodes of the first kind."

A number of metals, such as nickel, cobalt, chromium, and tungsten, do not give reproducible potentials when used as electrodes. Such metals are harder and more brittle than those which do behave satisfactorily. It is thought that crystal deformations and oxide coatings may account for this behavior.

The silver–silver chloride electrode, discussed previously as a reference electrode (page 252), is an example of an "electrode of the second kind." The potential is a function of the activity of chloride ions in the solution. The equilibrium can be written

$$AgCl(s) + e \rightleftharpoons Ag + Cl^- \qquad E^\circ = +0.22 \text{ V}$$

and the potential is given by

$$E = 0.22 - 0.059 \log \frac{1}{a_{Cl^-}}$$

In an electrode of the second kind, the ions in solution, Cl^- in this case, do not exchange electrons directly with the metal electrode. Rather, they regulate the concentration of silver ions which exchange electrons with the metal surface.

A widely used "electrode of the third kind" is the mercury–EDTA electrode. It was observed by Reilley and Schmid[2] that the potential of a mercury electrode responds reversibly to other metal ions in solution in the presence of the mercury–EDTA complex. We can represent the electrode system as follows (using M^{2+} as the metal of interest and Y^{4-} for the tetraanion of EDTA):

$$Hg|M^{2+} + MY^{2-} + HgY^{2-} + Hg^{2+}$$

A small amount of HgY^{2-} is added to the solution. Some of it dissociates to give Hg^{2+} ions, but since the complex is so stable, most of the mercury remains in the form of HgY^{2-}. The metal to be determined must form a weaker EDTA complex than does mercury. Hence M^{2+} will not remove an appreciable amount of Y^{4-} from Hg^{2+}. Letting K_{HgY} and K_{MY} represent the stability constants of the two complexes, we can derive a relation between the potential and the concentration

[2] C. N. Reilley and R. W. Schmid, *Anal. Chem.*, **30**, 947 (1958); R. W. Schmid and C. N. Reilley, *J. Am. Chem. Soc.*, **78**, 5513 (1956).

of M^{2+} as follows. First mercury(II) ions exchange electrons with the metal:

$$Hg^{2+} + 2e \rightleftharpoons Hg \qquad E° = +0.85 \text{ V}$$

$$E = 0.85 - \frac{0.059}{2} \log \frac{1}{[Hg^{2+}]}$$

If we substitute for $[Hg^{2+}]$ from the expression for K_{HgY} and for $[Y^{4-}]$ using the expression for K_{MY}, we obtain the equation

$$E = 0.85 - 0.03 \log \frac{K_{HgY}[MY^{2-}]}{K_{MY}[HgY^{2-}][M^{2+}]}$$

Let us assume we are titrating M^{2+} with Y^{4-}. Near the equivalence point the term $[MY^{2-}]$ becomes essentially constant. The terms K_{HgY} and K_{MY} are constants, and the concentration of the mercury complex, $[HgY^{2-}]$, is constant throughout the titration since the complex is so stable. Hence the equation above becomes

$$E = K - 0.03 \log \frac{1}{[M^{2+}]}$$

$$E = K - 0.03\, pM$$

The term K is a composite of all the constant terms.

The mercury electrode has proved applicable to the potentiometric determination of about 30 metal ions, either by direct or back-titration procedures. Potentials are established rather rapidly with the electrode since all the species except mercury itself are in solution.

An inert metal, such as platinum or gold, serves well as the indicator electrode for redox couples such as $Fe^{3+} + e \rightleftharpoons Fe^{2+}$ (page 252). The function of the metal is simply to transfer electrons; it does not itself enter into the redox reaction. The potential developed is a function of the ratio of the concentrations of the oxidized and reduced forms of the redox couple.

Care must be taken in the use of platinum in the presence of strong oxidizing agents, especially in solutions that contain a high concentration of chloride ions. Platinum is more readily oxidized in the presence of chloride ions because of the formation of the stable $PtCl_4^{2-}$ complex. There is also danger in using platinum in the presence of very strong reducing agents such as chromium(II) ion. The latter ion can reduce hydrogen ion in acid solution, and the reaction, which is normally slow, is catalyzed by platinum.

Membrane Electrodes

Membrane electrodes differ in principle from the metal electrodes we have just discussed. No electrons are given up by, or to, the membrane. Rather, a membrane allows certain kinds of ions to penetrate it, but excludes others. The glass electrode, used for determining pH, is the most widely known example of a membrane electrode. It was observed many years ago that a potential is developed at a thin glass membrane which separates two solutions of different hydrogen ion activity. Furthermore, with certain types of glass the potential developed is dependent on the difference in activities of hydrogen ions on either side of the membrane and is not affected, at least appreciably, by the presence of other ions in the solutions. The glass electrode has been widely studied, and the results have led to the development of glasses which respond selectively to ions other than hydrogen. In addition, other types of membranes have been proposed and are now commercially available for the analytical determination of a number of ions.

We shall consider the glass electrode in some detail since it has been so thoroughly studied. Then we shall look briefly at liquid membrane and solid-state electrodes.

The Glass Electrode for the Measurement of pH

The usual commercial glass electrode consists of a thin glass bulb containing an internal reference electrode, usually silver–silver chloride. The activity of hydrogen ion inside the bulb is constant. The bulb is immersed in the solution whose *p*H is to be determined, and electrolytic contact between the test solution and an external reference electrode is provided. (This contact is not necessarily a conventional salt bridge; sometimes it is simply a pinhole in the reference electrode tube or a small fiber which soaks up water and provides for electrolyte movement, as shown in Fig. 12.1.)

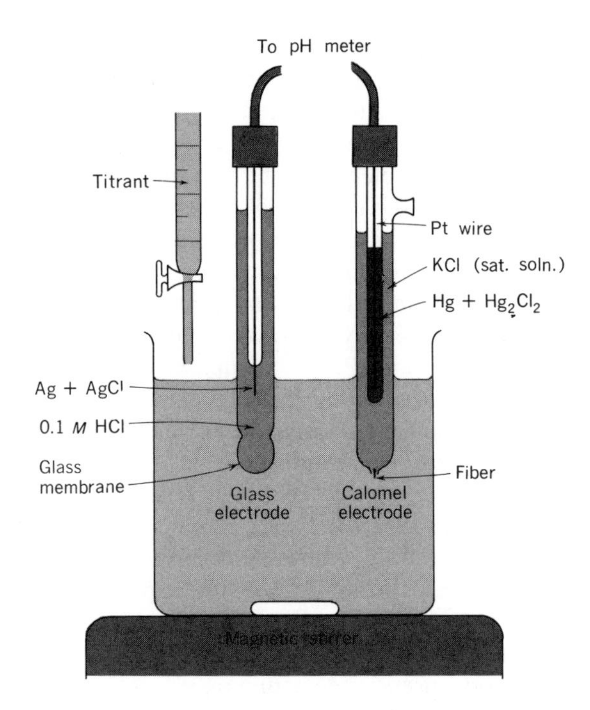

Figure 12.1 Apparatus for potentiometric acid-base titration with glass–calomel electrode pair.

The cell can be represented as follows:

$$\underbrace{Ag|AgCl,\ HCl\ (0.1\ M)|glass}_{\text{Glass electrode}}|\underbrace{H^+(x\ M)}_{\text{Test solution}}||\underbrace{KCl\ (sat.),\ Hg_2Cl_2|Hg}_{\text{Calomel electrode}}$$

Note that we have simply placed a reference electrode in each of the two solutions separated by the glass membrane. Since the potentials of the two reference electrodes remain constant, any change in the potential of the cell when we change test solutions must reflect a change in the potential developed across the glass membrane, provided that any change in the liquid junction potential between calomel electrode and test solution is quite small. It is found experimentally that the potential of this cell follows the relation

$$E = k + 0.0591\ p\text{H}$$

at 25°C and over a *p*H range of about 0 to 10 or 12, depending upon the composition of the glass and of the test solution.

Compared with other indicator electrodes for hydrogen ions, the glass electrode has several advantages. First, except for a very slight leakage of KCl from the calomel electrode, no foreign substance is added to the solution whose *p*H is being measured. Second, substances that are easily oxidized or reduced can be present in the solution without interfering. Such substances might react with hydrogen, for example, in the hydrogen electrode. Third, since potentials in general are independent of the physical sizes of electrodes, the electrode can be made quite small so that very small volumes of solutions can be measured. Fourth, there is no catalytic surface susceptible to poisoning as in the hydrogen electrode. Finally, sparingly buffered solutions can be measured accurately, and the electrode is well suited for continuous measurements.

The glass membrane is somewhat fragile, but this is not a serious disadvantage. One effect that can lead to a serious error, however, is the response of the electrode to ions other than hydrogen in solutions of high *p*H. In a sodium hydroxide solution of, say, *p*H 12, where $[H^+]$ is about $10^{-12}\ M$ and $[Na^+]$ about $10^{-2}\ M$, the potential of the glass electrode depends somewhat upon the activity of Na^+ and hence yields an error in *p*H despite a high inherent selectivity in responding to H^+ rather than to other ions.

If the membrane is made of soda-lime glass, this effect is sufficient to invalidate measurements in solutions of *p*H greater than about 10.0. It is possible to correct a *p*H reading for this effect. Manufacturers of glass electrodes furnish a chart of "sodium ion corrections" for use in solutions of *p*H greater than 10. In recent years other types of glass have been developed that do not show an appreciable error in solutions up to a *p*H of about 12.5.[3] These glasses contain a high percentage of lithium oxide in place of sodium oxide and are often referred to as "lithium glasses."

An error also occurs with the glass electrode in very acidic solutions (*p*H below about 0). This is thought to be caused by a change in the activity of water in the gel layer of the glass (see below) and the penetration of anions into the gel.

[3] G. A. Perley, *Anal. Chem.*, **21**, 391, 394, 559 (1949).

The high electrical resistance of the glass membrane (in the range of millions to hundreds of millions of ohms) used to be a serious disadvantage, because the traditional voltage measuring devices were unsuited for measurements involving high-resistance circuits. The development of electronic amplifiers led to the appearance of commercial *p*H meters just before World War II. These are now of several sorts, but for our purpose let us simply say that the *p*H meter has an amplifier that responds to the cell voltage by producing a proportional deflection in a meter or, more recently, a digital readout. Since the cell voltage varies linearly with *p*H, the meter scale is easily calibrated to read *p*H units directly. The *p*H meter can also be used as a voltmeter for measuring other sorts of electrochemical cells; usually, a millivolt scale is provided along with the *p*H scale.

Two different glass electrodes are not likely to give exactly the same potential reading when immersed in the same solution. In fact, a small potential may be developed at a glass membrane even if the same solution is on both sides of the membrane. It is thought that this effect results from "strains" in the glass, and it is often referred to as an "asymmetry potential." Because of this effect, it is necessary to calibrate the scale using a buffer, and it is necessary to repeat this calibration if one glass electrode is replaced by another.

Theory of the Glass-Membrane Potential

A number of proposals have been made to explain the mechanism by which the glass electrode functions. Considerable information has been gained from the efforts mentioned above to reduce the "sodium ion error" by changing the composition of the glass.

It was formerly believed that the potential of the glass electrode arose from the migration of H^+ through the glass. If H^+ could do this, and other ions could not, then a potential analogous to a liquid junction potential (Chapter 10) would develop across the membrane and a Nernstian response to H^+ would be expected. However, it has apparently been demonstrated that H^+ does not in fact traverse the membrane, and other ideas (which are unfortunately more difficult to explain to the beginning student) have been developed.

Glass consists of a negatively charged silicate network containing small cations, principally sodium ions. Other cations, such as hydrogen, can enter the glass by displacing sodium ions, but negative ions are repelled by the negatively charged silicate network. Not all glass membranes give the response to *p*H mentioned before; quartz and Pyrex glass are practically insensitive to changes in *p*H. The glass commonly used for many years in commercial glass electrodes is "Corning 015," which has the approximate composition (mole percent) Na_2O, 22, CaO, 6, and SiO_2, 72. The electrical resistance of this glass is relatively low and the glass is sensitive only to hydrogen ions up to a *p*H of about 10. The lithium glasses mentioned above were developed in an attempt to minimize the alkaline error.

It is known that a glass membrane must be soaked in water before it becomes operative. The outer surface of the glass becomes hydrated when the electrode is soaked; the inner surface is already hydrated. It is thought that the glass membrane has the following general structure:

	E_i		E_d		E_d		E_e	
Internal solution $a_{H^+} = a_1$		Inner hydrated gel $\leftarrow 10^{-4}$ mm $\rightarrow$		Dry glass layer $\leftarrow 0.1$ mm $\rightarrow$		Outer hydrated gel $\leftarrow 10^{-4}$ mm $\rightarrow$		External solution $a_{H^+} = a_2$

Surface occupied by H^+ Surface occupied by H^+

There is a center layer of dry glass sandwiched between two layers of silicic acid gel. These gels are formed by water molecules penetrating into the surface of the silicate lattice. In the process the following reaction occurs:

$$H^+(aq) + Na^+Gl^-(s) \rightleftharpoons Na^+(aq) + H^+Gl^-(s)$$

This is the type of reaction which occurs with cation exchange resins (Chapter 18); the protons are free to move and exchange with other ions, but the silicate lattice is fixed.

At the inner and outer surfaces of the glass membrane, *boundary* potentials E_i and E_e are developed. The explanation for the potential is similar to that of a liquid junction potential. At the interface between the gel and solution hydrogen ions tend to migrate in the direction of lesser hydrogen ion activity. The external and internal potentials are given by the expressions

$$E_e = k_1 + 0.059 \log \frac{(a_2)_1}{(a_2)_s}$$

$$E_i = k_2 + 0.059 \log \frac{(a_1)_1}{(a_1)_s}$$

where $(a_2)_1$ and $(a_1)_1$ are the hydrogen ion activities in the two solutions on either side of the membrane; $(a_2)_s$ and $(a_1)_s$ are the corresponding activities in the gels.

If the gel surfaces have the same number of sites from which protons can leave, then $k_1 = k_2$ and $(a_1)_s = (a_2)_s$. Hence the total boundary potential, E_b, is given by

$$E_b = E_e - E_i = 0.059 \log \frac{(a_2)_1}{(a_1)_1}$$

In the glass electrode $(a_1)_1$ is constant. Hence

$$E_b = k + 0.059 \log(a_2)_1$$

This is the same equation found experimentally for the glass electrode (above).

Potential differences are also developed at the boundaries between the dry glass layer and the inner and outer gel layers. The potentials are developed by the tendencies of sodium and hydrogen ions to migrate in the direction of lesser activity. Sodium ions are unhydrated and loosely bound in the dry glass layer; they can move from the silicate framework toward the gel layers. At the same time, hydrogen ions tend to move toward the dry glass layer, and since hydrogen and sodium ions have different mobilities, a potential difference is developed. It is thought that these *diffusion* potentials, designated E_d in the diagram above, are equal and opposite in sign; hence they cancel. The potential across the glass membrane is therefore dependent only on the difference in hydrogen ion activities in the inner and outer solutions,

$$E = k + 0.059 \log a_2$$

Ion-Selective Glass Electrodes

In recent years glass membranes have been developed which are selective for a given cation, just as the conventional glass electrode is selective for hydrogen ions. For example, glass electrodes are now available commercially for sodium, potassium, lithium, and silver ions. These electrodes are not *specific* for a given ion, but rather they possess a certain *selectivity* for a given ion and for that reason are called *ion-selective* electrodes.

We have already mentioned how the composition of the glass reduced the sodium ion error of the ordinary glass electrode at high *p*H. Eisenman and his collaborators[4] studied the influence of alkali metal ions and hydrogen ions on the glass membrane potential as a function of the composition of the glass. They found, for example, that a glass composed of 28% Na_2O, 5% Al_2O_3, and 68% SiO_2 shows a selective response to potassium ions down to an activity of 10^{-4}. A glass with a composition of 15% Li_2O, 25% Al_2O_3, and 60% SiO_2 can be used to determine lithium in the presence of both sodium and potassium ions.

Eisenman related the glass-electrode response to the ion-exchange properties of the glass surface. For example, suppose that the following exchange reaction occurs at a glass surface:

$$H^+Gl^-(s) + Na^+(aq) \rightleftharpoons Na^+Gl^-(s) + H^+(aq)$$

The boundary potential depends upon the activities of both sodium and hydrogen ions. In addition, sodium ions are able to penetrate the gel and influence the diffusion potential. The diffusion potentials do not cancel as in the case we discussed earlier. Instead, the total potential is given by the expression

$$E = k + 0.059 \log \Big\{ \underbrace{(a_{H^+})_{aq}}_{\text{Term } A} + \underbrace{K_{ex}\left(\frac{U_{Na}}{U_H}\right)(a_{Na^+})_{aq}}_{\text{Term } B} \Big\}$$

where U_{Na} and U_H are measures of the mobility of sodium and hydrogen ions in the gel, and K_{ex} is the equilibrium constant of the ion-exchange reaction above.

Note that if term *B* in the equation above is small (K_{ex} small), the potential is dependent mainly on the hydrogen ion activity. This is the case for the ordinary *p*H-sensitive glass electrode. On the other hand, if the term *B* is large (K_{ex} large) and term *A* is small, the potential is largely a function of the sodium ion activity. It can be seen that the selectivity of the glass electrode is related to the cation-exchange selectivity of the glass surface.

[4] See G. Eisenman, *Advances in Analytical Chemistry and Instrumentation*, Vol. 4, John Wiley & Sons, Inc., New York, 1965, p. 213, for a summary of this work.

Commercial ion-selective electrodes are very similar in construction to the glass electrode. The internal reference electrode is usually silver–silver chloride and the filling solution is the chloride salt of the cation to which the electrode is most responsive. It should be remembered that all these electrodes are responsive to some degree to hydrogen ions and must be used at sufficiently high *p*H that term *A* in the preceding equation is quite low.

Liquid-Membrane Electrodes

These electrodes employ as the membrane a water-immiscible liquid which will selectively bond the ion being determined. An example is the calcium ion electrode, which uses a cation exchanger that contains a phosphoric acid group. The exchanger has a greater tendency to bond calcium than to bond most other cations. A schematic representation and a drawing of the commercial electrode are shown in Fig. 12.2. The liquid exchanger is held between two porous glass or plastic membranes. It separates the test solution from the internal solution; the latter contains a fixed concentration of calcium chloride and a silver–silver chloride electrode.

At each interface the following reaction occurs:

$$Ca^{2+}(aq) + Ex^{2-} \rightleftharpoons CaEx$$

The potentials developed at the two interfaces are given by

$$E_e = k_1 + \frac{0.059}{2} \log \frac{(a_1)_{aq}}{(a_1)_{Ex}}$$

$$E_i = k_2 + \frac{0.059}{2} \log \frac{(a_2)_{aq}}{(a_2)_{Ex}}$$

where $(a_1)_{aq}$ and $(a_2)_{aq}$ are the activities of calcium ion in the aqueous solutions and $(a_1)_{Ex}$ and $(a_2)_{Ex}$ are the activities in the exchanger phase. Normally, the

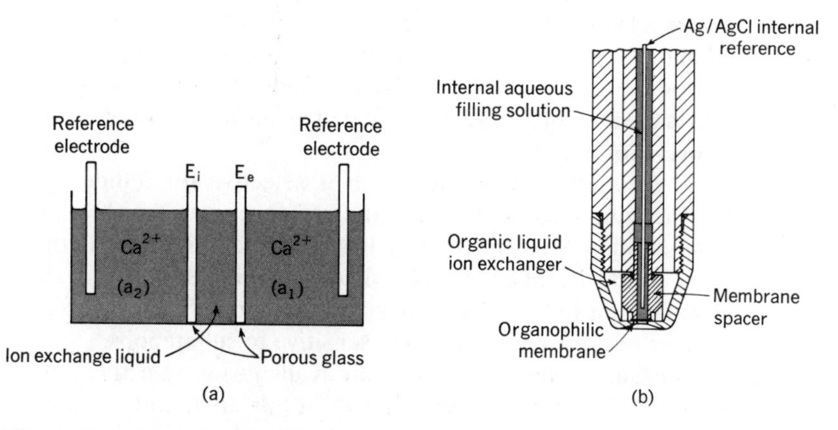

Figure 12.2 Liquid membrane electrode. (a) Schematic. (b) Commercial electrode. (Courtesy of Orion Research Inc., Cambridge, Mass.)

latter activities are equal. Hence the total potential is given by

$$E = E_e - E_i = k + \frac{0.059}{2} \log \frac{(a_1)_{aq}}{(a_2)_{aq}}$$

Since $(a_2)_{aq}$ is fixed this becomes

$$E = k + \frac{0.059}{2} \log (a_1)_{aq}$$

The calcium electrode is about 3000 times more selective for calcium than for sodium or potassium. Its selectivity ratio for calcium over magnesium is about 200, and for calcium over strontium about 70. It is independent of pH in the range 5.5 to 11, and it can be used to determine calcium ion concentrations as low as 10^{-5} M.

Several other ion-selective membrane electrodes are available, including anion-exchange membranes for anions such as nitrate, chloride, and perchlorate. An electrode which responds to several divalent cations is used to measure water hardness.

Solid-State and Precipitate Electrodes

Solid materials which are selective to anions, just as glass is selective to cations, have been used successfully as membrane electrodes in recent years. A *solid-state* electrode has a membrane in the form of a single crystal or a pellet of the compound which contains the anion to be determined. A *precipitate* electrode is prepared by suspending a finely divided insoluble salt in an inert semiflexible matrix from which a membrane can be fabricated.

A very successful solid-state electrode is that used to determine fluoride ion. A single crystal of lanthanum fluoride acts as the membrane. The crystal is doped with a rare earth element, europium (II), to increase its electrical conductivity. The electrode responds to fluoride ion down to concentrations of about 10^{-5} M. It is a thousand times more selective for fluoride than for the other common anions. Hydroxide ion does interfere and the electrode is limited in use over the pH range of about 0 to 8.5. Other successful solid-state electrodes are available for chloride, bromide, iodide, sulfide, cyanide, and thiocyanate ions. In these electrodes the membrane is a cast pellet of an insoluble salt of the anion, such as silver chloride for chloride ions.

A precipitate electrode which is selective for iodide can be prepared by polymerizing silicone rubber in the presence of an equal weight of finely divided silver iodide. After the mixture has solidified, it is sealed in the bottom of a glass tube. Inside the tube is placed a silver–silver iodide electrode dipping in a solution of potassium iodide. Barium sulfate suspended in paraffin is used in a similar manner to prepare an electrode sensitive to sulfate ions.

Some examples of commercially available ion-selective electrodes are given in Table 12.1. Research is very active in this area, and new electrodes and novel applications are appearing constantly. Developments in this field promise to be important in a number of areas, such as biology and medicine, agriculture, water pollution, and oceanography. Some of the ions most easily determined by potentiometry, for example, Na^+, K^+, and F^-, are among those very ions for which other analytical methods tend to be unsatisfactory. For instance, the initial impulse behind the development of electrodes for Na^+ and K^+ came from physiologists and clinicians, who were almost totally frustrated in studies of body electrolytes using other analytical techniques. In addition, the mechanisms by which these electrodes respond to various ions are far from totally clear, and one may expect that further studies will provide additional insight into the basic origins of the electrode potentials.

TABLE 12.1 Some Commercially Available Ion Selective Electrodes

Electrode	Type	Concentration range (M)	Some interferences
Bromide	Solid state	5×10^{-6} to 1	S^{2-}, I^-
Cadmium	Solid state	10^{-7} to 1	Ag^+, Hg^{2+}, Cu^{2+}
Calcium	Liquid membrane	10^{-5} to 1	Zn^{2+}, Fe^{3+}, Pb^{2+}
Chloride	Solid state	5×10^{-5} to 1	S^{2-}, Br^-, I^-, CN^-
Cyanide	Solid state	10^{-6} to 10^{-2}	S^{2-}, I^-
Fluoride	Solid state	10^{-6} to 1	OH^-
Iodide	Solid state	2×10^{-7} to 1	S^{2-}
Potassium	Liquid membrane	10^{-5} to 1	Cs^+
Sodium	Glass	10^{-6} to 1	Ag^+, H^+

DIRECT POTENTIOMETRY

Potentiometric Determination of *p*H

As previously mentioned, one of the principal applications of direct potentiometry is the determination of the *p*H of aqueous solutions. We wish now to examine briefly the meaning of the term *p*H as it is measured experimentally by direct potentiometry. A complete discussion of this topic has been given by Bates.[5]

The term *p*H was defined by Sorensen in 1909 as the negative logarithm of the hydrogen ion *concentration*. It was later realized that the emf of galvanic cells used to measure *p*H was dependent more upon the *activity* of hydrogen ions than upon the *concentration*. Hence the definition of *p*H was taken to be

$$pH = -\log a_{H^+}$$

This definition is satisfactory from a theoretical standpoint, but the quantity cannot be measured experimentally. There is no way to measure unambiguously the activity of a single ion species. (In thermodynamic terms it is said that *p*H is

[5] R. G. Bates, *Determination of pH: Theory and Practice*, 2nd ed., John Wiley & Sons, Inc., New York, 1973.

proportional to the work required to transfer hydrogen ions reversibly from the solution examined to one in which the activity of hydrogen ions is unity. No experiment can be performed in which hydrogen ions are transferred without at the same time transferring negative ions.)

The quantity measured potentiometrically is actually neither concentration nor activity of hydrogen ion. It is therefore preferable to define *p*H in terms of the emf of the cell employed for the measurement. For example, suppose that such a cell consisted of a suitable reference electrode connected by a salt bridge to the solution being treated, in which a hydrogen electrode was immersed:

$$\text{Reference} \| H^+(x\,M) | H_2, \text{Pt}$$

The equation relating the potential of this cell to *p*H is

$$E = E_{\text{ref}} + 0.059p\text{H}$$

Strictly speaking, this equation should also contain a term E_j, the liquid junction potential, which may be small with an approximate salt bridge, but not zero. Hence the equation should read

$$E = E_{\text{ref}} + 0.059p\text{H} + E_j$$

where E_{ref} is the potential of the reference electrode. Calling $E_{\text{ref}} + E_j$ a constant, k, this becomes

$$E = k + 0.059p\text{H}$$

or

$$p\text{H} = \frac{E - k}{0.059}$$

The preceding equation contains two unknowns, *p*H and k; hence it cannot be used to evaluate both quantities. It is necessary to assign arbitrarily a *p*H value to some standard buffer in order to fix a practical scale of *p*H.

The Bureau of Standards determines the *p*H of certain buffers by careful measurements of selected cells, using reasonable assumptions regarding activity coefficients.[6] Table 12.2 contains a few examples of the buffers recommended by the Bureau. It would be possible, of course, to define a *p*H scale, according to the equation

$$E = k + 0.059p\text{H}$$

using a single buffer. However, the term k is not strictly constant over a wide *p*H range largely because of changes in E_j with changes in composition of the solution. It is therefore recommended that a *p*H standard be chosen which is close to that of the unknown. Even better are two standards, one on either side of the unknown.

It is possible to arrange experimental conditions so that the term k is constant to about ±1 mV over the *p*H range 2 to 10. This leads to an uncertainty in the

[6] R. G. Bates, *op. cit.*; an excellent summary is given by H. F. Walton, *Principles and Methods of Chemical Analysis*, 2nd ed., Prentice-Hall, Inc., Englewood Cliffs, N.J., 1964, pp. 246–248.

TABLE 12.2 *p*H Values of NBS Standard Buffers*

Composition	25°C	*p*H 30°C	40°C
$KH_3(C_2O_4)_2 \cdot 2H_2O(0.05\ M)$†	1.68	1.69	1.70
KHC_4O_6(sat. at 25°C)	3.56	3.55	3.54
$KHC_8H_4O_4(0.05\ M)$	4.01	4.01	4.03
$KH_2PO_4(0.025\ M) + Na_2HPO_4(0.025\ M)$	6.86	6.85	6.84
$Na_2B_4O_7 \cdot 10H_2O(0.001\ M)$	9.18	9.14	9.07
$Ca(OH)_2$ (sat. at 25°C)†	12.45	12.30	11.99

* R. G. Bates in *Treatise on Analytical Chemistry*, Part I, Vol. I, I. M. Kolthoff and P. J. Elving, eds., Interscience Publishers, Inc., New York, 1959, p. 375.

† Secondary standards; the other four are primary standards.

value of *p*H of about ±0.01 to 0.02 unit. Hence it is not possible to obtain significant *p*H numbers to any greater precision than this by potentiometric measurement.

Most *p*H measurements made by analytical chemists and biochemists employ a glass electrode connected to a calomel reference electrode by a potassium chloride salt bridge. Assuming that one of the Bureau of Standards buffers is used to standardize the *p*H meter, just what does the measured, or practical, *p*H number mean? This question is thoroughly discussed in an article by Feldman.[7] The *p*H measured will not be exactly equal to $-\log a_{H^+}$, but under ordinary conditions it will be nearly so. By "ordinary conditions" we mean:

1. The ionic strength of the test solution is less than about 3.
2. No unusual ions of exceptional mobility (e.g., very large organic ions, highly hydrated lithium ion, etc.) are present in the solution.
3. The *p*H range is about 2 to 10.
4. There are no charged suspensions, such as clays, soils, or ion-exchange resins in the test solution. The meaning of *p*H measurements near cell surfaces in biological systems may be questionable. Measurements of *p*H on protein solutions apparently give reasonable results.

Care must be taken in calculating hydrogen ion concentration from a practical *p*H measurement. In order to make such a calculation it is necessary to know something about the activity coefficient of hydrogen ion:

$$pH \cong -\log a_{H^+} = -\log[H^+]\gamma_{H^+}$$

If the value of γ_{H^+} is taken as unity, very large errors may result, depending upon the ionic strength of the solution as well as other factors. It is estimated that, at an ionic strength of 0.16, the value of interest in biological chemistry, the magnitude of the error in determining the concentration of hydrogen ion is about 25% if it is assumed that $[H^+] = a_{H^+}$.

[7] I. Feldman, *Anal. Chem.*, **28**, 1859 (1956).

Determination of Concentrations of Other Ions

The potential of any indicator electrode is apparently dependent on the activity rather than the concentration of the ion to which it is sensitive. If the concentration of an ion is to be determined by direct potentiometry, the total ionic strength of the sample and standard solutions must be similar. For example, suppose calcium ion is being determined with a calcium ion-selective electrode. The potential is given by

$$E = k + \frac{0.059}{2} \log a_{Ca^{2+}}$$

$$E = k + \frac{0.059}{2} \log \gamma_{Ca^{2+}}[Ca^{2+}]$$

$$E = k + \frac{0.059}{2} \log \gamma_{Ca^{2+}} + \frac{0.059}{2} \log [Ca^{2+}]$$

If the ionic strength is held constant, the activity coefficient, $\gamma_{Ca^{2+}}$, remains constant for all concentrations of calcium ion. The second term on the right side of the equation is constant and can be combined with k, giving

$$E = k' + \frac{0.059}{2} \log [Ca^{2+}]$$

In practice it is customary to prepare a calibration curve of potential versus the logarithm of the concentration of the ion being determined, with the ionic strength held constant. A salt to which the electrode is not sensitive is added to both the standards and the test solution. The plot is linear if the ionic strength is constant; otherwise, it is nonlinear. The potential of the unknown solution, at the same ionic strength as the standards, is measured and the concentration determined from the calibration curve. In the determination of unbounded calcium in serum, the standards and unknowns are diluted with 0.15 M NaCl. The measurement gives only the unbound calcium, not that which is complexed.

POTENTIOMETRIC TITRATIONS

In a potentiometric titration, the end point is detected by determining the volume at which a relatively large change in potential occurs as the titrant is added. Figure 12.1 shows a schematic experimental setup for such a titration, using a glass electrode as an example of an indicator electrode. The method can be employed for all the reactions used for titrimetric purposes: acid-base, redox, precipitation, and complex formation. The proper indicator electrode is selected; a reference electrode, such as calomel, completes the cell. The titration may be performed manually, or the procedure may be automated. We shall consider the various techniques under three headings, recognizing that all variants cannot be discussed in a book of this sort: manual potentiometric titrations, automatic recording of the titration curve, and automatic titrant shutoff at the end point.

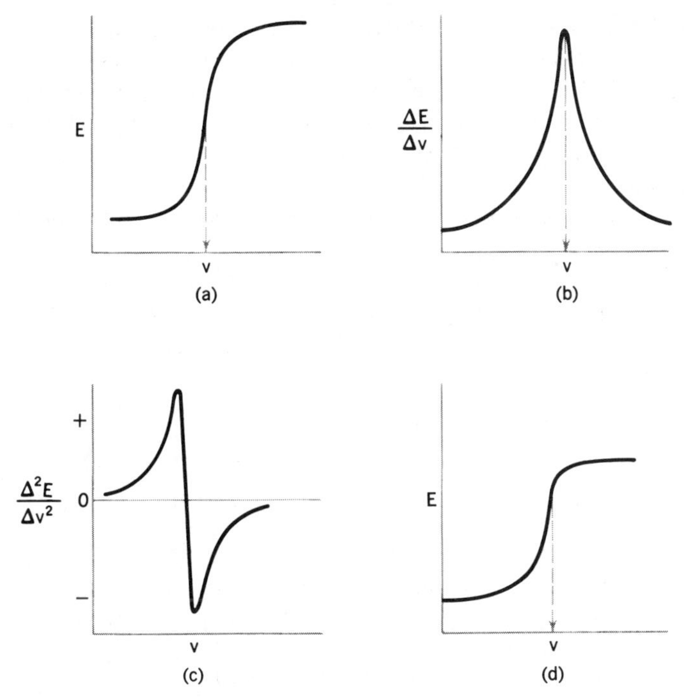

Figure 12.3 Methods of plotting potentiometric titration data.

In manual titrations the potential is measured after the addition of each successive increment of titrant, and the resulting readings are plotted on graph paper versus the volume of titrant to give a titration curve such as shown in Fig. 12.3(a). In many cases, a simple potentiometer could be used. However, if the glass electrode is involved, as in most acid-base titrations, a measuring device with a high input impedance is required because of the high resistance of the glass; typically, a commercial *p*H meter is employed. Because these *p*H meters have become so common, they are widely utilized for all sorts of titrations, even where their use is not obligatory.

Once the titration curve is at hand, a subjective element enters the procedure. The analyst must determine where the curve is steepest, normally by some sort of inspection. He may draw a vertical line through the steep portion of the curve and find the intersection of this line with the volume axis. There must be some uncertainty in this procedure, and this will, of course, be reflected in the ultimate volume reading. For a reaction which goes well to completion, the titration curve is so steep near the equivalence point that the uncertainty is small; for a reaction with a small equilibrium constant, the precision with which an end point may be reproduced becomes poorer.

Figure 12.3(b) shows a plot of the slope of a titration curve, that is, the change in potential with change in volume ($\Delta E/\Delta V$) against volume of titrant. The resulting curve rises to a maximum at the equivalence point. The volume at the equivalence point is determined by dropping a vertical line from the peak to the volume axis. There is some uncertainty, of course, in locating exactly the peak in the curve. The more complete the reaction, the sharper the peak, and hence the more accurate the location of the equivalence point.

Figure 12.3(c) shows a plot of the change in the slope of a titration curve ($\Delta^2 E/\Delta V^2$) against the volume of titrant. At the point where the slope $\Delta E/\Delta V$ is a maximum, the derivative of the slope is zero. The end point is located by drawing a vertical line from the point at which $\Delta^2 E/\Delta V^2$ is zero to the volume âxis. The portion of the curve joining the maximum and minimum values of $\Delta^2 E/\Delta V^2$ is steeper the more complete the titration reaction.

The curves shown in Fig. 9.3(a), (b), and (c) are for "symmetrical" reactions, that is, reactions in which 1 mol of titrant reacts with 1 mol of the substance titrated. Examples of acid-base, redox, and precipitation reactions that are symmetrical are

$$H_3O^+ + OH^- \longrightarrow 2H_2O$$

$$Ag^+ + Cl^- \longrightarrow AgCl(s)$$

$$Fe^{2+} + Ce^{4+} \longrightarrow Fe^{3+} + Ce^{3+}$$

For such reactions the midpoint of the steep portion of the curve in Fig. 12.3(a) corresponds to the equivalence point (see page 259). Likewise, the peak in the curve of Fig. 12.3(b) and the zero value of the second derivative in Fig. 12.3(c) occur at the equivalence point.[8]

For nonsymmetrical reactions such as

$$2Ag^+ + CrO_4{}^{2-} \longrightarrow Ag_2CrO_4(s)$$

and

$$Sn^{2+} + 2Ce^{4+} \longrightarrow Sn^{4+} + 2Ce^{3+}$$

the equivalence point does not occur at the midpoint of the curve of Fig. 12.3(d). The potential at the equivalence point in the titration of tin(II) (E_1°) with cerium(IV) ion (E_2°) is $(2E_1^\circ + E_2^\circ)/3$ (see page 261). Similarly, the maximal value of $\Delta E/\Delta V$ for a nonsymmetrical reaction does not coincide exactly with the equivalence point. Nevertheless, the maximum is usually taken as the end point of a titration. The error made by this procedure is quite small.

It is possible to locate the end point by a simple systematic method based upon the actual data without resorting to a graph. Only potential readings near the equivalence point need be recorded. Some definite increment of volume, say 0.10 ml, is selected and a number of readings is taken, 0.10 ml apart, on either side of the equivalence point. An example is given in Table 12.3, where the values of the first and second derivatives are included. It can be seen from the second derivative values that the slope changes sign, and hence goes through zero, between 25.00 and 25.10 ml of titrant. The volume at which the value of zero is reached is closer to 25.00 than to 25.10, since the reading of +120 is closer to zero than is −224. Since 0.10 ml caused a total change in the second derivative of

[8] A detailed mathematical analysis has shown that this statement, found in most textbooks, is not strictly true for certain cases where dilution by the titrant lowers the concentration of the reaction product. However, the error is negligible. See L. Meites and J. A. Goldman, *Anal. Chim. Acta*, **29**, 472 (1963), and W. Lund, *Talanta*, **23**, 619 (1976).

TABLE 12.3 Potential Readings Near the Equivalence Point

Titrant, ml	E, mV	$\Delta E/\Delta V/0.1$ ml	$\Delta^2 E/\Delta V^2$
24.70	210		
		12	+6
24.80	222		
		18	+102
24.90	240		
		120	+120
25.00	360		
		240	−224
25.10	600		
		16	−7
25.20	616		
		9	
25.30	625		

$120 - (-224) = 344$, the fraction $(120/344) \times 0.10$ ml is the approximate number of milliliters in excess of 25.00 necessary to bring the second derivative to the value zero. Hence the calculated volume at the equivalence point is

$$V = 25.00 + 0.10\left(\frac{120}{120 + 224}\right) = 25.035 \text{ ml}$$

This linear interpolation procedure is satisfactory because the center segment of the curve in Fig. 12.3(c) is practically linear near the equivalence point.

By a very simple experimental modification, values of $\Delta E/\Delta V$, the change in potential with change in volume of titrant, may be obtained directly. These values can be plotted as shown in Fig. 12.3(b), or the volume corresponding to the maximal value of $\Delta E/\Delta V$ may be obtained from inspection of the data. No reference electrode is required for this method. Two indicator electrodes are employed, but one is kept separated from the main body of solution until a potential reading has been taken. A simple device that can be used for performing such a titration is shown in Fig. 12.4. One electrode is in the body of the solution,

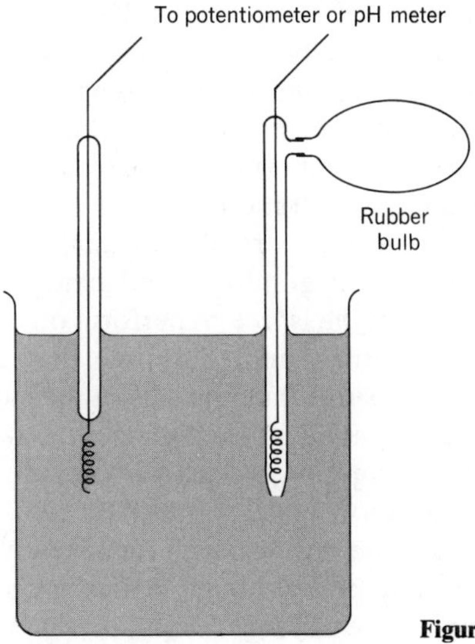

Figure 12.4 Apparatus for differential titration.

the other inside a small glass tube that dips into the same solution. When the solutions inside and outside the tube are uniform in composition, there is no difference in potential between the electrodes. When titrant is added to the main body of the solution, however, a difference in potential develops, since the solution inside the tube no longer has the same composition as that outside. The potential difference is noted, and then the solution is expelled from the tube by squeezing the rubber bulb so that the composition becomes uniform again. The potential difference falls to zero as bulk solution is sucked into the tube. The volume of solution in the tube must, of course, be kept small compared to the total volume of solution titrated. The method is rapid and the results can be accurate.

Automatic Recording of Titration Curves

Recording potentiometers have become commonplace in chemistry laboratories. With typical "recorders," as they are called, voltages from many different origins can be plotted with pen and ink as a function of time. The voltage developed in a circuit of inherently very high resistance, as enountered in glass-electrode setups, cannot be fed directly to a recorder for reasons that are beyond the scope of this book. Suffice it here to say that it is necessary to "match" the input impedance of the recorder with the impedance of the voltage source. The circuitry of most modern *p*H meters does this. Thus the student who is not interested in instrumentation per se can simply note that with relatively inexpensive commercially available circuits one may take the potential between the electrodes during a potentiometric titration and feed this to a recorder. Since the recorder plots voltage versus time, it is obvious that a potentiometric titration curve can be recorded provided the titrant is added at a constant rate. Thus, to record automatically a potentiometric titration curve, it is clear that one requires an electrode setup, as in manual titrations, plus an impedance matching device, a recorder, and a constant-flow buret. Commercial *p*H meters serve for the impedance matching and many are fitted with terminals for wires connecting with the recorder. Constant-flow burets usually rely, not upon gravity feed as with ordinary burets, but upon the action of a motor-driven plunger in a syringe type of assembly. Such devices are available commercially. Rapid stirring, usually accomplished with a magnetic stirrer, is important in order to maintain a uniform concentration in the solution during the titration.

After the titration curve has been recorded, the analyst faces the same problem of selecting the end point as with graphs plotted manually. On the other hand, while the titration is in progress, he is free to perform other tasks, perhaps preparing the next sample for titration, and thus time may be saved. The continuous nature of the recording insures that features of the curve will be seen which might be missed in a manual titration unless points were taken at very close intervals. A real disadvantage of recording, at least with the typical titrant flow rates, lies in the fact that the titration reaction must be rapid. In a manual titration, if the reaction is slow, the operator can wait for a steady reading before adding the next increment of titrant, and although it may try his patience, the titration will work. With automatic recording, the potentials may never reach equilibrium values at any stage of the titration, the recorded curve may be distorted, and a large potential break may appear before the equivalence point is actually reached.

Automatic Titrant Shut-Off

Finally, potentiometric titrations can be completely automated so that the buret is mechanically shut off at the end point. In some cases the buret is a conventional one and the meniscus is read in the usual manner. The drive mechanism of a syringe buret can be fitted with a revolution counter that provides direct digital readout.

One type of automatic titrator titrates the sample to a pre-set potential. The difference between the indicator and reference electrodes at the equivalence point is determined ahead of time, and the instrument is then set to shut off the buret at this potential. Even with good stirring, mixing is not instantaneous, and it is necessary to provide for restarting the titrant flow if the buret shuts off too soon due to a high local concentration of titrant around the indicator electrode. End-point "anticipation" is provided for by positioning the delivery tip of the buret very close to the indicator electrode.[9] Then the solution adjacent to the electrode is at a more advanced stage of titration than is the bulk of the solution. Thus the electrode reaches the equivalence potential somewhat early, and the delivery of titrant is stopped too soon. However, as the solution becomes uniform in concentration through further stirring, the potential drops back, and the buret is turned on again. This process is repeated until the entire solution reaches the equivalence potential. A human operator does the same sort of thing in a visual titration when he adds titrant rapidly in the early stages and then dropwise as fleeting color changes are seen near the equivalence point.

Another type of automatic titrator, developed by Malmstadt and Fett,[10] is based upon electronic differentiation. It is possible, by using a resistance-capacitance network with the right time constant, to construct an electronic circuit which produces a voltage that is proportional to the derivative (with respect to time) of a voltage which is fed into it. By combining two such circuits and including appropriate amplification, Malmstadt was able to trigger a relay system that shut off a buret when the second derivative of the voltage input from the electrodes changed sign. (Electronically, the second derivative is more suitable than the first derivative for actuating such a relay.) At constant titrant flow, differentiation with respect to time is equivalent to differentiation with respect to volume of titrant. Actually, the commercial instrument based upon Malmstadt's ideas employs an ordinary gravity-feed buret, but near the equivalence point (the only region of interest, really) the hydrostatic head is constant enough.

For a person with an occasional titration to perform, automatic titrators have little to offer. On the other hand, a laboratory with hundreds of samples to titrate

[9] J. J. Lingane, *Electroanalytical Chemistry*, 2nd ed., Interscience Publishers, Inc., New York, 1958, p. 159.

[10] H. V. Malmstadt and E. R. Fett, *Anal. Chem.*, **26**, 1348 (1954); **27**, 1757 (1955); **29**, 1901 (1957).

each week, a repetitive and boring task, may find them nearly indispensable. The titrator cannot do anything that people cannot do, but it is cheaper than people and it does not complain about its work provided it is properly serviced.

REFERENCES

R. G. BATES, "Concept and Determination of *p*H," Chap. 10 of I. M. Kolthoff, P. J. Elving, and E. B. Sandell, eds., *Treatise on Analytical Chemistry*, Part I, Vol. 1, Interscience Publishers, Inc., New York, 1959.

R. G. BATES, "Electrometric Methods of *p*H Determination," Vol. 3-A of F. J. Welcher, ed., *Standard Methods of Chemical Analysis*, D. Van Nostrand, Inc., New York, 1966.

R. A. DURST, ed., *Ion-Selective Electrodes*, (National Bureau of Standards Special Publication 314), U.S. Government Printing Office, Washington, D.C., 1969.

G. EISENMAN, ed., *Glass Electrodes for Hydrogen and Other Cations*, Marcel Dekker, Inc., New York, 1967.

J. J. LINGANE, *Electroanalytical Chemistry*, 2nd ed., Interscience Publishers, Inc., New York, 1958.

QUESTIONS

1. *Glass electrode.* Explain how the glass electrode functions as an indicator electrode for hydrogen ions.
2. *Membrane electrode.* Describe the various types of membrane electrodes. Explain what is meant by an "ion-selective" electrode.
3. *pH scale.* What two assumptions are made in the establishment of the conventional scale of *p*H? What is the uncertainty involved in determining the *p*H of a solution?
4. *Potentiometric titration.* What are the advantages of potentiometric titration over direct potentiometry?
5. *Concentrations of ions.* Explain how the molarity of an ion such as Na^+ is determined experimentally with an ion-selective electrode. Why is the ionic strength of the solution held constant?

Multiple-choice: In the following multiple-choice questions, select the *one best* answer.

6. What information *cannot* be obtained from the potentiometric titration curve of a weak acid titrated with a strong base? (a) The volume of base required to reach the equivalence point; (b) The pK_a of the acid; (c) The pK_b of the base; (d) The best buffering region for the acid-base system.
7. Which of the following electrodes is a membrane electrode? (a) glass electrode; (b) solid-state electrode; (c) precipitate electrode; (d) all of the above; (e) more than one but not all of the above.
8. Which indicator electrode would be best to use to follow the titration of Fe^{2+} with Ce^{4+}? (a) iron; (b) cerium; (c) glass; (d) platinum; (e) calomel.

9. Which of the following is an "electrode of the second kind"? (a) silver; (b) silver–silver chloride; (c) mercury–EDTA; (d) platinum.

10. Which of the following statements concerning the glass electrode is *false*? (a) It is a membrane electrode; (b) If made of soft glass, it gives low *p*H readings in solutions of high *p*H; (c) It has a relatively low electrical resistance; (d) It can be used in sparingly buffered solutions.

PROBLEMS

1. *Potentiometric titration.* A 40.0-ml sample of 0.100 *M* HCl is diluted to 100 ml and titrated with 0.100 *M* NaOH. (a) Calculate the *p*H of the solution after the addition of the following volumes of titrant: 0.00, 10.0, 20.0, 39.0, 39.9, 40.0, 40.1, and 50.0. (b) Calculate the potential readings one would obtain in a potentiometric titration using a cell which consists of a hydrogen-saturated calomel electrode pair. (c) Plot the curve: potential vs. milliters of titrant.

2. *Potentiometric titration.* Repeat Problem 1, substituting formic acid for hydrochloric acid.

3. *First and second derivatives.* From the data of Problems 1 and 2, plot $\Delta E/\Delta V$ and $\Delta^2 E/\Delta V^2$ vs. volume of titrant. You may wish to obtain several more values of the potential near the equivalence point.

4. *Potentiometric titration.* 2.50 mmol of Sn^{2+} is dissolved in 50 ml of solution and titrated with 0.10 *M* Ce^{4+} using a platinum-saturated calomel electrode pair. Calculate the potential after the addition of the following volumes of titrant: 0.00, 10.0, 25.0, 49.9, 50.0, 50.1, and 60.0. Plot the titration curve.

5. *Glass electrode.* A cell made up of a glass electrode and a saturated calomel electrode (as on page 307) gives a voltage reading of 0.614 when the test solution is a buffer of *p*H 7.00. Calculate the *p*H of solutions which give readings of (a) 0.531 V, and (b) 0.685 V in the same cell.

6. *Membrane electrode.* A liquid-membrane electrode is used to determine the concentration of calcium ion in a solution. The potential reading obtained with this electrode and a reference dipping into a solution which is 0.0010 *M* in Ca^{2+} is 0.260 V. In the same cell a solution of unknown concentration gives a reading of 0.272 V. The ionic strengths of the two solutions are the same. Calculate the concentration of calcium ion in the unknown.

7. *Equivalence point.* Given the following potential readings near the equivalence point of a titration:

Titrant, ml	*E, mV*
29.90	240
30.00	250
30.10	266
30.20	526
30.30	666
30.40	740
30.50	750

Calculate the volume at the equivalence point.

8. *Measurement of pH.* Derive an equation relating the potential of the following cell to pH:

$$Pt, H_2(1 \text{ atm})|H^+(x\,M)||HCl(0.10\,M), AgCl|Ag$$

9. *Buffer solution.* What should be the ratio of acetic acid to sodium acetate in a buffer so that the potential reading obtained in the cell in Problem 8 is 0.58 V? The Ag–AgCl electrode is positive.

10. *Measurement of pH.* Show that the potential of the cell

$$M|M^{2+}||KCl(\text{sat.}), Hg_2Cl_2|Hg$$

is related to the pH of the solution in the left-hand electrode by the equation

$$E = k + 0.059\,p\text{H}$$

where k is a constant. Assume that M^{2+} forms a slightly soluble hydroxide, $M(OH)_2$, and that the concentration of M^{2+} depends on the hydroxide ion concentration as expressed by the solubility product constant.

11. *Potential—pH.* How many grams of HCl should be dissolved in 1.0 liter of solution so that the potential reading obtained with a hydrogen-saturated calomel electrode pair is 0.350 V? The SCE is positive. (Use 0.246 V as the potential of the SCE.)

12. *Potential—pH.* A hydrogen-saturated calomel electrode pair is inserted in a solution of HCl and a potential reading of 0.287 V is measured. When these electrodes are inserted in a solution of NaOH, the reading is 1.049 V. The SCE is the positive electrode. How many milliliters of each solution should be mixed to prepare 100 ml of a solution that will give a reading of 0.990 V with the same electrodes? Use E of SCE = 0.246 V.

13. *Nernst equation.* A platinum-saturated calomel electrode pair is placed in a solution of pH 1.00 which contains TiO^{2+} and Ti^{3+} ions. The potential reading is 0.223 V with the SCE positive. Calculate the ratio of TiO^{2+} to Ti^{3+} ions in the solution. Use $E° = +0.100$ V for the couple

$$TiO^{2+} + 2H^+ + e \rightleftharpoons Ti^{3+} + H_2O$$

and $E = +0.246$ V for the SCE.

14. *Complex formation.* A sample of 4.00 mmol of M^{2+} is titrated with X^- according to the reaction

$$M^{2+} + X^- \rightleftharpoons MX^+$$

The potential of the cell

$$Hg|Hg_2Cl_2, KCl(\text{sat.})||M^{2+}|M$$

is 0.030 V at the equivalence point with the M electrode positive. Given that $M^{2+} + 2e \rightleftharpoons M$, $E° = 0.480$ V, and E of the SCE is 0.246 V, calculate the stability constant of the MX^+ complex. The volume at the equivalence point is 100 ml.

15. *Titrimetric analysis.* A 2.00-g sample of a weak acid, HA, MW 80.0, is dissolved in 50.0 ml of water and titrated with 0.200 M NaOH using a hydrogen-saturated calomel electrode pair (Problem 11). When half of the acid is neutralized the potential reading is 0.541 V; at the equivalence point the reading is 0.777 V. The SCE is positive. Calculate the percentage of HA in the sample.

16. *Redox titration.* A 50.0-ml sample of 0.100 M Fe^{2+} is titrated potentiometrically with 0.100 M Ce^{4+} using a platinum-saturated calomel (E = 0.25 V) electrode pair. When the potential reading is 0.58 V, how many milliliters of the Ce^{4+} solution has been added? The polarity of the platinum electrode is positive.

17. *Quinhydrone electrode.* The so-called "quinhydrone electrode" is an indicator electrode for hydrogen ions. A platinum electrode is inserted in the test solution which is saturated with quinhydrone, a 1 : 1 compound of quinone and hydroquinone. The redox couple is

$$\text{Q} + 2H^+ + 2e \rightleftharpoons \text{H}_2\text{Q} \qquad E° = 0.70 \text{ V}$$

(Q = quinone, $O{=}C_6H_4{=}O$; H_2Q = hydroquinone, $HO{-}C_6H_4{-}OH$)

Derive an equation relating the pH of a solution to the potential of the following cell:

$$\text{Pt}|\text{H}^+(x\,M) + \text{H}_2\text{Q} + \text{Q}||\text{KCl}(1M), \text{Hg}_2\text{Cl}_2|\text{Hg}$$

H_2Q stands for hydroquinone and Q for quinone. You can assume that the dissociation of quinhydrone produces an equal number of moles of quinone and hydroquinone.

18. *Quinhydrone electrode.* Answer the following questions about the quinhydrone electrode described in Problem 17: (a) What is the polarity of the calomel electrode at pH 4? (b) At what pH is the potential of the cell zero? (d) At high pH some H_2Q is neutralized and the ratio of H_2Q to Q is no longer unity. Suppose that the electrode is used to measure the pH of a solution which actually has a pH of 12.00. Will the pH value obtained with the electrode be above 12, below 12, or exactly 12?

19. *Mercury electrode.* Calculate the potential of the mercury indicator electrode at the equivalence point in the titration of 5.0 mmol of Pb^{2+} with EDTA at pH 5.00. Assume that the volume at the equivalence point is 100 ml and that the concentration of HgY^{2-} is 0.010 M.

20. *Precipitation titration.* A 50-ml portion of 0.10 M $AgNO_3$ is titrated with 0.10 M NaCl using a silver indicator electrode. (a) Calculate the potential of the indicator electrode after the addition of 49.9, 50.0, and 50.1 ml of titrant; (b) Repeat part (a), substituting NaBr for NaCl; (c) Repeat part (a), substituting NaI for NaCl. Calculate ΔE per 0.20 ml of titrant for each salt.

CHAPTER THIRTEEN

other electrical methods of analysis

Electrochemistry is a large field, including in its broadest aspects fuel cells; conversion of solar to electrical energy; corrosion of metals; batteries; commercial processes for the large-scale production of such materials as aluminum, chlorine, and sodium hydroxide; and electroorganic synthesis. Some electrochemical techniques have found extensive analytical application, and a recognizable specialty called electroanalytical chemistry has emerged. This area cannot be treated thoroughly in an introductory textbook. Rather, we have selected a limited number of the more classical electroanalytical techniques which the student will be able to understand with his present background. One such topic has already been discussed (potentiometry, Chapter 12). The groundwork for the present chapter was laid in Chapter 10; the student may find it useful to review electrode potentials, galvanic cells, and the Nernst equation.

ELECTROLYSIS

Introduction

Consider the following galvanic cell:

$$\mathrm{Cu|Cu^{2+}(1\ \mathit{M})||Ag^{+}(1\ \mathit{M})|Ag}$$

$$\begin{array}{lr}
\mathrm{2Ag^{+} + 2\mathit{e} \rightleftharpoons 2Ag} & E^\circ = +0.80\ \mathrm{V} \\
\mathrm{Cu^{2+} + 2\mathit{e} \rightleftharpoons Cu} & E^\circ = +0.34\ \mathrm{V} \\
\hline
\mathrm{2Ag^{+} + Cu = Cu^{2+} + 2Ag} & E^\circ_{\mathrm{cell}} = +0.46
\end{array}$$

The positive sign for E°_{cell} says, according to the convention described in Chapter 10, that the silver electrode is the positive one, that the copper electrode is negative, and that the direction of spontaneous reaction, if the cell is allowed to discharge, is from left to right as written above.

Now let us connect this cell in series opposition with an external voltage source. This source may be a battery or an electronic power supply. The positive side of the source is connected with the positive electrode of the galvanic cell, and negative to negative, as shown in Fig. 13.1. The arrow through the source in the figure means that we can vary the external voltage applied to the cell; the circles marked A and V represent an ammeter and a voltmeter, respectively.

Suppose that we adjust the external voltage source to exactly 0.46 V and close the switch. The ammeter will show no current, because the source and the galvanic cell, connected in opposition to each other, just balance; neither can send electrons through the other. If we apply a smaller voltage than 0.46 V, a current will be observed: electrons will flow from the copper electrode through the external circuit and into the silver electrode; the cell reaction is proceeding

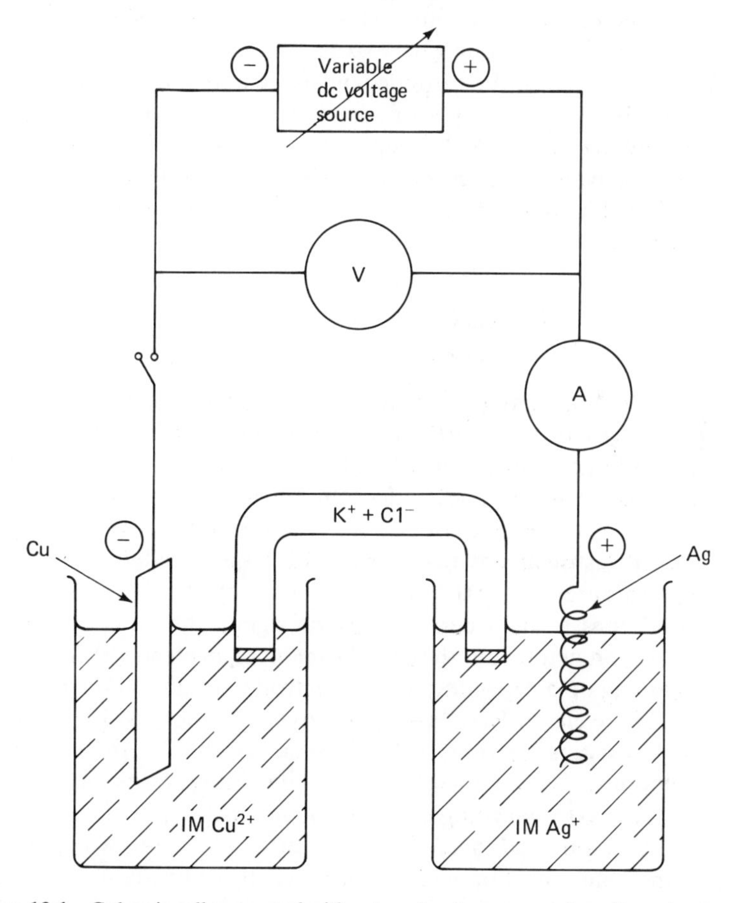

Figure 13.1 Galvanic cell connected with external voltage source (see discussion in text).

spontaneously from left to right, with Ag^+ being reduced to Ag and Cu oxidized to Cu^{2+}. That is, we observe the spontaneous discharge of the galvanic cell through an external resistance.

If, on the other hand, we apply against the galvanic cell a voltage greater than 0.46 V, again a current will flow, but it will be in the direction opposite to that above. Electrons will flow from the negative side of the external source into the copper electrode, and they will flow away from the silver electrode into the external circuit. Cu^{2+} will be reduced to Cu, and Ag will be oxidized to Ag^+, which is to say that the cell reaction has been reversed; it is now proceeding from right to left as written above. This process is called *electrolysis*, and we say that the cell, which was galvanic earlier in our discussion, is now an electrolytic cell.

In the spontaneous discharge of a galvanic cell, electrical energy is derived from the inherent tendency of a redox reaction to occur. In an electrolytic cell, on the other hand, an external source of electrical energy is used to force the chemical reaction to occur in the direction opposite to the spontaneous one.

Because of the opposite directions of electron flow in galvanic and electrolytic cells, students sometimes experience confusion regarding signs. In either kind of cell, the *cathode* is defined as the electrode at which reduction occurs, and the *anode* is the electrode where oxidation occurs. In the galvanic cell above, Ag^+ is reduced to Ag at the cathode, and as we noted, the silver electrode is the positive one. In the electrolytic mode, the silver electrode is still positive, but now oxidation of Ag to Ag^+ occurs here, and hence this electrode becomes the anode. In the galvanic case, the negative copper electrode is the anode (oxidation of Cu to Cu^{2+}) while in electrolysis, reduction occurs here and copper is the cathode. In summary:

	Cathode	Anode
Galvanic	+	−
Electrolytic	−	+

Although it may be obvious, we should note specifically that a current flowing in the circuit of Figure 13.1 involves electron flow in the external wiring and migration of ions in the cell solutions, including the salt bridge. This current is carried across the electrode–solution interfaces by electron transfer reactions involving the redox couples Cu^{2+}–Cu and Ag^+–Ag. Such a current, which is directly associated with electrode reactions, is called a *faradaic current.* (Faradaic currents are usually the major ones in electrolysis experiments, but under special conditions a small current may be significant which involves, not an electron transfer reaction, but the establishment of an electrical double layer, analogous to charging up a capacitor, at the electrode–solution interface.)

In the electrolysis described above, with the cell of Fig. 13.1, we initially faced a preexisting galvanic cell whose electrodes were in equilibrium with their solutions; this cell possessed a definite voltage of its own before we did anything with the external voltage source. Sometimes in practice, however, we place into a solution a pair of inert electrodes (often platinum) which presumably adopt the same potential. Suppose, for instance, that two platinum electrodes are inserted into an aqueous solution of copper sulfate, as shown in Fig. 13.2. No galvanic cell

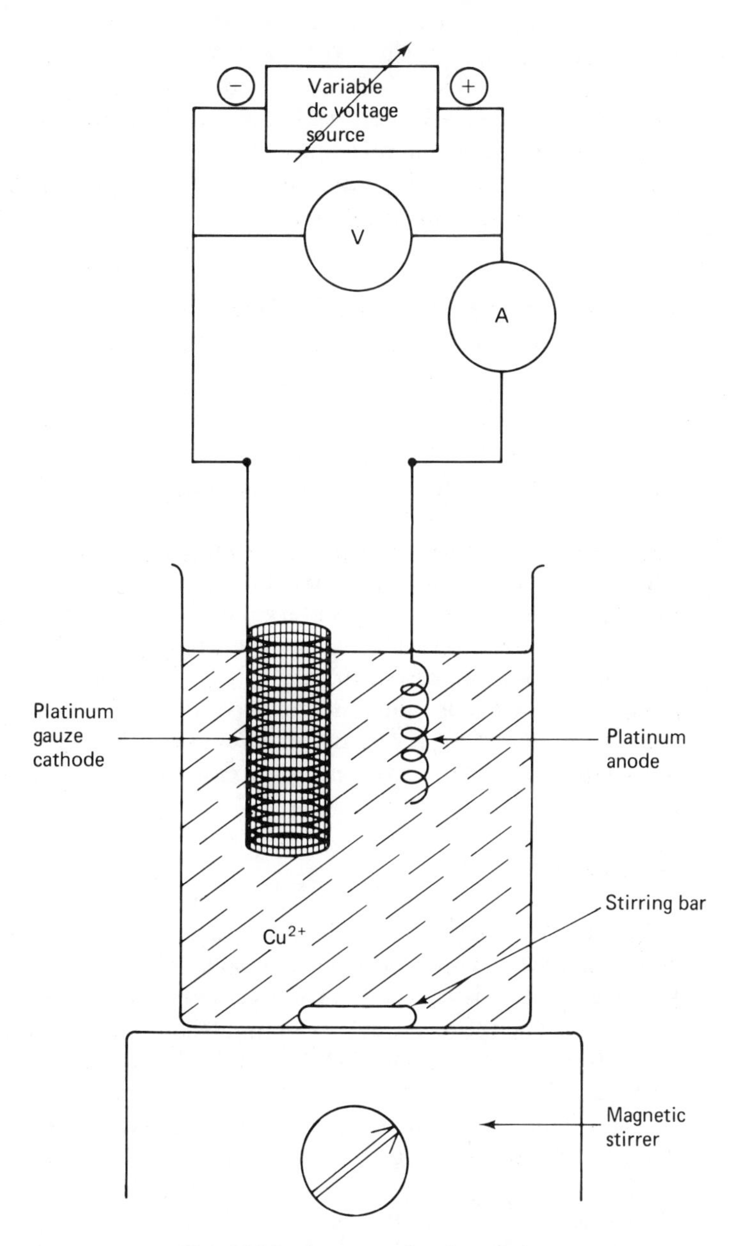

Figure 13.2 Apparatus for electrolysis.

really exists: if two identical electrodes are placed in the same solution, they should have the same potential, and there is no voltage developed between them.

Now let us close the switch and turn up the voltage source. When the applied voltage becomes high enough (as measured with voltmeter V in Fig. 13.2), we shall observe the appearance of a red–brown color due to copper metal on the negative electrode and bubbles of gas will be seen around the positive electrode.

We are reducing Cu^{2+} ($Cu^{2+} + 2e \rightarrow Cu$) and we are oxidizing water ($2H_2O \rightarrow O_2 + 4H^+ + 4e$). The platinum cathode has become copper-plated, and the solution sees it as a copper electrode; around the anode, where there was originally nothing of electrochemical interest, there are now molecules of O_2. In other words, when the electrolysis began, we created from its products a galvanic cell which we may write as follows:

$$Pt, O_2|H^+ + H_2O + Cu^{2+}|Cu$$

Supposing for simplicity that the ions are all at unit activity and that the partial pressure of oxygen above the solution is 1 atm, we can calculate the voltage of this galvanic cell from the standard potentials:

$$\begin{array}{ll} 2Cu^{2+} + 4e \rightleftharpoons 2Cu & E^\circ = +0.34 \text{ V} \\ O_2 + 4H^+ + 4e \rightleftharpoons 2H_2O & E^\circ = +1.23 \text{ V} \\ \hline 2Cu^{2+} + 2H_2O \rightleftharpoons 2Cu + O_2 + 4H^+ & E^\circ_{cell} = -0.89 \text{ V} \end{array}$$

The negative sign for E°_{cell} indicates spontaneity from right to left as the cell reaction is written above. To make the electrolysis go (i.e., for the reaction to go from left to right) we must apply a larger voltage than the galvanic value of 0.89 V.

The galvanic voltage which must be overcome by the external power source in order to initiate electrolysis (0.46 V in the cell on page 326, 0.89 V in the case immediately above) is called the *decomposition potential* or sometimes the "back emf," E_d. Values of E_d for particular cases are easily calculated using the methods of Chapter 10 once the products of the electrolysis have been decided upon (see below).

Products of Electrolysis

Sometimes several chemical substances are present in a solution, and we may wish to predict what the electrode reactions will be if the solution is electrolyzed. In principle, this is simple. For the cathode reaction, the rule is: of all those substances which have access to the cathode, that one which is most easily reduced will be the one that *is* reduced. We evaluate the ease of reduction, of course, from the single electrode potential. For reactions written $Ox + ne \rightleftharpoons Red$, using the sign convention of this text, the most positive (least negative) potential indicates the greatest tendency to go from left to right.

Example 1. A solution contains the following ions, each at a concentration of 1.0 M: Zn^{2+}, H^+, Cu^{2+}, Ag^+. Platinum electrodes are inserted and the applied voltage is increased until electrolysis begins. What product is formed at the cathode?

The standard potentials are:

$$\begin{array}{rl} Ag^+ + e \rightleftharpoons Ag & E^\circ = +0.80 \text{ V} \\ Cu^{2+} + 2e \rightleftharpoons Cu & E^\circ = +0.34 \text{ V} \\ H^+ + e \rightleftharpoons \tfrac{1}{2}H_2 & E^\circ = 0 \text{ V} \\ Zn^{2+} + 2e \rightleftharpoons Zn & E^\circ = -0.76 \text{ V} \end{array}$$

Using the rule above, we select the reduction of Ag^+ as the most facile cathode reaction, and the product will be a plate of silver metal on the surface of the platinum cathode.

Of course, if the concentrations had not all been unity, then the actual potentials to be compared would have been calculated from the Nernst equation. For example, if $[Ag^+]$ had been 0.1 *M*, then

$$E = +0.80 - 0.06 \log \frac{1}{0.01} = +0.68 \text{ V}$$

Since the concentrations appear in the log term of the Nernst equation, the $E°$ values are usually adequate for predicting electrode reactions unless they are quite close together or the concentrations are quite disparate.

One way of looking at the situation in the preceding example is the following. The external voltage source is pumping electrons onto the cathode. If these are sufficiently consumed in the reaction $Ag^+ + e \rightleftharpoons Ag$, then the cathode is prevented from becoming negative enough to reduce, say, Cu^{2+}, which has a lower electron affinity than Ag^+.

At the anode, that substance which gives up electrons most readily is the one which will be oxidized. Again, we judge the ease of oxidation from the potential, but here we look for the least positive (most negative) value, that is, the value indicating the greatest tendency of $Ox + ne \rightleftharpoons Red$ to go from right to left.

Example 2. A solution contains the following species, each at a concentration of 0.5 *M*: hydroquinone (H_2Q), Cl^-, Br^-, H^+. Platinum electrodes are inserted and a current is passed. What product is formed at the anode?

The standard potentials are:

$$PtBr_4^{2-} + 2e \rightleftharpoons Pt + 4Br^- \qquad E° = +0.58 \text{ V}$$

$$Q + 2H^+ + 2e \rightleftharpoons H_2Q \qquad E° = +0.70 \text{ V}$$

$$PtCl_4^{2-} + 2e \rightleftharpoons Pt + 4Cl^- \qquad E° = +0.73 \text{ V}$$

$$Br_2 + 2e \rightleftharpoons 2Br^- \qquad E° = +1.09 \text{ V}$$

$$O_2 + 4H^+ + 4e \rightleftharpoons 2H_2O \qquad E° = +1.23 \text{ V}$$

$$Cl_2 + 2e \rightleftharpoons 2Cl^- \qquad E° = +1.35 \text{ V}$$

Note that we must not forget any possible reductant: we include the oxidation of water as a possibility, and the oxidation of the platinum anode itself must also be considered. Platinum, which is sometimes called an inert electrode, is really only relatively so, and platinum may be oxidized in electrolysis, depending upon the potential it is allowed to reach and the composition of the solution. The oxidation is facilitated in this example by the presence of Br^-, which converts Pt^{2+} into a stable complex ion.

Inspecting the standard potentials, we see that +0.58 is the least positive one. Strictly, of course, in this example the standard potentials cannot be used directly. For instance, because $[H^+] = 0.5$ *M*, +1.23 V is not the correct value to use to determine how easily water is oxidized. But the student can easily verify, using the Nernst equation, that lowering $[H^+]$ from 1 *M* to 0.5 *M* will not shift this potential enough to make the oxidation of water a serious contender: a tenfold change in $[H^+]$ shifts the potential by only 0.06 V. Thus, in this example, the oxidation of the platinum anode is the best possibility. It may be noted that we cannot really calculate the potential for this, but it does not matter. The Nernst equation for the process is

$$E = +0.58 - \frac{0.06}{2}\log\frac{1 \times 0.5^4}{[PtBr_4^{2-}]}$$

Prior to the electrolysis, there is no $PtBr_4^{2-}$ in the solution according to the statement of the problem. But if we insert zero into the Nernst equation, we calculate that E is infinite, scarcely possible physically. Perhaps there are a few of these ions in the solution, formed in some manner such as a tiny amount of air oxidation of Pt. In any event, we do not know the concentration and hence cannot calculate E. But we do know that if $PtBr_4^{2-}$ is very small, E will be even less positive than +0.58 V, and we still confidently select this reaction for the anode reaction. If we were to start the electrolysis, the platinum anode would lose weight and $PtBr_4^{2-}$ would appear in the solution. Because platinum electrodes are expensive, such a situation is usually undesirable.

The student should be cautioned not to inspect potentials blindly without a physical picture in mind of the electrochemical cell. For example, electrolysis experiments are often performed with a silver anode in a chloride solution, where the following reaction proceeds from right to left:

$$AgCl + e \rightleftharpoons Ag + Cl^- \qquad E° = +0.22\text{ V}$$

AgCl is a solid which tends to adhere to the surface of the silver anode, and it cannot reach the cathode in appreciable quantity. Thus the reduction of AgCl is not a possible cathode reaction even if +0.22 V be the largest positive number among those which the student may be contemplating. As a second example of a common mistake, consider:

$$O_2 + 4H^+ + 4e \rightleftharpoons 2H_2O \qquad E° = +1.23\text{ V}$$

If, in a particular electrolysis experiment, the preceding reaction occurs at the anode, generating O_2, the student may seize upon the reduction of O_2 as the cathode reaction because +1.23 V is so large. But oxygen is not very soluble in water, and most of it bubbles out of the solution around the anode. In addition, special precautions are sometimes taken, such as a blanket of nitrogen over the cathode compartment or a porous diaphragm to minimize transport of O_2 to the cathode by currents in the solution. Thus the student is warned, in predicting electrolysis products, not to consider electrode reactions which are impossible because the appropriate electroactive species are not physically present at the electrodes.

Finally, there are sometimes problems in predicting electrolysis products arising from kinetic effects. Detailed discussion is beyond the scope of this text, but it should be mentioned that occasionally an electrode reaction which is possible thermodynamically (i.e., one that would be predicted to proceed on the basis of the potential) does not, in fact, occur during an electrolysis. The student should realize that the *tendency* for a reaction to occur and the *rate* at which it proceeds are two different things. Predicting electrolysis products from potentials alone will lead to an error once in a while; for our purposes here, we shall just have to live with this. Among examples which the student may encounter, oxyanions such as ClO_4^-, NO_3^-, and SO_4^{2-} are prominent for not reducing nearly so easily as the potentials would indicate.

It must be emphasized that in this section we have been talking about only the *initial* products of an electrolysis. As the electrode reactions proceed in a prolonged experiment, concentrations change, potentials shift, and new electrode reactions may become possible. Aspects of this problem are found later in the chapter.

Voltage Requirements

We saw earlier that in order to initiate an electrolysis we had to exceed the galvanic back emf or decomposition potential, E_d. This value, which is given by $E_d = E_{\text{anode}} - E_{\text{cathode}}$, is easily calculated, as we saw. Of course, E_d changes as the electrolysis proceeds, but it can still be calculated for any desired conditions, using the Nernst equations for E_{anode} and E_{cathode}. However, for an electrolysis actually to proceed, we must apply a larger voltage than E_d. The electrolytic cell offers a resistance, R, to the flow of current, and in order to pass a finite current, i, we know from Ohm's law that we must exceed E_d by a value $i \times R$. For example, to pass a current of 0.2 A through a cell whose resistance is 5 ohms, we require an additional $5 \times 0.2 = 1$ V above E_d.

There is also another problem to consider. Suppose that our goal is to reduce Cu^{2+} to Cu at a cathode. Now, before the electrolysis is started, the concentration of Cu^{2+} is presumably uniform throughout the solution, as shown by curve A in Fig. 13.3. As the electrolysis proceeds, however, it is those ions which are near the

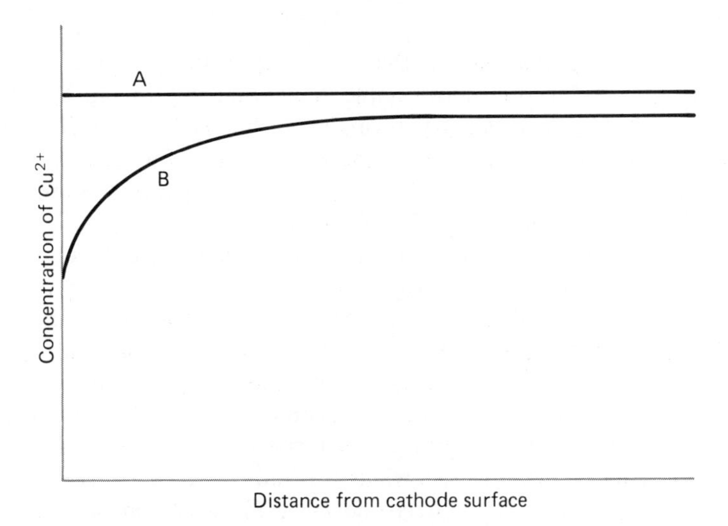

Figure 13.3 Concentration profiles. Curve A, before the start of the electrolysis. Curve B, after the electrolysis has proceeded for a certain length of time.

cathode surface which are reduced to copper metal, thereby lowering the concentration of Cu^{2+} at the interface below that in the body of the solution, as suggested in Fig. 13.3, curve B. Stirring the solution will help in this regard, but there exists at any such interface a stagnant film which is not well mixed with the bulk solution, even by vigorous stirring. Copper ions will tend to migrate into this depleted layer, of course, but some degree of depletion always persists if the electrolysis is being carried out at a finite rate. (The depleted layer may vary from a few angstrom units to an appreciable fraction of a millimeter, depending upon the rate of electrolysis and the stirring conditions.)

In general, we do not know the value of $[Cu^{2+}]$ at the electrode–solution interface, and in calculating E_d, we insert the bulk concentration into the Nernst equation for $E_{cathode}$. Thus, even if equilibrium between electrode and solution existed at the interface, our value for E_d would be too small. To take account of this in developing an equation for the required applied voltage, we shall add on a term designated $\omega_{cathode}$. It may be considered that this term arises from the slow diffusion of Cu^{2+} into the depleted layer; in this sense, $\omega_{cathode}$ represents a kinetic intrusion into the otherwise thermodynamic calculation of $E_{cathode}$. An ω term originating in a nonuniform concentration profile as in Fig. 13.3, curve B, is called a *concentration overpotential.*

Noninstantaneous diffusion, as in the example described above, is not the only reason for an electrode potential to depart from the Nernstian value calculated from bulk concentrations. A chemical step with slow kinetics may lead to the same result. For example, suppose that H^+ were being reduced at a cathode. Presumably the initial electron transfer reaction leads to hydrogen atoms:

$$H^+ + e \longrightarrow H$$

To form the final product, hydrogen gas, the atoms must combine:

$$2H \longrightarrow H_2$$

Now the *equilibrium* for this combination lies far to the right; under ordinary conditions, H_2 is far more stable than atomic hydrogen. But the *rate* of combination may be slow. Thus, to form the product H_2 at an appreciable rate, the cathode must be more negative than the Nernst equation suggests. This effect, which is often called *activation overpotential,* will be included in the ω term in writing an equation for the required applied voltage.

Activation overpotential arising from a slow kinetic step in the overall electrode process may occur in many cases, but it is particularly pronounced where a gas is liberated. The two common examples in electroanalytical chemistry are reduction of H^+ to form H_2, used as an illustration above, and oxidation of water to form O_2, which is a frequent anode reaction. The magnitude of this overpotential depends upon the chemical nature of the electrode material and its physical state (e.g., surface area), the temperature, and the rate at which the electrolysis is performed, which is often stated as a current density (amperes per square centimeter of electrode surface). Some measured values of hydrogen activation overpotentials are given in Table 13.1. The differences between smooth and platinized platinum electrodes reflect the surface area effect;

platinized platinum is smooth platinum upon whose surface has been deposited by rapid electrolysis a rough platinum layer of large surface area. Comparing platinum and mercury is interesting; because platinum catalyzes the combination of hydrogen atoms to form H_2, the overpotential is much lower than with mercury. The high overpotential for hydrogen evolution on mercury is exploited in polarography, as we shall see in a later section of this chapter.

TABLE 13.1 Some Overpotentials for Hydrogen Evolution (Volts)*

Current density, A/cm^2	Smooth platinum	Platinized platinum	Mercury
0.0001	—	0.0034	—
0.001	0.024	0.015	0.9
0.01	0.068	0.030	1.04
0.1	0.29	0.041	1.07
1.0	0.68	0.048	1.11

* *International Critical Tables,* Vol. VI, pp. 339–340. McGraw-Hill Book Co., Inc., New York, 1929.

To summarize overpotential for our purposes, then, let us say that in an electrolysis carried out at a finite rate, the electrodes are not in equilibrium with the bulk solution, and potentials calculated using bulk concentrations in the Nernst equation are incorrect. Where slow diffusion is the major effect, the electrode may be virtually in equilibrium with solution right at the interface, but in other cases chemical steps with slow kinetics may even preclude this. In any event, overpotential means that the cathode will be more negative than we would have predicted from the Nernst equation, and the anode more positive, if the electrolysis is actually proceeding at more than an infinitesimal rate.

Summarizing this section, we may write the following equation for the applied potential:

$$E_{\text{applied}} = E_{\text{anode}} - E_{\text{cathode}} + iR + \omega_{\text{cathode}} + \omega_{\text{anode}}$$

where $E_{\text{anode}} - E_{\text{cathode}} = E_d$ is the Nernstian decomposition potential.

Electrogravimetry

In its usual form, electrogravimetry involves plating a metal onto a weighed platinum cathode and then reweighing to determine the quantity of the metal. The experimental setup is similar to that shown in Fig. 13.2. The determination of copper serves as an example. The sample, perhaps a copper alloy, is dissolved in nitric acid. The platinum gauze cathode, having been cleaned in nitric acid, rinsed, dried in an oven, and weighed, is inserted into the solution and an electrical connection is made using some sort of clamp. The applied voltage is increased until the ammeter shows a current and the cathode develops a coppery appearance; bubbles will be seen rising from the anode. In practice, the terms in the equation for $E_{applied}$ in the section above cannot be calculated accurately. We simply start with a rough idea of what $E_{applied}$ should be and add some extra voltage to make certain that the electrolysis goes. Of course, for a case as common as the deposition of copper, directions can be found in the literature. After a reasonable time, one can test for completeness of deposition by lowering the cathode and observing whether the fresh platinum surface acquires a copper plate. At the end, the cathode is removed from the solution with the voltage still applied (to prevent redissolving the copper plate by galvanic action) while being rinsed with distilled water. The cathode is then dipped into ethanol or acetone to facilitate drying, dried quickly in an oven to avoid surface oxidation of the copper, and finally cooled and weighed.

The nitric acid medium is desirable for the experiment described above. As the concentration of Cu^{2+} is lowered by the electroreduction, the cathode becomes increasingly negative until the reduction of nitrate begins:

$$NO_3^- + 10H^+ + 8e \longrightarrow NH_4^+ + 3H_2O$$

This stabilizes the cathode potential, which does not then become sufficiently negative to reduce certain other metals such as nickel which may be present in the sample. (It also prevents the reduction of H^+, which is undesirable in this case because concurrent hydrogen evolution tends to cause a spongy and nonadherent copper deposit.) Since the supply of nitric acid is large, this control of the cathode potential is maintained for a sufficient time.

Simple electrogravimetric determinations as described above are widely used for metals. The technique works very well when a fairly noble metal such as silver or copper is to be determined in samples whose other constituents are less easily reduced than H^+. The intervention of hydrogen evolution prevents the cathode from becoming negative enough to reduce the other metals. Sometimes, as in the example with NO_3^- described above, other species (called depolarizers or potential buffers) are added to serve this same purpose. For example, the iron(III)–iron(II) system,

$$Fe^{3+} + e \rightleftharpoons Fe^{2+} \qquad E^\circ = +0.77\ V$$

limits the cathode potential to a value no more negative than +0.77 V if the concentrations of the two ions are equal. If no substance is present that is more easily oxidized than iron(II) ion, then iron(III) ion is formed at the anode; the reduction of the latter at the cathode serves to limit the cathode potential while a fairly constant ratio of the two ions is maintained in the solution. When H^+ serves as a potential buffer, there is a wide latitude for cathode potential control by adjusting the pH of the solution. The instrumentation for this type of electrolysis is relatively inexpensive. The sensitivity of the method is determined largely by the weighing process.

Controlled Potential Electrolysis

In classical electrolysis as described in the last section, it is easy to measure the applied voltage using voltmeter V in Fig. 13.2, and it is easy to determine whether anything is happening by watching the ammeter. In many cases, however, this is not enough. In an electrolytic reduction, it is the potential of the cathode that really determines what will happen there. The student may refer again to the equation given earlier for $E_{applied}$ to see that the applied voltage does not tell us much about the cathode potential itself because we probably do not know other terms in the equation, such as the iR drop or the anode overpotential. Thus it is often desirable to measure the cathode potential against a reference electrode which is not in the electrolysis circuit and whose potential remains constant throughout the electrolysis. Then, by adjustments to the applied voltage, we may control the cathode potential, that is, prevent its becoming more negative than desired. (We happen to use electrolytic reductions as examples in this chapter, thereby focusing attention on the cathode, but it should be noted that the same considerations would apply in a case where oxidation at an anode was of primary interest.) An example will clarify the need for cathode potential control.

Example. A solution whose volume is 100 ml is about 0.1 M in both Ag^+ and Cu^{2+}. It is desired to determine the exact quantity of silver by electrodeposition on a weighed platinum cathode; the results will be higher if copper also plates onto the cathode. Can this separation of silver from copper be done?

Suppose that the analytical balance is sensitive to 0.1 mg; then if 0.1 mg of Ag^+ is left in solution we shall consider the electrodeposition of silver to be "complete." The molar concentration will then be

$$[Ag^+] = \frac{0.1 \text{ mg}}{100 \text{ ml} \times 108 \text{ mg/mmol}} = 9 \times 10^{-6}\,M$$

The potential which the cathode must attain in order to lower $[Ag^+]$ to this value is

$$E_{cathode} = +0.80 - 0.06 \log \frac{1}{9 \times 10^{-6}} = +0.50 \text{ V}$$

The concentration of Cu^{2+} is 0.1 M. The potential of an electrode which can be in equilibrium with this ion is given by

$$E_{cathode} = +0.34 - \frac{0.06}{2} \log \frac{1}{0.1} = +0.31 \text{ V}$$

That is, for reduction of Cu^{2+} to copper metal to occur, the cathode must be more negative than +0.31 V.

Since silver reduction is "complete" at +0.50 V, we see that it is possible to plate silver on the cathode without interference from Cu^{2+}, provided that we monitor the cathode potential and prevent it from becoming more negative than some value between +0.50 and +0.31 V, say, +0.40 V.

A manually operated experimental setup for limiting the cathode potential to the desired value is shown in Fig. 13.4. The apparatus is seen to include the features required for any electrolysis: variable voltage source, anode, and cathode. But there is, in addition, a third electrode in the solution, the reference electrode, frequently a saturated calomel electrode. The potential difference between the cathode and the reference electrode is measured with a voltmeter of high input resistance (so as to draw little current).

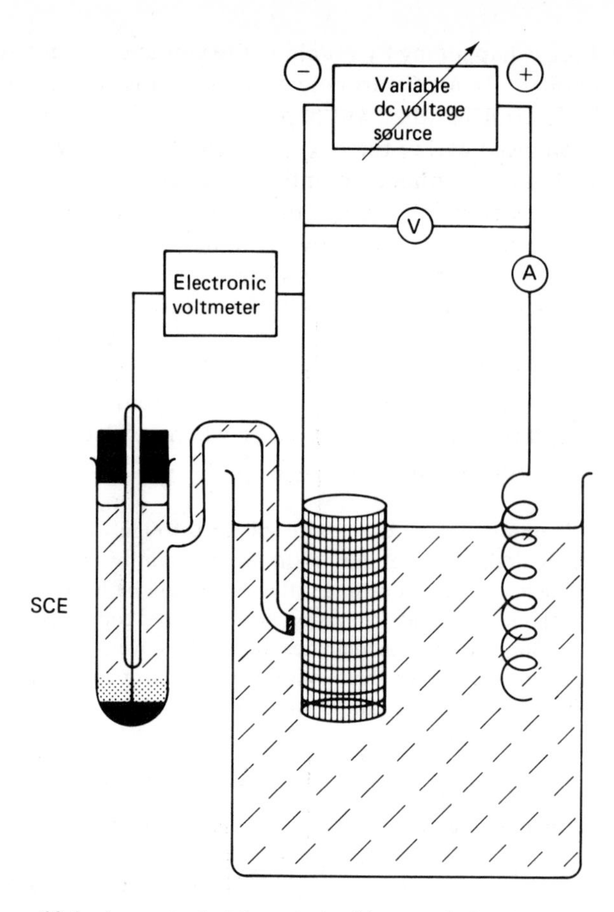

Figure 13.4 Apparatus for electrolysis with controlled cathode potential.

Returning to the example above, suppose that we decide to limit the cathode potential to +0.40 V. If the reference electrode is an SCE, its potential is +0.25 V. This means that the voltage between SCE and cathode is to be no smaller than (0.40 − 0.25) = 0.15 V, with the cathode positive. The desired potential relationships are shown in Fig. 13.5. (Students are sometimes confused by the signs, but a little thought shows that there is no reason to be: in this example, the cathode is more positive than the SCE, but it is still negative with respect to the anode, about which we have said little but which is perhaps platinum at which O_2 is being evolved at a potential above +1 V, depending upon the acidity and overpotential effects.)

Now to perform the electrolysis, we turn up the applied voltage until an adequate current flows, but we watch the high resistance voltmeter as the electrolysis proceeds. At first there is no problem: the high concentration of Ag^+ itself limits the cathode potential to a fairly positive value. But as $[Ag^+]$ decreases during the electrolysis, the SCE-cathode voltage will decrease. As it nears 0.15 V, the operator will cut back the applied voltage so that this value is not passed, and

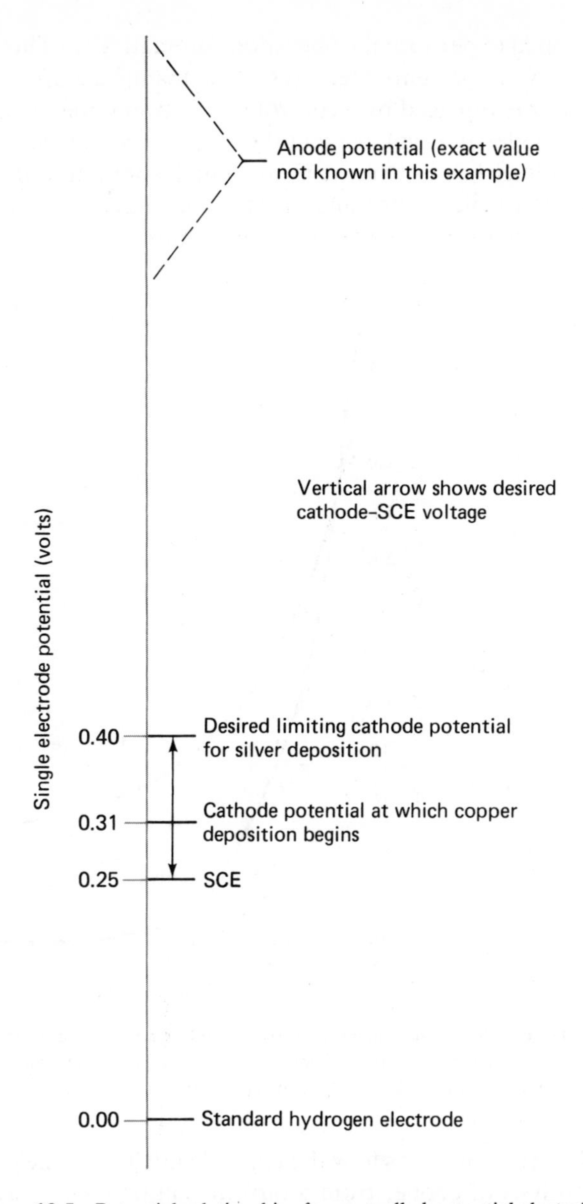

Figure 13.5 Potential relationships for controlled potential electrolysis.

he will need to make readjustments with increasing frequency as the electrolysis nears completion. Eventually, the current will decay to a very low value (because the potential control prevents the intrusion of other electrode reactions) and the run will be terminated.

Manual operation as described above is tedious; as the electrolysis proceeds, changes in concentrations, *iR* drop, and overpotentials require many adjustments to the applied voltage. As a result, instruments called *potentiostats* have been developed to perform the operation automatically. These commercially available units have the essential features of the manual setup in Fig. 13.4, but the human operator is replaced by a control link between the measuring device monitoring the cathode–SCE voltage and the variable voltage source. When the cathode–SCE voltage begins to move away from the predetermined value, an error signal is generated which, after amplification, changes the applied voltage via either a mechanical or totally electronic interaction.

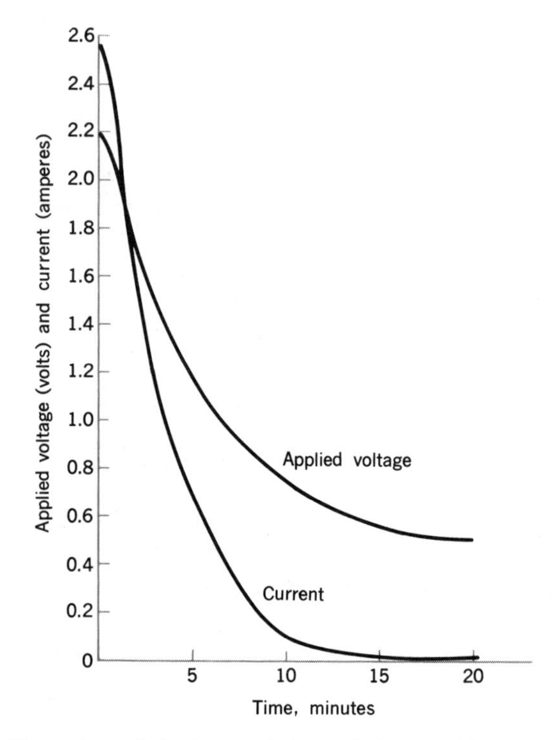

Figure 13.6 Change in applied voltage and current during controlled potential electrolysis of copper from a tartrate solution. (After J. J. Lingane, *Electroanalytical Chemistry*, 2nd ed., Interscience Publishers, Inc., New York, 1958.)

Figure 13.6 shows how the applied voltage and the current changed during an electrolysis of copper from a tartrate solution, using a potentiostat to limit the cathode potential to −0.36 V vs. SCE.[1] Ideally, the current falls off exponentially according to

$$i_t = i_0 \times 10^{-kt}$$

where i_t is the current at time t, i_0 the initial current, and k a constant, but slight departures from this behavior are common in practice.

[1] J. J. Ligane, *Electroanalytical Chemistry*, 2nd ed., Interscience Publishers, Inc., New York, 1958, p. 217.

Some examples of successful separations of metals by controlled potential electrolysis are given in Table 13.2. The table is only illustrative, not exhaustive. Sometimes a separation which appears nonfeasible upon inspection of potentials for the reduction of uncomplexed metal ions can be accomplished by choosing a complexing electrolyte which selectively changes the ease of reduction of some of the metal ions in the sample solution.

TABLE 13.2 Some Examples of Separations and Determinations by Controlled Potential Electrolysis

Metal determined	Metals from which separated	Electrolyte	$E_{cathode}$, V, vs. SCE
Ag	Cu and other, more active metals	Acetate buffer	+ 0.1
Cu	Ni, Sn, Pb, Sb, Bi	Tartrate with hydrazine and Cl^-	− 0.3
Bi	Pb, Sn, Sb, and other, more active metals	Tartrate with hydrazine and Cl^-	− 0.4
Sb	Sn	H_2SO_4 with hydrazine	− 0.2
Cd	Zn	Acetate buffer	− 0.8
Ni	Zn, Al, Fe	Ammoniacal tartrate	− 1.1

A mercury cathode, because it is a liquid, is not a convenient electrode for electrogravimetry, but it offers advantages for certain separations. Because of the high overpotential for hydrogen evolution, some of the metals more active than hydrogen are easily reduced at this electrode, and the solubility of a number of free metals in mercury to form dilute amalgams further facilitates the reductions by lowering the activity of the metal below unity. An excellent discussion of the mercury cathode and its applications is available.[2]

Applications of controlled potential electrolysis extend beyond analytical chemistry. The ability to control the potential of an electrode at any desired value corresponds to the possession of a large number of oxidizing and reducing agents, for which one may expect many uses in both inorganic and organic chemistry. The electrochemical preparation of adiponitrile (used in the manufacture of nylon) from acrylonitrile on a commercial basis is one example. The chemical problem of coupling two carbon atoms which possess positive character is circumvented by electrochemical generation of a radical which attacks another molecule forming an intermediate that further reduces to the product:

$$CH_2{=}CH{-}C{\equiv}N \xrightarrow[H^+]{1e} \dot{C}H_2{-}CH{=}C{=}N{-}H$$

$$\downarrow CH_2{=}CH{-}C{\equiv}N$$

$$\begin{matrix} CH_2{-}CH_2{-}C{\equiv}N \\ | \\ CH_2{-}CH_2{-}C{\equiv}N \end{matrix} \xleftarrow[H^+]{1e} \begin{matrix} CH_2{-}CH{=}C{=}N{-}H \\ | \\ CH_2{-}\dot{C}H{-}C{\equiv}N \end{matrix}$$

[2] J. A. Maxwell and R. P. Graham, *Chem. Rev.*, **46**, 471 (1950).

A second example is the reduction of aromatic nitro compounds to either hydroxylamines or amines by appropriate choice of cathode potential:

$$\begin{array}{ccc} \text{Ar-NO}_2 & \xrightarrow[\text{0.4 V vs. SCE}]{4e} & \text{Ar-NHOH} \\ 6e \downarrow {\scriptstyle -0.9\text{ V vs. SCE}} & & \\ \text{Ar-NH}_2 & & \end{array}$$

Other products are also obtainable under various experimental conditions.[3] Attractive aspects of synthetic steps based upon controlled potential electrolysis include frequent high yields and the fact that workup is easier because side reactions are often avoided and the products are not contaminated with excess chemical redox reagents. Application to organic chemistry goes back at least 35 years,[4] but the technique still has not been adopted in many laboratories. It has been employed by inorganic chemists to prepare intermediate oxidation states of several elements.

We shall refer to controlled potential electrolysis again. The technique may be adapted to coulometry (see the next section), and there is a relation to polarography (see below), which is often used as a pilot technique for selecting the potential for a controlled potential electrolysis.

COULOMETRY

Coulometry refers to the measurement of coulombs (i.e., quantity of electricity). In analytical chemistry, the term implies a measurement of coulombs under such conditions that the measured quantity is associated with a particular electrochemical reaction. This permits a simple, straightforward analytical calculation based on Faraday's law.

When a current of 1 ampere passes for 1 second, the quantity of electricity involved is 1 coulomb; that is,

$$\text{coulombs} = \text{amperes} \times \text{seconds}$$

Other units, such as millicoulombs, are sometimes convenient, just as we have millimoles, micrograms, etc. Now, according to Faraday's law, one equivalent of any electroactive substance requires 96,493 coulombs for complete reaction. The value 96,493 coulombs is called the Faraday, F (96,500 is close enough for many calculations, as in the problems at the end of this chapter).

Now, in a coulometric analysis, when conditions are such that all of the electricity passed through an electrolytic cell is associated with the desired electrode reaction, then

$$\begin{array}{c}\text{grams of substance} \\ \text{being determined}\end{array} = \frac{\text{number of coulombs} \times \text{MW}}{n \times F}$$

[3] A. J. Fry, *Synthetic Organic Electrochemistry*, Harper & Row, Publishers, New York, 1972, Table 6.3, p. 233.

[4] J. J. Lingane, C. G. Swain, and M. Fields, *J. Amer. Chem. Soc.*, **65**, 1348 (1943).

where n is the number of electrons transferred and F the Faraday. The student may note that the units in the equation above are correct:

$$\text{Grams} = \frac{\text{coulombs} \times \text{grams/mole}}{\text{equivalents/mole} \times \text{coulombs/equivalent}} = \text{grams}$$

Of course, if one began with a known quantity of material, then the equation could be turned around to calculate n for an unknown electrochemical process.

Coulometry is simple in principle and relatively uncomplicated in laboratory application. The major problem is to ensure that all the electricity passed through the cell is, in fact, associated with the desired electrode reaction. Referring back to the example on page 337, we might determine silver by measuring the coulombs required for the complete reaction:

$$Ag^+ + e \rightleftharpoons Ag$$

But we would have to make certain that the reduction of Cu^{2+} or of any other species did not intrude. Faraday's law would still hold, that is, the total coulombs would relate to the total equivalents of silver and copper, but we would not know how many of these coulombs to assign to silver. When we are certain that only one reaction is occurring at an electrode, we sometimes say that the reaction is proceeding at 100% *current efficiency.*

Controlled Potential Coulometry

An obvious way to prevent the intrusion of an undesired electrode reaction is to control the potential of the electrode, as described in the last section. This will require three electrodes in the solution (cathode, anode, and reference) and a potentiostat to control the potential of the electrode at which the desired reaction occurs. Sometimes the electrode where the reaction of interest occurs is called the *working electrode.* The second electrode in the electrolysis circuit is then called the *auxiliary* or *counter electrode*; some electrochemical process must occur here, of course, but we are not primarily concerned with it except to ensure that it does not introduce undesired substances into the test solution.

During a controlled potential coulometric experiment, the current will decay somewhat as shown in Fig. 13.6. Since the current is not constant, the total coulombs, Q, associated with the desired reaction is given by

$$Q = \int_0^t i\,dt$$

This integral represents the area under the current–time curve. There are many ways to obtain this area. In the past it was common to place a second electrolysis cell in series with the test cell, so that the current through the two was the same. In this second cell, a well-characterized electrode reaction occurred at 100% current efficiency which generated a conveniently measured product—a gas whose volume at known pressure could be determined, an ion which was easily titrated, or whatever.[5] Nowadays, the integration is commonly performed electronically.

Controlled potential coulometry has been applied to the determination of a number of metals, such as lead, copper, cadmium, silver, and uranium; certain organic compounds; and halide ions via the following reaction, proceeding from right to left at a silver anode:

$$AgX + e \rightleftharpoons Ag + X^-$$

The technique has also been used frequently to determine n values in the investigation of new electrochemical reactions.

Coulometric Titrations

This technique has been used more widely than controlled potential coulometry. In its basic form, the apparatus is relatively simple and inexpensive. Required is a constant-current source; essentially, this is a high dc voltage in series with such a large resistor that changes in the electrolytic cell resistance during operation are negligible and the current remains constant. Switching the current on and off is analogous to opening and closing a buret; we titrate an electroactive species by adding or removing electrons at a cathode or anode as the working electrode. The other requirement is a clock. The number of coulombs involved in the electrode reaction is then calculated from the value of the current and the time required for complete reaction.

Again, a major problem is to ensure that the desired electrode reaction proceeds at 100% current efficiency. From the discussion in previous sections, the student will appreciate the danger that the potential of the working electrode may shift sufficiently to permit the intrusion of an undesired electrode reaction which will contribute an unknown quantity to the total number of coulombs, thereby vitiating the analytical calculation for the desired constituent.

For this reason, *direct* coulometric titrations in which the substance being determined reacts exclusively at the working electrode are uncommon. More generally useful is an *indirect* coulometric titration in which the electrolysis generates a reagent, which in turn reacts chemically with the desired constituent. The potential of the working electrode is kept fairly constant by maintaining a high concentration of the substance which is undergoing the electrode reaction to generate the titrant. For example, in the coulometric titration of iron(II) ion, electrolytically generated cerium(IV) ion can be used.[6] Cerium(III) ion, from which cerium(IV) ion is generated by anodic oxidation, is present in large concentration, and hence the anode potential is kept from becoming sufficiently positive for oxygen to be evolved. At the start of the electrolysis, iron(II) ion is directly oxidized at the anode; then the potential becomes more positive, reaching a value sufficiently positive to oxidize cerium(III) to cerium(IV) ion. The cerium(IV) ion, in turn, oxidizes any remaining iron(II) ion in the body of the

[5] For a description of earlier coulometers, see J. J. Lingane, *Electroanalytical Chemistry*, 2nd ed., Interscience Publishers, Inc., New York, 1958, pp. 452–459.

[6] N. H. Furman, W. D. Cooke, and C. N. Reilley, *Anal. Chem.*, **23**, 945 (1951).

solution. The quantity of electricity used is the same, of course, as if the iron(II) ion alone were directly oxidized at the anode. If the direct oxidation of iron(II) ion were attempted, however, oxygen would be liberated before the oxidation of iron(II) ion were complete, and the analysis would not be valid.

It is not necessary, of course, that the substance being titrated be itself electroactive as it was in the example above. For instance, bromine generated by the anodic oxidation of bromide ion may be used to titrate phenol:

$$C_6H_5OH + 3Br_2 \longrightarrow C_6H_2Br_3OH + 3HBr$$

Since phenol does not react at either electrode, the process is totally indirect.

Figure 13.7 shows an example of a coulometric titration setup. The platinum generator electrode is placed directly in the test solution, but the auxiliary electrode is in a separate compartment, the bottom of which is a sintered glass filter disk. This separation of the latter electrode tends to prevent the convective transport of any undesirable electrode product into the test solution. The clock and the current source are wired together so that one switch turns on both. The current source may provide for the selection of several values for the constant

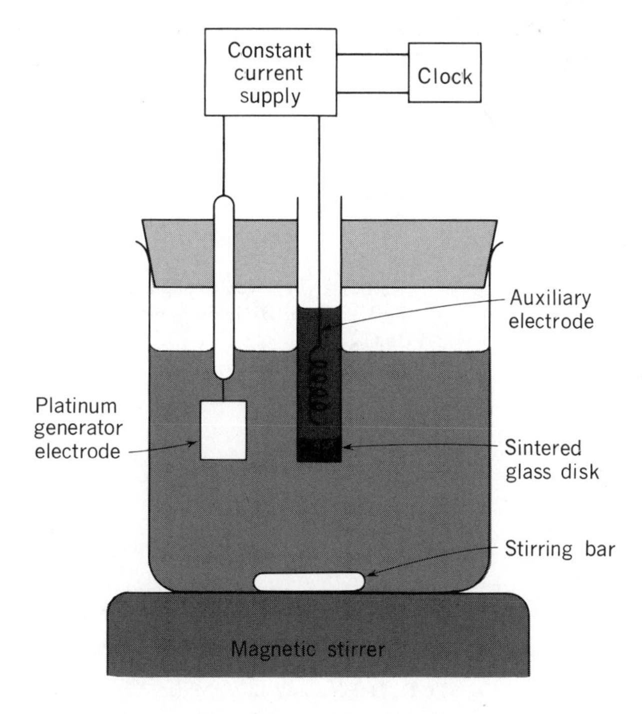

Figure 13.7 Cell for coulometric titration.

current, perhaps in the range of 1 to 200 mA. This is somewhat analogous to the availability of titrant solutions of various concentrations in classical titrimetry.

The titration is performed by operating the switch like a buret stopcock until the reaction of the desired constituent, directly at the electrode and/or indirectly with another substance formed at the electrode, is complete. The quantity is then calculated from the time required, the known constant current, and Faraday's law. Obviously this requires that we know when the reaction is complete; that is, we need to know, as in any titration, when to stop. In fact, any of the usual end-point techniques can be employed in coulometric titrations—visual, potentiometric, photometric, amperometric, etc. For example, in the titration of phenol with electrolytically generated Br_2 mentioned above, a dye such as indigo carmine which is destroyed by excess Br_2 may be added for a visual end point. For a coulometric titration with a potentiometric end point, there will be four electrodes in the solution: an anode and a cathode for the electrolytic generation of titrant, connected to the constant current source, and an indicator-reference electrode pair, connected with an electronic voltmeter or recording potentiometer for the potentiometric readout. The titration cell can be positioned in the sample compartment of a spectrophotometer for a photometric end point.

Conditions have been worked out for the electrolytic generation of a large number of titrants for coulometric analysis. Some examples are given in Table 13.3; more complete compilations may be found elsewhere.[7] These include H^+

TABLE 13.3 Some Examples of Electrolytically Generated Titrants for Coulometric Titrations

Titrant	Substance titrated
Acids and bases	
H^+	Bases
OH^-	Acids
Oxidants	
Br_2	As(III), U(IV), olefins, phenols, aromatic amines
Ce(IV)	Fe(II), I^-, Ti(III), hydroquinone, metol (*p*-methylaminophenol sulfate)
Mn(III)	H_2O_2, Fe(II)
Ag(II)	Ce(III), As(III)
$Fe(CN)_6^{4-}$	Tl(I)
Reductants	
Fe(II)	MnO_4^-, Ce(IV), VO_3^-, Cr(VI)
Ti(III)	Fe(III), U(VI)
Precipitants	
Ag(I)	Cl^-, Br^-, I^-, mercaptans
Ce(III)	F^-

[7] For example, see D. D. DeFord and J. W. Miller, "Coulometric Analysis," Chap. 49 of I. M. Kolthoff and P. J. Elving, eds., *Treatise on Analytical Chemistry*, Part I, Vol. 4, Wiley–Interscience, New York, 1963; see especially Table 49.1, p. 2516.

and OH^-, oxidizing and reducing agents, and reagents for precipitation and complexation.

It is advantageous that only the fundamental quantities current and time are required. Errors and inconveniences associated with the preparation and storage of standard solutions are eliminated. Since small currents and short times can be measured accurately with modern instruments, coulometric titrations are routinely applicable to smaller samples than can be titrated conventionally, perhaps by a factor of 10 to 100. Sample sizes from 100 mg down to a few hundredths of a microgram in volumes of 10 to 50 ml have been employed. Excellent results have been reported with solution volumes as small as 10 microliters.[8] The reason that the method can be employed for such small samples can be seen as follows:

$$96{,}500 \text{ coulombs} = 1 \text{ equivalent}$$

$$1 \text{ coulomb} = 0.00001036 \text{ equivalent}$$

$$= 10.36 \text{ microequivalents}$$

Since a quantity of electricity as small as 0.1 coulomb can be measured with a precision of about 1 part per thousand, 1.036 microequivalents can be determined with this precision. For a substance of equivalent weight 100, this corresponds to a precision of about 0.1 microgram.

Because the titrant is generated in the test solution where it is to react, it is possible to titrate coulometrically with reagents whose solutions are not sufficiently stable to be kept as standards in the laboratory. Examples are Cl_2 and Br_2, which volatilize from aqueous solutions, Ag(II), which oxidizes water upon standing, Mn(III), whose solutions tend to form Mn(II) and Mn(IV), and Cu(I), which disproportionates to Cu and Cu(II) in aqueous solution.

POLAROGRAPHY

Polarography is a form of electrolysis in which the working electrode is a special microelectrode, a dropping mercury electrode, and in which a current–voltage curve (voltammogram) is recorded. As most writers now use the term, polarography is a special case of voltammetry in which the microelectrode is dropping mercury. Because of the special properties of this electrode, polarography is much more widely employed than is voltammetry with other microelectrodes. The technique was invented by Heyrovsky in Czechoslovakia in 1922, who received a Nobel Prize for this work in 1959, partly for its own sake and partly because of its wide applicability to research efforts in many fields.

Figure 13.8 shows a typical experimental setup for polarography. We shall briefly explain the essential features of this technique using simple examples, and show a few of the many applications to analytical chemistry. The term *micro*electrode is somewhat misleading in that the drops of mercury are visible to the unaided eye, but they are much smaller than the electrodes used for ordinary

[8] R. Schreiber and W. D. Cooke, *Anal. Chem.*, **27**, 1475 (1955).

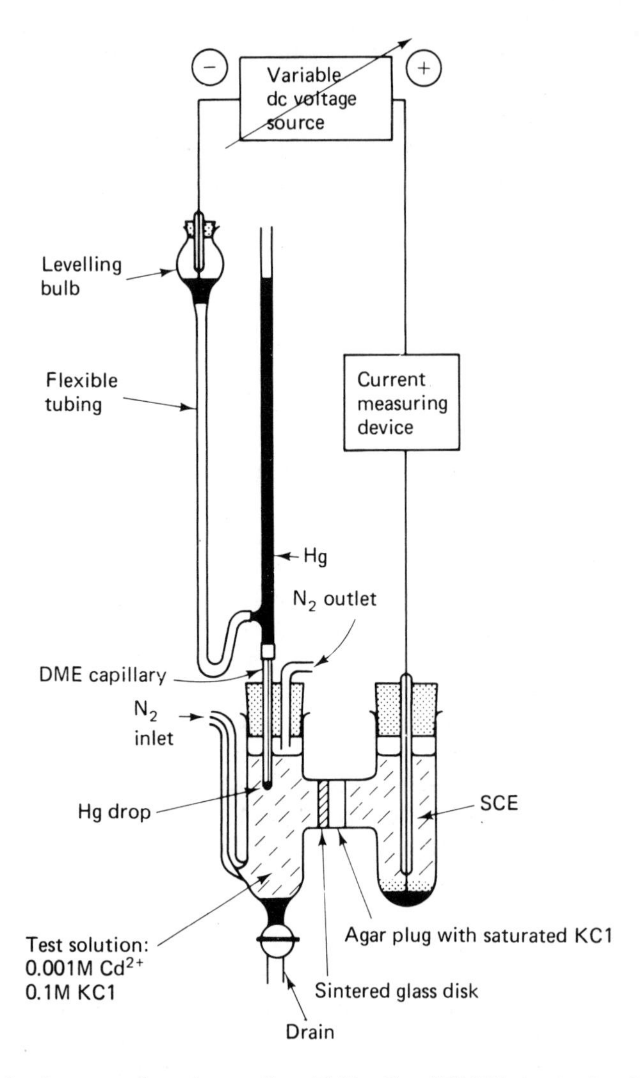

Figure 13.8 Apparatus for polarography with H-cell and DME in its simplest form. The apparatus is operated manually; the dc source may be a battery and voltage divider and the current measuring device, a galvanometer adapted as a microammeter. In a recording instrument the voltage source is programmed to vary E applied automatically with time and the current measuring device plots the polarogram with a pen-and-ink recorder.

electrolyses as described earlier. We shall refer to the dropping mercury electrode as the DME.

Relation of Applied Voltage and Cathode Potential

In practice, far more reductions than oxidations are studied in polarography; thus the DME is normally a cathode, as in the reduction of Cd^{2+} suggested in Fig.

13.8. We earlier saw the equation:

$$E_{\text{applied}} = E_{\text{anode}} - E_{\text{cathode}} + iR + \omega_{\text{cathode}} + \omega_{\text{anode}}$$

Let us consider the magnitudes of these terms. First, the current, i, will be very small. An electron transfer reaction is required to carry the current across the electrode–solution interface, and because the electrode has a very small surface area, the number of cadmium ions reducing per unit time will be small. Typically, currents in polarography range from fractions of a microampere up to a few μA ($1\ \mu A = 10^{-6}$ A). Since the value of R is also held down by the usual high concentration of strong electrolyte (KCl in the example of Fig. 13.8), the iR term in the preceding equation is usually negligible. For example, if $i = 1\ \mu A$ and $R = 100$ ohms, then $iR = 1 \times 10^{-4}$ V $= 0.1$ mV. (We are considering here a typical aqueous solution; in certain organic solvents where electrolytes are not very soluble, R may be quite high and the iR term appreciable.)

Second, it turns out that the anode potential is virtually constant. The SCE anode in Fig. 13.8 is *not* a microelectrode. Despite the fact that the anode reaction ($Hg_2Cl_2 + 2e \rightleftharpoons 2Hg + 2Cl^-$) proceeds from right to left during electrolysis, because the current is so small a negligible quantity of Cl^- is consumed, and even this can be replaced from solid KCl in equilibrium with the saturated solution. Thus E_{anode} remains constant at the SCE value of +0.25 V. Furthermore, with the tiny current and the large electrode (low current density), the half-reaction can furnish electrons rapidly enough to prevent the electrode from becoming more positive than +0.25 V (i.e., ω_{anode} is essentially zero).

Thus the equation above boils down to

$$E_{\text{applied}} = +0.25 - E_{\text{cathode}} + \omega_{\text{cathode}}$$

In other words, any change in the applied voltage is reflected totally in the potential of the cathode, E_{DME}, which we think of as comprising a Nernstian component and an overpotential term. Note that we do not need a three-electrode system, as we did in large-scale electrolyses, in order to monitor the cathode potential. The SCE, for the reasons given above, serves as both anode and reference electrode. Thus the applied voltage is the same as E_{DME} vs. SCE. (With organic solvents where the iR term may be appreciable as noted earlier, a three-electrode system is often employed in order to obtain directly values for E_{DME} without the necessity of correcting for the iR drop.) Incidentally, it should be noted that electrode potentials in polarography are customarily referred to the SCE rather than to the standard hydrogen electrode in reporting experimental results. As we saw in earlier chapters, it is the practice elsewhere to report single electrode potentials against hydrogen, even though an SCE or some other reference was actually employed in the laboratory.

The Polarogram

As a basis for the discussion which follows, the current–voltage curve, or polarogram, for the example being considered is shown in Fig. 13.9. Along LM, a small current, called the residual current, flows. Near M, the decomposition

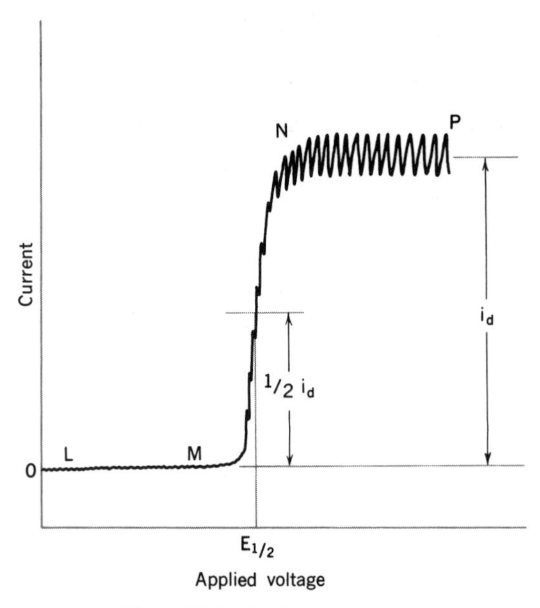

Figure 13.9 Typical polarogram.

voltage of the cell is reached, and the current increases rapidly from M to N, with further increase in applied voltage. The cell reaction is the following, proceeding from left to right:

$$Cd^{2+} + 2Hg + 2Cl^- \rightarrow Cd + Hg_2Cl_2$$

[The cadmium metal which is formed at the mercury cathode is soluble in mercury. The resulting solution, called an amalgam, is often symbolized by Cd(Hg). For this reason, some writers formulate the cathode reaction as follows:

$$Cd^{2+} + Hg + 2e \rightarrow Cd(Hg)$$

The student should not confuse Hg which is present in the SCE anode compartment with the mercury which flows from the DME cathode.]

Finally, as the voltage scan is continued, the current levels to a limiting value in the region NP. It is seen in the figure that the curve is not smooth. Rather, the current oscillates as the mercury drops grow and fall, reflecting the changes in surface area. Actually, at the instant a drop disengages, the electrode area decreases to the tiny cross-sectional area of the capillary bore and the faradaic current falls nearly to zero, but the actual polarogram will not show this because of inertia in current measuring devices. In working up polarographic data, a consistent method, using maximum currents, average currents, or whatever must be adopted.

Important Features of the DME

The DME is usually a 5- to 10-cm length of glass capillary tubing with an internal diameter of perhaps 0.04 to 0.08 mm. The combination of capillary length and mercury pressure is adjusted so that the small mercury drops grow and fall at regular intervals of about 2 to 5 sec. This must be reproduced in order to obtain comparable currents from one run to another.

The DME possesses several advantages over solid microelectrodes such as platinum. First, the surface exposed to the solution is reproducible, smooth, and continually renewed with fresh mercury. This leads to highly reproducible current–voltage curves, independent of the prior use of the electrode.

Second, the DME may be contrasted favorably with stationary electrodes in regard to the depleted layer adjacent to the electrode surface. With a stationary electrode, *peaks* rather than plateaus are found in the current–voltage curves. This results from the fact that the depletion extends progressively further from the electrode into the bulk solution as the electrolysis proceeds, and the electroactive species must diffuse further along shallower concentration gradients to reach the electrode as time goes by. Although there is nothing intrinsically bad about a peak as such, if, during the recording of a current–voltage curve, any jarring or vibration occurs, the diffusion setup is disturbed and erratic currents are observed. In principle, this problem can be countered by stirring the solution (which has the added advantage of yielding larger currents), but in practice it is difficult to reproduce the hydrodynamic variables in stirred solutions, and again, currents tend to be erratic. With the DME, the diffusion layer remains quite thin because of the periodic dropping of the mercury, and it is reestablished reproducibly at each new drop. The controlled, local stirring from the falling drops and the movement of the electrode surface toward the bulk solution as the drop grows, along with the increasing surface area, prevent the falling off with time. The oscillation of the current is a nuisance, but it is so regular that in practice it is not difficult to obtain an average value in a reproducible manner.

Third, a very important advantage of the DME results from the high activation overpotential for hydrogen evolution on mercury. Once reduction of hydrogen ion begins, the supply from an aqueous solvent is so abundant that the current rapidly attains an enormous (by polarographic standards) value, whereupon it becomes impossible to observe the small currents associated with the processes we desire to study. The potential at which reduction of H^+ becomes appreciable depends, of course, upon the pH of the solution, but in any case it is much more negative on mercury than on platinum, gold, or any other useful electrode material. This means that many reduction processes can be studied at the DME, even though thermodynamically hydrogen evolution should interfere.

Fourth, with many metals, amalgam formation lowers the activity of the metal, thereby facilitating the reduction of the metal ion. This, plus the overpotential effect mentioned above, means that many metals which are more active than hydrogen can still be dealt with in polarography. For example, polarograms have been obtained even for the alkali metals, although, to be sure, only in solutions of very high pH.

One of the major disadvantages of mercury as an electrode, when compared with more noble metals such as platinum, is the oxidation of the mercury itself at a potential which is not very positive. Once mercury oxidation begins, an enormous current results from the ample supply, and the oxidation of other species can no longer be observed. This means that the DME has a very limited use as an anode. The exact potential at which mercury oxidation becomes a problem depends upon the composition of the solution. For example, ions such as the halides, which form slightly soluble mercury(I) salts, facilitate the oxidation. Roughly, the DME is seldom useful at potentials more positive than about +0.4 V vs. SCE, while a platinum anode can be used up to the potential where water begins to oxidize, perhaps +1.4 V vs. SCE. Nevertheless, there is a wide potential range within which the DME is useful, and in this region it is unexcelled.

A second important disadvantage of the DME is the magnitude of the residual current. Part of this may sometimes be faradaic, resulting from electroactive trace impurities such as oxygen or easily reduced metal ions. However, cleaning up the solutions does not eliminate an appreciable residual current which is nonfaradaic in nature and which is much larger with the DME than with stationary microelectrodes of constant surface area. It is beyond the scope of this chapter to consider this charging or capacitance current in detail, but roughly we may view it as follows.

Consider a mercury droplet in an electrolyte solution, say KCl. At potentials more positive than about −0.5 V vs. SCE, the mercury surface is positively charged with respect to the adjacent solution, which contains an excess of Cl^- [Fig. 13.10(a)]. At potentials more negative than about −0.5 V, the surface double layer is reversed [Fig. 13.10(b)]. At about −0.5 V (the exact value depends upon the nature and concentration of ions in the solution) the electrode surface is uncharged with respect to the solution; in this condition, the mercury is said to be at its *electrocapillary maximum.* We may think of the charged double layer as the analog of a charged capacitor in electronics. Normally, once a capacitor is charged

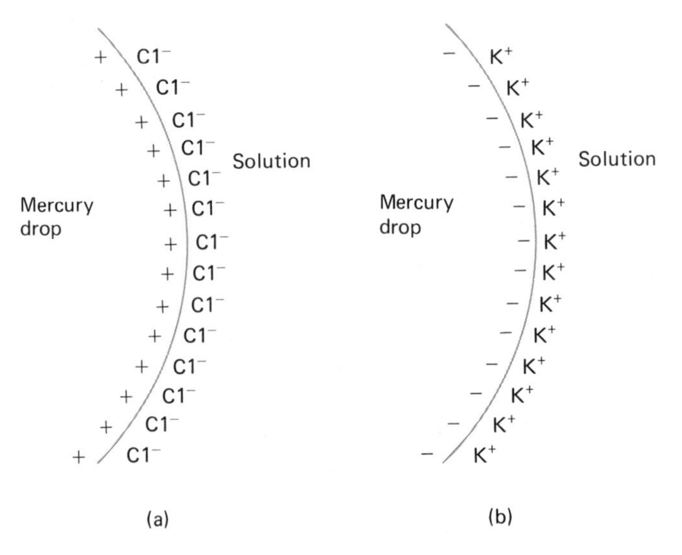

Figure 13.10 Schematic diagrams of the electrical double layers at the surface of a mercury drop in a potassium chloride solution at two potentials on opposite sides of the electrocapillary maximum.

up, no current need flow to keep it that way (unless the circuit provides a drain), but the growing mercury drop is like a capacitor whose plates are increasing in surface area; the result is a continual charging current at all potentials except at the electrocapillary maximum. The capacitance current changes sign at the electrocapillary maximum: at more positive potentials, electrons are flowing out of the DME into the external circuit, at more negative potentials, into it.

Figure 13.11 shows on an expanded scale a portion of the residual current region of Fig. 13.9 so that the shapes of the individual drop profiles can be seen. The current is maximal immediately after the disengagement of a drop, decaying toward zero as the drop continues to grow. The reason for this is that the rate of change of surface area with time, dA/dt, is greatest when the drop is smallest, thereby requiring the highest current flow to maintain the double-layer potential.

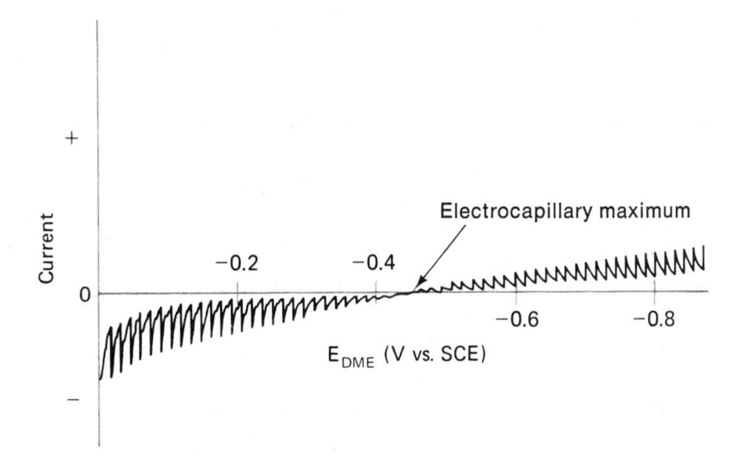

Figure 13.11 Portion of residual current from a polarogram.

The residual current represents a problem because it limits the analytical sensitivity of polarography. As the concentration of the desired constituent decreases, the limiting current (Fig. 13.9) decreases; this is the basis of polarography as an analytical technique. Now, if the current becomes too small, it can be amplified almost as much as we wish by electronic circuitry, but of course the residual current will also be amplified. The essential question is: how small a limiting current can be distinguished from the residual current? There is no hard and fast numerical answer that will serve all purposes and satisfy everyone, but it is clear that as the faradaic limiting current approaches the residual current in magnitude, we are getting into trouble. Very roughly, the lower limit for detection of most electroactive solutes by ordinary polarography is in the neighborhood of 10^{-6} *M*. There are special techniques which are beyond the scope of this text for pushing down the lower limit, for example, by sampling the current late in the drop life when the residual current is relatively small.

The Limiting or Diffusion Current

Returning to the reduction of Cd^{2+} at the DME, we may observe that there are three principal ways by which this ion could reach the electrode. First is convection: if the solution were vigorously stirred, Cd^{2+} would be swept toward the electrode by bulk flow of solution. In polarography, though, the solution is kept as quiet as possible; stirring, as noted earlier, is difficult to control and to reproduce.

Second, a cation like Cd^{2+} might feel the electrostatic pull of a negatively charged electrode. However, this is virtually eliminated by the large excess of "supporting electrolyte" (KCl in Fig. 13.9). The tendency of Cl^- to cluster about Cd^{2+} in the solution and the tendency of K^+ to huddle about the negative electrode tend to shield Cd^{2+} from the electrostatic attraction of the electrode. This lowers the limiting current below what it would otherwise be, but it is advantageous in the long run. The electrostatic migration of Cd^{2+} is subject to influence from other ions in the solution. Thus extraneous components of unknown samples being analyzed for cadmium might influence the limiting current due to Cd^{2+} in an uncontrolled manner. By the deliberate addition of a large excess of KCl, such effects are swamped out and minor differences from one sample to another may not affect the Cd^{2+} response. It may be mentioned again, as noted above, that the high concentration of KCl also serves to minimize the iR term. It is sometimes recommended that the supporting electrolyte be 50 to 100 times as concentrated as the electroactive species.

This illustrates, by the way, an important principle that is seen in many areas of analytical chemistry. The effects of other components of the sample upon the determination of the desired constituent are sometimes called *matrix effects*, particularly when major components influence the determination of minor or trace constituents. One solution to this problem is to create one's own reproducible matrix by the addition of a large quantity of material, KCl in the example above, which has the effect of making diverse samples more similar in gross composition. The material, of course, must be added to both calibration standards and unknowns.

Returning to the limiting current, the third way by which Cd^{2+} can reach the electrode is by diffusion. After the DME becomes sufficiently negative for reduction to begin (M in Fig. 13.9), solution adjacent to the electrode starts to become depleted in Cd^{2+}, and a concentration gradient is established leading to diffusion of Cd^{2+} from bulk solution to electrode surface. The rate of diffusion is proportional to the concentration difference:

$$\text{Rate of diffusion} \propto [Cd^{2+}]_{bulk} - [Cd^{2+}]_{interface}$$

On the limiting current plateau, NP in Fig. 13.9, the DME is sufficiently negative that virtually each Cd^{2+} ion reduces immediately upon reaching the electrode. Thus $[Cd^{2+}]_{interface}$ falls to practically zero. Now the faradaic current depends upon the rate of supply of Cd^{2+} to the electrode (i.e., upon the rate of diffusion), so we may write

$$i_d = k[Cd^{2+}]_{bulk}$$

where i_d is called the diffusion current. This, of course, shows the basis for analyses based upon the measurement of diffusion currents.

Most limiting currents in polarography are diffusion currents, but there are a number of exceptions. As an example, consider glucose. This sugar exists in aqueous solution as an equilibrium mixture of two forms, an open-chain aldehyde and a cyclic hemiacetal:

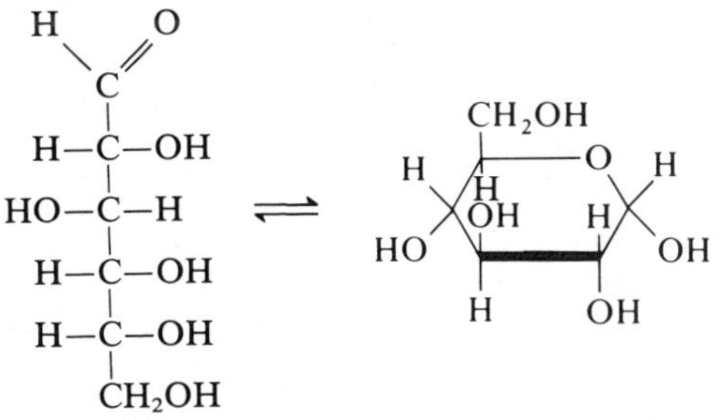

```
  H     O
   \   //
     C
     |
  H—C—OH                    CH2OH
     |                   H  |——————O  H
 HO—C—H       ⇌           \/H      \ /
     |                  HO \OH    H/ \OH
  H—C—OH                    |______|
     |                      H      OH
  H—C—OH
     |
    CH2OH
```

The ring form, which predominates in aqueous solution, is not reducible at the DME, while the free aldehyde group in the open-chain form is reducible to the alcohol. Studies of this system have shown that the limiting current is not diffusion-controlled but rather depends upon the rate of conversion of the nonreducible form into the aldehyde.[9] Such a current, sometimes very small, is said to be kinetically controlled, and polarography in such cases can be used to study the kinetics of conversion of nonreducible precursor to reducible form.

Ilkovic Equation

Because of mathematical problems in treating diffusion along the radii of an expanding, nearly spherical electrode, a totally rigorous equation for the diffusion current is difficult to derive, but the equation given by Ilkovic in 1934 is valid within a few percent. This equation for the average value of the diffusion current is

$$i_d = 607 n D^{1/2} C m^{2/3} t^{1/6}$$

The term i_d is the average diffusion current in microamperes, n the number of faradays per mole of reactant, D the diffusion coefficient (cm^2/s), C the concentration (millimolar), m the mass of mercury flowing (mg/s), and t the drop time (s). The constant 607 is a combination of natural constants including the faraday; the value is slightly temperature-dependent and the figure given here is for 25°C. The principal importance of the Ilkovic equation is that it shows in a quantitative manner the influence of many factors on the diffusion current.

It is convenient to divide the factors in the Ilkovic equation into two parts:

1. $nCD^{1/2}$, which is determined by the properties of the solution.
2. $m^{2/3}t^{1/6}$, which is dependent upon the characteristics of the capillary.

[9] P. Delahay and J. E. Strassner, *J. Am. Chem. Soc.*, **74**, 893 (1952).

The first part contains the concentration of the electroactive species. We have previously mentioned that the diffusion current is proportional to concentration, and this is the factor of primary interest to the analytical chemist. The term n, of course, is a property of the solute, the number of faradays per mole of reactant. The diffusion coefficient D is also a property of the solute in a given solvent. For simple ions this factor is related to the equivalent conductance and charge of the ion as well as to the temperature. The diffusion current increases with increase in temperature, this being about 2% per degree for the DME. Hence, to ensure an error no greater than ±1% in the diffusion current, the temperature should be controlled to within ±0.5°C.

The second part, $m^{2/3}t^{1/6}$, can be evaluated by measuring the mass of mercury flowing through the capillary per second and the drop time in the diffusion current region. Current obtained with different capillaries can be compared when this factor is known. The surface tension of mercury and hence the drop time varies with the potential of the microelectrode. For practical purposes the drop time can be considered approximately constant from 0 to about −1.0 V. At more negative potentials this term decreases much more rapidly, and this must be taken into account in calculating diffusion currents. All other factors being constant, the diffusion current is found to depend upon the effective pressure on the dropping mercury. The current should be proportional to the square root of the height of the mercury column if the process is diffusion controlled.

Polarographic Maxima

Polarograms are frequently seen where the wave is not a smooth, S-shaped curve as in Fig. 13.9. Sometimes the current rises sharply to a peak and then rapidly decreases to the plateau as shown in Fig. 13.12(a); in other cases, the

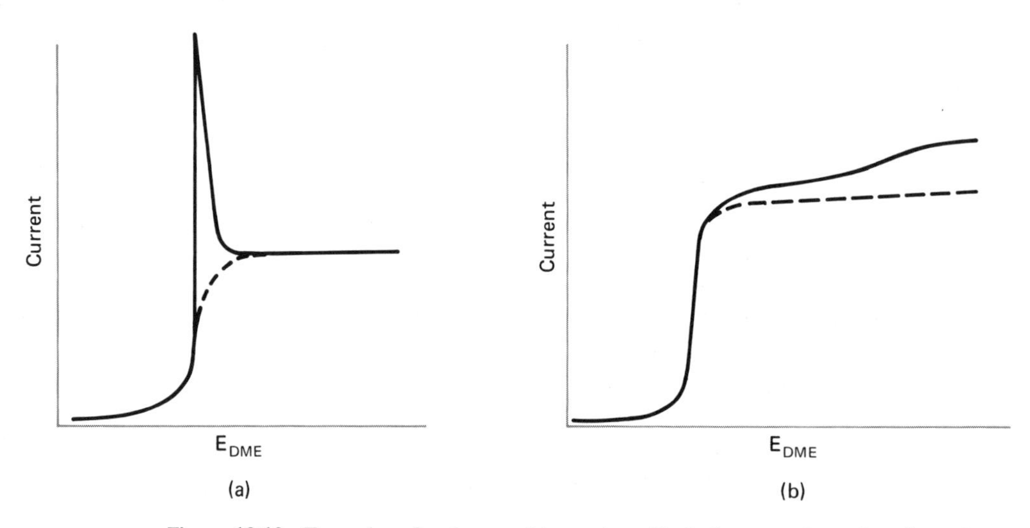

Figure 13.12 Examples of polarographic maxima. Dashed curves show the effect of maximum suppressors.

current rises gradually where it ought to be level and even gives the appearance of a second wave, as in Fig. 13.12(b). These anomalous currents are called *maxima*, and they are undesirable in most polarographic work. There is still no completely satisfactory explanation for their origin. It is clear that an abnormally high current must result from an enhanced delivery of reducible material to the electrode, but the mechanism for this is not totally clear. Adsorption of the electroactive species as a function of potential used to be invoked, but most investigators believe that this is not an adequate explanation. There is evidence for an unusual streaming of solution past the surface of the mercury drop at certain potentials which could supply extra materials to the electrode, but the cause of this is uncertain. It may be related to asymmetry in the electrical field around the drop arising from the fact that the drop is not a free sphere but rather a somewhat tear-shaped appendage to the capillary.

In any event, we know empirically how to eliminate maxima by adding so-called "maximum suppressors" to the solution. The commonest one used to be gelatin. More recently, the commercial surfactant Triton X-100 has become more widely used because it works as well as gelatin while its stock solutions are more stable. Typically, only very small concentrations of suppressors are required. For example, Triton X-100 is often used at a level of 0.002% in the final solution. The suppressors have some effect upon the diffusion current. In careful work, it is desirable to study the effect of suppressor concentration on the polarogram and then to employ the minimal concentration that is effective. Obviously, the suppressor concentration should be constant in all unknown and standard solutions which are to be compared.

The Half-Wave Potential

The potential at which the current is one-half the limiting current (i.e., where $i = i_d/2$) is called the *half-wave potential*, $E_{1/2}$ (see Fig. 13.9). At this potential, half of the Cd^{2+} ions that reach the cathode in a given time are reduced. $E_{1/2}$ is characteristic of the particular redox system and is independent of the concentration of the electroactive species in solution. This does not mean that it is very useful for qualitative identification except in very restricted circumstances, however, because too many systems exhibit $E_{1/2}$ values that are too close together.

For an electrode process which is fast enough for equilibrium to be established at the surface of the DME in the face of the changing potential, thermodynamic information is available; that is, the $E_{1/2}$ value is related to $E°$ for the redox couple. Although it is understood in physical chemistry that such true reversibility cannot be achieved in a process which is occurring at a finite rate there are many electrode processes that approach it so closely that we cannot tell the difference. For such a case, it can be shown that the following equation holds:

$$E_{\text{DME}} = E_{1/2} - \frac{0.059}{n} \log \frac{i}{i_d - i}$$

where the i values are currents at points on the polarographic wave corresponding to various values of E_{DME} and n is the number of electrons gained by the reducible species. A plot of E_{DME} vs. $\log[i/(i_d - i)]$ is a straight line whose slope is $-0.059/n$; $E_{1/2}$ is the value of E_{DME} when the log term is zero. Figure 13.13 shows such plots for two processes of differing n values.

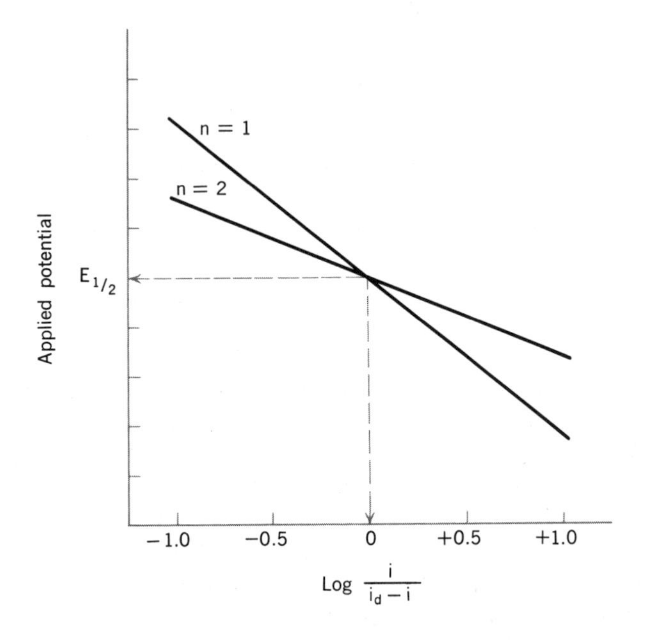

Figure 13.13 Plots of log $i/(i_d - i)$ vs. E_{app} for polarographic wave.

It should be emphasized that *many* polarographic processes are not reversible, even in the sense explained above. Those that depart appreciably from reversibility can be recognized quickly by the fact that the wave is "drawn out," the current not rising steeply as it does in Fig. 13.9. Thermodynamic conclusions about ease of reduction of chemical species cannot be drawn from such data; there is a slow kinetic step in the electrode process which prevents equilibration at any potential during the polarographic scan. In such cases, $E_{1/2}$ values reflect rate constants rather than positions of equilibrium.

Applications of Polarography

Because of the many substances that undergo redox reactions, a complete list of applications of polarography would be enormous; we can give only a brief sketch here. A polarographic scan is a quick way to determine the potential at which an ion or molecule will reduce, and it requires very little sample. Thus polarography is a useful pilot technique for establishing the conditions for a successful macroscale controlled potential electrolysis using a mercury cathode. Sometimes half-wave potentials for a series of organic compounds reflect electronic structure in a systematic manner to provide useful data on substituent effects that correlate nicely with other molecular properties.[10]

Applications to analysis include the determination of small quantities of metal ions—copper, lead, zinc, iron, cadmium, uranium, and many others. A number of inorganic anions yield useful waves; examples are iodate, bromate, selenite, and chromate.

Solutions containing O_2 exhibit two successive polarographic waves, the first representing a two-electron reduction to the level of H_2O_2 and the second the complete four-electron reduction to H_2O; $E_{1/2}$ values in acid solution are about -0.05 V and -0.8 V vs. SCE. Diffusion current measurements on one or the other of these waves provide a useful method for the determination of dissolved oxygen in aqueous solutions; this has been used, for example, in photosynthetic studies involving algal suspensions and isolated chloroplasts. Because these two oxygen waves extend over most of the useful range for the DME, they interfere with the polarographic determination of most other substances, and the solutions must normally be free of oxygen. This is usually accomplished by bubbling pure nitrogen through the test solution; this is terminated before the polarogram is recorded to avoid stirring of the solution by the bubbles. A nitrogen inlet is seen in the apparatus of Fig. 13.8.

Many organic compounds reduce at the DME to yield analytically useful polarographic waves. Examples are carbonyl, nitro, azo, and olefinic compounds; many halogenated compounds reduce according to

$$RCH_2X + H^+ + 2e \longrightarrow RCH_3 + X^-$$

Polarography has been applied to the determination of many compounds of biological interest, including drugs, hormones, and vitamins. It has also been used, along with other techniques, to elucidate the redox properties of biological compounds such as flavin derivatives and the pyridine coenzymes.

Table 13.4 lists some examples of polarographic analyses which have been reported, but it should be pointed out that this is by no means an exhaustive compilation.

In favorable cases, two or more components can be determined with a single polarogram; of course, the $E_{1/2}$-values must be sufficiently different for the individual waves to be distinguished. Figure 13.14 is a sketch from a polarogram obtained on a mixture of three reducible ions which illustrates this point. When the waves for two species in a sample are too close together, then a separation must precede the polarographic measurement step. Sometimes solution parameters such as the concentration of complexing agents or the pH can be manipulated in such a way as to improve the polarographic resolution of two components.

[10] See, for example, P. Zuman, *Substituent Effects in Organic Polarography*, Plenum Press, New York, 1967, and P. Zuman, *The Elucidation of Organic Electrode Processes*, Academic Press, Inc., New York, 1969.

TABLE 13.4 Some Examples of Polarographic Reductions

Ion or compound	Solvent* and supporting electrolyte	$E_{1/2}$, V, vs. SCE	Reduction product
Cd^{2+}	1 *M* HCl	− 0.642	Cd(Hg)
Cu^{2+}	1 *M* HCl	− 0.22	Cu(Hg)
Pb^{2+}	1 *M* HCl	− 0.435	Pb(Hg)
Sn^{4+}	1 *M* HCl–4 *M* NH_4Cl	− 0.25	Sn^{2+}
		− 0.52	Sn(Hg)
$CrO_4{}^{2-}$	2 *M* K_2CO_3	− 0.47	Cr^{3+}
Ni^{2+}	0.1 *M* KCl	− 1.1	Ni(Hg)
Zn^{2+}	1 *M* KSCN	− 1.05	Zn(Hg)
$MoO_4{}^{2-}$	1 *M* HCl	− 0.63	$MoCl_6{}^{3-}$
Azobenzene	Aqueous ethanol, *p*H 2–3	− 0.11	Hydrazobenzene
Benzaldehyde	50% ethanol, H_3PO_4, *p*H 1.3	− 0.98	$C_6H_5—CH(OH)—CH(OH)—C_6H_5$
1,4-Naphthoquinone	Aqueous ethanol, phosphate buffer, pH 3	+ 0.12	Naphthohydroquinone
Nicotinamide-adenine	0.1 *M* tetra-*n*-butylammonium carbonate buffer, *p*H 9	− 1.12	4,4′-Dimeric 1-electron product from electrogenerated free radical
Progesterone	50% ethanol buffered at *p*H 6	− 1.36	—
Quinine	Acetate buffer, *p*H 3	− 1.00	—
Riboflavin	H_3PO_4, *p*H 1.8	− 0.16	—

* Solvent is water unless otherwise noted.

Polarographic Techniques

Most polarographic work is performed with instruments which record the current–voltage curve automatically. The test solution is placed in the cell, nitrogen is bubbled through it for 10 to 20 min to remove oxygen, and finally the curve is recorded in a few minutes. The method is commonly employed to determine concentrations in the range 10^{-4} to 10^{-2} *M*, but under favorable conditions, concentrations as low as 10^{-6} *M* can be detected. Errors of perhaps 2 to 5% are fairly common, although these may be reduced by a factor of 10 or so in special cases.[11]

Since the concentration is proportional to the diffusion current, the main problem in polarographic analysis is to measure this current accurately. Correction must be made, of course, for the residual current. This can be measured

[11] Generations of analytical chemists grew up expecting errors of 1 part per thousand in classical volumetric and gravimetric analysis and errors of several percent with instruments such as spectrophotometers and polarographs. This was partly because we tackled tougher problems with the instruments, but in most cases it also reflected the capabilities of the instruments themselves. Technological developments are rapidly changing this situation. There is no reason other than perhaps cost for the chemist today not to see a wide variety of instruments which are just as accurate as the analytical balance and the buret.

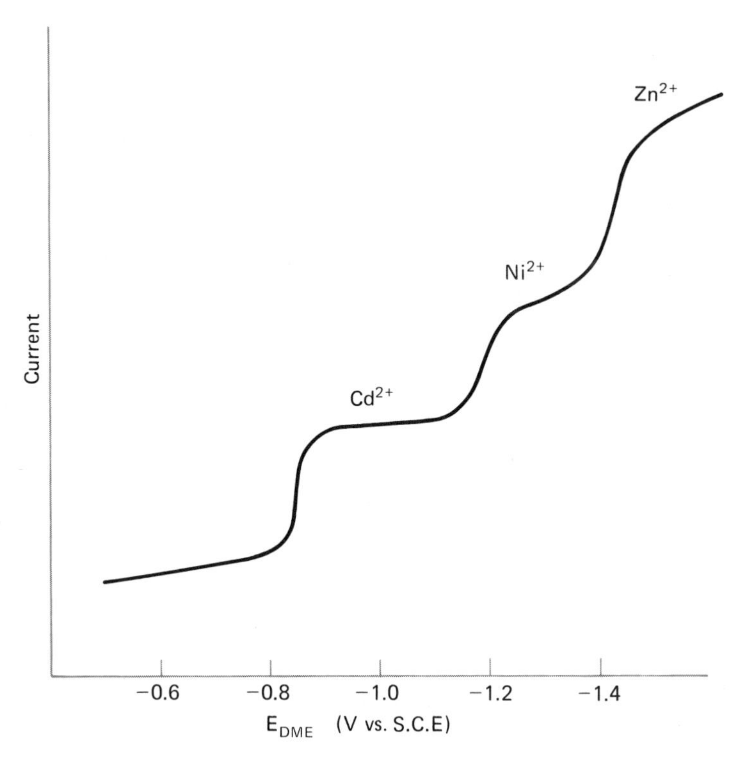

Figure 13.14 Sketch of the average current for a polarogram on a mixture of electroactive ions in an ammonia buffer.

separately on a solution that contains the supporting electrolyte alone, and subtracted from the current, which is measured on the diffusion plateau at the same voltage. With a recording instrument, the residual current shown by LM in Fig. 13.9 can be extrapolated to correct the current in region NP. Books on polarography among the references at the end of this chapter should be consulted for details; that by Meites is particularly valuable in helping inexperienced people attain a professional level.

In principle, once i_d is determined it is possible to calculate the concentration using the Ilkovic equation. The values of m and t for the particular capillary can be easily obtained by collecting and weighing mercury drops for a timed interval. Usually, though, the diffusion coefficient, D, is not accurately known for the experimental conditions that obtain, and this approach is not satisfactory.

It is possible, however, to hold D constant by controlling the overall composition of the solution; if m and t are also controlled, the Ilkovic equation for carefully specified conditions boils down to

$$i_d = kC$$

The constant k can then be evaluated from a polarogram on a standard solution of known concentration, whereupon unknown concentrations can be calculated from their i_d values. A careful worker will actually run several standards to assure himself that k is, in fact, constant (i.e., that a graph of i_d vs. C is linear).

A technique known as the *method of standard addition* is sometimes useful, especially in cases where variable compositions of unknown samples make it difficult to prepare standards with the same matrix as the unknowns. A polarogram is recorded for the unknown sample, and then a known volume of the unknown is spiked with a known quantity of a standard solution and a second polarogram is obtained. From the increase in i_d caused by the known addition of the desired constituent, the original unknown concentration can be calculated. If the spike is a small enough volume not to dilute the sample appreciably, then the added material is subject to the same matrix as existed in the original unknown solution. An example illustrates the calculation.

Example. A lead solution of unknown concentration yields a diffusion current of 1.00 μA. Then to 10.00 ml of the unknown solution is added 0.500 ml of a standard solution of lead whose concentration is 0.0400 M. The diffusion current with the spiked solution is 1.50 μA. Calculate the lead concentration of the unknown solution.

Let C be the unknown concentration. Then the concentration of lead in the spiked solution is given by

$$\frac{10.00C + (0.500)(0.0400)}{10.500} = \frac{10.00C + 0.0200}{10.500}$$

Since i_d is proportional to concentration,

$$\frac{1.00}{C} = \frac{1.50}{(10.00C + 0.0200)/10.500}$$

$$C = 0.00348\ M$$

Amperometric Titrations

Since the diffusion current is proportional to the concentration of the electroactive species, the polarographic technique can be used to follow the progress of a titration involving this substance. The voltage scan required for a complete polarogram is not needed here, of course. Once the region of the diffusion plateau is identified, the applied voltage is set at a value on this plateau and i_d values are measured as the titration proceeds. The essential apparatus is some sort of polarographic cell, perhaps similar to the one shown in Fig. 13.8, with provision for the penetration of a buret tip into the solution. The titration curve is a graph of i_d vs. volume of titrant. The shape of the curve depends upon whether the substance titrated, the titrant, or both, can be reduced at the applied voltage which is selected.

Consider the titration of substance S with titrant T to form reaction product P:

$$\mathrm{S + T \rightarrow P}$$

Suppose that P is not electroactive; perhaps it is a precipitate, as in the titration of Pb^{2+} with $CrO_4^{\ 2-}$ to form insoluble lead chromate, $PbCrO_4$. Figure 13.15(a)

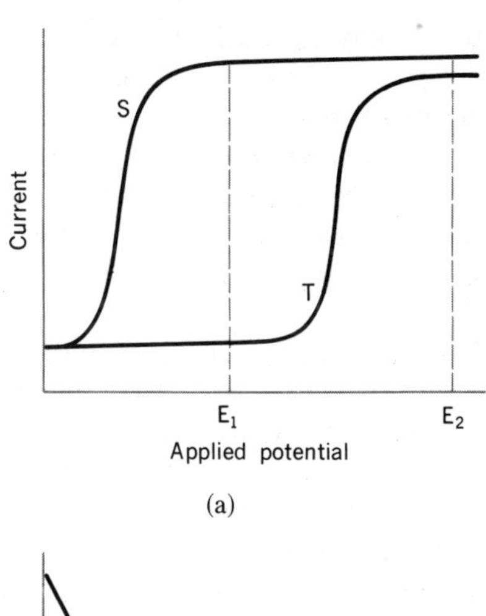

(a)

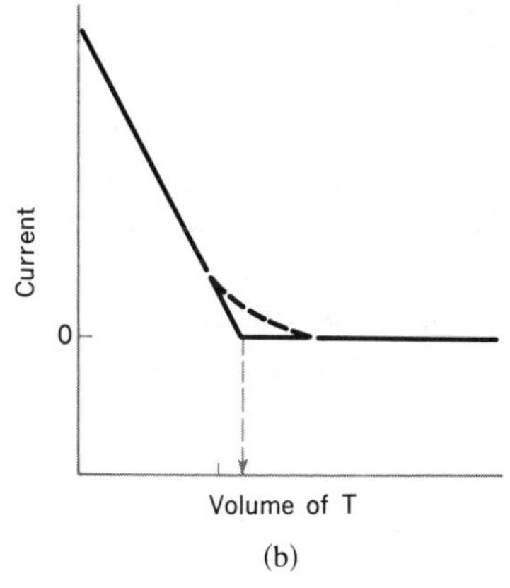

(b)

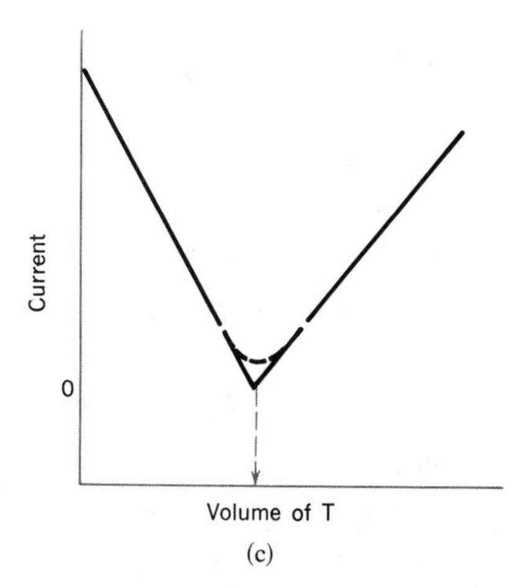

(c)

Figure 13.15 (a) Polarograms of substance titrated and of titrant. (b) Amperometric titration of S with T at E_1. (c) Amperometric titration of S with T at E_2.

shows polarograms of S and T obtained separately. Now suppose that we select the applied voltage E_1, where S yields a diffusion current but T does not. As the titration proceeds, consuming S, the value of i_d drops, as seen in Fig. 13.15(b). Past the equivalence point, the current will level off to some small residual value.

The break at the equivalence point may be sharp or more or less rounded, depending upon the extent to which the titration reaction goes to completion in a region where there is an excess of neither S nor T.

Suppose, on the other hand, that we perform the titration with an applied voltage E_2, where both S and T are reducible at the DME. The resulting titration curve is shown in Fig. 13.15(c). The first part of the curve is the same as in the previous case, but past the equivalence point the accumulation of excess titrant leads to increasing i_d values.

Because errors in individual i_d values tend to cancel when the "best" straight lines are drawn through a number of points, the error in locating the equivalence point is of the same order as in classical titrimetric analysis. Titrations can be performed rapidly, because there is no need for the operator to "hit" the end point, which can be found by extrapolation from points on either side. The extrapolation technique also permits the use of reactions which do not go well to completion near the equivalence point. Substances titrated are usually in the concentration range 10^{-4} to 10^{-1} *M*.

It should be remembered that the titrant not only removes S through reaction but also, by its very addition, dilutes the solution. Measured i_d values should be corrected for the dilution effect by multiplication by the factor $(v_0 + v)/v_0$, where v_0 is the initial volume of the solution and v is the volume of titrant added at the point where the correction is being applied. This correction can be minimized by using a very concentrated titrant solution, delivered from a good microburet to keep the volume error small.

REFERENCES

A. J. BARD, ed., *Electroanalytical Chemistry: A Series of Advances,* Marcel Dekker, Inc., New York, 1966– (a continuing series of volumes).

M. BREZINA and P. ZUMAN, *Polarography in Medicine, Biochemistry, and Pharmacy,* Interscience Publishers, Inc., New York, 1958.

A. J. FRY, *Synthetic Organic Electrochemistry,* Harper & Row, Publishers, New York, 1972.

I. M. KOLTHOFF and J. J. LINGANE, *Polarography,* Vols. I and II, Interscience Publishers, Inc., New York, 1952.

J. J. LINGANE, *Electroanalytical Chemistry,* 2nd ed., Interscience Publishers, Inc., New York, 1958.

L. MEITES, *Polarographic Techniques,* 2nd ed., Wiley–Interscience, New York, 1965.

C. N. REILLEY, "Fundamentals of Electrode Processes," and C. N. REILLEY and R. W. MURRAY, "Introduction to Electrochemical Techniques," Chaps. 42 and 43 of I. M. Kolthoff and P. J. Elving, eds., *Treatise on Analytical Chemistry,* Part I, Vol. 4, Wiley–Interscience, New York, 1963.

D. T. SAWYER and J. L. ROBERTS, Jr., *Experimental Electrochemistry for Chemists,* John Wiley & Sons, Inc., New York, 1974.

J. T. STOCK, *Amperometric Titrations,* Wiley–Interscience, New York, 1965.

QUESTIONS

1. *Products of electrolysis.* Predict the first products of electrolysis in the following cells. Each electrolyte is 0.10 *M* in an aqueous solution and the solution is 0.10 *M* in H^+.

Cathode	*Electrolyte*	*Anode*
Pt	NaBr	Pt
Pt	KI	Cu
Cu	$CuSO_4$	Cu
Pt	$FeCl_3$	Pt
Pt	$H_2C_2O_4$	Pt

2. *Terms.* Explain what is meant by the following terms: (a) charging current; (b) migration current; (c) activation overpotential; (d) concentration overpotential.

3. *Controlled potential electrolysis.* Explain why controlled potential electrolysis is needed to separate silver and copper, but constant potential electrolysis can be used to separate nickel and copper.

4. *Amperometric titrations.* Consider the amperometric titration of A with B, where

$$A + B \rightleftharpoons AB(s)$$

Sketch the titration curve at a potential where (a) B gives a diffusion current, A does not; and (b) A gives an anodic, and B a cathodic diffusion current.

5. *Amperometric titration.* Suppose that the reaction of A with B to form a precipitate of AB is essentially complete before the reaction of C with B to form CB begins. A solution containing both A and C is titrated with B. Sketch the shapes of the amperometric titration curves (one electrode polarizable) at a potential where (a) A and B give diffusion currents, C does not; (b) C gives a diffusion current, A and B do not; and (c) B and C give diffusion currents, A does not.

Multiple-choice: In the following multiple-choice questions, select the *one best* answer.

6. In the lead storage battery, one electrode is made of Pb, the other of PbO_2. The discharge reaction is

$$Pb(s) + PbO_2(s) + 2H_2SO_4 \rightleftharpoons 2PbSO_4(s) + 2H_2O$$

During discharge (a) PbO_2 is the positive electrode; (b) Pb is the anode; (c) reduction occurs at the PbO_2 electrode; (d) all of the above; (e) none of the above.

7. In the lead storage battery (Question 6), how many moles of $PbSO_4$ is formed when 96,500 coulombs of electricity is drawn from the battery? (a) 0.50; (b) 1.0; (c) 2.0; (d) 96,500/2.

8. Which of the following statements is *false*? (a) Relatively few substances have been determined by a primary coulometric titration; (b) Neutral molecules cannot be reduced polarographically; (c) Activation overpotential is caused by a slow electrode reaction; (d) In a completely polarized electrode, the reaction can still be reversible.

9. Which of the following statements is *false*? The DME has the following advantages over a solid electrode such as Pt. (a) The surface is reproducible; (b) The diffusion layer

remains thin; (c) The overpotential for H_2 is high; (d) The electrode is equally useful as an anode as well as a cathode.

10. Suppose that some H_2O_2 is formed at the anode of a H_2–O_2 coulometer and is reduced at the cathode. The reading on the coulometer is (a) high; (b) low; (c) correct; (d) no way to predict.

PROBLEMS

1. *Decomposition voltage.* Calculate the equilibrium decomposition voltages of the following electrolytes, all with platinum electrodes: (a) 1.0 *M* H_2SO_4; (b) 0.10 *M* H_2SO_4; (c) 0.010 *M* H_2SO_4.

2. *Decomposition voltage.* Calculate the equilibrium decomposition voltages of the following cells:

(a) $\overset{-}{\text{Pt}}|\text{Ag}^+(0.010\,M) + \text{H}^+(0.10\,M) + \text{H}_2\text{O}|\overset{+}{\text{Pt}}$

(b) $\overset{-}{\text{Pt}}|\text{Na}^+(0.10\,M) + \text{H}^+(0.050\,M) + \text{Cl}^-(0.10\,M), \text{AgCl}|\overset{+}{\text{Ag}}$

(c) $\overset{-}{\text{Pt}}|\text{Cu}^{2+}(0.10\,M) + \text{H}^+(0.010\,M)|\overset{+}{\text{Cu}}$

(d) $\overset{-}{\text{Pt}}|\text{Cu}^{2+}(0.20\,M) + \text{H}^+(0.010\,M)|\overset{+}{\text{Pt}}$

3. *Decomposition voltage.* Calculate the value of the equilibrium decomposition voltage of the cell in part (d) of Problem 2 after the electrolysis has proceeded until the concentration of Cu^{2+} has been reduced to 1.0×10^{-6} *M*. Note the increase in $[H^+]$.

4. *Applied voltage.* What voltage must be applied to a pair of smooth platinum electrodes immersed in a 0.020 *M* solution of Ag^+ in order for the initial current to be 0.80 A? The $[H^+]$ is 0.10*M*, the cell resistance is 0.40 ohm, and the overpotential terms at the anode and cathode are 0.75 V and 0.10 V, respectively.

5. *Electrolysis.* A solution which is 0.15 *M* in Cu^{2+} and 0.15 *M* in H^+ is electrolyzed between platinum electrodes. (a) Assuming that the overpotential term for hydrogen is negligible, what is the concentration of Cu^{2+} when H_2 begins to be liberated at the cathode? (b) Repeat part (a) but assume that the overpotential term for H_2 is 0.80 V.

6. *Faraday's law.* 100 ml of an aqueous solution of $CuSO_4$ is electrolyzed with Cu deposited at the cathode and O_2 liberated at the anode. The initial *p*H is 5.00; after a certain time the *p*H of the solution is 1.00. Calculate the number of coulombs of current which passed.

7. *Faraday's law.* An aqueous solution of $CuSO_4$ of *p*H 5.00 is electrolyzed, producing Cu at the cathode and O_2 at the anode. 50 ml of 0.10 *M* NaOH is required to titrate the acid produced during the electrolysis. Calculate the weight in grams of Cu deposited on the cathode.

8. *Overpotential—pH.* A solution which is 0.10 *M* in Zn^{2+} has a *p*H of 1.00. It is electrolyzed using Pt electrodes. (a) Assuming that the overpotential term for H_2 is zero, what is the first product at the cathode? (b) If the overpotential term for H_2 is 1.00 V, what is the first product at the cathode? (c) In part (b) calculate the $[Zn^{2+}]$ when H_2 begins to be liberated, noting that the $[H^+]$ increases as $[Zn^{2+}]$ decreases.

9. *Effect of pH.* (a) Theoretically, what must be the pH of a solution so that the concentration of Cr^{3+} can be reduced by electrolysis to 10^{-6} M before evolution of H_2? Assume that the overpotential for H_2 is zero. (b) Repeat part (a) for Mn^{2+}. (c) Repeat part (a) for Al^{3+}.

10. *Effect of overpotential.* Repeat Problem 9 except that the overpotential for H_2 is 0.80 V.

11. *Overpotential.* Calculate the overpotential term for H_2 to allow the concentration of Cd^{2+} to be reduced to 10^{-6} M by electrolysis before evolution of H_2 from a solution of pH 2.30.

12. *Mercury cathode.* Calculate the pH of a solution so that the concentration of Mn^{2+} can be reduced to 10^{-6} M by electrolysis before evolution of H_2 for the following conditions: (a) no overpotential; (b) 1.0 V overpotential; (c) mercury cathode, 1.0-V overpotential, activity of manganese in the amalgam = 10^{-6}.

13. *Activity.* If a cathode having a potential of +0.50 V is actually in equilibrium with a 0.050 M solution of Cu^{2+}, what is the activity of metallic copper on the surface of the electrode?

14. *Controlled potential electrolysis.* In a controlled potential electrolysis of a silver solution, the cathode potential reaches a value of + 0.20 V when the $[Ag^+]$ in the body of the solution is 0.010 M. (a) What is the overpotential term at the cathode? (b) What is the $[Ag^+]$ actually in equilibrium with electrode surface?

15. *Potential buffer.* What should be the ratio of the concentrations of Fe^{3+} to Fe^{2+} ions in a potential buffer in order to limit the cathode potential to + 0.80 V?

16. *Internal electrolysis.* An electrolysis carried out by short-circuiting a galvanic cell is called an *internal electrolysis.* Suppose that it is desired to separate copper from nickel in a solution which is 0.10 M in each Cu^{2+} and Ni^{2+}. The following cell can be used:

$$Cu|Cu^{2+}(0.10\ M) + Ni^{2+}(0.10\ M)||Pb^{2+}(0.10\ M)|Pb$$

The cell is short-circuited and allowed to run down. (a) Calculate the potential of the cathode at the end of the electrolysis. (b) What should the cathode potential be in order to deposit nickel? (c) What is the final concentration of Cu^{2+}?

17. *Coulometric titration.* The concentration of a hydrochloric acid solution is determined by coulometric titration using a platinum cathode and a silver anode. (a) What are the products at the two electrodes? (b) If 10.00 ml of acid is electrolyzed and a weight of 0.1587 g of silver is deposited in a silver coulometer, what is the molarity of the acid? (c) If the titration is done at constant current, what must the current be in order for the reaction to be complete in 10 min?

18. *Coulometer.* (a) Calculate the volume of hydrogen plus oxygen (STP) that should be produced per coulomb in a hydrogen–oxygen coulometer. (b) What volume would this be if the gas is saturated with water vapor at 25°C and a pressure of 740 mm of mercury?

19. *Microelectrolysis.* Suppose that a cadmium solution is electrolyzed in the cell shown in Fig. 13.8 for 5.00 min, the diffusion current being 2.00 μA. Calculate (a) the number of micrograms of cadmium deposited at the cathode, and (b) the number of micrograms of Hg_2Cl_2 formed at the anode. (c) Calculate the time in hours that would be

required to reduce the concentration of cadmium in 10 ml of solution from 0.0010 *M* to 0.0009 *M* by electrolysis at a current of 2.00 μA.

20. *Half-wave potential.* A metal ion is reduced polarographically, the diffusion current being 10 μA. The following currents were obtained at the indicated potentials (all negative): 0.444 V, 1.0 μA; 0.465 V, 2.0 μA; 0.489 V, 4.0 μA; 0.511 V, 6.0 μA; 0.535 V, 8.0 μA; 0.556 V, 9.0 μA. Calculate (a) the half-wave potential; (b) the value of n.

21. *Diffusion current.* The diffusion current constant of Zn^{2+} in 0.10 *M* KCl is 3.42. What diffusion current in microamperes is obtained with a 0.00100 *M* solution of Zn^{2+} using a capillary with a drop time of (a) 3.00 sec; (b) 4.00 sec; (c) 5.00 sec? Assume for each drop time that 1 drop of mercury weighs 5.00 mg.

22. *Pilot–ion method.* The ratio of diffusion current constants of cadmium to lead is 0.924. A polarogram is run on a solution of unknown lead concentration, the cadmium ion concentration being 0.0012 *M*. The diffusion currents are: lead 5.60 μA, cadmium 4.20 μA. Calculate the concentration of lead in the unknown.

23. *Method of standard addition.* A lead solution of unknown concentration gives a diffusion current of 5.00 μA. To 5.00 ml of this solution is added 10.0 ml of a 0.00150 *M* solution of Pb^{2+}, and the polarogram is run again, giving a diffusion current of 14.2 μA. Calculate the concentration of lead in the unknown.

24. *Method of standard addition.* The following is a modification of the method of standard addition. A 5.00-ml portion of a lead solution of unknown concentration is diluted to 25.0 ml and a polarogram run, the diffusion current being 0.40 μA. Another 5.00-ml portion of the same lead solution is mixed with 10.0 ml of a 0.00100 *M* solution of lead, the mixture diluted to 25.0 ml, the polarogram run, and the wave height is 2.00 μA. Calculate the concentration of lead in the unknown.

25. *Diffusion current.* A 0.0014 *M* solution of Pb^{2+} gives a diffusion current of 4.2 μA. What should be the concentration of another lead solution so that when equal volumes of the two are mixed, the diffusion current will be 12 μA?

26. *Polarographic analysis.* A sample containing 12.0% $CdCl_2$ is to be dissolved and the solution diluted to 100 ml in a volumetric flask. A polarogram is to be recorded using 10.0 ml of the solution. What is the largest sample that can be taken so that the diffusion current does not exceed 16.0 μA? The diffusion-current constant of cadmium is 3.50 and the value of $m^{2/3}t^{1/6}$ for the capillary is 1.20.

27. *Diffusion coefficient.* A certain metal undergoes reduction by taking up two electrons. The average diffusion current of a 0.00200 *M* solution of the metal is 10.8 μA and the value of $m^{2/3}t^{1/6}$ for the capillary is 1.60. Calculate the diffusion coefficient of the metal ion.

28. *Reduction of oxygen.* The diffusion coefficient of O_2 in dilute aqueous solution is 2.6×10^{-5} cm^2/sec. A 0.25 millimolar solution of O_2 in an appropriate supporting electrolyte gives a polarographic wave with a diffusion current of 5.8 μA. The DME constants are: $m = 1.85$ mg/sec; $t = 4.09$ sec. Calculate the value of n. To which product, water or hydrogen peroxide, is O_2 reduced under these conditions?

CHAPTER FOURTEEN

spectrophotometry

INTRODUCTION

Chemists have long used color as an aid in the identification of chemical substances. Spectrophotometry may be thought of as an extension of visual inspection in which a more detailed study of the absorption of radiant energy by chemical species permits greater precision in their characterization and quantitative measurement. Replacing the human eye with other detectors of radiation permits the study of absorption outside the visible region of the spectrum, and frequently spectrophotometric experiments can be performed automatically. In current usage, the term spectrophotometry suggests the measurement of the extent to which radiant energy is absorbed by a chemical system as a function of the wavelength of the radiation, as well as isolated absorption measurements at a given wavelength. In order to understand spectrophotometry, we need to review the terminology employed in characterizing radiant energy, consider in an elementary fashion the interaction of radiation with chemical species, and see in a general way what the instruments do. The student should understand that this chapter is only an introduction to the broad subject of spectrophotometry, and that it is possible to go much more deeply into nearly every topic that is mentioned here.

THE ELECTROMAGNETIC SPECTRUM

Various experiments in the physics laboratory are best interpreted in terms of the idea that light is propagated in the form of transverse waves. By appropriate

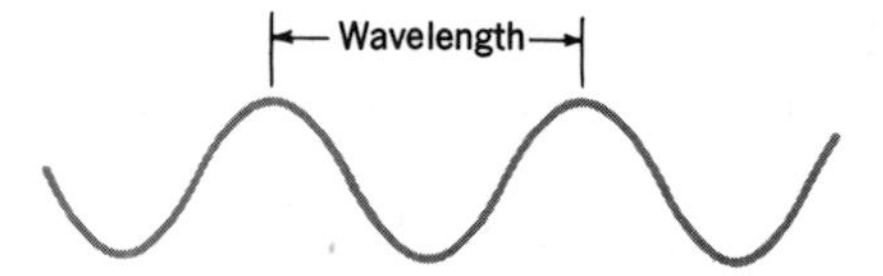

Figure 14.1 Transverse wave.

measurements, these waves may be characterized with regard to wavelength, velocity, and the other terms which may be used to describe any wave motion. In Fig. 14.1, it is indicated that the *wavelength* refers to the distance between two adjacent crests (or troughs) of the wave. The reciprocal of the wavelength, which is the number of waves in a unit length, is referred to as the *wave number.* The wave front is moving with a certain *velocity.* The number of complete cycles or waves passing a fixed point in a unit time is termed the *frequency.* The relationship of these properties is as follows, using the symbols λ for wavelength, $\bar{\nu}$ for wave number, ν for frequency, and c for the velocity of light:

$$\frac{1}{\lambda} = \bar{\nu} = \frac{\nu}{c}$$

The velocity of light is about 3×10^{10} cm/sec. Various units are employed for wavelength, depending upon the region of the spectrum: For ultraviolet and visible radiation, the angstrom unit and the nanometer are widely used, while the micrometer is the common unit for the infrared region. A micrometer, μm, is defined as 10^{-6} m, and a nanometer, nm, is 10^{-9} m, or 10^{-7} cm. One angstrom unit (Å) is 10^{-10} m or 10^{-8} cm. Thus there are 10 Å in 1 nm. Wave number is often used by chemists as a frequency unit because it has convenient numerical values ($\bar{\nu}$ and ν are related by a constant factor, c, the velocity of light); the common unit of wave number is the reciprocal centimeter, cm^{-1}.

Luminous bodies such as the sun or an electric bulb emit a broad spectrum comprising many wavelengths. Those wavelengths associated with *visible light* are capable of affecting the retina of the human eye and hence give rise to the subjective impressions of vision. But much of the radiation emitted by hot bodies lies outside the region where the eye is sensitive, and we speak of the *ultraviolet* and *infrared* regions of the spectrum which lie on either side of the visible. The entire electromagnetic spectrum is classified approximately as shown in Fig. 14.2.

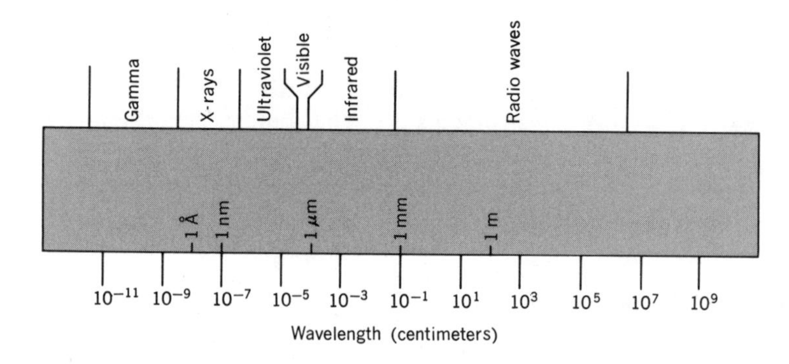

Figure 14.2 Approximate classification of electromagnetic spectrum.

Within the visible region of the spectrum, persons with normal color vision are able to correlate the wavelength of light striking the eye with the subjective sensation of color, and color is indeed sometimes used for convenience in designating certain portions of the spectrum, as shown in the rough classification in Table 14.1.

TABLE 14.1 Visible Spectrum and Complementary Colors

Wavelength, nm	Color	Complementary color
400–435	Violet	Yellow-green
435–480	Blue	Yellow
480–490	Green-blue	Orange
490–500	Blue-green	Red
500-560	Green	Purple
560-580	Yellow-green	Violet
580-595	Yellow	Blue
595-610	Orange	Green-blue
610-750	Red	Blue-green

We "see" objects by means of either transmitted or reflected light. When "white light," containing a whole spectrum of wavelengths, passes through a medium such as a colored glass or a chemical solution which is transparent to certain wavelengths but absorbs others, the medium appears colored to the observer. Since only the transmitted waves reach the eye, their wavelengths dictate the color of the medium. This color is said to be *complementary* to the color that would be perceived if the absorbed light could be inspected, because the transmitted and absorbed light together make up the original white light. Similarly, opaque colored objects absorb some wavelengths and reflect others when illuminated with white light.

THE INTERACTION OF RADIANT ENERGY WITH MOLECULES

The wave theory of light explains many optical phenomena, such as reflection, refraction, and diffraction, but there are other experimental results, such as the photoelectric effect, that are best interpreted in terms of the idea that a beam of light is a stream of particulate energy packets called photons. Each of these particles possesses a characteristic energy which is related to the frequency of the light by the equation

$$E = h\nu$$

where h is Planck's constant. Light of a certain frequency (or wavelength) is associated with photons, each of which possesses a definite quantity of energy. As explained below, it is the quantity of energy possessed by a photon which determines whether a certain molecular species will absorb or transmit light of the corresponding wavelength.

In addition to the ordinary energy of translational motion, which is not of concern here, a molecule possesses internal energy which may be subdivided into three classes. First, the molecule may be rotating about various axes, and possess a certain quantity of *rotational energy*. Second, atoms or groups of atoms within the molecule may be vibrating, that is moving periodically with respect to each other about their equilibrium positions, conferring *vibrational energy* upon the molecule. Finally, a molecule possesses *electronic energy*, by which we mean the potential energy associated with the distribution of negative electric charges (electrons) about the positively charged nuclei of the atoms.

$$E_{\text{int}} = E_{\text{elec}} + E_{\text{vib}} + E_{\text{rot}}$$

One of the basic ideas of quantum theory is that a molecule may not possess any arbitrary quantity of internal energy, but rather, it can exist only in certain "permitted" energy states. If a molecule is to absorb energy and be raised to a higher energy level, it must absorb a quantity appropriate for the transition. It cannot absorb an arbitrary quantity of energy determined by the experimenter and linger in an energy state intermediate between its permitted levels. This quantization of molecular energy, coupled with the concept that photons possess definite quantities of energy, sets the stage for selectivity in the absorption of radiant energy by molecules. When molecules are irradiated with many wavelengths, they will abstract from the incident beam those wavelengths corresponding to photons of energy appropriate for permitted molecular energy transitions, and other wavelengths will simply be transmitted.

The rotational energy levels of a molecule are quite closely spaced, as indicated schematically in Fig. 14.3. Thus pure rotational transitions require relatively little energy and are induced by radiation of very low frequency (long wavelength). It is in the far infrared and "microwave" regions of the spectrum (wavelengths of perhaps 100 μm to 10 cm) that absorption of radiation is correlated with changes in rotational energy alone. Studies of absorption in this region have contributed fundamental information regarding molecular structure but have found relatively little application in analytical chemistry.[1]

Vibrational energy levels are farther apart (Fig. 14.3), and more energetic photons are required if absorption is to increase the vibrational energy of a molecule. Absorption due to vibrational transitions is seen in the infrared region of the spectrum, roughly from 2 to 100 μm. Pure vibrational changes are not observed, however, because rotational transitions are superimposed upon them. Thus a typical vibrational absorption spectrum is composed of complex bands rather than single lines. In practice, an infrared absorption spectrum consists, not of discrete lines as might be supposed from Fig. 14.3, but rather of broad envelopes extending over wavelength spans, because of distortion of molecular energy levels by neighboring molecules and because of the inability of the instrument to resolve closely spaced lines.

[1] J. H. Goldstein, "Microwave Spectrophotometry," Chap. 62 of I. M. Kolthoff and P. J. Elving, eds., *Treatise on Analytical Chemistry*, Part I, Vol. 5, Wiley–Interscience, New York, 1964.

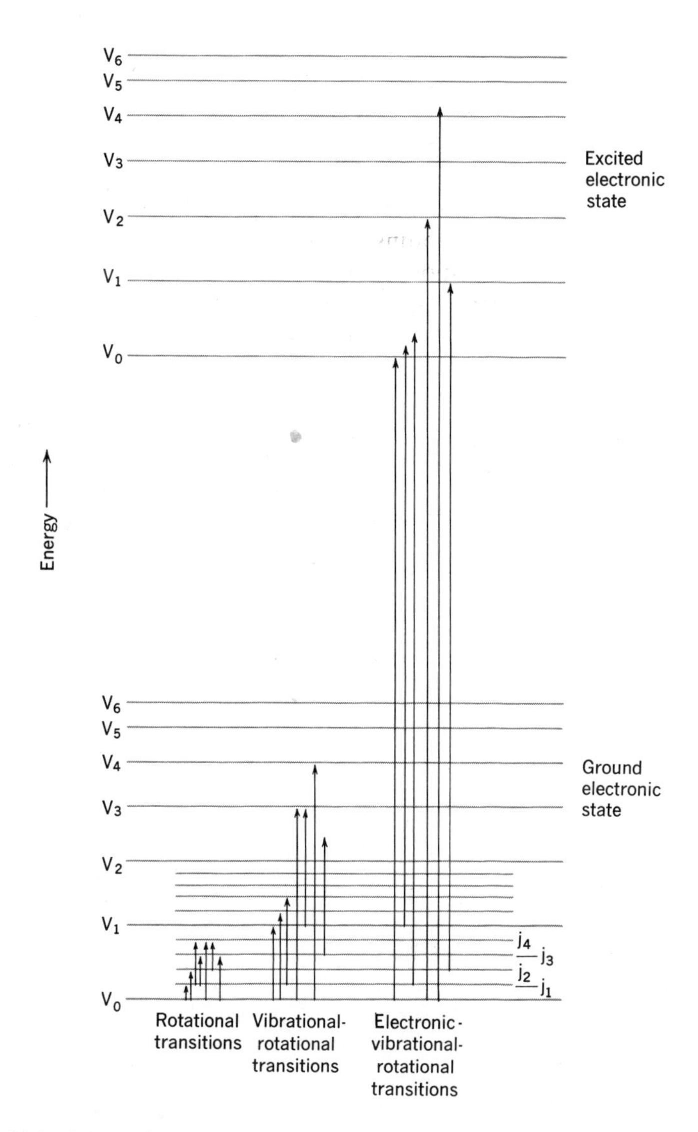

Figure 14.3 Schematic energy level diagram. Two electronic levels are shown: V_0, V_1, etc. are vibrational levels, and a few rotational levels represented by j values are shown.

Absorption of visible light and ultraviolet radiation[2] increases the electronic energy of a molecule. That is, the energy contributed by the photons enables electrons to overcome some of the restraint of the nuclei and move out to new orbitals of higher energy. Vibrational and rotational effects are superimposed upon the electronic change, but the region where the absorption is found is

[2] Many writers prefer that the word *light* be restricted to the visible region. It would then be improper to speak of "ultraviolet light" or "infrared light." For wavelengths to which the eye is unresponsive, *radiation* or *radiant energy* is preferred.

determined by the electronic energy levels of the molecule. The vibrational and rotational changes introduce "fine structure" into the spectrum, so that the absorption involves a band of wavelengths rather than a single line. The individual lines making up the band are usually not resolved under experimental conditions, and the observed visible or ultraviolet absorption spectrum generally consists of peaks exhibiting a smooth curvature.

Once the absorption has been recorded, the fate of the excited molecules is usually not of interest in ordinary spectrophotometry for analytical purposes, but we may briefly note, for the curious student, that the molecule tends not to remain in the excited state but rather to get rid of the excess energy. Commonly, the energy is degraded into heat by a stepwise process involving collisions with other molecules. (This heat is not noticeable in an ordinary spectrophotometric experiment.) Sometimes the energy is re-emitted as radiation, usually of longer wavelength than was originally absorbed; this phenomenon is known as *fluorescence* (if there is a certain time delay in reemission, the term *phosphorescence* is used). Fluorescence can lead to errors in absorption measurements if the reemitted radiation reaches the detector of the instrument. Finally, in some cases the absorbed energy may cause the molecule to dissociate into free radicals or ions, which may then proceed through a complicated series of reactions.

INFRARED SPECTROPHOTOMETRY

Infrared spectrophotometry is very important in modern chemistry, particularly (although not exclusively) in the organic area. It is a routine tool for detecting functional groups, identifying compounds, and analyzing mixtures. Instruments which record infrared spectra are commercially available and easy to use on a routine basis.

When we describe the structure of a molecule in terms of fixed bond lengths and bond angles, we are depicting a sort of average situation. Imagine a model of a complex molecule constructed of wooden balls connected by springs and suspended from a wire. Deliver a blow to the molecule, and it will become a quivering object with all its atoms in motion with respect to each other as perhaps dozens of springs stretch and compress and bend. This motion, which may at first appear hopelessly complicated, can be resolved into a series of individual vibratory modes whose natural frequencies depend upon the masses of the wooden balls and the characteristics of the springs. In a real molecule, analogous vibrations are occurring: pairs of atoms vibrating with respect to each other as the individual bonds lengthen and shorten, entire groups oscillating with respect to other atoms or groups, ring structures "breathing" (i.e., expanding and contracting, etc.). Now, if there is an oscillating electric dipole associated with a particular vibratory mode, then there will occur an interaction with the electrical vector of electromagnetic radiation of this same frequency, leading to the absorption of energy which shows up as an increased amplitude of vibration.

Most groups, such as C—H, O—H, C=O, and C≡N, give rise to infrared absorption bands which vary only slightly from one molecule to another depend-

TABLE 14.2 Some Infrared Group Frequencies

Group		Frequency, cm^{-1}	Wavelength, μm
OH	Alcohol	3580–3650	2.74–2.79
	H-bonded	3210–3550	2.82–3.12
	Acid	2500–2700	3.70–4.00
NH	Amine	3300–3700	2.70–3.03
CH	Alkane	2850–2960	3.37–3.50
	Alkene	3010–3095	3.23–3.32
	Alkyne	3300	3.03
	Aromatic	~3030	~3.30
C≡C	Alkyne	2140–2260	4.42–4.76
C=C	Alkene	1620–1680	5.95–6.16
	Aromatic	~1600	~6.25
C=O	Aldehyde	1720–1740	5.75–5.81
	Ketone	1675–1725	5.79–5.97
	Acid	1700–1725	5.79–5.87
	Ester	1720–1750	5.71–5.86
C≡N	Nitrile	2000–2300	4.35–5.00
NO_2	Nitro	1500–1650	6.06–6.67

ing upon other substituents. Examples of these are shown in Table 14.2. In addition to these group frequencies, which can usually be definitely assigned, complex molecules may exhibit a myriad of absorption bands whose exact origins are difficult to ascertain but which are extremely useful for qualitative identification. Many of these occur in what is called the "fingerprint" region of the spectrum (ca. 6.5 to 14 μm). Figure 14.4 shows a sample infrared spectrum; note

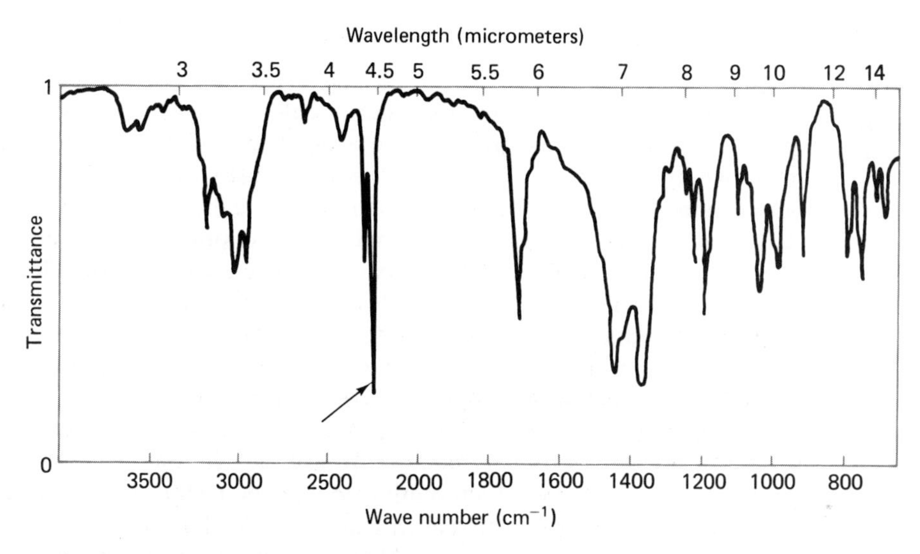

Figure 14.4 Infrared spectrum of acetonitrile ($CH_3C{\equiv}N$). Spectrum obtained on a thin film (0.005 mm) of the pure liquid. Arrow shows band due to the C≡N group.

the richness of detail found even for a very simple molecule, and note how easy it is to spot the absorption band associated with the presence of a particular functional group.

ULTRAVIOLET AND VISIBLE SPECTRA

Absorption spectra in the ultraviolet and visible regions generally consist of one or a few broad absorption bands, as shown in Fig. 14.5. All molecules can absorb radiation in the UV-visible region because they contain electrons, both shared and unshared, which can be excited to higher energy levels. The wavelengths at which absorption occurs depends upon how firmly the electrons are bound in the molecule. The electrons in a single covalent bond are tightly bound, and radiation of high energy, or short wavelength, is required for their excitation. For example, alkanes, which contain only C—H and C—C single bonds, show no absorption above 160 nm. Methane shows a peak at 122 nm (Table 14.3) which is designated a σ-σ^* transition. This means that an electron in a sigma-bonding orbital is excited to a sigma antibonding orbital.

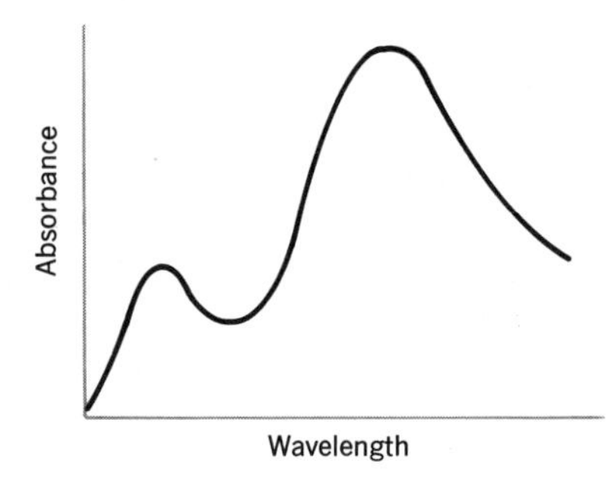

Figure 14.5 Typical visible or ultraviolet absorption spectrum.

If a molecule contains an atom such as chlorine which has unshared electron pairs, a nonbonding electron can be excited to a higher energy level. Since nonbonding electrons are not as tightly bound as are sigma-bonding electrons, the absorption takes place at longer wavelengths. Note that such a transition occurs in CH_3Cl at 173 nm (Table 14.3), and that the transitions in CH_3Br and CH_3I occur at even longer wavelengths. This is because the electrons are held less tightly by bromine and iodine. The transition described here is designated n-σ^* to indicate that a nonbonding electron is raised to a σ-antibonding orbital.

Electrons in double or triple bonds are rather easily excited to higher pi-orbitals. A transition is designated π-π^* when a pi-electron is raised from a pi-bonding orbital to a pi-antibonding orbital. The absorption of energy in such a transition is usually stronger than σ-σ^* transitions. In conjugated molecules (i.e., those containing a series of alternating double bonds) the absorption is shifted to longer wavelengths, as can be seen in Table 14.3. Such molecules are described by writing resonance structures, saying that the electron is more "delocalized" than if it were confined to one bond between two atoms. The shift to longer wavelengths

TABLE 14.3 Some Electronic Transitions in Organic Molecules

Compound	Wavelength, nm
Single bonds	
CH_4	122
$CH_3{-}CH_3$	135
CH_3Cl	173
CH_3Br	204
CH_3I	258
CH_3OH	184
CH_3OCH_3	184
Double bonds	
$CH_2{=}CH_2$	162
$-(CH{=}CH)_2-$	217
$-(CH{=}CH)_3-$	258
$-(CH{=}CH)_4-$	300
$-(CH{=}CH)_5-$	330
$(CH_3)_2C{=}O$	190, 280
$CH_3CH{=}CH{-}CHO$	217
$CH_2{=}CH{-}CH{=}CH{-}CHO$	263
Triple bonds	
$HC{\equiv}CH$	178
$HC{\equiv}N$	175

reflects the fact that the electron in a conjugated system is less tightly bound than one in a nonconjugated system.

It has been found that π-π^* transitions in molecules which contain unsaturated groups are very similar irrespective of the atoms which make up the double bonds. Note (Table 14.3) that the absorptions in acetylene, $H{-}C{\equiv}C{-}H$, and hydrogen cyanide, $H{-}C{\equiv}N$, occur at about the same wavelengths. The same is true for the two conjugated systems $-C{=}C{-}C{=}C-$ and $-C{=}C{-}C{=}O$, both giving peaks around 217 nm. Many years ago the association between unsaturation and the absorption of light was recognized by organic chemists, and the term *chromophore* was introduced to describe the role of such groups as $C{=}C$, $C{=}O$, and $N{=}N$ in shifting the absorption of light toward the visible region.

Most applications of ultraviolet and visible spectrophotometry to organic compounds are based on n-π^* or π-π^* transitions and hence require the presence of chromophoric groups in the molecule. These transitions occur in the region of the spectrum (about 200 to 700 nm) which is convenient to use experimentally. Commercial UV-visible spectrophotometers usually operate from about 220 to 1000 nm. The qualitative identification of organic compounds in this region is much more limited than in the infrared. This is because the absorption bands are broad and lacking in detail. However, certain functional groups, such as carbonyl, nitro, and conjugated systems, do show characteristic peaks, and useful information can often be obtained concerning the presence or absence of such groups in the molecule.

QUANTITATIVE ASPECTS OF ABSORPTION

Absorption spectra can be obtained using samples in various forms: gases, thin films of liquids, solutions in various solvents, and even solids. Most analytical work involves solutions, and we wish here to develop a quantitative description of the relationship between the concentration of a solution and its ability to absorb radiation. At the same time, we must realize that the extent to which absorption occurs will depend also upon the distance traversed by the radiation through the solution. As we have seen, absorption also depends upon the wavelength of the radiation and the nature of the molecular species in solution, but for the time being we may suppose that we can control these.

Bouguer's (Lambert's) Law

The relationship between the absorption of radiation and the length of the path through the absorbing medium was first formulated by Bouguer (1729), although it is sometimes attributed to Lambert (1768). Let us subdivide a homogeneous absorbing medium such as a chemical solution into imaginary layers, each of the same thickness. If a beam of monochromatic radiation (i.e., radiation of a single wavelength) is directed through the medium, it is found that each layer absorbs an equal fraction of the radiation, or each layer diminishes the radiant power of the beam by an equal fraction. Suppose, for example, that the first layer absorbed half of the radiation incident upon it. Then the second layer would absorb half of the radiation incident upon *it*, and the radiant power emerging from this second layer would be one-fourth that of the original power; from the third layer, one-eighth, etc.

Bouguer's finding may be formulated mathematically as follows, where P_0 is the incident radiant power and P is the power emergent from a layer of medium b units thick:

$$-\frac{dP}{db} = k_1 P$$

The minus sign indicates that the power decreases with absorption. For the student who is unfamiliar with calculus, we may express this equation verbally as: The decrease in radiant power per unit thickness of adsorbing medium is proportional to the radiant power. For the student who has studied calculus, let us rearrange the preceding equation to

$$-\frac{dP}{P} = k_1 db$$

and integrate between limits P_0 and P and 0 and b:

$$-\int_{P_0}^{P} \frac{dP}{P} = k_1 \int_0^b db$$

$$-(\ln P - \ln P_0) = k_1 b$$

$$\ln P_0 - \ln P = k_1 b$$

$$\ln \frac{P_0}{P} = k_1 b$$

Usually the equation is written with base-10 logarithms, which simply changes the constant:

$$\log \frac{P_0}{P} = k_2 b$$

A verbal statement of this equation might be: The power of the transmitted radiation decreases in an exponential fashion as the thickness of the absorbing medium increases arithmetically. Some writers consider this integration step to be a "derivation" of Bouguer's law, but actually the two formulations are equivalent representations of what we are here taking as an experimental finding.

Bouguer's law appears to describe correctly, without exception, the absorption of monochromatic radiation by various thicknesses of a homogeneous medium. The student can convince himself that the law applies strictly only with monochromatic radiation by considering an extreme case. Pass two wavelengths through a medium, one of which is absorbed appreciably and the other not at all. According to Bouguer's law, if we allow the thickness of the medium to increase indefinitely, then the transmitted radiant power should approach zero. But it cannot fall to zero if an appreciable fraction is not absorbed at all.

It may be noted that Bouguer's law takes the same form as other familiar functions such as the rate expression for first-order kinetics or radioactive decay, and the compound interest law.

Beer's Law

The relationship between the concentration of an absorbing species and the extent of absorption was formulated by Beer in 1859. Beer's law is analogous to Bouguer's law in describing an exponential decrease in transmitted radiant power with an arithmetic increase in concentration. Thus

$$-\frac{dP}{dc} = k_3 P$$

which upon integration and conversion to ordinary logarithms becomes

$$\log \frac{P_0}{P} = k_4 c$$

Beer's law is strictly applicable only for monochromatic radiation and where the nature of the absorbing species is fixed over the concentration range in question. We shall comment further on this point in connection with so-called "deviations" from Beer's law.

Combined Bouguer-Beer Law

Bouguer's and Beer's laws are readily combined into a convenient expression. We note that, in studying the effect of changing concentration upon absorption, the path length through the solution would be held constant, but the measured results would depend upon the magnitude of the constant value. In other words, in Beer's law as written above, $k_4 = f(b)$. Similarly, in Bouguer's law, $k_2 = f(c)$. Substitution of these fundamental relationships into Bouguer's and Beer's laws gives

$$\log \frac{P_0}{P} = f(c)b \quad \text{(Bouguer)} \qquad \text{and} \qquad \log \frac{P_0}{P} = f(b)c \quad \text{(Beer)}$$

The two laws must apply simultaneously at any point, so that

$$f(c)b = f(b)c$$

or, separating the variables,

$$\frac{f(c)}{c} = \frac{f(b)}{b}$$

Now, the only condition under which two functions of independent variables can be equal is that they both equal a constant:

$$\frac{f(c)}{c} = \frac{f(b)}{b} = K$$

or

$$f(c) = Kc \qquad \text{and} \qquad f(b) = Kb$$

Substitution into either the Bouguer or the Beer expression yields the same result:

$$\log \frac{P_0}{P} = f(c)b = Kbc$$

$$\log \frac{P_0}{P} = f(b)c = Kbc$$

Nomenclature and Units

Unfortunately, the development of the nomenclature regarding the Bouguer-Beer law has not been systematic, and a confusing array of terms appears in the literature. In analytical chemistry, the tendency in the United States has been to adopt recommendations of a Joint Committee on Nomenclature in Applied Spectroscopy, established by the Society for Applied Spectroscopy and the American Society for Testing Materials (ASTM).[3]

[3] For the report of this committee, see H. K. Hughes et al., *Anal. Chem.*, **24**, 1349 (1952).

The symbols P_0 and P as used here are recommended for the incident and transmitted radiant powers, respectively.[4] The term $\log(P_0/P)$ is called the *absorbance* and given the symbol A. Other terms which have been used synonymously with absorbance and which the student may encounter in the literature are *extinction*, *optical density*, and *absorbancy*.

The symbol b is accepted for the length of the path through the absorbing medium; it is ordinarily expressed in centimeters. Other writers have used the letter l for the same quantity, and, more rarely, the letters d or t.

Two different units for c, the concentration of absorbing solute, are often used, grams per liter and moles per liter. It is apparent that the value of the constant (designated K above) in the Bouguer-Beer law will depend upon which concentration system is used. When c is in grams per liter, the constant is called the *absorptivity*, symbol a. When c is in moles per liter, the constant is the *molar absorptivity*, symbol ε. Thus, in the recommended system, the Bouguer-Beer law may take two forms:

$$A = abc_{\text{g/liter}} \quad \text{or} \quad A = \varepsilon bc_{\text{mol/liter}}$$

It is apparent that $\varepsilon = a \times \text{MW}$, where MW refers to the molecular weight of the absorbing substance in the solution. Other designations for a are *specific extinction*, *extinction coefficient*, *Bunsen coefficient*, and *specific absorption*. Similarly, some writers call ε the *molar extinction coefficient*, *molecular extinction*, and various other names.

The *transmittance*, $T = P/P_0$, is simply the fraction of the incident power which is transmitted by a sample. The *percent transmittance*, $\%T = P/P_0 \times 100$, is also encountered. If $A = \log(P_0/P)$ and $T = P/P_0$, then $A = \log(1/T)$. Since, from Beer's law, absorbance is directly proportional to concentration, it is clear that transmittance is not; $\log T$ must be plotted vs. c to obtain a linear graph.

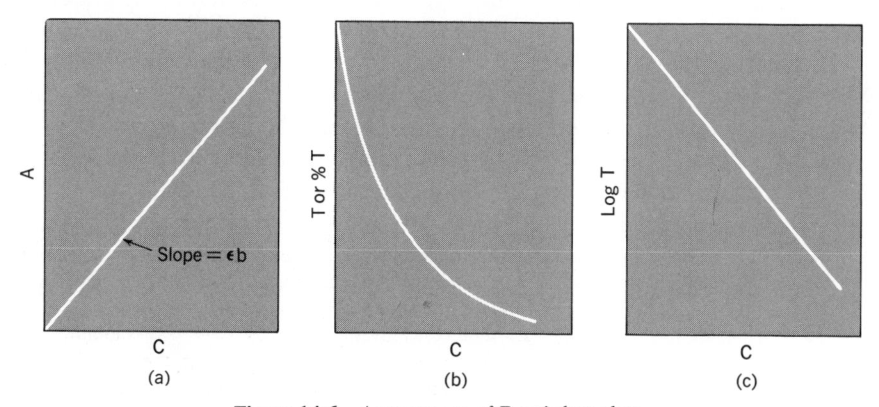

Figure 14.6 Appearance of Beer's law plots.

[4] Many writers use I_0 and I for these terms, standing for *intensity* of the beam, but in ordinary spectrophotometers the quantity actually measured is the rate at which radiant energy is absorbed at the detector; this is best called "radiant power." Such units as watts or ergs per second might be employed, but in spectrophotometry we deal with a ratio (P/P_0) or the logarithm of a ratio $(\log P_0/P)$, and the units cancel.

Figure 14.6 shows the situation. Analytical chemists prefer absorbance plots, but the student should be familiar with transmittance because it is encountered frequently. The detectors of most instruments generate a signal which is linear in transmittance, because they respond linearly to radiant power. Thus, if an instrument is to be read in absorbance units, there must be a logarithmic scale on the readout device or the signal must be altered logarithmically by an electronic circuit or in some mechanical fashion.

Deviations from the Bouguer-Beer Law

According to the Bouguer-Beer law (or, as many writers say, simply Beer's law), a plot of absorbance vs. molar concentration will be a straight line of slope εb. Frequently, however, measurements on real chemical systems yield Beer's law plots which are not linear over the entire concentration range of interest. Such curvature suggests that ε is not a constant, independent of concentration, for such systems, but closer consideration leads to a somewhat more sophisticated view. The value of ε is expected to depend upon the nature of the absorbing species in solution and upon the wavelength of the radiation. Most deviations from Beer's law encountered in analytical practice are attributable to failure or inability to control these two aspects, and hence may be called apparent deviations because they reflect experimental difficulties more than any inadequacy of Beer's law itself.[5]

Consider, for example, absorbance measurements on a series of solutions of a weak acid, HB. The degree of dissociation of HB (fraction ionized) varies with the quantity of HB introduced into each solution if the final volumes are the same. Under this circumstance, it is possible to encounter either positive or negative Beer's law deviations, depending upon the ε-values of the two species, HB and B^-, at the wavelength employed. Since the fraction of the material present as B^- decreases with increasing analytical concentration of HB, a negative deviation from Beer's law will be seen if $\varepsilon_{B^-} > \varepsilon_{HB}$. On the other hand, if $\varepsilon_{HB} > \varepsilon_{B^-}$, a positive deviation will result. The system should follow Beer's law at a wavelength[6] where $\varepsilon_{HB} = \varepsilon_{B^-}$. These deviations from Beer's law may be circumvented, not only by performing measurements at the isosbestic wavelength (which lowers the sensitivity because ε-values generally are not maximal here),

[5] There is another class of deviations which may be considered *real* rather than apparent, but they are not likely to be encountered in analytical chemistry. For example, it is shown in the theory of optics that ε for a substance in solution will change with changes in the refractive index of the solution. Since changes in refractive index attend concentration changes, Beer's law should not hold, even ideally. However, this effect is very small and is generally well within the experimental errors of spectrophotometry. Another real deviation from Beer's law sometimes occurs when relatively strong radiation passes through a medium containing only a few absorbing molecules. Under these conditions, all the molecules may be elevated to higher energy states by only a fraction of the available photons, and hence there will be no opportunity for further absorption regardless of how many more photons may be available. This situation, known as *saturation*, is ordinarily not encountered in analytical practice.

[6] A wavelength where two or more species in equilibrium with one another have the same ε-value is called an *isosbestic point.*

but by adjusting all the solutions to a very low pH by addition of strong acid so as to repress the ionization of HB, or by addition of sufficient strong alkali to transform all the material into B^-.

Many examples of this sort of Beer's law deviation are known. The general viewpoint here is that there is nothing wrong with Beer's law, that ε-values for individual species are constant over a wide concentration range, and that the deviations are predictable from a knowledge of the equilibria in which these species participate. Equilibria involving ions are often sensitive to added electrolytes, and failure to control the ionic strength may create problems in spectrophotometry. Temperature and various other factors may further complicate the situation.

Even with systems that are "well-behaved" chemically, deviations from Beer's law may occur because of characteristics of the instruments used in measuring absorbance values. In days past, such deviations sometimes resulted from fatigue effects in detectors, nonlinearity in amplifiiers and readout devices, and instability in the sources of radiant energy. These problems have largely been solved in modern spectrophotometric instruments.

We pointed out earlier that the Bouguer-Beer law demands monochromatic radiation. Because ε values depend upon wavelength, measured absorbance values reflect the wavelength distribution in the radiation, which, in a practical spectrophotometer, is never strictly monochromatic. Think again of an absorbing solution as a series of imaginary layers of equal thickness. Now if heterochromatic radiation passes through the first layer, the more strongly absorbed wavelengths are abstracted from the beam to a greater extent than the others. Thus the radiation impinging upon the second layer will be richer in the less strongly absorbed wavelengths, and the second layer will not absorb the same fraction of the radiation incident upon it as did the first layer. Since the Bouguer-Beer law states that each layer will absorb an equal fraction, deviation from the law will clearly result.[7]

Although it must be pointed out that instrumental characteristics can lead to deviations from the Bouguer-Beer law, it is a practical fact that the better modern spectrophotometers are capable of performing well in this regard. This is not the case with the colorimeters or filter photometers which employ broad band-pass filters to isolate the desired radiation and which are still widely used in clinical and control laboratories. Further, good spectrophotometers can be operated in such a way as to lose some of their fine characteristics. Thus the student should file away in the back of his mind the warning to check before assuming the Bouguer-Beer law to hold for a particular chemical system with a particular instrument.

INSTRUMENTATION FOR SPECTROPHOTOMETRY

A spectrophotometer is an instrument for measuring the transmittance or absorbance of a sample as a function of wavelength; measurements on a series of

[7] A mathematical analysis of this type of deviation may be found in L. Meites and H. C. Thomas, *Advanced Analytical Chemistry*, McGraw-Hill Book Company, New York, 1958, p. 255.

samples at a single wavelength may also be performed. Such instruments may be classified as manual or recording, or as single or double beam. In practice, single-beam instruments are usually operated manually and double-beam instruments generally feature automatic recording of absorption spectra, but it is possible to record a spectrum with a single-beam instrument. An alternative classification is based upon the spectral region, and we speak of infrared or ultraviolet spectrophotometers, etc. A complete understanding of spectrophotometers requires a detailed knowledge of optics and electronics which is far beyond the scope of this book. It is possible, though, for the student at this stage to understand what the instruments do. By combining this general, fundamental understanding with detailed instructions furnished in the form of "manuals" by the manufacturers, the chemist can obtain good data with modern spectrophotometers. Manually operated, single-beam spectrophotometers will be discussed first, because this provides the background for appreciating the capabilities of the more complex instruments.

Single-Beam Spectrophotometers

The essential components of a spectrophotometer, which are shown schematically in Fig. 14.7, are the following:

1. A continuous source of radiant energy covering the region of the spectrum in which the instrument is designed to operate.
2. A monochromator, which is a device for isolating a narrow band of wavelengths from the broad spectrum emitted by the source (of course, strict monochromaticity is not attained).

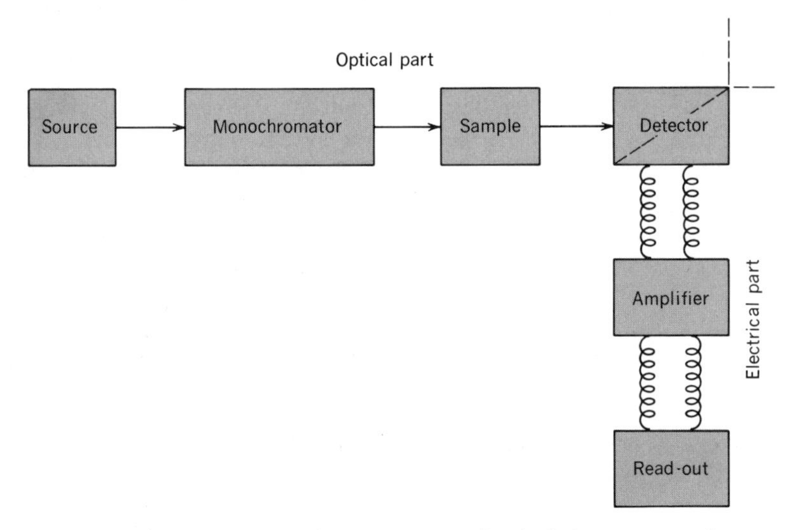

Figure 14.7 Block diagram showing components of a single-beam spectrophotometer. Arrows represent radiant energy, coiled lines electrical connections. The optical part and the electrical part of the instrument meet at the detector, a transducer which converts radiant energy into electrical energy.

3. A container for the sample.
4. A detector, which is a *transducer* that converts radiant energy into an electrical signal.
5. An amplifier and associated circuitry which renders the electrical signal appropriate for readout.
6. A readout system on which is displayed the magnitude of the electrical signal.

Both single- and double-beam spectrophotometers, and instruments which operate in various regions of the spectrum, all have these essential components, although the details are quite different in the several cases. In accord with the goal set forth above, we shall discuss these components briefly.

Source

The usual source of radiant energy for the visible region of the spectrum as well as the near infrared and near ultraviolet is an incandescent lamp with a tungsten filament. Under ordinary operating conditions, the output of this tungsten lamp is adequate from about 325 or 350 nm to about 3 μm. The energy emitted by the heated filament varies greatly with wavelength, as shown in Fig. 14.8. The energy distribution is a function of the temperature of the filament, which depends in turn upon the voltage supplied to the lamp; an increase in operating temperature increases the total energy output and shifts the peak of Fig. 14.8 to shorter wavelength. Thus the voltage to the lamp should be a stable one; a regulated power supply is incorporated into the instrument. The heat from a tungsten lamp may be troublesome; often the lamp housing is water-jacketed or cooled with a blower to prevent warming of the sample or of other instrument components.

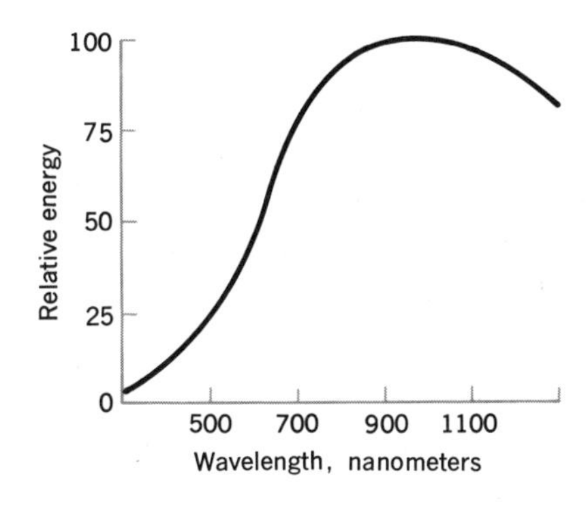

Figure 14.8 Relative output of a typical tungsten lamp as a function of wavelength.

Below about 325 to 350 nm, the output of a tungsten lamp is inadequate for spectrophotometers, and a different source must be used. Most common is a hydrogen (or deuterium) discharge tube, which is used from about 175 to 375 or 400 nm. When a discharge between two electrodes excites emission by a sample of a gas such as hydrogen, a discontinuous line spectrum characteristic of the gas is obtained provided the pressure is relatively low. As the hydrogen pressure is increased, the lines broaden and eventually overlap, until at relatively high pressures, a continuous spectrum is emitted. The pressure required in a hydrogen discharge tube is lower than that with certain other gases; also the tube runs cooler. Such tubes are conveniently small—about the size of common radio tubes. The envelope is usually glass, but a quartz window is provided to pass the ultraviolet radiation. A high-voltage power supply is required for gaseous discharge tubes. In some spectrophotometers, provision is made for interchanging tungsten and hydrogen discharge sources in order to cover the visible and ultraviolet regions through which the instruments operate.

The source for infrared spectrophotometers, which commonly operate from about 2 to 15 μm, is usually the Nernst glower. This is a small rod of ceramic appearance fabricated from a special mixture of metal oxides, with platinum leads sealed into the ends. The rod is nonconducting at room temperature, but it is brought into a conducting state when heated, after which a flow of current maintains a glow which is rich in infrared radiation.

Monochromator

This is an optical device for isolating from a continuous source a beam of radiation of high spectral purity of any desired wavelength. The essential components of a monochromator are a slit system and a dispersive element. Radiation from the source is focused upon the entrance slit, then collimated by a lens or mirror so that a parallel beam falls upon the dispersing element, which is either a prism or a diffraction grating. By mechanically turning the prism or grating, various portions of the spectrum produced by the dispersive element are focused on the exit slit, whence, by a further optical path, they encounter the sample.

The student may recall from elementary physics the action of a prism in dispersing white light into a spectrum. When a beam of light passes through the interface between two different media, such as air and glass, bending takes place which is called *refraction*. The extent of the bending depends upon the index of refraction of the glass. This index of refraction varies with the wavelength of the light; the blues are bent more than the reds, as shown in Fig. 14.9. As a result of the variation of refractive index with wavelength, the prism is able to disperse or spread out a beam of white light into a spectrum, in which the various colors

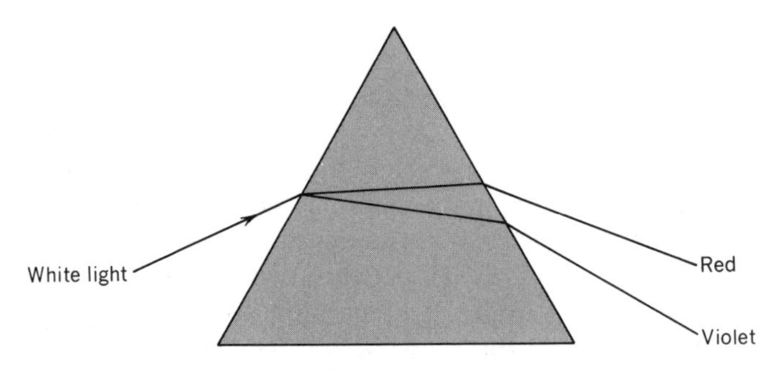

Figure 14.9 Dispersion of white light by a prism.

making up the white light may be recognized separately. Infrared and ultraviolet radiation are dispersed in the same manner, but here the words *light* and *color* are not used and the prism material is not glass. The material of choice represents a compromise between dispersive power and transparency in the desired wavelength region, along with several other factors. Spectrophotometers covering mainly the visible region of the spectrum have glass prisms, whereas quartz is the prism material for instruments covering the ultraviolet and near infrared as well as the visible; infrared spectrophotometers commonly have prisms of rock salt.

The spectral purity of the emergent radiation from the monochromator depends upon the dispersive power of the prism and the width of the exit slit. At first thought, one might suppose that monochromaticity could be approached as closely as desired by merely decreasing the slit width sufficiently, but this is not the case. Eventually, the slit becomes so narrow that diffractive effects at its edges only create a loss of radiant power with no increase in spectral purity (this is the so-called "Rayleigh diffraction limit"); actually, before this limit is approached in a typical spectrophotometer, the narrowed slit is passing insufficient energy to activate the detector.

With prism monochromators, a given slit width does not yield the same degree of monochromaticity throughout the spectrum. The wavelength dependence of the dispersion of a prism is such that the wavelengths in the spectrum are not spread out uniformly. The dispersion is greater for the shorter wavelengths, and hence wider slits may here achieve the same degree of spectral purity as would narrower slits at longer wavelengths.

Figure 14.10 is a schematic diagram of the optical system of a particular single-beam spectrophotometer with a quartz prism. The back of the prism is

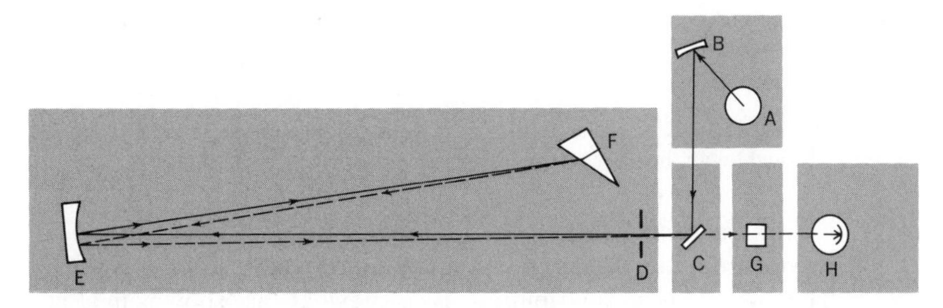

Figure 14.10 Schematic diagram of optical system of Beckman model DU spectrophotometer. *A*, light source; *B*, *C*, mirrors; *D*, slit; *E*, collimating mirror; *F*, quartz prism with reflecting back surface; *G*, cell; *H*, phototube. (Courtesy of Beckman Instruments, Inc.)

coated with a reflective metallized surface so that the radiation actually passes twice through the dispersive element. This not only enhances the dispersion but also is of great geometric convenience.

A diffraction grating (reflection) is made by ruling on a polished metal surface, such as aluminum, a large number of parallel lines. For the infrared region there are about 1500 to 2500 lines/inch; for the ultraviolet and visible regions, about

15,000 to 30,000 lines/inch. When light is reflected from this surface, that which strikes the rulings is dissipated by scattering; the unruled portions reflect regularly, acting as individual light sources. Overlapping of the waves from these sources establishes an interference pattern which results in the dispersion of the reflected light into its component wavelengths. The student is referred to elementary physics texts for a full explanation of this phenomenon.

The machines for ruling the lines on a grating must be constructed to very close tolerances, and original gratings are expensive. Much cheaper, and much more widely used, are *replica gratings*, large number of which can be prepared from a single master grating. The original is coated with a plastic material, which, after hardening, is stripped off to yield a replica. The plastic is made reflective by evaporating a film of metal, generally aluminum, onto the ruled face; the grating is mounted in the monochromator in such a way that rotation allows various portions of the spectrum to illuminate the exit slit.

Gratings differ from prisms in rendering a uniform dispersion throughout the entire spectrum; in other words, a single slit width yields the same degree of monochromaticity of the emergent radiation throughout the spectrum. Figure 14.11 shows the optical path through a widely used grating instrument.

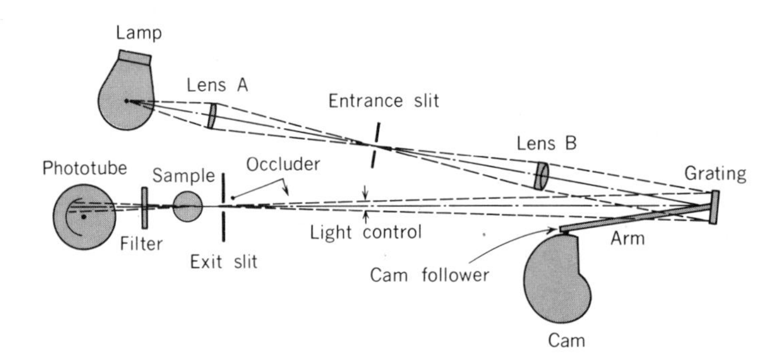

Figure 14.11 Schematic diagram of optical system of Spectronic 20 (Courtesy of Bausch and Lomb, Inc.)

As mentioned earlier, the radiation emergent from a monochromator is not monochromatic, although it is much more nearly so than is the original source. The wavelength distribution is somewhat as shown in Fig. 14.12. The terminology employed in describing the width of the band shown in the figure is not entirely standard, and advertisements for instruments often quote figures for "band width" without specifying what is meant. A particular terminology which is widely understood is shown in the figure.

A problem in monochromators is so-called "stray light," by which is meant radiation of unspecified wavelengths which is reflected about inside the monochromator and which may find its way to the exit slit. In good instruments, this is minimized by using dull black surfaces and by inserting baffles in appropriate positions. It is cut to an extremely low level in the finer instruments which employ double monochromators, generally combining both prism and grating. With

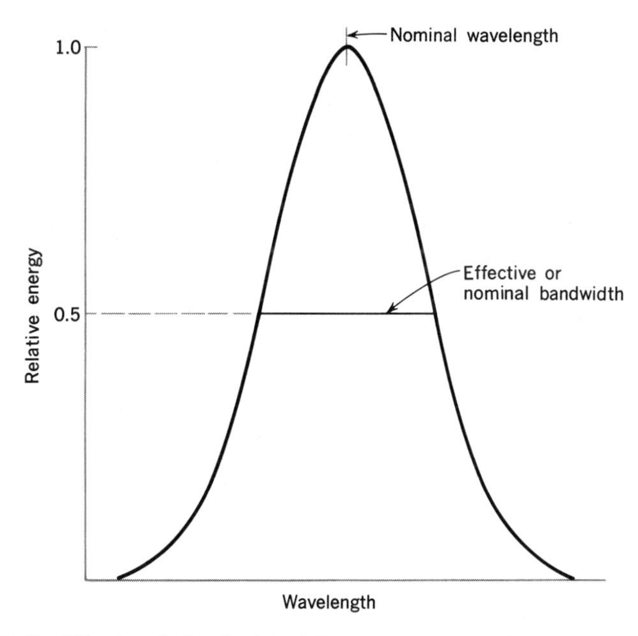

Figure 14.12 Wavelength distribution of the energy emergent from a monochromator.

ordinary instruments, spurious absorbance readings due to stray light may be obtained in spectral regions where very little energy of the desired wavelengths is available.

Until quite recently, instruments without true monochromators were widely used for absorbance measurements, mainly in the visible region, in laboratories where a low initial investment, simplicity, and speed were more important than the quality of the results. These instruments, designated *filter photometers*, utilized colored glass filters to isolate fairly broad wavelength bands from the source. They served admirably for many routine analyses, but they have been largely displaced by inexpensive grating spectrophotometers.

Sample Container

Most spectrophotometry involves solutions, and thus most sample containers are cells for placing liquids in the beam of the spectrophotometer. The cell must transmit radiant energy in the spectral region of interest; thus glass cells serve in the visible region, quartz or special high-silica glasses in the ultraviolet, and rock salt in the infrared. It must be remembered that the cell, which in a sense is merely a container for the sample, is actually more than this; when in position, it becomes part of the optical path through the spectrophotometer, and its optical properties are important. In less expensive instruments, cylindrical test tubes are sometimes used as sample containers. It is important that such tubes be positioned reproducibly by marking one side of the tube and facing the mark in the same direction whenever it is placed in the instrument. The better cells have flat optical surfaces. The cells must be filled so that the light beam goes through the solution, with the meniscus entirely above the beam. Cells are generally held in position by kinematic design of the holder or by spring clips which ensure reproducible positioning in the cell compartment of the instrument.

Typical visible and ultraviolet cells have path lengths of 1 cm, but a wide variety is available, ranging from very short paths, fractions of a millimeter, up to 10 cm or even more. Special microcells may be obtained, by means of which minute volumes of solution yield an ordinary path length, and adjustable cells of variable path length are also available, particularly for infrared work. The variety of infrared cells currently on the market is beyond the scope of this discussion. Problems in the infrared are different from those in the ultraviolet and visible regions. Because solvents which are infrared-transparent are not available, the tendency is to run concentrated solutions at short path lengths (0.1 mm or even less) to minimize absorption by the solvent, and the cells are thus quite different from those employed at shorter wavelengths.

Detector

In a detector for a spectrophotometer, we desire high sensitivity in the spectral region of interest, linear response to radiant power, a fast response time, amenability to amplification, and high stability or low "noise" level, although in practice it is necessary to compromise among these factors. Higher sensitivity, for example, can be bought only at the expense of increased noise. The types of detection that have been most widely used are based upon photochemical change (mainly photographic), the photoelectric effect, and the thermoelectric effect. Photography is no longer used in ordinary spectrophotometry; generally speaking, photoelectric detectors are employed in the visible and ultraviolet regions and detectors based upon thermal effects are used in the infrared.

The commonest photoelectric detector is the *phototube*. This is an evacuated envelope, with a transparent window, containing a pair of electrodes across which a potential is maintained. The surface of the negative electrode is photosensitive; that is, electrons are ejected from this surface when it is irradiated with photons of sufficient energy. The electrons are accelerated across the potential difference to the positive electrode, and a current flows in the circuit. Whether or not electrons are emitted depends upon the nature of the cathode surface and the frequency of the radiation; the number of electrons emitted per unit time, and hence the current, depends upon the radiant power. A variety of phototubes are available which differ in the material of the cathode surface (also in the transparent window) and hence in their response to radiation of various frequencies. A number of spectrophotometers provide for interchanging detectors so as to maintain a good response over a broad wavelength range.

Photomultiplier tubes are more sensitive than ordinary phototubes because of high amplification accomplished with the tube itself. Such a tube has a series of electrodes, each at a progressively more positive potential than the cathode. The geometry of the tube is such that the primary photoelectrons are focused into a beam and accelerated to an electrode which is, say, 50 to 90 V more positive than the cathode. The bombardment of this electrode (or dynode, as it is called) releases many more secondary electrons which are accelerated to a third, more positive, electrode, etc., for perhaps 10 stages. A regulated high-voltage power supply, furnishing about 500 to 900 V, is required to operate the tube. The output of the photomultiplier is still further amplified with an external electronic amplifier. The enhanced sensitivity of this detector permits narrower slit widths in the monochromator and hence better resolution of spectral fine structure.

The common infrared detector is the thermocouple. The student may recall the thermoelectric effect: If two dissimilar metals are joined at two points, a potential is developed if the two junctions are at different temperatures. Heating of one of the junctions by the infrared radiation is thus the basis of detection. This junction is specially designed to have a low heat capacity so that it will be warmed appreciably by radiant energy of the low power encountered in the instrument.

Amplification and Readout

It is beyond the scope of this text to discuss the detailed electronics of amplification and readout as they are accomplished in various spectrophotometers. To give an idea of what may be involved, we may briefly consider one possibility. Let us place a large load resistor in series with a phototube, as shown schematically in Fig. 14.13. Suppose the radiant power supplied to the cathode is such that a current of 1 microampere (10^{-6} A) flows in the circuit. If the resistor has a value of 1 megohm (10^{6} ohms) as shown in the figure, then according to Ohm's law the voltage across the resistor, $E = iR$, is $10^{-6} \times 10^{6} = 1$ V. Although 1 V is a fair voltage, it cannot be measured by connecting an ordinary voltmeter across the resistor. As soon as such a connection is made, the meter becomes part of the circuit, establishing a parallel shunt around the resistor. Since the resistance of a typical voltmeter is very low as compared with 10^{6} ohms, most of the current bypasses the large resistance and flows through the meter, and the voltage across the resistor, although measured correctly as it is, is no longer 1 V, but perhaps only a few millivolts.

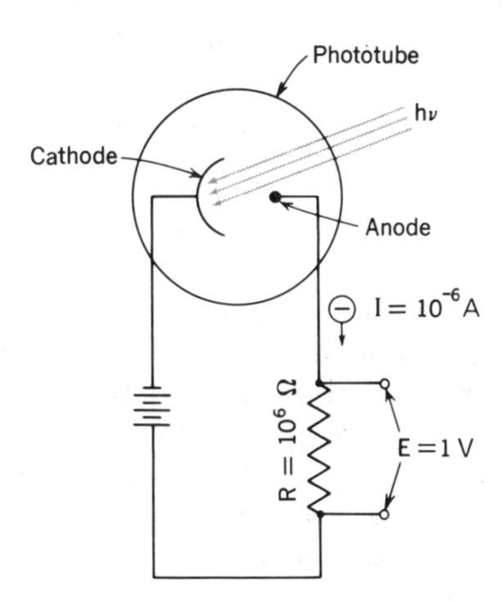

Figure 14.13 Simple phototube circuit (see text).

This problem is solved by an amplifier of high input resistance so that the phototube circuit is not drained. Rather, the voltage across the load resistor is used to *control* a circuit which draws its power from an independent source and which has an output large enough to operate a meter or other readout device.

Operation of a Single-Beam Spectrophotometer

We shall first describe the usual mode of operation of a typical manually operated, single-beam spectrophotometer, and then examine briefly some possible variations on the common procedure.

Ordinary Operation

Typically, there is an opaque shutter, controlled by the operator, which may be placed in front of the phototube so that the tube is in darkness. With this shutter in position, a small current ("dark current") flows in the phototube circuit due to thermal emission of electrons by the cathode or perhaps a small leakage in the tube. By means of a knob on the instrument, the operator cancels out the dark current and sets the scale on the instrument to read infinite absorbance (zero transmittance). Next, with the wavelength set at the desired value and a cell containing a reference solution in the beam (the reference may be the pure solvent, a "blank" from an analytical procedure, etc.), the shutter is removed to expose the detector. Now, by adjusting the radiant power to the detector by means of the monochromator slit control, and/or by changing electronically the gain of the amplifier, the instrument scale is set to read zero absorbance (100% transmittance). With a scale thus established, the sample solution is placed in the beam and its absorbance or transmittance is read off. (The scale is generally linear in transmittance, but most instruments have an absorbance scale alongside the transmittance scale and either one can be read.)

The scale, set up as described above, must be reestablished whenever the wavelength is changed, in order to compensate for the variation of source output with wavelength and the wavelength dependence of the detector response, as well as any absorption by the reference solution or the cell. It is good practice to check the dark current and reference solution settings frequently because of possible drift in the circuit and in the output of the source. Usually two cells are used, one for the reference solution and one for the samples to be measured; it is obvious that these cells should be matched with regard to path length and optical qualities.

Differential Measurements

Ordinarily, the reference solution in spectrophotometry is the pure solvent or a "blank" solution of some kind which contains little or none of the substance being determined. Using this reference and the dark current adjustment, a scale is set up as described above and shown in the upper part of Fig. 14.14. Also shown along the scale are the absorbance and transmittance values of two solutions, one an unknown which is to be measured and the other a standard solution containing a known quantity of the substance being determined.

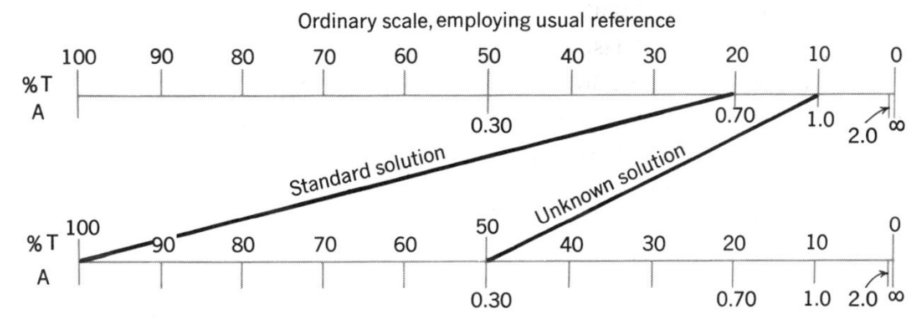

Figure 14.14 Scale expansion in differential spectrophotometer.

Now we must recognize that the instrument does not know anything about the sort of solution that is in the beam. All it is capable of is to produce a reading of zero absorbance (100% T) when a certain radiant power falls upon the detector and the amplification of the electronic circuit is appropriate. Thus the instrument can be set to read zero absorbance with a strongly absorbing solution, instead of the usual reference in the beam, by opening the slits of the monochromator and/or increasing the gain of the amplifier. Suppose, then, that we place the standard solution shown in Fig. 14.14 in the beam and set the absorbance reading at zero. As we have attempted to show in the figure by the lower scale and the lines tying it to the upper scale, we have now accomplished essentially a scale expansion. What was only a small portion of the upper scale becomes a much larger portion of the lower scale. Note that the difference in absorbances of the two solutions is 0.30 unit in each case, but 0.30 unit is a much large portion of the lower scale than the upper. Hence we may expect that a constant instrumental error will result in a smaller relative concentration error. Actually, in some cases the error can be reduced to as little as a part per thousand, and the spectrophotometric measurement, normally not this good, can compete with ordinary titrimetric and gravimetric techniques, which are usually considered more precise. Several applications of differential spectrophotometry have been given by Bastian,[8] and a detailed mathematical analysis of the technique and its errors has been developed by Hiskey.[9]

To see more clearly how the error is reduced in differential spectrophotometry, let us consider an example. Suppose that we wish to determine copper by measuring the absorbance of blue copper(II) solutions. Let us simply assume, to establish a specific basis for our discussion, that a solution with a copper concentration of 2 mg/ml (solution A) can be measured against a pure water reference with an error of 1%. Now consider the error if a solution containing 20 mg of copper/ml (solution B) were measured, using as a reference, not water, but another cupric solution containing 18 mg of copper(II)/ml (solution C). The concentration difference between B and C is the same as the difference between

[8] R. Bastian et al., *Anal. Chem.*, **21**, 972 (1949); **22**, 160 (1950); **23**, 580 (1951).

[9] C. F. Hiskey et al., *Anal. Chem.*, **21**, 1440 (1949); **22**, 1464 (1950); **23**, 506 (1951); **23**, 1196 (1951); **24**, 342 (1952).

solution A and water. Thus, if Beer's law is obeyed, the foregoing two measurements will give rise to the same absorbance value and the same error. In other words, supposing solution B were our unknown solution, we could determine how much it differed from solution C with a 1% error. But we know *accurately* the concentration of solution C, because it is a carefully prepared reference solution (by *accurately*, we mean that no spectrophotometric error is involved). Thus, so far as errors in spectrophotometry are concerned, the concentration of solution B can be determined with an error of only 0.1%. (In measuring solution A vs. water, the error is $0.02/2 \times 100 = 1\%$; if B is measured vs. C, the error is still $0.02/2 \times 100 = 1\%$, but the error in the absolute concentration of B is only $0.02/20 \times 100 = 0.1\%$.)

The differential approach not only leads to lower errors, as explained above, but also permits the extension of spectrophotometry to the analysis of solutions which would be too highly absorbing for ordinary measurements.

It is possible to achieve even greater precision by setting both ends of the scale with standard solutions in the beam. In other words, the 100% T is set as described above using a standard solution more dilute than the unknown, but the 0% T is set, not with a shutter in the beam, but with a more concentrated standard solution. Reilley and Crawford, whose definitive paper[10] should be consulted for further details, refer to this as the "method of ultimate precision."

Double-Beam Spectrophotometers

Recording spectrophotometers which automatically plot the absorbance of a sample as a function of wavelength are almost always double-beam instruments. It is far beyond the scope of this text to present a full discussion of these devices which have practically revolutionized the taking of absorption spectra for modern chemists, but it is obligatory at least to give an idea of what they do. The student should realize that, easy as they may be to operate, these instruments are extremely complicated. We shall discuss here one type of instrument, the optical null type, and that only briefly and in very general and schematic terms. Reference should be made to Fig. 14.15 throughout this discussion. The figure is not intended to represent any particular real instrument, but it is presented only to give the student an idea of the sort of thing that the instrument makers have been able to do for chemists. We have chosen to describe the optical null instrument, not because it is the best or even the most common, but because it is the easiest type for beginning students to understand.

Radiation from the source passes through a monochromator as in a single-beam instrument and encounters a chopper. The chopper, driven by a synchronous motor (not shown in the figure), is a rotating mirror of such shape and such placement that it permits the beam to pass straight through during half of its period of rotation. During the other half the beam encounters a reflective surface that turns it through a right angle, directing it upward in the figure. The direction may be changed again by other stationary mirrors as desired. Thus we now have

[10] C. N. Reilley and C. M. Crawford, *Anal. Chem.*, **27**, 716 (1955).

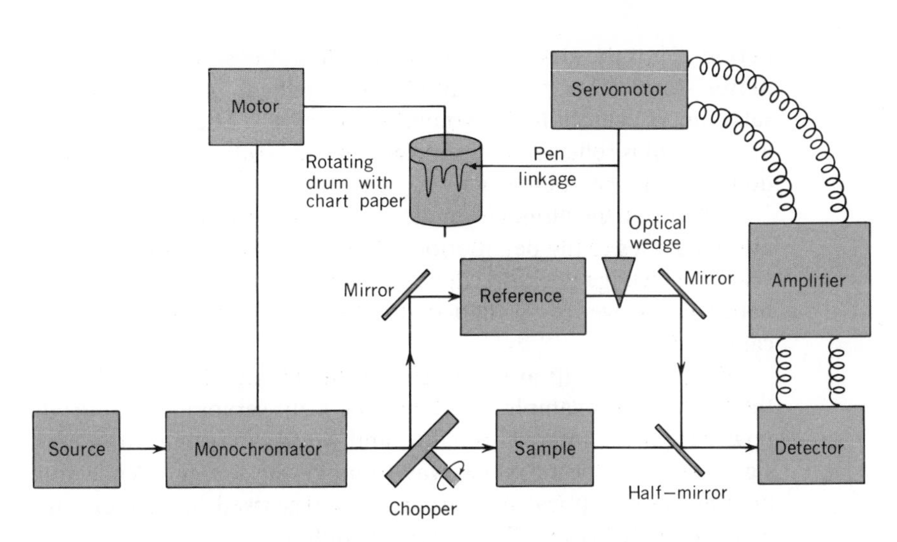

Figure 14.15 Schematic diagram of a possible double-beam optical null recording spectrophotometer.

two beams, from the same source, which are not steady but pulsate at a frequency determined by the chopper. One beam is passing while the other is blocked, and they alternate perhaps many times a second. One beam next passes through the sample while the other encounters a reference solution. The beams are then recombined so that they both fall upon a single detector.

Now as the chopper rotates, the detector "sees" first one beam, then the other. Suppose that the monochromator is set at a wavelength where the sample does not absorb. Then the sample beam and the reference beam are of equal power, and the detector sees the same thing, regardless of the position of the chopper. A steady, constant power impinges upon the detector, and the electrical output of the detector circuit is a voltage which does not vary with time, in other words, a dc voltage. Now the signal from the detector is fed to an amplifier, and the important characteristic of this amplifier is that it is an ac amplifier which is tuned to the frequency of the chopper. The dc voltage from the detector is not amplified, and the output of the amplifier is inadequate to do anything in the instrument.

Let us now change the wavelength so that the sample absorbs. Now the radiant power in the reference beam is greater than in the sample beam, and the detector, looking first at one and then the other, generates an electrical signal which reflects this pulsation in radiant power. This signal is an ac voltage, a square wave, actually, superimposed upon the dc signal mentioned above. The amplifier amplifies this ac voltage, which is then fed to a motor (a so-called "servo").

The motor drives an "optical wedge" or "variable-density wedge" into the reference beam. This wedge is a special device which blocks part of the beam and diminishes the radiant power in a smoothly progressive fashion as it moves. Now, when the wedge has moved so as to attenuate the reference beam to the same extent as the sample absorption has attenuated the sample beam, then once again the detector sees the same thing regardless of where the chopper is in its rotational period. Thus the electrical signal is again dc, the amplifier output falls off, and the motor stops. This is an example of "feedback"; the imbalance which set the servomotor in motion has "committed suicide." The position of the optical wedge at this point is reflected mechanically by the position of a pen which moves up and down on a piece of chart paper.

If we drive the monochromator with a motor which also moves the chart paper at right angles to the pen motion, we are then able to obtain a plot of wavelength vs. optical wedge position which becomes wavelength vs. transmittance or absorbance if the wedge is shaped properly and the chart paper is appropriately calibrated.

Real instruments are more complicated than we have suggested above and in the figure. For example, there may be a slit servomotor controlling the radiant power from the monochromator, and we have not indicated how an imbalance signal tells the motor which way to move the wedge. While some instruments operate on an optical null principle as discussed here, there are others which employ an electrical null that is quite different.

These instruments are extremely complicated, but they have been engineered so well that anyone can operate them by merely inserting samples and pushing buttons. Obtaining the top performance of which the instrument is capable, however, requires more knowledge than this. Most chemists must strike a compromise between the extremes of total ignorance and devoting a lifetime to the study of instrumentation. Unfortunately, the compromise in many cases falls so far toward the side of ignorance that data are obtained which are no better than a less expensive instrument could have furnished, and the instrument may even be needlessly damaged.

ERRORS IN SPECTROPHOTOMETRY

Errors in spectrophotometric measurements may arise from a host of causes, some of which have been anticipated in the discussion of instrumentation above. Many can be countered by care and common sense. Sample cells should be clean. Certain substances (e.g., proteins) sometimes adsorb very strongly in the cell and are washed out only with difficulty. Fingerprints may absorb ultraviolet radiation. The positioning of the cells in the beam must be reproducible. Gas bubbles must not be present in the optical path. The wavelength calibration of the instrument should be checked occasionally, and drift or instability in the circuit must be corrected. It must not be assumed that Beer's law holds for an untested chemical system. Sample instability may lead to errors if the measurements are not carefully timed.

The concentration of the absorbing species is of great importance in determining the error after all other controllable errors have been minimized. It is intuitively reasonable that the solution being measured should not absorb practically all of the radiation nor should it absorb hardly any. We might expect, then, that the error in a spectrophotometric determination of concentration would be minimal at some intermediate absorbance value away from the extreme ends of the scale. An expression can be derived from Beer's law which shows where this minimum error occurs.

Recall that

$$A = \log \frac{P_0}{P} = \frac{1}{2.3} \ln \frac{P_0}{P} = \varepsilon bc$$

Let the relative error in concentration be $dc/c = dA/A$. We want to obtain an expression for dc/c and then inquire where this expression has a minimum. Differentiating Beer's law, $A = (1/2.3) \ln (P_0/P)$, we obtain

$$dA = \frac{1}{2.3} d \ln \frac{P_0}{P} = \frac{(-P_0/P^2)\, dP}{2.3(P_0/P)}$$

Dividing numerator and denominator by P_0/P yields

$$dA = -\frac{(1/P)\, dP}{2.3} = -\frac{dP}{2.3P}$$

Dividing both sides by A, we obtain

$$\frac{dA}{A} = -\frac{dP}{2.3PA}$$

From Beer's law, $P = P_0 \times 10^{-A}$; substitution into the above equation gives

$$\frac{dA}{A} = \frac{dP}{2.3AP_0 \times 10^{-A}} = \frac{dc}{c}$$

It is convenient to normalize the equation by setting $P_0 = 1$, corresponding to the customary actual operation of setting the instrument to 100% T or zero absorbance with a reference solution in the beam. This gives

$$\frac{dc}{c} = -\frac{dP}{2.3A \times 10^{-A}}$$

Now the minimum in dc/c occurs when the term, $A \times 10^{-A}$, is at a maximum. To find this maximum, we differentiate and set the derivative equal to zero:

$$\frac{d(A \times 10^{-A})}{dA} = 10^{-A} - 2.3A \times 10^{-A} = 0$$

or

$$10^{-A}(1 - 2.3A) = 0$$

If 10^{-A} is zero, A is infinite, and the error is infinite. Setting the other term equal to zero yields

$$1 - 2.3A = 0$$

$$2.3A = 1$$

$$A = \frac{1}{2.3} = 0.43$$

An absorbance value of 0.43 corresponds to 36.8% transmittance. (The student who is familiar with calculus may notice that the absorbance of 0.43 could, so far as we have actually shown above, represent either a maximum or a minimum error. Such a student will know that the first derivative may be tested for this; it turns out that we are dealing with a minimum error.)

The term dP in the equations above, following the usual practice in calculus, may be taken as an approximation of ΔP, the error in P. This is often called the *photometric error*, and for our purposes simply represents the uncertainty in reading the instrument scale. This uncertainty is considered constant in the present discussion, and it is probably roughly so with so many actual instruments. To find the relative error in concentration as a function of the photometric error at the optimal concentration, substitute 0.43 for A in the preceding equation for dc/c:

$$\frac{dc}{c} = -\frac{dP}{2.3 \times 0.43 \times 10^{-0.43}} = -2.72\,dP$$

Thus, if a 1% error were made in reading the instrument, the relative error in c would be 2.72% at best; with absorbance values above and below 0.43, it would be even larger. Photometric errors may range from 0.1% to several percent, depending upon the instrument employed.

The relative error in concentration resulting from a 1% photometric error is plotted against percent transmittance in Fig. 14.16. The curve approaches infinity at both 0 and 100% T, passes through a minimum at 36.8% T, but is actually not very far from minimal over a fair range, say, 10 to 80% T (absorbance values of about 0.1 to 1.0).

The student who has difficulty in visualizing why there should be a minimum in the error curve, or in following the calculus treatment above, could think of the situation this way: Refer to Fig. 14.6(b). A given error in measuring % T gives a small error, Δc, in concentration at very low concentrations, because the % T vs. c curve is very steep. But, where c is very low, a small error in c becomes a large

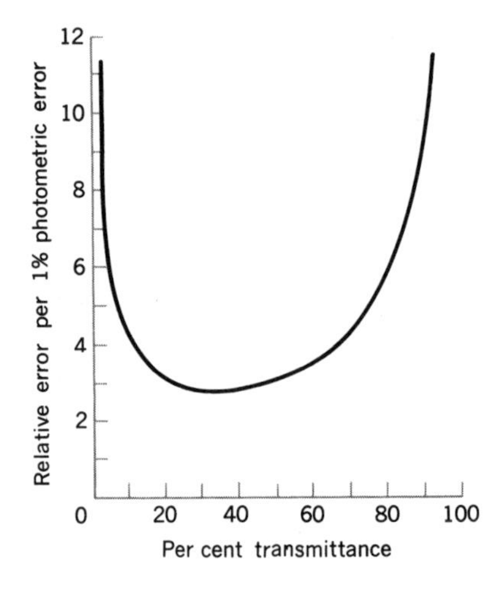

Figure 14.16 Error curve.

relative error, $\Delta c/c$. At high concentrations, the same error in % T represents a much larger absolute error in c, because the % T vs. c curve is much flatter. Somewhere in between, there will be a point where these two effects meet to give a minimal relative error, $\Delta c/c$. This happens to fall at 36.8% T.

In the error treatment above, it was supposed that the error in measuring transmittance was constant, independent of the value of the transmittance; the error was considered to arise entirely from uncertainty in reading the instrument scale. In some of the best modern instruments, on the other hand, the limiting factor in the accuracy lies elsewhere, usually in the "noise" level of the detector circuit. In such cases, dP is not constant, and a different error function is obtained which is minimal, not at 36.8% T but at a lower T value. Actually, with a complex instrument it may be difficult to decide which of several factors limits the accuracy. Thus the way in which dP varies with P may not be clear, and it may not be legitimate to calculate a % T value corresponding to minimal error.[11] With one of the well-known, high-quality, modern instruments, the Cary Model 14 recording spectrophotometer, the minimal error is said to occur at about 10% T ($A = 1$).[12]

APPLICATIONS OF SPECTROPHOTOMETRY

Plotting Spectrophotometric Data

Absorption spectra are most frequently plotted as % T vs. wavelength (λ), A or ε vs. λ, and log A or log ε vs. λ. Comparison of these plots may be made clear by reference to Figs. 14.17, 14.18, and 14.19. Analytical chemists generally

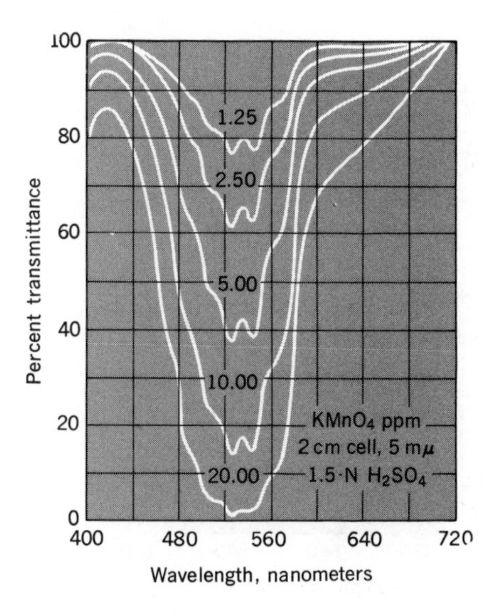

Figure 14.17 Transmittance-wavelength curves for solutions of potassium permanganate. (M. G. Mellon, ed., *Analytical Absorption Spectroscopy*, John Wiley & Sons, Inc., New York, 1950. Used by permission of the author and publisher.)

[11] R. P. Bauman, *Absorption Spectroscopy*, John Wiley & Sons, Inc., New York, 1962, p. 376.

[12] *Instructions for Cary Recording Spectrophotometer Model 14*, Applied Physics Corporation, Monrovia, Calif., p. 5.

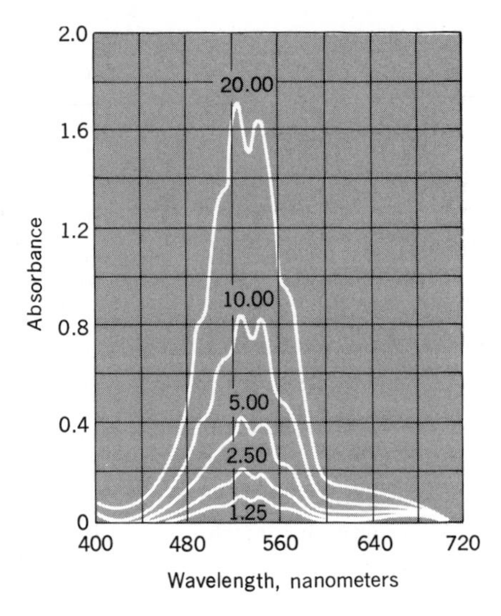

Figure 14.18 Absorbance-wavelength curves for solutions of potassium permanganate. (M. G. Mellon, ed., *Analytical Absorption Spectroscopy*, John Wiley & Sons, Inc., New York, 1950. Used by permission of the author and publisher.)

prefer absorbance to % T for the ordinate. Note that a minimum in % T corresponds to a maximum in A. The two curves are not mirror images, however, because A and % T are related logarithmically [$A = \log(1/T)$]. Sometimes ε-values are calculated from absorption data and plotted against λ.

It may be seen from Fig. 14.18 that the shape of the absorption spectrum depends upon the concentration of the solution if the ordinate is linear in absorbance. That is, the curves in Fig. 14.18 are not superimposable by simple vertical displacement. This is clear from Beer's law, $A = \varepsilon bc$, which shows that

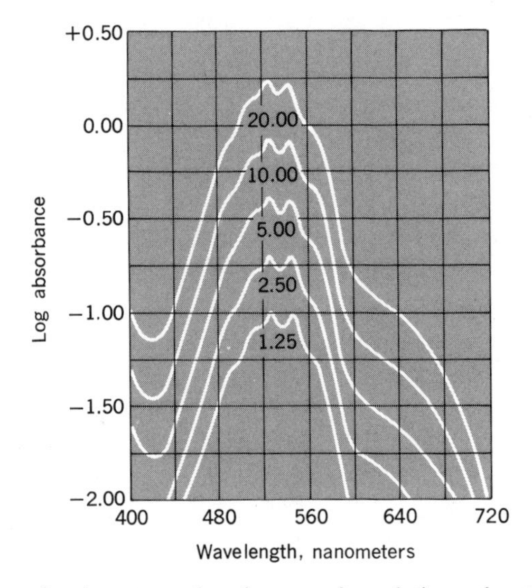

Figure 14.19 Log absorbance-wavelength curves for solutions of potassium permanganate. (M. G. Mellon, ed., *Analytical Absorption Spectroscopy*, John Wiley & Sons, Inc., New York, 1950. Used by permission of the author and publisher.)

changing the concentration changes the absorbance at each wavelength by a constant *multiple*. On the other hand, as seen in Fig. 14.19, the shape of the curve is independent of concentration if the ordinate is log A. That this should be the case is seen by taking logarithms of both sides of the Beer's law equation:

$$\log A = \log(\varepsilon bc) = \log \varepsilon + \log b + \log c$$

Now the concentration term is *added* rather than multiplied, and hence increasing the concentration adds a constant increment to log A at each wavelength across the spectrum. The curve for the higher concentration is thus displaced upward, but could be superimposed upon the lower one by a simple vertical movement. The same ε vs. λ plot should be obtained regardless of concentration provided the system follows Beer's law at all wavelengths. It is common practice, particularly among organic chemists, to plot log ε vs. λ.

Identification of Chemical Substances

The student is familiar with simple color tests which are used for identification purposes. The purple color of permanganate solutions, the blue of copper, the yellow of chromate, and many others might be mentioned. The absorption spectrum of a compound, determined with a spectrophotometer, may be considered as a more elegant, objective, and reliable indication of identity. The spectrum is another physical constant, so to speak, which, along with melting point, refractive index, and other properties, may be used for characterization. Like the others, absorption spectra are not infallible proof of identity, but simply represent another tool available for intelligent application.

It must be remembered that an absorption spectrum depends upon not only the chemical nature of the compound in question but also other factors. Changing the solvent often results in shifts in absorption bands. The shape of a band and particularly the appearance of "fine structure" may well depend upon instrument characteristics such as the resolution of the monochromator, the amplifier gain, and the rate of scan as it relates to inertia in the recorder. Treating a recording spectrophotometer as a "black box" can lead to peculiar absorption spectra.

Spectra of many thousands of compounds and materials have been recorded, and locating the proper ones for comparison in connection with a particular problem may be extremely difficult. Several catalogs and compilations are available.[13] Increasingly, large laboratories are employing machine data-handling techniques to store and retrieve spectra as well as other important information.

Multicomponent Analysis

A spectrophotometer cannot *analyze* a sample. It becomes a useful tool only after the sample has been treated in such a way that the measurement is

[13] For example, H. M. Hershenson, *Ultraviolet and Visible Absorption Spectra: Indexes for 1930–1954 and 1955–1959,* Academic Press, Inc., New York, 1956 and 1961; R. A. Friedel and M. Orchin, *Ultraviolet Spectra of Aromatic Compounds,* John Wiley & Sons, Inc., New York, 1951.

interpretable in unambiguous terms. In many cases, however, it is not necessary that each individual component of a complex sample be isolated from all others. In spectrophotometry, for example, it is sometimes possible to measure more than one constituent in a single solution. Let us suppose a solution to contain two absorbing constituents, X and Y. The complexity of the situation depends upon the absorption spectra of X and Y.

Case 1

The spectra do not overlap, or at least it is possible to find a wavelength where X absorbs and Y does not and a similar wavelength for measuring Y. Figure 14.20 shows such a situation. The constituents X and Y are simply measured at the wavelengths λ_1 and λ_2, respectively.

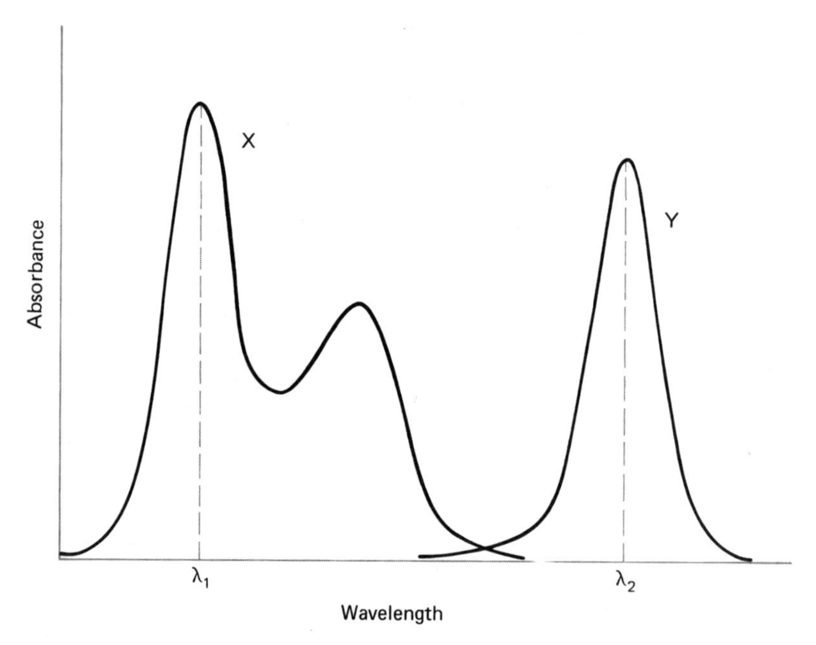

Figure 14.20 Absorption spectra of compounds X and Y. (There is no overlap at the two wavelengths which are to be employed.)

Case 2

One-way overlap of the spectra: As shown in Fig. 14.21, Y does not interfere with the measurement of X at λ_1, but X does absorb appreciably along with Y at λ_2. The approach to this problem is simple in principle. The concentration of X is determined directly from the absorbance of the solution at λ_1. Then the absorbance contributed at λ_2 by this concentration of X is calculated from the previously known molar absorptivity of X at λ_2. This contribution is subtracted from the measured absorbance of the solution at λ_2, yielding the absorbance due to Y, whose concentration is then calculated in the usual manner.

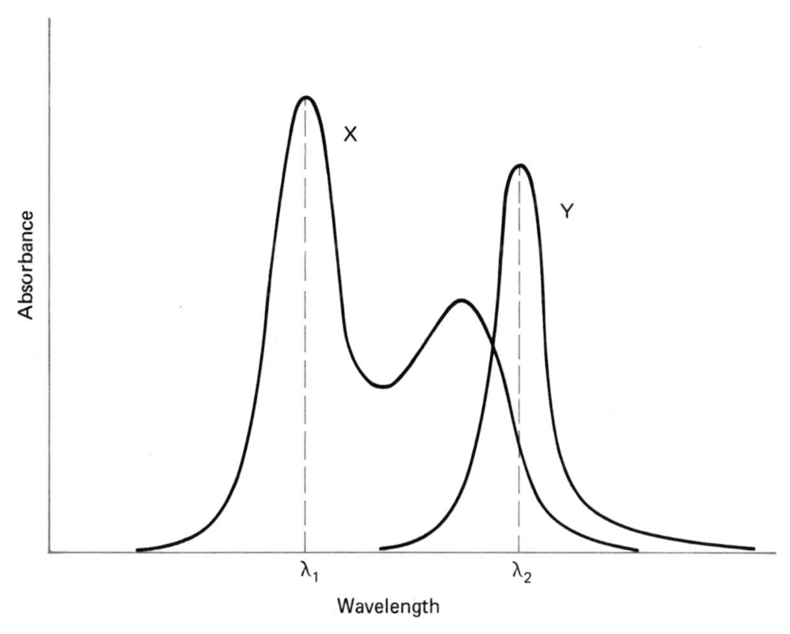

Figure 14.21 Absorption spectra of compounds X and Y. (One-way overlap: X can be measured with no interference from Y, but X interferes with the direct measurement of Y.)

Case 3

Two-way overlap of the spectra: When no wavelength can be found where either X or Y absorbs exclusively, as suggested in Fig. 14.22, it is necessary to solve two simultaneous equations in two unknowns. Let

$$A_1 = \text{measured absorbance at } \lambda_1$$

$$A_2 = \text{measured absorbance at } \lambda_2$$

$$\varepsilon_{X_1} = \text{molar absorptivity of X at } \lambda_1$$

$$\varepsilon_{X_2} = \text{molar absorptivity of X at } \lambda_2$$

$$\varepsilon_{Y_1} = \text{molar absorptivity of Y at } \lambda_1$$

$$\varepsilon_{Y_2} = \text{molar absorptivity of Y at } \lambda_2$$

$$C_X = \text{molar concentration of X}$$

$$C_Y = \text{molar concentration of Y}$$

$$b = \text{path length}$$

Since the total absorbance is the sum of the contributions of the individual absorbing constituents of the solution:

$$A_1 = \varepsilon_{X_1} b C_X + \varepsilon_{Y_1} b C_Y$$

$$A_2 = \varepsilon_{X_2} b C_X + \varepsilon_{Y_2} b C_Y$$

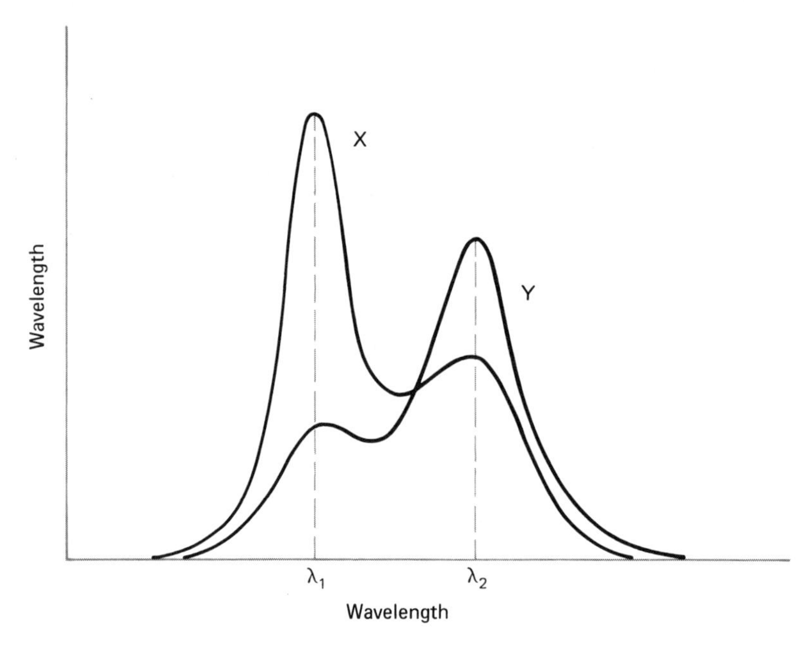

Figure 14.22 Absorption spectra of compounds X and Y. (Two-way overlap: there is no wavelength where either compound can be measured without interference by the other.)

C_X and C_Y are the only unknowns in these equations and hence their values can be readily determined. The ε values must be known, of course, from measurements on pure solutions of X and Y at the two wavelengths.

Equations can be set up in principle for any number of components provided that absorbance values are measured at as many wavelengths. However, the importance of small errors in measurement is magnified as the number of components increases, and in practice this approach is generally limited to two- or possibly three-component systems. An exception to this is possible if a computer is available. Then, particularly if the spectrum is recorded, it becomes not too difficult to "overdetermine" the system (i.e., take absorbance values at many more wavelengths than there are components) and by a rapid series of successive approximations obtain reliable values for a large number of components.

Preparation of Samples for Spectrophotometric Analysis

So far we have said little about chemistry in this chapter. But seldom will the analyst receive a sample which is ready to be measured without some sort of pretreatment. Often separations of interfering substances are necessary; some of the available techniques are considered in later chapters.

Many organic compounds absorb in the ultraviolet region of the spectrum, and pretreatment then involves only separation of interferences. Some elements in the periodic table absorb strongly in the visible or ultraviolet, at least in certain oxidation states, and the preliminary steps may involve redox reactions as well as separations. Manganese, for example, if often determined spectrophotometrically after oxidation to Mn(VII) by means of persulfate or periodate:

$$2Mn^{2+} + 5S_2O_8^{2-} + 8H_2O \longrightarrow 2MnO_4^- + 10SO_4^{2-} + 16H^+$$

The purple MnO_4^- solution is measured at about 525 nm. Chromium is determined similarly after oxidation to Cr(VI).

Development of absorption by means of inorganic reagents is occasionally possible. For example, iron may be determined by means of the red color obtained by treating iron(III) solutions with thiocyanate:

$$Fe^{3+} + SCN^- \longrightarrow (FeSCN)^{2+}$$

The system is complicated by the tendency to form higher complexes such as $[Fe(SCN)_2]^+$. Other examples of colored complexes formed with inorganic reagents are the blue tetraammine copper complex, $[Cu(NH_3)_4]^{2+}$, and the several complex heteropoly acids such as phosphomolybdic, $H_3P(Mo_3O_{10})_4 \cdot 29H_2O$, which are used to determine elements such as phosphorus and silicon. Iodide ion forms yellowish complex ions which exhibit absorption maxima in the ultraviolet region with several metals, including bismuth, antimony, and palladium (e.g., PdI_4^{2-}).

The colored complexes formed by metal ions with organic reagents offer the most impressive variety of spectrophotometric methods, and they are especially useful in the field of trace analysis. Most of these complexes are of the chelate type which were discussed more fully elsewhere (pages 87 and 189). We mention here only a few points of special interest regarding their adaptation to spectrophotometric analysis.

In some regards, the low aqueous solubility of many of the metal chelate compounds is disadvantageous, but on the other hand, extraction of metals into nonaqueous solvents by means of chelating agents may lead to very powerful analytical methods. In favorable cases, it may be possible to concentrate the metal, separate it from interferences, and develop the absorbing system in a single step. For example, the chelates of 8-hydroxyquinoline (oxine, page 88) with such metals as aluminium, iron(III) ion, cadmium, gallium, lead, and copper are soluble in chloroform, and extractions with this solvent generally precede the spectrophotometric determination. By controlling the *p*H of the aqueous phase and by adding complexing agents which mask certain metal ions, the extraction can be made quite selective. Reasonable absorbance values are generally obtained with chloroform solutions whose metal concentrations are of the order of a few micrograms per milliliter.[14]

The solvent used in spectrophotometric procedures poses a problem in some regions of the spectrum. The solvent must not only dissolve the sample but also must not absorb appreciably in the region in which the determination is made. Water is an excellent solvent in that it is transparent throughout the visible region

[14] See E. B. Sandell, *Colorimetric Determination of Traces of Metals*, 3rd ed., Interscience Publishers, Inc., New York, 1959, for spectrophotometric methods for most inorganic ions.

and down to a wavelength of about 200 nm in the ultraviolet. However, since water is a poor solvent for many organic compounds, organic solvents are commonly employed for these substances. The transparency cutoff points in the ultraviolet region of a number of solvents are listed in Table 14.4. Aliphatic hydrocarbons, methanol, ethanol, and diethyl ether are transparent to ultraviolet radiation and are frequently employed as solvents for organic compounds.

TABLE 14.4 Solvents for Ultraviolet and Visible Regions

Solvent	Approximate transparency minimum, nm	Solvent	Approximate transparency minimum, nm
Water	190	Chloroform	250
Methanol	210	Carbon tetrachloride	265
Cyclohexane	210	Benzene	280
Hexane	210	Toluene	285
Diethyl ether	220	Pyridine	305
p-Dioxane	220	Acetone	330
Ethanol	220	Carbon disulfide	380

There is no single solvent which is transparent throughout the infrared region. Carbon tetrachloride is useful up to 7.6 μm, and carbon disulfide up to 15 μm. Not all substances are soluble in these solvents, however, and other liquids with more restricted ranges must be used. Water exhibits strong absorption in the infrared and must be avoided. Rock-salt cells are frequently employed in the infrared region, and water would dissolve the salt.

Because of the problems encountered in working with solutions in the infrared, various other techniques have been employed. Liquids can be measured directly, using a very thin film placed between rock-salt plates. Another technique employs a liquid *mull*, made by dispersing the sample in a viscous hydrocarbon, such as the mineral oil Nujol. Still another employs solid potassium bromide which is highly transparent to infrared radiation. The finely ground sample is mixed homogeneously with potassium bromide and pressed into a small disk. The disk can be placed directly in the beam of a spectrophotometer and absorption measurements made.

PHOTOMETRIC TITRATIONS

Various properties of a solution may be measured in order to assess the progress of a titration toward the equivalence point. We have seen, for example, in Chapter 12 that the potential of an indicator electrode may be used for this purpose, and we have described another end-point technique, amperometric, in Chapter 13. The absorbance of a solution may likewise be measured during the course of a titration; we have available, then, still another end-point technique which may be useful in certain circumstances. Our discussion of photometric titrations will be brief; a more complete treatment may be found in two reviews which also provide a guide to the literature.[15,16]

As a matter of fact, visual titrations are really photometric in nature. "The color change reflects a change in the absorption of light by the solution, accompanying changes in the concentrations of absorptive species. In a visual titration, one actually employs all the features of an automatic photometric titrator: Light passes through the solution to the eye, which is a photosensitive transducer responding with a signal to the brain. The brain is analogous to the circuitry of an instrument which amplifies the signal and otherwise renders it appropriate for transmission to an electromechanical shutoff system; traversing a motor neuron, the signal triggers a muscular response that closes the buret to terminate the titration. In visual titrations, the most complicated and expensive instruments of all—people—act as automatic photometric titrators."[16]

Photometric titrations often possess advantages of sensitivity of end-point detection and circumvention of interferences over visual titrations. Further, they are not restricted to the wavelength region where the human eye responds, and they are fairly easily automated. In comparison with potentiometric titrations, the photometric approach is often advantageous for borderline cases of titrations which are approaching nonfeasibility. While the potential of an indicator electrode responds to the logarithm of a concentration (or a concentration ratio), the absorbance of a solution is directly proportional to concentration. If the equilibrium constant of a titration reaction is undesirably small, concentrations will not change as rapidly as one would like in the vicinity of the equivalence point, and the potential, because of the logarithmic compression, will change even less rapidly. The absorbance, on the other hand, will change just as rapidly as the concentration. This is not a unique advantage of photometric titrations, of course. The same consideration applies to amperometric and other end points which are obtained by linear extrapolation.

Just as with amperometric titrations (Chapter 13, page 362), it is necessary to correct measured absorbance values for dilution if the volume of titrant is appreciable compared with the initial volume of the solution.

Titrations without Indicators

Sometimes a substance directly involved in the titration reaction absorbs appreciably at an accessible wavelength, and the titration can be followed spectrophotometrically without adding an indicator. The shapes of the titration curve are predictable from the ε values of the chemical species concerned. Some typical photometric titration curves of this type are shown in Fig. 14.23. If the titration reaction is appreciably incomplete in the vicinity of the equivalence point, the curve will become rounded, as shown by the dashed portion of curve A

[15] J. B. Headridge, *Photometric Titrations*, Pergamon Press, Inc., New York, 1961.

[16] A. L. Underwood, "Photometric Titrations," a chapter in *Advances in Analytical Chemistry and Instrumentation*, Vol. 3, C. N. Reilley, ed., Wiley–Interscience, New York, 1964, p. 31.

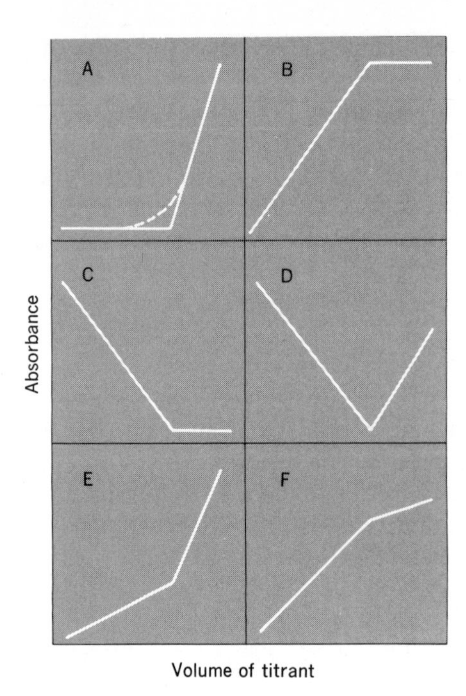

Figure 14.23 Typical photometric titration curves: (A) $\varepsilon_t > \varepsilon_s = \varepsilon_p$ (usually $\varepsilon_s = \varepsilon_p = 0$); (B) $\varepsilon_p > \varepsilon_s$, $\varepsilon_t = 0$; (C) $\varepsilon_s > \varepsilon_p$; $\varepsilon_t = 0$; $(D) \varepsilon_s > \varepsilon_p$, $\varepsilon_t > 0$; (E) $\varepsilon_p > \varepsilon_s$; $\varepsilon_t > \varepsilon_p$; (F) $\varepsilon_p > \varepsilon_s$, $\varepsilon_t < \varepsilon_p$. ε_s, ε_p, and ε_t are the molar absorptivities of the substance titrated, reaction product, and titrant, respectively.

in Fig. 14.23. The end point is then located by the intersection of extrapolated straight lines drawn through points taken sufficiently before and after the rounded portion. Titration curves of this sort are easily calculated: One simply computes the concentrations of absorbing species at any point using the equilibrium constant of the reaction; then the contribution of each species to the absorbance of the solution is calculated from Beer's law, using known ε values and path length.

Titrations with Indicators and the Titration of Mixtures

In cases where none of the species involved in the titration reaction absorbs sufficiently, an indicator may be added to the solution. Figure 14.24 shows an example of an indicator titration, a chelometric titration of copper(II) ion with EDTA using the metallochromic indicator pyrocatechol violet. In this titration, a wavelength was selected where the free indicator absorbs more strongly than the copper-indicator complex. We see in the figure, first the reaction of free copper(II) ion with EDTA, which does not affect the absorbance of the solution at this wavelength. Then, as the end point is approached, copper is pulled away from the indicator by the titrant, and the absorbance rises as free indicator accumulates, until finally all of the copper has been titrated and the absorbance becomes constant again.

Figure 14.25 shows a photometric titration of a mixture of bismuth and copper with EDTA. At the wavelength selected, the copper(II)–EDTA chelate absorbs strongly, while the other species (Bi^{3+}, bismuth–EDTA chelate, and EDTA) have ε-values of zero. The bismuth chelate is much more stable than the copper(II)

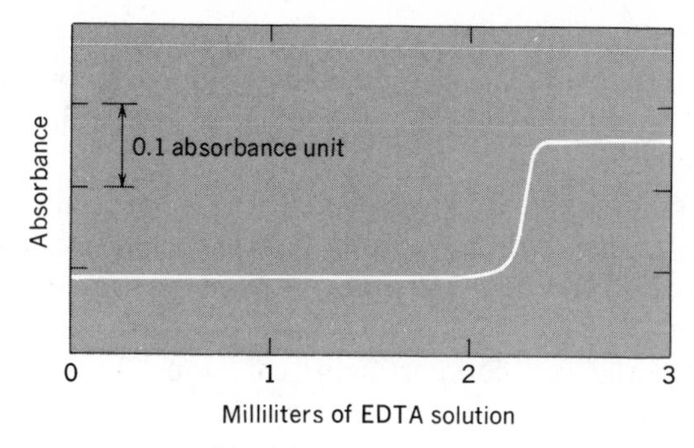

Figure 14.24 Titration of copper with EDTA, using pyrocatechol violet indicator: 14.01 mg of copper in 150 ml titrated with 10^{-1}-*M* EDTA at 440 nm (Data of T. M. Robertson.)

one. Thus, as EDTA is added to the Bi^{3+}–Cu^{2+} mixture, the bismuth chelate is formed first. When $[Bi^{3+}]$ has been reduced to a very low value, the copper(II)–EDTA chelate begins to form, and, because this is the strongly absorbing species, the absorbance begins to rise. After the copper end point, the curve levels off as excess, nonabsorbing EDTA is added.

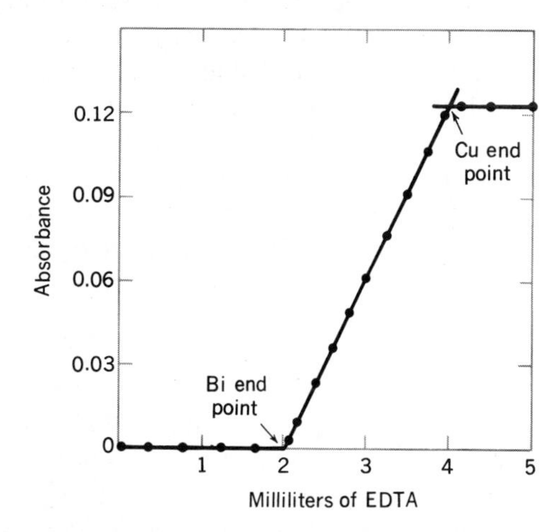

Figure 14.25 Titration of bismuth-copper mixture with EDTA; 41.8 mg of bismuth and 13.1 mg of copper in 100 ml (each 2×10^{-3}-*M*) buffered at *p*H 2, titrated at 745 nm with 10^{-1}-*M* EDTA.

Instrumentation

The simplest approach to a photometric titration is to titrate in a flask or beaker on the laboratory bench, taking samples out of the titration vessel for absorbance measurements as the titration proceeds. Of course, the samples must be returned each time, and this technique is inconvenient for more than an occasional single titration. On the other hand, any good spectrophotometer can be used without modification, and it is more obvious to a beginner exactly what is going on than might be the case if more elaborate instrumentation were employed.

Sometimes it is possible to fit a spectrophotometer with a modified cell compartment so that a titration vessel such as a beaker can be positioned in the light beam. It is convenient to stir by means of a magnetic stirrer underneath the compartment. The buret tip is introduced into the solution through a hole in the cover of the cell compartment; care must be taken that this arrangement be light-tight.

With a recording spectrophotometer, it is possible to record absorbance vs. time at a constant wavelength. If titrant is introduced into a titration vessel in the sample beam at a constant flow rate, if adequate stirring is provided, and also if the titration reaction is rapid, then the plot of absorbance vs. time readily becomes a photometric titration curve. With relatively simple on-the-spot modification, the output signal of a manual spectrophotometer can be recorded; thus an expensive double beam recording instrument is not really required for this application.

Finally, photometric titrators which terminate titrant flow at the end point are on the market. The operator merely sets up and starts the titration, and then later reads a buret. We described briefly in Chapter 12 an automatic potentiometric titrator based upon double electronic differentiation of the voltage from a pair of electrodes. Now the first derivatives of photometric titration curves like those in Fig. 14.23 exhibit the sigmoid shape associated with a typical potentiometric titration. By differentiating electronically a signal arising from a photodetector and then feeding this into the circuit of the existing potentiometric titrator, an instrument was devised for automatic photometric titration. Of course, it was necessary to add to the potentiometric titrator not only an additional differentiating circuit but also optical components, source, photodetector, etc.[17]

FLUORESCENCE

As mentioned early in this chapter (page 374), once absorption of radiation has occurred, the energy may be reemitted as radiation usually of longer wavelength than was originally absorbed. This phenomenon, known as *fluorescence*, can be used for the analysis of many inorganic and organic species. It offers advantages in its high sensitivity and specificity, and it has found many applications in the field of biochemistry.

Principles

The absorption process which leads to fluorescence usually involves a π-π^* transition (page 376) in a complex organic molecule. The process is represented schematically in Fig. 14.26. The molecule absorbs radiation, labeled $h\nu_{ex}$, raising

[17] H. V. Malmstadt and C. B. Roberts, *Anal. Chem.*, **28**, 1408 (1956).

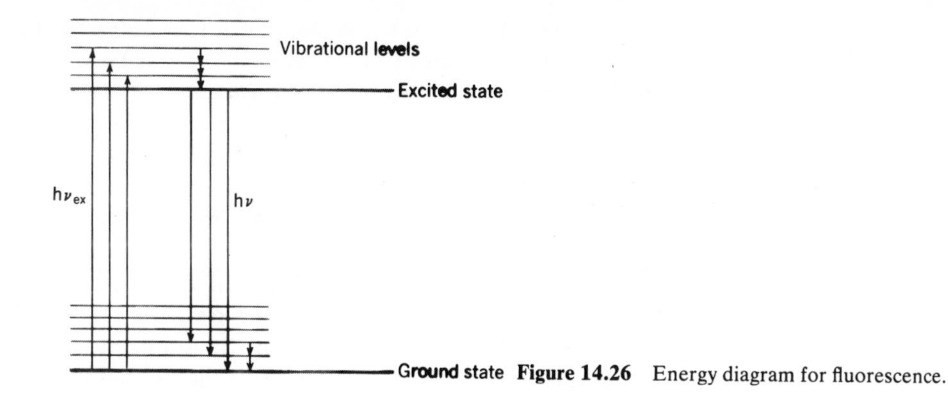

Figure 14.26 Energy diagram for fluorescence.

the electron from the ground state to an excited state. In this higher-energy level, the molecule can be in one of several vibrational states. The molecule loses vibrational energy through collisions with solvent molecules, falling to the lowest vibrational level in the excited state. In this lowest level the probability of return to the ground state by emission of a photon is greatest. Fluorescence always involves a transition from the lowest vibrational level of the excited state; the electron can return to any of several vibrational levels of the ground state. In the ground state, the molecule loses energy and returns to the lowest vibrational level by collisions with solvent molecules. Both the absorption and emission of radiation occurs over a band of wavelengths, just as in ultraviolet and visible spectrophotometry.

Generally, the exciting radiation is in the ultraviolet region; the emitted radiation is usually in the visible region, although it may be in the ultraviolet region. The lifetime in the excited state is about 10^{-8} s and emission of fluorescent radiation occurs with practically no time delay. The ground and excited states are "singlets," meaning that the electron pair in the molecular orbital have paired spins. In certain cases the electron in the excited state can cross over to a "triplet" state (electron spins not paired) before the transition back to the ground state occurs. In this case the transition is much less probable than before, and the lifetime of the excited triplet state may be of the order of 10^{-4} to several seconds. Emission of radiation may persist for some time after the exciting radiation is discontinued, and the phenomenon is then called *phosphorescence*.

The molecular structure of a compound strongly influences its fluorescent behavior. Aromatic and heterocyclic compounds, particularly those which contain electron-donating groups, such as —OH and —NH_2 tend to fluoresce strongly. Other groups, such as —Br, —I, —NO_2, and —COOH, tend to inhibit fluorescence. It is thought that heavy atoms such as bromine and iodine cause a greater chance of a singlet excited state changing over to a triplet, thus decreasing the probability of fluorescence occurring.

If the concentration of the fluorescent compound is small, the power (intensity), F, of the fluorescent light is directly proportional to concentration, c,

$$F = kc$$

A plot of the fluorescent power vs. concentration gives a straight line if the concentration is sufficiently low that the absorbance ($A = abc$) is less than about 0.01 to 0.05. The linear relationship usually holds for concentrations up to a few parts per million, depending upon the emitting species. At higher concentrations the fluorescent power lies below an extrapolation of the straight-line plot. One reason for this behavior is that at high concentrations the distribution of exciting radiation is not uniform. The first layers of solution may absorb a much larger fraction of the radiation than layers at greater depth in the solution. In dilute solutions, where, for example, at least 90% of the radiation is transmitted, the distribution of the absorbed light is much more uniform.

Two other factors contribute to the negative deviation from the linear plot. One is called *self-quenching*; at high concentrations the excited molecules may lose energy by collisions among themselves, rather than by fluorescence. The other is *self-absorption*; if the fluorescent radiation is absorbed by a molecule in its ground state, the intensity of the fluorescent light is decreased as it passes through the solution.

Measurement of Fluorescence

A number of instruments have been described for measuring fluorescence. All of these operate on the same principle; they differ in types of components, degree of sophistication, and performance characteristics. The basic design is shown in Fig. 14.27. The source is usually a mercury vapor lamp or a xenon arc lamp. The mercury vapor lamp produces an intense line spectrum. Since most fluorescent molecules absorb a variety of wavelengths of ultraviolet light, a line source is often suitable. The xenon arc lamp produces a continuous spectrum over the range of about 250 to 600 nm, with the peak intensity at about 470 nm.

The ultraviolet light from the source passes through a filter or monochromator to allow selection of the wavelength for excitation. Instruments called *fluorometers* employ filters at this point as well as in the fluorescent beam.

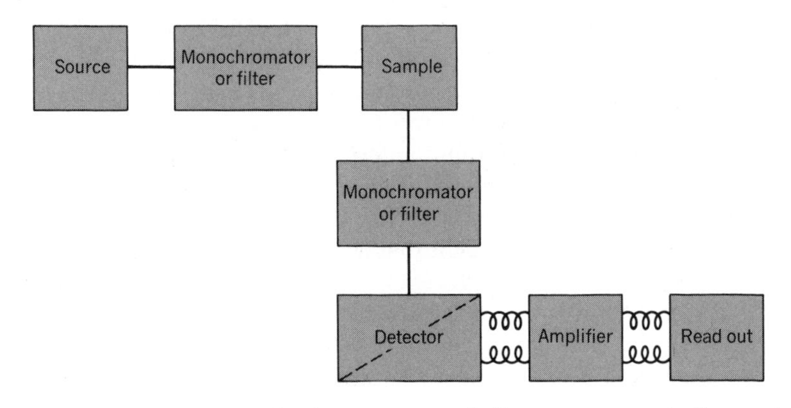

Figure 14.27 Block diagram showing components of a fluorometer or spectrofluorometer.

Spectrofluorometers employ a monochromator, usually a grating, to select the desired wavelength.

The exciting radiation then enters the sample solution which is held in a glass of quartz cell. Fluorescent radiation is emitted in all directions and must be separated from the incident radiation for measurement. This is done most conveniently by measuring the fluorescence at right angles to the incident beam. A spectrofluorometer contains a second monochromator which permits the selection of the wavelength of maximum emission.

The fluorescent radiation next reaches a detector. Since the fluorescent signal is of low intensity, it must be amplified. Usually a photomultiplier tube (page 390) is used, and its output is further amplified with an external electronic amplifier. Finally, some readout device is employed, generally either a voltmeter or a pen-and-ink recorder.

Applications

Fluorescent procedures have been used successfully for the determination of both organic and inorganic substances and have proved especially useful for biological systems. The fluorometric method is inherently more sensitive than methods based on the absorption of radiation. The reason for this is the fact that the power of the fluorescent light, F, can be measured independently of the power of the incident radiation, P_0. The sensitivity of a fluorescent measurement can be increased by increasing P_0, or by amplifying the signal produced by the fluorescent radiation. In spectrophotometry the absorbance is dependent on the ratio of P_0 and P. Increasing P_0 increases P proportionately and does not affect the absorbance. Amplification of the detector signal also affects P_0 and P and does not affect the absorbance. Determinations at the part per billion level are fairly common with fluorescent methods.

Inorganic substances can be measured by converting them to species which are fluorescent. For example, the chelating agent, 8-hydroxyquinoline (page 87), forms fluorescent compounds with metals such as beryllium, aluminum, and zinc and has been used in their determinations. Inorganic anions have been determined by their "quenching" effect, that is, their ability to decrease fluorescence. Iodide ion is an effective quencher, apparently competing with the fluorescent material for the excitation energy.

There are many applications of fluorometric methods to organic compounds, including such substances as enzymes, vitamins, and steroids. A summary of over 100 different substances has been given by Weissler and White.[18]

A number of factors other than concentration may affect the intensity of fluorescent light. Among these are pH, temperature, and the concentration of foreign ions. Calibration curves are normally prepared using the same conditions which are employed in the determination of an unknown sample.

[18] A. Weissler and C. E. White in L. Meites, ed., *Handbook of Analytical Chemistry*, McGraw-Hill Book Company, New York, 1963.

REFERENCES

R. P. BAUMAN, *Absorption Spectroscopy*, John Wiley & Sons, Inc., New York, 1962.

J. R. DYER, *Applications of Absorption Spectroscopy of Organic Compounds*, Prentice-Hall, Inc., Englewood Cliffs, N.J., 1965.

G. G. GUILBAULT, *Practical Fluorescence*, Marcel Dekker, Inc., New York, 1973.

D. M. HERCULES, ed., *Fluorescence and Phosphorescence Analysis*, Wiley–Interscience, New York, 1966.

E. J. MEEHAN, "Optical Methods: Emission and Absorption of Radiant Energy," Chap. 53; "Fundamentals of Spectrophotometry," Chap. 54; and "Spectroscopic Apparatus and Measurements," Chap. 55, of I. M. Kolthoff and P. J. Elving, eds., *Treatise on Analytical Chemistry*, Part I, Vol. 5, Wiley–Interscience, New York, 1964.

A. A. SCHILT and B. JASELSKIS, "Ultraviolet and Visible Spectrophotometry," Chap. 58 of I. M. Kolthoff and P. J. Elving, eds., *Treatise on Analytical Chemistry*, Part I, Vol. 5, Wiley–Interscience, New York, 1964.

L. A. SMITH, "Infrared Spectroscopy," Chap. 66 of I. M. Kolthoff and P. J. Elving, eds., *Treatise on Analytical Chemistry*, Part I, Vol. 6, Wiley–Interscience, New York, 1965.

S. UDENFRIEND, *Fluorescence Assay in Biology and Medicine*, Academic Press, Inc., New York, 1962.

C. E. WHITE and R. J. ARGAUER, *Fluorescence Analysis: A Practical Approach*, Marcel Dekker, Inc., New York, 1970.

QUESTIONS

1. *Terms.* Explain clearly the meaning of the following terms: (a) complementary color; (b) chromophore; (c) isosbestic point; (d) fine structure; (e) wave number; (f) nanometer; (g) n-π^* transition.

2. *Beer's law.* A plot of absorbance vs. concentration at constant b gives a straight line if Beer's law is obeyed. (a) What property of the line is represented by the absorptivity? (b) Why are photometric measurements of concentration usually carried out at the wavelength where the absorptivity is a maximum?

3. *Beer's law.* Distinguish between real and apparent deviations from Beer's law. Under what circumstances does one get positive deviations from Beer's law in the case of a weak acid?

4. *Errors.* (a) Explain how a differential measurement reduces the error in spectrophotometry. (b) Criticize the statement "A substance can be determined spectrophotometrically in the range 0 to 15 ppm with an accuracy of ±0.08 ppm."

5. *Fluorescence.* (a) What are the advantages of the fluorescent method for analytical purposes? (b) Why does a plot of fluorescent power vs. concentration vary from linearity at high concentrations of solute?

Multiple-choice: In the following multiple-choice questions, select the *one best* answer.

6. Which of the following relations is *not* correct? (a) $A = abc_{\mathrm{g/liter}}$; (b) $A = \varepsilon bc_{\mathrm{m/liter}}$; (c) $\varepsilon = a \times \mathrm{MW}$; (d) $A = \log P/P_0$; (e) $A = \log 1/T$.

7. The term $A \times 10^{-A}$ has a maximum value when A has a value of (a) 2.72; (b) 2.30; (c) 1.00; (d) 0.43; (e) 0.00.

8. Which of the following statements about monochromators is correct? (a) Prisms disperse shorter wavelengths better than longer wavelengths; (b) Gratings disperse all wavelengths uniformly; (c) For the ultraviolet region a prism should be made of quartz; (d) all of the above.

9. The detector used in infrared spectrophotometers is (a) a phototube; (b) photographic film; (c) a photomultiplier tube; (d) a thermocouple.

10. The use of beam chopping and tuned amplifiers in double-beam, optical null spectrophotometers (a) largely eliminates problems caused by fluctuations in the output of the source; (b) permits the operator to obtain spectra of two samples at the same time; (c) eliminates the need for reference solutions or solvent blanks; (d) makes it possible to record absorbance as a function of wavelength.

PROBLEMS

1. *Absorbance.* The scale on a spectrophotometer extends from 1 to 100% transmittance. (a) What are the values of the absorbances at these two extremes? (b) If the percent transmittance is actually zero, what is the value of the absorbance?

2. *Absorbance.* If the absorbance were defined in terms of natural rather than common logarithms, what would be the values of the absorbance in part (a) of Problem 1? What would be the absorbance of a solution which has a transmittance of 36.8%?

3. *Absorbance.* Convert the following values of percent transmittance to absorbance: (a) 98; (b) 80; (c) 50; (d) 10; (e) 1.0.

4. *Transmittance.* Convert the following values of absorbance to percent transmittance: (a) 0.03; (b) 0.10; (c) 0.50; (d) 1.00; (e) 1.70; (f) 2.00.

5. *Transmittance.* The percent transmittance of a solution in a 2.0-cm cell is 50. Calculate the percent transmittance of this solution in cells of the following lengths: (a) 4.00 cm; (b) 1.0 cm; (c) 0.20 cm.

6. *Absorptivity.* The absorbance of a solution containing 2.0×10^{-3} g/liter of a solute, MW 180 g/mol, in a 2.0-cm cell is 0.800. Calculate (a) the absorptivity; (b) the molar absorptivity.

7. *Molar absorptivity.* A solution contains 3.0 mg of Fe^{2+}/liter. The Fe^{2+} is converted into a complex with 1,10-phenanthroline and the absorbance of the solution in a 1.0-cm cell is 0.60. The MW of the complex is 596. Calculate (a) the absorptivity; (b) the molar absorptivity of the complex.

8. *Absorbance.* 5.00 ml of a permanganate solution of unknown concentration has an absorbance of 0.356 at a certain wavelength. Then 1.00 ml of a standard permanganate solution containing 5.00 μg of Mn/ml is added to the first solution and the absorbance becomes 0.333. Calculate the concentration of the unknown permanganate solution in μg of Mn/ml.

9. *Dilution.* A compound of MW 150 has a molar absorptivity of 4.0×10^5. How many grams of this compound should be dissolved in exactly 1 liter of solution so that after a

200-fold dilution the resulting solution will have an absorbance of 0.60 in a 1.0-cm cell?

10. *Dilution.* A substance S has a molar absorptivity of 1.0×10^5. A chemist wishes to dilute 1.00 ml of a 0.0100 M solution of S so that the resulting solution will have an absorbance of 0.200 in a 1.0-cm cell. To what volume (ml) should he dilute the 1.00 ml of S?

11. *Dilution.* Water quality engineers are often interested in the dilution of polluting materials as they flow downstream in a river. An engineer added a 1.00-liter sample of a dye to a river at the sewage disposal plant just below a large city. When the "spill" reached a point 100 miles downstream, he took a 1.00-liter sample from the river, filtered, it, and measured the absorbance due to the dye to be 0.128. In the laboratory he diluted a 1.00-ml sample of the original dye solution to 1.00 liter, diluted 1.00 ml of the resulting solution to 1.00 liter, and made one further 1000-fold dilution of the dye. The absorbance of the final solution was 1.024. Calculate the degree of dilution, expressed as final volume to original volume, of the dye as it moved downstream.

12. *Blood volume.* The blood volume of a man can be measured by injecting a known amount of a harmless dye into a vein and determining the concentration of the dye after it has been well mixed by circulation. The blood volume is obtained by dividing the plasma volume by the fractional volume of plasma in the blood.

In a certain determination, 1.00 ml of Evans blue is injected into a 75-kg man. Ten minutes later a blood sample is withdrawn from the man. The blood is centrifuged to separate the plasma from the cells, and it is found that the plasma makes up 53% of the blood volume. The absorbance of the plasma in a 1.00-cm cell, using a blank as a reference, is 0.380.

Another 1.00-ml sample of the same Evans blue solution is diluted to exactly 1 liter in a volumetric flask. Then 10.0 ml of this solution is further diluted to 50.0 ml in a volumetric flask. The absorbance of the fully diluted solution is 0.200, against the same reference as above. Calculate the man's blood volume in liters.

13. *Molecular weight.* A sample of an amine of unknown molecular weight is converted into the amine picrate (a 1:1 addition compound) by treatment with picric acid (MW 229). Most amine picrates have about the same molar absorptivities, $\log \varepsilon = 4.13$ at 380 nm in 95% ethanol. A solution of the amine picrate was prepared by dissolving 0.0250 g of the sample in exactly 1 liter of 95% ethanol. The absorbance of the solution in a 1.00-cm cell at 380 nm is 0.760. Estimate the molecular weight of the unknown amine.

14. *Analysis.* A 1.00-g sample of steel is dissolved in nitric acid and the manganese oxidized to permanganate with potassium periodate. The solution is made up to 250 ml in a volumetric flask and found to have an absorbance which is 1.50 times as great as that of a 0.00120 M solution of $KMnO_4$. Calculate the percentage of Mn in the steel.

15. *Analysis.* A 0.425-g sample of steel is dissolved in acid, and the manganese oxidized to permanganate. The solution is made up to 100 ml in a volumetric flask and the absorbance at 520 nm in a 2.0-cm cell is 0.80. The molar absorptivity of permanganate at 520 nm is 2235. Calculate the percentage of Mn in the steel.

16. *Photometric error.* If the error in determining concentration is to be 0.50% in a solution which gives an absorbance reading of 0.60, what must the photometric error be?

17. *Relative error.* Calculate the error in determining concentration per 1% photometric error for the following values of percent transmittance: (a) 1.0; (b) 10; (c) 50; (d) 80; (e) 99.

18. *Relative error.* The molar absorptivity of the compound A (MW 125) at 480 nm is 2500. What weight of sample containing about 1.5% of A should be taken for analysis so that maximum accuracy may be obtained in the photometric determination of A? The solution is finally diluted to 100 ml and the cell used is 1.00 cm in length.

19. *Relative error.* Suppose that the Environmental Protection Agency suspects that the wastewater of a paper pulp mill contains 1.0% SO_2 by weight. The SO_2 can be determined by reacting it with a dye, 1 mol of SO_2 reacting with 1 mol of the dye to form a colored species which has a molar absorptivity of 4300 at a certain wavelength. What volume of the wastewater should be taken for analysis so that the photometric error will be minimal? The water is treated with the dye, diluted to 100 ml, and the absorbance measured in a 1.0-cm cell. Assume that the density of the wastewater is 1.0 g/ml.

20. *Differential measurement.* In a conventional spectrophotometric measurement a standard solution which is 0.0010 M in compound X gives an absorbance reading of 0.699. An unknown solution of X reads $A = 1.000$. If the standard solution is used as the reference with $A = 0.000$, answer the following: (a) What is the absorbance of the unknown? (b) What are the differences in % T of the two solutions in the two methods of measurement?

21. *Differential measurement.* Solution A, concentration c, gives an absorbance reading against a blank of 0.30. (a) Calculate the absorbance readings of solutions B, C, and D of concentrations $2c$, $3c$, and $4c$, respectively. (b) Calculate the percentage transmittances of the four solutions and the differences between A and B, B and C, and C and D. (c) If A is used as the reference and is set to give an absorbance of 0.000, repeat the calculations of parts (a) and (b).

22. *Differential measurement.* A solution of concentration c has an absorbance of 0.4343 when the reference is the pure solvent. (a) What is the relative error in c if the photometric error is 0.20%? (b) If a solution whose concentration is $3c$ is used as the reference and one of $4c$ is measured, what is the relative error in the latter concentration, the photometric error being the same as in part (a)?

23. *Multicomponent analysis.* The absorption spectra of two colored substances A and B are determined and the following data obtained in a 1.00-cm cell:

Solution	Concentration	A at 450 nm	B at 700 nm
A alone	5.0×10^{-4} M	0.800	0.100
B alone	2.0×10^{-4} M	0.100	0.600
A + B	Unknown	0.600	1.000

Calculate the concentrations of A and B in the unknown solution.

24. *Absorbance—emf.* The compound nicotinamide-adenine dinucleotide (NAD^+) is an important coenzyme, participating in enzyme-catalyzed redox reactions. The redox couple may be written

$$NAD^+ + H^+ + 2e \rightleftharpoons NADH \qquad E_f^\circ = -0.31\text{ V}$$

where E_f° is the potential when the molarities of NAD^+ and NADH are each unity, but the pH is 7.00.

NAD^+ and NADH both absorb in the ultraviolet region of the spectrum. Some molar absorptivities are:

$$260 \text{ nm:} \quad \varepsilon \text{ of } NAD^+ = 18{,}000; \ \varepsilon \text{ of NADH} = 14{,}400$$

$$340 \text{ nm:} \quad \varepsilon \text{ of } NAD^+ = 0; \ \varepsilon \text{ of NADH} = 6{,}000$$

NAD^+ oxidizes ethyl alcohol as follows:

$$NAD^+ + C_2H_5OH \rightleftharpoons NADH + C_2H_4O + H^+$$

A chemist takes exactly 0.0010 mmol of NAD^+, some alcohol, and some enzyme catalyst. He dilutes the solution to 10.0 ml and buffers the solution at pH 7.00. After the solution has equilibrated he measures the absorbance of the solution in a 1.0-cm cell at 340 nm and finds it to be 0.480. Assuming that the absorbance is due only to the NADH, calculate (a) the potential of the NAD^+–NADH couple in the solution. (b) What is the potential of the couple

$$C_2H_4O + 2H^+ + 2e \rightleftharpoons C_2H_5OH$$

in the same solution?

25. *Stability of a complex.* When a large excess of the ion X^- is added to a solution of the metal M^{2+} the complex MX_3^- is formed. The concentrations of other complex species can be considered negligible. The MX_3^- complex absorbs strongly at 350 nm, where other species absorb negligibly. A solution which is $5.0 \times 10^{-4}\,F$ in M^{2+} is made 0.20 F in X^-, and the absorbance at 350 nm is found to be 0.80 in a 1.00-cm cell. Another $5.0 \times 10^{-4}\,F$ solution of M^{2+} is made 0.0025 F in X^- and the absorbance at 350 nm found to be 0.64 in the same cell. Assuming that all the M^{2+} is converted into the complex in the first solution but not in the second, calculate the stability constant of MX_3^-.

26. *Method of continuous variations.* The formula of a complex formed by a reaction such as $M + nX \rightleftharpoons MX_n$ can be determined by the method of continuous variations. The absorbances of a series of solutions of varying composition are measured at the wavelength at which the complex shows its maximum absorptivity. The number of moles of M and X is kept constant, while the mole fractions of the reactants are varied.

The following data were obtained for the reaction of M and X. Solution A is 0.0100 F in M; solution B is 0.0100 F in X. The absorbances can be assumed to be a measure of the concentration of MX_n.

Solution	A, ml	B, ml	Absorbance
1	10.00	0.00	0.000
2	9.00	1.00	0.133
3	7.00	3.00	0.400
4	5.00	5.00	0.667
5	3.00	7.00	0.932
6	2.00	8.00	0.800
7	1.00	9.00	0.400
8	0.00	10.00	0.000

Plot the absorbance against mole fraction X, and calculate the formula of the complex from the mole fraction at which the maximum absorbance is obtained.

27. *Mole-ratio method.* The formula of the complex MX_n can also be determined by the mole-ratio method. Solutions are prepared in which the concentration of M is held constant and that of X is varied. The absorbance of MX_n is measured.

The following data were obtained by mixing solution A, 0.0100 F in M, solution B, 0.0100 F in X, and solution C, 0.100 F in $HClO_4$.

A, ml	B, ml	C, ml	Absorbance
4.00	0.00	16.00	0.000
4.00	1.00	15.00	0.125
4.00	3.00	13.00	0.375
4.00	4.00	12.00	0.500
4.00	5.00	11.00	0.625
4.00	6.00	10.00	0.750
4.00	10.00	6.00	1.000
4.00	12.00	4.00	1.000

Plot the absorbances against the ratio of moles of X to moles of M. Calculate the formula of the complex from the mole ratio at which the absorbance reaches a maximum and levels off.

28. *Photometric titration.* Lead forms a chelate PbY^{2-} with EDTA, the log of whose stability constant is 18.3. This chelate has an absorption maximum at 240 nm, with a molar absorptivity of 6500. Suppose that 100 ml of a 5.0×10^{-5} F solution of a lead salt is titrated with a 1.0×10^{-3}-F EDTA solution. The lead solution is buffered at pH 3.00. The length of the path through the titration cell is 3.0 cm. Calculate the absorbance of the solution at the following volumes of titrant, correcting for dilution: (a) 1.00, (b) 2.00, (c) 4.00, (d) 5.00, (e) 6.00, and (f) 8.00 ml. Plot the titration curve.

29. *pK of an indicator.* Exactly 1.00-mmol portions of an indicator, a weak acid, HIn, and its salt NaIn are dissolved in 1.00-liter volumes of various buffer solutions. The absorbances of the resulting solutions are measured at 650 nm in a 1.00-cm cell and found to be:

*p*H	*Absorbance*
12.00	0.840
11.00	0.840
10.00	0.840
7.00	0.588
2.00	0.000
1.00	0.000

Calculate the molar absorptivities of HIn and In^- at 650 nm and calculate the pK_a of the indicator.

30. *pK of an indicator.* A solution of a weak acid indicator, HA, is prepared by dissolving 4.0 mmol of HA in 10.0 liters of solution. The following values of absorbances were measured in a 1.0-cm cell in various buffers:

*p*H	*Absorbance*
1.00	1.20
2.00	1.20
7.30	0.40
11.00	0.20
12.00	0.20

(a) Assuming that at low *p*H the principal species in solution is HA and at high *p*H the principal species is A^-, calculate the pK_a of the indicator. (b) Calculate the absorbance at *p*H 6.10.

31. *pK of an indicator.* A solution of a weak-base indicator, BOH, is prepared by dissolving 5.0 mmol of BOH in 10.0 liters of solution. The following values of absorbances were measured in a 1.0-cm cell in various buffers.

*p*H	*A at 450 nm*	*A at 580 nm*
1.00	1.20	0.600
2.00	1.20	0.600
6.30	0.40	0.600
11.00	0.20	0.600
12.00	0.20	0.600

(a) Assuming that at low *p*H the principal species in solution is B^+ and that high *p*H the principal species is BOH, calculate the absorbance at 450 nm and a *p*H of 5.10. (b) Another solution of BOH of different concentration is prepared which gives absorbance readings as follows: $A = 0.960$ at 580 nm, and $A = 1.12$ at 450 nm. Calculate the *p*H of this solution.

CHAPTER FIFTEEN

flame emission and atomic absorption spectrophotometry

Flame emission spectroscopy (or flame photometry, as it is often called) involves the measurement of radiant energy emitted by an excited atom population. In atomic absorption spectroscopy, the absorption of radiant energy by atoms in the ground state is measured. The two methods are often discussed together for convenience, despite their fundamental difference, because the flame is common to both: whether emission or absorption is to be measured, the required atom population is usually obtained from the sample by means of the flame.

FLAME EMISSION SPECTROSCOPY (FLAME PHOTOMETRY)

Emission Spectroscopy

Flame photometry is a branch of emission spectroscopy, a major field which represents, in a sense, the other side of the coin from spectrophotometry or absorption spectroscopy. In absorption studies, at least for analytical purposes, the fate of the excited species is not of interest (unless it involves a process that interferes with the absorption measurement, as fluorescence might). In emission spectroscopy, the sample is excited by various means, and the emission of radiation by excited species as they revert to lower energy states is measured.

Types of Emission Spectra

There are three major types of spectra emitted by excited substances. Incandescent solids emit *continuous spectra* which are characterized by the absence of discrete lines. That is, a graph showing emitted energy as a function of frequency or wavelength does not exhibit sharp peaks. An example is the radiation from the hot tungsten filament in a light bulb, as shown in Fig. 14.8; the spectral distribution of the energy approximates that of a black-body radiator.

Excited molecules in the gas phase emit *band spectra.* When a molecule in an excited electronic state of energy, E_2, undergoes a transition to a state of lower energy, E_1, a photon of energy $h\nu$ is emitted, where

$$h\nu = E_2 - E_1$$

But within each electronic state, the molecule may exist in a number of vibrational-rotational substates of slightly different energies. Hence the radiation from a large assembly of excited molecules comprises a number of frequencies which are grouped into bands. Each band corresponds to a transition from one excited electronic state to another electronic state of lower energy. The slightly different frequencies within each band result from the multiplicity of vibrational and rotational states within each electronic energy level. We saw a similar situation with regard to absorption in Chapter 14. The individual frequencies which make up a band may not be resolved by the particular instrument employed to examine the spectrum, in which case only an envelope over the individual lines is actually recorded.

Excited atoms or monatomic ions in the gaseous state emit *line spectra.* Vibrational and rotational "fine structure" is absent from the spectra of monatomic species, and the emission spectrum is a series of individual frequencies or lines corresponding to transitions between various electronic energy levels.

A spectral line is not a line in the mathematical sense of having no width. There is a *natural width,* of the order of 10^{-4} Å, which results from the probability distribution associated with each electronic energy level of an atom. Observed lines, in both emission and absorption, are considerably wider than this as a result of additional factors. *Doppler broadening,* of the order of 10^{-2} Å at typical flame temperatures, reflects the different velocity components of the atoms along the line of observation; the Doppler effect is discussed in elementary physics textbooks. *Pressure broadening,* which in flames is typically of the same order of magnitude as Doppler broadening, results from perturbations of the energy levels of an absorbing or emitting atom by neighboring atoms or molecules. Lines observed in flame work are generally between 10^{-2} and 10^{-1} Å in width. Measuring an actual line profile requires spectroscopic apparatus of higher resolving power than would be used for analytical work.

Methods of Excitation

In this chapter we are concerned with excitation in a flame, but it may be mentioned that other methods are used as well. The dc arc, for example, is very common in emission spectroscopy. An electrical arc is struck between two electrodes, using a voltage of perhaps 200 to 300 V. The electrodes are often graphite rods, to one of which the sample is applied in some manner. For instance, a depression may be machined into one end of a rod to hold a drop of sample solution. The rod is dried in an oven or under an infrared lamp. When an arc is formed between this rod and a second electrode, sufficient energy is provided to volatilize the sample residue, dissociate it into atoms, and excite these atoms, at least for many of the elements. Examination of the radiant energy emitted from the arc will disclose that it comprises discrete frequencies (or wavelengths) that serve to identify components of the sample. Under very carefully standardized conditions, quantitative analysis may be possible by measuring the radiant power at appropriate wavelengths, but there are many problems in quantitative emission spectroscopy with arc sources. Although very high temperatures are attained in the arc (say, 4000 to 8000°C), all the sample components may not be volatilized uniformly, and the time dependence of the emission spectrum may be tricky. Further, an arc tends to wander about as it plays upon an electrode surface, and much flickering is observed as well. Ac arcs, operated at 1000 V or more, are somewhat steadier than dc arcs. High-voltage sparks are also used to excite spectra in some cases.

Flame Sources

Of the common sources used in emission spectroscopy, the flame is the least energetic and excites the fewest elements—about 50 of the metallic ones. But where it is applicable, the flame has notable advantages over arc or spark excitation. A well-regulated flame is a much more stable source than an arc or spark. Further, the emission spectrum of an element in a flame is relatively simple. Only a few of the lines that are seen in an arc spectrum are found in flame emission. This places a much lighter burden upon the resolving power of the monochromator with regard to interferences; that is, it is easier to find an emission line for a particular element which does not have lines of other elements as near neighbors.

Temperature

The temperature of the flame is obviously one of the most important variables in flame photometry. This is determined by the nature of the fuel and its rate of supply, the supply of air or oxygen, and the design of the burner. Some of the less expensive instruments employ a Meker burner in which ordinary city gas burns in air. Temperatures are not much over 1500°C in the hottest portion of the flame, and relatively few elements such as the alkali metals are excited. Hydrogen–oxygen flames are widely employed in better instruments; temperatures well over 2000°C are attainable, providing sufficient energy for most purposes, and the flame itself produces very little background radiation to interfere with the observation of desired spectral lines.

Sample Introduction

Several arrangements are possible for introducing the sample solution into the flame. Instruments that employ a Meker burner generally have a separate atomization chamber in which the sample solution, introduced through a funnel, is broken up into fine droplets by the action of a blast of air. The geometry of the chamber is such that large drops are removed by the walls or by special baffles. The fine mist is swept by the air stream into the base of the burner, where it mixes with the burner gases and is carried into the flame. This type of atomizer is uneconomical with regard to sample consumption, which may be as high as 20 or 30 ml/min; much of the sample never reaches the flame.

The other common arrangement, which is better for many purposes, employs an integral aspirator-burner with which the sample solution is sprayed directly into the flame. Such a burner is depicted schematically in Fig. 15.1.

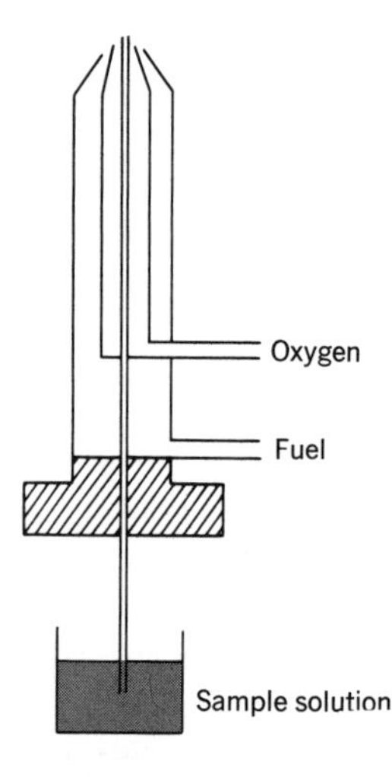

Figure 15.1 Schematic diagram of an integral aspirator-burner.

The flow of oxygen through the constricted tip of the inner annulus draws sample solution from a small beaker up through the capillary. At the tip of the capillary, the column of liquid encounters strong shear forces which disperse it into droplets that are carried directly into the flame by the rushing gases. The fuel, most commonly hydrogen, is supplied through an outer annulus. Sample consumption is comparatively low, perhaps 0.5 to 2 ml/min. Over a period of time, there may be a tendency for an encrustation of solid salts to form near the capillary tip. This, of course, may affect the rate of sample feed through the capillary. This rate is also sensitive to variations in the viscosity of the sample solutions. Thus, for the most accurate work, it may be helpful to force the solution through the capillary at a constant flow rate. This is easily accomplished using a motor-driven hypodermic syringe connected to the bottom of the capillary with a section of plastic tubing.

Instrumentation

Figure 15.2 shows in block form the components of a basic, research-level flame photometer. Commercial instruments over a wide price range represent compromises with regard to one or another of the components in order to provide

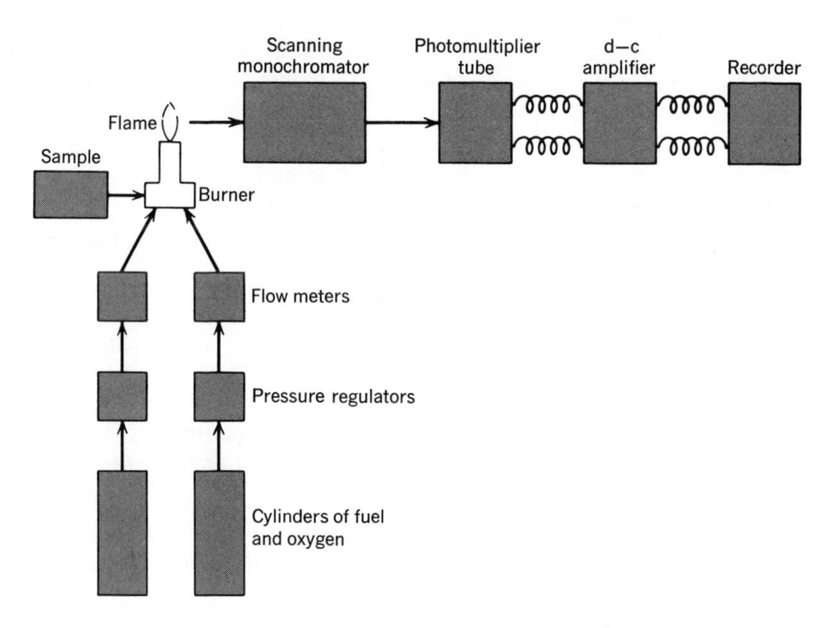

Figure 15.2 Block diagram of a flame photometer.

adequate capability under marketable conditions for applications of varying difficulty and sophistication. For example, in determining sodium using a low-temperature flame, only a few lines of the alkali and alkaline earth metals appear in the emission spectrum. Thus an inexpensive filter of colored glass could serve to transmit the desired sodium frequency to the detector while blocking the radiation emitted by, say potassium or calcium, and an expensive monochromator would not be needed.

The scanning monochromator and recorder are convenient for evaluating baseline effects arising from the flame background emission and for examining lines from several elements. But a less expensive arrangement is possible in which a monochromator is manually set for a certain wavelength. A background reading is obtained by introducing distilled water or some sort of analytical "blank" into the flame, and then the sample is measured at the same wavelength. A simple electrical meter would then be used in place of the recorder.

If sensitivity is not a problem, the photomultiplier tube may be replaced by an ordinary phototube. This simplifies the circuitry because a high-voltage power supply is no longer required to operate the detector. Flow meters for the fuel and oxygen are not essential if the pressures are well regulated, the aspirator is clean, and standards are run frequently along with the unknown samples.

Some instruments provide for a direct comparison of the radiant power emitted by two elements in the sample, thereby permitting the use of an *internal standard* to minimize the effects of variation in sample feed and fluctuations in the flame. For example, suppose sodium analyses were being performed. A constant quantity of a lithium salt might be added to all the standard and unknown sodium solutions. The choice of lithium would be based upon its similarity to sodium in its response to excitational variations and the unlikelihood of its occurrence in the particular set of unknowns at hand. The radiation emitted in one direction from the flame passes to a filter that transmits only the yellow sodium line to a detector. Radiation emitted in another direction goes to a filter that transmits only the lithium line, and thence to a second detector. A measuring circuit compares the two detector signals. The calibration curve for such a case would be a graph with the ratios of sodium-to-lithium emissions plotted against the sodium concentrations of the standard solutions.

Problems

The sequence of events required to produce an emission spectrum in the flame is almost forbidding enough to suggest that quantitative flame photometry should not work. Droplets of sample solution must evaporate; the resulting tiny solid particles must dissociate into atoms; the atoms must be excited. All these processes occur within a distance of a few cm as the sample droplets are being carried upward at a high velocity by the flame gases. Actually, it is more critical than this, because the optical system does not examine the whole flame but sees only the radiation from a region lying a certain distance above the burner tip. With the proper illumination, one may see unevaporated sample droplets issuing in profusion from the top of the flame. The flame gases are diluted by inrushing air as a result of the low pressure created by high velocity. There is at no point a stable, equilibrium atom population of the sort that might be obtained by holding the sample in a furnace for a sufficient time.[1] Thus the observed emission intensity depends upon a series of kinetic effects—rate of evaporation, rate of dissociation, etc.

It turns out, however, that flame photometry does work provided conditions are very carefully controlled. Under the best circumstances, deviations of the order of perhaps 2% may be seen in replicate analyses. But there are many traps for the unwary, and errors of 50 or 100% are probably far more common than many people realize.

Interferences

We shall list briefly some of the commonest problems in quantitative flame photometry.

1. *Radiation from other elements.* Perhaps no two spectral lines have exactly the same wavelength, but some are very close together. Whether emission by one metal will interfere in determining another will depend upon how close the lines are and the quality of the monochromator. With a good instrument, it is usually possible to measure an element without interference of this type, but with inexpensive filter instruments this is often not the case. Band spectra emitted by

[1] The furnace as a stable source for emission spectroscopy has in fact been considered by various people; technical problems—materials of construction, techniques for sample introduction, contamination from one sample to another—remain to be overcome.

excited molecules formed in the combustion process may also represent a problem in some cases.

2. Cation enhancement. In high-temperature flames, some of the metal atoms may ionize; for example,

$$Na \rightleftharpoons Na^+ + e$$

The ion has an emission spectrum of its own, with different frequencies from those of the atomic spectrum. Thus ionization decreases the radiant power of atomic emission. Sometimes a second metal, say potassium in this example, represses by its own ionization the ionization of the first (sodium). It is as though the partial pressure of free electrons had been increased in the equilibrium. The sodium atom emission is thereby enhanced. Sodium analysis on unknown samples containing varying quantities of potassium are thus subject to errors. One solution to this problem would be addition of a large quantity of a potassium salt to all the solutions—unknowns and standards—so as to swamp out variations from one sample to another.

3. Anion interference. Many examples of this are known. For instance, phosphate and sulfate ions lower the emission of calcium well below the level found for, say, calcium chloride solutions. Little is known of the detailed mechanism; presumably the solid residue resulting from solvent evaporation is less readily dissociated into atoms than is calcium chloride. Again, the swamping technique is often useful. For example, addition of 0.1 M EDTA to the solutions overcomes the effects of at least 0.01 M phosphate and sulfate on the emission from 6×10^{-4} M calcium.[2] Addition of a large excess of phosphate or sulfate would also overcome the effects of variations in these ions from sample to sample, but in this case EDTA is more satisfactory. For some reason which is not known, EDTA not only counters the effect of the other anions, but also enhances calcium emission, leading to improved sensitivity.[3]

Because our goal in this chapter is to provide a brief introduction to flame photometry without unduly lengthening the book, there are many aspects which are not considered here. Anyone undertaking a flame photometric analysis should do a great deal of reading and exercise the utmost attention to detail in preparing standard and unknown solutions in the laboratory. Matrix effects are very important; that is, the overall composition of a solution may have a large effect upon the emission intensity of a particular element.

Applications

The important applications of flame photometry involve analysis that are difficult or impossible to perform in any other way, at least where speed is much more important than accuracy. For many years the main forte of flame photometry was analysis for alkali metals, mainly sodium and potassium, and to a lesser extent, the alkaline earths, primarily calcium. Analyses for these ions

[2] A. C. West and W. D. Cooke, *Anal. Chem.*, **32**, 1471 (1960).
[3] *Ibid.*

became important in connection with electrolyte balance studies in physiology and in clinical chemistry laboratories. The alkali metals form few compounds containing chromophores to provide a basis for ultraviolet or visible spectrophotometry, they are not electroactive at reasonable potentials for electroanalytical techniques such as polarography, and they form few insoluble compounds. Thus flame photometry has been exceedingly useful in studies involving these elements. It is too early to state whether it will be supplanted in the future by other methods such as potentiometry with ion-selective electrodes, but at least for some time to come, flame photometry will probably continue to be important in biomedical research, agronomy, water analysis, nutrition studies, and other areas where alkali metals must be determined.

It would be incorrect to leave the student with the impression that flame photometry has been applied only to the alkali metals. Many other examples may be found in the literature involving perhaps 50 of the metallic elements. Lead in gasoline samples can be determined by spraying the gasoline directly into the flame. In many cases, metals have been extracted as chelates into organic solvents which are then subjected to flame photometry. Matrix effects are limited by the extraction process, and thus interferences are often minimal with this technique. Sensitivity is often greater with organic solvents than with water. The enhancement of metal emission probably results largely from a smaller droplet size and more rapid evaporation of the solvent, with a small contribution from the slightly higher flame temperature.

ATOMIC ABSORPTION SPECTROSCOPY

In a typical flame, most of the atoms by far are in the ground electronic state rather than in an excited state. For example, with regard to the transition that leads to the yellow sodium line at 589 nm, the ratio of excited atoms to ground-state atoms at 2700°C is about 6×10^{-4}. Further, the number of excited atoms varies exponentially with temperature, whereas, with so few atoms excited, the ground-state population is practically constant over a reasonable temperature range. (Note that this argument neglects the effect of temperature on evaporation and dissociation.) In 1955, Walsh pointed out these facts and suggested that improved analytical methods might be possible based upon absorption of radiation by ground state atoms in the flame.[4] This technique was developed rapidly during the 1960s. Unhappily, some of the early workers as well as the instrument manufacturers propounded overly optimistic views concerning interferences in atomic absorption. While it is true that the flame is no longer an excitation source, one may still expect kinetic effects related to evaporation and dissociation in the formation of the ground state atom population. Nevertheless, experience has proved that atomic absorption is a very useful technique for determining small quantities of a number of metals.

[4] A. Walsh, *Spectrochim. Acta*, **7**, 108 (1955).

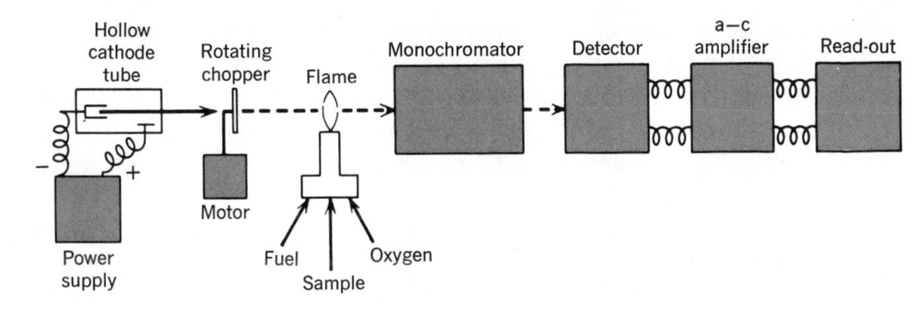

Figure 15.3 Components of an atomic absorption spectrophotometer.

Instrumentation

Figure 15.3 shows in schematic form the basic components of an atomic absorption spectrophotometer. It might be supposed that a continuous source such as those discussed in Chapter 14 could be used, but in practice this is not done. Absorption lines due to atomic species are much narrower than the bands encountered in ordinary spectrophotometry. If the band of radiation provided by a monochromator is much wider than the absorption line, then much of the radiation has no chance of being absorbed, and sensitivity will suffer. If the band pass of the monochromator is diminished enough to yield reasonable absorbance values, then an extremely intense continuous source is required in order to furnish enough energy in the very narrow wavelength region passed by the monochromator to operate the detector system. Thus the common source in atomic absorption is the one originally suggested by Walsh,[4] the hollow-cathode discharge tube. This tube contains an anode and a hollow, cylindrical cathode in an inert gas atmosphere (often argon) at low pressure. The tube is operated with a power supply that furnishes several hundred volts. Atoms of the gas are ionized in the electrical discharge, and collisions of the energetic ions with the cathode surface dislodge atoms of the metal which are excited. This results in the emission of a line spectrum of the metal which appears as a glow within the hollow cathode space.

A suitable line in the emission spectrum of the source is selected for the analysis. This line, a so-called resonance line, represents a transition from an excited atomic state to the ground state, and it thus represents the correct frequency for absorption by ground-state atoms in the flame. Hollow-cathode emission lines are very sharp—narrower in general than absorption lines in the flame—and are thus well suited to atomic absorption work. The only requirement of the monochromator is that other lines in the spectrum of the source, arising from the metal cathode and the inert gas, not be transmitted to the detector. This permits the use of a fairly inexpensive monochromator. Perhaps the major disadvantage of the hollow-cathode source is that a different tube is required for each metal determined.[5] Demountable tubes can be constructed in which various

[5] This is not strictly correct; in some cases, the cathode can be fabricated from an alloy so that emission lines of several metals may be obtained from a single tube.

metal cathodes can be interchanged, but the nuisance of pumping out the air each time the tube is opened has precluded their extensive use. After evacuation, bleeding-in the inert gas to the correct pressure is an additional problem. Thus sealed tubes, a different one for each metal, are commonly used. Some commercial instruments provide for a rapid and convenient switch from one tube to another.

The flame in atomic absorption is itself an emitting source just as it is in flame photometry, and the detector responds to radiation from the flame as well as from the hollow-cathode source. Yet, if uncorrected, this would obviously interfere with absorption measurements: if radiation emitted from the flame and unabsorbed radiation from the hollow-cathode source were not distinguished, then absorbance values would be too low and would be subject to the annoying fluctuations of flame emission. This problem is eliminated by the combination of the beam chopper with the tuned amplifier shown in Fig. 15.3. Only an ac electrical signal whose frequency is the same as that of the pulsating beam created by the chopper is amplified; unamplified signals are below the sensitivity of the readout device and go unnoticed. If emission from the flame were steady, it would give rise to only a dc signal at the detector, and even if not, it is improbable that more than a tiny component would have the correct frequency to be amplified.

Most atomic absorption spectrophotometers are single-beam instruments. Double-beam design compensates for source and detector fluctuations but not for the major source of noise and instability, the flame itself. Single-beam operation is the same as described in Chapter 14: P_0 is determined with distilled water or an analytical blank sprayed into the flame, and then P is measured with the sample. The Bouguer-Beer law applies as in any spectrophotometry. Special burners which provide a long path length through the flame have been devised to improve the sensitivity via Bouguer's law. Radiation may be passed several times through the flame by means of a mirror arrangement, or a fishtail burner shape may provide a long path.

Applications

Atomic absorption has gained very rapid acceptance and is now a very widely used analytical tool. There are several advantages as compared with flame emission:

1. Because absorption depends upon the ground state population, the sensitivity may be higher, especially for elements that are difficult to excite (zinc, for example, may be determined down to 0.5 ppm, while the lower limit in emission is perhaps 500 ppm or so).
2. Ground-state population is much less sensitive to flame temperature than excited population.
3. Interference from spectral lines of other elements and flame background emission is minimized by the beam-chopping technique.

But it should be remembered that the same sequence of events as in flame emission—atomization, evaporation, and dissociation—occurs in atomic absorption, and many interferences operate at these steps, particularly dissociation. The gross composition of the sample may have a great effect upon the ground-state population of a particular trace metal in the flame. Proper attention to such matrix effects in preparing standard solutions as well as unknown samples is very important.

Atomic absorption spectrophotometry is too widely used to permit a survey of applications in this chapter. Procedures may be found in the literature for determining 40 or more metals in a wide variety of samples that are important in industry, agriculture, medicine, oceanography, and many other fields. Basic studies, for example, on the distribution of lead in the environment, involving plants, soils, animal tissues, and urine samples, depend very heavily upon atomic absorption. Mercury, which represents an unusual example, is discussed more fully later.

Applications of atomic absorption may also be found in which the primary goal is not an immediate analysis but rather a better understanding of flame processes. For example, Cook et al. measured ground-state atom populations by atomic absorption at various heights above the burner tip. These data, along with temperature and emission measurements, provided a clear explanation for many reported discrepancies involving enhancement and suppression effects in flame photometry.[6]

Flameless Atomic Absorption

The determination of mercury at low concentrations represents an important exception to the use of the flame to obtain an atom population for atomic absorption. Mercury ions in solution are easily reduced to metallic mercury, and the element is sufficiently volatile to yield an atomic vapor without recourse to high temperature. The solution containing mercury salts is treated with a reducing agent such as tin(II) chloride, and the finely divided metallic mercury is volatilized by bubbling air through the solution. The vapor is swept by the air stream into an absorption cell with quartz windows where the attenuation of a mercury line (253.7 nm) is measured. The source is an ordinary mercury vapor lamp. As the air bubbling is continued, the absorbance rises to a maximum value and then diminishes. It is convenient to record absorbance as a function of time so that the maximum is easily found. Although there is no obvious theoretical reason to expect it, the maximal absorbance is proportional to the mercury concentration of the original solution. The method is effective well into the nanogram level.

Because much of the sample which is aspirated into a flame is not converted to atomic vapor during its brief transit through the hot zone, there have been many efforts in recent years to replace the flame with a high-temperature source which is less subject to kinetic factors. Furnace atomizers are currently the most prominent of these. The sample is placed in a graphite tube which is heated electrically to produce the atomic vapor. An inert gas is present to prevent oxidation of the

[6] J. H. Gibson, W. Grossman, and W. D. Cooke, *Anal. Chem.*, **35**, 266 (1963).

graphite. Higher sensitivities are typically obtained than with flames; with some elements, as little as 10^{-12} or 10^{-13} g can be detected. A totally satisfactory arrangement is still to be developed, however. The sudden heating produces a burst of ground-state atoms whose absorption must be caught on the fly and whose level is subject to matrix effects in an undesirable degree, and residues in the tube may influence later samples. Sooner or later this problem will be solved; it probably requires no new science, but rather some combination of technological inspiration and space-age materials.

REFERENCES

G. D. CHRISTIAN AND F. J. FELDMAN, *Atomic Absorption Spectroscopy: Applications in Agriculture, Biology, and Medicine*, Wiley–Interscience, New York, 1970.

J. A. DEAN, *Flame Photometry*, McGraw-Hill Book Company, New York, 1960.

J. A. DEAN and T. C. RAINS, eds., *Flame Emission and Atomic Absorption Spectrometry*. Vols. 1–3. Marcel Dekker. Inc., New York, 1969.

W. T. ELWELL and J. A. F. GIDLEY, *Atomic Absorption Spectrophotometry*, Pergamon Press, Inc., New York, 1966.

R. HERRMANN and C. Th. J. ALKEMADE, *Chemical Analysis by Flame Photometry*, 2nd ed., trans. by P. T. Gilbert, Wiley–Interscience, New York, 1963.

J. RAMIREZ-MUNOZ, *Atomic Absorption Spectroscopy*, Elsevier Publishing Co., New York, 1968.

J. W. ROBINSON, *Atomic Absorption Spectroscopy*, Marcel Dekker, Inc., New York, 1966.

B. L. VALLEE and R. E. THIERS, "Flame Photometry," Chap. 65 of I. M. Kolthoff and P. J. Elving, eds., *Treatise on Analytical Chemistry*, Part I, Vol. 6, Wiley–Interscience, New York, 1965.

QUESTIONS

1. *Terms.* Explain the meaning of the following terms: (a) continuous spectrum; (b) Doppler broadening; (c) internal standard; (d) cation enhancement; (e) resonance line.
2. *Excitation.* Point out the methods used for excitation in flame photometry.
3. *Instrumentation.* Describe the basic components and the operation of (a) a flame photometer, and (b) an atomic absorption spectrometer.
4. *Applications.* What are the principal applications of flame photometry and atomic absorption spectroscopy in analytical chemistry?
5. *Interferences.* What are the most common problems encountered in the application of flame photometry to analytical determinations?

Multiple-choice: In the following multiple-choice questions, select the *one best* answer.

6. Line spectra are emitted by (a) incandescent solids; (b) excited molecules; (c) excited atoms and monatomic ions; (d) molecules in the ground electronic state.

7. Band spectra are emitted by (a) tungsten lamps and Nernst glowers; (b) excited atoms and monatomic ions; (c) excited molecules; (d) incandescent solids.

8. Which of the following statements is false? (a) A well-regulated flame is a much more stable source than an arc or a spark; (b) The natural width of a spectral line is about 0.01 Å; (c) A flame is a less energetic source than an arc; (d) Ionization decreases the radiant power of atomic emission.

9. Which of the following best explains why atomic absorption spectrophotometry is sometimes more sensitive than flame emission spectroscopy? (a) An absorption line in a flame is always much broader than an emission line because of the Doppler effect; (b) Hollow cathode discharge tubes have a much greater radiant power output than do ordinary flames; (c) Detectors employed in absorption work are inherently more sensitive than those used to measure emission; (d) At the temperature of a typical flame, the population of ground-state atoms is much greater than the population of excited atoms.

10. The method of standard addition compensates for matrix effects provided that (a) the addition contains a large enough quantity of some substance to swamp out sample variations; (b) the addition contains none of the substances being determined; (c) the addition dilutes the sample enough that the concentrations of interfering substances are lowered to negligible values; (d) the addition does not dilute the sample appreciably and does not itself introduce appreciable quantities of interfering substances.

CHAPTER SIXTEEN

solvent extraction

INTRODUCTION

The partition of solutes between two immiscible liquids offers many attractive possibilities for analytical separations. Even where the primary goal is not analytical but rather preparative, solvent extraction may be an important step in the sequence that leads to a pure product in the organic, inorganic, or biochemical laboratory. Although complicated apparatus is sometimes employed, frequently only a separatory funnel is required. Often a solvent extraction separation can be accomplished in a few minutes. The technique is applicable over a wide concentration range, and has been used extensively for the isolation of extremely minute quantities of carrier-free isotopes obtained by nuclear transmutation as well as industrial materials produced by the ton. Solvent extraction separations are usually "clean" in the sense that there is no analog of coprecipitation with such systems. Aside from its intrinsic interest, there is an important reason for discussing solvent extraction in this text: We shall use a particular approach to solvent extraction, the Craig pseudocountercurrent technique, as a model to aid our understanding of chromatographic processes in Chapter 17.

DISTRIBUTION LAW

When a solute distributes itself between two immiscible liquids, there is a definite relationship between the solute concentrations in the two phases at equilibrium. Nernst gave the first clear statement of the distribution law when he pointed out in

1891 that a solute will distribute itself between two immiscible liquids in such a way that the ratio of concentrations at equilibrium is constant at a particular temperature:

$$\frac{[A]_1}{[A]_2} = \text{constant}$$

$[A]_1$ represents the concentration of a solute A in the liquid phase 1.

Although this relationship holds fairly well in certain cases, in reality it is inexact. Strictly, in thermodynamic terms, it is the activity ratio rather than the concentration ratio that should be constant. The activity of a chemical species in one phase maintains a constant ratio to the activity of the same species in the other liquid phase:

$$\frac{a_{A1}}{a_{A2}} = K_{D_A}$$

Here a_{A_1} represents the activity of solute A in phase 1. The true constant K_{D_A} is called the *distribution coefficient* of species A. In approximate calculations, which are adequate for many purposes, concentrations rather than activities may be employed in problems involving K_D values such as those at the end of this chapter.

Sometimes it is necessary or desirable to take into account the chemical complications in extraction equilibria. For example, consider the distribution of benzoic acid between the two liquid phases benzene and water. In the aqueous phase, benzoic acid is partly ionized,

$$HBz + H_2O \rightleftharpoons H_3O^+ + Bz^-$$

In the benzene phase, benzoic acid is partially dimerized by hydrogen bonding in the carboxyl groups,

$$2HBz \rightleftharpoons (HBz)_2$$

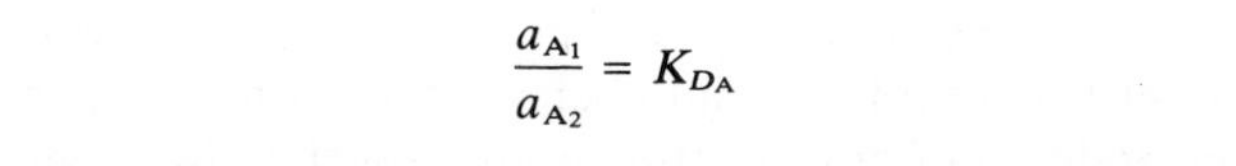

O
C
OH
O····HO
C
C
OH····O

Each particular species, HBz, Bz^-, $(HBz)_2$, will have its own particular K_D value. The system water, benzene, and benzoic acid may then be described by three distribution coefficients:

$$K_{D_{HBz}} = \frac{a_{HBz_{org}}}{a_{HBz_{aq}}}$$

$$K_{D_{Bz}} = \frac{a_{Bz^-_{org}}}{a_{Bz^-_{aq}}}$$

$$K_{D(HBz)_2} = \frac{a_{(HBz)_{2org}}}{a_{(HBz)_{2aq}}}$$

Now it happens that the benzoate ion, in fact, remains almost totally in the aqueous phase, and the benzoic acid dimer exists only in the organic phase.

Further, in a practical experiment, the chemist will usually want to know where the "benzoic acid" *is*, not whether part of it is ionized or dimerized. Also, he will be more interested in how much is there than in its thermodynamic activity. He would be better served, then, by an expression combining the concentrations of all the species in the two phases:

$$D = \frac{\text{total benzoic in organic phase}}{\text{total benzoic in aqueous phase}}$$

$$= \frac{[\mathrm{HBz}]_{\mathrm{org}} + 2[(\mathrm{HBz})_2]_{\mathrm{org}}}{[\mathrm{HBz}]_{\mathrm{aq}} + [\mathrm{Bz}^-]_{\mathrm{aq}}}$$

The ratio D is called the *distribution ratio.*

It is clear that D will not remain constant over a range of experimental conditions. For example, raising the pH of the aqueous phase will lower D by converting benzoic acid into benzoate ion, which does not extract into benzene. The addition of any electrolyte may affect D by changing activity coefficients. However, the distribution ratio is useful when its value is known for a particular set of conditions.

EXAMPLES OF SOLVENT EXTRACTION EQUILIBRIA

In this section we shall consider examples illustrating the manner in which equilibrium expressions may be manipulated to obtain equations which show the factors upon which D values depend.

Partition of a Weak Acid

Consider a weak acid, HB. Assume for simplicity that the acid is monomeric in both solvent phases, and that the anion of the acid does not penetrate the organic phase. The pertinent equilibrium expressions then are

$$D = \frac{[\mathrm{HB}]_{\mathrm{org}}}{[\mathrm{HB}]_{\mathrm{aq}} + [\mathrm{B}^-]_{\mathrm{aq}}} \tag{1}$$

$$K_{D_{\mathrm{HB}}} = \frac{[\mathrm{HB}]_{\mathrm{org}}}{[\mathrm{HB}]_{\mathrm{aq}}} \tag{2}$$

$$K_a = \frac{[\mathrm{H_3O^+}]_{\mathrm{aq}}[\mathrm{B}^-]_{\mathrm{aq}}}{[\mathrm{HB}]_{\mathrm{aq}}} \tag{3}$$

Rearranging (3) gives

$$[\mathrm{B}^-]_{\mathrm{aq}} = K_a \frac{[\mathrm{HB}]_{\mathrm{aq}}}{[\mathrm{H_3O^+}]_{\mathrm{aq}}}$$

and substitution into (1) yields

$$D = \frac{[\mathrm{HB}]_{\mathrm{org}}}{[\mathrm{HB}]_{\mathrm{aq}} + (K_a[\mathrm{HB}]_{\mathrm{aq}}/[\mathrm{H_3O^+}]_{\mathrm{aq}})}$$

or

$$D = \frac{[\mathrm{HB}]_{\mathrm{org}}}{[\mathrm{HB}]_{\mathrm{aq}}\{1 + (K_a/[\mathrm{H_3O^+}]_{\mathrm{aq}})\}}$$

Referring to (2), we see that

$$D = \frac{K_{D_{\mathrm{HB}}}}{1 + (K_a/[\mathrm{H_3O^+}]_{\mathrm{aq}})}$$

Thus we have derived an expression showing explicitly the dependence of the distribution ratio upon the distribution coefficient of the weak acid, its ionization constant, and the *p*H of the aqueous phase. It might well be that we could capitalize upon inherent differences in the values of the appropriate constants to effect the separation of a mixture of acids by regulating the *p*H of the aqueous phase.

Extraction of a Metal as a Chelate Compound

Many important separations of metal ions have been developed around the formation of chelate compounds with a variety of organic reagents. As an example, consider the reagent 8-quinolinol (8-hydroxyquinoline), often referred to by the trivial name "oxine,"

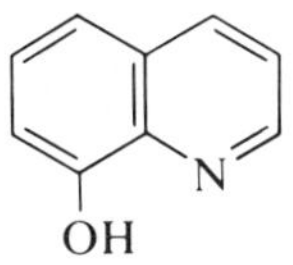

This reagent forms neutral, water-insoluble, chloroform- or carbon tetrachloride-soluble molecules with metal ions; the cupric oxinate chelate may be depicted as follows:

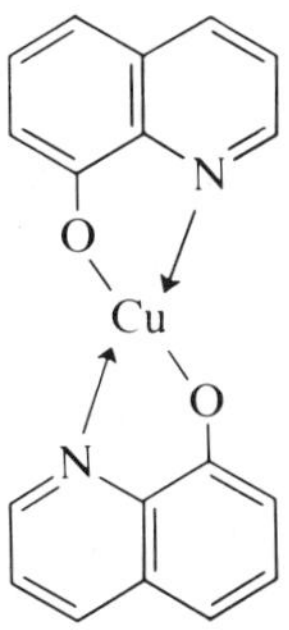

If we abbreviate oxine as HOx, we may write the chelation reaction as

$$\mathrm{Cu^{2+} + 2HOx \rightleftharpoons Cu(Ox)_2 + 2H^+}$$

Another very important chelating agent for the solvent extraction of metal ions is diphenylthiocarbazone or "dithizone,"

$$\begin{array}{l} \qquad\qquad C_6H_5 \\ \qquad\qquad | \\ \quad NH{-}NH \\ \;\,/ \\ S{=}C \\ \;\,\backslash \\ \quad N{=}N \\ \qquad\;\; | \\ \qquad\; C_6H_5 \end{array}$$

The chelation reaction may be written as

$$M^{n+} + n\text{HDz} \rightleftharpoons \text{M(Dz)}_n + nH^+$$

Consider the extraction of an aqueous solution of the metal ion M^{n+} with an organic solvent containing a chelating agent HX. A distribution ratio may be written for the metal:

$$D = \frac{C_{M_{org}}}{C_{M_{aq}}}$$

where C_M is the total metal concentration regardless of what form it is in. Let us assume for simplicity (but not a bad assumption in many actual cases) that the only metal in the organic phase is present as the chelate, MX_n (i.e., $C_{M_{org}} = [MX_n]_{org}$). Also, let us ignore possible lower complexes such as MX or MX_{n-1} which may exist in the aqueous phase. (A large excess of HX would perhaps ensure the validity of this assumption.) Thus

$$C_{M_{aq}} = [M^{n+}]_{aq} + [MX_n]_{aq}$$

If the partition of the chelate greatly favors the organic phase, which is frequently the case, we may neglect $[MX_n]_{aq}$ as compared with $[M^{n+}]_{aq}$, and thus we may write

$$D = \frac{[MX_n]_{org}}{[M^{n+}]_{aq}}$$

Rearranging the distribution coefficient expression for the chelate gives

$$[MX_n]_{org} = K_{DMXn}[MX_n]_{aq}$$

and substitution into the expression for D yields

$$D = \frac{K_{DMXn}[MX_n]_{aq}}{[M^{n+}]_{aq}} = \frac{K_{DMXn}}{[M^{n+}]_{aq}/[MX_n]_{aq}}$$

Next, consider the formation constant of the chelate in the aqueous phase:

$$M^n + nX^- \rightleftharpoons MX_n$$

$$K_f = \frac{[MX_n]_{aq}}{[M^{n+}]_{aq}[X^-]^n_{aq}} \quad \text{or} \quad [M^{n+}]_{aq} = \frac{[MX_n]_{aq}}{K_f[X^-]^n_{aq}}$$

and the acid dissociation of the chelating agent:

$$HX + H_2O \rightleftharpoons H_3O^+ + X^-$$

$$K_a = \frac{[H_3O^+]_{aq}[X^-]_{aq}}{[HX]_{aq}} \quad \text{or} \quad [X^-]_{aq} = \frac{K_a[HX]_{aq}}{[H_3O^+]_{aq}}$$

Thus

$$[M^{n+}]_{aq} = \frac{[MX_n]_{aq}[H_3O^+]^n_{aq}}{K_f K^n_a [HX]^n_{aq}}$$

and

$$\frac{[M^{n+}]_{aq}}{[MX_n]_{aq}} = \frac{[H_3O^+]^n_{aq}}{K_f K^n_a [HX]^n_{aq}}$$

Returning to the expression for D, we obtain

$$D = \frac{K_{D_{MX_n}} K_f K^n_a [HX]^n_{aq}}{[H_3O^+]^n_{aq}}$$

Next consider the distribution coefficient for the chelating agent:

$$K_{D_{HX}} = \frac{[HX]_{org}}{[HX]_{aq}} \quad \text{or} \quad [HX]_{aq} = \frac{[HX]_{org}}{K_{D_{HX}}}$$

Then

$$D = \frac{K_{D_{MX_n}} K_f K^n_a [HX]^n_{org}}{K^n_{D_{HX}} [H_3O^+]^n_{aq}}$$

This equation gives the distribution ratio for the metal in terms of (a) constants $K_{D_{MX_n}}$, $K_{D_{HX}}$, K_f, and K_a which are properties of the particular compounds involved in the selected system, and (b) variables $[H_3O^+]_{aq}$ and $[HX]_{org}$ which are subject to experimental manipulation for a particular system. Let us lump together the constants into an *extraction constant*, K_{ex}:

$$\frac{K_{D_{MX_n}} K_f K^n_a}{K^n_{D_{HX}}} = K_{ex}.$$

Then

$$D = \frac{K_{ex}[HX]^n_{org}}{[H_3O^+]^n_{aq}}$$

Taking logarithms,

$$\log D = \log K_{ex} + n \log [HX]_{org} - n \log [H_3O^+]_{aq}$$

or

$$\log D = \log K_{ex} + n \log [HX]_{org} + n\,pH$$

Thus a plot of $\log D$ vs. pH should be a straight line with a slope of n and an intercept on the $\log D$ axis of $\log K_{ex} + n \log [HX]_{org}$. The higher is the charge on the metal ion, the steeper is the slope of the line. Varying the concentration of the chelating agent shifts the curves along the pH axis.

Note that no term involving metal concentration appears in the final equation. One of the attractive aspects of solvent extraction (unlike, say, precipitation) is the fact that it works all the way from tracer levels up to macro quantities.

Of course, the constants K_f and $K_{D_{MX_n}}$ vary from one metal ion to another. This is the basis for separation by extraction of aqueous solutions with organic solvents containing chelating agents.

EXTRACTION SYSTEMS INVOLVING ION PAIRS AND SOLVATES

Generally, simple metal salts tend to be more soluble in a highly polar solvent like water than in organic solvents of much lower dielectric constant. Many ions are solvated by water, and the energy of solvation contributes to the disruption of the crystal lattice of the solid salt. Furthermore, less work is required to separate ions of opposite charge in a high-dielectric solvent. Usually, then, the formation of an uncharged species is necessary if an ion is to be extracted from water into an organic solvent. We have seen an example of this in the extraction of metals converted into neutral chelates of 8-quinolinol. The metal ion is bound in the chelate by definite chemical bonds, often largely covalent in character.

Sometimes, on the other hand, an uncharged species extractable into an organic solvent is obtained through the association of ions of opposite charge. In point of fact, it must be admitted that it is difficult to draw a line between an ion pair and a neutral molecule. Probably, if the components stay together in water, it will be called a molecule; if the components are separated in water sufficiently that an entity cannot be detected, this entity will be called an ion pair if it does show up in a nonpolar solvent.

A common example of an extraction system involving ion-pair formation in the organic phase is found in the use of tetraphenylarsonium chloride to extract permanganate, perrhenate, and pertechnetate from water into chloroform. The species which passes into the organic phase is an ion pair, $[(C_6H_5)_4As^+, ReO_4^-]$. Similarly, the extraction of uranyl ion, UO_2^{2+}, from aqueous nitrate solutions into solvents such as ether (an important process in uranium chemistry) involves an association of the type $[UO_2^{2+}, 2NO_3^-]$. It is believed that the uranyl ion is solvated by ether as well as by water, a fact which doubtless facilitates penetration of the organic phase by an ion pair which then takes on more of the character of the solvent.

It has been known for many years that iron(III) ion can be extracted into ether from strong hydrochloric acid solution. This process is useful for the separation of bulk quantities of iron prior to the determination of other elements in ferrous alloys. Despite extensive study, the system water–ether–HCl–Fe^{3+} is still not completely understood. There is evidence that the extractable species is an ion pair of the type $[H_3O^+, Fe(H_2O)_2Cl_4^-]$; other equilibria may intrude, such as solvation of both proton and iron(III) ion by ether:

$$H_3O^+ + C_4H_{10}O \rightleftharpoons C_4H_{10}OH^+ + H_2O$$

$$Fe(H_2O)_2Cl_4^- + 2C_4H_{10}O \rightleftharpoons Fe(C_4H_{10}O)_2Cl_4^- + 2H_2O$$

Thus under certain conditions the species in the ether phase may be [$C_4H_{10}OH^+$, $Fe(C_4H_{10}O)_2Cl_4^-$]. The system is undoubtedly complicated, and mixtures of various solvated ion pairs probably participate under the usual conditions of the extraction.

Multiple Extractions

In an ideal separation by solvent extraction, all of the desired substance would end up in one solvent and all the interfering substances in the other. Such all-or-none transfer from one solvent to another is rare, and we are much more likely to encounter mixtures of substances that differ only somewhat in their tendencies to pass from one solvent into another. Thus one transfer does not lead to a clean separation. In such cases, we must consider how best to combine a number of successive partial separations until we eventually achieve the desired degree of purity.

In considering how two phases may be brought together repetitively, four levels of complexity may be distinguished.[1] First would be the simple, one-shot contact as mentioned above. Second, one phase could be brought repeatedly into contact with fresh portions of a second phase. This would be applicable where one substance remained quantitatively in one phase, while another substance was distributed between the two phases. An example might be repeated extraction of an aqueous solution with successive portions of an organic solvent. The Soxhlet extractor would fall in this category, as would the technique of reprecipitation in gravimetric analysis.

Third, one phase may move while in contact with a second phase which remains stationary. The moving phase may move continuously, as in the various chromatographic techniques, or in a series of equilibrium steps, as in the Craig apparatus described below. Some techniques of this type have been designated "countercurrent," but this is not really the case, since only one phase moves. The term "pseudocountercurrent" is sometimes applied to such processes.

Fourth, we list true countercurrent methods, in which both phases move, continually in contact with each other, in opposite directions. Fractional distillation is an example of a true countercurrent process: Refluxing liquid runs continuously down the distilling column in contact with rising vapors. Countercurrent processes are extensively employed by chemical engineers in large-scale plant operations. Because of experimental difficulties as well as problems in the theoretical treatment, however, they are not nearly so common in the research laboratory as techniques in the third category.

In this section we shall discuss the second type, multiple extractions of one solvent batch with successive portions of another. The Craig process is described in a later section.

Suppose that we have a weight W of solute A dissolved in water. For simplicity, let the extraction with an organic solvent be uncomplicated by other

[1] H. A. Laitinen and W. E. Harris, *Chemical Analysis*, 2nd ed., McGraw-Hill Book Company, New York, 1975, p. 407.

equilibria (i.e., A has the same chemical structure in both solvents). Let

$$K_{D_A} = \frac{[A]_{org}}{[A]_{aq}}$$

Let us extract the aqueous solution with an equal volume of organic solvent, perhaps ether. In the extraction, suppose that the weight w of A moves from the aqueous solution into the organic solvent. Then

$$K_{D_A} = \frac{w}{W - w}$$

whence

$$WK_{D_A} - wK_{d_A} = w$$

$$w + wK_{D_A} = WK_{D_A}$$

$$w(1 + K_{D_A}) = WK_{D_A}$$

$$w = \frac{WK_{D_A}}{1 + K_{D_A}}$$

Now the fraction, f_{org}, extracted into the organic solvent is w/W. Dividing both sides of the equation above by W gives

$$f_{org} = \frac{w}{W} = \frac{K_{D_A}}{1 + K_{D_A}}$$

Since the fraction of A extracted into the organic solvent plus the fraction remaining in the aqueous solution must be equal to 1, the fraction of A remaining in the aqueous solution is

$$f_{aq} = 1 - \frac{K_{D_A}}{1 + K_{D_A}} = \frac{1 + K_{D_A} - K_{D_A}}{1 + K_{D_A}} = \frac{1}{1 + K_{D_A}}$$

Now let us separate the two solvent layers, set aside the organic solution, and extract the aqueous solution with a second, equal volume of fresh organic solvent. Of that fraction of A remaining in the aqueous solution after the first extraction, the fraction $K_{D_A}/(1 + K_{D_A})$ will be removed in the organic phase and the fraction $1/(1 + K_{D_A})$ will remain in the water. Thus, after two extractions,

$$f_{aq} = \left(\frac{1}{1 + K_{D_A}}\right)\left(\frac{1}{1 + K_{D_A}}\right) = \left(\frac{1}{1 + K_{D_A}}\right)^2$$

More generally, after n extractions of an aqueous solution with successive equal portions of organic solvent, the fraction of the original solute A unextracted (i.e., remaining in the water) is given by

$$f_{aq} = \left(\frac{1}{1 + K_{D_A}}\right)^n$$

The fraction extracted can always be obtained, of course, by subtracting the unextracted fraction from 1.

If the volume of organic solvent used each time is not the same as that of the water solution, it is easily shown that the expression for the unextracted fraction becomes

$$f_{aq} = \left(\frac{V_{aq}}{V_{aq} + K_{DA} V_{org}}\right)^n$$

where V_{aq} and V_{org} are the volumes of aqueous and organic phases, respectively.

Example. Suppose that K_D (org/aq) for a certain solute in a water–chloroform system is 10. Calculate the percentage of solute extracted from 50 ml of water by 100 ml of chloroform where (a) the chloroform is used all at once, and (b) the 100 ml of chloroform is divided into five 20-ml portions which are employed one after the other.

(a) $f_{aq} = \dfrac{50}{50 + (10)(100)} = 0.04762$

$f_{org} = 1 - 0.04762 = 0.9524$

$\%$ extracted $= 95.24$

(b) $f_{aq} = \left(\dfrac{50}{50 + (10)(20)}\right)^5 = 0.00032$

$f_{org} = 1 - 0.00032 = 0.9997$

$\%$ extracted $= 99.97$

Note that the extraction was more effective in removing the solute from the water when the chloroform was divided into several portions than it was when the same total volume of chloroform was employed in a one-shot process.

CRAIG PSEUDOCOUNTERCURRENT EXTRACTION

Basic Idea of the Craig Experiment

Consider an aqueous solution containing 1000 mg of some solute in a separatory funnel. Let it be a simple solute whose partition is uncomplicated by ionization, dimerization, etc. Add to the funnel an equal (for convenience) volume of an immiscible organic solvent. Also, for simplicity, suppose that the distribution coefficient of the solute is unity. Now, after equilibration, there will be 500 mg of the solute in the aqueous phase and 500 mg in the organic phase.

Next, take a second funnel and transfer the lighter liquid (say, the organic phase) from the first funnel into it. Then add fresh aqueous solvent to this organic phase in the second funnel and add fresh organic phase to the first funnel. Now shake both funnels until equilibration is achieved. There will then be 250 mg of the solute in each layer in each funnel.

As the student probably suspects by now, we next secure a third funnel, and transfer the organic solution from the second funnel into the third and also introduce a fresh portion of the aqueous solvent. We replace the organic layer in the second funnel using the organic phase from the first one, and we add fresh organic phase to the latter. Then all three funnels are shaken to secure equilibrium.

After two or three transfers, it becomes easier to keep track of what is going on by introducing a simple schematic representation. Portions of aqueous and organic solvent are represented by boxes, and where an aqueous box adjoins an organic box we have equilibration of the phases. Within this box, we give the weight of solute present in that phase. Thus the first step is depicted as follows:

		Funnel 0		
fresh org	fresh org	fresh org		
		1000 mg	fresh aq	fresh aq

The two phases in funnel 0 then equilibrate:

		Funnel 0		
. . . fresh org	fresh org	org 500 mg		
		⇅ aq 500 mg	fresh aq	fresh aq . . .

In the second step noted above, we have performed one transfer ($n = 1$), and the resulting situation may be represented by

	Funnel 0	Funnel 1		
fresh org	fresh org	500 mg		
	500 mg	fresh aq	fresh aq	fresh aq

Equilibration then takes place, giving

	Funnel 0	Funnel 1		
. . . fresh org	org 250 mg	org 250 mg		
	⇅ aq 250 mg	⇅ aq 250 mg	fresh aq	fresh aq . . .

In effect, we are pushing the top row of boxes toward the right with each transfer.

For the third step, where n, the number of transfers, is 2, the resulting situation is

	Funnel 0	Funnel 1	Funnel 2		
fresh org	fresh org	250 mg	250 mg		
	250 mg	250 mg	fresh aq	fresh aq	fresh aq

After equilibration the distribution is as follows:

	Funnel 0	Funnel 1	Funnel 2		
. . . fresh org	org 125 mg	org 250 mg	org 125 mg		
	⇅	⇅	⇅		
	aq 125 mg	aq 250 mg	aq 125 mg	fresh aq	fresh aq . . .

Note that the number of funnels is one more than the number of transfers, and hence the first funnel is labeled number 0. Let us write diagrams for two more steps, after equilibration takes place: fourth step, $n = 3$:

	Funnel 0	Funnel 1	Funnel 2	Funnel 3	
fresh org	org 62.5 mg	org 187.4 mg	org 187.5 mg	org 62.5 mg	
	⇅	⇅	⇅	⇅	
	aq 62.5 mg	aq 187.5 mg	aq 187.5 mg	aq 62.5 mg	fresh aq

fifth step, $n = 4$:

	0	1	2	3	4	
	31.25	125	187.5	125	31.25	
	⇅	⇅	⇅	⇅	⇅	
	31.25	125	187.5	125	31.25	

Perhaps we have carried this far enough to see what is happening. As the number of transfers is increased, the solute spreads out through more and more funnels, but it is "bunching up" toward the center (because $K_D = 1$) and the fraction of the solute in the extreme funnels is decreasing. It may also be surmised that for a different solute with a distribution coefficient, not 1, but favoring the aqueous phase, the peak concentration would not appear in the middle funnel but rather toward the left in our diagram. Likewise, a solute relatively more soluble in the organic layer would peak toward the right of the center.

Binomial Distribution in the Craig Extraction

Let us now formulate a more general mathematical treatment of the Craig countercurrent distribution. Actually, the mathematics is not difficult. In the first step of the treatment above, we distributed the solute between the two phases according to the distribution coefficient, K_D. Exactly as shown on page 442, the fractions of solute in the two phases are given by

$$f_{\text{org}} = \frac{K_D}{1 + K_D}$$

and

$$f_{\text{aq}} = \frac{1}{1 + K_D}$$

In the apparatus that Craig developed, and in the formulations encountered in the literature, the lighter phase (here, organic) is transferred from vessel 0 to vessel number 1. Fresh organic phase is then introduced into 0, and fresh aqueous phase into 1. After equilibration of the two vessels,

$$f_{\text{org}_0} = \left(\frac{1}{1 + K_D}\right)\left(\frac{K_D}{1 + K_D}\right)$$

(The fraction $1/1 + K_D$ was present in the aqueous layer in the vessel after the first equilibration, as shown above, and the fraction $K_D/1 + K_D$ passed over into the fresh organic solvent upon equilibration. Thus the product of the two fractions gives the fraction of the original W now present in the organic layer of vessel 0.) Likewise,

$$f_{\text{aq}_0} = \left(\frac{1}{1 + K_D}\right)\left(\frac{1}{1 + K_D}\right)$$

(At the beginning of this step, the fraction $1/1 + K_D$ was present in the aqueous layer, as shown above, and the fraction $1/1 + K_D$ *of that* is what remains after equilibration with fresh organic phase.) Also,

$$f_{\text{org}_1} = \left(\frac{K_D}{1 + K_D}\right)\left(\frac{K_D}{1 + K_D}\right)$$

and

$$f_{\text{aq}_1} = \left(\frac{K_D}{1 + K_D}\right)\left(\frac{1}{1 + K_D}\right)$$

We have gone through these steps to make certain that the student understands what is happening. Actually, it is easier finally to consider the total solute in each vessel instead of focusing upon the two phases separately, although we may add what is in the two phases to get the total. Let us work this out for vessels 0, 1, and 2 where $n = 2$, including for practice all stages up to $n = 2$.

$n = 0$:

$$f_0 = \underbrace{\left(\frac{K_D}{1 + K_D}\right)}_{\text{Organic phase}} + \underbrace{\left(\frac{1}{1 + K_D}\right)}_{\text{Aqueous phase}} = \frac{1 + K_D}{1 + K_D} = 1$$

Since, where $n = 0$, all of the solute is in this vessel, the fraction contributed by the organic phase and the fraction in the aqueous phase must add up to 1.

$n = 1$:

$$f_0 = \left(\frac{1}{1 + K_D}\right)\left(\frac{K_D}{1 + K_D}\right) + \left(\frac{1}{1 + K_D}\right)\left(\frac{1}{1 + K_D}\right) = \frac{1}{1 + K_D}$$

$$f_1 = \left(\frac{K_D}{1 + K_D}\right)\left(\frac{K_D}{1 + K_D}\right) + \left(\frac{K_D}{1 + K_D}\right)\left(\frac{1}{1 + K_D}\right) = \frac{K_D}{1 + K_D}$$

Here, $f_0 + f_1 = 1$, and we may confirm our result:

$$\frac{1}{1 + K_D} + \frac{K_D}{1 + K_D} = \frac{1 + K_D}{1 + K_D} = 1$$

$n = 2$:

$$f_0 = \left(\frac{1}{1 + K_D}\right)\left(\frac{1}{1 + K_D}\right) = \left(\frac{1}{1 + K_D}\right)^2$$

$$f_1 = \left(\frac{1}{1 + K_D}\right)\left(\frac{K_D}{1 + K_D}\right) + \left(\frac{K_D}{1 + K_D}\right)\left(\frac{1}{1 + K_D}\right) = 2\left(\frac{1}{1 + K_D}\right)\left(\frac{K_D}{1 + K_D}\right)$$

$$f_2 = \left(\frac{K_D}{1 + K_D}\right)\left(\frac{K_D}{1 + K_D}\right) = \left(\frac{K_D}{1 + K_D}\right)^2$$

Now, examine the foregoing fractions:

$$f_0 = \left(\frac{1}{1 + K_D}\right)^2$$

$$f_1 = 2\left(\frac{1}{1 + K_D}\right)\left(\frac{K_D}{1 + K_D}\right)$$

$$f_2 = \left(\frac{K_D}{1 + K_D}\right)^2$$

The alert student will note that these three terms are the terms in the expansion of the binomial

$$\left(\frac{1}{1 + K_D} + \frac{K_D}{1 + K_D}\right)^2$$

In general, for any number of transfers, n, it may be shown that the fractions of the total solute to be found in the various vessels 0, 1, 2, . . . , n are given by the terms in the expression of the binomial

$$\left(\frac{1}{1 + K_D} + \frac{K_D}{1 + K_D}\right)^n$$

In working with the formulas above, we assumed equal volumes of the two phases. In general, where the volumes are not necessarily equal (but the same in all vessels), it may be shown that the fractions of solute in the various vessels are given by the terms in the expansion of the binomial

$$\left(\frac{1}{1 + E} + \frac{E}{1 + E}\right)^n$$

where

$$E = K_D \times \frac{V_{\text{upper}}}{V_{\text{lower}}}$$

and V_{upper} and V_{lower} are the volumes of upper and lower phases, respectively.

If n is not small, expansion of the binomial becomes tedious. Fortunately, mathematical tables are available where it is worked out. Any single term in the binomial expansion can be obtained directly, using the formula

$$f_{n,r} = \left(\frac{n!}{r!(n - r)!}\right)\left(\frac{1}{1 + K_D}\right)^n K_D^r$$

where $f_{n,r}$ is the fraction of solute in the rth tube after n transfers. Some writers use the form

$$f_{n,r} = \frac{n!K_D^r}{r!(n - r)!(1 + K_D)^n}$$

During a series of successive transfers, a solute moves through the vessels of a Craig apparatus as a sort of wave of diminishing amplitude. The solute spreads through more and more vessels as n increases, but at the same time the fraction of the vessels which contain the solute decreases. Figure 16.1 shows theoretical distributions of solute for various numbers of transfers, n, for the case where $K_D = 1$ and $V_{\text{org}} = V_{\text{aq}}$.

For a given number of transfers, substances with different distribution coefficients are distributed differently. Figure 16.2 shows theoretical distributions after 16 transfers for various values of the distribution coefficient. We see in the figure that solutes with different K_D values are beginning to separate after the particular number of transfers indicated, but that separation is not complete in any case. This means that for these solutes more stages would be required in order to obtain pure components in quantitative yield.

It may be seen from the mathematical treatment given above why it is said that a Craig distribution is a *binomial* one. (Other types commonly encountered are *Gaussian*, as we saw in Chapter 3 for the distribution of random errors, and Poisson distributions.) However, as the number of transfers increases, the binomial distribution approximates more and more closely the Gaussian. (Poisson distributions also tend to Gaussian as n increases.) When n becomes large,

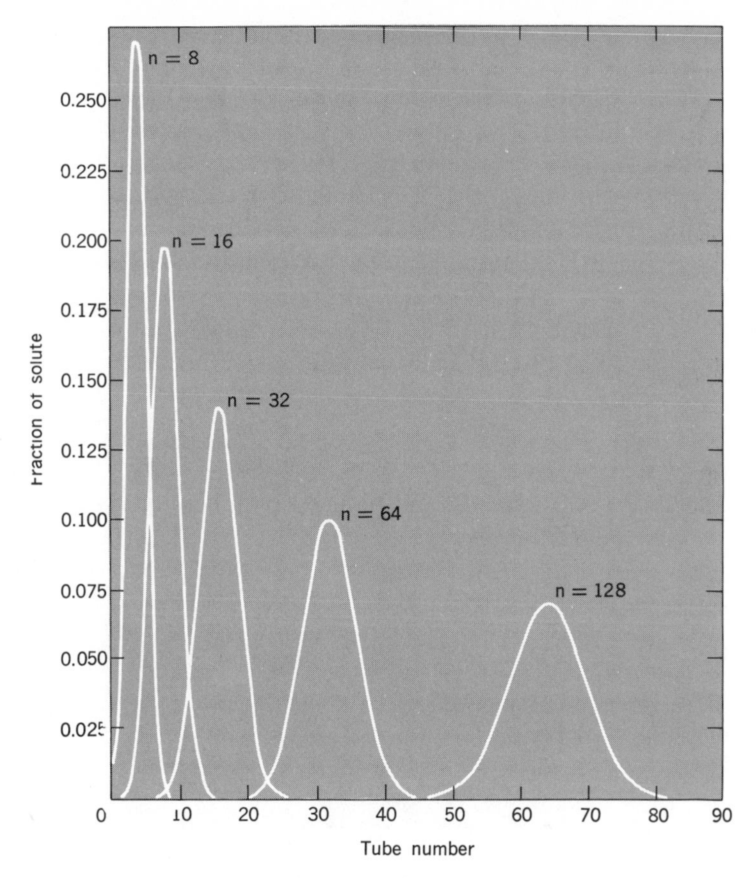

Figure 16.1 Theoretical distributions for solute with distribution coefficient $K_D = 1$ after various numbers of transfers (n) in a Craig experiment.

say 50, although there is no definite demarcation, a Gaussian treatment is sufficiently accurate for describing a Craig distribution, and it is more convenient than the more cumbersome binomial theorem for such cases. The approximate equations then are

$$r_{\max} = \frac{nK_D}{K_D + 1}$$

$$f_{\max} = \frac{1}{\sqrt{2\pi nK_D/(1 + K_D)^2}}$$

$$f_x = f_{\max} \times e^{-[x^2/[2nK_D/(1+K_D)^2]]}$$

where $r_{\max}$ is the number of the tube containing the maximal quantity of the solute, $f_{\max}$ is the fraction of solute in this tube, f_x is the fraction of solute in a tube x tubes distant from $r_{\max}$, and e is the base of natural logarithms. To use these equations for the case of unequal volumes of the two phases, simply replace K_D by E as defined earlier in the chapter.

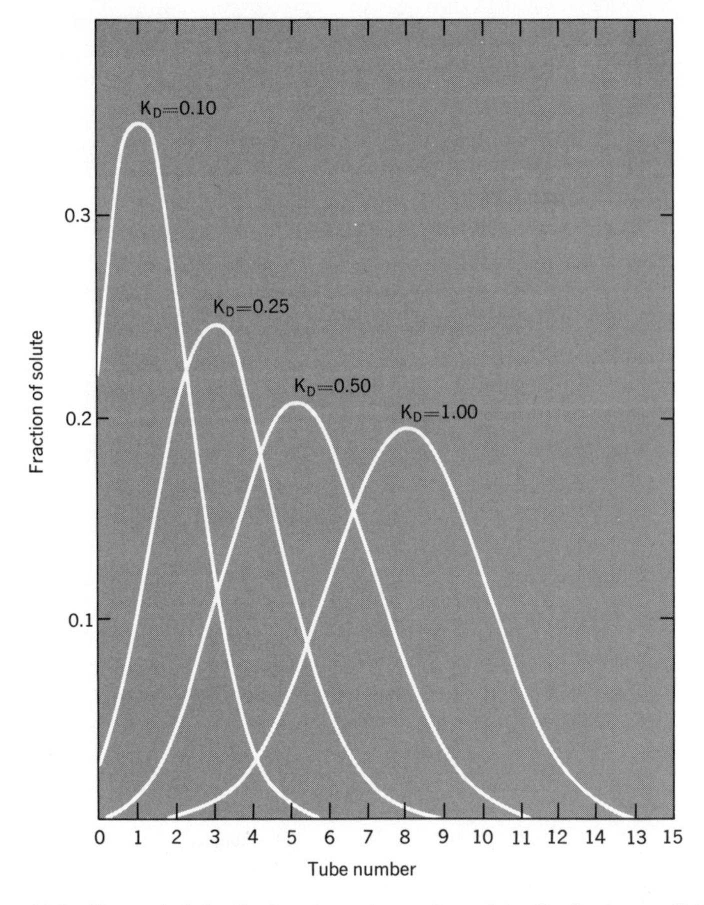

Figure 16.2 Theoretical distributions for various values of the distribution coefficient K_D, in a 16-transfer Craig experiment.

Apparatus for Craig Extraction

For a large number of transfers, say 100 or 1000, manual operation with separatory funnels would be impossible in a practical sense. Craig, who expounded both the theory and practice of this type of extraction process, developed apparatus to take much of the labor out of the procedure. The typical Craig apparatus is based upon glass units shaped as shown in Fig. 16.3. Although most people think in terms of vigorous shaking to equilibrate two liquid phases, Craig showed that a gentle sloshing of one liquid over another was actually more effective; further, troublesome emulsions were often avoided by the less vigorous technique. Thus it was unnecessary to adhere to the usual separatory funnel shape when designing a more automated apparatus. In the unit shown in Fig. 16.3, the two liquids are equilibrated by gently rocking the apparatus about 20 times, and the phases are allowed to separate. The apparatus is then rotated as indicated in the figure so that the cell is in a vertical position, whereupon the upper phase runs

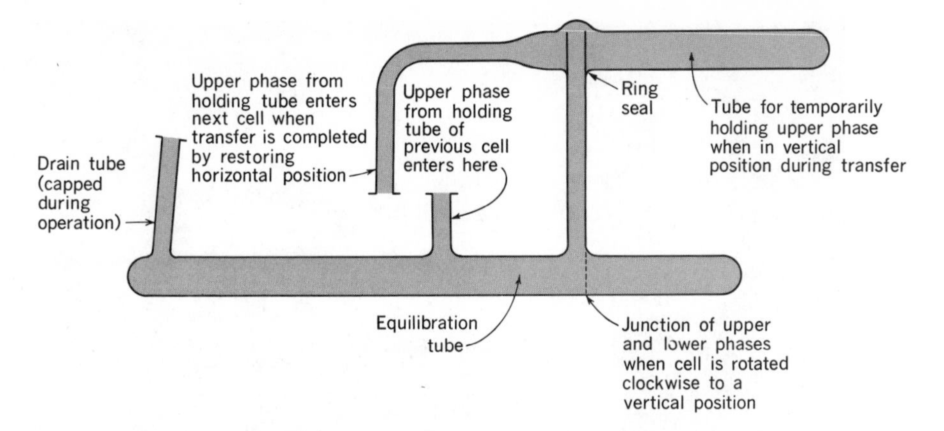

Figure 16.3 Typical glass cell for Craig apparatus.

out of the equilibrium tube and into a temporary holding tube. Obviously, the volume of the lower phase must be such that the solvent interface occurs at the level of the runoff side arm. When the cell is returned to a horizontal position, the lighter phase runs out of the holding tube and into the equilibration tube of the next cell in line. (The liquid is prevented from returning to the equilibration tube whence it came by the design of the cell in the area labeled "ring seal" in the figure.)

A battery of cells is clamped firmly to a metal frame so that all may be rocked and tilted together. The liquid-tight joints between adjacent cells are held together by spring clamps. In practice, the lower phase is introduced into all of the cells at the beginning of the experiment. The solute mixture to be separated is placed in the first cell (number 0) in either phase. A solvent reservoir and metering device introduces the appropriate voume of the lighter phase into the first cell after each transfer. In the simpler instruments, the rocking and tilting are done manually by the operator. Larger outfits with as many as 1000 cells are operated by a motorized robot so that the entire distribution experiment requires no attention once it is set in motion. Two commercial Craig machines are shown in Figs. 16.4 and 16.5.

After the distribution has been completed, the bank of cells is tilted so that the liquids in the cells may be collected from the drain tubes shown in Fig. 16.3. If the experiment is a preparative one, the pooled solvents from tubes containing the desired solute may be evaporated to obtain the desired material. In an analytical run, the solutions from the individual cells may be analyzed by appropriate means, for example, by spectrophotometry, titrimetry, or measurement of refractive index.

Applications of Craig Extraction

Many of the applications of the Craig extraction process are found in the biochemical area. It has proved particularly useful for separating peptides in the

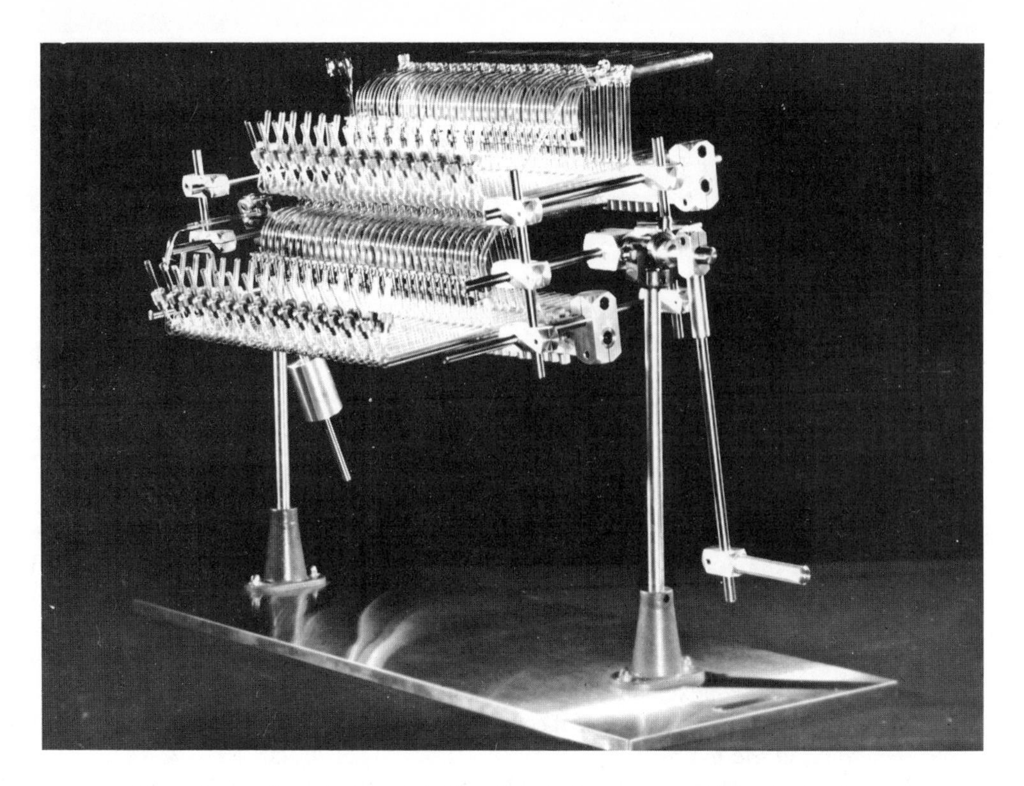

Figure 16.4 Manually operated 60-cell Craig Countercurrent Distribution Apparatus. The cells are arranged in two banks of thirty each. (Courtesy of H. O. Post Scientific Instrument Co., Inc.)

molecular weight range of about 500 to 5000. Some antibiotics and certain hormones are polypeptides, and several examples of successful purification by the Craig method may be found. Crude tyrocidin from a bacterial source was separated after 673 transfers into three major components (A, B, and C) and several minor ones; 1600 transfers on component A led to a crystallizable material, and 2140 more transfers eliminated the last trace of a particularly troublesome impurity. Peptide hormones from the posterior lobe of the pituitary gland have been fractionated by Craig extraction; distribution between 2-butanol and 0.1 M aqueous acetic acid was employed in the final purification of synthetic oxytocin and in comparing this product with the naturally occurring hormone, and similar work was done on vasopressin. Craig distribution has been employed for isolating and characterizing ACTH (adrenocorticotrophic hormone) from the anterior pituitary. Although proteins present many difficulties, some have been successfully handled by the Craig technique, notably insulin, ribonuclease, lysozyme, and the serum albumins. The technique is used for purification of many pharmaceutical preparations. Chemists other than biochemists have been slow to adopt the technique. Only a few inorganic applications appear in the literature.

Figure 16.5 1000-Cell Craig Countercurrent Distribution Instrument for automatic operation. The instrument can also be operated as two 500-cell units running concurrently. Driving mechanism is on the right-hand side under the table, with a driving shaft extending through the table top to the instrument. The unit is enclosed in a fume hood with glass doors and removable sides and top to afford access. (Courtesy of H. O. Post Scientific Instrument Co., Inc.)

Craig Extraction as a Model for Continuous Separation Processes

In the Craig extraction experiment, the heavier liquid phase remains stationary, while the lighter phase is transported down the series of cells, carrying solutes with it to various extents in accord with their partition properties. In chromatography, which we shall consider in Chapter 17, there is likewise a stationary and a moving phase (not necessarily both liquid). But here the moving phase flows continuously, and equilibrium is not actually attained at any time during the experiment. This leads to great difficulty in formulating the separation process mathematically, and the difficulty appears in the form of complicated equations with many correction factors and adjustable parameters. It is easy for the student to lose sight of what is going on in such processes. In the Craig experiment, on the other hand, the intermittent flow of the moving phase permits equilibration in each step of the overall process, and as we have seen above, the theoretical treatment is not unusually difficult. The Craig extraction process then, while it is by no means real chromatography, is useful as a teaching tool in explaining chromatography to beginners. We shall try to show the similarities and differences in more detail in the next chapter.

REFERENCES

D. DYRSSEN, J. O. LILJENZIN, and J. RYDBERG, eds., *Solvent Extraction Chemistry,* Wiley–Interscience, New York, 1967.

H. IRVING and R. J. P. WILLIAMS, "Liquid–Liquid Extraction," Chap. 31, p. 1309, of I. M. Kolthoff and P. J. Elving, eds., *Treatise on Analytical Chemistry,* Part I, Vol. 3, Interscience Publishers, Inc., New York, 1961.

C. J. O. R. MORRIS and P. MORRIS, *Separation Methods in Biochemistry,* Wiley-Interscience, New York, 1963.

G. H. MORRISON and H. FREISER, *Solvent Extraction in Analytical Chemistry,* John Wiley & Sons, Inc., New York, 1957.

J. STARY, *The Solvent Extraction of Metal Chelates,* Macmillan Publishing Co., Inc., New York, 1964.

QUESTIONS

1. *Terms.* Explain clearly the meaning of the following: (a) distribution coefficient; (b) distribution ratio; (c) countercurrent extraction; (d) pseudocountercurrent extraction; (e) ion pair.

2. *Distribution ratio.* Consider a substance which does not behave in a simple fashion, but rather participates in equilibria such as dissociation or dimerization, so that its distribution between two liquid phases is not adequately described by a K_D value. We might then employ a distribution ratio, D, in describing its behavior in a Craig experiment. If D were not independent of the analytical concentration of the substance (Why might it not be?), what would be the effect upon the shape of the Craig distribution curve?

3. *Binomial distribution.* Expansion of the binomial $(x + y)^4$ gives

$$x^4 + 4x^3y + 6x^2y^2 + 4xy^3 + y^4$$

   Assume that these terms represent the fractions of solute in tubes 0, 1, 2, 3, and 4 of a Craig extraction experiment where $x = y = \frac{1}{2}$. Show that this expression gives the same fraction of solute in tube 2 ($6x^2y^2$) as the expression on page 448 for $n = 4$ and $r = 2$. Recall that $K_D = 1$.

4. *Stability constants.* The stability constants of complex ions can be determined by a solvent extraction method. A chelating agent is employed and the distribution ratio D is determined as a function of the concentration of the complexing agent in the aqueous phase. Consider a metal ion, M^{2+}, which forms a chelate MT_2 with HT, and a complex ion MX^+ with X^-. Derive an expression relating D to $[X^-]$, involving the stability constant K_s, for the reaction

$$M^{2+} + X^- \rightleftharpoons MX^+$$

   You can assume that MT_2 is the only metal-containing species in the organic phase and that it exists predominantly in that phase. Also assume that no appreciable complexing occurs between M^{2+} and HT in the aqueous phase.

5. *Stability constants.* Repeat Question 4, except that the complexing agent is a weak acid, HX. The constant desired is for the reaction

$$M^{2+} + HX \rightleftharpoons MX^{+} + H^{+}$$

PROBLEMS

1. *Extraction.* The distribution coefficient (organic/aqueous) for the partition of molecular iodine between water and carbon tetrachloride is 85. 50 ml of a 0.010 *M* aqueous solution of I_2 is extracted with 50 ml of CCl_4. Calculate (a) the number of millimoles of I_2 extracted; (b) the concentration of I_2 in the aqueous phase after extraction.

2. *Extraction.* The distribution ratio (organic/aqueous) for the extraction of solute A from water into ether is 10.0. (a) If 100 ml of water containing 1.00 g of A is shaken with 100 ml of ether, what percentage of A is extracted into ether? (b) If the aqueous phase is extracted twice with 50-ml portions of ether, what percentage of A is extracted?

3. *Extraction.* The distribution ratio of iodine between an organic solvent and water is 8.00. If 50.0 ml of 0.100 *M* aqueous iodine is shaken with 200 ml of the organic solvent until equilibrium is reached, how many milliliters of 0.0600 *M* $Na_2S_2O_3$ is required to titrate a 25.0-ml aliquot of the organic phase?

4. *Distribution ratio.* 100 ml of a solution which is 0.200 *M* in the weak acid HA is extracted with 25.0 ml of ether. After the extraction a 25.0-ml aliquot of the aqueous phase required 30.0 ml of 0.0400 *M* NaOH for titration. Calculate the distribution ratio (organic/aqueous) of HA.

5. *Distribution ratio.* When an aqueous solution of $FeCl_3$ in concentrated HCl is shaken with twice its volume of ether containing HCl, 99% of the iron is extracted. Calculate the distribution ratio (organic/aqueous) of the compound.

6. *Separation of iron and chromium.* Iron(III) can be separated from other metals, such as chromium, by extraction into ether from a strong aqueous HCl solution. If 50 ml of aqueous HCl containing 0.25 g of iron(III) is treated with 150 ml of ether in a one-shot process, how many milligrams of iron is left in the aqueous phase? Use a value of 50 for *D*.

7. *Distribution ratio.* Two extractions with 20-ml portions of an organic solvent removed 89% of a solute from 100 ml of an aqueous solution. Calculate the distribution ratio (organic/aqueous) of the solute.

8. *Multiple extractions.* The distribution coefficient (aqueous/organic) of a solute S between water and benzene is 8.0. Calculate the number of mmol of S extracted from 100 ml of a 0.12 *M* aqueous solution of S if the extraction is carried out with the following quantities of benzene: (a) one 100-ml portion; (b) two 50-ml portions; (c) four 25-ml portions; (d) ten 10-ml portions.

9. *Multiple extractions.* 100 ml of 0.10 *M* S (Problem 8) is to be extracted with benzene. It is desired to reduce the concentration of S to 0.0010 *M* or less. What total volume of benzene will be required if the extraction is done with (a) 25-ml, and (b) 10-ml portions of the solvent?

10. *Multiple extractions.* 50 ml of a 0.050 M aqueous solution of a solute X is to be extracted with chloroform. Calculate the minimum value of the distribution coefficient that would allow the concentration of X to be lowered to 1.0×10^{-3} M by extraction with (a) two 25-ml, and (b) five 10-ml portions of chloroform.

11. *Craig extraction.* The distribution coefficient for the extraction of solute A from water into an organic solvent is 10 (organic/aqueous). 1.00 g of A is dissolved in water and placed in tube 0 of a Craig extraction apparatus and extracted with the organic solvent. The organic phase is transferred from tube 0 to tube 1, and so on. (a) Calculate the fraction of A remaining in tube 0 after five transfers. (b) Calculate the fraction of A in tubes 1, 2, 3, 4, and 5.

12. *Craig extraction.* Repeat Problem 11 except that the value of K_D is 100.

13. *Extraction of a metal chelate.* A chelating agent, HT, dissolved in an organic solvent extracts a metal, M^{2+}, from an aqueous solution according to the reaction

$$M^{2+}(aq) + 2HT(org) \rightleftharpoons MT_2(org) + 2H^+(aq)$$

The equilibrium constant for this reaction is 0.010. (a) Identify this equilibrium constant in terms of other constants (page 439). (b) Calculate the pH values at which 1, 25, 50, 75, and 99.9% of the metal is extracted into the organic phase. 10 ml of the aqueous solution is shaken with 10 ml of a 0.010 M solution of HT. Assume that the concentration of the metal is so small that the concentration of HT in the aqueous phase is negligible. (c) Plot the results, percent extracted vs. pH.

14. *Extraction of a metal chelate.* Repeat Problem 13 for a metal, N^{2+}, where the K of the extraction reaction is 1.0×10^{-6}. Suggest a pH at which metals M^{2+} and N^{2+} could be separated quantitatively by solvent extraction.

15. *Extraction of a metal chelate.* A certain metal ion is extracted by a chelating agent as in Problems 8 and 9. The concentration of the chelating agent in the organic phase is 0.010 M. The following data are obtained:

pH	1	2	3	4	5
D	10^{-8}	10^{-4}	1	10^4	10^8

Make a plot of log D vs. pH (page 439) and evaluate n and K.

16. *Separation of acids.* Given two acids, HA and HB, with the following distribution coefficients and dissociation constants:

	K_D	K_a
HA	10	1×10^{-5}
HB	1000	1×10^{-10}

Calculate the distribution ratios of the two acids at pH values 4, 5, 6, 7, 8, 9, and 10. Assuming that the ratio of the two D's needs to be 10^6 to 1 for a quantitative extraction of HB without extracting HA appreciably, what is the lowest pH (roughly) at which such a separation can be accomplished?

17. *Acid dissociation constant.* An acid, HX, has a distribution ratio of 10 between an organic solvent and water. At pH 5.0, half of the acid is extracted into the organic solvent. Calculate the dissociation constant of HX.

CHAPTER SEVENTEEN

gas-liquid chromatography

INTRODUCTION

The resolution of mixtures into their components is important in all the branches of chemistry and no less so in the many other fields where chemical techniques are employed in solving a wide variety of problems. Thus the impact of a powerful and versatile separation technique will be felt throughout much of modern science. In this connection, the significance of chromatography can scarcely be overstated. Utilizing chromatographic methods, separations in many cases are accomplished much more rapidly and effectively than before, and many separations are routinely successful which would never have been attempted by other techniques. Unparalleled breakthroughs in biochemistry—for example, in our understanding of the structure and function of enzymes and other proteins—have stemmed directly from the application of chromatography to biological research. Evaluating air and water pollution, determining pesticide residues on fruits and vegetables, identifying and classifying bacteria, monitoring respiratory gases during anesthesia, searching for organic compounds and living organisms on other planets, determining metabolic pathways and mechanisms of action of drugs—a list of all such studies based upon chromatography would be a lengthy one indeed.

Although forerunners can be found in the nineteenth century, it is generally considered that a paper published in 1906 by Michael Tswett, a lecturer in botany at the University of Warsaw, provided the first description in nearly modern terms of a chromatographic separation.[1] Tswett described the resolution of the

[1] M. Tswett, *Ber. Deut. Botan. Ges.*, **24**, 235 (1906).

chlorophylls and other pigments in a plant extract as follows:

> If a petroleum ether solution of [crude] chlorophyll is filtered through a column of adsorbent (I use mainly calcium carbonate which is packed firmly into a narrow glass tube), then the pigments are resolved from top to bottom into various colored zones according to an adsorption sequence, where the more strongly adsorbed pigments displace the more weakly adsorbed ones and force them further downward. This separation becomes practically complete if, following the pigment solution, a stream of pure solvent is passed through the column. Like a spectrum of light rays, the different components of the pigment mixture are systematically resolved on the calcium carbonate column and can be identified and also determined quantitatively. Such a preparation I term a chromatogram, and the corresponding method the chromatographic method.

Although the term chromatography is derived from Greek words meaning "color" and "write," the color of the compounds is obviously incidental to the separation process; Tswett himself anticipated applications to a wide variety of chemical systems. Had his work been immediately seized upon and extended, several sciences might have progressed more rapidly. As it was, chromatography remained dormant until about 1931, when separations of plant carotene pigments were reported by the prominent organic chemist Kuhn. This research attracted more attention, and adsorption chromatography became widely used in the field of natural product chemistry.

More recently, there have been four major developments: ion-exchange chromatography in the late 1930s, partition chromatography in 1941, gas chromatography in 1952, and gel-filtration chromatography in 1959. In addition to these major advances, which provided additional mechanisms to adsorption for distributing solutes between stationary and mobile phases, there have also appeared modifications in the geometry of the chromatographic system, as in paper and thin-layer chromatography.

Theoretical developments which permit a thorough understanding of the chromatographic process and hence clarify the factors which determine column performance appeared first in connection with gas chromatography. But certain of these insights have proved, with suitable adjustments, to be equally helpful in understanding chromatography where the moving phase is a liquid. Thus there began about 1968 a revolution in liquid chromatography which promises new speed and efficiency in the separation of non-volatile compounds which do not lend themselves to the gas chromatographic approach.

Because of its great practical importance in many research areas, chromatography is a fast-moving field. Efforts continue along many lines, of which we may mention a few: better detectors, new column-packing materials, improved interfacing with other instruments (such as the mass spectrometer) which serves to identify the separated components, new data-processing techniques based upon computers, and new mathematical models which provide additional insight into the nature of the process. Our goal in this book is to present a basic introduction which will acquaint the student with the nature of the chromatographic process, explain in simple terms what the instruments fundamentally do, and show some of the applications which have made chromatography indispensible in so many fields.

DEFINITION AND CLASSIFICATION OF CHROMATOGRAPHY

Although the meaning of the term is largely understood by chemists, a good definition of chromatography is difficult to formulate. It is a collective term applied to methods which appear diverse in some regards but share certain common features. A definition should emphasize that components of the sample are distributed between two phases, but this alone is inadequate because we do not wish the term to embrace all separation processes. Keulemans' definition serves as well as any:

> Chromatography is a physical method of separation, in which the components to be separated are distributed between two phases, one of these phases constituting a stationary bed of large surface area, the other being a fluid that percolates through or along the stationary bed.[2]

The stationary phase may be either a solid or a liquid, and the moving phase may be either a liquid or a gas. Thus all the known types of chromatography fall into the four categories that are shown in Table 17.1: liquid–solid, gas–solid, liquid–liquid, and gas–liquid.

TABLE 17.1 Summary of Types of Chromatography

Stationary phase	SOLID		LIQUID	
Moving phase	Liquid	Gas	Liquid	Gas
Examples	Tswett's original chromatography with petroleum ether solutions and $CaCO_3$ columns Ion-exchange chromatography	Gas–solid chromatography or GSC	Partition chromatography on silica gel columns Paper chromatography	Gas–liquid chromatography or GLC

In all the chromatographic techniques, the solutes to be separated migrate along a column (or, as in paper or thin-layer chromatography, the physical equivalent of a column), and of course the basis of the separation lies in different

[2] A. I. M. Keulemans, *Gas Chromatography*, 2nd ed., Reinhold Publishing Corp., New York, 1959, p. 2.

rates of migration for the different solutes. We may think of the rate of migration of a solute as the result of two factors, one tending to move the solute and the other to retard it. In Tswett's original process, the tendency of solutes to adsorb on the solid phase retarded their movement, while their solubility in the moving liquid phase tended to move them along. A slight difference between two solutes in the firmness of their adsorption and in their interaction with the moving solvent becomes the basis of a separation when the solute molecules repeatedly distribute between the two phases over and over again throughout the length of the column.

In this chapter we consider gas–liquid chromatography, or GLC, by far the more important form of gas chromatography. In the next chapter we shall describe chromatographic techniques where the moving phase is a liquid.

BASIC APPARATUS FOR GLC

To orient readers who are totally unfamiliar with gas chromatography, we shall first describe the apparatus and technique of GLC briefly and in general terms. Then we shall consider the theory, next indicate more fully the functions of the components of the apparatus, and finally give some illustrative applications which show the power and versatility of the method.

Figure 17.1 is a schematic diagram of a common type of basic GLC instrument. Although gas chromatographs can become very complicated if additional features are included, the basic instrument is a fairly simple one. The moving phase in GLC is a gas, most commonly helium, hydrogen, or nitrogen. The choice of carrier gas depends primarily upon the characteristics of the detector, as we shall see later. The user buys a cylinder of the compressed gas and attaches his own reducing value to it. Commercial gas chromatographs usually provide an additional regulating valve for good control of the pressure at the inlet of the column. With instruments of the type shown, employing the thermal conductivity detector, the carrier gas passes through one side of the detector and then enters the column. Near the column inlet is a device whereby samples may be introduced into the carrier gas stream. The samples may be gases or volatile liquids. The injection port is heated so that liquid samples are quickly vaporized. Samples of a few microliters of liquid or a few milliliters of gas are commonly introduced through a rubber septum by means of a hypodermic syringe.

The gas stream next encounters the column, which is mounted in a constant-temperature oven. This is the heart of the instrument, the place where the basic chromatographic process takes place. Columns vary widely in size and packing material. A common size is 6 ft long and $\frac{1}{4}$ inch internal diameter, made of copper or stainless steel tubing; to save space, it may be U-shaped or coiled into a spiral. The tubing is packed with a relatively inert, pulverized solid material of large surface area, frequently diatomaceous earth or firebrick. The solid, however, is actually only a mechanical support for a liquid; before it is packed into the column, it is impregnated with the desired liquid which serves as the real stationary phase.

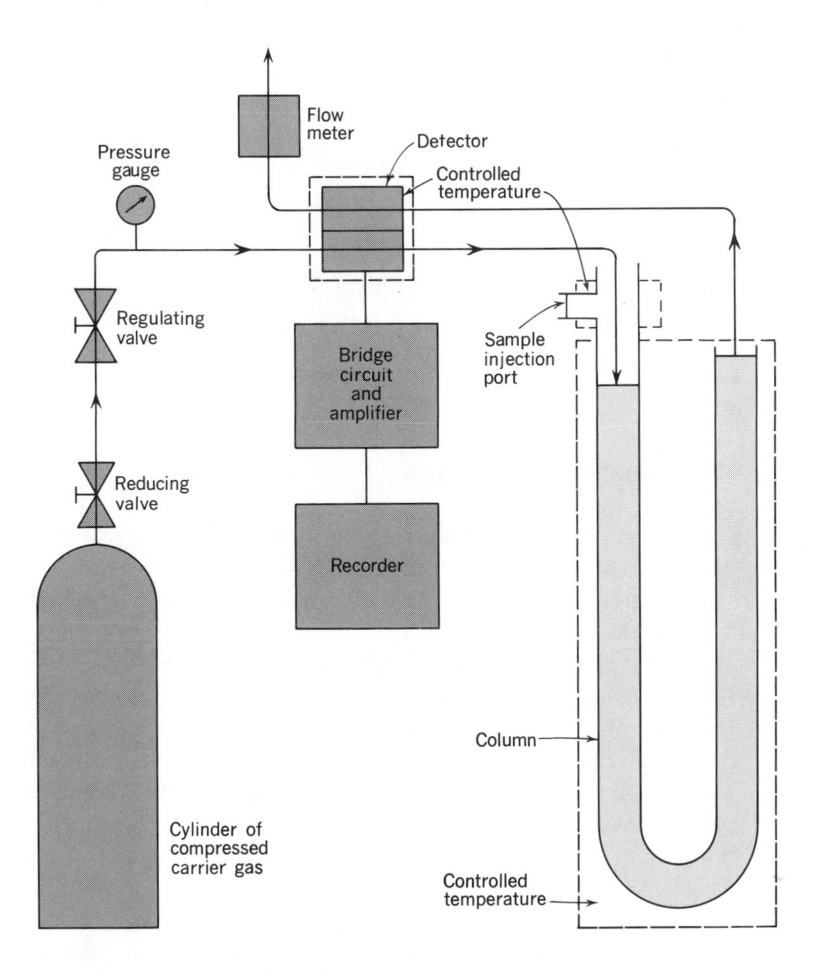

Figure 17.1 Schematic diagram of a gas chromatograph with thermal conductivity detector. Large arrows indicate direction of gas flow.

This liquid must be stable and nonvolatile at the temperature of the column, and it must be appropriate for the particular separation.

After emerging from the column, the gas stream passes through the other side of the detector. Elution of a solute from the column thus sets up an imbalance between the two sides of the detector which is recorded electrically. The carrier gas flow rate is important, and usually a flow meter of some sort is provided. There may be another regulating valve at the outlet end of the system, although normally the emerging gases are vented at atmospheric pressure. Because continual exposure of laboratory workers to the vapors of chromatographed compounds may be unwise even though the samples are usually small, attention should be paid to ventilation at the outlet of the instrument. Provision can be made to trap separated solutes as they emerge from the column if this is required for further investigation.

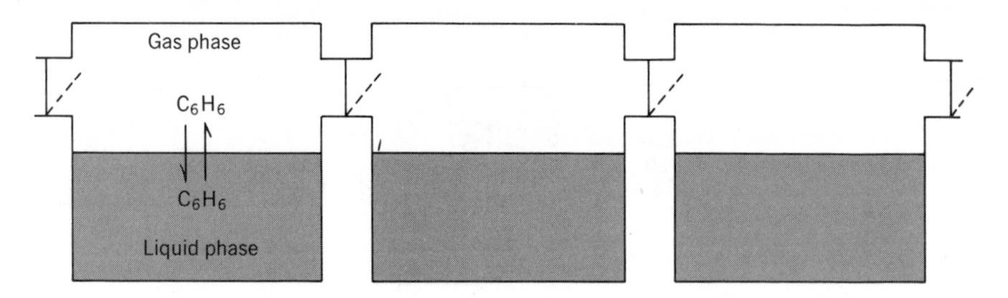

Figure 17.2 Imaginary chambers for a Craig model of a GLC experiment.

THEORY OF GLC

The Theoretical Plate Concept

It is suggested that the reader review the Craig countercurrent solvent extraction distribution in Chapter 16; we shall approach GLC in essentially the same manner. Suppose that we have a series of small chambers as shown in Fig. 17.2, each containing a portion of a nonvolatile liquid, which serves as the stationary phase. Let us introduce into the first chamber a sample of the mobile phase, a gas such as nitrogen, containing vapor of an organic compound, say benzene. If the liquid is a suitable one for our purpose, some of the benzene will dissolve in it, and some will remain in the space above it. Now Henry's law, in its usual form, states that the partial pressure exerted by a solute in dilute solution is proportional to its mole fraction. Thus, for the equilibrium distribution of benzene between liquid and vapor phases in our chamber, we may write

$$p_{\text{benzene}} = kX_{\text{benzene}}$$

where p_{benzene} is the partial pressure of benzene in the vapor phase, X_{benzene} the mole fraction of benzene in the liquid, and k a constant. In gas chromatography, partial pressure and mole fraction are often replaced by concentration terms which yield a dimensionless *distribution coefficient, K*:

$$K = \frac{\text{concentration of benzene in liquid phase, wt/ml}}{\text{concentration of benzene in gas phase, wt/ml}} = \frac{C_L}{C_G}$$

It is customary for the liquid term to be the numerator. K is also called a *partition coefficient* by many writers.

Now let us perform the same sort of operation that we employed in the Craig extraction. Transfer the gas (nitrogen plus benzene vapor) from the first chamber into the second one, where it encounters fresh liquid that contains no benzene. Introduce fresh nitrogen containing no benzene into the first chamber. Wait until equilibrium has been established in both chambers. Then transfer the gas from the second chamber into the third and from the first into the second, and introduce fresh nitrogen into the first. As the transfers proceed, the benzene will become distributed through the chambers in the same manner as a solute in the Craig solvent extraction system, illustrated in Fig. 16.1. The position of the benzene band on the horizontal axis will depend upon the number of transfers, the value of K, and the relative volumes of vapor and liquid in the individual chambers (cf. the discussion accompanying Fig. 16.1). Just as we saw in the Craig experiment, a second compound, say toluene, could be separated from benzene in this gas-liquid system if the K values for the two compounds were different, provided sufficient stages were employed (cf. Fig. 16.2). After a large number of transfers, the binomial distribution will become virtually the symmetrical Gaussian one.

There is a major difference in practice between Craig solvent extraction and chromatography which should be pointed out, although it is not important with regard to the principle involved. With Craig apparatus, it is customary to stop when the desired separation has been achieved and drain the solutions from the tubes containing the solutes. In modern chromatography, on the other hand, the flow of the mobile phase is continued until the solutes have migrated the entire length of the column whence they emerge, one after the other, to enter the detector. The practice of terminating the Craig process as soon as possible simply reflects the more cumbersome nature of the apparatus and the time-consuming character of the process.

The equilibrium chambers in the apparatus described above are called *theoretical plates*, a term which originated in distillation theory and was later carried over into chromatography. Each cell in the Craig apparatus is a plate in this sense. Now a chromatographic column operates under conditions of continuous flow of the moving phase, and equilibrium is not attained at any point in the column. However, after traversing a certain length of column, a mixture will have been subjected to the same degree of fractionation as would have been achieved in one equilibrium step. That length of column which accomplishes this is called the *height equivalent of a theoretical plate* or HETP. The total length of a column divided by HETP is the number, n, of theoretical plates in the column,[3] and it is customary to rate column performance in terms of number of plates. A good column will have more plates than a poor column of the same length. The great efficiency of GLC for performing difficult separations lies in the fact that large numbers of plates are fairly easily obtained with columns of reasonable length; columns with a couple of thousand plates may be only 5 or 6 ft long. (To give an idea of the power of the chromatographic technique, it may be noted that columns for fractional distillation are likewise rated in terms of theoretical plates; a fairly good 6-ft conventional distilling column may have something of the order of 20 or 30 plates.[4])

[3] The reader should note that n in this chapter is the number of *plates*, while in Chapter 16, n was the number of *transfers* in the Craig solvent extraction experiment. Sometimes our goal of avoiding confusion in the use of symbols conflicts with our desire to conform to the commonest usage as found elsewhere in the literature.

[4] The direct comparison of numbers of plates in distillation and chromatography is not quite fair unless it is qualified as follows. In a fractional distillation, all of the plates in the column are utilized throughout the experiment. In chromatography, on the other hand, after the sample components have migrated along the column, that portion of the column already traversed might just as well not be there any longer. Thus chromatography basically requires more plates than distillation. But these plates are so readily obtained that, even taking this factor into account, chromatographic columns are ordinarily far more efficient than distilling columns.

In gas chromatography, samples are injected as rapidly as possible, so that a substance is placed on the column as a narrow "plug." However, just as in the Craig extraction technique, when the substance moves along it spreads out through more and more plates but occupies a progressively smaller fraction of the total number of plates which it has encountered. In other words, as we increase the number of plates, the absolute width of the elution band increases, but there is a decrease in width relative to the total base of the operation. In this perspective, elution bands look narrow as they emerge from a good column, broad from a poor column.

Calculation of the Number of Theoretical Plates

It is often desirable to evaluate a chromatographic column by measuring n, the number of theoretical plates. With a Craig apparatus, the plates may simply be counted. But this cannot be done with a column, because the plates are now imaginary; a theoretical plate is a mental concept, part of a model developed to explain the chromatographic process in familiar terms. We can determine the apparent number of plates, however, because the model implies a relationship between the characteristics of a chromatographic elution band and the number of plates in the column. The mathematical derivation of an equation for calculating n from a chromatogram is beyond the scope of this book,[5] but the general idea is intuitively reasonable, and the result is a very simple formula. In the Craig extraction scheme, we were able to calculate how a solute distributed itself through a known number of tubes. We may readily imagine that, if we had known instead the characteristics of the distribution, we might have turned our thinking around and calculated the number of tubes. Similarly, as implied in the preceding paragraph, the time required to elute a solute from a column and the width of the elution band should enable us to calculate n.

Figure 17.3 shows a Gaussian elution band and the parameters which are used to calculate n. The time from injection of the sample to the appearance of the peak of the elution band at the detector is called the *retention time*, t_R. The formula for calculating n from t_R and the bandwidth depends upon where the width is measured. The two commonest ways are given here. If the width is measured halfway between the baseline and the top of the band, we may designate it as $w_{1/2}$. Then the formula for obtaining n is

$$n = 5.54\left(\frac{t_R}{w_{1/2}}\right)^2$$

If the width is measured at the baseline using the construction shown in the figure,

[5] The interested reader may find a derivation in O. E. Schupp, III, *Gas Chromatography*, Vol. 13 of E. S. Perry and A. Weissberger, eds., *Technique of Organic Chemistry*, Wiley–Interscience, New York, 1968, pp. 39–46.

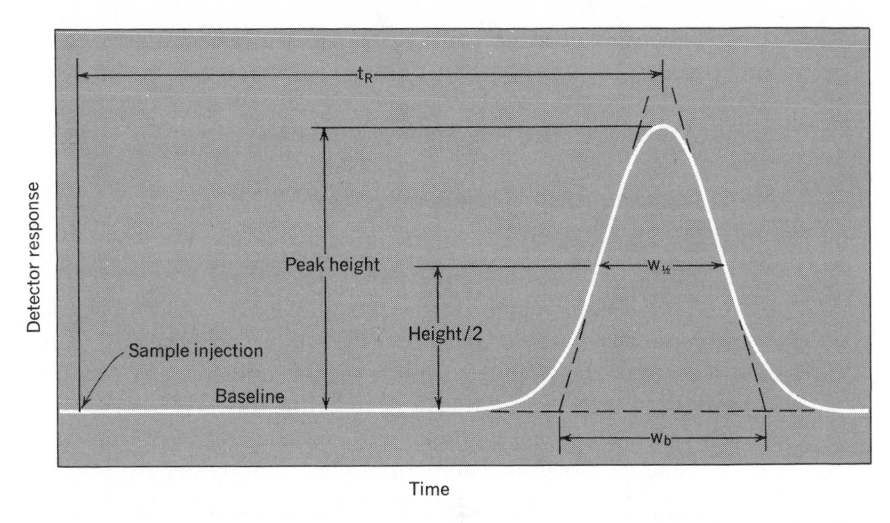

Figure 17.3 Chromatographic elution band showing measurement of t_R and w for estimating n, the number of theoretical plates.

we designate it w_b; the formula then becomes

$$n = 16\left(\frac{t_R}{w_b}\right)^2$$

In the latter case, tangents to the band are drawn at the two inflection points; the width, w_b, is the distance between the intersections of these tangents with the baseline. The two formulas give comparable results, and the choice is a matter of personal preference. Although there may be some difficulty in locating accurately the inflection points of an elution band, the slopes of the tangents are not highly sensitive to this error.

If a mixture of several compounds with different K values is injected into the chromatograph, then of course several elution bands will be recorded. These will show different retention times, but the bandwidths will also be different; thus comparable (although probably not identical) n values will be calculated from the several bands. A component with a larger K value will spend more time in the liquid phase and hence take longer to be eluted. Referring to a Craig type of model with discrete plates, we should say that more transfers would be required to elute such a compound. But the band is also broadened thereby, so that theoretically the n value should be the same as that obtained for an earlier band.

For the purpose of determining n, the units for the abscissa in Fig. 17.3 make no difference. Thus, although t_R is defined as a time, we may actually measure distances on the recorder chart paper in centimeters or millimeters; of course, t_R and w must be measured in the same units.

The number of plates in a given column is found to vary with sample size in a fairly regular way. Overloading of any column leads to a deterioration of performance and poor separations. To obtain a value for n, sometimes extrapolation to zero sample size is performed on a graph of measured n-values vs. sample size.

Nonideal Behavior: The van Deemter Equation

GLC as Linear Nonideal Chromatography

The plate model based upon a Craig type of solute distribution between gas and liquid phases, with equilibrium attained prior to each transfer from one plate to the next, represents what is sometimes called *linear ideal chromatography*. Linear in this connection means that the distribution coefficient, K, is independent of solute concentration; thus a graph of concentration in the liquid phase vs. concentration in the gas phase is a straight line (Fig. 17.4, curve 1a). Such a graph is often called an *isotherm*. In general, a linear isotherm leads to a symmetrical elution band as shown by curve 1b in Fig. 17.4. A departure from Henry's law behavior, shown by a nonlinear isotherm (Fig. 17.4, curves 2a and 3a), leads to a skewed elution band (Fig. 17.4, curves 2b and 3b). Referring to curves 2a and 2b, we may interpret the asymmetric elution band as follows: Where the solute concentration is high, a greater fraction of solute remains in the gas phase than is the case at lower concentrations. As a result, solute at the peak will move faster through the column, and hence the peak will tend to catch up to the leading edge of the elution band and leave a long tail at the trailing edge.

Nonlinear isotherms of the type shown in Fig. 17.4, curve 2a, are common in adsorption equilibria, and thus skewed elution bands with tailing as shown in curve 2b are often encountered in gas–solid chromatography and in liquid adsorption chromatography. At the low concentrations normally employed in

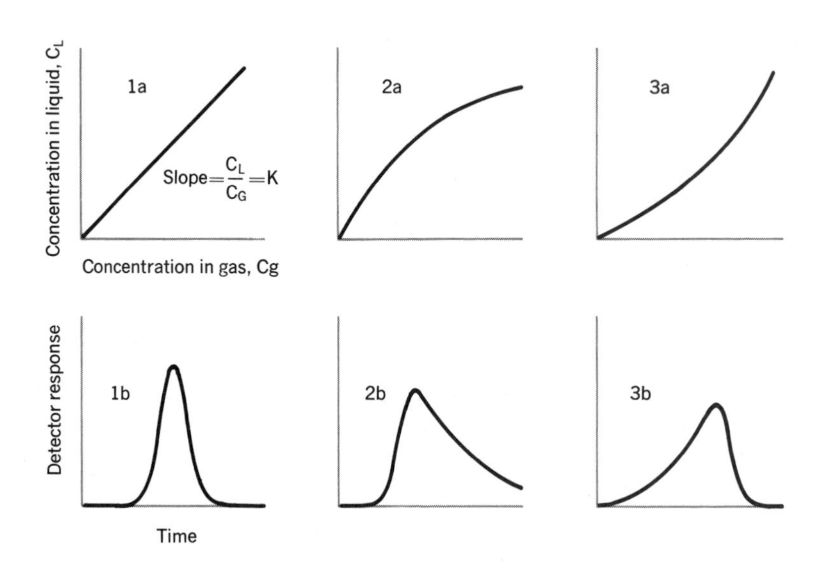

Figure 17.4 Linear and nonlinear isotherms (a) and the shapes of the corresponding elution bands (b).

GLC, deviations from Henry's law do not generally occur, and thus GLC commonly provides an example of linear chromatography.

Ideal behavior, on the other hand, is not attainable in any actual chromatographic process. First, ideality would require that the sample be placed initially in only the first plate. This can be done in the Craig apparatus, but in a column, where HETP may be only a fraction of a millimeter, it is impossible with a finite sample. Second, ideal chromatography would require a column whose packing was perfectly uniform in regard to particle size and shape, liquid loading, and geometric array; further, the velocity of the moving phase would have to be the same everywhere in the column. Third, in ideal chromatography, equilibrium would always exist at all points in the column between the stationary and moving phases with regard to the solute distribution; because the flow of the moving phase is continuous, this means that equilibration would have to be instantaneous. Finally, ideal chromatography would require that the solute move along the column as a result only of the motion of the moving phase; that is, the solute could not spread out in the column by its own tendency to diffuse.

Thus, under the usual conditions, GLC provides an example of *linear nonideal chromatography*. Departures from the requirements for the ideal process as listed in the paragraph above result in elution bands which are broader than the hypothetical bands for an ideal case. A detailed theoretical consideration of the kinetic factors which lead to this band broadening has given considerable insight into the nature of the chromatographic process in real columns. Not only has this been intellectually satisfying; practical improvements have resulted in terms of better separations in less time, and these have spilled over from GLC to revitalize liquid chromatography as well. A mathematical analysis of chromatographic theory is far beyond the scope of this text, but the principal factors are understandable in qualitative terms. We may summarize these for what insights they provide and leave the details to the experts.

The pioneering treatment of nonideal behavior in GLC was presented in 1956 by van Deemter et al.[6] These workers considered three major factors that caused band broadening when a solute, initially a narrow plug, migrated along a column. An equation was derived containing three terms which represented the contributions of these three factors to HETP.

Eddy Diffusion

The first factor, which is usually called eddy diffusion, arises from the multiplicity of pathways for a gas flowing through a column which is packed with particles of various sizes and shapes arranged in an irregular manner. As it flows through the channels among the packing particles, the carrier gas divides into many streams which may merge and again split in a complex manner as the gas follows a myriad of routes through the column. Likewise, solute molecules, as they move along with the flowing carrier gas, will follow many paths, some shorter and some longer than the average distance. This means that the original solute plug will spread out, some molecules reaching the detector sooner, some later,

[6] J. J. van Deemter, E. J. Zuiderwig, and A. Klinkenberg, *Chem. Eng. Sci.*, **5**, 271 (1956).

with many at an average time. The magnitude of the contribution of eddy diffusion to band broadening ought to depend upon the size of the packing particles, their shape, and the uniformity of their distribution in the column.

There has been some dispute about the importance of eddy diffusion which appears now to be settled. Perhaps this factor was more severe with the column packing materials and techniques of the earlier years of GLC. Measurements with very carefully packed columns have suggested that it is less important than early workers supposed. Apparently, in a good column most of the pathways actually followed by the gas are about the same length. Also, it has been proposed that lateral diffusion of solute from one portion of the gas stream to an adjacent one might counter the effect of eddy diffusion upon the solute profile. Furthermore, it appears that some of the band broadening formerly attributed to eddy diffusion actually occurred outside the column, for example in the glass wool that was sometimes used to plug the end of the column, in the tubing that led from the column to the detector, and in the detector itself. It now seems clear that eddy diffusion may be only a very small factor in a well-designed gas chromatograph with a good column, and indeed it may be practically negligible as compared with other band broadening factors. Nevertheless, eddy diffusion is worth noting because it is negligible only if rendered so by proper attention to details in the design and fabrication of the chromatograph and the column.

Longitudinal Diffusion

The second factor contributing to band broadening is longitudinal diffusion of the solute in the gas phase. Solute molecules tend to diffuse along concentration gradients, and thus a solute band moving along a column will broaden as molecules spread into the regions of lower concentration ahead of the band and behind it. Solute molecules spend part of their time in the gas phase and part in the liquid phase, but diffusion is much faster in gases than in liquids (diffusion coefficients in the gas phase are typically of the order of 10^5 times as great as in liquids). Thus diffusion in the liquid phase is generally considered negligible in GLC. Diffusion takes time, and thus the contribution of diffusion to band broadening will increase with the length of time required to elute the band from the column. Hence the diffusion term in the van Deemter equation involves the velocity of the carrier gas, becoming smaller as the velocity increases.

Nonequilibrium in Mass Transfer

The last factor in nonideality arises from the fact that equilibrium cannot be attained for the distribution of the solute between stationary and mobile phases in the face of continuous flow of the carrier gas. Using C_L and C_G for solute concentrations in the liquid and gas phases, respectively, we may write Henry's law as

$$C_L = KC_G$$

But Henry's law describes an equilibrium situation, and for cases where there is insufficient time for equilibration, we must write

$$C_L = KC_G \times f(t)$$

where $f(t)$ is some function of time which reflects the kinetics of the mass transfer process between the two phases. When t is large (i.e., when equilibrium is approached), then, of course, $f(t)$ must approach unity so as to yield Henry's law.

The band broadening that arises from the fact that equilibration is not instantaneous is depicted in Fig. 17.5, where both ideal and nonideal solute distributions are shown. The ideal case (upper portion of Fig. 17.5) is simply a Craig distribution, where equilibrium is attained in all of the plates before transfers are made. The lower part of Fig. 17.5 illustrates the case where the gas phase is moving continuously, equilibrium is never attained at any point, and hence in general $C_L \neq KC_G$. Where the front of the solute zone in the gas phase encounters fresh liquid, some of the solute dissolves in the liquid but, because we do not wait for equilibrium, $C_L < KC_G$. The result is that some of the solute

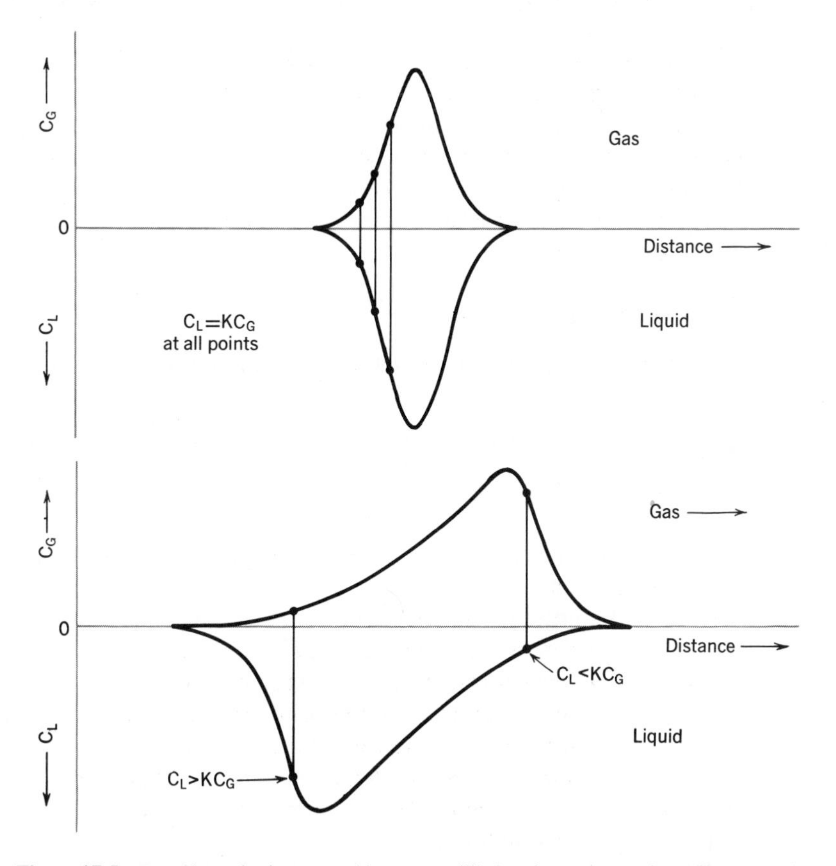

Figure 17.5 Band broadening caused by nonequilibrium in mass transfers: Concentration profiles in gas and liquid phases for a solute zone within a column. Upper: ideal distribution assuming instantaneous equilibration. Lower: actual distribution resulting from finite rate of mass transfer.

which would have been retarded by the liquid phase if equilibrium had been attained actually continues to migrate along the column with the carrier gas. At the other extreme, the rear of the gas-phase solute zone, solute cannot leave the liquid rapidly enough to equilibrate with the fresh carrier gas, and $C_L > KC_G$. In other words, part of the solute is retarded to a greater degree by the stationary liquid than would have been the case if equilibration had been instantaneous. This nonideal behavior continues as the solute traverses the entire length of the column, with the obvious result that the elution band is broader than it would otherwise have been.

The extent of band broadening arising from the slow mass transfer kinetics depends upon the time available for the process to move toward equilibrium. Thus the nonequilibrium term in the van Deemter equation ought to involve the carrier gas velocity and, in contrast with the longitudinal diffusion term considered above, it should become larger as the velocity increases. Furthermore, we might expect that solute in a thin film of liquid would be closer to equilibrium with the moving gas stream than would be the case for solute which had penetrated deep pools of liquid. Thus the magnitude of the nonequilibrium term should depend upon the quantity of liquid with which the solid support is impregnated and the manner of its distribution within the nonuniform pores of the solid. In addition, the nonequilibrium term contains the diffusion coefficient of the solute in the liquid phase. The faster the solute molecules can move along concentration gradients in the liquid to and from the interface with the gas, the closer will be the approach to equilibrium in the mass transfer process, and the less the band spreading attributable to this factor.

The van Deemter Equation

Consideration of the three factors discussed above led to the following equation:

$$\text{HETP} = 2\lambda d_p + \frac{2\gamma D_G}{u} + \frac{8kd_f^2}{\pi^2(1+k)^2 D_L}u$$

where

λ = dimensionless parameter measuring the irregularity of the column packing
d_p = diameter of the packing particles
γ = correction factor accounting for the irregularity of diffusion pathways through the packing material
D_G = diffusion coefficient of the solute in the gas phase
u = linear velocity of the carrier gas
k = constant for a particular solute and a particular column
d_f = "effective film thickness," a measure of the liquid loading of the packing material
D_L = diffusion coefficient of the solute in the liquid phase

Theoretically minded people have scrutinized the three terms in the van Deemter equation very carefully and have proposed extensions of these as well as additional factors in order to gain a better understanding of the chromatographic process. Such work, which involves sophisticated mathematics, is far beyond the scope of an introductory textbook. The van Deemter equation as given here focuses attention upon the major factors which cause band broadening and serves quite well, for the interested reader, as a takeoff point for understanding the further refinements that have followed.

The van Deemter equation is often seen in the abbreviated form

$$\text{HETP} = A + \frac{B}{u} + Cu$$

where

$$\begin{aligned} A &= \text{eddy diffusion term} \\ B/u &= \text{longitudinal diffusion term} \\ Cu &= \text{nonequilibrium in mass transfer term} \end{aligned}$$

A graph of HETP vs. u is a branch of a hyperbola, as shown in Fig. 17.6. The dashed construction on the graph depicts the contributions of the A, B/u, and Cu terms to HETP at various carrier gas velocities. The A term remains constant, independent of velocity. At very low velocities, most of the band broadening is due to longitudinal diffusion, while at high velocities, the increasing departure from equilibrium in the mass transfer process becomes dominant. There is an optimal velocity where the best balance of these factors is obtained, that is where

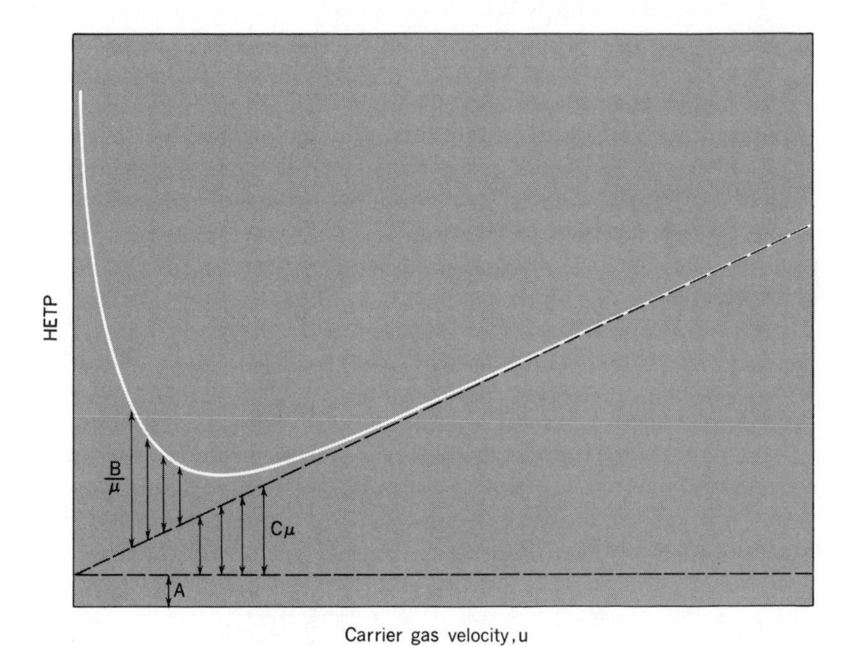

Figure 17.6 Schematic depiction of the van Deemter equation HETP $= A + B/u + Cu$. Note that the contribution of the A term to HETP is independent of velocity, that B/u increases as velocity decreases, and that Cu predominates at high velocities.

HETP is a minimum or the number of plates in the column is a maximum. The minimum in the graph is a rather shallow one, and in practice it is not necessary to locate it exactly; it is obviously advantageous, however, to be in the right neighborhood.

Although more recent work has led to a more sophisticated understanding of the processes that occur in the column, the van Deemter equation as presented here is fairly good, at least with regard to predicting the shape of a graph of HETP vs. u. Curves very similar to the one shown in Fig. 17.6 are in fact obtained in the laboratory. Such curves are not calculated because values are generally not at hand for the parameters such as λ and γ in the van Deemter equation. A, B, and C are easily obtained from experimental data by measuring HETP at three carrier gas velocities and setting up three equations in three unknowns. The values will vary greatly from one sort of column to another; the following have been presented as typical for packed columns of the sort commonly used in analysis:[7]

$$A \cong 0 - 1 \text{ mm}$$

$$B \cong 10 \text{ mm}^2/\text{s}$$

$$C \cong 0.001 - 0.01 \text{ s}$$

$$\text{HETP}_{\text{min}} \cong 0.5 - 2 \text{ mm}$$

$$u_{\text{opt}} \cong 1 - 10 \text{ cm/s}$$

In summary, the plate theory of chromatography enables us to calculate a very useful measure of column performance, the number of theoretical plates or, if we wish, HETP. This is very simply done using one of the formulas on page 464. But the plate theory in itself does not suggest how the performance of a column may be improved. The so-called rate theory, as exemplified by the van Deemter equation, on the other hand, gives definite factors such as particle size, liquid loading, and carrier gas velocity over which we have some control by which improved performance may be obtained.

Resolution

In general, the positions of elution bands on the horizontal axis of the chromatogram and their widths will determine how complete a separation of the starting mixture has been accomplished. Although samples may have many components, we suppose that one pair of these will be the most difficult to separate and confine our discussion of resolving mixtures to two-component systems.

The *resolution*, R, of two components is often defined as follows, using the terms shown in Fig. 17.7:

$$R = \frac{2(t_{R_2} - t_{R_1})}{w_{b_1} + w_{b_2}}$$

[7] A. B. Littlewood, *Gas Chromatography: Principles, Techniques, and Applications*, 2nd ed., Academic Press, Inc., New York, 1970, p. 202.

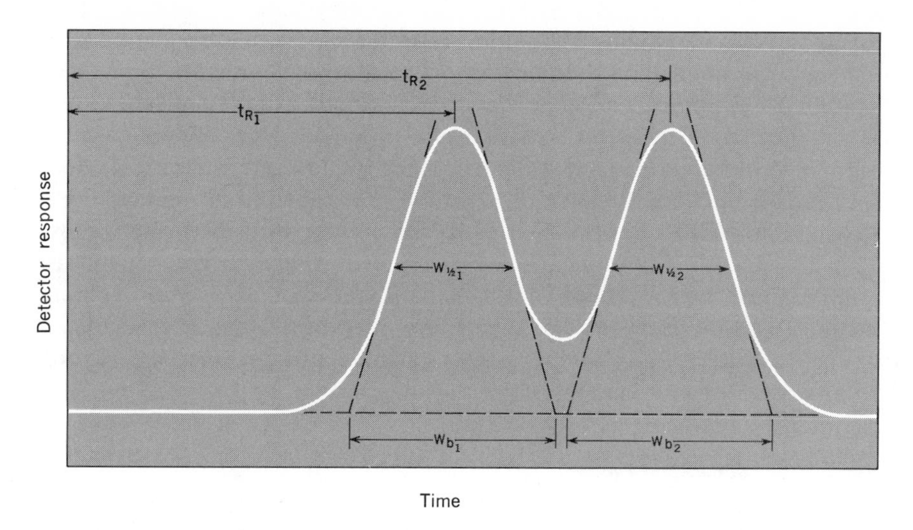

Figure 17.7 Measurements used to calculate the resolution, R, of two peaks.

Alternatively, if the widths are measured halfway between the baseline and the tops of the bands, the equation becomes

$$R = \frac{2(t_{R_2} - t_{R_1})}{1.699(w_{1/2_1} + w_{1/2_2})}$$

Often, for two bands which are close together, the widths are about the same; in that case, the 2 can be removed from the numerator and one of the w terms dropped from the denominator.

If $R = 1.5$, the two solutes are virtually completely separated; there will be only 0.3% overlap of the two elution bands. If $R = 1$, the separation is adequate for most analyses; the overlap is about 2%. As R decreases below 1, the overlap becomes progressively more severe until, at about $R = 0.75$ (50% overlap), the separation becomes unsatisfactory for most purposes. For quantitative analysis by GLC, the area under a solute peak is the best measure of the quantity of that solute in the sample. Complete resolution of a mixture is not required for this. If overlap is not too great and if the peaks are symmetrical, fairly good estimates of the areas can be made. This can be extended by computer methods for analyzing a complex shape into a family of discrete peaks. Of course, whatever technique is used, some point will be reached beyond which good results cannot be obtained. It then becomes necessary to consider the factors that determine resolution in order to improve the situation.

Length of Column

The number of theoretical plates in a column, everything else being the same, is proportional to its length, and hence one of the obvious ways to improve resolution is to employ a longer column. The separation of two peaks, $t_{R_2} - t_{R_1}$, is directly proportional to the distance that the two solutes migrate, whereas the width of an elution band increases directly with the square root of the distance. Thus, as we lengthen a column, two bands will separate faster than they broaden, so to speak, and resolution will improve. There are limitations, however, on column length, of which we may mention two. If a column is too long, the pressure required to give a reasonable carrier gas flow rate is excessive. Second, the longer the column, the longer will be the time required for elution. The efficiency of a busy laboratory handling many samples might be better served by improving the resolution in some manner that did not lead to great increases in t_R values. Nevertheless, if preliminary experiments with an ordinary column, perhaps 4 to 6 ft long, showed a fair separation with a particular sort of sample, then it might be sensible to employ a somewhat longer column, say 8 or 10 ft, in order to achieve a really good separation.

Separation Factor

If a satisfactory separation is not obtained with a good column of reasonable length after careful attention to operating parameters such as temperature (see below) and carrier gas flow rate, then the best approach is usually to try a different stationary liquid phase. In other words, if logical attempts to achieve resolution by narrowing the solute bands fail, then we must move the peaks farther apart by changing the K values for the solutes.

The ratio of retention times, t_{R_2}/t_{R_1}, is called the *separation factor*, S (some writers use S.F. and some α for this ratio). Usually, the ratio of retention times is about the same as the ratio of the K values for the two solutes. Thus

$$S = \frac{t_{R_2}}{t_{R_1}} \cong \frac{K_2}{K_1}$$

Note that S is not the same thing as the resolution, R. The ratio of retention times, measured at the peaks of the elution bands, does not in itself describe the effectiveness of the separation, because it tells nothing about the widths of the bands. However, it may be shown, as we might expect intuitively, that there is a relation between R and S if the number of theoretical plates in the column is taken into account:

$$R = \frac{n^{1/2}(S - 1)/S}{4}$$

Taking $R = 1$, which as noted earlier represents a fairly good separation of two solutes, we may plot the number of plates required vs. the separation factor, S, which yields the curve shown in Fig. 17.8. The curve approaches the ordinate axis asymptotically, reflecting the fact that, if $S = 1$, no separation is possible however long the column. As S increases above 1, the number of plates required decreases rapidly. In other words, if we can promote a rather small increase in S by changing the liquid phase, it may be far more effective in improving the resolution than even a large increase in the length of the column.

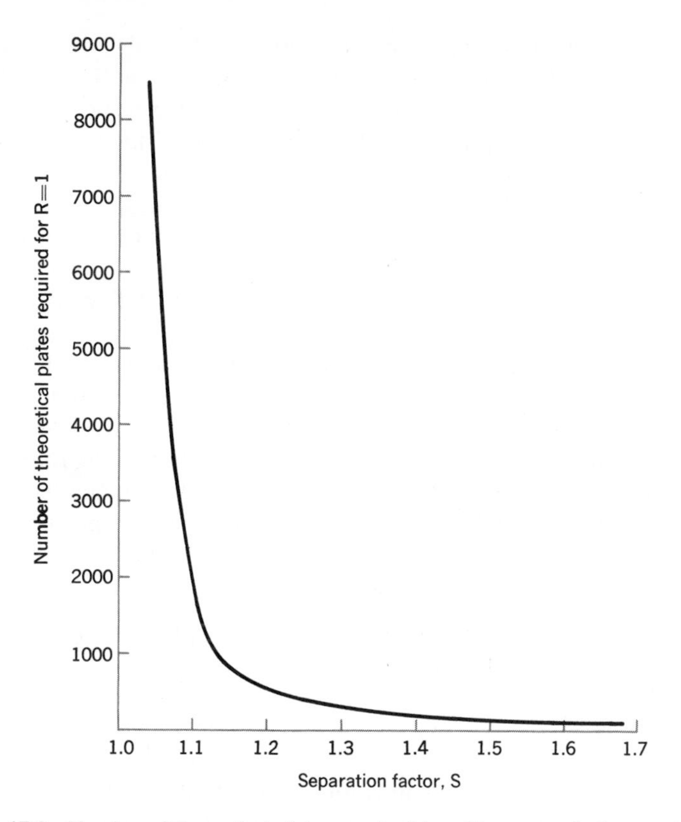

Figure 17.8 Number of theoretical plates required to achieve a resolution equal to one plotted as a function of the separation factor.

Factors in Retention

For many purposes, the *retention volume* of a solute, V_R, is more convenient than the retention time. The retention volume is the product of the retention time and the carrier gas flow rate; since there is an inverse relation between flow rate and retention time, V_R is independent of flow rate. The flow rate is ordinarily measured at a point beyond the column, as shown in Fig. 17.1. This measured value, F, should be corrected to account for the fact that the column is at a different temperature than the flow meter; a further correction is required if the flow meter introduces moisture into the gas or creates an appreciable pressure drop of its own. If F_c is the corrected flow rate, then

$$V_R = t_R \times F_c$$

Because the gas is compressible, its velocity is not uniform throughout the column—the gas moves faster near the outlet end—and V_R should be corrected to an average column pressure to yield the *corrected retention volume*, V_R°. The corrections are not discussed in detail in this text, but they may be found in the books on gas chromatography listed in the references.

Consider a compound which does not interact with the stationary phase (i.e., $K = 0$). This compound is not retarded by the liquid in the column, but its retention time is not zero; some time, which we may call t_G, will be required simply to wash the compound through the column. Using the same corrections as above, we may calculate a corrected retention volume, V_G°, for such a compound. This volume should amount to the same thing as the space within the column which is available to the gas, that is, the portion of the column which is not occupied by the packing material and its liquid load. Other terms, such as "interstitial volume" and "void volume," have been used for the analogous space in other forms of chromatography such as ion exchange.

The fixed gases of the air are not appreciably soluble in most organic liquids that are used as stationary phases. Thus it is easy to obtain t_G by injecting a little air along with the sample. The thermal conductivity detector responds to air, and a small blip called the "air peak" appears on the recorder chart, from which V_G° can be calculated. (The other common detector, the flame ionization detector, does not respond to the gases of the air; another approach is then required.)

Distribution Coefficient and Liquid Load

Let V_L be the volume of the liquid phase in the column, and recall the distribution coefficient:

$$K = \frac{C_L, \text{wt/ml}}{C_G, \text{wt/ml}}$$

Now when the peak of an elution band for some solute appears at the detector, we suppose that half of that solute is still in the column and half has been eluted in a volume of $V_R^\circ - V_G^\circ$. Then we may cancel the weight terms in the preceding equation and write

$$K = \frac{1/V_L}{1/(V_R^\circ - V_G^\circ)} = \frac{V_R^\circ - V_G^\circ}{V_L}$$

whence

$$KV_L = V_R^\circ - V_G^\circ$$

and

$$V_R^\circ = V_G^\circ + KV_L$$

More rigorous derivations of this equation and a discussion of the assumptions involved may be found in monographs on gas chromatography.

V_G° is often quite small compared with V_R°. Thus it is frequently the case that the retention volume for a solute is almost directly proportional to the quantity of liquid phase in the column. Similarly, with a given liquid loading, the retention volume will vary directly with K. In other words, the experimenter has considerable control over retention in his choice of the liquid phase and the quantity used in preparing the column.

It may be noted that the time corresponding to a given retention volume is also under the control of the operator via the flow rate, but, as we have seen, the analysis time cannot be shortened at will in this manner without the penalty of poorer resolution. With certain types of mixtures, it may be advantageous to vary the flow rate during the chromatographic run, starting with a low value for optimal resolution of solutes whose K values are small and close together, and continuously increasing the flow rate to accelerate the elution of laggard solutes which are easily resolved but are spending more time than necessary in the column. This technique is known as "programmed flow gas chromatography."

Temperature

Virtually every aspect of GLC is sensitive to temperature to some extent. The volume of the liquid phase and hence also the gas space in the column, the viscosity of the carrier gas and therefore the inlet pressure required for a given flow rate, diffusion coefficients of solutes—these are a few examples of factors which may be affected to some degree by the column temperature. But we are particularly interested here in the often very pronounced direct effect of temperature upon solute retention. The usual effect of a temperature increase is to lower the value of the distribution coefficient K; in other words, at the higher temperature, a solute is driven out of the liquid phase in accord with the general rule that an increase in temperature lowers the solubility of a gas in a liquid. Decreasing K in turn decreases the retention time and the retention volume. The magnitude of the effect depends upon the nature of the solute and of the liquid phase and the temperature region investigated, but roughly the change in retention volume is of the order of a few percent, say 3 to 12%, per degree. An analysis, then, is completed most rapidly at the highest column temperature compatible with the desired separation and sample stability.

On the other hand, the separation factor for a pair of solutes is generally larger the lower the temperature. As a crude rule of thumb, the higher the temperature the more similar the behavior of two compounds in a GLC column. Thus the column temperature selected for an analysis ought to be low enough to achieve the necessary separation, but no lower than this so as not to waste time.

For a series of solutes which interact in the same manner with the liquid phase, for instance an homologous series, the distribution coefficients generally bear an inverse relation to the vapor pressures; the larger the vapor pressure, the smaller is K. In general, the lower the boiling point of a solute, the greater will be its vapor pressure at a given temperature. Thus the components of a mixture of such solutes will emerge from a column in order of increasing boiling points. (If some of the solutes interact in specific ways with the stationary liquid, e.g., by hydrogen bonding, then this simple rule may not hold.)

Figure 17.9 (a) shows a chromatogram of a hydrocarbon mixture obtained in the ordinary manner with the column held at a certain temperature. The temperature selected was a compromise: it was too high to yield an optimal separation of the lower compounds in the hydrocarbon series and too low for the higher-molecular-weight compounds. A much nicer chromatogram of the same mixture is seen in Fig. 17.9(b): resolution is better, and in fact a number of impurities in the hydrocarbons that were mixed to prepare the sample may be seen

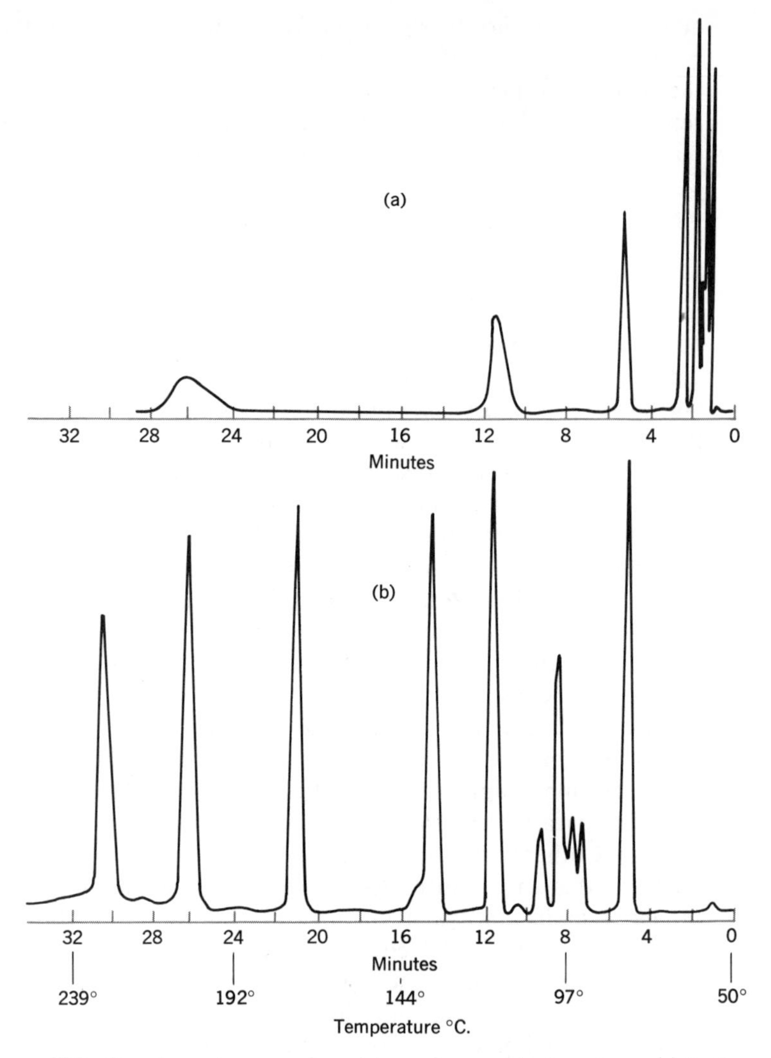

Figure 17.9 Gas chromatograms of a mixture of normal hydrocarbons. (a) Isothermal chromatogram of the following mixture at 168°C: (1) pentane, (2) hexane, (3) heptane, (4) 1-octene, (5) decane, (6) 1-dodecene, (7) 1-tetradecene. (b) Programmed temperature chromatogram of the same mixture. (Reprinted from *Anal. Chem.*, **30**, 1157 (1956); copyright by the American Chemical Society. Reprinted by permission of the copyright owner.)

which do not show up in curve (a). All the bands have about the same shape, which facilitates quantitative measurements; peak number 7 which was very low in curve (a) is now higher above the baseline, representing a better signal-to-noise ratio for improved quantitative accuracy. Curve (b) was obtained by the technique of *programmed temperature GLC.* Here the temperature of the column was raised during the chromatographic run, starting at a temperature that was suitable for the lower members of the series, and finishing at a higher temperature where the elution of the higher boiling components was more satisfactory. The effect is rather similar to that of programming the flow rate as mentioned in the section above. Various temperature–time functions have been studied, but the commonest by far for ordinary work is a linear program: the temperature increases linearly with time at so many degrees per minute. Modern chromatographs often provide for this capability, and the operator can select on the panel an initial temperature, a rate of increase, and a final temperature.

EXPERIMENTAL ASPECTS OF GLC

Having seen what gas–liquid chromatography is, and after considering the theory, we may now discuss briefly some of the more practical aspects which will make the gas chromatograph less of a mysterious "black box."

Carrier Gas

Various gases have been used in GLC, for example, hydrogen, helium, nitrogen, argon, carbon dioxide, and even water vapor. The lighter gases, hydrogen and helium, permit more longitudinal diffusion of solutes, which tends to lower column efficiency, particularly at lower flow rates. Thus nitrogen might be a better choice of carrier gas in order to accomplish a really difficult separation. In addition, it is cheaper than helium and safer in the laboratory than hydrogen. However, there is another consideration, the characteristics of the detector. It is obviously desirable that the response of the detector to the components of the sample differ greatly from its response to the ever-present carrier gas. In the case of the thermal conductivity cell, which is one of the most widely used detectors, this requirement is much better met by the lighter gases, hydrogen and helium, as we shall see below. Thus, with instruments employing this detector, helium is by far the commonest carrier gas in the United States, while in Europe, where helium is very expensive, hydrogen is more widely used. With the flame ionization detector, which has become a rather common one in recent years, nitrogen is probably the most widely used carrier gas.

Sampling System

Liquid samples, typically ranging from a small fraction of 1 μl to perhaps 25 μl or more, are usually injected through a rubber septum by means of a hypodermic syringe. Special syringes delivering various volumes in the microliter range are on the market, sometimes equipped with mechanical devices that aid in reproducing sample size. Gaseous samples may also be injected, or they may be introduced by means of various gas-sampling devices designed for commercial chromatographs. The injection technique is important: the sample should be introduced as a sharp "plug" rather than being slowly bled into the carrier gas stream. Slow injection leads to much more band spreading than is necessary; actually, HETP calculated from the elution peak as described above is a function of the injection rate. Good injection technique requires practice.

It is important that the size of the sample not be too large for the apparatus. Overloading has an extremely deleterious effect upon column efficiency. The lower limit of sample size is determined by the detector: so far as the column is concerned, the smaller the sample the better. The sensitivity of the detector determines how small a sample can be handled.

The temperature at the injection port is very important. If a liquid sample evaporates slowly, the result is similar to that caused by injecting too slowly. The injection port is usually heated independently of the heating unit surrounding the column, and generally it should be held at a temperature above the boiling points of the sample components. On the other hand, the temperature should be below a level where the compounds would decompose.

Column

The stationary phase in GLC is a liquid but it cannot be allowed simply to slosh around inside a tube. The liquid must be immobilized, preferably in the form of a thin layer of large surface area. This is most commonly accomplished by impregnating a ground-up solid material with the liquid phase before the column is packed. The solid should be chemically inert toward the substances which will be chromatographed, stable at the operating temperature, and of large surface area per unit weight. The pressure drop required for desirable gas flow rates should not be excessive. Mechanical strength is desirable so that the particles will not break and alter the particle size distribution with handling. Most of the solids employed as supports in GLC are highly porous, but the characteristics of the pores are very important. For example, the pores in silica gel tend to be narrow; they fill up with the liquid and provide insufficient area of gas–liquid interface. The active adsorbents such as activated charcoal and silica gel are poor solid supports. Even when coated with the liquid film, these solids adsorb sample components, causing "tailing" of the elution bands as shown in curve 2b of Fig. 17.4. The commonest solid support materials are diatomaceous earth (a deposit formed on ocean bottoms from the siliceous shells of a certain type of algae) and firebrick. The materials are ground and carefully graded with respect to particle size and often subjected to various chemical pretreatments to improve their surface qualities. The preparation of the solid, its impregnation with the liquid phase, and the final packing into copper, stainless steel, or glass columns used to be an art that was cultivated by chromatographers. Today, it is much more common to buy ready-made columns from the manufacturers who have made available a wide variety of very good columns.

There is another type of column for GLC called the open-tubular or capillary column. This is a long, thin tube of glass or other material such as stainless steel, perhaps 0.1 to 1 mm in diameter, sometimes several hundred feet long, coiled up to save space. The inner surface is coated with a very thin layer of the stationary liquid phase, just that quantity which will adhere as a film on the glass or metal; there is no column packing in the usual sense. The pathways through the column are practically the same length for all molecules of the sample, and hence eddy diffusion is virtually zero in open-tubular columns. The very thin liquid film, containing no deep, stagnant pools, promotes a rapid approach to equilibrium in the partition process. The columns are very long, and it has been argued that they are not much more efficient than packed columns of comparable length would be. On the other hand, packed columns of that length would require relatively enormous pressure drops to attain reasonable flow rates. Because of the very light liquid loading, open-tubular columns can handle only very small samples, and their widespread use awaited the development of very sensitive detectors. Very impressive separations are often obtained with these columns. Open-tubular is perhaps preferable to capillary in designating such columns; the tubing is not fine enough to be considered a capillary in the original sense of hair-like.

The stationary liquid phase must be selected with the particular separation problem in mind. The liquid should have a very low vapor pressure at the column temperature; a common rule of thumb suggests a boiling point at least 200°C above the temperature to which the liquid will be subjected. The two important reasons for desiring low volatility are, first, loss of liquid will eventually destroy the column, and second, the detector will respond to the vapor of the stationary phase with resulting drift of the recorder baseline and lowered sensitivity toward the components of the analytical sample. Some commercial gas chromatographs have dual columns to compensate for the effect of liquid bleed on the detector, but even here, excessive volatility is undesirable.

Obviously, the liquid phase should be thermally stable at the column temperature, and, except in special cases, it should not react chemically with the sample components. The liquid must have an appreciable solvent power for the sample. Recalling the old rule that "like dissolves like," it may be stated rather crudely that generally there should be some chemical resemblance between the liquid substrate and the solutes to be separated. Thus the saturated hydrocarbon squalane ($C_{30}H_{62}$, MW 423, boiling point about 350°C) is a good liquid phase for the separation of low-molecular-weight alkanes on a column that will not be heated above about 150°C. For the separation of aromatic hydrocarbons, the aromatic liquid benzyldiphenyl, useful up to about 120°C, is sometimes recommended. A polyglycol column might be used to separate a mixture of alcohols. Of course, it is not required that the stationary liquid match the solutes functional group for functional group, and sometimes one liquid phase will serve for the separation of a variety of mixtures. This has led to the designation "general purpose" for some columns, although this is misleading in that no liquid phase provides completely general effectiveness for separating all classes of solutes. Examples of general purpose liquids are silicone oils and greases, useful for a wide variety of nonpolar solutes, and polyethyleneglycols (Carbowaxes), widely used for mixtures of polar solutes. Lists of liquid phases, recommended temperature limits, and the types of compounds for whose separation they are useful may be found in monographs on GLC.

The quantity of liquid substance applied to the solid support is important. If too much liquid is present, solutes spend too much time diffusing through the liquid phase, and the separation efficiency is lowered. Too little liquid allows solutes to interact with the solid itself, in which case adsorption may cause "tailing" and consequent overlapping of elution bands.

Detector

The separation process occurs in the column, and hence this component must be considered the heart of the instrument. On the other hand, the separation would be of little value without some way to detect and measure the separated solutes as they emerge from the column.

Two types of detectors are commonly distinguished, integral and differential. An integral detector provides at any instant a measure of the total quantity of eluted material which has passed through it up to that time. The first paper on GLC[8] described an example of an integral detector. A mixture of fatty acids was chromatographed, and the effluent gas from the column was bubbled through an aqueous solution containing a *p*H indicator. When an acid emerged from the column, the *p*H of the solution dropped, the indicator changed color, and a light beam of appropriate wavelength passing through the solution was attenuated by absorption. The resulting change in the electrical signal from a photodetector activated a relay which turned on a buret containing sodium hydroxide. The addition of the base restored the *p*H of the solution to its original value, and hence the indicator to its original color, whereupon the buret was automatically shut off. The volume of titrant was recorded as a function of time, and the resulting chromatogram, of the type shown in Fig. 17.10 (a), consisted of a series of steps, each representing the titration of one of the acids in the original mixture. The quantity of each acid was easily found by measuring the height of the corresponding step. This detector, which seems crude and clumsy by hindsight, was useful at the time because it was easily assembled from available components, and it served its purpose very nicely at the birth of GLC. But integral detectors have been largely supplanted by differential detectors; the latter are found on the overwhelming majority of modern gas chromatographs, and it is these that we wish to emphasize in this chapter.

Differential detectors yield the familiar chromatograms consisting of peaks rather than steps, as shown in Fig. 17.10 (b). Two major classes may be distinguished: first, those which measure the *concentration* of a solute by means of some physical property of the effluent gas stream, and second, those which respond to the solute directly and hence measure its *mass flow rate.* This distinction will be clarified as we examine one example of each type of differential detector. First, though, we list some general detector characteristics which are useful in evaluating any detector.

1. Sensitivity. As explained below, the sensitivity of the detector represents an important limitation upon the smallest quantity of a solute that can be determined by GLC, and the increased demand for trace analyses in many different fields has stimulated the development of more sensitive detectors. For example, our growing awareness of the impact of trace contaminants upon biological ecosys-

[8] A. T. James and A. J. P. Martin, *Biochem. J.*, **50**, 679 (1952).

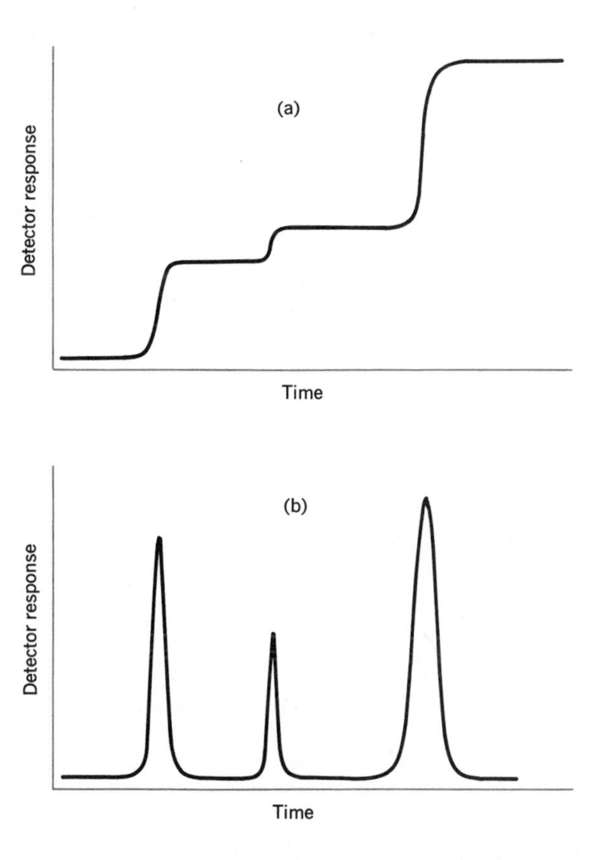

Figure 17.10 Comparison of chromatograms obtained with (a) integral and (b) differential detectors.

tems has provided a market for very sensitive detectors in studies of water pollution and pesticide residues in food products. It is possible for detector sensitivity to have a direct economic, political, or legal implication. For example, certain federal agencies are empowered to establish permissible levels of various poisons in foods. However, an important exception was legislated: the so-called Delaney clause requires a level of zero for any compound which is known to be carcinogenic in man or experimental animals. Now zero means zero to a politician, but to an analytical chemist it means a quantity smaller than he can detect with available methodology. It is perhaps not coincidental that the growing agitation by spokesmen for the food industry to repeal the Delaney clause has paralleled the increasing sensitivity of GLC and other analytical techniques. It may be argued that a level of zero means nothing outside the context of a particular analytical method employing a specified instrument.

Various measures of detector sensitivity are found in the literature, but basically, for our purpose, we may consider the sensitivity as the slope of a curve showing detector response as a function of the quantity measured, as shown in Fig.

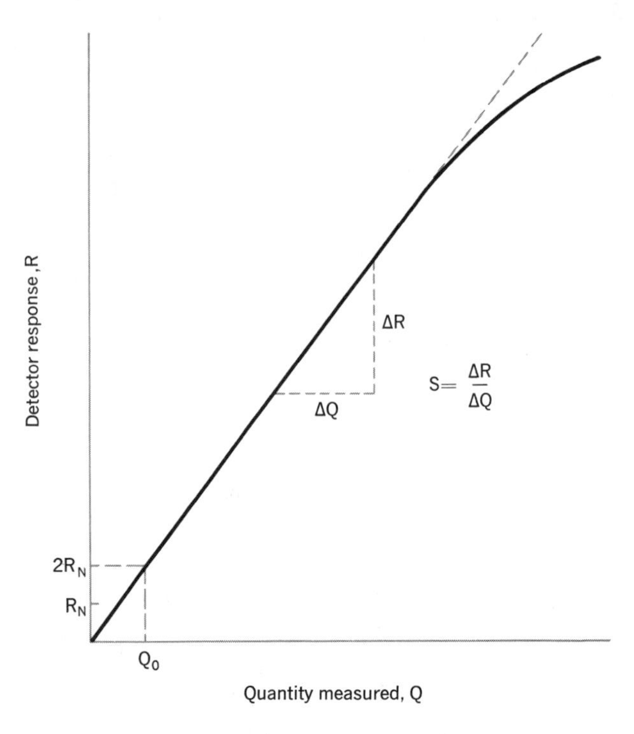

Figure 17.11 Detector response curve showing the definition of the sensitivity, S, and the relationship of peak-to-peak noise level, R_N, to limit of detection, Q_0.

17.11. A general expression for the sensitivity, then, is

$$S = \frac{\Delta R}{\Delta Q}$$

as shown in Fig. 17.11.

2. Stability. The baseline of a chromatogram is subject to short-term fluctuations of a largely random nature which are called "noise." A longer range upward or downward trend in the baseline is called "drift." Noise and drift, illustrated in Fig. 17.12, may originate in various instrument components such as amplifiers or recorders and in fluctuations of the carrier gas flow rate. Drift is seen in programmed temperature operation if the column reaches a temperature where the stationary liquid volatilizes. Much of the problem is eliminated by good circuit design, high-quality components, and proper operation of the chromatograph. There will always be, however, an inherent detector noise level which, along with the sensitivity, sets the lower limit for the quantity of a solute that can be detected.

The basic problem can be easily understood in qualitative terms. Suppose that a solute elution band is too small to be measured accurately. Perhaps the first thing to come to mind is this: why not increase the amplification in order to enlarge the elution band on the recorder chart? The answer is, this can be done but there is a limitation upon the benefit to be derived from it because the noise is amplified too. The smallest elution band that can be distinguished from noise peaks

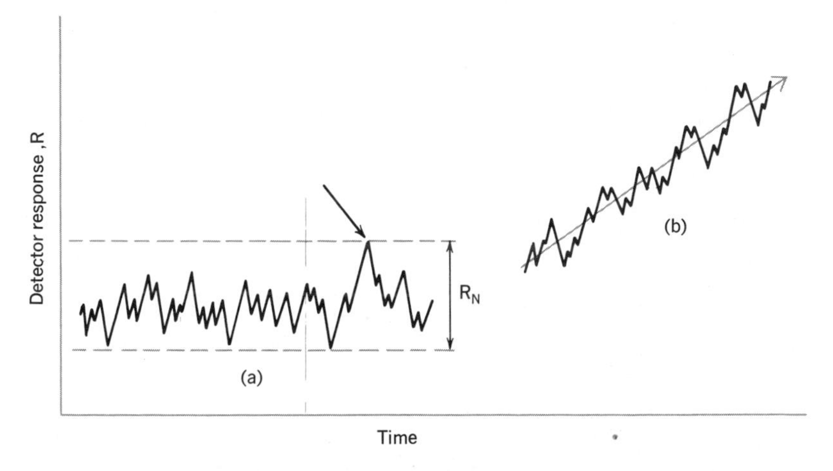

Figure 17.12 Expanded baselines. Left-hand portion illustrates noise and shows the measurement of the peak-to-peak noise level, R_N; the peak marked by the arrow might be ignored in estimating R_N (see text). Right-hand portion illustrates short-term noise superimposed upon upward drift.

corresponds to the limit of detection for a solute. This becomes essentially a statistical problem: how much larger than the random baseline fluctuations must an elution band be in order to yield acceptable odds that we shall identify it correctly as an elution band and not confuse it with noise. This is a problem not unlike some of those encountered in Chapter 2; we need to estimate the level beyond which a recorder deflection is probably not noise but rather is due to a definite cause, the sample. Various analyses of this problem have been presented in the literature, and various recommendations may be found. Perhaps the commonest advice is to take, as the limit of detection, that quantity of a solute which gives an elution band whose height is twice the peak-to-peak noise level. The peak-to-peak noise level, R_N, is explained in Fig. 17.12.

The relationship among sensitivity, noise level, and limit of detection may be formulated as follows. Recalling the definition of sensitivity,

$$S = \frac{\Delta R}{\Delta Q}$$

if we associate the limit of detection, Q_0, with twice the peak-to-noise level, $2 \times R_N$, then we may write

$$S = \frac{2R_N}{Q_0}$$

or

$$Q_0 = \frac{2R_N}{S}$$

In other words, low noise level and high sensitivity are favorable detector attributes with regard to the limit of detection.

Unusually large noise peaks such as the peak marked by the arrow in Fig. 17.12 occur infrequently. If such peaks are excluded in estimating R_N, then the limit of detection will appear to be better. Along with this goes, of course, an increased risk of reporting an analytical result for a solute when, in fact, a noise peak was measured.

3. Linearity. The ideal detector response would be linear with respect to the quantity measured, Q. This is the case with commonly used detectors within certain concentration limits, but eventually, as shown in Fig. 17.11, the response generally falls off.

4. Versatility. It is obviously advantageous that a detector respond to a wide variety of chemical compounds. None of the components of a sample would then be overlooked, nor would it be necessary to change detectors in order to handle various types of samples.

5. Response time. The detector should respond rapidly to the presence of the solute, or, as it is sometimes said, there should be a small "time constant." The total response time for a chromatograph is a function not only of the detector itself, but also of inertia in other components, for example, the recorder.

6. Chemical activity. In many cases, this is not an important factor, but sometimes solutes which have been separated by GLC are subjected to further study. For example, a mass spectrum or an infrared spectrum may be desired in order to identify the solute with certainty. In that event, it is obviously important that the solute not be decomposed in the detection process.

There are additional desiderata which may be classed as nonfunctional but which may be important in certain circumstances, such as low cost, simplicity, safety, and ability to withstand abuse.

The geometry of the detector and the pathway to it are very important. Solutes which have been separated in the column must not remix in the tubing leading to the detector nor inside the detector itself. A small volume within the detector is also conducive to a fast response. Stagnant pockets of gas must be avoided, and the dead volume between the column and the detector should be as small as possible.

Thermal Conductivity Detector

One of the most widely used detectors for general purpose GLC is the thermal conductivity cell. This device contains either a heated metal filament (generally platinum, a platinum–rhodium alloy, or tungsten) or a thermistor. Thermistors are small beads prepared by fusing a mixture of metal oxides, generally of manganese, cobalt, nickel, and traces of other metals. There is usually a thin protective layer of glass on the surface, and fine platinum alloy wires provide electrical connections. The important property of thermistors in the present context is an unusually large temperature coefficient of electrical resistance.

The heated detector element, filament or thermistor, under steady-state conditions, adopts a certain temperature determined by the heat supplied to it and the rate at which it loses heat to the walls of the chamber which surrounds it. Although a small amount of heat is lost through radiation and by conduction through the metal electrical leads, the temperature of the element is determined primarily by the thermal conductivity of the gas in the space between the element and the walls. Detection is based upon the fact that different gases have different thermal conductivities. When the composition of the gas changes, the temperature of the element changes, and this is reflected by a change in the electrical resistance of the element.

As shown schematically in Fig. 17.1 the detector generally has two sides, each with its own element. The pure carrier gas traverses one side of the detector, which is ahead of the sample injection port, while the column effluent flows through the other side. This is seen in more detail in Fig. 17.13, where one type of detector employing thermistors is illustrated schematically.

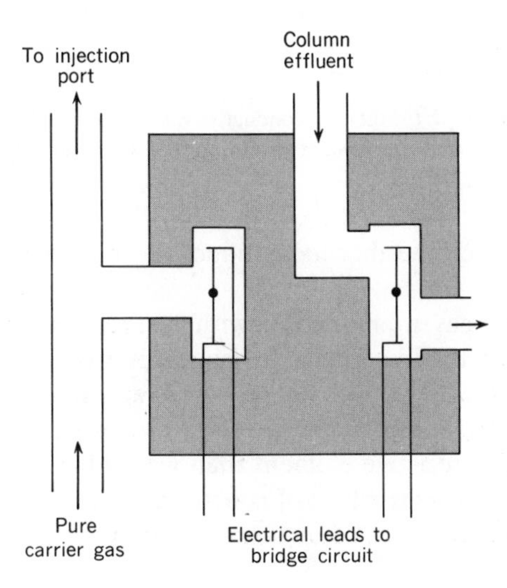

Figure 17.13 Schematic diagram of a thermal conductivity cell. The black dots are thermistor beads.

As we said, the detector elements are simply electrical resistances selected for their unusually large temperature coefficients of resistance. Thus the circuitry associated with the thermal conductivity detector is exactly what one would expect from elementary physics regarding resistance measurements. The two resistances in the two sides of the detector are two arms of a Wheatstone bridge circuit, as shown in Fig. 17.14. Before the injection of the sample into the chromatograph, pure carrier gas is flowing through both sides of the detector; the adjustable resistors are set so that the bridge is balanced, which establishes the baseline on the recorder chart. Now, after injection, when a solute emerges from the column, the value of R_s in Fig. 17.14 changes, while the other resistances remain the same. The bridge goes out of balance, and a voltage appears across the leads labeled "to recorder" in the figure. After the solute has passed through the detector, the bridge returns to its original balance. Thus a record of the voltage across the bridge vs. time will exhibit a peak as shown in Fig. 17.3 for the elution of each separated component of the sample. Basically, the thermal conductivity detector responds to changes in the concentration of a solute in the carrier gas

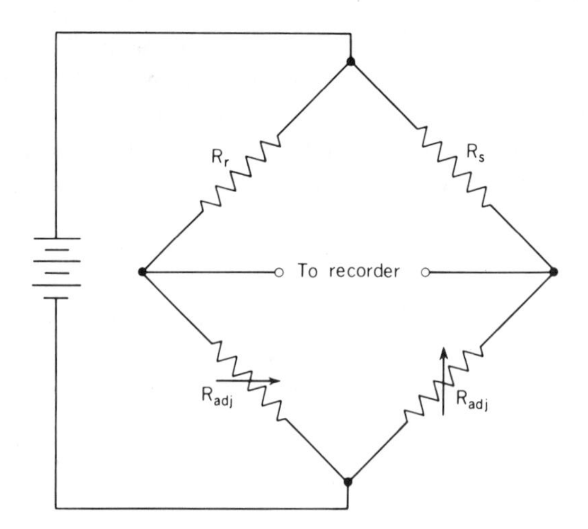

Figure 17.14 Wheatstone bridge circuit for thermal conductivity detector. R_r and R_s are the resistive elements in the reference and sample sides of the detector. R_{adj} is adjustable by the operator in order to balance the bridge.

stream, reflecting the way in which the thermal conductivity of the gas mixture depends upon the concentration.

Helium is an attractive carrier gas in conjunction with the thermal conductivity cell because its thermal conductivity, like that of hydrogen, is much greater than that of most organic compounds, while it does not represent an explosion hazard. Thus the appearance of an eluted solute at the detector causes a much greater change in the temperature of the resistive element than would be the case, say, with nitrogen as the carrier gas. This implies, of course, a greater sensitivity in detection, or a lower limit of detection. A few thermal conductivity values are given in Table 17.2.

TABLE 17.2 Thermal Conductivities of Some Gases and Organic Vapors*

Hydrogen	5.34
Helium	4.16
Methane	1.09
Nitrogen	0.75
Ethane	0.73
n-Butane	0.56
Ethanol	0.53
Benzene	0.44
Acetone	0.42
Ethyl acetate	0.41
Chloroform	0.25
Carbon tetrachloride	0.22

* Values are calories per second conducted through a 1-cm layer of gas 1 m^2 in area at 100°C, with a temperature gradient of 1°C/cm.

The thermal conductivity detector is relatively simple and inexpensive, rugged, and reliable. Its sensitivity is adequate for many purposes. The sensitivity may be increased by operating the elements at a higher temperature by furnishing a larger bridge current, but this involves a trade-off with regard to the life expectancy of the elements. This detector, in general, is nondestructive; that is, solutes may be recovered unchanged and subjected to further investigation. As may be inferred from Table 17.2, the response is not the same for all compounds, and accurate quantitative work requires calibration with known quantities of the various solutes.

Flame Ionization Detector

This detector was developed in response to the need, in certain applications, for higher sensitivity and faster response time than are provided by the thermal conductivity cell. The sensitivity of a detector depends upon not only its type but also upon the specific design and the manner in which it is operated. Thus a definite numerical comparison is difficult, but very roughly we may state that the flame ionization detector is several hundred to a thousand times as sensitive as the thermal conductivity detector. This detector is in very wide use, although perhaps it still runs second to thermal conductivity in actual numbers. The circuitry associated with the flame ionization detector is more complicated than the simple bridge circuit just discussed and gas chromatographs equipped with this detector are more expensive. Not only is the detector fast and sensitive, it is fairly stable, linear over a wide solute range, and responsive to almost all organic compounds. It is unresponsive to many inorganic compounds, including water.

The general principle of the flame ionization detector is as follows. The thermal energy in a hydrogen flame is sufficient to cause many molecules to ionize. The effluent gas from the column is mixed with hydrogen and burned at the tip of a metal jet in an excess of air. A potential is applied between the jet itself and a second electrode located above or around the flame. Ordinarily, the jet is the positive electrode. When ions are formed in the flame, the gas space between the two electrodes becomes more conducting, and an increased current flows in the circuit. This current passes through a resistor, the voltage across which is amplified to yield a signal which is fed to a recorder. Hydrogen may serve as the carrier gas, although it is more common to use nitrogen, in which case hydrogen is fed into the gas stream just ahead of the burner. Major aspects of the setup are shown in schematic form in Fig. 17.15.

With the flame ionization detector, the concentration of ions in the space between the electrodes and hence the magnitude of the current depends upon the rate at which solute molecules are delivered to the flame. A given weight of solute reaching the flame in unit time will yield the same detector response regardless of the degree of dilution by the carrier gas. This is the basis for the statement that this detector responds, not to solute concentration, but rather to the mass flow rate of solute. It should also be noted that the flame ionization detector is destructive of the sample components, in contrast with detectors such as the thermal conductivity cell which respond to some physical property of the gas related to solute concentration.

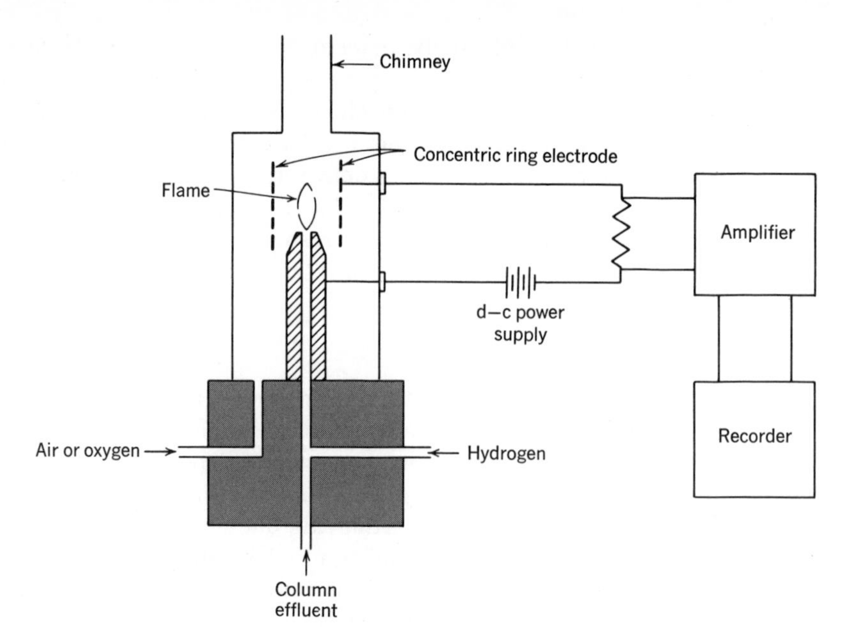

Figure 17.15 Schematic diagram of a flame ionization detector and the associated circuitry.

Although flame ionization is the most common, it may be mentioned that detectors are available in which solute molecules are ionized by radioactive sources. In one of these, the β-ray ionization detector, the source is a β-emitter such as tritium (^{3}H) or ^{99}Sr. The kinetic energy of the β-particle is much greater than that required to ionize a solute molecule; thus a series of ions is produced by one collision after another as each β-particle travels through the gas flowing through the detector. Isotopes which emit α-particles have also been employed as ionization sources. Attention must be given to safety with these detectors: along with the dangerous voltages associated with ionization detectors, the radioactive source represents a potential health hazard in the event of leakage or inadequate shielding.

In recent years, the electron-capture detector has become important for certain purposes. The column effluent passes through a cell containing a source of β-particles (usually a metal foil containing ^{63}Ni). Collisions of carrier gas molecules (often a mixture of N_2 and CH_4) with these particles produce ions and secondary electrons, which migrate to a positive electrode, yielding a certain current. When a solute which is capable of capturing electrons elutes from the column, there is a drop in this current which serves as the basis for detection. Very roughly, the electron-capture detector may be about 1000 times as sensitive as the flame ionization detector, but another important advantage in certain applications is selectivity. The detector is relatively insensitive to many hydrocarbons, alcohols, amines, and other compounds while responding 100,000 to 1 million times more strongly to certain other compounds such as heavily halogenated species. It has proved very useful for detecting certain pesticides (e.g., DDT, aldrin, and dieldrin) in samples where large excesses of other compounds tax the resolving ability of the column.

Still other types of detectors may be found in monographs on GLC and in the current research literature.

APPLICATIONS OF GLC

In discussing the theory of GLC, we emphasized that the column was a separation tool, but the gas chromatograph as a whole, because of the detection and recording of elution bands, is an analytical instrument which provides both qualitative and quantitative information about the components of a sample.

Identification of Compounds

With a particular column, and with all the variables, such as temperature and flow rate, carefully controlled, the retention time or retention volume of a solute is a property of that solute, just as its boiling point or refractive index is a property. This implies that retention behavior could be used to identify a compound. It must be stated, however, that this is not the forte of the gas chromatograph. Instruments such as the mass spectrometer, the infrared spectrophotometer, and the nmr spectrometer provide far more information about the nature of an unknown compound. In fact, for an analyst starting from scratch, with no information about the sample or whence it came, it would be virtually impossible to identify the components by their retention times alone. The thousands of known compounds simply provide too many possibilities from which to choose. In such a case, the best approach utilizes the capabilities of two instruments. For example, the chromatograph is used to separate the components of the sample mixture, and these are then introduced consecutively into the mass spectrometer. Various interfacing devices for accomplishing this automatically have been described. Frequently, the mass spectral data are handled by a computer which prints out possible identities of the compounds.

On the other hand, the analyst is not always faced with a totally unknown sample. The source of the sample and its history may permit reasonable guesses regarding some of the components. In such a case, a comparison of retention times with those of known compounds may confirm the identities of some of the components. Such an identification is quite likely to be correct if spiking the sample with a known compound does not lead to an additional elution band with several different columns at several temperatures.

Within an homologous series, say a series of normal alkanes, the logarithm of the retention volume is a linear function of the number of carbon atoms if the experiment is very carefully performed. Such a plot is shown in Fig. 17.16 (a). Graphs of the sort shown in Fig. 17.16 (b) may be useful in establishing the series in which an unknown compound belongs.

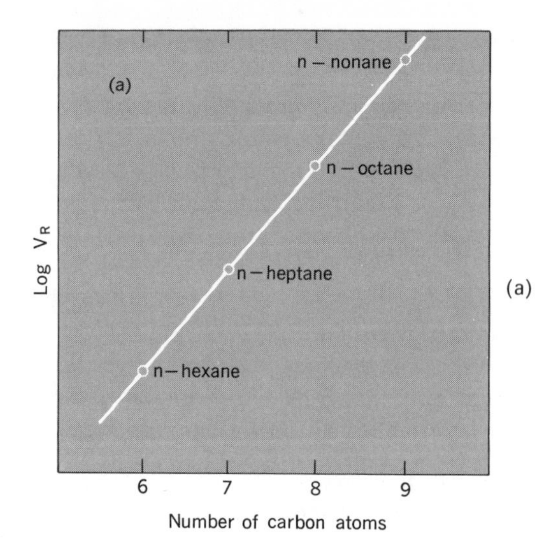

(a)

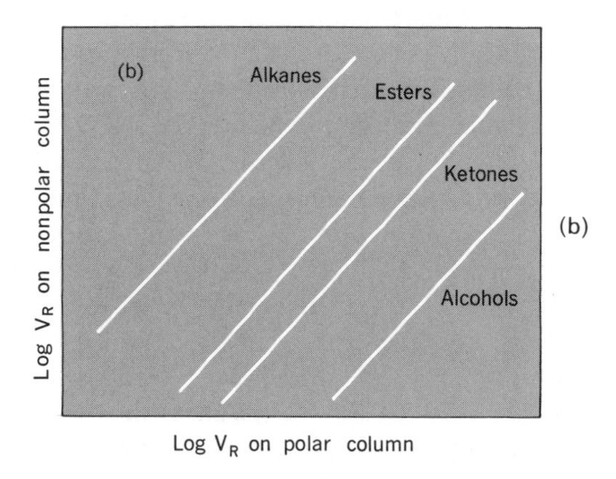

(b)

Figure 17.16 Identification of organic compounds by retention volume measurements.

Quantitative Analysis

Quantitative analysis by GLC depends upon the relationship between the quantity of a solute and the size of the resulting elution band. In general, with differential detectors (which are almost always employed), the best measure of the quantity of a solute is the area under the elution band. Solutes with very low retention times yield sharp, narrow bands, in which case the height of the band may be an adequate measurement. Otherwise, some sort of integration is required to obtain the area. The detector sensitivity is different for various compounds; this may be inferred from Table 17.2 for the thermal conductivity cell, and the same is true for other types of detectors. Thus there is no way to relate the area of an elution band to quantity of solute other than by calibration with known samples. Once this is done, we may write

$$\text{quantity of a solute} = \text{calibration factor} \times \text{area under elution band}$$

The units in which the area is measured make no difference provided the calibration factor is appropriate.

Various integrative methods may be applied to the measurement of the area under an elution band, of which we may mention a few examples. A planimeter may be employed. The baseline is extrapolated under the peak, and a stylus on a movable arm is traced around the enclosed area; a dial with a vernier device yields a number that is proportional to the area. The elution band may be cut out of the chart paper with scissors and weighed on an analytical balance. The weight will be proportional to the area provided the paper is uniform. Still another method involves construction of a triangle by extending the baseline and drawing tangents through the inflection points as shown in Fig. 17.3. The area of the triangle is found by measuring its base and its height and applying the usual formula, area = base × height/2. It is assumed that the area of the triangle is the same as the area under the elution band.

Various devices which provide an integral readout may be obtained as accessories to laboratory recorders, in some cases from the recorder manufacturers. The ball-and-disk integrator is an example of these.

Although in many cases the analyst may wish to examine a recorded chromatogram, it is unnecessary to have such a record so far as integration per se is concerned. Thus electronic devices have been developed which accept the detector signal directly and provide an integrated output. By an analog-to-digital conversion, this output may be transformed into a suitable input for a digital computer. The computer may then print out a complete record of the analysis including retention times, areas under elution bands, percentages of the components in the sample, and other information of interest.

Applicability of GLC

There are far too many applications of GLC to permit, in the available space, anything more thorough than listing a few examples. Each year thousands of papers are published in which GLC is at least mentioned, and the technique has spread outside of chemistry into many other fields.

The major limitation is volatility. The sample must have an appreciable vapor pressure at the temperature of the column, and this immediately eliminates many kinds of samples. An actual count is impossible, but it has been estimated that perhaps 20% of the known chemical compounds can be handled directly in a gas chromatograph. Biologists use GLC extensively, but unfortunately many of the most important biological compounds are insufficiently volatile, including amino acids, peptides, proteins, vitamins, coenzymes, carbohydrates, and nucleic acids. Sometimes, however, it is possible to convert nonvolatile sample components into volatile derivatives which can then be chromatographed. Obviously a difficult or time-consuming preliminary chemical step would nullify the speed and simplicity of the chromatographic analysis; hence there is a continuous search for reagents and reaction conditions which will derivatize all of the components of the sample quickly, cleanly, and quantitatively. For example, many studies have been directed toward the preparation of volatile amino acid derivatives. These involve both reactions of the carboxyl group, such as the formation of methyl or other alkyl esters, and reactions of the amino group, such as the formation of the trifluoroacetyl derivative. In recent years, volatile trimethylsilyl derivatives of acids, alcohols, amines, monosaccharides, and many other compounds have been studied extensively.

The technique called "pyrolysis gas chromatography" has become quite common in studies of nonvolatile polymeric materials such as tars, paints, rubber, synthetic films and fibers, and various sorts of plastics. The sample is heated very rapidly to a high temperature in a nonoxidizing atmosphere, and the gaseous products of the thermal decomposition are swept into the chromatographic column. In some cases the products are known compounds, but frequently the chromatogram is complex with many bands which are not identified. Despite the difficulty of a complete chemical interpretation, such a chromatogram may be reproducible and highly characteristic of the starting material, and it may be used in a manner somewhat analogous to the use of fingerprints. An interesting application of this technique is the identification of bacteria. In one example, the characterization of Salmonella organisms was described.[9] The cultured cells were harvested, washed free of culture medium, and centrifuged. The wet, packed cells were freeze-dried, and a sample of about 80 μg of the dried bacteria was subjected to pyrolysis GLC. Forty-seven species of Salmonella were correctly classified by examination of the chromatograms. Correlations were observed between GLC bands and groupings based upon traditional serological and biochemical classification tests. It is thought that some of the characteristic GLC band patterns arise from species differences in the hexose sugars of the bacterial cell wall.

Most inorganic samples are not sufficiently volatile to permit the direct application of GLC, although some work has been done at very high temperatures using molten salts or eutectic mixtures as the stationary liquid phase. Halides of some elements such as tin, titanium, arsenic, and antimony are fairly volatile and have been separated by GLC. A number of metals such as beryllium, aluminum, copper, iron, chromium, and cobalt have been subjected to GLC in the form of fairly volatile chelate compounds with acetylacetone and its fluorinated derivatives.[10] For example, aluminum, iron, and copper have been determined in alloys by dissolution of the sample followed by extraction of the metals into a chloroform solution of trifluoroacetylacetone which is then chromatographed.[11] Relative errors of the order of 0.2 to 3% were reported.

GLC figures prominently in efforts to monitor and control the distribution of pollutants in the environment. For example, the U.S. Environmental Protection Agency operates an extensive program for monitoring pesticide levels in soils

[9] E. Reiner et al., *Anal. Chem.*, **44**, 1058 (1972).

[10] R. W. Moshier and R. E. Sievers, *Gas Chromatography of Metal Chelates*, Pergamon Press, Inc., New York, 1965.

[11] R. W. Moshier and J. E. Schwarberg, *Talanta*, **13**, 445 (1966).

throughout the country; the goal is to establish a baseline showing exactly where we are at the present time so that trends in the future can be interpreted meaningfully. Details vary, but the general approach involves extracting the soil sample with an organic solvent and chromatographing the extract in an instrument with a very sensitive detector. Similar programs are under way for crops, water, fish, and wildlife, and GLC is important in all of these.

Although it is perhaps more fun to discuss unusual applications and those which relate to popularly relevant subjects, the chemistry student should also keep in mind that thousands of chromatograms are recorded every day in laboratories where GLC is a routine analytical tool.

REFERENCES

C. E. BENNETT, S. DAL NOGARE, and L. W. SAFRANSKI, "Chromatography: Gas," Chap. 37 of I. M. Kolthoff and P. J. Elving, eds., *Treatise on Analytical Chemistry*, Part I, Vol. 3, Interscience Publishers, Inc., New York, 1961.

H. P. BURCHFIELD and E. E. STORRS, *Biochemical Applications of Gas Chromatography*, Academic Press, Inc., New York, 1962.

S. DAL NOGARE and R. S. JUVET, JR., *Gas–Liquid Chromatography*, Wiley–Interscience, New York, 1962.

R. A. JONES, *An Introduction to Gas–Liquid Chromatography*, Academic Press, Inc., New York, 1970.

A. I. M. KEULEMANS, *Gas Chromatography*, 2nd ed., Reinhold Publishing Corp., New York, 1959.

A. B. LITTLEWOOD, *Gas Chromatography: Principles, Techniques, and Applications*, 2nd ed., Academic Press, Inc., New York, 1970.

J. H. PURNELL, *Gas Chromatography*, John Wiley & Sons, Inc., New York, 1962.

O. E. SCHUPP, 3, *Gas Chromatography*, Vol. 13 of E. S. Perry and A. Weissberger, eds., *Technique of Organic Chemistry*, Wiley–Interscience, New York, 1968.

QUESTIONS

1. *Theoretical plate.* Explain clearly what is meant by "height equivalent of a theoretical plate" in connection with a continuous-flow separation process such as gas–liquid chromatography.

2. *Theoretical plates.* Using peaks numbered 5, 6 and 7 in the chromatogram shown in Fig. 17.9(a), obtain three estimates of the number of theoretical plates in the column. The column was 4 ft long. Obtain estimates of HETP, in mm, for this column under the operating conditions that were used. (1 inch = 2.54 cm.)

3. *Band spreading.* By discussing the factors that contribute to band spreading in a GLC column, show that it is reasonable to expect that there will be an optimum carrier gas velocity.

4. *Detector.* Suppose that a manufacturer redesigned his thermal conductivity detector, taking advantage of the latest improvements in materials and fabrication methods. In

the new detector, the peak-to-peak noise level was found to be half of what it was in the old one, and the sensitivity had been doubled. What was the effect of these improvements upon the limit of detection for a certain organic compound?

5. *Rate of sample injection.* Let us say that you are a skilled chromatographer and you have just hired an inexperienced person to work in your laboratory. The newcomer runs a chromatogram on a sample that you yourself have run many times. What in the appearance of the chromatograms might suggest to you that he had injected the sample too slowly?

Multiple-choice: In the following multiple-choice questions, select the *one best* answer.

6. Which of the following would have no effect upon the resolution of a pair of solutes in GLC? (a) increasing the length of the column; (b) changing to a more sensitive detector; (c) injecting the sample more slowly; (d) changing the chemical nature of the stationary liquid.

7. Which of the following would have practically no effect upon the retention volume of a solute in GLC? (a) changing the carrier gas flow rate; (b) increasing the stationary liquid loading of the column packing from 5 to 10% by weight; (c) increasing the column temperature; (d) changing the chemical nature of the stationary liquid.

8. As a thermal conductivity detector deteriorated with age, the peak-to-peak noise level doubled and the sensitivity decreased by a factor of 3. The limit of detection for an organic compound (a) remained the same; (b) increased by a factor of 6; (c) decreased by a factor of 6; (d) increased by a factor of 2.

9. Increasing the quantity of stationary liquid phase applied to the column packing will in general, everything else the same, (a) increase t_R for a solute; (b) decrease t_R for a solute; (c) not influence t_R for a solute; (d) increase or decrease t_R for a solute, depending upon the boiling point of the solute.

10. The open-tubular GLC column is a long, thin tube of glass of stainless steel, perhaps 0.1 mm internal diameter and a couple of hundred feet long, with the inner surface coated with a very thin film of stationary liquid phase. There is no packing material in the usual sense. What can be predicted about this column? (a) Eddy diffusion should be very small, perhaps almost zero; (b) Equilibration of solutes between gas and liquid phases should be rapid; (c) The very light liquid loading might lead to adsorption of some solutes unless the inner surface of the tube had very little polar character; (d) All of the above predictions are reasonable.

11. In GLC, interaction of solutes with the solid support will often cause (a) unusually narrow elution bands; (b) excessive eddy diffusion; (c) asymmetric elution bands with "tailing"; (d) decreased detector sensitivity.

12. The main reason for developing more sensitive detectors for gas chromatography is that (a) smaller quantities of sample components can be determined; (b) the time required for analysis will be shortened; (c) it will be possible to employ much longer columns; (d) it will be possible to employ much shorter columns.

13. The separation factor, S, in chromatography (a) depends upon the length of the column; (b) depends upon the square root of the length of the column; (c) depends upon the number of theoretical plates in the column; (d) has nothing to do with any of the answers above.

14. When solutes are subjected to Craig-type countercurrent extraction, they form bands which broaden as the number of stages increases. Which factor below, operative in chromatography, does not contribute to band spreading in the Craig experiment? (a) eddy diffusion; (b) longitudinal diffusion; (c) nonequilibrium in mass transfer; (d) none of the above factors contributes.

15. The main advantage of ionization detectors over the thermal conductivity cell for GLC is that (a) the ionization detectors are less sensitive; (b) the ionization detectors are more sensitive; (c) many organic compounds can be detected with ionization detectors, while thermal conductivity cells respond to only a few inorganic ions; (d) shorter columns can accomplish the same separations when ionization detectors are employed.

CHAPTER EIGHTEEN

liquid chromatography

INTRODUCTION

In this chapter we shall consider several forms of chromatography where the moving phase is a liquid. Perhaps 80% of all chemical compounds are better handled in solution than in the gas phase, largely because of volatility limitations. Thus liquid chromatography is potentially more important than gas chromatography, although the latter will retain its dominant position where volatile compounds are involved. Historically, liquid chromatography was invented first, but the theoretical ideas and the modern instrumentation that led to highly efficient performance were developed first in GLC. In 1968 or so, an effort began to upgrade liquid chromatography utilizing concepts, particularly the kinetic approach, which had been successful in GLC. This process continues, and its rapid pace makes liquid chromatography today one of the most exciting aspects of analytical chemistry.

We shall first examine the phase distribution processes that are involved in the common forms of liquid chromatography. Next, to provide perspective in which to view recent developments, we shall briefly consider the conventional mode of operation. Finally, we shall describe modern practice as it has evolved so far. Concepts presented in Chapter 17 will be useful in this one, and we shall refer to these rather than presenting a self-contained theoretical treatment here.

Adsorption

Imagine a solid material with a clean, dry surface. Now if this surface is exposed to a fluid—gas, pure liquid, or solution—there is generally a tendency for molecules of gas, solvent, or solute to interact with the surface. If the solid material is very finely divided or is highly porous, in other words, if there is a large surface area, then the extent of adsorption may be appreciable. For example, if a good adsorbent is introduced into a vessel containing a gas, the decrease in pressure as the surface attracts gas molecules is easily measurable. An atom, ion, or molecule in the surface layer of a solid, unlike its counterparts in the interior, does not have neighboring particles on all sides. Thus residual attractive forces are exerted upon components of the fluid which bathes the surface, and the free energy of the system may be minimized if such components concentrate at the interface. In certain systems under certain conditions, the adsorbed layer may be only one molecule thick; frequently, however, the adsorbed molecules can in turn hold others so that a multimolecular layer is built up. In the case of the adsorption of ions on the surfaces of ionic solids, the force involved is an obvious electrostatic one. In other cases, van der Waals forces (dipole–dipole, dipole–induced dipole, and London forces) are responsible for adsorption.

If a solution containing a solute is placed in contact with a solid adsorbent, that solute will distribute itself between the two phases, with the position of equilibrium determined by the affinity of the surface for the solute, the affinity of the surface for the solvent (i.e., the solvent competes with the solute for available surface sites), and the affinity of the solvent for the solute.

Adsorption Equilibrium

The isotherm which describes an adsorption equilibrium is usually nonlinear. Many systems follow the Freundlich equation, at least if the concentration is not too high; this may be given in the form

$$C_S = KC_L^{1/n}$$

where C_S is the concentration of an adsorbed solute on a solid phase in equilibrium with a solution of solute concentration C_L. Typical units for C_S are millimoles of solute per gram of adsorbent, and for C_L, molarity. K and n are constants. If n were equal to 1, the Freundlich equation would resemble other equilibrium expressions such as Henry's law or the equation for the distribution coefficient in solvent extraction. But in general $n > 1$, and hence a graph of C_S vs. C_L, which is called an adsorption isotherm, resembles curve 2a in Fig. 17.4. Taking logarithms of both sides of the Freundlich equation yields

$$\log C_S = \log K + \frac{1}{n} \log C_L$$

which suggests that a graph of $\log C_S$ vs. $\log C_L$ is a straight line of slope $1/n$ and has an intercept of $\log K$ on the $\log C_S$ axis.

K and n are constants only for a given system and, of course, only for a stated temperature. They vary with the nature of the adsorbent and its surface character and with the solvent and the solute. K is more sensitive than is n to changes in the nature of the solute, and separations based upon adsorption depend largely upon differences in the K values for various solutes. As we saw in the last chapter, nonlinear isotherms are associated with skewed chromatographic elution bands; frequently the bands obtained with adsorption columns resemble curve 2b in Fig. 17.4, although at very low concentrations, where the isotherm may approximate linearity, fairly symmetrical bands are sometimes seen.

Variability in the surface properties of commercial adsorbents has been a problem for chromatographers. In some cases, simply washing an adsorbent such as alumina with acid or alkali considerably modifies its behavior, and the temperature at which it is dried may also be very important. Different batches of adsorbents, even those from the same producer, may exhibit troublesome variability. In recent years, the marketing of specially prepared adsorbents which are tested and labelled "for chromatography" has been helpful.

A few general rules regarding the adsorbability of a solute may be stated. Everything else the same, the more polar a compound, the more strongly will it be adsorbed from solution. Other factors equal, high molecular weight favors adsorption. The nature of the solvent from which adsorption occurs is very important: the more polar the solvent, the stronger is its tendency to occupy surface sites in competition with the solute, and hence the less the adsorption of the solute.

A wide variety of adsorbents have been employed in various applications. Among the more common ones are sucrose, starch, cellulose, calcium carbonate, magnesium carbonate, silica gel, alumina, and charcoal.

Ion Exchange

Ion-Exchange Resins

A wide variety of materials, both natural and synthetic, organic and inorganic, exhibit ion-exchange behavior, but in the research laboratory, where uniformity from one batch to another is important, the preferred ion exchangers are usually synthetic materials known as ion-exchange resins. The resins are prepared by introducing ionizable groups into an organic polymer matrix. The commonest matrix is a copolymer of styrene and divinylbenzene.

The polymerization of styrene yields a linear polymer:

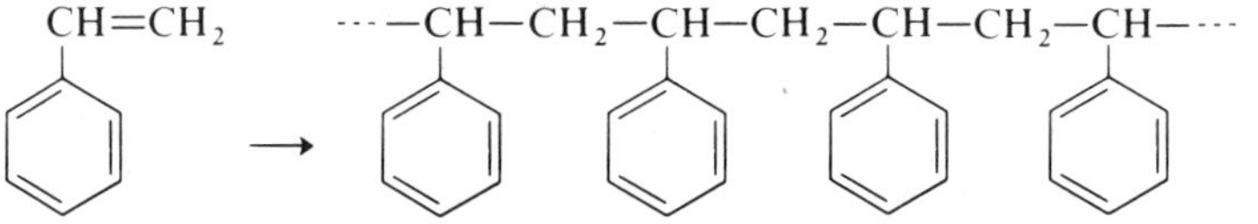

$$CH{=}CH_2(C_6H_5) \longrightarrow \cdots-CH(C_6H_5)-CH_2-CH(C_6H_5)-CH_2-CH(C_6H_5)-CH_2-CH(C_6H_5)-\cdots$$

Styrene Polystyrene

The addition of the bifunctional monomer divinylbenzene to the polymerization mix, on the other hand, links together the polystyrene chains and yields a material with a three-dimensional network structure:

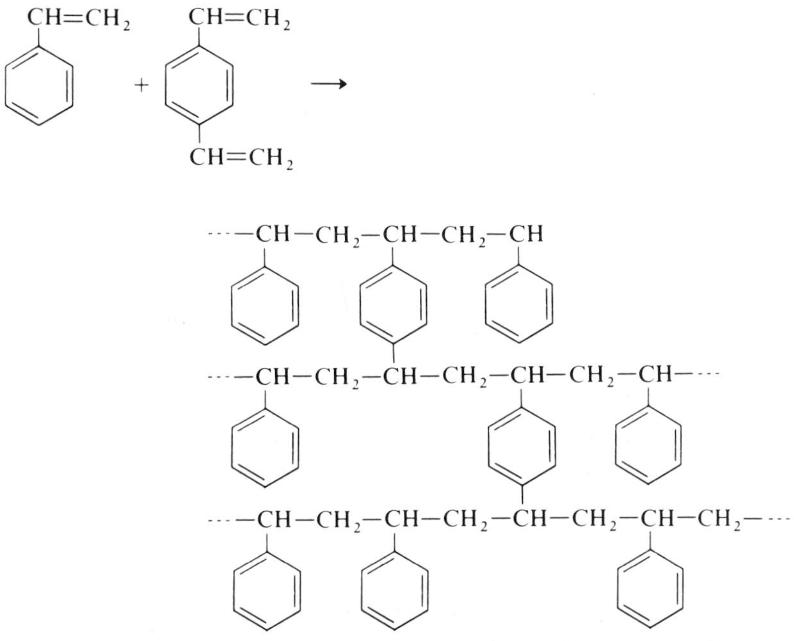

$$\underset{\text{styrene}}{C_6H_5{-}CH{=}CH_2} + CH_2{=}CH{-}C_6H_4{-}CH{=}CH_2 \longrightarrow$$

$$\cdots{-}CH{-}CH_2{-}CH{-}CH_2{-}CH$$

$$\cdots{-}CH{-}CH_2{-}CH{-}CH_2{-}CH{-}CH_2{-}CH{-}\cdots$$

$$\cdots{-}CH{-}CH_2{-}CH{-}CH_2{-}CH{-}CH_2{-}CH{-}CH_2{-}\cdots$$

By varying the divinylbenzene content, the degree of cross-linking can be controlled quite reproducibly. General-purpose resins usually contain 8 to 12% divinylbenzene; a resin with, say, 8% divinylbenzene is said to be "8% cross-linked." The resins are made in the form of spherical beads by the process of emulsion polymerization. The bead diameter is controlled; this is usually in the range of 0.1 to 0.5 mm, although other sizes can be made for special purposes.

To prepare a typical cation-exchange resin, the polymer is sulfonated to introduce $—SO_3H$ groups into the aromatic rings. Because these sulfonic acid groups are highly polar, the polymer thus acquires a high affinity for water. When the resin particles are suspended in water, they increase in size because of the water uptake. This swelling is limited, of course, by the cross-linking; a linear (non-cross-linked) polymer would swell indefinitely until it finally yielded a molecular dispersion in solution.

The arylsulfonic acids are strong acids. Thus these groups are ionized when water penetrates the resin beads:

$$R{-}SO_3H + H_2O \rightleftharpoons R{-}SO_3^- + H_3O^+$$

But, unlike ordinary electrolytes, here the anion is permanently attached to the immovable polymer matrix, and it cannot migrate through the aqueous phase within the pores of the resin. This fixation of the anion in turn restricts the mobility of the cation, H_3O^+. Electrical neutrality is maintained within the resin, and H_3O^+ will not leave the resin particle unless it is replaced by some other cation. The exchange is stoichiometric; that is, one H_3O^+ is replaced by one Na^+, two H_3O^+ by one Ca^{2+}, etc. As we discuss below, ion exchange is an equilibrium process, and seldom does it go to completion; but regardless of the extent to which it proceeds, the stoichiometry is exact in that one positive charge leaves the resin for each one which enters.

The introduction of basic groups into the polymer yields anion exchange materials. One of the common strong base anion exchangers may be represented as

$$R-C_6H_4-CH_2-\overset{\displaystyle CH_3}{\underset{\displaystyle CH_3}{\overset{|}{\underset{|}{N^+}}}}-CH_3 \qquad X^-$$

where X^- is an anion such as OH^-, Cl^-, or NO_3^-. The exchangeable ion (i.e., the ion which is not fixed to the polymer matrix) is called the "counter ion."

Ion-Exchange Equilibrium

Suppose that a resin containing the exchangeable counter ion B is placed in a solution containing ion A of the same charge. The exchange reaction

$$\underset{\text{Solution}}{A} + \underset{\substack{\text{Resin}\\\text{phase}}}{RB} \rightleftharpoons \underset{\substack{\text{Resin}\\\text{phase}}}{RA} + \underset{\text{Solution}}{B}$$

takes place, and equilibrium will be attained with some of each ion in the resin phase and some in solution, for which we may write an equilibrium constant,

$$K = \frac{a_{A_r} \times a_{B_s}}{a_{B_r} \times a_{A_s}}$$

where a_{A_r} represents the activity of ion A in the resin phase and a_{A_s} its activity in the solution outside the resin pores. This expression can be written

$$K = \frac{X_{A_r} \times (B)_s}{X_{B_r} \times (A)_s} \times \frac{\gamma_{A_r} \times \gamma_{B_s}}{\gamma_{B_r} \times \gamma_{A_s}}$$

or

$$K = Q \times \frac{\gamma_{A_r}\gamma_{B_s}}{\gamma_{B_r}\gamma_{A_s}}$$

where X_{A_r} and X_{B_r} are the concentrations in mole fraction in the resin phase, parentheses mean molal concentrations in the external solution, and γ is the activity coefficient. The term Q is called the concentration quotient, or "practical selectivity coefficient." Some workers use the term "selectivity coefficient" to refer to the product of Q and the activity coefficient ratio of the ions in solution:

$$Q_\gamma = Q\frac{\gamma_{B_s}}{\gamma_{A_s}} = K\frac{\gamma_{B_r}}{\gamma_{A_r}}$$

At low concentrations of the ions in solution, $Q_\gamma \cong Q$, since the activity coefficients approach unity.

It is clear why Q_γ values are called selectivity coefficients: If Q_γ is large, the resin is showing a preference for ion A; if Q_γ is small, the selectivity of the resin favors ion B. We may never speak of the tendency of a resin to pick up a certain ion without noting that there is already another ion in the resin, that is, we should consider, not the tendency for the resin to pick up ion A in an absolute sense, but rather the tendency to pick up A at the expense of ion B. The tendency to pick up ion A will be different if the resin phase contains some other ion C instead of B as the counter ion.

However, we can put a certain ion on a resin and then compare a series of other ions using this as a reference. For the ions in this series we may simply write distribution ratios:

$$D = \frac{\text{concentration of an ion in the resin}}{\text{concentration of the same ion in solution}}$$

The conventional units of D are

$$\frac{\text{amount/kg of dry resin}}{\text{amount/liter of solution}}$$

The "amount" term may be in milligrams, moles, or whatever, since its units cancel in the D ratio.

A distribution ratio with different units is sometimes used, with the symbol D_V.

$$D_V = \frac{\text{amount/liter of resin bed}}{\text{amount/liter of solution}}$$

The conversion factor for D to D_V is the so-called bed density, ρ,

$$D_V = D \times \rho$$

where ρ is in kilograms of dry resin per liter of resin bed.

The significant aspect of ion exchangers is, of course, their selectivity; that is, D values are different for various ions, and hence separations may be accomplished by ion exchange.

Neutral molecules can find their way into the pores of an ion-exchange resin, but they are not subject to forces so strong as those acting on ions, and in general they can be washed out by water or some other solvent. Solutes in the resin which are not so strongly held as ions are said to be "sorbed"; the pickup of such solutes by the resin is called "sorption." Sorption can sometimes be used to effect separations, but in this discussion we are concerned only with legitimate ion exchange. Of course, the sign of the charge on an ion is important in selectivity, but this is so obvious as to be trivial. A cation cannot participate in exchange on an anion exchange resin; it might find its way into the resin pores by some sort of general electrolyte sorption, but it would not be strongly held and could be washed out with water.

In a series of ions which have the proper sign to act as true counter ions, the magnitude of the charge is important. Normally, the resin prefers the ion of higher charge. Thus the extent of exchange with, say, H_3O^+, would decrease in the order

$$Th^{4+} > Al^{3+} > Ca^{2+} > Na^+$$

provided proper allowance was made for other factors such as concentration.

There are exceptions to this, but it is a good rule of thumb under ordinary conditions.

With a series of ions of the same charge, the resin still shows selectivity. For example, with the alkali metals, the following order is generally found with cation-exchange resins:

$$Cs^+ > Rb^+ > K^+ > Na^+ > Li^+$$

The important factor here is probably the radius of the ion; the smaller an ion of given charge, the more strongly it will be held by the resin. At first glance, the above order may not appear consistent with this statement, which would imply that Cs^+ is a smaller ion than Li^+. The usual values of ionic radii, however, are obtained by X-ray diffraction studies of solid crystals, and these "crystallographic" or "naked" radii are not the right ones to use here. The ions in solution are hydrated, and it is the radius of the hydrated ion that determines the ion exchange behavior. Such hydrated radii are much more difficult to measure, but estimates are available. While the naked radius of Li^+ is 0.68 Å, the hydrated radius is about 10 Å; the naked and hydrated radii for Cs^+ are 1.65 Å and 5.05 Å, respectively.

Liquid–Liquid Partition

If a solution of sodium silicate is acidified under the proper conditions, the precipitated silicic acid takes the form of a gel, a hydrophilic network structure which contains a large quantity of water. If this water is then driven out by heating the gel to an appropriate temperature, the silica which remains is a hard solid with a highly porous structure of very large surface area which is known as silica gel. Silica gel has a high affinity for water and is widely used as a desiccant. In 1941, Martin and Synge employed silica gel as a solid support to immobilize water as the stationary phase in a chromatographic column.[1] With a mixture of *n*-butanol and chloroform as the moving phase, acetylated amino acids were separated on this column. The basis for the separation was considered to be partition of the solutes between the stationary water phase and the mobile organic phase, the same process utilized in Craig's countercurrent solvent extraction technique adapted to the chromatographic mode. This particular form of chromatography was not widely adopted because shortly after it was proposed, a variant known as paper chromatography appeared which was more convenient in view of the technological limitations upon chromatography in the 1940s. Liquid–liquid partition chromatography in columns is returning, however, in a newer form which will be mentioned later.

Gel Filtration

The process of gel filtration is also known as gel permeation and molecular exclusion. Chromatography based upon this process is widely used in biochemis-

[1] A. J. P. Martin and R. L. M. Synge, *Biochem. J.*, **35**, 91 (1941).

try to separate proteins and other macromolecules, and in industrial laboratories to characterize synthetic polymers. The most important column-packing materials are cross-linked dextrans and polyacrylamides, sold under the trade names Sephadexes and Bio-Gels, respectively.

When certain bacteria grow on substrates containing sucrose, they synthesize polysaccharides called dextrans. Dextrans are linear polymers of D-glucose with a small amount of cross-linking between adjacent chains; a typical molecular weight is of the order of 75,000. Additional cross-links can be introduced artificially by means of various chemical reagents to yield a network structure in which the pore size is controlled. The commercial materials are offered in the form of small beads of stated diameters with designations that reflect the pore sizes.

The glucose residues in the cross-linked dextrans are hydrophilic, and when the beads are placed in an aqueous solution, there is an appreciable uptake of water which is called the "water regain." This is accompanied by the swelling of the particles to form the Sephadex gel. The process with which we are now concerned is the distribution of solutes between an aqueous phase within the gel particles and an external aqueous phase. At first glance, transferring a solute from water into water is no process at all. But selectivity with regard to different solutes is achieved on the basis of the pore size. A small molecule may freely enter the gel phase, in other words, the water within the particles is available. In a column operation, this will have a retarding effect because molecules which are able to penetrate the gel will spend part of their time sheltered from the moving phase. On the other hand, the internal water phase will not be available for a large molecule that cannot penetrate the pores, and such a molecule will not experience retardation by the stationary phase. In between, there will be molecules of intermediate size which can penetrate some of the larger pores and will then be retarded in some degree.

Let V_R be the retention volume for a solute in a chromatographic experiment with a Sephadex column. Let V_0 be the interstitial volume or void volume, that is, the volume within the column which is available for the moving phase (this is analogous to V_G in Chapter 17), and let V_L be the volume of the water within the gel particles which is available for accepting solutes. Then we can write an equation of the same sort that we had in GLC:

$$V_R = V_0 + KV_L$$

where K has the form of a distribution coefficient. If a solute is completely excluded from the interior of the gel particle, then $K = 0$ and $V_R = V_0$; colored marker substances which approximate this behavior are available from the Sephadex supplier. For a solute which can freely enter the gel particle, there should be no preference for water inside or outside the gel, and hence $K = 1$ and $V_R = V_0 + V_L$. For molecules of intermediate size which can penetrate the gel to some extent but not freely, K values should fall between 0 and 1. If sieving based upon molecular size were the only phenomenon occurring, then K values greater than 1 would never be encountered. In fact, however, such values are sometimes obtained, suggesting that solutes may interact with the dextran matrix itself; effects such as adsorption, hydrogen bonding, and ion exchange would

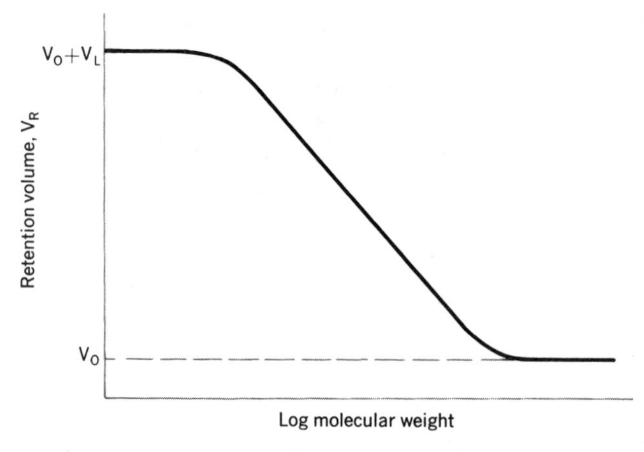

Figure 18.1 Relation between retention volume and molecular weight for solutes on a gel filtration column.

explain such behavior. Figure 18.1 shows the manner in which solute elution volume typically varies with molecular weight.

The polyacrylamide gels are prepared by copolymerizing acrylamide, $H_2N{-}CO{-}CH{=}CH_2$, with methylenebisacrylamide,

$$H_2N{-}CO{-}CH{=}CH{-}CH_2{-}CH{=}CH{-}CO{-}NH_2$$

the latter providing cross-linking. The polymer, which is produced in the form of spherical beads, hydrates readily when soaked in water because of the polar amide groups. Bio-Gel columns are similar to Sephadex with regard to their general applicability.

Recently, gel filtration materials into which ionizable groups have been deliberately introduced have appeared on the market. Presumably, the ion-exchange aspect offers additional selectivity over that provided by molecular exclusion alone for the separation of solutes which are ionic. Materials are also now available which are suitable for use with certain organic solvents rather than water, extending gel filtration further into organic chemistry.

CONVENTIONAL LIQUID CHROMATOGRAPHY TECHNIQUES

We saw in Chapter 17 that the gas chromatograph is a complete analytical instrument in the sense that components of the sample are not only separated but are measured as well. Liquid chromatography developed earlier, in a period when speed and automation were not major concerns in the laboratory and before the "systems" approach had been applied to analytical instrumentation. The great power of liquid chromatography as a separation tool was recognized 40 years ago, but the idea of building an efficient analytical instrument around it goes back only about 10 years. These recent developments will be better appreciated if we examine briefly the conventional approach, which, as a matter of fact, is still widely used.

Various sizes of columns may be employed, the major consideration being adequate capacity to accept the sample without overloading the stationary phase. It is a common rule of thumb that the length of the column should be at least 10 times its diameter. For a typical case, let us say that we have a column 20 cm long and 1 or 2 cm in diameter, something like the one shown in Fig. 18.2. The packing material, an adsorbent like alumina or perhaps an ion-exchange resin, had probably been added in the form of a suspension in a portion of the moving phase and allowed to settle into a wet bed with a little liquid remaining above the surface. Now by opening the stopcock, the liquid level is allowed to fall just to the top of the bed, and then a small portion of sample solution (a few tenths of a milliliter up to perhaps a couple of milliliters) is carefully pipetted onto the top of the bed. The liquid reservoir is positioned, and the flow of the mobile phase is started. The desired flow rate is obtained by gravity alone, by inserting the outlet end of the column into an evacuated vessel as shown in the figure, or by pumping liquid in at the top of the column. A typical flow rate might be a few tenths of a milliliter per minute, possibly faster if the separation were not a difficult one.

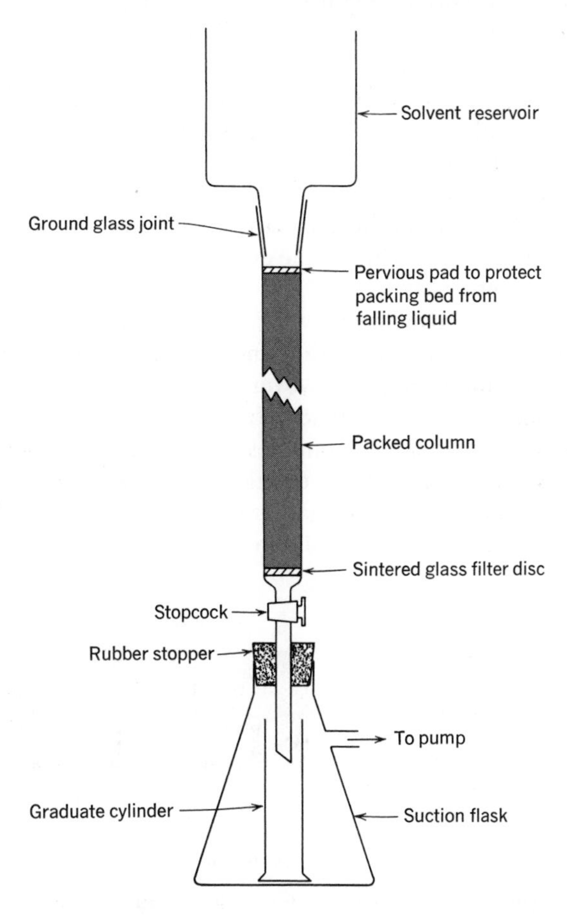

Figure 18.2 Possible setup for conventional liquid chromatography.

The effluent solution is collected in a series of fractions of convenient volume. The solution may drip into a graduated cylinder which the operator dumps into a beaker or test tube each time a certain volume, say 5 or 10 ml, has accumulated. It is not uncommon for the chromatographic elution process to require several hours, even all day or overnight. In such a case, a mechanical device called a fraction collector is convenient. The operator sets up a series of tubes on a turntable which positions a new tube under the column when the desired volume has been collected in the previous one; activation can be based upon time, drop counting, or the deflection of a light beam by the rising meniscus in the tube.

It sometimes happens that no single moving phase is well suited for the elution of all the components of a sample. In adsorption, for example, a fairly nonpolar solvent may be ideal for eluting some of the less polar solutes, whereas the more polar solutes may then show an inordinately long retention. In such a case, the technique of gradient elution may be useful. The composition of the mobile phase is changed continuously by allowing a more polar solvent to flow into the reservoir containing the less polar one, whence the mixture flows to the column. The reservoir is stirred. Figure 18.3 shows such an arrangement schematically. Now the laggard solutes will move along faster as the eluting power of the solvent mixture increases. Bands with serious "tailing" may be sharpened, since the tail sees a stronger solvent than the front of the band. The result of gradient elution is rather similar to that of temperature programming in GLC, which was mentioned in the last chapter.

The individual fractions of the effluent solution are examined by whatever means is appropriate—spectrophotometry, polarography, radioactive assay, titrimetry, etc.—to locate the desired components of the sample and to determine their quantities.

As seen above, the conventional column operation is slow and tedious, and it is poorly suited to automation and modern methods of data processing. It is not totally without merit, however; for example, fairly large columns can be used which have sufficient capacity for preparative-scale work. But for handling large numbers of analytical samples, recent developments enable us to do better. Before describing these developments, we shall complete our discussion of conventional liquid chromatography by considering briefly two widely used forms which do not employ a column in the usual sense.

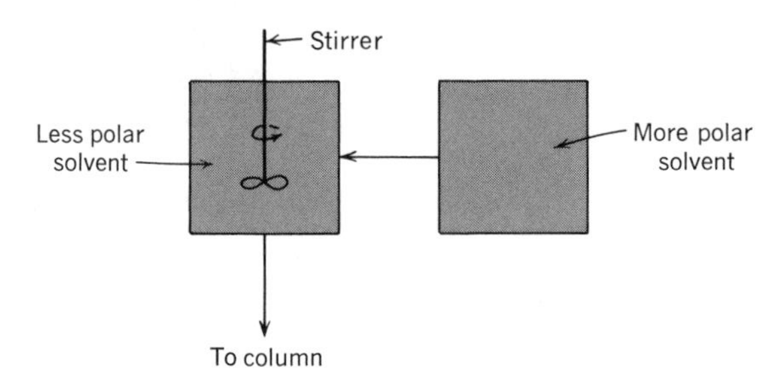

Figure 18.3 Arrangement for gradient elution in liquid chromatography based upon adsorption. Large arrows show solvent flow.

Paper Chromatography

In 1944, again from Martin's laboratory, the separation of a mixture of amino acids was reported using paper chromatography.[2] In this technique, a small volume of sample solution is applied near one end of a strip of filter paper and the spot is allowed to dry (blowing on it with a hair dryer is convenient). The end of the strip is then placed in a trough containing a suitable solvent within a closed chamber. In ascending paper chromatography, the paper is suspended from the top of the chamber so that it dips into the solvent at the bottom, and the solvent creeps up the paper by capillarity. In the descending form, the paper is anchored in a solvent trough at the top of the chamber, and the solvent migrates downward by capillarity assisted by gravity. After the solvent front has moved almost the length of the paper, the strip is removed, dried, and examined. In a successful case, solutes from the original mixture will have migrated along the paper at different rates, forming a series of separated spots. If the compounds are colored, of course the spots can be seen. If not, they must be found in some other way. Some compounds fluoresce, in which case glowing spots may be seen when the paper is held under an ultraviolet lamp. For amino acids, the paper is usually sprayed with a solution of ninhydrin, a reagent which reacts with the amino group to yield a purple compound. For quantitative analysis, the spots may be cut out with scissors, the solutes leached from the paper by appropriate solvents, and the solutions examined by a suitable technique, often spectrophotometry.

For identification purposes, spots are often characterized by their R_f values. An R_f value is the ratio of the distance moved by a solute to the distance that the solvent front moved during the same time. Identical R_f values for a known and an unknown compound using several different solvent systems provides good evidence that the two are identical, especially if they are run side by side along the same strip of paper.

It sometimes happens that all the components of a sample cannot be separated using any one solvent system; some components separate better in one system, some in another. Two-dimensional paper chromatography may then be employed. The sample is spotted near one corner of a square filter paper sheet. After migration of the solutes parallel to one edge of the paper using one solvent system, the paper is turned 90°, and a second solvent system carries the solutes into the unused portion of the paper. The pattern of ninhydrin-stained spots that results from applying this technique to the amino acids in a protein hydrolyzate is often called a "fingerprint" of the protein.

It was considered at first that paper chromatography was simply a form of liquid–liquid partition. The hydrophilic cellulose fibers of the paper can bind water; after exposure to a humid atmosphere, filter paper that appears dry may actually contain a large percentage of water, say 20% or more by weight. Thus the paper was considered to be the analog of a column containing a stationary aqueous phase. Solutes then were partitioned between this water and the moving immiscible organic solvent. It was soon realized, however, that this model was too

[2] R. Consden, A. H. Gordon, and A. J. P. Martin, *Biochem. J.*, **38**, 224 (1944).

simple. Separations were obtained where the moving phase was miscible with water or in some cases was itself an aqueous solution. Thus, although liquid–liquid partition may indeed play a role in some cases, the mechanism of paper chromatography is often more complicated than that. Interactions between solutes and the cellulose support are probably involved, for example, adsorption and hydrogen bonding. Carboxyl and other ionizable groups are introduced into cellulose during the pulping and bleaching operations of paper manufacture, and hence the paper may also act as an ion exchanger.

Thin-Layer Chromatography

Thin-layer chromatography, or TLC, like paper chromatography, is inexpensive and simple to perform. It has an advantage of speed over paper chromatography: the process may require only a half hour or so, whereas a typical separation on paper requires several hours. TLC is very popular and is used routinely in many laboratories.

The separation medium is a layer perhaps 0.1 to 0.3 mm thick of an adsorbent solid on a glass, plastic, or aluminum plate. A typical plate is 8 × 2 inches. Typical solids are alumina, silica gel, and cellulose. Workers used to prepare their own plates by coating the glass with an aqueous suspension of the solid, which usually contained a binder such as plaster of paris, and then drying the plates in the oven. Precoated glass plates and sheets of plastic and aluminium foil which can be cut to size with scissors are commercially available, and probably the majority of workers use these today.

The sample, generally a mixture of organic compounds, is applied near one end of the plate as a small volume of solution, usually a few microliters containing microgram quantities of the compounds. A hypodermic syringe or a small glass pipet may be used. The sample spot is dried, and then the end of the plate is dipped into a suitable moving phase. The solvent moves up the thin layer of solid on the plate, and as it moves, sample solutes are carried along at rates which depend upon their solubilities in the moving phase and their interactions with the solid. After the solvent front has migrated perhaps 10 cm, the plate is dried and examined for solute spots as in paper chromatography. A two-dimensional run using two different moving phases is often performed; here a square plate is used rather than a narrow one. The separation may be followed by a quantitative determination; where a spot is located, the adsorbent can be scraped from the plate with a spatula, the solute eluted from the solid material with a suitable solvent, and the concentration of the solution determined by a technique such as spectrophotometry.

MODERN LIQUID CHROMATOGRAPHY

Although a theoretical consideration of liquid chromatography can become very complicated, our purpose here requires a brief and qualitative treatment. Fortunately, it is possible to understand recent developments in these terms. Let us look at the situation in the following manner. We have been "spoiled" by GLC. We have become accustomed to chromatography where excellent separations required only a few minutes; perhaps integrations were performed electronically and a computer printed out complete analyses. After this experience, we are no longer content with the conventional technique when we encounter nonvolatile samples which require liquid moving phases. Several hours to elute a sample, assaying eluate fractions one at a time, one or two samples a day—out of the question! Yet we need the power of chromatography for increasingly difficult separations. So we must speed it up and at least partly automate it. How shall we do this?

The first thing that comes to mind is to increase the flow rate; solutes will elute faster. Gravity will no longer suffice, and we shall need a pump, along with plumbing connections that do not leak under high pressure. But there is more to consider than this. It has been well known for years that high flow rates in liquid chromatography are incompatible with good separations. Recalling the last chapter, we suspect that kinetic factors are responsible for this. So we need more than a pump: we must consider how to overcome the increasing departure from equilibrium that accompanies the increased flow rate, and this leads into the area of column technology.

In comparing liquid chromatography with GLC, it is instructive to consider the three terms in the van Deemter equation:

$$\text{HETP} = A + \frac{B}{u} + Cu$$

A, the eddy diffusion term, is probably small, and we shall not expect to benefit greatly by improvements here. Smaller diameter of the column packing particles tends to lessen A, but smaller particles are more difficult to pack uniformly, a factor which operates in the opposite direction. In seeking a big improvement in column efficiency, we must turn to factors other than eddy diffusion. (But this is not the whole story on particle size, because the size may also affect the nonequilibrium term in the van Deemter equation, as we shall note below.)

In GLC, the B/u term was very important at low carrier gas flow rates. This is not the case in liquid chromatography because of the fact that diffusion coefficients of solutes in the liquid phase are very much smaller than in gases. Thus low flow rates in liquid chromatography are not deleterious so far as the separation itself is concerned, although we take no comfort in this. We are interested in faster flow rates, not slower ones, in our quest for high performance which includes speed as well as good separations. For all practical purposes, we may neglect the effect of longitudinal diffusion in liquid chromatography.

There is only one factor left: If we want high-speed liquid chromatography that is also highly efficient with regard to the resolution of solutes, it becomes crucial to obtain faster mass transfer of solutes between stationary and moving phases. One possibility is to raise the temperature. If the sample is stable at, say 80°, we may operate the column at this temperature and get somewhat faster mass transfer kinetics than at room temperature. The higher temperature also lowers the viscosity of the liquid, with the result that the desired flow rate may be obtained with a smaller pressure drop. But only small gains are made in this way. The major breakthrough to date has been the development of new column packing materials.

The old packing materials are porous solids—silica gel, diatomaceous earth, ion-exchange resin beads, etc.—which have deep pores that fill with stagnant portions of mobile phase. Diffusion of solutes through such pores is slow and represents the major obstacle to high-speed operation. So far, there have been two basic approaches to the solution of this problem. At the present time, the best performance available is provided by packing materials known as "superficially porous supports" or "porous layer beads." These are commercially available under tradenames such as Zipax (E. I. du Pont de Nemours and Co., Inc.) and Corasil (Waters Associates). The porous layer beads represent an ingenious combination of large overall particle size for easy column packing and low pressure drop with very small pore depth for rapid equilibration. Each bead is an impervious sphere of glass or siliceous material of some sort, 30 to 50 μm in diameter, with a thin surface coating of porous material, as shown schematically in Fig. 18.4. With Zipax, the surface layer is made up of bonded spherical microbeads, each about 200 nm in diameter. The interstices in this partly fused bead structure are large enough to permit solute molecules free access to a very thin film of stationary liquid. The Zipax surface is a very weak adsorbent; this material is intended primarily as a support for a stationary liquid phase. Corasil I has a very thin layer of silica gel bonded to the surface of a glass bead. It has been used as a support for a liquid phase as well as an adsorbent for liquid–solid chromatography. Corasil II has a thicker silica gel layer and was developed primarily as an adsorbent. Materials are also available for ion-exchange chromatography; these

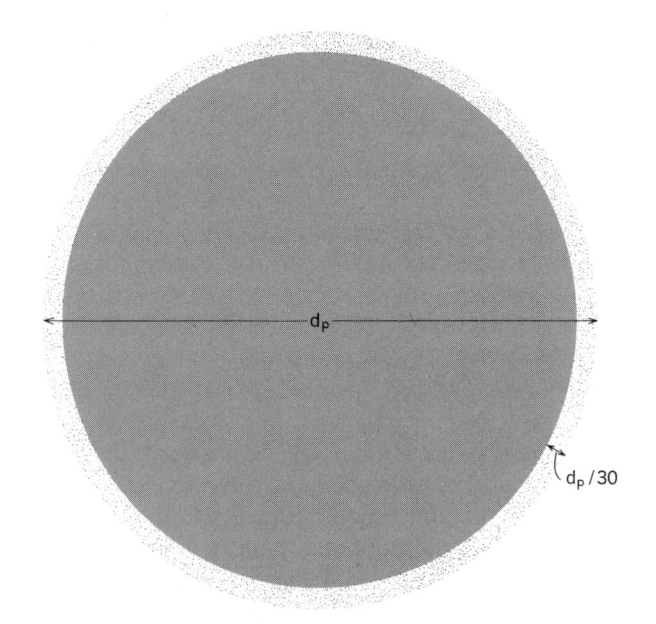

Figure 18.4 Schematic section through the center of a porous layer bead.

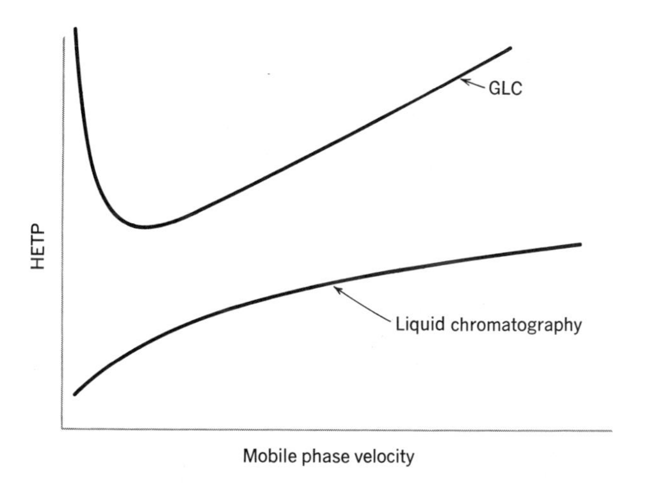

Figure 18.5 Typical graphs of HETP vs mobile phase velocity for gas and liquid chromatography.

comprise a thin, porous outer shell of an ion-exchange medium over a solid, nonporous core. Interesting packing materials have been developed where molecules which play somewhat the role of a stationary liquid are chemically bonded to a silica matrix; the purpose is to eliminate deterioration of liquid loaded columns resulting from solubility of the stationary phase in the mobile liquid and from stripping due to shear forces at high carrier velocities.

Figure 18.5 shows how HETP varies with mobile phase velocity, with the typical behavior in GLC also shown for contrast. The point noted earlier, that longitudinal diffusion is not a problem at low velocities in the liquid case, may be seen in the figure. It may also be seen that the price for a fast flow rate, in terms of increased HETP, is not so bad as might have been expected; the curve almost levels off at very fast velocities. Possibly this relates to the occurrence of turbulent flow at high carrier velocities.

The new packing materials have led to the development of liquid chromatographs which are operated in much the same manner as gas chromatographs. Small samples are injected, picked up by the flowing liquid carrier stream, and carried to columns which are typically 50 to 100 cm long and 2 or 3 mm in internal diameter. Typical flow rates are 1 or 2 ml/min, requiring pressures of 500 to several thousand psi, depending upon the column. The column effluent passes through a detector, the output of which is recorded as in GLC. With superficially porous supports, sample capacities are small, and very sensitive detectors are required. Two types are in common use, one based upon the absorption of ultraviolet radiation by sample components, the other upon measurement of refractive index. The former is generally the more sensitive but not universally applicable, while the latter is less sensitive but responds to a larger number of compounds. Figure 18.6 shows a chromatogram obtained with such an instrument.

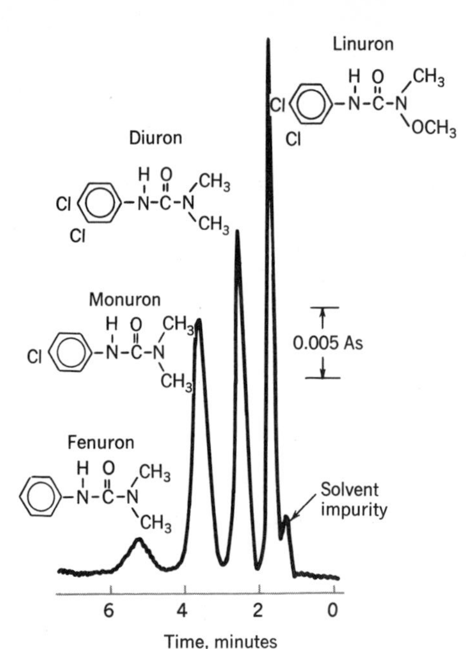

Figure 18.6 Separation of substituted urea herbicides. Column = 50 cm, 2.1 mm i.d., 1.0% β, β'-oxydipropionitrile on 37–44 μ controlled surface porosity support; carrier = di-n-butyl ether; flow rate = 1.14 cc/min; sample 1 μl of a solution of 67 μg/ml each in carrier. (Courtesy of the publisher of the *Journal of Chromatographic Science*.)

The other approach to the problem of slow diffusion involves simply grinding the traditional sorts of porous packing materials to a very small particle size, say 2 to 5 μm in diameter. This obviously leads to shorter pores and faster equilibration of stationary and mobile phases. There have been experimental problems in packing columns uniformly with these very small particles. In addition, achieving desired flow rates through such finely divided materials requires high inlet pressure, perhaps of the order of 10,000 psi; pumps which provide this without leaking are being developed. One may expect that liquid chromatographs accepting columns with such packing materials will be marketed before long.

Modern liquid chromatography is in its infancy, but because it promises so much, it is an active field and rapid advances may be expected. Particularly important would be detectors of high sensitivity and broader applicability than the present ultraviolet detector. Many chemists would like to see high-efficiency operation on a preparative scale, but basically there is a conflict between capacity for large samples and optimization for high-speed separations, and there may be disappointments. However, some degree of scale-up is certainly possible.

REFERENCES

L. FISCHER, *An Introduction to Gel Chromatography*, Part II of T. S. Work and E. Work, eds., *Laboratory Techniques in Biochemistry and Molecular Biology*, Vol. I, North-Holland Publishing Co., Amsterdam, 1970.

F. HELFFERICH, *Ion Exchange*, McGraw-Hill Book Company, New York, 1962.

J. G. KIRCHNER, *Thin-Layer Chromatography*, Vol. XII of E. S. Perry and A. Weissberger, eds., *Technique of Organic Chemistry*, Wiley–Interscience, New York, 1967.

J. J. KIRKLAND, ed., *Modern Practice of Liquid Chromatography*, Wiley–Interscience, New York, 1971.

I. M. KOLTHOFF and P. J. ELVING, eds., *Treatise on Analytical Chemistry*, Part I, Vol. 3, Interscience Publishers, Inc., New York, 1961, Chaps. 33, 34, 35, 36.

H. A. LAITINEN and W. E. HARRIS, *Chemical Analysis*, 2nd ed., McGraw-Hill Book Company, New York, 1975, Chaps. 24, 25.

L. R. SNYDER and J. J. KIRKLAND, *Introduction to Modern Liquid Chromatography*, Wiley–Interscience, New York, 1974.

QUESTIONS

1. *Adsorption.* Suppose that a particular solution–solid adsorption process is correctly described by an equation of the Freundlich type and that a particular gas–liquid interaction follows Henry's law. Compare the shapes of the elution curves obtained in chromatographic experiments based upon these two processes and explain your comments.

2. *Ion exchange.* Explain the difference between the practical selectivity coefficient of an ion-exchange resin with regard to two ions, A and B, and the distribution coefficient for an individual ion.

3. *Terms.* Explain clearly the meaning of the following terms: (a) cross-linking; (b) gradient elution; (c) porous layer beads; (d) R_f value; (e) gel filtration.

Multiple-choice: In the following multiple-choice questions, select the *one best* answer.

4. A neutral molecule such as ethanol or sugar which found its way into the pores of a typical anion exchange resin could be eliminated (a) only by replacement with a cation; (b) only by replacement with an anion; (c) only if replaced by another organic molecule on a one-for-one exchange basis; (d) by flushing out with water.

5. In chromatography, a substance for which the distribution coefficient, K, is zero may be used to estimate (a) the volume within the column available to the moving phase; (b) the volume within the column occupied by the packing; (c) the volume within the pores of the packing material; (d) the total volume of the column.

6. Which of the following statements is false? (a) The more polar a compound the more strongly it will be adsorbed from a solution; (b) A high molecular weight favors adsorption, other factors being equal; (c) The more polar is the solvent, the stronger is the adsorption of the solute; (d) The adsorption isotherm is usually nonlinear.

PROBLEMS

1. *Resin capacity.* A 1.016-g sample of an anion-exchange resin in the chloride form lost 0.3012 g of moisture when heated overnight at 60°C under vacuum. A sample of the undried resin weighing 0.6630 g was washed with nitric acid to remove the chloride. If 15.25 ml of 0.1000 M $AgNO_3$ was required to titrate the chloride, calculate the capacity of the dry resin in milliequivalents per gram.

2. *Distribution coefficient.* A 5.00-ml solution containing a radioactive element is shaken with 100 mg of a cation-exchange resin. The original activity of the solution is 80,000 counts/ml, and after equilibration with the resin, the activity is 35,000 counts/ml. Calculate the distribution coefficient (weight) for this element.

3. *Water analysis.* A 10.00-ml portion of seawater was passed through a column packed with an ample quantity of a strong-acid cation-exchange resin in the hydrogen form. Distilled water was then passed through the column to flush out any unbound ions. The column effluent, including the wash, was then titrated with 0.1000 *M* NaOH using phenolphthalein indicator. The required volume of NaOH was 62.25 ml. Calculate the total milliequivalents of cations (Na^+, Mg^{2+}, K^+, etc.) per liter of seawater.

4. *Gel filtration.* A gel filtration column is prepared using a Sephadex gel packing which excludes molecules whose molecular weights are larger than about 500,000. The volume of water within the gel particles available for accepting solutes is 200 ml. The interstitial or void volume is 50 ml. (a) What is the expected retention volume, V_R, for a protein of molecular weight 700,000? (b) What is the expected retention volume for a small protein such as cytochrome *c*, MW = 12,400, which can freely penetrate even the smallest pores of the gel phase? (c) What can be said about the behavior of a compound of molecular weight 15,000 which elutes with a retention volume of 400 ml?

CHAPTER NINETEEN

real analytical chemistry

For students who are completing introductory courses in analytical chemistry it may be wise to provide a few words of encouragement and of caution and to place what has been accomplished in perspective relative to the field as a whole. You have seen, not real analytical chemistry as it is currently practiced, but rather a series of topics and exercises selected for their suitability for beginners and for the manner in which they fit into the undergraduate chemistry curriculum. (This is pretty much the case for all your courses: in organic chemistry, for example, you do not synthesize novel compounds nor develop new theories; rather, you learn accepted theories and standard reactions and, in the laboratory, you learn to perform a number of routine operations.) One has to start somewhere.

In many cases, depending upon ultimate career goals, undergraduate chemistry courses provide essential background. For example, organic chemistry, again, is important to students who plan to study biochemistry, pharmacology, or other fields in the biomedical area. But such students may well not expect to work professionally in organic chemistry as such. In this regard, analytical chemistry, if not unique, is at least remarkable for the frequency with which it crops up in other fields in a manner requiring virtually professional competence on the part of people whose major interests may lie elsewhere. Thus we approach the answer to the first question you might ask about real analytical chemistry.

WHO DOES IT?

The answer is "all sorts of people do it." In the mid-1960s, a rough survey by one of the authors disclosed that about 60% of the research papers in the leading professional journal *Analytical Chemistry* were authored by people who were not themselves analytical chemists. And some of them were not even primarily chemists. Browsing through recent issues of the same journal shows that "outsiders" still do analytical chemistry. Table 19.1 illustrates this point. It

TABLE 19.1 Origins of Some of the Research Papers in *Analytical Chemistry* for December, 1977

Veterans Administration Hospital and Department of Radiology, University of Florida, Gainesville
Department of Electrical Engineering and Computer Sciences, University of California, Berkeley
National Institute of Public Health, Laboratory of Toxicology, Bilthoven, The Netherlands
Israel Institute for Biological Research, Ness-Ziona
Eastern Regional Research Center, Agricultural Research Service, U.S. Department of Agriculture, Philadelphia
Research Institute on Alcoholism, New York State Department of Mental Hygiene, Buffalo
Chemistry Division, Naval Weapons Center, China Lake, California
Mineral Engineering Department, University of Wyoming, Laramie
National Bureau of Standards Institute for Materials Research, Washington
Grand Forks Energy Research Center, Energy Research and Development Administration, North Dakota
Department of Microbiology, Colorado State University, Fort Collins
Laboratoire de Toxicologie Générale et Biotoxicologie, Faculté de Pharmacie, Marseilles, France
Chemical Engineering Department, University of Southern California, Los Angeles
Departments of Medicinal Chemistry and Pharmacy, University of Illinois at the Medical Center, Chicago
Department of Health and Human Ecology, California State College, San Bernardino
U.S. Geological Survey, Reston, Virginia
Division of Clinical Pharmacology, University of Colorado Medical Center, Denver

should be noted that the table does not represent a thorough study; for example, there are doubtless analytical chemists working in some of the listed laboratories. But it does show the importance of analytical work in a wide variety of scientific endeavors. (To emphasize this, we biased the table by omitting chemistry departments where analytical research might be pursued for its own sake.) Thus, depending upon his career, the student who is now reading this book could find himself using analytical results to reach decisions in his own field or, in some cases, performing or supervising the analytical work himself. An excellent example is the premedical student who becomes a pathologist. It is not at all unusual for such a person to acquire responsibility for a clinical laboratory where he will supervise both analytical and microbiological work.

Having seen that a wide variety of people may become involved in analytical work, we may ask the next question.

WHAT DO THEY DO?

At the outset we should note that the actual laboratory work may be performed by anyone who can learn certain mechanical operations; much of it can even be automated to eliminate human labor altogether. But behind this operation there must be someone to select the analytical methods, to purchase the equipment, to hire and train the people, to see that standards are maintained, to trouble-shoot when problems arise, and to stand as a responsible person at the interface between the analytical laboratory and the ultimate users of the results. Such a person is going to read widely to keep up with what other people with similar problems are doing, he is going to look constantly for ways to improve his operation from the standpoints of reliability and cost, and when new situations arise he is going to be a researcher. In fact, he probably uses some of his time for research in any event, just because he likes to.

Undergraduates often ask what is done in research or how one chooses a research problem. Because analytical chemistry provides needed answers in so many different fields, the problems, particularly in industrial, government, or medical laboratories, are often brought to the analytical chemist ready-made. Creativity is then directed toward finding the simplest and least expensive analytical method that will provide the necessary reliability.

Samples that are totally unknown and that might contain any of the millions of chemical compounds seldom come to the analytical laboratory. Usually, the sample has a history that provides some idea of what is likely to be in it, and the person who wants it analyzed will provide important information. Is a total analysis required or need only one or two components be determined? Must errors be held to a few parts per thousand or will 5 or 10% be all right? Can the desired constituent be measured directly or must interfering substances be dealt with first? Is this the first of many such samples or is it a one-shot deal? These and other questions must be considered before the method is selected. Then comes library work: How have other people handled similar problems? Finally, decisions are made that lead to a step-by-step prescription for performing the analysis based upon the sample, the capabilities of the laboratory, and the uses to which the results will be put.

Academic research in analytical chemistry is harder to characterize because it involves highly individualistic people who enjoy a great deal of freedom. But, very roughly, we may note two broad approaches in choosing research problems, one relating to the technique and the other to the sample. One may become, for instance, a spectroscopist or an electroanalytical chemist or an expert in mass spectrometry or chromatography or solvent extraction or whatever. He will devote himself to very basic work in improving the technique—new sources of excitation, better detectors, new column packings, new methods of data processing. To demonstrate that he has accomplished something, he will then report some analytical applications to any kinds of samples that are convenient.

In the other approach, the analytical chemist devotes himself to a certain element or group in the periodic table, or certain classes of organic compounds, or samples from certain sources. He will then apply any analytical techniques that show promise of improving the methodology in his chosen area. An analytical chemist interested in clinical chemistry provides an example of this approach to the selection of research problems. He will deal with blood or urine samples, let us say, and with what he perceives as the present or future needs of clinicians for laboratory results as diagnostic aids. He may concern himself with improving current methods in regard to reliability, speed, or cost, or he may develop methods for determining sample components which are presently neglected.

It is impossible to describe in a few pages the professional activity of any modern scientist, and this is no less true of the analytical chemist. Chemistry has been called the handmaiden of the sciences; this is to say that chemistry provides materials, techniques, and viewpoints for deeper understanding in physics, geology, biology, agriculture, and medicine. If this be true, then analytical chemistry is the handmaiden of chemistry, for if we do not know what is in our samples then we do not know what we are doing. The student who remains in touch with science throughout his life will see increasing numbers of important conclusions founded upon analytical work in whatever fields he chooses to observe.

LABORATORY DIRECTIONS AND PROCEDURES

CHAPTER TWENTY

general laboratory directions

INTRODUCTION

In determining the constitution of an unknown sample, a sound method must first be selected. It then remains a matter of technique alone whether the ultimate measurement is performed upon *all* or part of the desired constituent. The heart of *quantitative technique* is simply to carry a sample through a number of manipulations without accidental losses and without introducing foreign material. Since every conceivable fortuity cannot be anticipated in writing a text, the student must develop independent judgment in connection with his laboratory work. Common sense plus awareness of the danger spots are the main requirements of the beginning student in this regard.

Neatness and Cleanliness

Good analysts are usually scrupulously neat. The student with an orderly laboratory bench is not likely to mix up samples, add the wrong reagents, or spill solutions and break glassware. Neatness in the laboratory must extend, of course, from the student's own bench to the shelves where materials are available for the whole class. Much time can be wasted in searching for a small item in a jumble of glassware or in finding a certain reagent bottle that has been misplaced on the side-shelf. Neatness also includes stewardship over the more permanent laboratory fixtures, such as ovens, hot plates, hoods, sinks, and the benches themselves. Corrosive materials that are spilled must be cleaned up immediately from equipment, benches, or floors. It is important that plumbing be conserved by washing acids and bases down the drains with copious volumes of water.

No analysis should ever be performed using anything but clean glassware. Glassware that looks clean may or may not be clean as an analyst understands the term. Surfaces on which no visible dirt appears are often still contaminated by a thin, invisible film of greasy material. When water is delivered from a vessel so contaminated, it does not drain uniformly from the glass surface, but leaves behind isolated drops that are troublesome or sometimes impossible to recover. Glassware into which a brush can be inserted, such as beakers or Erlenmeyer flasks, are best cleaned with soap or synthetic detergent. Pipets, burets, and volumetric flasks may require hot detergent solutions for thorough cleaning. If the glass surface still does not drain uniformly, it may be necessary to use *cleaning solution*,[1] whose strong oxidizing properties ensure clean glass surfaces in most cases. After cleaning, apparatus should be rinsed several times with tap water, then with small portions of distilled water, and finally allowed to drain. Cleaning solution is usually avoided in biological work because many microorganisms are sensitive to traces of chromium which remain on the glass even after thorough rinsing.

Planning and Efficiency

The beginning student will find that time spent in planning his work will save him many wasted hours during the laboratory periods. Before arriving at the laboratory, he should be familiar with the experiment to be performed. Some of the operations of analytical chemistry are very time-consuming. Fortunately, however, some of these operations require very little attention when once set in motion. For example, it is sometimes necessary to dry primary standards or unknown samples in an oven before they are used. It is foolish for a student to come to the laboratory, place material in the oven, and then sit around for an hour or more while it dries. Depending upon the laboratory regulations in his particular school, it may be possible for him to place a sample in the oven before the regular laboratory period, or, perhaps, to work on another experiment while his sample is drying during the laboratory period. Filtration and ignition of precipitates are examples of lengthy operations that do not need constant attention. The student should plan his work, considering more than one experiment at a time (with the instructor's help at first, if necessary) in such a way that he need not be idle for long periods.

[1] Many different formulas are available for the preparation of cleaning solution, and the choice seems largely personal. A satisfactory solution can be made as follows: In a 600-ml beaker place 20 to 25 g of technical-grade sodium dichromate and 15 ml of water. Then add slowly and carefully about 450 ml of technical-grade sulfuric acid with occasional stirring. Cool the solution and store it in a glass-stoppered bottle. The appearance of the green chromium(III) ion indicates that the solution is exhausted. Cleaning solution is very corrosive; take special precautions not to get it on your skin or clothing. If any is spilled on the skin, rinse the affected parts quickly and thoroughly with tap water. Cleaning solution is most effective when heated to about 60 to 70°C. Consult your instructor for directions before using this solution. The solution should not be discarded after use. It is returned to the storage bottle until the color turns green as mentioned above.

Reagents

The purity of the reagents used in analytical chemistry is a matter of utmost importance. Fortunately, reagents sufficiently pure for most analytical work are now available commercially, and even primary standards can be purchased in some cases. In work where a certain impurity might be very deleterious, however, it is best to test each batch of reagents rather than to rely on the manufacturer. Except in special cases as noted in the laboratory directions, the student in the beginning course may assume that the reagents provided him are of adequate purity.

Reagents of high purity cost more, of course, than reagents that need not be so carefully manufactured. For some uses, less expensive grades of chemicals may be employed. For example, the most costly grades of sulfuric acid and alkali dichromate would be extravagant for the preparation of "cleaning solution." Also, many examples can be found of cheaper grades of reagents which do not contain the particular impurities that imperil certain analyses. However, since a reagent is often used for a number of different analyses, it is generally most practical to stock the analytical laboratory with only the better grades of reagents.

Reagents are roughly classified as follows, although some of the designations have not been precisely defined:

1. *Technical* (or *commercial*)-grade chemicals are used mainly in industrial processes on a large scale, and are seldom employed in the analytical laboratory except for such purposes as the preparation of cleaning solution.
2. *U.S.P.* reagents meet purity standards that can be found in the *United States Pharmacopoeia.* These standards were established primarily for the guidance of pharmacists and the medical profession, and in many cases, impurities are freely tolerated that are not incompatible with the use to which these persons will put the compounds. Thus U.S.P. reagents are not usually suitable for analytical chemistry.
3. *C.P.* reagents are often much more pure than U.S.P. reagents. On the other hand, the designation C.P. (standing for "chemically pure") has no definite meaning; standards of purity for reagents of this class have not been established. Thus C.P. reagents may often be used for analytical purposes, but there are many situations where they are not sufficiently pure. In many analyses it is necessary to test these reagents for certain impurities before they may be used.
4. *Reagent-grade* chemicals conform to the specifications established by the Committee on Analytical Reagents of the American Chemical Society.[2] The label on such a reagent bears a statement such as "meets A.C.S. specifications," and generally furnishes information regarding the actual percentages of various impurities or at least maximal limits of impurities. Many reagent-grade chemicals are used as primary standards in the introductory laboratory. A label from a bottle of potassium acid phthalate (potassium

[2] *Reagent Chemicals, American Chemical Society Specifications,* 5th ed., Washington, D.C., 1974.

biphthalate) is shown in Table 20.1. Note that the purity is 99.95 to 100.05%, well within the usual requirements for a primary standard.

TABLE 20.1 Label of a Reagent-Grade Chemical

Potassium Biphthalate, Crystal, Primary Standard	
Meets A.C.S. Specifications	
Assay	99.95–100.05%
Insoluble matter	0.005%
pH of a 0.05 M solution at 25°C	4.00
Chlorine compounds (as Cl)	0.003%
Sulfur compounds (as S)	0.002%
Heavy metals (as Pb)	0.0005%
Iron (Fe)	0.0005%
Sodium (Na)	0.0005%

Substances of even higher degrees of purity are available for research and work in such areas as solid-state physics. These ultrapure materials may be obtained from the National Bureau of Standards and various commercial supply houses.

Certain acids and bases are provided in the laboratory as concentrated solutions. Table 20.2 gives the composition of most of the common reagents. Dilute solutions are prepared as needed by adding the concentrated solution to water. The degree of dilution is often indicated by the ratio of the volume of concentrated solution to that of water. For example, 1:4 nitric acid means that 1 volume of nitric acid is added to 4 volumes of water.

TABLE 20.2 Composition of Concentrated Common Acids and Bases

Reagent	Density, g/ml	Percent by weight	Approximate molarity
Acetic acid, $HC_2H_3O_2$	1.057	99.5	17
Ammonium hydroxide, NH_4OH	0.90	28 (as NH_3)	15
Hydrochloric acid, HCl	1.18	36	12
Nitric acid, HNO_3	1.42	72	16
Perchloric acid, $HClO_4$	1.68	71	12
Phosphoric acid, H_3PO_4	1.69	85	15
Sodium hydroxide, $NaOH$	—	50	16
Sulfuric acid, H_2SO_4	1.83	85	15

The chances of contamination increase enormously when a reagent bottle is placed in the laboratory for the use of a large number of people. Thus it is most important that students carefully adhere to certain rules governing the use of the reagent shelf. In addition to the following instructions, any further sugrestions of the instructor should be heeded. (1) The reagent shelf should be clean and orderly. (2) Any spilled chemicals must be cleaned up immediately. (3) The stoppers of reagent bottles should not be placed on the shelf or laboratory bench. Stoppers may be placed on clean towels or watch glasses, although it is best to hold them between two fingers while reagents are being withdrawn. (4) The mouths of reagent bottles should be kept clean. (5) Pipets, droppers, or other instruments should never be inserted into reagent bottles. Rather, a slight excess of reagent should be poured into a clean beaker from which the pipetting is done and the excess discarded, not returned to the bottle. (6) Fingers, spatulas, or other implements should not be inserted into bottles of solid reagents.

APPARATUS

In addition to the usual equipment found in any chemistry laboratory, there are certain items that are of special interest to the analytical chemist. Some of the more important items are described in this section, and advice is given regarding their use.

Wash Bottle

Each student should have a wash bottle of reasonable capacity capable of delivering a stream of distilled water (about 1 mm in diameter) from a tip connected flexibly to the main part of the bottle. A convenient type, constructed from a 1-liter Florence flask, glass tubing, a short section of rubber tubing, and a two-hole rubber stopper, is shown in Fig. 20.1. Other types are also available, including a polyethylene bottle (Fig. 20.1) whose body is squeezed to force water from the tip. It is sometimes advisable to have additional wash bottles available

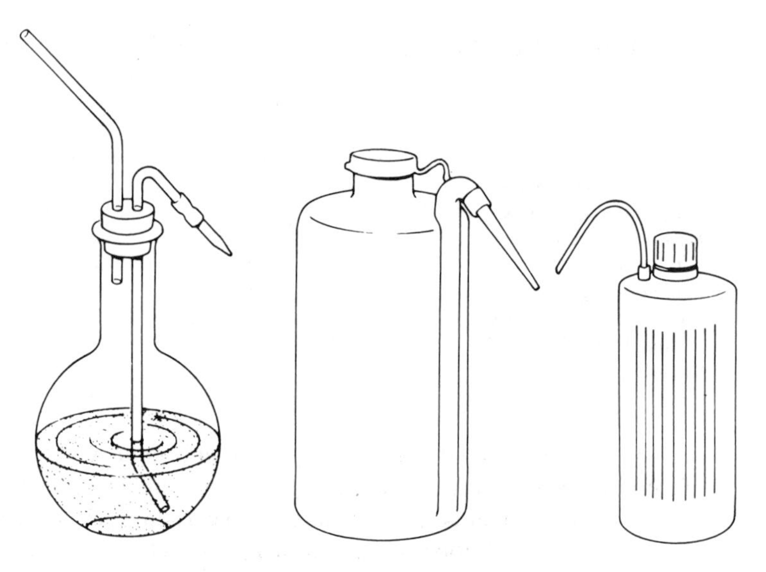

Figure 20.1 Wash bottles.

for hot water and special solvents. The wash bottle is used whenever a fine, directed stream of distilled water is needed, as when rinsing down the sides of a glass vessel to ensure that no droplets of sample solution are lost.

Stirring Rods

As the name implies, stirring rods are used for stirring solutions or suspensions, generally in beakers. The rods are cut from a length of solid glass rod, generally 3 or 4 mm in diameter, so as to extend about 6 or 8 cm from the top of the beaker. The ends should be fire-polished. In addition to their stirring function, stirring rods have other useful purposes. For example, they are used in transferring solutions from one vessel into another. When an aqueous solution is poured from the lip of a vessel such as a beaker, there is a tendency for some of the liquid to run down the outside surface of the glass. This is prevented by pouring the solution down a stirring rod, the rod being held in contact with the lip of the vessel and directing the flow of liquid into the receptacle (see Fig. 20.8). Stirring rods also serve as handles for "rubber policemen" (sections of rubber tubing sealed together at one end, with the other end slipped over a stirring rod, used to salvage small quantities of precipitates from the walls of beakers).

Desiccator

A desiccator is a vessel, usually of glass but occasionally of metal, which is used to equilibrate objects with a controlled atmosphere. Since the desiccator usually stands in the open, the temperature of this atmosphere generally approaches room temperature. It is normally the humidity of this atmosphere which is of interest. Objects such as weighing bottles or crucibles, and chemical substances, tend to pick up moisture from the air. The desiccator provides an opportunity for such materials to come to equilibrium with an atmosphere of low and controlled moisture content so that errors due to the weighing of water along with the objects can be avoided. A common type of desiccator is shown in Fig. 20.2.

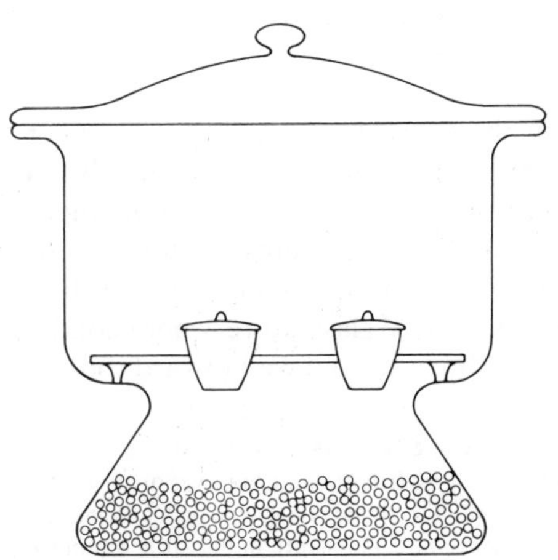

Figure 20.2 Desiccator.

The nature of the drying agent placed in the bottom of the desiccator determines the equilibrium partial pressure of water vapor in the desiccator space. Table 20.3 contains the results of Trusell and Diehl,[3] who studied the efficiency of various chemical desiccants. These workers studied 21 common drying agents by passing known volumes of wet nitrogen over the material, condensing the residual water in a liquid nitrogen trap, and weighing the water. The most powerful desiccant is not necessarily the best for a given application. Phosphorus pentoxide, for example, has a tendency to acquire a surface glaze as it picks up water, which prevents the bulk of the material from being effective. Calcium chloride, while rather poor with regard to equilibrium water vapor pressure, is inexpensive, has a fairly high capacity,[4] and is adequate for much analytical work.

TABLE 20.3 Efficiency of Chemical Desiccants*

Material	Residual water (γ/liter)†
$Mg(ClO_4)_2 \cdot 0.12H_2O$	0.2
$Mg(ClO_4)_2 \cdot 1.48H_2O$	1.5
BaO (96.2%)	2.8
Al_2O_3 (anhydrous)	2.9
P_2O_5	3.5
Molecular Sieve 5A (Linde)	3.9
$Mg(ClO_4)_2$ (88%) + $KMnO_4$ (0.86%)	4.4
$LiClO_4$ (anhydrous)	13
$CaCl_2 \cdot 0.18\ H_2O$	67
$CaSO_4 \cdot 0.02\ H_2O$ (Drierite)	67
Silica gel	70
NaOH (91%, Ascarite)	93
$CaCl_2$ (anhydrous)	137
$CaSO_4 \cdot 0.21\ H_2O$ (Anhydrocel)	207
$NaOH \cdot 0.03\ H_2O$	513
$Ba(ClO_4)_2$ (anhydrous)	599
CaO	656

* Data of F. Trusell and H. Diehl, *Anal. Chem.*, **35**, 674 (1963).
† $1\gamma = 10^{-6}$ g.

After reagents or objects such as crucibles have been dried in the oven, or perhaps at even higher temperatures, they are usually cooled to room temperature in the desiccator prior to weighing. When a hot object cools in the desiccator, a partial vacuum is created, and care must be taken in opening the vessel lest a sudden rush of air blow material out of a crucible or disturb the desiccant itself. For this reason, and also because glass is a very poor conductor of heat, it is usually best to allow a very hot object to cool well toward room temperature before it is

[3] F. Trusell and H. Diehl, *Anal. Chem.*, **35**, 674 (1963).

[4] Capacity must be distinguished from equilibrium vapor pressure in describing desiccants. A desiccant may have a high capacity (i.e., it may be able to pick up a large weight of water vapor per unit weight of desiccant), but still leave much moisture in the air at equilibrium.

placed in the desiccator. After a hot object has been placed in the desiccator, it is well to cover the vessel in such a way as to leave a small opening at one side. This allows air displaced by the warm object to reenter as the object cools, and hence minimizes the tendency to form a vacuum. The desiccator is completely closed during the final stages of cooling.

The desiccator cover should slide smoothly on its ground-glass surface. This surface should be lightly greased with a light lubricant such as Vaseline (never stopcock grease!) Needless to say, the desiccator should be scrupulously clean and should never contain exhausted desiccant. After filling the desiccant chamber, beware of dust from the desiccant in the upper part of the desiccator.

Pipets

Some common types of pipets are shown in Fig. 20.3. The *transfer pipet* is used to transfer an accurately known volume of solution from one container to another. The pipet should be cleaned if distilled water does not drain uniformly, but leaves droplets of water adhering to the inner surface. Cleaning can be done with a warm solution of detergent or with cleaning solution (consult instructor).

The pipet is filled by gentle suction to about 2 cm above the etch line [Fig. 20.4(a)], using an aspirator bulb. Alternatively, a water aspirator can be used to apply suction. A long rubber tube is attached from the top of the pipet to the trap shown in Fig. 20.11. The tip of the pipet should be kept well below the surface of the liquid during the filling operation. The forefinger is then quickly placed over

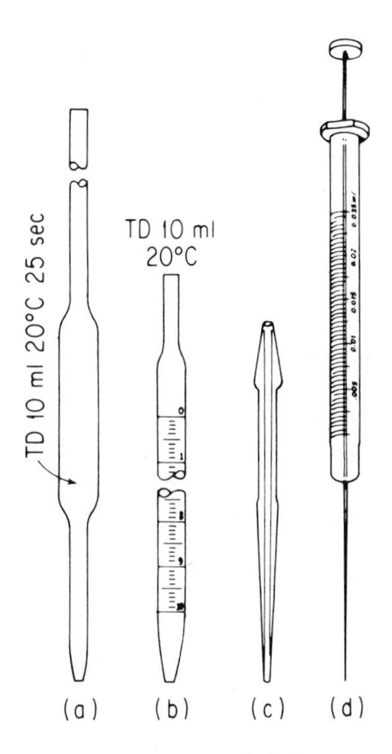

Figure 20.3 Pipets: (a) transfer pipet, (b) measuring pipet, (c) lambda pipet, and (d) microliter syringe.

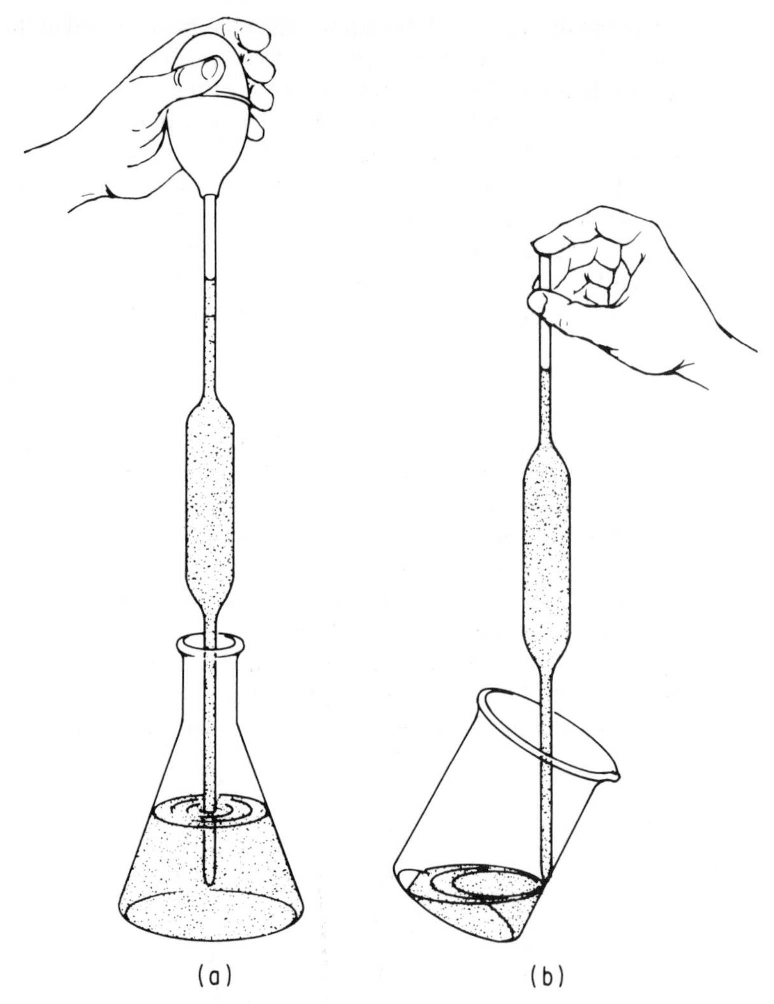

Figure 20.4 (a) Filling pipet—liquid drawn above graduation mark, and (b) use of forefinger to adjust liquid level in pipet.

the top of the pipet [Fig. 20.4(b)], and the solution is allowed to drain out until the bottom of the meniscus coincides with the etched line. Any hanging droplets of solution are removed by touching the tip of the pipet to the side of the beaker, and the stem is wiped with a piece of tissue paper to remove drops of solution from the outside surface. The contents of the pipet are then allowed to run into the desired container, with care being taken to avoid spattering. With the pipet in a vertical position, allow the solution to drain down the inner wall for about 30 sec after emptying, and then touch the tip of the pipet to the inner side of the receiving vessel at the liquid surface. A small volume of solution will remain in the tip of the pipet, but the pipet has been calibrated to take this into account; thus this small final quantity of solution is *not* to be blown out or otherwise disturbed. Pipets with damaged tips are not to be trusted.

Measuring pipets are graduated much like burets and are used for measuring volumes of solutions more accurately than could be done with graduated cylinders. However, measuring pipets are not ordinarily used where high accuracy is required.

Two types of micropipets are also shown in Fig. 20.3. The so-called *lambda* pipets are available in capacities of 0.001 to 2 ml, where 0.001 ml = 1 lambda. They are filled and emptied using a syringe. Those calibrated to *contain* a certain volume are rinsed with a suitable solvent. Those calibrated to *deliver* are not rinsed, but the last drop is forced out of the pipet with the syringe. Microliter syringes are widely used for delivering small volumes in such operations as gas chromatography. They can be bought equipped with stainless steel tips for use in injecting a sample into a closed system. The syringe shown in Fig. 20.3 has a capacity of 0.025 ml (25 microliters, or 25 lambdas) and the smallest divisions correspond to 0.0005 ml.

The National Bureau of Standards specifies 20°C as the standard temperature for calibration of volumetric glassware. The use of such glassware at other temperatures leads to errors. However, the errors are normally small, and pipets can be used at "room temperature" without special precautions except in work of highest accuracy.

Burets

A common form of buret is shown in Fig. 20.5(a). The buret is used to deliver accurately known but variable volumes, most in titrations. The stopcock plug is made of either glass or Teflon. The Teflon stopcock requires no lubrication, but the glass plug should be lightly greased with stopcock grease (not one containing silicones). If too heavy a coating is applied the stopcock may leak and also some of the grease may plug the buret tip. To lubricate a stopcock, remove the plug and wipe old grease away from both plug and barrel with a cloth or paper tissue. Make sure the small openings are not plugged with grease (pipe cleaners are helpful in this event). Then spread a thin, uniform layer of stopcock grease over the plug, keeping the application especially thin in the region near the hole in the plug. Finally, insert the plug in the barrel and rotate it rapidly in place, applying a slight inward pressure. The lubricant should appear uniform and transparent, and no particles of grease should appear in the bore.

Burets must be cleaned carefully to assure a uniform drainage of solutions down the inner surfaces. A hot, dilute detergent solution may be used for this purpose, especially if used in conjunction with a long-handled buret brush. Cleaning solution may also be used, applied hot for a few minutes or overnight at room temperature. The instructor should be consulted for directions on the proper use of cleaning solution. When not in use, the buret should be filled with distilled water and capped (paper cups or small beakers are convenient) to prevent the entry of dust.

It is poor practice to leave solutions standing in burets for long periods. After each laboratory period, solutions in burets should be discarded, and the burets rinsed with distilled water and stored as suggested above. It is especially

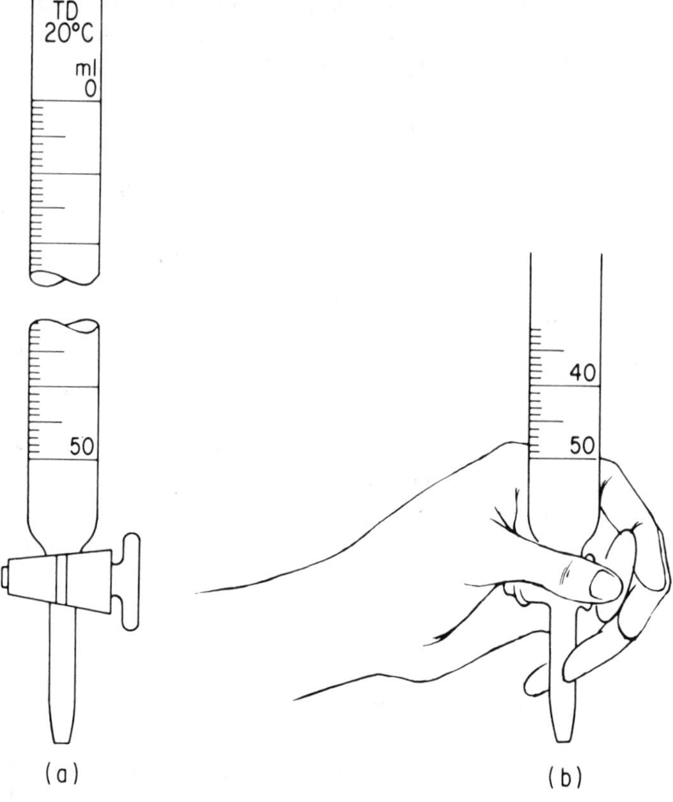

Figure 20.5 (a) Buret and (b) method of grasping stopcock.

important that alkaline solutions not stand in burets for more than short periods of time. Such solutions, which attack glass, cause stopcocks to "freeze," and the burets may be ruined.

The beginner must be very cautious in reading burets. In order to become familiar with the graduations and adept at estimating between them, much practice is needed early in the laboratory work. An ordinary 50-ml buret is graduated in 0.1-ml intervals and should be read to the nearest hundredth of a milliliter. An aqueous solution in a buret (or any tube) forms a concave surface referred to as a *meniscus*. In the case of solutions that are not deeply colored, the position of the bottom of the meniscus is ordinarily read (the top is taken if the solution is so intensely colored that the bottom cannot be seen, e.g., with permanganate solutions). It is most helpful to cast a shadow on the bottom of the meniscus by means of a darkened area on a paper or card held just behind the buret with the dark area slightly below the meniscus. Great care must be taken to avoid parallax errors in reading burets: the eye must be on the same level with the meniscus. If the meniscus is near a graduation that extends well around the buret, the right eye-level can be found by seeking a position so that the graduation mark seen at the back of the buret merges with the same line at the front. A loop of paper encircling the buret just below the meniscus serves the same purpose.

Before a titration is started, it must be ascertained that there are no air bubbles in the tip of the buret. Such bubbles register in the graduated portion of the buret as liquid is delivered if they escape from the tip during a titration, and hence cause errors. When a solution is delivered rapidly from a buret, the liquid running down the inner wall is somewhat detained. After the stopcock has been closed, it is important to wait a few seconds for this "drainage" before taking a reading.

In performing titrations, the student should develop a technique that permits both speed and accuracy. The solution being titrated, generally in an Erlenmeyer flask, should be gently swirled as the titrant is delivered. One way to accomplish this, while retaining control of the stopcock and permitting ease of reading the buret, is to face the buret, with the stopcock on the right, and operate the stopcock with the left hand from behind the buret while swirling the solution with the right hand [Fig. 20.5(b)]. The thumb and forefinger are wrapped around the handle to turn the stopcock, and inward pressure is applied to keep the stopcock seated in the barrel. The last two fingers push against the tip of the buret to absorb the inward pressure.

Volumetric Flasks

A typical volumetric flask is shown in Fig. 20.6. The flask contains the stated volume when filled so that the bottom of the meniscus coincides with the etched line. If the solution is poured from the flask, the volume delivered is somewhat less than the stated volume, and volumetric flasks are never used for measuring out solutions into other containers. They are used whenever it is desired to make a solution up to an accurately known volume.

When solutions are made up in volumetric flasks, it is important that they be well mixed. This is accomplished by repeatedly inverting and shaking the flask. Some analysts make a practice of mixing the solution thoroughly before the final volume has been adjusted, and mixing again after the flask has been filled to the mark: it is easier to agitate the solution vigorously when the narrow upper portion of the flask has not been filled.

Solutions should not be heated in volumetric flasks, even those made of Pyrex glass. There is a possibility that the flask may not return to its exact original volume upon cooling.

Most volumetric flasks have ground-glass or polyethylene stoppers, screw caps, or plastic snap caps. Alkaline solutions cause ground-glass stoppers to "freeze" and thus should never be stored in flasks equipped with such stoppers.

When a solid is dissolved in a volumetric flask, the final volume adjustment should not be made until all the solid has dissolved. In certain cases marked volume changes accompany the solution of solids, and these should be allowed to take place before the volume adjustment is made.

Funnels and Filter Paper

In gravimetric procedures the desired constituent is often separated in the form of a precipitate. This precipitate must be collected, washed free of undesirable contaminants from the mother liquor, dried, and weighed, either as such or

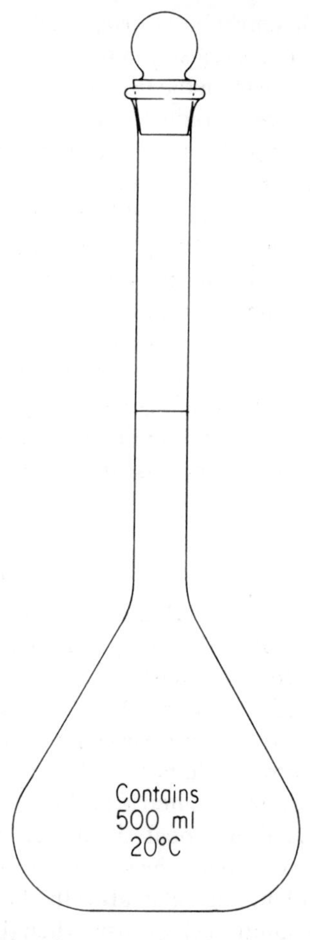

Figure 20.6 Volumetric flask.

after conversion into another form. Filtration is the common way of collecting precipitates, and washing is often accomplished during the same operation. Filtration is carried out with either funnels and filter paper or filtering crucibles. The important factors in choosing between the two are the temperature to which the precipitate must be heated to convert it into the desired weighing form and the ease with which the precipitate may be reduced.

The cellulose fibers of filter paper have a pronounced tendency to retain moisture, and a filter paper containing a precipitate cannot be dried and weighed as such with adequate accuracy. It is necessary to burn off the paper at a high temperature. During the burning, reducing conditions due to carbon and carbon monoxide prevail in the vicinity of the precipitate. Thus precipitates that cannot be heated to high temperatures or that are sensitive to reduction are normally not filtered using filter paper; filtering crucibles of the types described in a later section are employed. Some of the techniques given here, however, will apply to all types of filtration.

Various types of filter paper are available. For quantitative work, only paper of the so-called "ashless" quality should be used. This paper has been treated with hydrochloric and hydrofluoric acids during its manufacture. Thus it is low in inorganic material and leaves only a very small weight of ash when it is burned. (A typical figure for the ash from one circular paper 11 cm in diameter is 0.13 mg.) The weight of ash is normally ignored; for very accurate work, a correction can be applied, since the weight of ash is fairly constant for the papers in a given batch.

Within the ashless group, there are further varieties of paper that differ in porosity. The nature of the precipitate to be collected dictates the choice of paper. "Fast" papers are used for gelatinous, flocculent precipitates such as hydrous iron(III) oxide and for coarsely crystalline precipitates such as magnesium ammonium phosphate. Many precipitates that consist of small crystals (e.g., barium sulfate), will pass through the "fast" papers. "Medium" papers require a longer time for filtration, but retain smaller particles and are the most widely used. For very fine precipitates such as silica, "slow" paper is employed. Filtration at best is rather time-consuming, and the analyst should use the fastest paper consistent with retention of the precipitate.

Filter paper is normally folded so as to provide a space between the paper and the funnel, except at the top of the paper, which should fit snugly to the glass. The procedure is shown in Fig. 20.7. The second fold is made so that the ends fail to match by about $\frac{1}{8}$ inch. Then the paper is opened into a cone. The corner of the outside fold on the thicker side is torn off in order to fit the paper to the funnel more easily and to break up a possible air passage down the fold next to the funnel. With the paper cone held in place in the funnel, distilled water is poured in. A clean finger, applied cautiously to prevent tearing the fragile wet paper is used to smooth the paper and obtain a tight seal of paper to glass at the top. Air does not enter the liquid channel with a properly fitted paper, and thus the drainage from the stem of the funnel establishes a gentle suction which facilitates filtration. A malfunctioning filter can seriously delay an analysis; it is preferable by far to reject such a filter and prepare a new one.

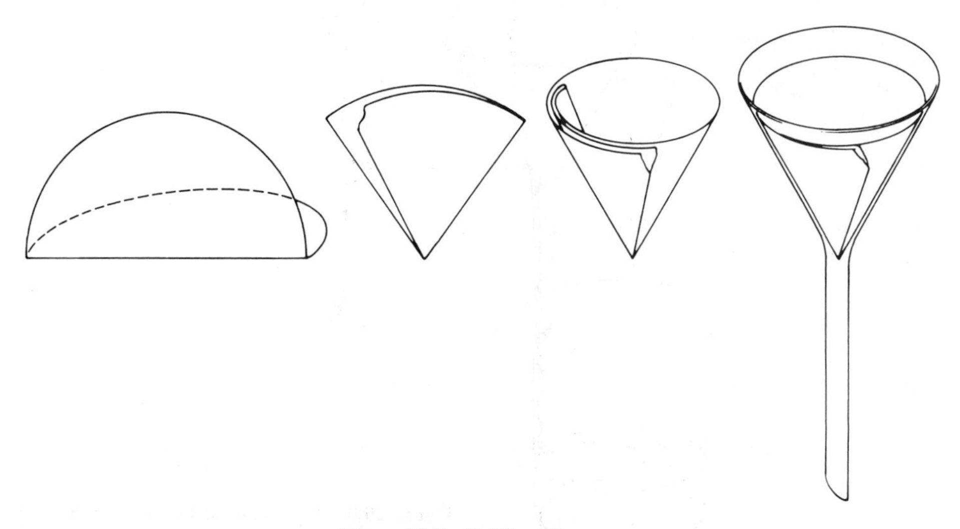

Figure 20.7 Folding filter paper.

Filter paper circles are available in various diameters. The size to be used depends upon the quantity of precipitate, not the volume of solution to be filtered. Larger paper than necessary should be avoided: the paper and the funnel should match with regard to size. It is especially important that the paper not extend above the edge of the glass funnel, but come within 1 or 2 cm of this edge. The precipitate should occupy about one-third of the paper cone and never more than one-half.

Technique of Washing and Filtering a Precipitate

Usually a precipitate is washed, either with water or a specified wash solution, before it is dried and weighed. The washing is generally carried out in conjunction with the filtration step (Fig. 20.8), wherein the precipitate is separated from its mother liquor in compact form. Once the precipitate is in the filter, it can be

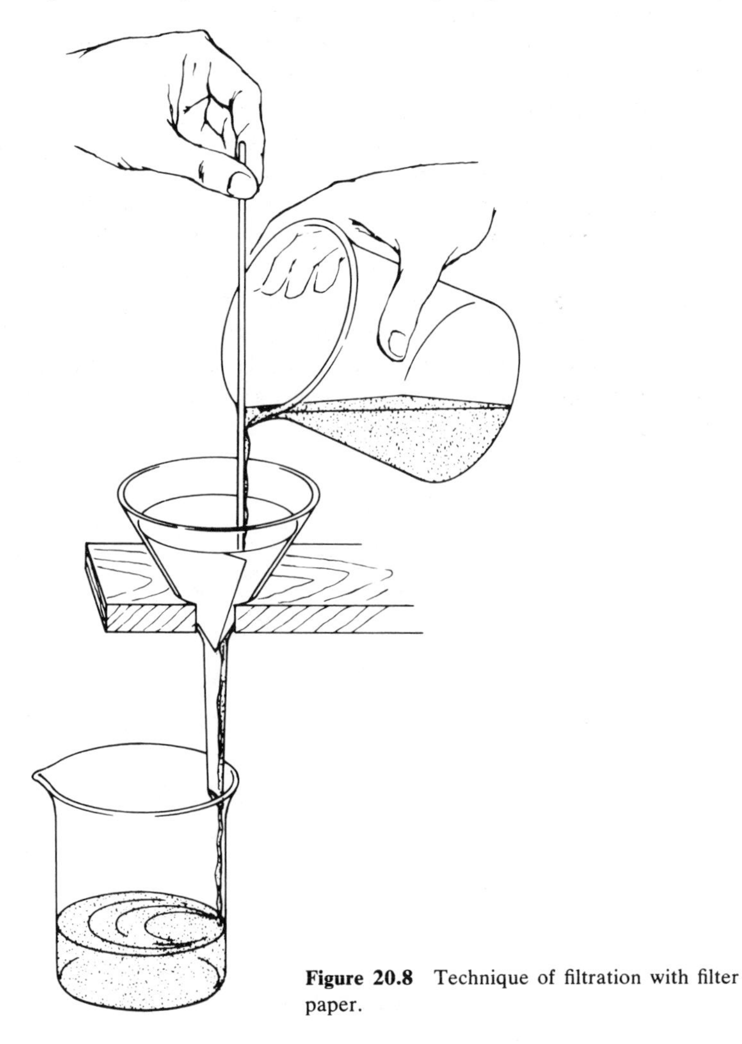

Figure 20.8 Technique of filtration with filter paper.

washed by passing wash solution through the filter. However, this technique is often rather inefficient; the wash solution does not penetrate uniformly into the compact mass of precipitate. It is usually preferable to wash the precipitate by decantation, at least in cases where the precipitate settles rapidly from suspension. The supernatent mother liquor is carefully poured off through the filter while as much of the precipitate as possible is retained in the beaker. The precipitate is then stirred with wash solution in the beaker, and the washings are decanted through the filter. This washing is repeated as often as desired,[5] until, in the final instance, the precipitate is not allowed to settle but is poured into the filter along with the wash solution. Residues of precipitate remaining in the beaker are usually transferred to the filter by a directed jet from a wash bottle, as shown in Fig. 20.9. If the precipitate tends to adhere to the glass, the last traces may be scavenged by means of a rubber policeman. The precipitate is then wiped from the policeman with a small bit of filter paper, which is added to the paper in the funnel for ignition.

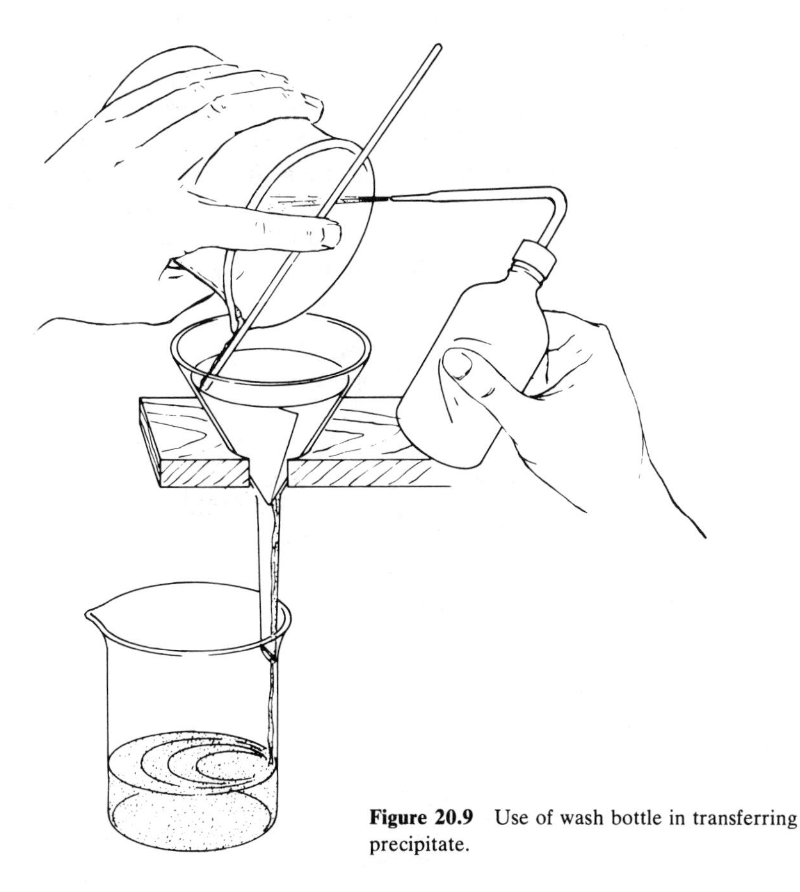

Figure 20.9 Use of wash bottle in transferring precipitate.

[5] It should be noted that several washings with small volumes of solution are more effective than one washing with the same total volume of wash solution.

The stem of the funnel should extend well into the vessel receiving the filtrate, and the tip of the stem should touch the inner surface of the vessel to prevent spattering of the filtrate. All transfers into the funnel should be made with the aid of a stirring rod, and care must be taken that no drops of solution are lost. The filtrate should be examined for turbidity; sometimes small amounts of precipitate run through the filter early in the filtration, but can be caught by refiltering through the same filter after its pores have been somewhat clogged by collected precipitate.

Ignition of Precipitate with Filter Paper

After the filter paper has drained as much as possible in the funnel, the top of the paper is folded over to encase the precipitate completely. Using great care to avoid tearing the fragile, wet paper, the bundle of paper is transferred from the funnel to the prepared crucible (see the discussion of crucibles below). It is better to handle the paper where it is three layers thick rather than by the other side. The next steps in the ignition of the material in the crucible are generally as follows:

1. Drying the Paper and Precipitate. This may be done in an oven at temperatures of 100 to 125°C if the schedule permits setting aside the experiment at this stage. If the ignition is to be followed through immediately, the drying may be accomplished with a burner. Place the covered crucible in a slanted position in a clay or silica triangle, and place a small flame beneath the crucible at about the middle of the underside. Too strong heating must be avoided; the flame should not touch the crucible, and the drying should be leisurely.

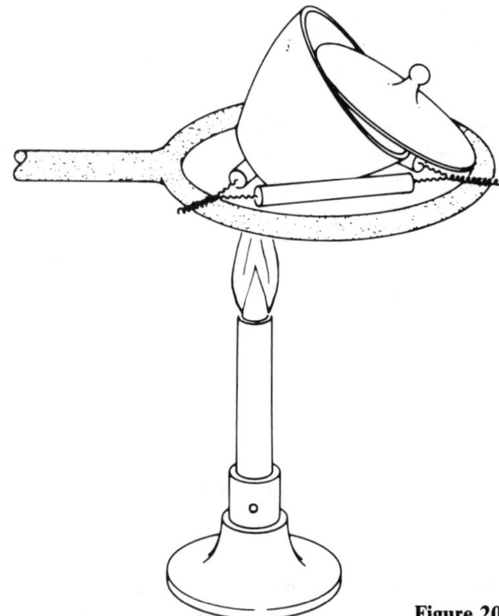

Figure 20.10 Ignition of precipitate.

2. Charring the Paper. (See Fig. 20.10). After the precipitate and paper are entirely dry, the crucible cover is set ajar to permit access of air, and the heating is increased to char the paper. Increase the size of the flame slightly and move it back under the base of the crucible. The paper should smoulder, but must not burn off with a flame. If the paper bursts into flame, cover the crucible immediately to extinguish it. Small particles of precipitate may be swept from the crucible by the violent activity of escaping gases; also, under these conditions, carbon of the paper may reduce certain precipitates which can be safely handled in filter paper under less vigorous conditions. Care must be taken that reducing gases from the flame are not deflected into the crucible by the underside of the cover. During the charring, tarry organic material distills from the paper, collecting on the crucible cover. This is burned off later at a higher temperature.

3. Burning Off the Carbon from the Paper. After the paper has completely charred and the danger of its catching fire is past, the size of the flame is increased until the bottom of the crucible becomes red. This should be done gradually. The carbon residue and the organic tars are burned away during this stage of the ignition. The heating is continued until this burning is complete, as evidenced by the disappearance of dark-colored material. It is well to turn the crucible from time to time so that all portions are heated thoroughly. Sometimes it is necessary to direct special attention to the underside of the cover to remove the tarry material collected there.

4. Final Stage of Ignition. To conclude the ignition, place the crucible upright, removing the cover to admit air freely, and heat at the temperature recommended for the particular precipitate. A Tirrill burner will heat a covered porcelain crucible to about 700°C, and a Meker burner will give a temperature roughly 100°C higher. With platinum crucibles, temperatures about 400°C higher can be obtained. The ignition is continued until the crucible has reached constant weight, that is, until the difference between two weighings with a heating period in between is less than about 0.5 mg.

Filtering Crucibles

Certain precipitates cannot be handled with filter paper, either because they are too easily reduced or because they cannot be heated to a temperature adequate to burn off the paper. Such precipitates are filtered by means of *filtering crucibles*, several types of which are pictured in Fig. 20.11. The *Gooch*, which is seldom used today, is a porcelain crucible with a number of small holes in the bottom. A mat of asbestos is formed on the inside of the bottom when the crucible is prepared for use, and this mat is the filtering medium. It is prepared by forming a suspension of specially cut and treated asbestos fibers in water, and pouring the suspension through the crucible under suction. Gooch crucibles can be ignited at high temperatures and are adequate for many different precipitates.

For precipitates that need not be heated above 500°C or so, *sintered* or *fritted glass crucibles* may be used. These crucibles are made of glass, and have a bottom

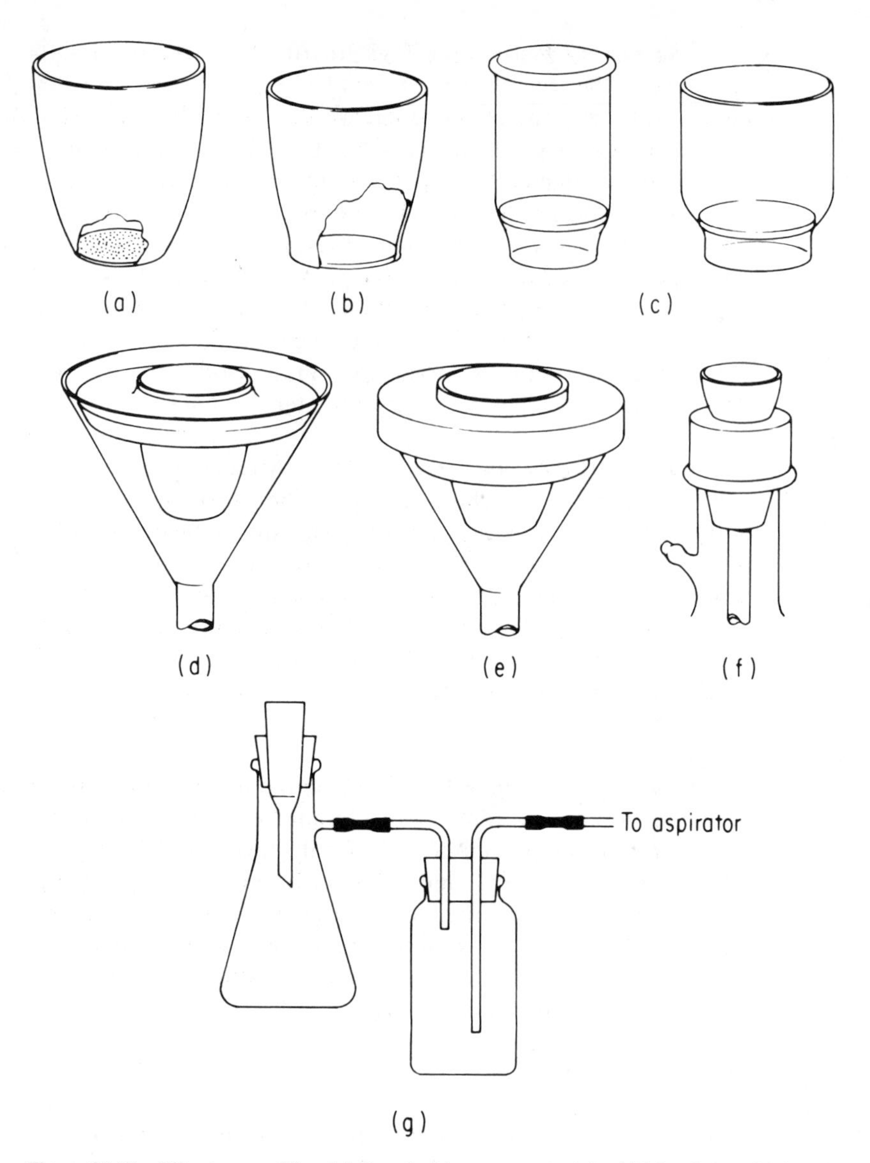

Figure 20.11 Filtering crucibles: (a) Gooch, (b) porous porcelain, (c) fritted glass, high and low forms, (d) Sargent holder, (e) Bailey holder, (f) Walter holder, and (g) filtration with suction, Walter holder.

of sintered ground glass fused to the body of the crucible. They are mounted in a suitable holder (as shown in Fig. 20.11) and suction is applied. Sintered glass crucibles are available in varying porosities for handling various types of precipitates. In the Corning Glass Co. Pyrex line, the three porosities designated coarse, medium, and fine (the letters C, M, and F appear near the upper edge of the glass crucible) serve most analytical usages. Although temperatures up to

500°C are said to be safe, sintered glass crucibles must be heated very gradually if such a temperature is to be approached without damage. It must be kept in mind that strong alkalies attack the crucibles, especially the sintered filtering disks. The crucibles are cleaned with solvents appropriate to the particular contamination at hand. For three reasons sintered glass crucibles are never heated directly over a flame. First, they may be broken; second, carbon, which is very difficult to remove, may be deposited in the fritted disk; and third, reducing gases from the flame may penetrate to the precipitate via the porous bottom. If a burner must be used to attain the desired temperature, the sintered glass crucible is placed within an ordinary porcelain crucible.

For precipitates which must be ignited at very high temperatures, *porous porcelain crucibles* may be employed. These are porcelain crucibles with unglazed, porous bottoms through which precipitates can be filtered off under suction. As with sintered glass crucibles, the porous porcelain crucibles must not be heated with an open flame. They are ignited within an ordinary porcelain crucible unless a furnace is used. The porous bottom is readily attacked by a strong alkali.

Crucibles for Ignition of Precipitates in Filter Paper

Different kinds of crucibles have been suggested for one purpose or another, but we may confine our discussion to the two most widely used, *porcelain* and *platinum*. Platinum is usually preferred, although porcelain, which is inexpensive but adequate for many purposes, is also widely used. There are certain cases where platinum must not be used (see below).

Porcelain crucibles are resistant to attack by many reagents; however, they are attacked by some, notably alkaline substances and hydrofluoric acid. They can be heated to high temperatures (about 1200°C), and their weight changes very slightly with strong and prolonged heating. It must be remembered that, although they can be heated safely to about 1200°C, porcelain crucibles do not attain this temperature over a burner. A platinum crucible over the same burner reaches a much higher temperature. In a furnace, on the other hand, porcelain and platinum come to the same temperature. As an example of the effect of this temperature difference between porcelain and platinum when a burner is used, note that in porcelain calcium oxalate cannot be ignited satisfactorily to the oxide, while in platinum the conversion is completed readily.

For many purposes, platinum crucibles are superior to porcelain. Platinum crucibles reach a higher temperature over a burner (but not in a furnace), and they cool much more rapidly because of the high thermal conductivity of the metal. Platinum is a very inert metal, and it is resistant to certain reagents that attack porcelain. A notable example of this is the fact that a precipitate can be treated with hydrofluoric acid in a platinum crucible. This is used, for example, in the accurate determination of silica by treating the impure SiO_2 precipitate with hydrofluoric acid, volatilizing silicon tetrafluoride, and obtaining the pure silica by difference.

On the other hand, platinum is attacked by certain substances, and because of its great expense, care must be exercised in its use. Your instructor will caution you as necessary on the use of platinum.[6]

The idea of *constant weight* should be explained more thoroughly. The weight of a precipitate in a crucible is normally obtained by difference: first the empty crucible is weighed, and then the crucible plus ignited precipitate. The ignition of the precipitate sometimes involves a chemical conversion—for example,

$$CaC_2O_4 \cdot 2H_2O \longrightarrow CaO \qquad \text{or} \qquad MgNH_4PO_4 \cdot 6H_2O \longrightarrow Mg_2P_2O_7$$

—into the final weighing form, or sometimes merely drying. This change can be assumed to be complete when the weight of the precipitate no longer changes with further heating. The precipitate is heated at the desired temperature for a reasonable length of time, and then cooled and weighed. It is then reheated, usually for a shorter period of time, and weighed again. If the two weighings agree (within, say, 0.2 to 0.5 mg) the ignition is considered complete. The precipitate has been ignited to "constant weight." Now, since the weight of precipitate is obtained by difference, the weight of the crucible alone must also be accurately known. Further, it is sometimes found that the weight of a crucible varies somewhat with the conditions under which it was prepared and dried. Thus the empty crucible is usually ignited *exactly* as the crucible plus precipitate will be ignited later. This ignition is continued until the crucible itself is at constant weight.

Special Instruments

Not too many years ago, the analytical chemist used only the apparatus described above, plus the analytical balance, for almost all determinations. A striking change has taken place, and nowadays *p*H meters, colorimeters, spectrophotometers, polarographs, and still more elaborate and complex instruments are found in most analytical laboratories. Directions for the use and care of instruments are necessarily specialized, and are best obtained from the manufacturer's bulletins and from personal instruction by experienced people. We include here only a few general remarks to fill out our discussion of analytical apparatus.

The rule of greatest importance is that no instrument should ever be touched by a person unfamiliar with the directions for its proper use and the precautions against damaging it. Some instruments contain fragile components which may be injured by improper handling, and sometimes a carefully worked out calibration may be ruined by manipulation of the wrong knobs.

The other rule that must always be remembered is that an instrument should never be used by a person who has not thought through its advantages and limitations for the job at hand, who does not have a proper estimate of the reliability of the data obtained, and who cannot interpret correctly the significance of the instrumental measurement and apply it with intelligence. Meaningless

[6] If platinumware is employed, a detailed discussion on the proper care of platinum should be studied. A good discussion is given in "Notes on the Care and Use of Platinum Crucibles and Dishes," a pamphlet circulated by Matthey Bishop, Inc., Malvern, Pa. 19355.

measurements are made every day by imposters masquerading as chemists. Anyone can learn to turn knobs and read galvanometers, but the assurance that a measurement has been made on the best possible system must come from a well-trained chemist.

CALIBRATION OF VOLUMETRIC GLASSWARE

Introduction

Table 20.4 shows some of the tolerance values established for volumetric glassware by the National Bureau of Standards.[7] It may be noted that glassware meeting these specifications is adequate for all but the most exacting work of the analytical laboratory. Such glassware, stated by the manufacturer to conform to NBS standards, may be purchased.[8] For an extra fee, it is also possible to obtain glassware that has actually been tested by the NBS. Because of the expense, however, the beginning analytical laboratory will rarely be equipped with glassware guaranteed to meet NBS tolerance specifications. Less expensive glassware stated by the manufacturer to meet tolerances about double those of the NBS is available,[9] but even this glassware may not be furnished the student. For this reason, and also because the history of a particular item involving possible damage may be obscure, it is often advisable for the student to calibrate his volumetric glassware. Of course, the individual instructor will make the decision in any case.

TABLE 20.4 Tolerances for Volumetric Glassware (ml)

Capacity, less than and including	Volumetric flasks	Transfer pipets	Burets
2		0.006	
5		0.01	0.01
10		0.02	0.02
25	0.03	0.03	0.03
50	0.05	0.05	0.05
100	0.08	0.08	0.10
200	0.10	0.10	
500	0.15		
1000	0.30		

Since most analytical work involves dilute, aqueous solutions, water is generally used as the reference material in the calibration of volumetric glassware. The general principle in calibration is to determine the weight of water contained

[7] Nat. Bur. Standards Circ. 434 (1941).

[8] For example, in the Kimball Glass Co., Kimax Class A Inc line and the Corning Glass Works Pyrex line of volumetric glassware.

[9] Kimball Glass Co. Kimax volumetric glassware.

in or delivered by a particular piece of glassware. Then, with the density of water known, the correct volume is found.

The units of volume commonly employed in analytical chemistry are the *liter* and the *milliliter.* The liter was formerly defined as the volume occupied by 1 kg of water at the temperature of its maximum density (about 4°C) under a pressure of 1 atm. In 1964 the Twelfth General Conference on Weights and Measurements, meeting in Paris, France, abolished this definition and instead made the liter a special name for the cubic decimeter. This new definition eliminates the previous discrepancy of 28 parts per million between the milliliter and cubic centimeter (1 ml was 1.000028 cc), and these two units are now equivalent.

The National Bureau of Standards has specified 20°C as the temperature at which glassware is calibrated. Since the laboratory temperature will usually not be exactly 20°C, glassware must, strictly speaking, be corrected when used at other temperatures because of errors due to expansion (or contraction) of both the glass vessel itself and the solution contained therein. The coefficient of expansion of glass is sufficiently small that the correction required for this factor is negligible for most work (it amounts to the order of 1 part per 10,000 for a change of 5°C). The change in the volume of the solution itself, on the other hand, is more important, but it can still be ignored in many cases if the working temperature is not far removed from 20°C (the volume change is of the order of 1 part per 1000 over a 5°C range).

As noted above, calibration data are secured by converting weight of water into volume via the density. Tables showing the density of water at various temperatures are available in handbooks. However, the data in such tables are usually given on the basis of weights *in vacuo*, while the actual weighings are made in air. Since the water being weighed generally displaces more air than do the weights, it is necessary to correct the weighings for the buoyancy effect of the air if

TABLE 20.5 Volume of 1 g of Water Weighed in Air with Brass Weights at Various Temperatures*

°C	ml	°C	ml
10	1.0016	21	1.0030
11	1.0017	22	1.0032
12	1.0018	23	1.0034
13	1.0019	24	1.0036
14	1.0020	25	1.0038
15	1.0021	26	1.0041
16	1.0022	27	1.0043
17	1.0023	28	1.0046
18	1.0025	29	1.0048
19	1.0026	30	1.0051
20	1.0028		

* In many single-pan balances (such as Mettler's), the weights are a chrome-nickel steel rather than brass. However, the density of this alloy is close enough to that of brass for the student to use the values in this table in ordinary calibration work as described in this chapter.

such handbook tables are to be used (see Chapter 21 for a fuller discussion of this buoyancy effect). On the other hand, we may change *in vacuo* densities into densities that would be obtained in air with brass weights, and use these directly with our weighings obtained in air. The values in Table 20.5, which are the reciprocals of such adjusted densities, may be used directly by the student without taking the buoyancy effect into consideration.

Methods of Calibration Commonly Employed

There are three general approaches to the calibration of volumetric glassware that are in wide use and with which the student should be familiar.

1. The first method, which we may designate as a *direct, absolute calibration*, is based on the principles outlined above. The volume of water delivered by a buret or pipet, or contained in a volumetric flask, is obtained directly from the weight of the water and its density. Directions are given below for the calibration of a buret, a pipet, and a volumetric flask using this method.
2. Volumetric glassware is sometimes calibrated by comparison with another vessel previously calibrated directly. We may refer to this as an *indirect, absolute calibration*, or *calibration by comparison*. This method is especially convenient if many pieces of glassware are to be calibrated, and it is sufficiently accurate for all ordinary usages provided the reference vessel itself has been accurately calibrated. Calibrating bulbs are not available in many student laboratories, and specific directions for their use are not included in this chapter. They are not difficult to use if proper care is taken. The student may obtain directions from the instructor if he is to use such equipment.
3. Sometimes it is necessary to know only the relationship between two items of glassware without knowing the absolute volume of either one. This situation arises, for example, in taking an aliquot portion of a solution. Suppose that it is desired to titrate one-fifth of an unknown sample. The unknown might be dissolved, appropriately treated preparatory to the titration, and diluted to volume in a 250-ml volumetric flask. A 50-ml pipet would then be used to withdraw an aliquot for titration. For the calculations in this analysis, it would not be necessary to know the exact volume of the flask or the pipet, but it would be required that the pipet hold exactly one-fifth as much solution as the flask. The method used for a *relative calibration* of this sort simply involves discharging the pipet five times into the flask and marking the level of the meniscus on the flask.

Calibration of Buret

The buret should be thoroughly cleaned so that it drains well, and the stopcock should be properly lubricated. Fill the buret with water and test for leakage, making a reading to the nearest 0.01 ml and repeating the reading after waiting at least 5 min. No noticeable change should have occurred. During the waiting period, weigh a stoppered 125-ml Erlenmeyer flask (or other suitable container) to the nearest milligram. Record this weight.

Fill the buret with distilled water that is at the temperature of the laboratory. This temperature must be measured and recorded. Then sweep any air bubbles from the tip of the buret by opening the stopcock to permit rapid outflow. Now withdraw water more slowly until the meniscus is at, or slightly below, the zero mark on the buret. After drainage is complete (at least 30 sec), read the buret to the nearest 0.01 ml. Record this "initial" reading. Remove any hanging drop of water from the tip of the buret by touching it lightly to the side of a vessel such as a beaker. Now run about 10 ml of water from the buret into the weighed flask. The tip of the buret should extend somewhat into the mouth of the flask, and care should be taken against spattering. The neck of the flask where contact with the stopper is made should not be wet. Quickly stopper the flask. Read the buret after allowing time for drainage and record this "final" reading. Then weigh the stoppered flask to the nearest milligram. Record this weight. Tabulation of the data as suggested in Table 20.6 is convenient.

TABLE 20.6 Sample Data for Calibration of Buret

Initial buret reading	Final buret reading	Apparent volume, ml	Initial weight of flask	Final weight of flask	Weight of water, g	Temperature, °C	Actual volume, ml	Correction, ml
0.38	10.16	9.78	62.576	72.311	9.735	24	9.77	−0.01
0.49	20.16	19.67	72.311	91.832	19.521	24	19.59	−0.08

Now refill the buret and obtain another "initial" reading. Run about 20 ml of water into the flask, obtain the "final" buret reading, and reweigh the flask. Note that we are calibrating the buret in 10-ml intervals, but starting each time from an initial reading near zero since in titrations we generally start near this point.

This process is repeated for the 30-, 40-, and 50-ml volumes. The flask, of course, should never be allowed to exceed the capacity of the balance. It should first be emptied, the neck dried with a clean towel, and reweighed.

For each calibration interval, multiply the weight of water by the appropriate value from Table 20.5 to obtain the actual volume of the water. The difference between this actual volume and the apparent volume (from the buret readings) is, of course, the correction. If the actual volume is larger than the apparent volume, the correction obtained by subtracting in this way will be positive, which means that it is to be added to the buret reading in future work with the buret.

The calibration should be repeated as a check on the work. Duplicate results should agree within 0.04 ml. With the weighings carried out to the nearest milligram, as directed above, errors in weighing should not affect the results because the weight is actually needed only to the nearest 0.01 g. Hence failure of duplicate results to agree is generally due to (1) leakage around the buret stopcock, (2) failure to wait for drainage before reading the buret, or (3) careless technique in collecting or handling the container and stopper.

Since our calibration is performed over 10-ml intervals, students often ask about the correction to be applied to a buret reading of an intermediate value such as 46 ml. Ordinarily, applying the correction of the nearest interval, in this case 50 ml, will be sufficient. It is perhaps somewhat better to interpolate between the known corrections. This is done most conveniently by a graphical method. The student may plot his corrections against the intervals 10, 20 ml, . . . , connecting the points with straight lines so as to obtain linear interpolations by simple inspection of the graph. Such an interpolation obviously is not likely to reflect the true situation within the buret, but it will usually be more than sufficiently accurate for the need at hand. The student must use good judgment in consideration of the precision desired in the buret reading. For example, an error of 0.04 ml in a 40-ml buret reading represents only 1 ppt.

Calibration of Pipet

The method is essentially the same as that described above for a buret. The pipet should naturally be carefully cleaned and rinsed before calibration. Weigh to the nearest milligram a clean, stoppered Erlenmeyer flask of adequate capacity. Now fill the pipet to a level above the etched line, using a beaker of distilled water which is at the temperature of the laboratory. This temperature must be measured, of course. Dry the outside of the pipet with a clean towel, and then release the finger pressure to allow the liquid level to fall to the etched line. The pipet should be held in a vertical position with the etched line at eye level for this operation. Hold the liquid level so that the bottom of the meniscus coincides with the etched line, and touch the tip of the pipet to the side of the beaker to remove any hanging drop. Then discharge the contents of the pipet into the weighed container. It is best to insert the tip of the pipet well into the container and to keep the tip touching the inner wall of the container during the delivery. With the tip still touching the container, allow the pipet to drain for 20 to 30 sec after the flow ceases. Do not disturb the last portion of water remaining in the tip. Then stopper the container and reweigh it. Calculate the volume of water delivered by the pipet from the weight and the appropriate value taken from Table 20.5. The calibration should be repeated, and the duplicate results should not differ by more than 1 ppt. Common errors are (1) warming the contents of the pipet by holding the enlarged portion in the palm of the hand, (2) failure to allow sufficient drainage time, (3) disturbing the residue of water that should remain in the tip of the pipet, and (4) general carelessness in handling the weighed container and stopper.

Calibration of Volumetric Flask

For large volumetric flasks, an ordinary analytical balance cannot be used. For example, a 250-ml flask may weigh as much as 100 g, and may weigh 350 g or so when filled with water. This, of course, exceeds the capacity of an analytical balance, and a large-capacity balance must be used. Because of possible inequality of the balance arms of such balances, weighing by the method of substitution, as described below, is recommended.

The flask should be thoroughly cleaned and rinsed, and then clamped in an inverted, vertical position until dry. Now stopper the flask and place it on the left-hand pan of the balance. Add copper shot to a beaker on the right-hand pan of the balance until the rest point of the balance comes on scale. Note the position of the rest point, and with the copper shot undisturbed, replace the flask with weights just appropriate to give the same rest point. Record this weight.

Now, using a small funnel, add distilled water at room temperature to the flask until the flask is nearly filled. Remove the funnel, being careful to avoid leaving drops of water on the neck of the flask above the mark. (If such drops do appear, remove them with a length of filter paper.) Very carefully complete the filling of the flask up to coincidence of the meniscus with the etched mark by means of a pipet or dropper. Reweigh the flask by substitution as above. From the weight of water, the temperature, and the appropriate value from Table 20.5 calculate the volume of water that the flask contains. The calibration should be checked, of course, by repeating the procedure. Duplicate results should agree within 0.1 ml for a 250-ml flask.

Relative Calibration of 50-ml Pipet and 250-ml Volumetric Flask

The pipet and flask should be carefully cleaned and rinsed. The flask should be dried. Using an approved technique, introduce into the flask five 50-ml portions of distilled water from the pipet. The tip of the pipet should extend well into the flask to avoid splashing, and the pipet should be operated carefully with regard to drainage and the other points noted previously in this chapter. Finally, mark the position of the bottom of the meniscus on the neck of the flask by means of the upper edge of a gummed label. The calibration should be repeated.

RECORDING LABORATORY DATA

There are three main requirements for the recording of data obtained in the analytical laboratory. These may be briefly expressed as follows: The record (1) should be complete, (2) should be intelligible to any reasonably competent chemist, and (3) should be easy to find on short notice. These requirements may be met by adherence to the following rules.

1. The student should have a bound notebook for recording his laboratory data, calculations, results, and all other matters pertinent to the analysis of a sample. The pages of the notebook should be numbered, and a table of contents should be developed so that any given experiment can be quickly found.
2. All data obtained in the laboratory should be recorded directly in the notebook at the time the work is performed. Especially forbidden is the recording of data on loose paper with the idea of copying it into the notebook later. While neatness may be sacrificed somewhat by taking the notebook directly into the laboratory, the prevention of loss of data and errors in transcribing them more than counterbalances this.

3. Entries should be recorded in ink. If a mistake is made and a recorded value is invalidated, it is not to be erased, but is crossed out so as to be still legible. A notation as to why it is rejected is made in the notebook.
4. The data in the notebook should be organized and recorded in a systematic way. This benefits the student because it is then relatively easy to locate errors in the analytical calculations; the student may thus be saved repeating an entire determination in order to obtain satisfactory results. To facilitate an orderly presentation of the data, the student should plan how best to record it before the experiment is actually begun. It is especially helpful to arrange beforehand a table in which the experimental data, the calculations, and the final results can be entered systematically.
5. The rules regarding significant figures (see Chapter 2) should be followed in recording data in the notebook.

Table 20.7 shows an example of a satisfactory laboratory record for a typical volumetric analysis. The opposite page in the notebook may be used for arithmetic calculations and other material which does not fit well into the tabulation. The student should, of course, follow any style suggested by the individual instructor. Some instructors grade unknowns directly from the notebook, while others prefer that the student summarize his results on a 3- by 5-inch card somewhat as shown in Table 20.8.

There are few situations where the graduate chemist (indeed, any professional) will not be expected to be neat, efficient, systematic, and able to locate quickly the results of his work and interpret them effectively for his superiors. The notebook in analytical chemistry should provide a good breeding ground for desirable habits.

SAFETY IN THE ANALYTICAL LABORATORY

It is unfortunate that laboratory safety is not usually emphasized sufficiently in college courses. Generally speaking, our colleagues in industry are much more safety conscious. It is largely up to the individual instructor to see that safety regulations are enforced as he may wish, but a few general remarks are quite in order in this book.

Injuries in the laboratory are usually due to one of the following (although we realize, of course, that there is overlap among these categories in many accidents): (1) fire, (2) poisons, (3) broken glass, and (4) explosions. Fire is not a common danger in the beginning analytical laboratory because inflammable substances (organic solvents, for example) are not used extensively. Whenever a solvent such as ether or alcohol is used, care must be taken that no open flames are in the vicinity. Similarly, because of the nature of his work, the beginning analytical student is not very likely to experience an explosion. Great care must be exercised in dealing with substances or apparatus where an explosion may occur. Perchloric acid comes to mind as an example of a widely used substance that can be very dangerous if not used properly. The possibility of an explosion must always be

TABLE 20.7 Sample Notebook Page for Soda Ash Analysis

Soda Ash Analysis		p. 25	
		5/18/80	
		unknown no. 186	

Method: sample dissolved in water and titrated with standard HCl solution, using modified methyl orange indicator

Reaction: $Na_2CO_3 + 2HCl \rightarrow CO_2 + H_2O + 2NaCl$

	I	II	III
Wt. of sample	0.3276 g	0.3342 g	0.3128 g
Final buret reading HCl	35.86 ml	36.32 ml	34.56 ml
Initial buret reading, HCl	0.18 ml	0.10 ml	0.38 ml
Volume HCl	35.68 ml	36.22 ml	34.18 ml
Initial buret reading, NaOH	0.37 ml	0.57 ml	0.92 ml
Final buret reading, NaOH	0.06 ml	0.37 ml	0.57 ml
Volume NaOH	0.31 ml	0.20 ml	0.35 ml

Volume relation of NaOH to HCl:
1.00 ml NaOH = 0.95 ml HCl

	I	II	III
ml HCl equiv. to NaOH	0.29 ml	0.19 ml	0.33 ml
Total HCl req. for sample	35.39 ml	36.03 ml	33.85 ml

Normality of HCl: 0.1076

$$\%Na_2CO_3 = \frac{\text{ml HCl} \times N \times EW}{\text{mg sample}} \times 100$$

	I	II	III
	61.60%	61.49%	61.70%
Average % Na_2CO_3		61.60%	
Average deviation, ppt		1.2 ppt	

considered whenever a container is at hand whose contents are under pressure, such as cylinders of compressed gases. Cuts from broken glass are common when glass tubing is improperly pushed into a tight hole in a rubber stopper and when unusual pressure is applied to a thin glass vessel such as a beaker. Where there is a possibility that necessary hand pressure will cause glass to break, the hand should be protected with a glove or towel.

Perhaps the greatest danger in the analytical laboratory is from poisons, if we include in the class corrosive substances such as strong acids and bases that readily

TABLE 20.8 Sample Report Card for Soda Ash Unknown

5/18/80 Name ______________________ Soda Ash Unknown No. 186

Normality of HCl: 0.1076

	Wt sample	ml HCl	% Na_2CO_3	Deviation
1	0.3276 g	35.39 ml	61.60	0.00
2	0.3342 g	36.03 ml	61.49	0.11
3	0.3128 g	33.85 ml	61.70	0.10

Average % Na_2CO_3: 61.60

Average deviation: 0.07 in 61.60 or 1.2 ppt

attack human tissues. Such reagents should be handled with the greatest care. When a substance such as sulfuric acid is spilled on the skin or splashed into an eye, the severity of the resulting burn may depend upon the speed with which the situation is handled. There may be no time in which to look for expert help, and thus each student should know beforehand the emergency treatment to be undertaken. With acid or alkali burns, the first step, which is to be taken *immediately*, ignoring the common courtesies and forgetting any possible embarrassment, is to wash the affected area with copious quantities of cool water. This should be followed by washing with a solution of a weak base such as bicarbonate in the case of acid burns, or a dilute solution of a weak acid such as acetic for alkali burns. The instructor should see that each student knows exactly where they are located. Prevention is far superior to any treatment, of course, and proper caution will prevent most acid burns. Under no circumstances should corrosive solutions be sucked into a pipet by mouth. In many laboratories it is required that the eyes be protected by prescription glasses or special safety glasses at all times.

While it is tempting to be most impressed by strong acids and bases, it should be borne in mind that nearly all the chemicals encountered in the laboratory are poisons. "Familiarity breeds contempt," and we often forget that common substances like hydrogen sulfide, benzene, carbon tetrachloride, and the vapor from mercury can be fatal to human beings. We cannot discuss the toxicology of all these poisons individually, but we must warn the student, and especially the instructor, to be alert to the dangers involved in laboratory work.

Safety charts, posters, and signs for the laboratory may be obtained from a number of sources such as (1) Manufacturing Chemists' Assoc., Inc., 1825 Connecticut Avenue, N.W., Washington, D.C. 20009, and (2) Fisher Scientific Co., 711 Forbes Avenue, Pittsburgh, Pa. 15219. A three-volume book entitled *Safety in the Chemical Laboratory*, by Norman V. Steere, is available from the Journal of Chemical Education, Office of Publications Coordinator, 238 Kent Road, Springfield, Pa. 19064.

CHAPTER TWENTY-ONE

the analytical balance

The measurement of weight is a fundamental operation of quantitative analysis, and it is extremely important for the beginning student to understand the principles of the analytical balance which is used in performing this operation. Until recent years, most balances found in the introductory laboratory were *two-pan* balances, also referred to as *equal-arm* balances. Today the so-called *single-pan* or *unequal-arm* (also sometimes called *constant-load*) balance is very common and is found in many introductory laboratories.

We shall describe first the determination of mass and weight and then consider these two types of balances in some detail.

MASS AND WEIGHT

The analytical balance is a lever of the first class; that is, the fulcrum (B in Fig. 21.1) lies between the points of application of forces (A and C in Fig. 21.1). In the equal-arm balance $l_1 = l_2$. Pans are suspended from A and B, and the object to be weighed (mass M_1) is placed on the left-hand pan while known weights (mass M_2) are placed on the right-hand pan. Both M_1 and M_2 are attracted by the earth (gravity), the forces being, according to Newton's second law,

$$F_1 = M_1 g$$

$$F_2 = M_2 g$$

where g is the acceleration of gravity. The operator adjusts the values of M_2 until

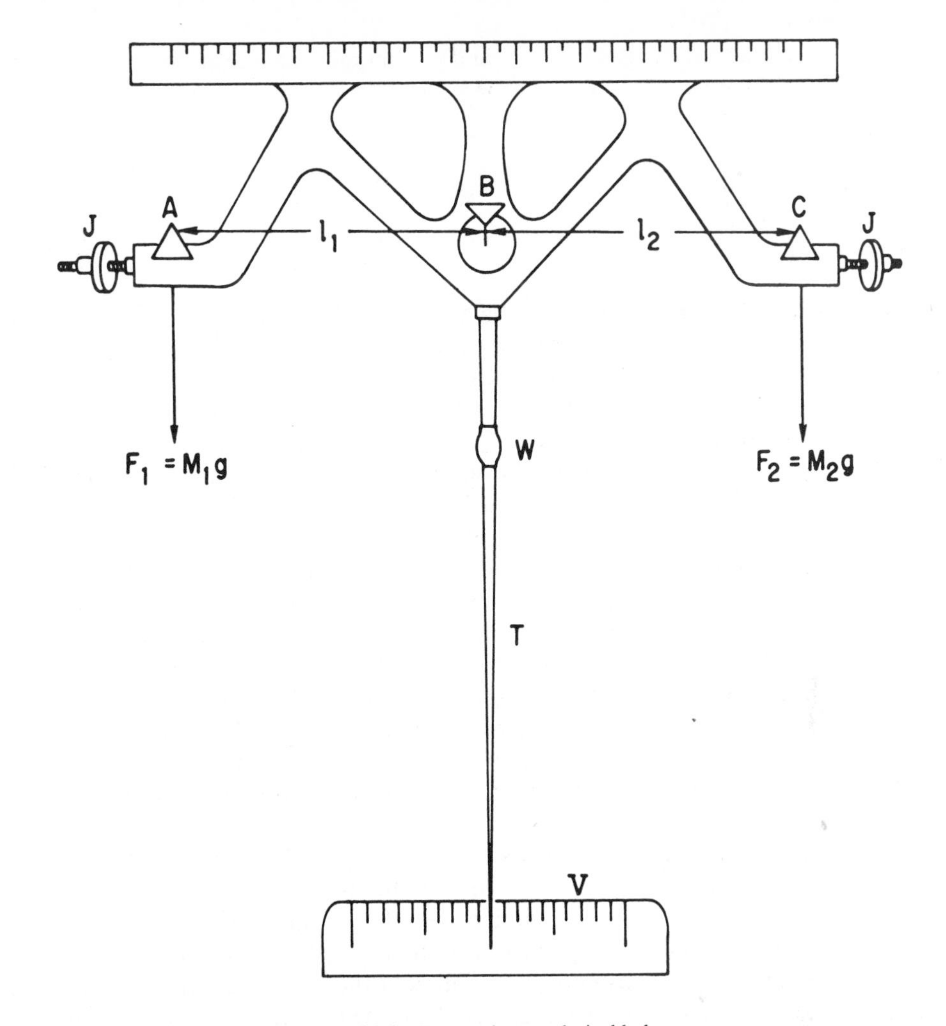

Figure 21.1 Beam of an analytical balance.

the long pointer returns to its original position in the center of the scale. Then, according to the principle of moments,

$$F_1 l_1 = F_2 l_2$$

and since $l_1 = l_2$, $F_1 = F_2$, and hence

$$M_1 g = M_2 g \qquad \text{or} \qquad M_1 = M_2$$

Since M_2 is known, M_1 is determined.

It should be noted that the *weight* of an object is the force exerted on the object by gravitational attraction (F_1 above). The *mass* is the quantity of matter of which the object is composed. The weight of an object is different at different locations on the earth's surface, whereas the mass is invariant. It is mass that the analyst determines, but since the term g cancels in the case of the equal-arm balance, the ratio of weights is the same as the ratio of masses. Hence it is customary to use the term *weight* instead of mass, and we commonly speak of the process of determining the mass of an object as *weighing.*

TWO-PAN BALANCES

General Features

The beam of a two-pan balance contains three prism-form "knife edges," A, B, and C of Fig. 21.1. These knife edges are mounted so as to lie in a plane, and the two terminal ones must be equidistant from the central knife edge. The latter knife edge, B, is above the center of gravity of the beam system and rests on a smooth plate at the top of a cylindrical column. The terminal knife edges, A and C, support *stirrups*, which in turn support the two pans. Each stirrup contains a smooth plate that rests on the terminal knife edge.

The knife edges are made of agate, a very hard and brittle material. They are ground to a very sharp edge and are slightly hollowed so that only the two ends touch the plates, which are also agate, when the balance is in operation. This results in a nearly frictionless bearing on which the beam can swing back and forth.

When the knife edges are in contact with the agate plates, they may be damaged by accidental jarring of the balance. Hence a mechanism to lift the beam and stirrups is provided so that the knife edges can be separated from the plates when the balance is not in operation or when objects are placed on or removed from the pans. This device is called the *beam arrest* and is operated by turning a milled knob at the front of the balance case. A *pan-arrest* mechanism is also provided to keep the pans from swinging when the beam is lowered. The mechanism is operated by a knob on the front of the balance case.

A long pointer, T, is attached to the center of the beam, and the tip of the pointer swings directly in front of the *pointer scale*, V, located at the base of the column. Observation of the position of the pointer on the pointer scale tells the analyst the extent of unbalance of the two sides. Attached to the pointer is a movable weight, W, which is used to adjust the sensitivity of the balance. At each end of the beam are adjusting screws, J, with nuts that can be moved in or out to decrease or increase the effective weight of either side of the beam.

The beam is either notched or graduated so that small weight adjustments can be made by placing a metal *rider* in different positions on the beam. The so-called *rider* balance has the beam graduated so that weight adjustments as small as 0.1 mg can be made by changing the position of the small wire rider. The rider can be moved from one position to another by a *rider carrier*, which can be operated from outside the balance case. *Chainomatic* balances normally have a notched beam in order to eliminate entirely the use of separate weights below 1 g. V-shaped cuts are made in the beam at regularly spaced intervals starting over the left-hand knife edge. A cylindrical rider is placed in the notches by a rider carrier, thereby adding units of 0.1 g (up to 1 g) to the right-hand side of the beam.

With a chainomatic balance, weights below 0.1 g are added by use of a fine gold chain. One end of the chain is attached to the right-hand side of the beam and the other to a movable support that can be raised or lowered by a control outside the balance case. By raising or lowering the support the fraction of the weight of the chain carried by the beam is varied. The effective weight of the chain on the right-hand side of the beam is read from a movable tape or dial against a stationary vernier, or by means of a vernier that moves up and down a graduated pillar.

Any balance can be equipped with a *damping* device. Such a device opposes the normal motion of the beam of the balance, causing it to cease swinging rather quickly. The equilibrium position can then be easily read from the pointer scale. Two methods are commonly used: air damping and magnetic damping. In the first case, a cylinder, closed at one end, is suspended from each stirrup and is hung concentrically within another cylinder. The latter cylinder is closed at the bottom and is mounted in a fixed position. As the beam moves, one cylinder moves up and down inside the other. The cylinders fit tightly enough so that the friction of the displaced air effects the desired slowing of the beam motion.

Magnetic damping is usually effected by attaching a small aluminum plate to one end (or to both) of the balance beam. This plate is mounted between the poles of a magnet, and as the beam swings it moves up and down in the magnetic field. This vertical motion of the nonmagnetic metal induces a current which produces a magnetic field that opposes the motion of the plate. Hence the oscillations of the beam are slowed down.

The entire balance is placed in a wooden, metal, or plastic case with glass windows on all four sides. The case is mounted on three or four legs, at least two of which are adjustable in length so that the balance can be leveled. A spirit level is usually provided to indicate whether the balance is level or not.

SINGLE-PAN BALANCES

There are two sources of error which are inherent in the two-pan balance but are eliminated in the single-pan balance. These are the lever arm error and the scale deflection, or sensitivity, error. For a two-pan balance to give accurate results it is necessary for the two balance arms to be equal in length. In practice, it is very difficult to make beams in which the balance arms are of identical length. If one arm is only 1/100,000 longer than the other, an error of 1/100,000 of the weight applied is introduced. With a load of 100 g, this amounts to an error of 1 mg. The error can be eliminated by weighing the object first on one pan, then on the other, and averaging the two results. It can also be eliminated by the *method of substitution*. The object is first counterbalanced with any suitable material and then it is replaced with the weights required to obtain the same balance rest point.

The scale deflection, or sensitivity, error results from a slight bending of the beam as the load is increased. For constant scale deflection per unit of weight added, the three knife edges must lie exactly on a straight line. An increasing load on a balance tends to bend the beam slightly, bringing the terminal knife edges below the plane of the central knife edge. This results in a smaller scale deflection per unit weight added at higher balance loads. The error can be corrected by using the method of substitution mentioned above.

The single-pan balance employs the method of substitution and hence eliminates the two errors inherent in the two-pan system. The essential differences in the two systems are illustrated in Fig. 21.2. The single-pan balance employs two knife edges rather than three, and the balance arms are unequal in length. A full complement of weights is suspended from the short arm, and the longer arm has a constant counterweight (plus a damping device) rigidly attached to the beam. Hence the empty balance is fully loaded and weighing is done by substitution, leaving the same load on the beam at all times. The object to be weighed is placed on the pan and weights are removed from the shorter arm by turning knobs on the outside of the balance case. A set of dials shows the sum of the weights removed. A partial release mechanism allows this adjustment of weights without danger of damaging the knife edges.

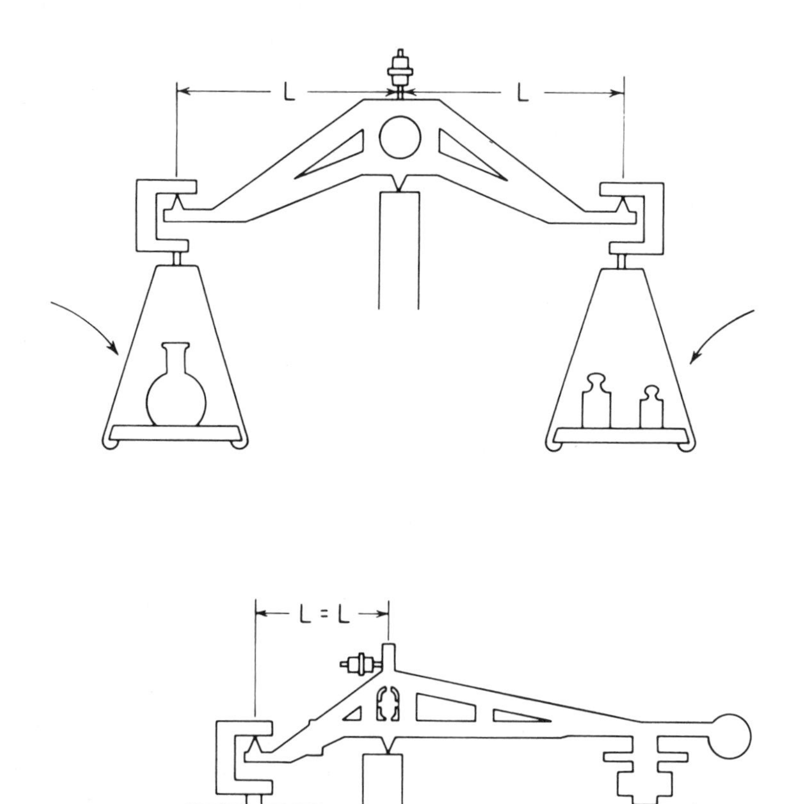

Figure 21.2 Comparison of single-pan and two-pan balances. (Courtesy of Mettler Instrument Corporation, Princeton, New Jersey.)

Once the sum of the weights removed is within 0.1 g (but on the low side) of the weight of the object, the beam is fully released and allowed to come to rest. A reticle, consisting of a scale inscribed on a glass plate, is attached to the longer arm of the balance. The deflection of the beam from its original point of rest is greatly magnified by projecting optically the image of this scale onto a glass plate on the front of the balance. The scale displacement in milligrams can be read directly and some device, such as a vernier, is employed to read to the nearest 0.1 mg. Most of these features can be seen in Fig. 21.3, which shows a "see-through" view of a standard Mettler analytical balance. Figure 21.4 shows a picture of the Mettler Model H78AR, which is designed for student use. The AR in the model designation refers to an air-release feature which causes the beam of the balance to be lowered slowly and uniformly onto the sapphire knife edges. This feature protects the knife edges from accidental damage which might occur if the beam is lowered too rapidly.

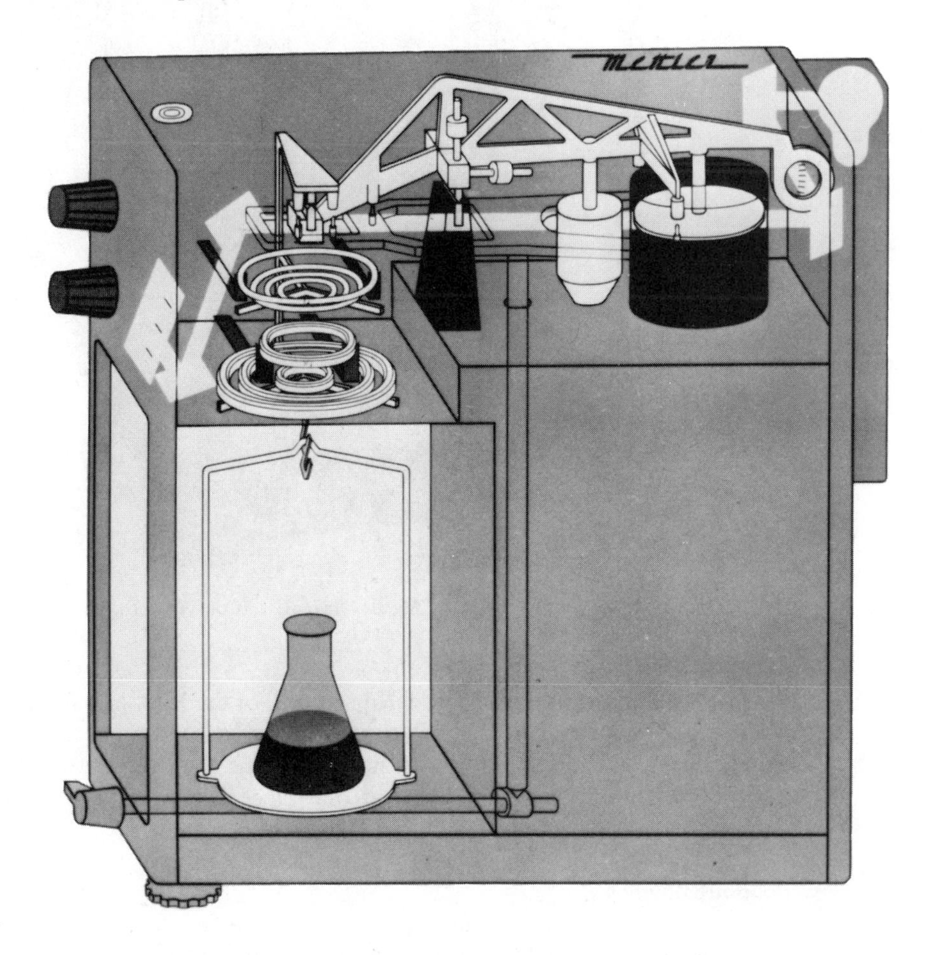

Figure 21.3 "See-through" view of a standard Mettler analytical balance. (Courtesy of Mettler Instrument Corporation, Princeton, New Jersey.)

Figure 21.4 Single-pan balance, Mettler H78AR. (Courtesy of Mettler Instrument Corporation, Princeton, New Jersey.)

The knife edges and planes (bearing stones) of the Mettler balances are made of synthetic sapphire (Al_2O_3). This material has proved more satisfactory than agate, since it is harder and more homogeneous than the natural stone. The built-in weights are concentric, solid one-piece ring weights of nonmagnetic chrome-nickel steel.

Readout Devices

The most common devices used on commercial balances to read to the nearest 0.1 mg are the vernier and the "full-digital readout." Figure 21.5 shows a reading scale in which the grams and tenths of a gram appear on the digital register and the

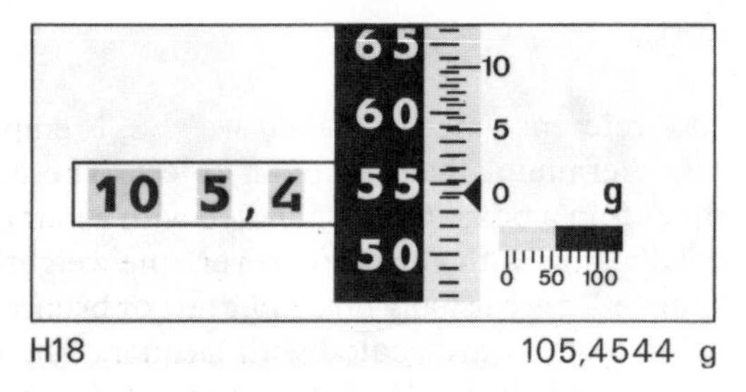

Figure 21.5 Scale of a single-pan balance employing vernier. (Courtesy of Mettler Instrument Corporation, Princeton, New Jersey.)

milligrams appear on the optical scale. Reading from the zero on the vernier gives a value of 54.4 mg or a total weight of 105.4544 g. A digital readout is shown in Fig. 21.6, all four decimals appearing as digits on the optical scale. A device such as this eliminates subjective weighing errors.

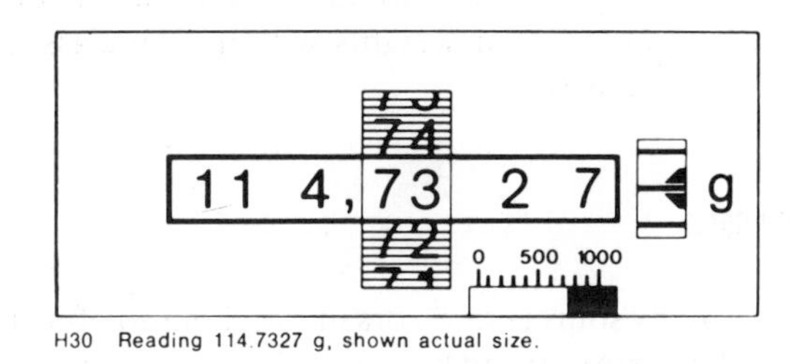

Figure 21.6 Digital readout. (Courtesy of Mettler Instrument Corporation, Princeton, New Jersey.)

Taring Devices

A number of commercial models contain so-called *tare* devices, which enable the operator to set the balance at zero with an empty weighing bottle, or beaker, on the pan. The operator can then weigh the sample by pouring it directly into the container and avoid the operation of subtracting the weight of the container from that of the container plus sample.

Preweighing Devices

Some models contain a built-in automatic "preweighing" system. Such a balance actually contains two weighing mechanisms, which use the same optical scale to provide coarse readings in grams and then fine readings in milligrams. The object to be weighed is placed on the pan, a knob turned to the proper position, and the weight to the nearest gram appears immediately on the optical scale. The coarse weight is then dialed in and the knob turned to another position in which the weight can be read to the nearest 0.1 mg. Such a device eliminates the trial-and-error dialing of weight knobs to find the coarse weight of the object.

ANALYTICAL WEIGHTS

A series of reference masses, called *weights*, is employed along with a two-pan balance is determining the mass of an object. These weights are kept in separate compartments in a box, and are handled with a pair of ivory-tipped forceps. The box is usually lined with velvet to prevent the weights from being scratched. The weights above 1 g are usually made of brass or bronze, and are plated with a metal such as chromium or are coated with lacquer. The weights less than 1 g, called *fractionals*, are usually made of aluminum. Riders are normally of aluminum and usually weigh 10 mg.

The Bureau of Standards classifies analytical weights in several categories. Class M weights are of one-piece construction and are generally reserved for reference standards or for work requiring high precision. They are frequently used to check the calibrations of the built-in weights of a single-pan balance. Class S weights may consist of a base into which the top or handle is screwed, and are used for most calibration work. Class S-1 weights are generally used for routine analytical work and class P weights are used for most student work. A list of acceptable tolerances for weights within each class has been established by the Bureau of Standards.[1]

ERRORS IN WEIGHING

There are many sources of errors in weighing an object on a single-pan or two-pan balance. The error caused by unequal balance arms in the two-pan balance can be eliminated by the method of transposition or substitution. Such an error is not encountered in the single-pan balance since the method of substitution is employed. Errors in analytical weights can be corrected by calibration against known weights. Obviously, the operator must avoid careless mistakes such as misreading the swing of a pointer, the value of a weight or a dial, or the position of a vernier. In addition, the following sources of error must be guarded against.

Changes in Moisture or Carbon Dioxide Content. Samples to be analyzed are usually dried before weighing, and in gravimetric procedures a final precipitate is usually heated at high temperature. Such materials may readily increase in weight by taking up water vapor or carbon dioxide from the air during the weighing process. For this reason, closed containers or weighing bottles (Fig. 21.7) are usually employed to hold a sample, and ignited precipitates are usually weighed in covered crucibles.

Electrification. A glass vessel should not be wiped with a dry cloth before the vessel is weighed. The object may acquire a charge of static electricity, making the swings of the beam erratic. The charged container is attracted to various parts of

[1] "Precision Laboratory Standards of Mass and Laboratory Weights," Nat. Bur. Standards Cir. 547 (1962).

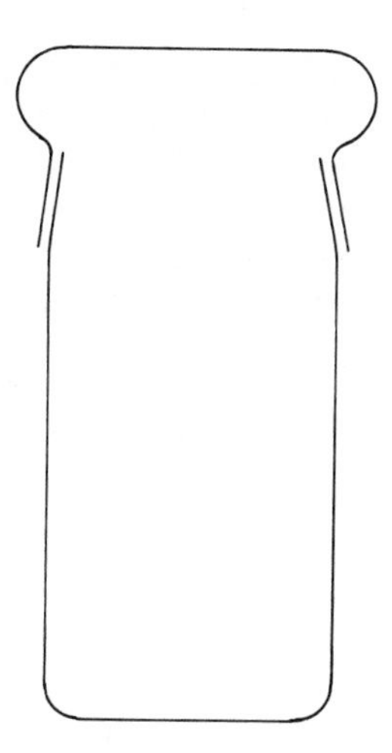

Figure 21.7 Weighing bottle.

the balance, and an error in weight may occur. This effect is not usually noticeable if the humidity is high.

Temperature. The temperature of the object, the balance, and the weights should be the same. For example, if the object is warmer than the weights, the pan is buoyed up by convection currents, and the apparent weight is less than the true weight. Crucibles that have been heated and samples that have been dried should always be cooled to room temperature before weighing.

Buoyancy. In the normal weighing process, both the object and the weights are buoyed up by the weight of air displaced. This is in accordance with the principle of Archimedes. If the object and weights displaced the same amount of air, no error would be introduced by this effect. This is not usually the case, however, since the density of the weights is normally different from that of the object. In quantitative analysis the density of the weights is usually larger, and hence the object displaces a greater volume of air than do the weights. The weight of the object is therefore less in air (the apparent weight) than it would be in a vacuum (the true weight). The true weight, W_v, is given by

$$W_v = W_a + (V_o - V_w)d_a$$

where W_a is the weight in air, V_o and V_w are the volumes of the object and weights respectively, and d_a is the density of air (about 0.0012 g/ml under usual conditions).

Since $V_o = W_v/D_o$ or approximately W_a/D_o, and $V_w = W_a/D_w$, where D_o and D_w are the densities of the object and weights, respectively,

$$W_v = W_a + \left(\frac{W_a}{D_o} - \frac{W_a}{D_w}\right)0.0012$$

or

$$W_v = W_a\left\{1 + \left(\frac{1}{D_o} - \frac{1}{D_w}\right)0.0012\right\}$$

In the usual weighings made in the analytical laboratory, the errors caused by buoyancy are quite small. Most analytical results are expressed in terms of the ratio of two weights. If the densities of the sample and final precipitate are nearly equal, no appreciable error is introduced by using weights in air. In such operations as the calibration of volumetric apparatus by weighing large volumes of water, the error is appreciable since the density of water is much less than that of the weights.

GENERAL RULES FOR USE OF BALANCES

The following points should be noted about the use of any analytical balance:

1. The balance should be kept clean. Remove dust from the pan or pans and from the floor of the balance with a camel-hair brush before starting a weighing.
2. Learn the capacity of your balance and never weigh an object heavier than this amount.
3. The object to be weighed should be at the same temperature as the balance. It is preferable not to handle the object with the fingers, and glass objects should not be rubbed with a dry cloth prior to weighing. A weighing bottle can be handled (but not rubbed) with a clean cloth or a strip of paper folded around it.
4. Chemicals should not be placed directly on the metal pan but should be weighed in weighing bottles. The desired quantity of sample is poured out into a suitable container and the bottle reweighed. The difference in the two weights is the weight of sample taken from the bottle.
5. At the completion of weighing make a final check to make sure that (a) the weight is recorded in your notebook, (b) the balance is clean, (c) no objects have been left on the pans, (d) the beam and pans have been arrested, and (e) the dust cover has been replaced.
6. If the balance does not behave properly, report this fact to the instructor.

Use of a Two-Pan Balance

There are several methods which can be employed when weighing an object on a two-pan balance. In any of these methods it is necessary to determine first the equilibrium position of the beam of the balance (zero point). The object is then placed on the left-hand pan, the weights on the right-hand pan, and the equilibrium position again determined (rest point). If the balance is damped, the analyst waits until the beam ceases swinging and reads the position of the pointer on the pointer scale. For an undamped balance the equilibrium position is calculated from observation of the swings of the pointer across the pointer scale. The final weight adjustment required to make the rest point coincide with the zero point is either calculated (rider balance) or made with the chain of a chainomatic balance.

The following rules should be observed:

1. To set the beam in motion, release the beam support slowly and then the pan arrests. To arrest the beam, engage the pan arrests as the pointer crosses the center of the scale and then raise the beam carefully.
2. While observing the swings of the pointer, sit directly in front of the balance to avoid errors due to parallax.
3. Always arrest the beam whenever weights are changed or an object is added to or removed from the pan. With a rider balance, fractional weights and the rider can be changed while the beam is released provided the pans are arrested. With a chainomatic balance, weights can be changed with the chain while the beam is in motion.
4. Always handle the weights with forceps, never with the fingers. Place the weights near the center of the pan. Use only the set of weights assigned to you.
5. At the completion of a weighing, remove all the weights from the balance pan and return them to the box.

Use of a Single-Pan Balance

Since the single-pan balances are damped, the empty balance is allowed to come to rest after the beam is fully released. The zero adjustment knob is then turned until the illuminated scale indicates a reading of zero; on a balance equipped with a vernier the zero lines of the projected scale and the vernier are made to coincide. The beam is arrested and the object to be weighed is placed on the pan. Then with the beam partially released, weights are removed by turning the knobs on the outside of the balance case. Once the sum of the weights removed is within 0.1 g (but on the low side) of the weight of the object, the beam is fully released and allowed to come to rest. The weight can then be read directly to four decimal places.

The following observations should be made:

1. Note the arrestment control knob. The beam can be partially released, fully released, or arrested. In the partial release position, weights can be changed without damaging the knife edges. The beam is fully released when a final weighing is made.
2. Note the zero adjust control. With the beam fully released, this knob can be turned to make the optical scale read zero with the pan empty.
3. Note the knobs which control the weights. One usually removes weights in 10-g increments; the other removes weights in 1-g units up to 9 g. A third knob may remove 0.1-g units.
4. At the completion of a weighing, arrest the beam and turn all the weight knobs until each dial reads zero.

CHAPTER TWENTY-TWO

laboratory procedures

This chapter contains directions for a number of experiments which illustrate a wide variety of analytical principles. Most of the exercises are from the classical areas of titrimetric and gravimetric analysis; a few illustrations of what might be called "simple instrumental" methods are included. The student is urged to acquaint himself fully with the theoretical principles discussed in earlier chapters before attempting the laboratory exercises.

ACID-BASE TITRATIONS

Directions are given for the preparation and standardization of hydrochloric acid and sodium hydroxide solutions, and for these solutions to be used in several analyses. The theoretical principles of acid-base titrations are discussed in Chapters 3 and 6.

EXPERIMENT 1. PREPARATION OF 0.1 N SOLUTIONS OF HYDROCHLORIC ACID AND SODIUM HYDROXIDE

Procedure

(*a*) ***Hydrochloric Acid.*** Measure into a clean, glass-stoppered bottle approximately 1 liter of distilled water. With a graduated cylinder or measuring pipet, add to the water about 8.5 ml of concentrated hydrochloric acid. Stopper the bottle, mix the solution well by inversion and shaking, and label the bottle.

(b) Sodium Hydroxide. Carbonate-free sodium hydroxide can be prepared most readily from a concentrated solution of the base, because sodium carbonate is insoluble in such a solution. A 1 : 1 solution of sodium hydroxide in water is available commercially, or may have been prepared by the instructor (Note 1). Carefully add 6 to 7 ml of this solution to approximately 1 liter of distilled water (Note 2) in a clean bottle, using a graduated pipet and rubber bulb. Close the bottle with a rubber stopper (Note 3), shake the solution well, and label the bottle.

Notes

1. If such a solution is not available, dissolve about 50 g of sodium hydroxide in 50 ml of water in a small, rubber-stoppered Erlenmeyer flask. Be careful in handling this solution, as considerable heat is generated. Allow the solution to stand until the sodium carbonate precipitate has settled. If necessary, the solution can be filtered through a Gooch crucible. Alternatively, carbonate-free base can be prepared by dissolving 4.0 to 4.5 g of sodium hydroxide in about 400 ml of distilled water and adding 10 ml of 0.25 *M* barium chloride solution. The solution is well mixed and then allowed to stand overnight so that barium carbonate will settle out. The solution is then decanted from the solid into a clean bottle and diluted to 1 liter.

2. Some directions call for boiling the water for about 5 min to remove carbon dioxide. If this is done (consult instructor), be sure to protect the water from the atmosphere as it cools.

3. Glass-stopped bottles should not be used since alkaline solutions cause the stoppers to stick so tightly that they are difficult or impossible to remove. Polyethylene bottles, if available, are excellent for storing dilute base solutions. It may be desirable to protect the solution from atmospheric carbon dioxide (consult instructor). This can be done by fitting a two-holed rubber stopper with a siphon and soda-lime tube, as shown in Fig. 22.1.

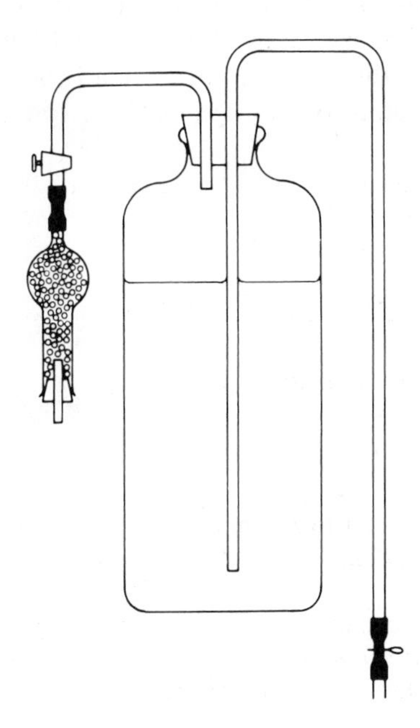

Figure 22.1 Bottle for storing carbonate-free base.

EXPERIMENT 2. DETERMINATION OF THE RELATIVE CONCENTRATIONS OF THE HYDROCHLORIC ACID SODIUM HYDROXIDE SOLUTIONS

In this experiment the ratio of the concentrations of the acid and base solutions is determined. Following standardization of either solution, the normality of the other can be calculated from this ratio.

Procedure

Rinse two clean burets and fill one with the hydrochloric acid and the other with the sodium hydroxide solution prepared in Experiment 1. Remove any air bubbles from the tips, lower the liquid level to the graduated portions, and record the initial reading of each buret.

Now run about 35 to 40 ml of the hydrochloric acid solution into a clean 250-ml Erlenmeyer flask and record the buret reading (Note 1). Add 2 drops of phenolphthalein indicator (Note 2) and about 50 ml of water from a graduated cylinder, rinsing down the walls of the flask. Now run into the flask the sodium hydroxide solution from the other buret, swirling the flask gently and steadily to mix the solutions. As an aid in preventing overrunning of the end point, notice the transient, local, pink coloration as it becomes more persistent with the progress of the titration. Finally, when the color first pervades the entire solution even after thorough mixing, stop the titration and record the buret reading. The color should persist for at least 15 sec or so, but may gradually fade because of the absorption of atmospheric carbon dioxide. It is well to rinse down the inside of the flask and also the buret tip with distilled water just before the termination of the titration so that stray droplets will not escape the reaction. If the end point is accidentally overrun, the titration can still be salvaged: run enough hydrochloric acid solution into the flask to turn the phenolphthalein indicator colorless, record the buret reading again, and then approach the end point once more with the sodium hydroxide solution.

Repeat the titration at least two more times (Note 3). Finally calculate the volume of acid equivalent to 1 ml of base:

$$1.000 \text{ ml of base} = \frac{\text{volume of acid}}{\text{volume of base}}$$

Use buret corrections if necessary (consult instructor).

Notes

1. This volume is recommended in order to minimize errors in reading a buret. The acid is usually titrated with base instead of base with acid to minimize absorption of carbon dioxide during the titration.

2. Methyl red, bromthymol blue, and other indicators can also be employed. Solutions are prepared as follows: *Phenolphthalein*: Dissolve 2 g of phenolphthalein per liter of 95% ethanol. *Methyl red*: Dissolve 1 g of the sodium salt of methyl red per liter of water.

Bromthymol blue: Dissolve 0.1 g of bromthymol blue in 8 ml of 0.02 *M* NaOH and dilute to 100 ml with water.

3. Do not allow the sodium hydroxide to remain in the buret any longer than necessary. As soon as the titrations are finished, drain the base from the buret and rinse thoroughly, first with dilute hydrochloric acid and then with water.

EXPERIMENT 3. STANDARDIZATION OF SODIUM HYDROXIDE SOLUTION WITH POTASSIUM ACID PHTHALATE

A number of good primary standards are available for standardizing base solutions. Directions are given here for the use of potassium acid phthalate, but these can be readily modified to suit another standard.

Procedure

Place about 4 to 5 g of pure potassium acid phthalate in a clean weighing bottle and dry the sample in an oven at 110°C for at least 1 hr. Cool the bottle and its contents in a desiccator (Note 1). Weigh accurately into each of three clean, numbered Erlenmeyer flasks about 0.7 to 0.9 g of the potassium acid phthalate (Note 2). Record the weights in your notebook.

To each flask add 50 ml of distilled water (Note 3) from a graduated cylinder, and shake the flask gently until the sample is dissolved. Add 2 drops of phenolphthalein to each flask. Rinse and fill a buret with the sodium hydroxide solution. Titrate the solution in the first flask with sodium hydroxide to the first permanent pink color. Your hydrochloric acid solution, in a second buret, may be used for back-titration if required. Repeat the titration with the other two samples, recording all data in your notebook.

Calculate the normality of the sodium hydroxide solution obtained in each of the three determinations. Average these values and compute the average deviation in the usual manner. If the average deviation exceeds about 2 parts per thousand, consult the instructor. Finally, calculate the normality of the hydrochloric acid solution from the normality of the sodium hydroxide and the volume ratio of the acid and base obtained in Experiment 2.

Notes

1. Potassium acid phthalate is relatively nonhygroscopic, and the drying process may be omitted (consult instructor).

2. Since 4 meq of potassium acid phthalate weighs $4 \times 204.2 = 816.8$ mg, the quantity recommended should require 35 to 45 ml of a 0.1 *N* base solution for titration.

3. Some directions call for boiling the water for about 5 min to remove carbon dioxide before use. If this is done (consult instructor), the water should be cooled to room temperature before the titration, and it should be protected from the atmosphere while cooling.

EXPERIMENT 4. STANDARDIZATION OF THE HYDROCHLORIC ACID SOLUTION WITH SODIUM CARBONATE

The hydrochloric acid solution can be standardized against a primary standard if so desired. Sodium carbonate is a good standard and is particularly recommended if the acid solution is to be used to titrate carbonate samples.

Procedure

Accurately weigh three samples (about 0.20 to 0.25 g each) of pure sodium carbonate (Note 1), which has been previously dried, into three Erlenmeyer flasks. Dissolve each sample with about 50 ml of distilled water and add 2 drops of methyl red[1] or methyl orange (see Note 2 and consult instructor).

(a) Methyl Red. Titrate each sample with the hydrochloric acid solution. Methyl red is yellow in basic and red in acid solution. As soon as the solution is distinctly red, add 1 additional ml of hydrochloric acid and remove the carbon dioxide by boiling the solution gently for about 5 min (Note 3). Cool the solution to room temperature and complete the titration. Back-titration can be done with the sodium hydroxide solution previously prepared. If the color change is not sharp, repeat the heating to remove carbon dioxide.

(b) Methyl Orange. Prepare a solution of *p*H 4 by dissolving 1 g of potassium acid phthalate in 100 ml of water. Add 2 drops of methyl orange to this solution and retain it for comparison purposes. Now titrate each sample with hydrochloric acid until the color matches that of the comparison solution.

Calculate the normality of the hydrochloric acid solution obtained in each of the three titrations. Average these values and compute the average deviation in the usual manner. If this figure exceeds 2 to 3 parts per thousand, consult the instructor. The normality of the sodium hydroxide solution can be calculated from the normality of the acid and the relative concentrations of acid and base obtained in Experiment 2.

Notes

1. Analytical-grade sodium carbonate (assay value 99.95%) can be used after drying for about $\frac{1}{2}$ hr at 270 to 300°C.

2. Various indicators and mixed indicators have been suggested for this titration. The *p*H at the equivalence point of the reaction

$$CO_3^{2-} + 2H^+ \rightleftharpoons H_2CO_3$$

[1] The color changes shown by these indicators can be modified by addition of a suitable dye. Methyl red is frequently mixed with methylene blue, the color change then being red-violet (acid) to green (base) with an intermediate shade of gray. The dye xylene cyanole FF is added to methyl orange to give "modified methyl orange." The color change is then pink (acid) to green (base) with a gray intermediate. See I. M. Kolthoff and C. B. Rosenblum, *Acid-Base Indicators*, Macmillan Publishing Co., Inc., New York, 1937.

is about 4, and methyl orange changes color near this *p*H. The titration curve is not very steep, however, and hence it is often suggested that excess acid be added and carbon dioxide removed by boiling or vigorous shaking. The subsequent titration of excess acid with base involves only strong electrolytes, and a sharp end point is obtained if the carbon dioxide is completely removed. An indicator blank must then be determined since methyl orange changes color at a *p*H appreciably different from 7. If methyl red (*p*H 5.4) is employed, no indicator blank is necessary. Directions are given here for the titration to the methyl red end point with removal of carbon dioxide and for the titration to the methyl orange end point without removal of carbon dioxide. The latter procedure is more rapid and is recommended where a high degree of accuracy is not required.

3. If insufficient acid is present to completely convert bicarbonate into carbonic acid, the indicator will turn back to its basic color as the carbon dioxide is expelled and the *p*H rises. The titration is then continued with acid. If excess acid is added, the indicator will retain the acid color and the titration is continued by addition of base.

EXPERIMENT 5. DETERMINATION OF THE PURITY OF POTASSIUM ACID PHTHALATE

Directions are given for the titration of a solid acid sample. Commercial unknowns usually contain potassium acid phthalate or sulfamic acid, either of which can be titrated with phenolphthalein indicator.

Procedure

Dry the unknown sample in a weighing bottle (unless otherwise directed) at 110°C for at least 1 hr, and cool in a desiccator. Weigh three samples of appropriate size (Note 1) into clean Erlenmeyer flasks and dissolve in 75 to 100 ml of distilled water (Note 2). Add 2 drops of phenolphthalein indicator to each flask.

Titrate the contents of the first flask with the standard sodium hydroxide solution to the first permanent pink color. Record all buret readings as usual and then titrate the other two samples in the same manner.

Calculate the percentage of potassium acid phthalate in each sample and obtain the average purity and average deviation (Note 3). Report the percentage purity as suggested on page 550.

Notes

1. The instructor will specify the size sample required to use 30 to 45 ml of a 0.1 *N* base for titration. If this weight is greater than 1 g, as it may be with potassium acid phthalate samples of low percentage purity, the samples need be weighed only to the nearest milligram. Why?

2. Some directions call for boiling the water for about 5 min to remove carbon dioxide before use. If this is done (consult instructor), the water should be cooled to room temperature before the titration, and it should be protected from the atmosphere while cooling.

3. The precision that can be obtained depends upon the homogeneity of the sample as well as the technique of the student. Consult the instructor for the expected precision.

EXPERIMENT 6. DETERMINATION OF ACETIC ACID CONTENT OF VINEGAR

The principal acid in vinegar is acetic acid, and federal standards require at least 4 g of acetic acid per 100 ml of vinegar. The total quantity of acid can be readily determined by titration with standard base using phenolphthalein indicator. Although other acids are present, the result is calculated as acetic acid.

Procedure

Pipet 25 ml of vinegar into a 250-ml volumetric flask, dilute to the mark, and mix thoroughly (Note). Pipet a 50-ml aliquot of this solution into an Erlenmeyer flask and add 50 ml of water and 2 drops of phenolphthalein indicator. Titrate with standard base to the first permanent pink color. Repeat the titration on two additional aliquots.

Assuming all the acid to be acetic, calculate the number of grams of acid per 100 ml of vinegar solution. Assuming that the density of vinegar is 1.000, what is the percentage of acetic acid by weight in vinegar? Average your results in the usual manner.

Note

This quantity should require a reasonable volume of 0.1 *N* base for titration. If a more concentrated base solution is employed, a 100-ml volumetric flask can be used. The dilution with water prevents the color of the vinegar solution from interfering in the detection of the end point.

EXPERIMENT 7. DETERMINATION OF THE ALKALINITY OF SODA ASH

Crude sodium carbonate, called soda ash, is commonly used as a commercial neutralizing agent. The titration with standard acid to the methyl orange end point gives the total alkalinity, which is mainly due to sodium carbonate. Small amounts of sodium hydroxide and sodium bicarbonate may also be present. The results are usually expressed as percentage of sodium carbonate or sodium oxide.

Since the samples are frequently nonhomogeneous, the method of aliquot portions is employed. Either methyl red or methyl orange can be employed as the indicator.

Procedure

Weigh accurately into a clean 250-ml beaker a sample of the dried unknown of appropriate size (Note). Dissolve the sample in about 125 ml of distilled water. Place a clean funnel in a 250-ml volumetric flask and transfer the solution from the beaker to the flask. Rinse the beaker, add the rinsings to the flask, and finally dilute to the mark. Mix the contents of the flask thoroughly by inversion and shaking.

Pipet a 50-ml aliquot into an Erlenmeyer flask and add 2 drops of methyl red or methyl orange. Titrate with standard acid according to procedure (a) or (b), page 568. Repeat the titration with two other 50-ml aliquots. At the end of the titrations, be sure to empty and thoroughly rinse the volumetric flask. An alkaline solution should not be left in a volumetric flask for a long period of time.

Report the percentage of sodium carbonate or sodium oxide (consult instructor) in the sample. A precision of 3 to 5 parts per thousand is not unusual for the titration.

Note

The instructor will specify the size of sample required to use 30 to 45 ml of 0.1 *N* acid for titration.

EXPERIMENT 8. TITRATION OF ANTACIDS

The purpose of this experiment is to determine the amount of acid neutralized by various commercial antacids. These products contain bases such as calcium carbonate, magnesium carbonate, and magnesium hydroxide. The latter compounds are not very soluble in water, but a direct titration can be carried out with hydrochloric acid if sufficient time is allowed for reaction between the solid and the titrant. A recurring end point may be obtained because this reaction is rather slow. The *p*H of a solution made by dissolving the antacid in water is in the range 8 to 9, indicating the presence of bicarbonate ion. Hence methyl orange should be used as the indicator in a direct titration.

In the procedure below, excess acid is added to react with the antacid, the solution is heated to remove CO_2, and the excess acid is titrated with standard base. Phenolphthalein can be used as the indicator and a reasonably sharp end point is obtained.

Procedure

Cut a tablet of the antacid in half and weigh one piece on the analytical balance (Note 1). Transfer the sample to a 250-ml Erlenmeyer flask; add 75 ml of standard 0.1 *N* HCl solution using a pipet. Heat the solution to boiling and boil it gently for about 3 min (Note 2). Cool the solution to room temperature, add 4 drops of phenolphthalein indicator, and titrate with standard base to the first permanent pink color.

(a) Calculate the grams of HCl neutralized by 1 g of the antacid. (b) Calculate the grams of 0.1 *N* hydrochloric acid solution neutralized by 1 g of the antacid. (Assume that 0.1 *N* HCl has a density of 1.00 g/ml.) You may have heard advertising claims about how many times its weight of "stomach acid" an antacid neutralizes. Which weight is referred to in such claims, the one calculated in (a) or in (b) above?

Notes

1. Tablets of such products as Tums and Rolaids weigh about 1.3 to 1.4 g and require 80 to 110 ml of 0.1 *N* HCl for direct titration. If 0.2 *N* standard acid and base are at hand the entire tablet can be titrated.

2. There will probably be a small amount of white solid (filler) which does not dissolve even after heating.

EXPERIMENT 9. TITRATIONS IN NONAQUEOUS MEDIA

The principles of titrations in media other than water are discussed in Chapter 6. Such titrations are rather widely used today, and it is appropriate to include one or two in a beginning course in volumetric analysis. The procedures given below employ a solution of perchloric acid in glacial acetic acid. This solution can be standardized against pure potassium acid phthalate which acts as a base in acetic acid solvent. The solution is then used to titrate an amine which is too weak a base to be titrated in water. Methyl violet is used as the indicator, or the end point can be detected potentiometrically using a glass and calomel electrode pair. These directions are based upon the recommendations of Fritz.[2]

Procedure

Preparation of Solutions

(a) 0.1 *N* $HClO_4$. Add 4.3 ml of 72% perchloric acid to 150 ml of glacial acetic acid, mix well, add 10 ml of acetic anhydride, and allow the solution to stand for 30 min (Note 1). Dilute to 500 ml with glacial acetic acid and allow the solution to cool to room temperature.

(b) 0.1 *N sodium acetate.* Dissolve 4.1 g of anhydrous sodium acetate in glacial acetic acid and dilute to 500 ml with the acid.

(c) *Methyl violet indicator.* This is prepared by dissolving 0.2 g of methyl violet in 100 ml of chlorobenzene.

Relative Concentrations. Fill two burets with the perchloric acid and sodium acetate solutions (Note 2). Withdraw about 35 to 40 ml of perchloric acid, add 2 drops of methyl violet indicator, and titrate with the sodium acetate solution. Take the first permanent violet tinge as the end point. Repeat the titration with two additional samples and calculate the relative concentrations of the two solutions.

Standardization. Weigh samples of about 0.5 to 0.6 g of pure potassium acid phthalate into three Erlenmeyer flasks and add about 60 ml of glacial acetic acid to each flask. Heat the first flask cautiously until the sample is in solution. Then cool and add 2 drops of methyl violet. Titrate with perchloric acid solution to the

[2] J. S. Fritz, *Acid-Base Titrations in Nonaqueous Solutions*, G. F. Smith Chemical Co., Columbus, Ohio, 1952.

disappearance of the violet tinge. Use the sodium acetate solution for back-titration if needed.

Repeat the titration with the other two samples and then calculate the normality of the perchloric acid solution. From the relative concentration obtained above, calculate the normality of the sodium acetate solution (Note 3).

Analysis. Amino acids or amines can be titrated with perchloric acid in acetic acid. If the sample is an amino acid, it is recommended that excess perchloric acid be added and the excess titrated with sodium acetate. Amines can be titrated directly with perchloric acid.

Weigh three samples of the substance to be determined, taking about 3 meq for each sample (consult instructor). If the sample is an amino acid, dissolve in exactly 50 ml (pipet) of standard perchloric acid. Add 2 drops of methyl violet indicator and back-titrate with sodium acetate, taking the first permanent violet tinge as the end point. If the sample is an amine, dissolve it in about 50 ml of acetic acid and titrate with standard perchloric acid to the first appearance of the violet color.

Calculate the number of milliequivalents of amino acid or amine in the sample, and report as desired by the instructor (Note 4).

Notes

1. The acetic anhydride is added to react with water in the perchloric acid.
2. Avoid contact of acetic acid solutions with the skin. If these solutions are spilled on the hands, rinse the hands immediately with tap water.
3. Room temperature should be noted when these titrations are carried out. If there is a large change in temperature between the time of standardization and analysis, a correction should be applied to the volume of acetic acid solution.
4. The instructor may prefer to give pure samples of the amine or amino acid as unknowns. The student then reports the equivalent weight of the unknown.

PRECIPITATION AND COMPLEX FORMATION TITRATIONS

The theoretical principles of complex formation and precipitation titrations are discussed in Chapters 8 and 9. In this section, laboratory directions are given for precipitation titrations involving the reaction of silver cation with chloride and thiocyanate anions. The Mohr, Volhard, and Fajans methods are illustrated. Titrations involving complex formation include the reaction of silver ion with cyanide ion and the use of EDTA in determining the hardness of water.

EXPERIMENT 10. PREPARATION OF 0.1 *M* SOLUTIONS OF SILVER NITRATE AND POTASSIUM THIOCYANATE

Silver nitrate can be weighed as a primary standard. however, a solution of approximately the desired molarity is usually prepared and then standardized using the indicator which will be employed in the analysis.

Procedure

(*a*) ***Silver Nitrate.*** Weigh on a trip balance about 8.5 g of silver nitrate (Note 1), dissolve the salt in distilled water in a beaker, and transfer the solution to a clean bottle (Note 2). Dilute the solution to 500 ml with distilled water, mix thoroughly, and label the bottle. Protect the solution from the sunlight as much as possible.

(*b*) ***Potassium Thiocyanate.*** If the Volhard method is to be used, prepare a 0.1 *M* thiocyanate solution by weighing about 4.9 g of potassium thiocyanate (Note 3) and dissolving the salt in water. Dilute the solution to about 500 ml and store in a clean bottle.

Notes

1. The molecular weight is 169.87. To prepare 500 ml of a 0.1 *M* solution, 8.5 g are needed which is ample for standardization and determination of an unknown. If 1 liter of solution is desired, take 17 g.
2. A brown-glass bottle is recommended to protect the solution from sunlight.
3. The molecular weight is 97.18. If 1 liter of the solution is desired, take 9.7 g of the salt.

EXPERIMENT 11. STANDARDIZATION OF SILVER NITRATE AND POTASSIUM THIOCYANATE SOLUTIONS

Sodium or potassium chloride can be used as a primary standard for a silver nitrate solution. Either potassium chromate (Mohr Method) or dichlorofluorescein (Fajans, or adsorption indicator, method) can be used as the indicator. Directions are given for both indicators below.

After standardization of the silver nitrate, this solution can be used to standardize the potassium thiocyanate solution. The two solutions are titrated directly, using iron(III) ion as the indicator.

Procedure

(*a*) ***Mohr Method.*** Weigh accurately three samples of pure, dry sodium chloride of about 0.20 to 0.25 g each (Notes 1 and 2) into 250-ml Erlenmeyer (Note 3) flasks. Dissolve each sample in 50 ml of distilled water and add 2 ml of 0.1 *M* potassium chromate solution (Note 4). Titrate the first sample with silver nitrate, swirling the solution constantly, until the reddish color of silver chromate begins to spread more widely through the solution, showing that the end point is almost reached. The formation of clumps of silver chloride is also an indicator that the end point is near. Continue the addition of silver nitrate dropwise until there is a permanent color change from the yellow of the chromate ion to the reddish color of silver chromate precipitate. Run an indicator blank (Note 5) if desired (consult instructor).

Titrate the other two samples in the same manner. Calculate the molarity of the silver nitrate solution and its chloride titer.

(*b*) ***Fajans Method.*** Weigh accurately three samples of pure, dry sodium chloride of about 0.20 to 0.25 g each (Notes 1 and 2) into 250-ml Erlenmeyer flasks. Dissolve each sample in 50 ml of water and add 10 drops to dichlorofluorescein indicator (Note 6) and 0.1 g dextrin (Note 7). Titrate the first sample with silver nitrate to the point where the color of the dispersed silver chloride changes from yellowish white to a definite pink (Note 8). The color change is reversible, and back-titration can be done with a standard sodium chloride solution. Titrate the other two samples in the same manner. Calculate the molarity of the silver nitrate solution and its chloride titer.

Notes

1. The molecular weight is 58.44. If potassium chloride, molecular weight 74.55, is used, take 0.25 to 0.30 g. The method of aliquot portions may be used if desired. Weigh accurately 1.1 to 1.2 g sodium chloride (1.5 to 1.6 g potassium chloride), dissolve the salt in a 250-ml volumetric flask, and withdraw 50-ml aliquots for titration.
2. Dry the salt at 120°C for at least 2 hr. For complete removal of the last traces of water, these salts should be heated to about 500 to 600°C in an electric furnace. This is not necessary except for very precise work.
3. Porcelain casseroles are sometimes recommended. The reddish color of silver chromate is more readily distinguished against a white background.
4. Dissolve 19.4 g in 1 liter of water.
5. To determine the indicator blank, add 2 ml of the indicator to about 100 ml of water to which is added a few tenths of a gram of chloride-free calcium carbonate. This gives a turbidity similar to that in the actual titration. Swirl the solution and add silver nitrate dropwise until the color matches that of the solution that was titrated. The blank should not be larger than about 0.05 ml.
6. This solution is prepared by dissolving 0.1 g of sodium dichlorofluoresceinate in 100 ml of water (or 0.1 g of dichlorofluorescein in 100 ml of 70% alcohol). Fluorescein, or its sodium salt, can be used in place of dichlorofluorescein.
7. The function of dextrin is to prevent coagulation of colloidal silver chloride.
8. The end point is easier to detect in diffuse light. Avoid direct sunlight.

(*c*) ***Titration of Silver Nitrate with Potassium Thiocyanate.*** Pipet 25 ml of standard silver nitrate solution into a 250-ml Erlenmeyer flask. Add 5 ml of 1 : 1 nitric acid (Note 1) and 1 ml of iron(III) alum solution as the indicator (Note 2). Titrate with thiocyanate, swirling the solution constantly, until the reddish-brown color begins to spread throughout the solution. Then add the thiocyanate dropwise, shaking the solution thoroughly between addition of drops. The end point is marked by the permanent appearance of the reddish color of the iron-thiocyanate complex.

Titrate two additional portions of the silver nitrate solution with thiocyanate. Calculate the molarity of the thiocyanate solution.

Notes

1. If the nitric acid has a yellow tinge indicating the presence of oxides of nitrogen, boil the solution until the oxides are expelled.
2. This is a saturated solution of iron(III) ammonium sulfate in 1 *M* nitric acid.

EXPERIMENT 12. DETERMINATION OF CHLORIDE

Chloride can be determined by titration with standard silver nitrate. A direct titration using either chromate ion or dichlorofluorescein as the indicator can be employed. Alternatively, excess standard silver solution can be added to the unknown and the excess titrated with standard potassium thiocyanate. Directions are given below for all three procedures. It is assumed that the sample is water soluble.

Procedure

(*a*) *Mohr Method.* Accurately weigh three samples of the dried material (Note 1) into three 250-ml Erlenmeyer flasks (Note 2). Dissolve each sample in about 50 ml of distilled water (Note 3) and add 2 ml of 0.1 *M* potassium chromate solution. Titrate the first sample with standard silver nitrate as directed in Experiment 11(a).

Titrate the other two samples in the same manner. Calculate the percentage of chloride in the sample.

(*b*) *Fajans Method.* Accurately weigh three samples of the dried material of appropriate size (Note 1) into three 250-ml Erlenmeyer flasks. Dissolve each sample in about 50 ml of distilled water (Note 3) and add 10 drops of dichlorofluorescein indicator and 0.1 g of dextrin. Titrate the first sample with standard silver nitrate as directed in Experiment 11(b).

Titrate the other two samples in the same manner. Calculate the percentage of chloride in the sample.

Notes

1. Consult the instructor regarding the size of sample. A large sample may be dissolved in a 250-ml volumetric flask and aliquot portions titrated, if desired.
2. Porcelain casseroles are sometimes recommended.
3. The solution should be nearly neutral. Dissolve a small portion of the material in about 10 ml of water and place a drop of the solution on a piece of litmus paper. If the solution is basic, add 1 drop of phenolphthalein to each solution and then add dropwise dilute nitric acid (about 1 to 150 ml of water), until the pink color of the indicator is just discharged. If the solution is acidic add 1 drop of phenolphthalein to each solution and then add dropwise 0.1 *N* sodium hydroxide until the solution is barely pink. Then add 1 or 2 drops of dilute nitric acid until the pink color of the indicator is just discharged.

(*c*) *Volhard Method.* Accurately weigh three samples of appropriate size (Note 1) into three 250-ml Erlenmeyer flasks. Dissolve each sample in about 50 ml of distilled water and then add 5 ml of 1 : 1 nitric acid (Note 2). To the first sample add standard silver nitrate solution from a buret until an excess of about 5 ml is present (Note 3). Add 1 to 2 ml of nitrobenzene (Note 4), stopper the flask with a rubber stopper, and shake the flask vigorously until the silver chloride is well coagulated (about 30 sec). Now add 1 ml of iron(III) alum indicator (Note 5) and titrate the excess silver nitrate with standard potassium thiocyanate solution. The end point is marked by the permanent appearance of the reddish color of the iron–thiocyanate complex.

Titrate the other two samples in the same manner. Calculate the percentage of chloride in the sample.

Notes

1. Consult the instructor. If desired, a larger sample may be taken, dissolved in a 250-ml volumetric flask, and aliquot portions titrated.
2. See Note 1 of Experiment 11(c).
3. The silver chloride coagulates near the equivalence point. Shake the solution well, allow it to stand for a few moments for the precipitate to settle, and then add a few drops of silver nitrate to the supernatant liquid. If no precipitate forms, silver nitrate is in excess.
4. Instead of adding nitrobenzene, one may filter off the silver chloride with a Gooch or sintered glass crucible. The precipitate is then washed with 1% nitric acid solution, the washings being added to the filtrate. The indicator is then added to the filtrate and the latter titrated with potassium thiocyanate.
5. See Note 2 of Experiment 11(c).

EXPERIMENT 13. DETERMINATION OF SILVER IN AN ALLOY

Silver can be determined by direct titration with potassium thiocyanate using iron(III) ion as the indicator.

Procedure

Accurately weigh three samples of a silver alloy of appropriate size (consult intructor) and place each in a 250-ml Erlenmeyer flask. Dissolve the first sample in 15 ml of 1:1 nitric acid. Boil the solution until all the oxides of nitrogen are removed and then dilute the solution to about 50 ml. Add 1 ml of iron(III) alum indicator (Note) and titrate with standard thiocyanate solution to the appearance of the reddish color of the iron–thiocyanate complex.

Dissolve and titrate the other two samples in the same manner. Calculate the percentage of silver in the alloy.

Note

See Note 2 of Experiment 11(c).

EXPERIMENT 14. DETERMINATION OF CYANIDE BY THE LIEBIG METHOD

The standard silver nitrate solution can be used to determine cyanide according to the Liebig method. There is no interference from chloride, bromide, or iodide ions since the silver salts of these anions are soluble in excess cyanide. Iodide ion is used as the indicator in the procedure given below.

Procedure

Secure a sample from the instructor (Note 1). Weigh accurately three 125-ml Erlenmeyer flasks that are clean, dry, and fitted with cork stoppers. Then add to each flask about 8 or 10 ml of the cyanide solution (Note 2) and reweigh the flask and contents. Add 3 ml of concentrated ammonia and about 0.1 g of potassium iodide to each flask and dilute to about 50 ml with distilled water. Titrate the first sample with silver nitrate, swirling the flask constantly, until 1 drop of titrant produces a faint permanent opalescence in the solution (Note 3).

Titrate the other two samples in the same manner. Calculate the percentage by weight of cyanide in the solution.

Notes

1. Since sodium and potassium cyanides are difficult to dry, the unknown sample may be dispensed as a solution. The solution should be titrated relatively soon since it slowly decomposes. *Caution:* Do not pipet any of this solution, because it is extremely poisonous. Also, do not acidify the solution under any circumstances.
2. This volume should contain 7 to 8 mmol of cyanide. Remember that each millimole of silver reacts with 2 mmol of cyanide.
3. After finishing the titration, add ammonia to the flask and pour the solution down a drain. Rinse the flask thoroughly with a large quantity of water.

EXPERIMENT 15. PREPARATION AND STANDARDIZATION OF SODIUM–EDTA SOLUTION

Titrations involving the use of the chelating agent EDTA are described in Chapter 8. Directions are given below for the preparation of a 0.01 *M* solution of sodium–EDTA and the standardization against calcium chloride.

Procedure

Weigh about 4 g of disodium dihydrogen EDTA dihydrate and about 0.1 g of $MgCl_2 \cdot 6H_2O$ into a clean 400-ml beaker. Dissolve the solids in water, transfer the solution to a clean 1-liter bottle, and dilute to about 1 liter (Note 1). Mix the solution thoroughly and label the bottle.

Prepare a standard calcium chloride solution as follows. Weigh accurately about 0.4 g of primary standard calcium carbonate that has been previously dried at 100°C. Transfer the solid to a 500-ml volumetric flask, using about 100 ml of water. Add 1:1 hydrochloric acid dropwise until effervescence ceases and the solution is clear. Dilute with water to the mark and mix the solution thoroughly.

Pipet a 50-ml portion of the calcium chloride solution into a 250-ml Erlenmeyer flask and add 5 ml of an ammonia–ammonium chloride buffer solution (Note 2). Then add 5 drops of Eriochrome Black T indicator (Note 3). Titrate carefully with the EDTA solution to the point where the color changes from wine-red to pure blue. No tinge of red should remain in the solution.

Repeat the titration with two other aliquots of the calcium solution. Calculate the molarity of the EDTA solution and the calcium carbonate titer.

Notes

1. If the solution is turbid, add a few drops of 0.1 *M* sodium hydroxide solution until the solution is clear.

2. Prepare this solution by dissolving about 6.75 g of ammonium chloride in 57 ml of concentrated ammonia and diluting to 100 ml. The *p*H of the buffer is slightly above 10.

3. Prepare by dissolving about 0.5 g of reagent-grade Eriochrome Black T in 100 ml of alcohol. If the solution is to be kept, date the bottle. It is recommended that solutions older than 6 weeks to 2 months not be used. Alternatively, the indicator may be used as a solid, which has a much longer shelf life. It is prepared by grinding 100 mg of the indicator into a mixture of 10 g of NaCl and 10 g of hydroxylamine hydrochloride. A small amount of the solid mixture is added to each titration flask with a spatula.

Alternatively, Calmagite may be used as the indicator. A solution is prepared by dissolving 0.05 g of the indicator in 100 ml of water. Add 4 drops of the indicator to each flask. The color change is from red to blue, as with Eriochrome Black T.

EXPERIMENT 16. DETERMINATION OF THE TOTAL HARDNESS OF WATER

The standard EDTA solution can be used to determine the total hardness of water.

Procedure

Obtain the water to be analyzed from the instructor and pipet a portion into each of three 250-ml Erlenmeyer flasks (Note 1). To the first sample add 1.0 ml of the buffer solution (Note 2) and 5 drops of indicator solution (Note 3). Titrate with the standard EDTA solution to a color change of wine-red to pure blue.

Repeat the procedure on the other two portions of water. Calculate the total hardness of the water as parts per million (ppm) of calcium carbonate. This is done as follows:

$$\text{Volume EDTA (ml)} \times \text{CaCO}_3 \text{ titer (mg/ml)} = \text{mg CaCO}_3$$

$$\frac{1000 \text{ ml/liter} \times \text{mg CaCO}_3}{\text{ml sample}} = \text{mg CaCO}_3\text{/liter, or ppm}$$

Notes

1. The volume of water titrated should be chosen so that 40 to 50 ml of the EDTA solution will be used for titration.

2. See Note 2 of Experiment 15.

3. See Note 3 of Experiment 15.

GRAVIMETRIC METHODS OF ANALYSIS

Gravimetric methods of analysis and the principles of separation by precipitation are discussed in Chapters 4 and 9. In this section directions are given for the precipitation of silver chloride, barium sulfate, and hydrous iron(III) hydroxide. These three compounds illustrate the common types of precipitates discussed in Chapter 4: curdy, crystalline, and gelatinous. An example of the use of an organic precipitant, 8-hydroxyquinoline, is also included.

EXPERIMENT 17. DETERMINATION OF CHLORINE IN A SOLUBLE CHLORIDE

The usual samples given to students are readily soluble in water, and no interfering ions are present. Hence this is a very simple determination that requires a relatively short time, and it is quite suitable for the introduction to gravimetric techniques. See Chapter 4 for a discussion of the properties of silver chloride and the possible errors encountered in its use in gravimetric analysis.

Procedure

The sample should be dried for 1 to 2 hr in an oven at 100 to 120°C. Clean three (Note 1) porous-bottom filtering crucibles (either sintered glass or porous porcelain) of "medium" porosity. Much of the dirt may be removed with detergents, although acids or other "strong" reagents may be necessary at times, depending upon the nature of the contamination. It is often well to draw a little concentrated nitric acid slowly through the porous filtering disc to eliminate certain material with which the disk may be impregnated. It is not wise to use "cleaning solution" on a porous-bottom crucible. The cleaned and thoroughly rinsed crucibles should be dried to constant weight at the same temperature at which the silver chloride is later to be dried. An oven is recommended, and temperatures of 100 to 150°C are suitable for ordinary student work.

Weigh out accurately three portions of the dried sample of about 0.5 to 0.7 g each (Note 2). Dissolve each portion in a 250-ml beaker, using 100 to 150 ml of water to which about 1 ml of concentrated nitric acid has been added. Prepare an approximately 0.1 M solution of silver nitrate (15 to 20 mg of $AgNO_3$/ml). Now heat the first of the chloride solutions to boiling, and with constant stirring add silver nitrate slowly to precipitate silver chloride. Obviously, in the case of an unknown chloride solution, the quantity of silver nitrate to be added cannot be specified. The student must determine when precipitation is complete. This is done by adding the silver nitrate in small portions, stirring vigorously, allowing the precipitate to settle somewhat, and noting whether a new cloud of precipitate appears upon further addition. After precipitation is complete, add about 10% more silver nitrate solution.

After the precipitate has coagulated well, remove the beaker from the heat, cover it with a watch glass, and set it aside to cool in the laboratory bench (for protection from light) for at least 1 hr. The other two samples are precipitated in the same way. After the first one has been done, the student will know the approximate amount of silver nitrate solution to add to the others and hence he can proceed more rapidly.

After the solution has cooled, it is filtered through a weighed crucible with suction, retaining the bulk of the precipitate in the beaker. It is wise to test the filtrate once again for completeness of precipitation, using a few drops of silver nitrate solution. The precipitate in the beaker is then washed by decantation with three 25-ml portions of 0.01 M nitric acid (about 2 drops of concentrated nitric acid in 100 ml of water). The washings, of course, are poured through the filter. Finally, the precipitate is stirred up in a small portion of wash solution and transferred into the crucible. Any precipitate remaining in the beaker is carefully rinsed out into the crucible, using a rubber policeman if necessary to remove precipitate adhering to the walls of the beaker. Now wash the precipitate in the crucible three or four times more with small portions of wash solution, allowing it to drain each time. Collect the last portion of wash solution separately and test it for the absence of silver ion with a drop of hydrochloric acid. Finally, drain the crucible completely with strong suction and place it in a covered beaker for drying in the oven. The watch glass covering the beaker should be raised with small glass hooks so as to permit the circulation of air over the precipitate.

After the samples have dried for about 2 hr, cool them in a desiccator and weigh. Return them to the oven for about 30 min, cool them in a desiccator, and reweigh. Drying may be considered complete if no more than 0.4 mg was lost during the second drying period.

Calculate the percentage of chlorine in the unknown sample. Report the result in the manner prescribed by the instructor.

Notes

1. Consult the instructor concerning the number of replicates to be run. Ordinarily, it is considered adequate to perform the analysis in triplicate.

2. The sample size should be such that a convenient quantity of precipitate is obtained. Consult the instructor. Portions of 0.5 to 0.7 g are appropriate for most student unknowns purchased currently.

EXPERIMENT 18. DETERMINATION OF SULFUR IN A SOLUBLE SULFATE

See Chapter 4 for a discussion of the properties of barium sulfate and the possible errors encountered in its use in gravimetric analysis.

Procedure

The sample should be dried in an oven at 100 to 120°C (Note 1). Weigh out accurately three portions of about 0.5 to 0.8 g each (Note 2) and dissolve each sample in about 200 ml of distilled water and 1 ml of concentrated hydrochloric acid. Prepare a 5% solution of barium chloride by dissolving 5 g of $BaCl_2 \cdot 2H_2O$ in 100 ml of water. Calculate the volume of this solution which will be required to precipitate the sulfate in each sample, including a 10% excess (Note 2). Measure this volume of solution into a clean beaker using a graduated cylinder.

Now heat both the sample solution and the solution of barium chloride nearly to boiling. Pour the hot barium chloride solution quickly but carefully into the hot sample solution and stir vigorously. Allow the precipitate to settle and test the supernatant liquid for completeness of precipitation by adding a few more drops of the barium chloride solution.

After precipitation is complete, cover the beaker with a watch glass and allow the precipitate to digest for 1 to 2 hr, keeping the solution hot (80 to 90°C) on a steam bath, hot plate, or using a low flame.

Either a fine porous porcelain filtering crucible or a funnel with filter paper can be used to collect the precipitate. The ignition temperature is too high to permit the use of sintered glass crucibles. If filter paper is used, the slow type such as Whatman No. 42 should be selected unless the precipitate has been unusually well digested, in which case No. 40 may suffice.

Three crucibles, porous porcelain (or, if filter paper is to be used, ordinary porcelain), and lids are cleaned, rinsed, and heated to constant weight. Use the highest temperature of the Tirrill burner for an ordinary crucible. Remember that a porous porcelain crucible must be heated within another crucible to prevent damage and the admission of gases from the flame through the porous bottom. Thus the full temperature of a Meker burner is required in this case.

The directions for washing and filtering the precipitate apply whichever type of filter is used. The solution is to be hot at the time of filtration. Decant the clear supernatant solution through the filter. Discard the clear filtrate so that if precipitate later runs through the filter only a small volume need be refiltered. Then rinse the precipitate into the funnel or filtering crucible with hot water. Remove any precipitate from the walls of the beaker with a rubber policeman and rinse such particles into the filter with hot water. If the filtrate is cloudy it must be refiltered, in which case the second passage generally clears it up. Continue to rinse the precipitate in the filter with hot water until a drop of silver nitrate solution added to a test portion of the washings collected in a test tube shows that chloride is absent.

After washing is complete, if filter paper was used, transfer the paper and precipitate carefully to the previously prepared crucible. Dry the precipitate slowly over a low flame (an oven may be used if desired). Then, increasing the heat, char the paper carefully and finally burn it off completely. Ignite the precipitate for about 20 min at the highest temperature of the Tirrill burner. The crucible should be uncovered and in a slanted position for free access of air to prevent reduction of barium sulfate to barium sulfide. (If such reduction is suspected, moisten the cooled precipitate with a little concentrated sulfuric acid and carefully raise the temperature, finishing the ignition once again at the highest temperature of the burner.) Cool and weigh the crucible and its contents, then reignite for a second 20-min period. This is repeated, of course, until constant weight is attained.

In the event that a filtering crucible was used instead of filter paper, ignition is the same except that there is no paper to be burned off; the crucible is heated inside an ordinary crucible, and a Meker burner is used. Be sure the precipitate is dried at a low temperature because steam formed upon strong ignition of wet material may carry precipitate out of the crucible.

Sulfur is usually reported as sulfur trioxide. Calculate the percentage of sulfur trioxide in the unknown sample and report the results in the style required by the instructor.

Notes

1. Consult the instructor. Certain samples of hydrated salts should not be dried in the oven. The common samples purchased from commercial suppliers of student unknowns are usually to be dried.
2. Consult the instructor regarding the weight of sample.

EXPERIMENT 19. DETERMINATION OF IRON IN AN OXIDE ORE

See Chapter 4 for a discussion of the properties of iron(III) hydroxide and the possible errors encountered in its use in gravimetric analysis.

Procedure

Accurately weigh out three samples of about 0.5 g each (Note 1) of the dried iron oxide unknown. Treat each sample in a clean 250-ml beaker with about 20 ml of 1 : 1 hydrochloric acid. Cover the beaker with a watch glass and heat the solution nearly to boiling. Continue this careful heating until solution of the sample is complete, or, if there is a persistent insoluble residue, until this residue (generally silica) shows no red-brown color.

Rinse down the watch glass and sides of the beaker with water from the wash bottle. Add 10 ml of saturated bromine water to the solution and boil gently until the bromine vapors are completely expelled. (If desired, the oxidation of iron(II) ion may be accomplished with nitric acid rather than bromine: add 1 to 2 ml of concentrated nitric acid dropwise to the solution and boil for 1 or 2 min to expel the nitrogen oxides.)

If there is an undissolved siliceous residue in the sample, the solution should be filtered at this point. A "fast" paper is adequate. Collect the filtrate in a clean 400-ml beaker. Wash the filter paper with dilute hydrochloric acid until no yellow remains, collecting the washings in the same beaker.

Now dilute the solution to about 200 ml, bring it nearly to a boil, and slowly add 1:2 ammonium hydroxide solution until a definite odor of ammonia persists in the vapors (Note 2). Hold the solution at the near-boiling temperature for another minute or so and then allow the precipitate to settle. Test for completeness of precipitation with a drop or two of ammonium hydroxide.

The precipitate is now to be separated from the mother liquor, washed, redissolved, and reprecipitated. In the first filtration, decantation is used to the greatest extent possible, retaining the bulk of the precipitate in the original beaker for redissolving. Much time is saved by this technique. Carefully decant the supernatant solution through a "fast" filter paper (Whatman No. 41, for example), retaining the precipitate in the beaker as much as possible. Then wash this precipitate with about 25 ml of 0.1 M ammonium nitrate wash solution (8 or 10 g of NH_4NO_3/liter). Allow the precipitate to settle and decant the washings through the filter. Repeat this washing a second time. Discard the filtrate and washings, and place the beaker containing the precipitate under the funnel. Now pour through the funnel about 50 ml of 1:10 hydrochloric acid solution. This should completely dissolve any precipitate in the funnel and the bulk of the precipitate in the beaker. Make certain that the acid washes the filter paper thoroughly. This can be accomplished by adding the acid from a pipet, directing a slow stream near the upper edge of the paper all the way around its circumference. No trace of yellow should remain on the paper.

The iron is now reprecipitated with ammonium hydroxide exactly as before. Allow the precipitate to settle, test for completeness of precipitation, and then decant the supernatant liquid through the same filter paper as used above after the first precipitation. Wash the precipitate by decantation three times with hot 0.1 M ammonium nitrate wash solution, finally bringing the precipitate onto the filter paper. Use a rubber policeman and rinsing to effect a quantitative transfer. (Alternatively, the inside of the beaker may be wiped with a small piece of ashless filter paper which is then added to the precipitate in the funnel.) The precipitate is now washed in the funnel with the hot ammonium nitrate solution until the filtrate shows only a faint test for chloride with silver nitrate.

After the precipitate has drained as much as possible, fold down the edges of the filter paper, and transfer paper with contents to a previously prepared porcelain crucible. Dry and char the paper in the usual manner and finally ignite the precipitate at bright red heat (a Meker burner is recommended, although a Tirrill burner may be adequate). There is some tendency for reduction of iron(III) oxide to the magnetic oxide, Fe_3O_4. Hence the ignition should be performed with plentiful access of air into the crucible. After ignition for about 45 min, cool the crucible in the desiccator and weigh. Reignite for 20-min periods until constant weight is attained.

Calculate the percentage of iron in your unknown sample. Report the results as requested by the instructor.

Notes

1. Consult the instructor regarding recommended weight of sample.

2. Unless it was perfectly clear, the ammonium hydroxide should have been previously filtered to remove silica, which is often suspended in the solution as a result of attack on the glass container.

EXPERIMENT 20. DETERMINATION OF NICKEL IN STEEL

One of the most common applications of organic precipitants is the precipitation of nickel by dimethylglyoxime (Chapter 4). In the determination of nickel in steel, an acid solution containing iron in the +3 oxidation state is treated with tartaric acid and an alcoholic solution of dimethylglyoxime. The solution is made slightly basic with ammonia, precipitating the nickel quantitively. Iron is not precipitated because it forms a soluble complex with tartrate ion.

Procedure

Weigh accurately three samples of about 1 g each (Note 1) of the steel into 400-ml beakers. Dissolve each sample by warming it with about 60 ml of 1:1 hydrochloric acid. Add cautiously 10 ml of 1 : 1 nitric acid and boil the solution gently to expel oxides of nitrogen. Dilute the solution to 200 ml, heat it nearly to boiling, and add 25 ml of a 20% solution of the tartaric acid (Note 2).

Now neutralize the solution with concentrated ammonia until a definite odor of ammonia persists in the vapors. Then add 1 ml of excess ammonia. If any insoluble material is evident, remove it by filtration (filter paper), and wash it with a hot solution containing a little ammonia and ammonium chloride. Combine the washings with the remainder of the solution.

Make the solution slightly acidic with hydrochloric acid and heat it to about 70°C. Add 20 ml of a 1% solution of dimethylglyoxime in alcohol (Note 3). Make the solution slightly alkaline with ammonia (note odor) and add 1 ml of excess. Allow the precipitate to digest for 30 min at 60°C and then cool the solution to room temperature.

Filter the solution through a weighed fritted-glass or Gooch crucible, and wash the precipitate with cold water until the washings are free of chloride ion. Dry the precipitate by heating the crucible in an oven at 110 to 120°C until a constant weight is obtained. Calculate the percentage of nickel in the steel.

Notes

1. Consult the instructor. The sample should contain about 25 to 35 mg of nickel.
2. The solution is prepared by dissolving 25 g of tartaric acid in 100 ml of water. It should be filtered before use if it is not clear.
3. Dissolve 10 g of dimethylglyoxime in 1 liter of 95% ethanol. For best results, about 10 ml of this solution should be used for each 1% of nickel present. If too much solution is added, dimethylglyoxime itself may precipitate.

OXIDATION-REDUCTION TITRATIONS

The principles of oxidation-reduction titrations are discussed in Chapter 10, and applications to analyses are treated in Chapter 11. In this section directions are given for the preparation and standardization of solutions of potassium permanganate, cerium(IV) sulfate, potassium dichromate, iodine, potassium bromate, and sodium thiosulfate. A number of experiments illustrating the applications of these reagents to analysis are included.

EXPERIMENT 21. PREPARATION OF A 0.1 *N* POTASSIUM PERMANGANATE SOLUTION

The permanganate titrations described here will be carried out in acid solution. Hence the equivalent weight of potassium permanganate is one-fifth the molecular weight, or 31.61. Directions are given for the preparation of 1 liter of a 0.1 *N* solution, thereby requiring 31.61 × 0.1 or about 3.2 g of the salt.

Procedure

Since potassium permanganate solutions are susceptible to decomposition (page 278), special precautions are recommended for preparing the solution if it is to be used over a period of several weeks. If the solution is to be prepared, standardized, and used the same day, the special precautions are not necessary (Note).

Weigh approximately 3.2 g of a good grade of potassium permanganate and place it in a clean 250-ml beaker. Dissolve the salt by adding 50 ml of water and stirring. Decant the solution into a large beaker and add 50 ml of additional water to dissolve any crystals remaining in the first beaker. Repeat this procedure until all the crystals are dissolved. Dilute the solution to about 1 liter, transfer to a glass-stoppered bottle and label properly.

If the instructor recommends removal of manganese dioxide, proceed as follows: Before transferring the solution to the bottle, heat it just to boiling and keep it slightly below the boiling point for 1 hr. Then allow the solution to cool, and filter it through a sintered glass crucible using suction. Transfer the solution to a glass-stoppered bottle and label properly.

Note

Consult the instructor. The instructor may have prepared in advance a large quantity of stock solution which has stood for a week or so. If the manganese dioxide has settled, the clear solution can be withdrawn through an all-glass siphon, avoiding filtration.

EXPERIMENT 22. STANDARDIZATION OF POTASSIUM PERMANGANATE SOLUTION

Procedure

(*a*) *Sodium Oxalate. Fowler-Bright method.* Weigh accurately three samples of about 0.25 to 0.30 g each (Note 1) of the dried salt into clean 500-ml Erlenmeyer flasks. Add 250 ml of dilute sulfuric acid (12.5 ml of concentrated acid diluted to 250 ml) which has previously been boiled 10 to 15 min and then cooled to 24 to 30°C. Swirl the flask until the solid dissolves and then titrate with permanganate. Steadily add the permanganate directly into the oxalate solution (not down the walls of the flask) and stir slowly. Add sufficient permanganate (35 to 40 ml) to come within a few milliliters of the equivalence point, adding this at a rate of about

25 to 35 ml/min. Let the solution stand until the pink color disappears (Note 2) and then heat the solution to 55 to 60°C. Complete the titration at this temperature, adding permanganate slowly until 1 drop imparts to the solution a faint pink color that persists at least 30 sec. The last milliliter should be added slowly, allowing each drop to be decolorized before adding another. The solution should be as warm as 55°C at the end of the titration.

To about 300 ml of previously boiled dilute sulfuric acid, add permanganate solution dropwise until the color matches that of the titrated solution. This volume (usually about 0.03 to 0.05 ml) is subtracted from the volume used in the titration.

After titration of three samples, calculate the normality obtained in each titration and average the results. With care, the average deviation should be as small as 2 parts per thousand (consult instructor).

Notes

1. The equivalent weight of sodium oxalate is 67.00. This weight is sufficient to react with about 35 to 45 ml of a 0.1 *N* solution.

2. This may take 30 to 45 sec, since the reaction is not instantaneous. If the color does not disappear, indicating excess permanganate, discard the solution and add less permanganate to the next sample.

McBride Method. In place of the more lengthy Fowler-Bright procedure, the McBride procedure may be preferred for beginning students.

Weigh accurately three samples of about 0.25 to 0.30 g each of dried sodium oxalate into clean 250-ml Erlenmeyer flasks. Dissolve each sample in about 75 ml of 1.5 *N* sulfuric acid (20 ml of concentrated sulfuric acid added to 400 ml of water). Then heat the first solution almost to boiling (80 to 90°C) and titrate slowly with the permanganate with constant swirling. The end point is marked by the apparance of a faint pink color that persists at least 30 sec. The temperature should not drop below 60°C during the titration. Titrate the other two solutions in the same manner.

To about 100 ml of the 1.5 *N* sulfuric acid, add permanganate solution dropwise until the color matches that of the titrated solution. This volume should be subtracted from the volume used in the titration.

Calculate the normality of the permanganate solution. The average deviation should be as small as about 2 parts per thousand (consult instructor).

(b) Arsenic(III) Oxide. Weigh accurately three portions of about 0.2 g each of pure arsenic(III) oxide, previously dried in the oven, into each of three 250-ml Erlenmeyer flasks. Add to each flask 10 ml of a cool sodium hydroxide solution made by dissolving 20 g of sodium hydroxide in 80 ml of water. Allow the flask to stand for 8 to 10 min, stirring the solution occasionally until the sample has completely dissolved. Then add 100 ml of water, 10 ml of concentrated hydrochloric acid, and 1 drop of 0.0025 *M* potassium iodide solution as a catalyst (Note). Titrate with the permanganate solution until a single drop imparts to the solution a faint pink color which persists for at least 30 sec after the liquid is swirled. The last milliliter should be added slowly, allowing each drop to be decolorized before adding another. Finally, run a blank to determine the volume of permanganate required to color the solution and to react with any reducible material in the reagents. This should amount to no more than 1 drop of permanganate.

Titrate the other two samples and calculate the normality of the permanganate solution. The average deviation should be about 1 to 2 parts per thousand.

Note

Potassium iodate or iodine monochloride can also be used as catalysts if desired. The potassium iodide solution is prepared by dissolving about 0.4 g in 1 liter of water.

(*c*) ***Iron Wire.*** Pure iron wire, free of rust, can be used as a primary standard for permanganate. Weigh accurately three samples of wire of about 0.2 g each into 150-ml beakers. Add 20 ml of 1 : 1 hydrochloric acid to the first sample and warm the solution on a steam bath or over a low flame until all the iron dissolves.

Since the sample is in hydrochloric acid solution, it is convenient to employ tin(II) chloride as the reducing agent and then add preventive solution before titration with permanganate. Directions for this procedure are given under Experiment 25.

Repeat the procedure on the other two samples and calculate the normality of the permanganate solution. The average deviation should be about 2 parts per thousand, or less.

EXPERIMENT 23. DETERMINATION OF AN OXALATE

Procedure

Weigh accurately three portions of the dried material of appropriate size (Note). If the Fowler-Bright procedure is followed, dissolve each sample in 250 ml of dilute sulfuric acid in a 500-ml Erlenmeyer flask. The acid (12.5 ml of concentrated acid diluted to 250 ml with water) should have been previously boiled and then cooled to 24 to 30°C. Titrate as directed on page 586.

If the McBride procedure is used, dissolve each sample in 75 ml of 1.5 *N* sulfuric acid (20 ml of concentrated acid to 400 ml of water) in a 250-ml Erlenmeyer flask. Heat the solution almost to boiling and titrate as directed on page 587.

Calculate and report the percentage of sodium oxalate in the sample in the usual manner.

Note

Consult the instructor. The equivalent weight of sodium oxalate is 67.00. For material containing about 25% sodium oxalate, a 1-g sample is a convenient amount.

EXPERIMENT 24. DETERMINATION OF HYDROGEN PEROXIDE

Procedure

Pipet 25 ml of the peroxide solution (Note 1) into a 250-ml volumetric flask, dilute the solution to the mark, and mix thoroughly. Transfer a 25-ml aliquot of this solution to a 250-ml Erlenmeyer flask to which has been added 5 ml of concentrated sulfuric acid and 75 ml of water. Titrate with standard permanganate solution to the first appearance of a permanent pink tinge.

Repeat the titration on two additional aliquots of the solution. Calculate the weight of hydrogen peroxide in the original 25-ml sample, and assuming that the weight was exactly 25 g, calculate the percentage by weight of hydrogen peroxide (Note 2).

Notes

1. If the density of the solution is about 1, this volume gives $25 \times 0.03 = 0.750$ g, or 750 mg H_2O_2. Since the eq wt of H_2O_2 is 17.01, this is about 44 meq. Thus one-tenth of this solution will use about 44 ml of a 0.1 N permanganate solution for titration.
2. Commercial hydrogen peroxide often contains small amounts of organic compounds such as acetanilide, which are added to stabilize the peroxide. These compounds may react with permanganate to some extent causing incorrect results.

EXPERIMENT 25. DETERMINATION OF IRON IN AN ORE

The determination of iron in iron ores by titration with permanganate has been discussed on page 280. Directions are given here for dissolving the sample in hydrochloric acid, and then for reduction of iron(III) ion, prior to titration, by both tin(II) chloride and amalgamated zinc (Jones reductor). When tin(II) chloride is used, Zimmermann-Reinhardt preventive solution is added to prevent oxidation of chloride ion by permanganate. When the Jones reductor is employed, hydrochloric acid is first removed by evaporation with sulfuric acid before reduction with zinc. No preventive solution is then required.

Procedure

Dissolving Sample. Weigh three samples of iron ore (Notes 1 and 2) of appropriate size (consult instructor) into three 150-ml beakers. Add 10 ml of concentrated hydrochloric acid and 10 ml of water to each beaker. Cover the beakers with watch glasses and keep the solution just below the boiling point on a steam bath, hot plate, or wire gauze until the ore dissolves (hood) (Note 3). This may require 30 to 60 min. At this point the only solid present should be a white residue of silica. If an appreciable amount of colored solid remains in the beaker, the sample must be fused to bring the remainder of the iron into solution (Note 4). Reduce the sample according to (a) or (b) below.

Notes

1. Some commercial samples are made from iron oxide and are easily soluble in acid. Such samples will not require lengthy heating to effect solution and will not leave a residue of silica. The instructor will alter the directions if such samples are used.

2. If the ore is not finely ground, it should be ground in an agate mortar before drying. If it is suspected that the ore contains organic matter, the sample, after weighing, should be ignited in an uncovered porcelain crucible for 5 min. This oxidizes organic matter.

3. If tin(II) chloride is to be used to reduce the iron, add successive portions of tin(II) chloride until the solution changes from yellow to colorless. This will aid in dissolving the sample. Avoid an excess of tin(II) chloride.

4. If fusion is required (consult instructor), dilute the solution with an equal volume of water and then filter. Wash the residue with 1% hydrochloric acid and then wash with water to remove the acid. Transfer the paper to a porcelain crucible and burn off the carbon. If the residue is white it may be disregarded, as the color was probably caused by organic matter. If the color remains, add about 5 g of potassium pyrosulfate and heat carefully until the salt just fuses. Maintain this temperature for 15 to 20 min or until all the iron reacts. Then cool and dissolve the residue in 25 ml of 1 : 1 hydrochloric acid and add this solution to the main filtrate.

(a) Reduction with Tin(II) Chloride. Adjust the volume of the solution to 15 to 20 ml by evaporation or dilution. The solution should be yellow in color because of the presence of iron(III) ion (Note 1). Keep the solution hot and reduce the iron in the first sample (Note 2) by adding tin(II) chloride (Note 3) drop by drop, until the color of the solution changes from yellow to colorless (or very light green). Add 1 or 2 drops excess tin(II) chloride. Cool the solution under the tap and rapidly pour in 20 ml of saturated mercury(II) chloride solution (Note 4). Allow the solution to stand for 3 min and then rinse the solution into a 500-ml Erlenmeyer flask. Dilute to a volume of about 300 ml and add 25 ml of Zimmermann-Reinhardt solution (Note 5). Titrate slowly with permanganate, swirling the flask constantly. The end point is marked by the first appearance of a faint pink tinge which persists when the solution is swirled (Note 6).

Reduce and titrate the second and third samples in the same manner (Note 7). Calculate and report the percentage of iron in the sample in the usual manner.

Notes

1. If tin(II) chloride has been used during dissolving, and the solution is colorless or almost so at this point, add a small crystal of potassium permanganate and heat a little longer until the yellow color is distinct. The reduction can then be followed more readily.

2. Reduce only one sample and finish the titration before reducing the second. Why?

3. This is prepared by dissolving 113 g of $SnCl_2 \cdot 2H_2O$ (free of iron) in 250 ml of concentrated hydrochloric acid, adding a few pieces of mossy tin, and diluting to 1 liter with water.

4. If the mercury(II) chloride were added slowly, part of it would be temporarily in contact with an excess of tin(II) chloride, which might reduce the substance to metallic mercury. Also, if the solution were hot, there would be a danger of forming mercury. The precipitate here should be white and silky and not large in quantity. If the precipitate is gray or black, indicating the presence of mercury, the sample should be discarded. If no

precipitate is obtained, indicating insufficient tin(II) chloride, the sample should be discarded.

5. This is prepared as follows: Dissolve 70 g of $MnSO_4$ in 500 ml of water and add slowly, with stirring, 110 ml of concentrated sulfuric acid and 200 ml of 85% phosphoric acid. Dilute to 1 liter.

6. The color may fade slowly because of oxidation of mercury(I) chloride or chloride ion by permanganate.

7. If desired (consult instructor), a blank can be determined by carrying a mixture of 10 ml of concentrated hydrochloric acid and 10 ml of water through the entire procedure. The blank normally will be about 0.03 to 0.05 ml of 0.1 *N* potassium permanganate.

(b) Jones Reductor. Add 5 ml of concentrated sulfuric acid to each beaker and evaporate the solution carefully until heavy white fumes appear (hood). This expels hydrochloric acid. Cool the solution and dilute to about 100 ml.

Arrange a Jones reductor as shown in Fig. 11.1, page 276. The column is about 2 cm inside diameter and about 45 cm long. In the lower end of the column is a perforated porcelain plate covered with glass wool. The column is filled with amalgamated zinc (Note 1), the zinc column being about 35 to 40 cm long. The tube below the stopcock is connected through a rubber stopper to a 250-ml suction flask. The side arm of the suction flask is connected to a second suction flask (as shown in Fig. 11.1), which acts as a water trap, and suction is applied to the arm of this second flask.

Wash the zinc column first with several portions of water, and then with 200 to 250 ml of 1 *N* sulfuric acid (about 30 ml of concentrated acid to 1 liter of water). The rate of flow of the solution through the column should be about 50 to 75 ml/min, and the liquid level should always be left just above the top of the zinc column (Note 2). Add to the solution collected in the receiving flask (about 250 ml) 0.1 *N* potassium permanganate solution dropwise (Note 3). If no more than 2 drops are required to impart a faint pink tinge to the solution, the reductor is ready for use. Otherwise, continue the washing with 250-ml portions of acid until no more than 2 drops of permanganate is required. Once a satisfactory blank has been obtained, place a clean receiving flask under the column and proceed to reduce the first sample.

Carefully pour the first sample through the reductor, controlling the rate of flow with the stopcock to about 50 ml/min. Stop the flow of the column when the solution stands just above the top of the zinc. Rinse the beaker that contained the sample with five 10-ml portions of 1 *N* sulfuric acid, and pour these washings through the column at the same rate of flow as before. Then wash the column with 100 ml of water (Note 4).

Remove the receiving flask from the column, rinse the end of the delivery tube, and add 5 ml of 85% phosphoric acid to the flask. Titrate immediately with standard permanganate, adding the solution rapidly at first, but dropwise near the end point, so that the end point will not be overstepped.

Reduce and titrate the other two samples in the same manner. Calculate the percentage of iron in the sample, remembering to subtract from the total volume of permanganate the volume used by the blank. The average deviation should be about 1 to 2 parts per thousand.

Notes

1. Prepare the amalgamated zinc as follows: To 300 ml of pure 20- to 30-mesh zinc in a beaker, add about 300 ml of a 2% mercury(II) nitrate (or chloride) solution, acidified with 2 ml of nitric acid. Stir for 5 min and then wash the zinc several times by decantation. The zinc should now have a bright silvery luster from the coating of mercury. Fill the reductor tube with water and then add the zinc slowly to the column.

2. From this point on, do not allow the top of the column to be exposed to the air. Hydrogen peroxide may be formed by the reaction of atmospheric oxygen and hydrogen liberated in the reductor. Permanganate oxidizes hydrogen peroxide, hence this effect will cause high results for iron.

3. The blank is necessary because the zinc may contain materials, particularly iron, that are oxidized by permanganate. New reductors frequently require three or four treatments before giving a satisfactory blank. If additional treatments are indicated, replace the zinc with a better grade.

4. The reductor should be left filled with water after use to prevent formation of basic salts, which tend to clog the column. No acid should be left in the column since it will eventually dissolve the zinc. For storage, the amalgam can be washed, transferred to a stoppered bottle, and covered with water.

EXPERIMENT 26. DETERMINATION OF OXYGEN IN PYROLUSITE

Procedure

Weigh accurately three samples of about 0.5 g of the finely ground and dried ore into 250-ml Erlenmeyer flasks. Calculate approximately the number of milliequivalents of oxygen in the sample (Note 1) and then the weight of sodium oxalate required to react with the sample. Add to each flask about 0.25 g of pure sodium oxalate in excess of the calculated amount, weighing this accurately on the balance. Add to each flask about 100 ml of sulfuric acid (10 ml of concentrated acid to 100 ml of water), cover the flask with a watch glass, and heat each flask on a steam bath or gently over a low flame. Shake the flask occasionally. Keep the temperature just below the boiling point and avoid allowing the solution to evaporate very much. After the samples have completely digested, as indicated by the disappearance of all the black or brown particles (Note 2) and the cessation of evolution of carbon dioxide, rinse the watch glass and rinse down the walls of the flask with distilled water. Titrate the hot solution with standard permanganate to the first appearance of a faint pink tinge.

After digestion and titration of all three samples, calculate the percentage of oxygen in the sample. The average deviation may be as high as 4 parts per thousand (consult instructor). Remember that the milliequivalents of oxygen are given by the relationship

$$\text{meq oxygen} + \text{meq } KMnO_4 = \text{meq } Na_2C_2O_4$$

Notes

1. Consult the instructor for approximate percentage of oxygen in the sample. For example, if the sample is about 10% oxygen, a 500-mg sample would contain 50 mg of

oxygen, or about 6 meq. Six milliequivalents of sodium oxalate (eq wt of 67) weigh about 0.4 g. Hence one should take about 0.65 g of sodium oxalate.

2. There may be a residue of white or light-brownish silica. This should be disregarded.

EXPERIMENT 27. PREPARATION OF A 0.1 *N* CERIUM(IV) SOLUTION

Procedure

(a) Cerium (IV) Sulfate. Weigh on a trip balance about 33 g of cerium(IV) sulfate or 63 g of $Ce(SO_4)_2 \cdot 2(NH_4)_2SO_4 \cdot 2H_2O$ (Note 1). Add this solid with stirring to a sulfuric acid solution made by adding 28 ml of concentrated sulfuric acid to 500 ml of water. Stir until the solid dissolves and dilute the solution to 1 liter in a clean bottle (Note 2). Label the bottle appropriately.

Notes

1. If cerium(IV) oxide is used, weigh 21 g of the solid into a 1500-ml beaker. Add to the solid with stirring a hot sulfuric acid solution made by adding 78 ml of concentrated sulfuric acid to 300 ml of water. After the oxide has dissolved, dilute the solution to 1 liter and transfer to a clean bottle.

2. If the solution is at all turbid, it should be filtered before use. The precipitation of cerium(IV) phosphate is quite slow in acid solution, and if the solution is to be used soon after standardization, the filtration may be omitted. However, if the solution is to be kept for some time, it is preferable to allow it to stand for 1 or 2 weeks and then filter before standardization.

(b) Cerium (IV) ammonium nitrate. Weigh accurately 54.83 g of primary standard grade cerium(IV) ammonium nitrate into a 1-liter beaker (Note 1). Then pour 56 ml of concentrated sulfuric acid over the salt and stir for about 2 min. Add carefully 100 ml of water and stir well (Note 2). Continue the addition of water slowly with stirring until the volume is about 600 ml. Then transfer the cool solution to a 1-liter volumetric flask, dilute to the mark, and mix the solution thoroughly.

Notes

1. The equivalent weight of the salt is 584.23. It is sufficiently accurate to weigh to only the second decimal place. Why? If the ordinary grade of salt is used, weigh about 56 g on a trip scale and dissolve as indicated. Dilute to approximately 1 liter and then standardize the solution against a primary standard.

2. Ordinarily, of course, it is not advisable to add water to concentrated sulfuric acid. In the present case, however, this is recommended (see H. Diehl and G. F. Smith, *Quantitative Analysis*, John Wiley & Sons, Inc., New York, 1952, p. 274). Perform the operation very cautiously to prevent spattering of the acid solution.

EXPERIMENT 28. STANDARDIZATION OF THE CERIUM(IV) SOLUTION

Procedure

(*a*) *Arsenic(III) Oxide.* Weigh accurately three portions of about 0.2 g each of pure arsenic(III) oxide, previously dried in an oven, into each of three 500-ml Erlenmeyer flasks. Add to each flask 10 ml of a cool solution of sodium hydroxide made by dissolving 20 g of sodium hydroxide in 80 ml of water. Allow the flask to stand 8 to 10 min, stirring occasionally until the sample has completely dissolved. Then add 100 ml of water and make the solution acidic by adding 25 ml of 1 : 10 sulfuric acid. Add 2 drops of 0.01 *M* osmium tetroxide solution (Note 1) as a catalyst and 2 drops of ferroin [iron(II) 1,10-phenanthroline sulfate] as the indicator (Note 2).

Titrate the pale pink solution with cerium(IV) solution. The pink color becomes more pronounced as the titration proceeds. The end point is approached without warning, at which point the solution changes from pink to a very faint blue (or colorless). The color change is very sharp.

Titrate the other samples in the same manner and calculate the normality of the cerium(IV) solution.

Notes

1. This solution is made by dissolving 0.125 g of OsO_4 in 50 ml of 0.1 *N* sulfuric acid.
2. This solution is prepared by dissolving 1.5 g of 1,10-phenanthroline monohydrate, $C_{12}H_8N_2 \cdot H_2O$, in 100 ml of 0.025 *M* iron(II) sulfate (freshly prepared). The indicator can be purchased from the G. F. Smith Chemical Co., Columbus, Ohio.

(*b*) *Iron Wire.* Weigh accurately three samples of pure iron wire, free of rust, of about 0.2 g each. Place each sample in a 150-ml beaker and add 20 ml of 1 : 1 hydrochloric acid to each beaker. Warm the solution on a steam bath or over a low flame until all the iron dissolves.

Adjust the volume of the solution to 15 to 20 ml by evaporation or dilution. The solution may not be intensely yellow if very little of the iron has been oxidized to the +3 state during the dissolving process. Add tin(II) chloride (Note 1) drop by drop to the first solution (hot) until the solution is colorless or a very light green, and then add 1 or 2 drops excess. The first drop or so may be sufficient to decolorize the solution. Now cool the solution under the tap and rapidly pour in 20 ml of saturated mercury(II) chloride solution (Note 2). Allow the solution to stand for 3 min and then rinse it into a 500-ml Erlenmeyer flask. Add 10 ml of concentrated hydrochloric acid, dilute the solution to about 300 ml, and add 2 drops of ferroin indicator (Note 3). Titrate with the cerium(IV) solution until the color changes from pink (or reddish orange) to a pale yellow.

Reduce and titrate the other two samples in the same manner. Calculate the normality of the cerium(IV) solution.

Notes

1. See Note 3, page 590.
2. See Note 4, page 590.
3. See Note 2 of part (a) above.

EXPERIMENT 29. DETERMINATION OF IRON IN AN ORE

Procedure

Weigh and dissolve the samples as directed in Experiment 25. Then reduce the first sample with tin(II) chloride or the Jones reductor as directed by the instructor. Directions are given in Experiment 25.

(a) Tin(II) Chloride. If tin(II) chloride is used, cool the solution under the tap after reduction is complete and rapidly add 20 ml of saturated mercury(II) chloride solution. Allow the solution to stand for 3 min and then rinse into a 500-ml Erlenmeyer flask. Dilute to a volume of about 300 ml with 1 *M* HCl and add 2 drops of ferroin indicator. Titrate with the standard cerium(IV) solution until the color changes from pink (or reddish orange) to a pale yellow.

(b) Jones Reductor. If the Jones reductor is used, add 5 ml of phosphoric acid, 2 drops of ferroin indicator, and dilute to about 300 ml with 1 *M* sulfuric acid. Titrate with the standard cerium(IV) solution until the color changes from pink (or reddish orange) to a pale yellow.

Reduce and titrate the other two samples in the same manner. Calculate and report the percentage of iron in the sample.

EXPERIMENT 30. PREPARATION AND STANDARDIZATION OF A 0.1 *N* POTASSIUM DICHROMATE SOLUTION

Procedure

Weigh accurately a sample of about 4.9 g of pure potassium dichromate (Note 1) that has been previously dried in the oven. Dissolve the sample in water in a 400-ml beaker and transfer the solution quantitatively to a 1-liter volumetric flask. Dilute to the mark and mix the solution thoroughly. Calculate the normality of the solution.

If the solution is to be standardized against iron (consult instructor), proceed as follows.

Weigh and dissolve samples of iron wire as directed under Experiment 21. Reduce the iron(III) iron with tin(II) chloride as directed on page 590, and rinse the solution into a 500-ml Erlenmeyer flask. Now add 250 ml of water, 5 ml of concentrated sulfuric acid, 5 ml of 85% phosphoric acid, and 8 drops of sodium diphenylaminesulfonate indicator (Note 2). Titrate slowly with the dichromate solution, swirling the flask constantly. The solution becomes green since the reaction produces chromium(III) ions. The oxidized form of the indicator is purple or violet. As the end point is approached, indicated by a transient purple color, add the dichromate dropwise until the color changes permanently to purple.

Calculate the normality of the dichromate solution. Compare this with the normality calculated from the weight of dichromate dissolved in the solution.

Notes

1. The equivalent weight is 294.18/6 or 49.03.

2. The indicator solution is made by dissolving 0.3 g of the sodium salt in 100 ml of water. If the barium salt is used, dissolve 0.3 g in 100 ml of hot water, add an excess of 0.1 *M* sodium sulfate solution, and let the solution stand until barium sulfate settles. Either filter or decant the solution from the precipitate.

EXPERIMENT 31. DETERMINATION OF IRON IN AN ORE, USING DICHROMATE

Procedure

Weigh and dissolve the first sample and reduce it with tin(II) chloride as directed under Experiment 25. Then cool the solution and rapidly add 20 ml of saturated mercury(II) chloride solution. Allow the solution to stand for 3 min and then rinse into a 500-ml Erlenmeyer flask. Now add 250 ml of water, 5 ml of concentrated sulfuric acid, 5 ml of 85% phosphoric acid, and 8 drops of sodium diphenylaminesulfonate indicator (see Note 2 of Experiment 30). Titrate with dichromate solution as directed in Experiment 30.

Reduce and titrate the other two samples in the same manner. Calculate and report the percentage of iron in the ore.

EXPERIMENT 32. PREPARATION AND STANDARDIZATION OF A 0.1 *N* IODINE SOLUTION

Procedure

Weigh on a trip balance about 12.7 g of reagent-grade iodine (Notes 1 and 2) and place this in a 250-ml beaker. Add to the beaker 40 g of potassium iodide, free of iodate (Note 3), and 25 ml of water. Stir to dissolve all the iodine and transfer the solution to a glass-stoppered bottle. Dilute to about 1 liter (Note 4).

Standardize the iodine solution as follows: weigh accurately a sample of reagent-grade arsenic(III) oxide (Note 5) of about 1.25 g into a 250-ml beaker.

Add a solution made by dissolving 3 g of sodium hydroxide in 10 ml of water. Allow the beaker to stand, swirling it occasionally, until the solid has completely dissolved. Then add 50 ml of water, 2 drops of phenolphthalein indicator, and 1:1 hydrochloric acid until the pink color of the indicator just disappears. Then add 1 ml of hydrochloric acid in excess. Transfer the solution to a 250-ml volumetric flask, dilute to the mark, and mix thoroughly.

Transfer with a pipet 25 ml of the arsenite solution to a 250-ml Erlenmeyer flask and dilute with 50 ml of water. Then carefully add small portions of sodium bicarbonate to neutralize the acid. When vigorous effervescence ceases, add 3 g of additional sodium bicarbonate to buffer the solution. Add 5 ml of starch indicator (Note 6) and titrate with iodine until the first appearance of the deep-blue color, which persists for at least 1 min. If the end point is overrun, some of the standard arsenic(III) solution can be used for back-titration.

Titrate two other aliquots in the same manner. Calculate the normality of the arsenic(III) solution from the weight of arsenic(III) oxide taken. Then calculate the normality of the iodine solution from the volumes of the two solutions used.

Notes

1. If the solution is to be standardized and only one unknown analyzed, 500 ml of solution will suffice. The quantities of iodine and potassium iodide can be halved (consult instructor).

2. If desired, the iodine can be weighed accurately and the solution made up to a definite volume. To prepare 500 ml of solution, first place about 20 g of potassium iodide in a large weighing bottle and dissolve this in 10 ml of water. After the solution has come to room temperature, weigh the bottle accurately (to the nearest milligram is sufficient). Take the bottle to a trip balance and there add about 6.4 g of iodine and stopper the bottle tightly. Do not open the bottle near the analytical balance as iodine fumes are corrosive. Reweigh the bottle and record the weight of iodine taken. When the iodine has dissolved, transfer the solution to a 500-ml volumetric flask, dilute to the mark, and mix thoroughly. Store the solution in a glass-stoppered bottle.

3. Test the salt for iodate as follows: Dissolve about 1 g in 20 ml of water and add 1 ml of 6 *N* sulfuric acid and 2 ml of starch solution. No blue color should develop in 30 sec.

4. A bottle of brown glass is preferable for storing this solution. In any event, the solution should be kept from the light as much as possible.

5. Pure arsenic(III) oxide is not hygroscopic and may not need drying unless it has been exposed to air for some time. If drying is necessary, it is usually preferable to place the material in a desiccator over sulfuric acid for about 12 hr. The oxide (octahedral variety) tends to sublime when heated as high as 125 to 150°C.

6. This solution is prepared as follows: Make a paste of 2 g of soluble starch and 25 ml of water and pour this gradually (with stirring) into 500 ml of boiling water. Continue the boiling for 1 or 2 min, add 1 g of boric acid as a preservative, and allow the solution to cool. Store the solution in a glass-stoppered bottle. Alternatively, a solid complex of starch and urea can be prepared by melting urea in a small beaker and adding soluble starch. The ratio of weights of urea to starch should be 4:1. The solution is mixed, allowed to cool and solidify, and then the solid is ground in a mortar. A small scoopful is used in place of starch solution. The complex dissolves readily in water and is quite stable.

EXPERIMENT 33. DETERMINATION OF THE PURITY OF ARSENIC(III) OXIDE

Procedure

Weigh three samples of the unknown of appropriate size (Note) into 500-ml Erlenmeyer flasks. Add to each flask 50 ml of water and 1 g of sodium hydroxide and warm until the sample dissolves. Cool the solution, add 2 drops of phenolphthalein indicator, and add 1 : 1 hydrochloric acid until the pink color of the indicator just disappears. Then add small portions of sodium bicarbonate to neutralize the acid. When vigorous effervescence ceases, add 3 g of additional bicarbonate to buffer the solution. Dilute to about 150 ml with water and add 5 ml of starch solution (Note 6, Experiment 32). Titrate the first sample with iodine solution to the first appearance of the deep-blue color, which lasts for at least 1 min.

Titrate the other two samples in the same manner. Calculate and report the percentage of arsenic(III) oxide in the sample.

Note

Consult instructor as to size of sample and method of drying. Commercial unknown samples are usually about 3 to 15% arsenic(III) oxide.

EXPERIMENT 34. PREPARATION AND STANDARDIZATION OF A 0.1 *N* SODIUM THIOSULFATE SOLUTION

Procedure

Dissolve about 25 g of sodium thiosulfate pentahydrate crystals in 1 liter of water that has been recently boiled and cooled. Add about 0.2 g of sodium carbonate as a preservative and store in a clean bottle.

Standardization

(a) Potassium Dichromate. Weigh three portions of pure, dry potassium dichromate (Note 1) of about 0.2 g each into 500-ml Erlenmeyer flasks. Dissolve each sample in about 100 ml of water and add 4 ml of concentrated sulfuric acid. To the first sample carefully add 2 g of sodium carbonate (Note 2) with gentle swirling to liberate carbon dioxide. Then add 5 g of potassium iodide dissolved in about 5 ml of water (Note 3), swirl, cover the flask with a watch glass, and allow the solution to stand for 3 min (Note 4). Dilute the solution to about 200 ml and titrate with thiosulfate solution until the yellowish color of iodine has nearly disappeared (Note 5). Then add 5 ml of starch solution and continue the titration until 1 drop of titrant removes the blue color of the starch–iodine complex. The final solution will be clear emerald green, the color imparted by chromium(III) ion.

Treat the other samples in the same manner and calculate the normality of the thiosulfate solution. The average deviation should be about 1 to 3 parts per thousand.

Notes

1. The material can be obtained from the Bureau of Standards, but the best grade available from commercial supply houses is sufficiently pure for most purposes. Dry at 150°C if necessary.

2. The purpose of this is to remove air from the flask and lessen the danger of air oxidation of iodide. Do not add too much carbonate, as this will use up too much acid. The final concentration of acid is about 0.4 *M* if these quantities are used.

3. Do not allow this solution to stand as the iodide may be oxidized by air. The iodide should be free of iodate (see Note 3, Experiment 32) or a blank must be determined.

4. The reaction is somewhat slow but should be complete within this time.

5. Starch should not be added to a solution that contains a large quantity of iodine. The starch may be coagulated and the complex with iodine may not easily break up. A recurring end point will then be obtained.

(b) Copper. Secure some clean copper wire or foil, weigh three pieces of about 0.20 to 0.25 g each (Note 1), and place then in 250-ml Erlenmeyer flasks. Add to each flask 5 ml of 1 : 1 nitric acid and dissolve the copper by warming the solution on a steam bath or over a low flame. Add 25 ml of water and boil the solution for about 1 min. Then add 5 ml of urea solution (1 g in 20 ml of water) and continue boiling for another minute (Note 2). Cool the solution under the tap and neutralize the acid with 1 : 3 ammonia, adding the ammonia carefully until a pale blue precipitate of copper(II) hydroxide is obtained (Note 3). Now add 5 ml of glacial acetic acid and cool the solution if it is warm. To the first sample, add 3 g of potassium iodide, cover the flask with a watch glass, and allow to stand for 2 min. Then titrate with thiosulfate solution until the brownish color of iodine is almost gone (Note 4). Add 5 ml of starch solution and 2 g of potassium thiocyanate (Note 5). Swirl the flask for about 15 sec and complete the titration, adding thiosulfate dropwise. At the end point the blue color of the solution disappears and the precipitate appears white, or slightly gray, when allowed to settle (Note 6).

Treat the second and third samples in the same manner and titrate with thiosulfate. Calculate the normality and copper titer of the thiosulfate.

Notes

1. The equivalent weight of copper is the atomic weight, 63.546.

2. Boiling removes oxides of nitrogen which result from the following reactions:

$$3Cu + 8HNO_3 \longrightarrow 3Cu(NO_3)_2 + 2NO + 4H_2O$$

$$2NO + O_2 \longrightarrow 2NO_2$$

Nitrous acid is formed by the reaction

$$NO_2 + NO + H_2O \longrightarrow 2HNO_2$$

and is decomposed by urea, according to the equation

$$2HNO_2 + (NH_2)_2CO \longrightarrow CO_2 + 2N_2 + 3H_2O$$

3. If excess ammonia is added, copper(II) hydroxide redissolves, forming the deep blue copper(II) ammonia complex, $Cu(NH_3)_4^{2+}$. If the excess of ammonia is large, a large quantity of ammonium acetate will be produced later; this may keep the reaction between copper(II) and iodide ions from being complete. The excess ammonia can be removed by boiling, and the precipitate will re-form.

4. After the first titration, calculate the approximate volume required for the other samples. Then titrate to within 0.5 ml of this volume before adding starch and thiocyanate.

5. Thiocyanate displaces adsorbed iodine from the precipitate.

6. The precipitate is seldom completely white. Do not continue addition of thiosulfate until the precipitate is white, since this will require considerable excess titrant.

EXPERIMENT 35. DETERMINATION OF COPPER IN AN ORE

Procedure

Accurately weigh three samples of appropriate size (consult instructor) of the finely ground, dried ore into 250-ml beakers. Add 5 ml of concentrated hydrochloric acid and (slowly) 10 ml of concentrated nitric acid (Note 1). Cover the beaker with a watch glass and heat over a low flame until only a white residue of silica remains. Remove the watch glass and add 10 ml of 1 : 1 sulfuric acid. Then evaporate the solution (hood) until white fumes of sulfur trioxide appear (Note 2). Cool the solution and add carefully 25 ml of water.

Next add 5 ml of saturated bromine water and boil the solution gently for several minutes to expel excess bromine (Note 3). Then add 1:3 ammonium hydroxide carefully until a slight precipitate of iron(II) hydroxide is formed, or, if no iron is present, the deep-blue color of the copper-ammonia complex is just formed. Now add 5 ml of glacial acetic acid, 2 g of ammonium acid fluoride (Notes 4 and 5), and stir until the iron(III) hydroxide redissolves.

Dissolve about 3 g of potassium iodide in 10 ml of water and add to the copper solution. Titrate the liberated iodine at once with thiosulfate until the brownish color of the iodine is almost gone (Note 6). Then add 5 ml of starch solution and 2 g of potassium thiocyanate (Note 7). Swirl the flask gently for about 15 sec and complete the titration, adding thiosulfate dropwise. At the end point the blue color of the solution disappears, and the precipitate appears white, or slightly gray, when allowed to settle (Note 8).

Treat the other two samples in the same manner and titrate with thiosulfate. Calculate the percentage of copper in the ore.

Notes

1. Some commercial unknowns are prepared from copper oxide and are easily dissolved by warming for 10 to 15 min with 15 ml of 1 : 2 sulfuric acid. No evaporation is then required. Continue the procedure with the addition of bromine water.

2. This removes nitric acid, which would react with iodide ion.

3. Bromine is added to ensure that arsenic and antimony are in the +5 oxidation state. These elements may be reduced back to the +3 state by sulfur in a sulfide ore. Unless the

excess bromine is removed, it will, of course, oxidize iodide ion. A test to see if bromine has been completely expelled is to hold in the vapors a piece of filter paper moistened with a starch solution which contains some potassium iodide. If the paper darkens, bromine is still present. If arsenic and antimony are absent, the bromine treatment can be omitted. Consult the instructor.

4. The container will be etched by hydrofluoric acid. It is preferable to transfer the solution at this point to a flask or beaker which is already etched or chipped. This container can be used for each sample after addition of fluoride and then discarded.

5. Samples of copper oxide (Note 1) do not normally contain iron, and it may not be necessary to add fluoride. Add glacial acetic acid and continue the procedure. Consult the instructor.

6, 7, 8. See Notes 4, 5, and 6 of Experiment 34(b).

EXPERIMENT 36. IODOMETRIC DETERMINATION OF HYDROGEN PEROXIDE

Procedure

Pipet 25 ml of the peroxide solution into a 250-ml volumetric flask, dilute to the mark, and mix thoroughly. Transfer a 25-ml aliquot of this solution to a 250-ml Erlenmeyer flask and add 8 ml of 1:6 sulfuric acid (about 6 *N*), 3 g of potassium iodide in 10 ml of water, and 3 drops of 3% ammonium molybdate solution. Titrate with thiosulfate until the brown color of iodine has almost disappeared. Then add 5 ml of starch solution and finish the titration to the disappearance of the deep blue color.

Titrate two other portions of the peroxide solution in the same manner. Calculate the weight of hydrogen peroxide in the original sample. Assuming that the weight of the sample was exactly 25 g, calculate the percentage of hydrogen peroxide by weight.

EXPERIMENT 37. DETERMINATION OF BLEACHING POWER BY IODOMETRY

Commercial bleaching products contain oxidizing agents such as hypochlorites or peroxides. The oxidizing power can be determined by the same procedure used in Experiment 36.

Procedure

Place an accurately measured sample of the bleach in a 250-ml Erlenmeyer flask. Liquid bleaches, such as Chlorox and Purex, contain sodium hypochlorite, and 2.00 ml is a convenient sample if 0.1 *N* thiosulfate is used as the titrant. Solid products, such as Chlorox II and Snowy Bleach, contain peroxides. A sample of 0.7 to 0.8 g is usually a convenient size for titration (Note 1). Add to the Erlenmeyer flask 75 ml of distilled water, 3 g of KI, 8 ml of 1:6 sulfuric acid, and 3 drops of 3% ammonium molybdate solution (Note 2). Titrate the liberated iodine with 0.1 N thiosulfate until the brown color of iodine has almost disappeared. Then add 5 ml of starch solution and finish the titration to the disappearance of the deep blue color.

Titrate at least two portions of each bleaching product being compared. The oxidizing ability of a bleach is usually reported as percent chlorine. That is, the calculation assumes that chlorine is the oxidizing agent, although in fact it may not be. Report the percent by weight of chlorine in each product, assuming that the liquid bleaches have a density of 1.000 g/ml. The equivalent weight of chlorine is the atomic weight.

Notes

1. The solids may not be homogeneous. If duplicate samples do give widely different results, a large sample can be taken, dissolved in a 500-ml volumetric flask, and 50-ml aliquots titrated.

2. Some of the bleaches react slowly with iodide ion. Molybdate ions catalyze the reaction. Alternatively, the solutions can be heated to about 60°C to speed up the reaction.

EXPERIMENT 38. DETERMINATION OF GLYCEROL BY OXIDATION WITH PERIODATE[3]

The periodate oxidation of alcohols containing hydroxyl groups on adjacent carbon atoms is discussed in Chapter 11. Excess periodate is added to the polyhydric alcohol and the oxidation takes place at room temperature. The excess periodate can be determined by reduction to iodate with an excess of standard arsenic(III) solution. The excess arsenic is titrated with a standard iodine solution. Potassium iodide can also be used to reduce the excess periodate. Alternatively, the formic acid produced in the reaction can be titrated with standard base.

Directions are given below for the determination of glycerol by oxidation with periodate. In this reaction 1 mol of glycerol produces 1 mol of formic acid plus 2 mol of formaldehyde (page 293). The excess periodate is destroyed by allowing it to oxidize ethylene glycol to formaldehyde (page 293), and the formic acid is titrated with standard base using bromthymol blue indicator.

Procedure

Prepare a dilute solution of sulfuric acid by adding 30 ml of 0.2 N H_2SO_4 to about 300 ml of distilled water in a 600-ml beaker. Then dilute the solution to about 500 ml. Now dissolve 15 g of sodium metaperiodate, $NaIO_4$, in 250 ml of the dilute sulfuric acid solution. It should not be necessary to use heat to make the solid dissolve.

Prepare a 0.1% solution of bromthymol blue by dissolving 0.1 g of the indicator in 8 ml of 0.02 M NaOH and diluting to 100 ml with distilled water.

[3] V. C. Mehlenbacher in *Organic Analysis*, Vol. 1, Interscience Publishers, Inc., New York, 1953.

Weigh the sample containing glycerol (Note 1) and place it in a 500-ml Erlenmeyer flask. Add about 50 ml of distilled water and 6 drops of bromthymol blue to the flask. Prepare a blank in the same manner. Now carefully add 0.2 *N* H_2SO_4 to the sample and blank until the color of each solution is definitely green or greenish yellow. Then neutralize each solution with 0.05 *N* NaOH to a definite blue color free of green. Add with a pipet 50 ml of the periodate solution to each flask, cover with a watch glass, and allow the solutions to stand at room temperature for 1 hr. Prepare a solution of ethylene glycol by mixing 30 ml of the glycol with 30 ml of water. Add to each the sample and the blank 10 ml of this solution and allow the solutions to stand for 20 min. Dilute each solution to about 250 ml with distilled water and titrate with standard 0.1 *N* NaOH (Note 2). Repeat the determination on two additional samples (consult instructor).

Subtract the volume of base used by the blank from that used by the sample. Then calculate the percentage of glycerol in the sample, noting that 1 mol of glycerol produces 1 mol of formic acid.

Notes

1. If the sample contains from 60 to 70% glycerol, the amount weighed should be from 0.55 to 0.75 g; if 50 to 60%, from 0.65 to 0.85 g; if 40 to 50%, from 0.80 to 1.00 g; if 30 to 40%, from 0.90 to 1.30 g. See the reference in footnote 3 for a complete table of recommended sample sizes.

2. If a *p*H meter is available, it is recommended[4] that the sample be titrated to a *p*H of 8.1 and the blank to a *p*H of 6.5.

OPTICAL METHODS OF ANALYSIS

The principles of spectrophotometry are discussed in Chapter 14. The experiments given in this section illustrate some of these principles as they are applied to chemical analysis. All the experiments can be carried out with a simple spectrophotometer which covers the visible region of the spectrum.

EXPERIMENT 39. DETERMINATION OF MANGANESE IN STEEL

Manganese in steel can be determined spectrophotometrically after oxidation to the purple permanganate ion.[5] The steel is dissolved in nitric acid and the oxidation is effected with potassium periodate:

$$2Mn^{2+} + 5IO_4^- + 3H_2O \longrightarrow 2MnO_4^- + 5IO_3^- + 6H^+$$

Since the yellow iron(III) ion is present, phosphoric acid is added to form the colorless iron–phosphate complex.

[4] *Ibid.*

[5] H. H. Willard and L. H. Greathouse, *J. Am. Chem. Soc.*, **39**, 2366 (1917).

If the steel contains chromium or nickel, the color of these ions interferes in the manganese determination. This interference can be canceled by the addition of about the same amounts of these elements to the standards as are in the unknown. Alternatively, a sample can be carried through the entire procedure except for the periodate oxidation and then used as a blank in setting the spectrophotometer to read zero absorbance. In the directions below, a steel of known manganese content which contains about the same quantities of nickel and cobalt as the unknown is used as the standard.

The experiment first calls for determining a spectral-transmittance curve to find the wavelength at which to perform the analysis. A check of Beer's law is then made by measuring the absorbances of permanganate solutions at several concentrations. The unknown is determined by comparison with the Beer's law plot.

Procedure

(*a*) *Spectral-Transmittance Curve.* Weigh accurately a sample of steel of known manganese content (Note 1). Dissolve the sample in 50 ml of 1 : 3 nitric acid, using heat and finally boiling gently for 1 to 2 min to remove oxides of nitrogen. Then remove the flask from the burner, add about 1 g of ammonium persulfate, and boil the solution gently for 10 to 15 min (Note 2). If a precipitate of an oxide of manganese forms or if the permanganate color develops, add a few drops of sodium sulfite or sulfurous acid (Note 3) and boil the solution a few minutes more to expel sulfur dioxide.

Dilute the solution with water to about 100 ml and add about 10 ml of 85% phosphoric acid and 0.5 g of potassium periodate. Boil the solution gently for about 3 min to effect oxidation to permanganate. Then cool the solution and dilute to 250 ml in a volumetric flask.

Measure the absorbance of the prepared solution or a suitable dilution thereof (Note 4) using the directions given for the spectrophotometer employed. Cover the range from about 440 to 700 nm,[6] taking readings at 20-nm intervals. In the region of maximum absorbance, take readings at intervals of 5 nm. Plot the absorbance vs. wavelength and connect the points to form a smooth curve. Select the proper wavelength to use for the determination of manganese (Note 5).

(*b*) *Beer's Law Check.* Secure four clean, dry 100-ml beakers or Erlenmeyer flasks and number them 1 to 4. In beaker 1 place some of the solution prepared in part (a) and do not dilute it. In beaker 2 place 30 ml of the standard solution (use a 10-ml pipet) and then add 10 ml of water. In beaker 3 place 10 ml of the standard solution and 10 ml of water, and in beaker 4 place 10 ml of the standard and 30 ml of water. Mix the solutions which were diluted and then measure the absorbance of each against distilled water at the wavelength of maximum absorbance. Plot the absorbance vs. concentration for the four solutions. Draw the best straight line between the points. Is Beer's law obeyed?

[6] The abbreviation nm is for a *nanometer*, 10^{-9} m or 10^{-7} cm. It is synonymous with millimicron, mμ, a term still widely used by chemists.

(c) *Analysis of Sample.* Weigh accurately a sample of steel whose manganese content is to be determined, dissolve it, and oxidize the manganese to permanganate as directed in part (a). The solution is finally diluted to 250 ml in a volumetric flask. Measure the absorbance of some of the solution against distilled water at the same wavelength used in part (b) to check Beer's law. Using the Beer's law plot, read off the concentration of permanganate and calculate the percentage of manganese in the sample.

Notes

1. Consult the instructor for the rough manganese content of the sample. Calculate the size sample needed to give a final solution whose absorbance falls in the range 0.7 and 0.8 if the molar absorptivity of permanganate is 2360 liters/mol-cm at 525 nm.
2. The ammonium persulfate is added to oxidize carbon or carbon compounds. The excess persulfate is destroyed by boiling the solution.
3. This reduces manganese to the bivalent state. The solution should be clear.
4. Any spectrophotometer or filter photometer can be used. Follow precisely the operating directions given by the instructor or in the operation manual.
5. This wavelength is 525 nm, but the wavelength calibration of the spectrophotometer is often not reliable and the value found with the instrument should be used.

EXPERIMENT 40. DETERMINATION OF IRON WITH 1,10-PHENANTHROLINE

The reaction between iron(II) ion and 1,10-phenanthroline to form a red complex serves as a good sensitive method for determining iron. The molar absorptivity of the complex, $[(C_{12}H_8N_2)_3Fe]^{2+}$, is 11,100 at 508 nm. The intensity of the color is independent of *p*H in the range 2 to 9. The complex is very stable and the color intensity does not change appreciably over very long periods of time. Beer's law is obeyed.

The iron must be in the +2 oxidation state, and hence a reducing agent is added before the color is developed. Hydroxylamine, as its hydrochloride, can be used, the reaction being

$$2Fe^{3+} + 2NH_2OH + 2OH^- \longrightarrow 2Fe^{2+} + N_2 + 4H_2O$$

The *p*H is adjusted to a value between 6 and 9 by addition of ammonia or sodium acetate. An excellent discussion of interferences and of applications of this method is given by Sandell.[7]

Procedure[8]

Prepare the Following Solutions:

(a) Dissolve 0.1 g of 1,10-phenanthroline monohydrate in 100 ml of distilled water, warming to effect solution if necessary.

[7] E. B. Sandell, *Colorimetric Determination of Traces of Metals*, 3rd ed., Interscience Publishers, Inc., New York, 1959.

[8] H. Diehl and G. F. Smith, *Quantitative Analysis*, John Wiley & Sons, Inc., New York, 1952.

(b) Dissolve 10 g of hydroxylamine hydrochloride in 100 ml of distilled water.
(c) Dissolve 10 g of sodium acetate in 100 ml of distilled water.
(d) Weigh accurately about 0.07 g of pure iron(II) ammonium sulfate, dissolve in water, and transfer the solution to a 1-liter volumetric flask. Add 2.5 ml of concentrated sulfuric acid and dilute the solution to the mark. Calculate the concentration of the solution in mg of iron per ml.

Into five 100-ml volumetric flasks, pipet 1-, 5-, 10-, 25-, and 50-ml portions of the standard iron solution. Put 50 ml of distilled water in another flask to serve as the blank and a measured volume of unknown in another (Note). To each flask add 1 ml of the hydroxylamine solution, 10 ml of the 1,10-phenanthroline solution, and 8 ml of the sodium acetate solution. Then dilute all the solutions to the 100-ml marks and allow them to stand for 10 min.

Using the blank as the reference and any one of the iron solutions prepared above, measure the absorbance at different wavelengths in the interval 400 to 600 nm. Take readings about 20 nm apart except in the region of maximum absorbance where intervals of 5 nm are used. Plot the absorbance vs. wavelength and connect the points to form a smooth curve. Select the proper wavelength to use for the determination of iron with 1,10-phenanthroline.

Using the selected wavelength, measure the absorbance of each of the standard solutions and the unknown. Plot the absorbance vs. the concentration of the standards. Note whether Beer's law is obeyed. From the absorbance of the unknown solution, calculate the concentration of iron (mg/liter) in the original solution.

Note

Prepared solutions may be used as unknowns. Consult the instructor concerning size of sample to be used. If a natural water is used be sure that it is colorless and free of turbidity.

EXPERIMENT 41. DETERMINATION OF NITRITE IN WATER[9]

The determination of nitrite ion in water is important in assessing the degree of pollution. The efficiency of a water purification process can be judged by the amount of nitrite ion in the water.

Nitrite ion can be determined in water by utilizing the reaction of this ion with amines (diazotization). The compound 4-aminobenzenesulfonic acid is diazotized according to the reaction

$$HSO_3-C_6H_4-NH_2 + NO_2^- + 2H^+ \rightarrow HSO_3-C_6H_4-\overset{+}{N}{\equiv}N + 2H_2O$$

[9] M. G. Mellon, *Quantitative Analysis,* Thomas Y. Crowell Co., New York, 1955, p. 512. See also *Standard Methods for the Examination of Water and Sewage,* 10th ed., American Public Health Association, New York, 1955, p. 153.

The diazonium salt is then coupled with 1-naphthylamine to form the colored product:

$$HSO_3-C_6H_4-\overset{+}{N}\equiv N + C_{10}H_7-NH_2 \longrightarrow HSO_3-C_6H_4-N\equiv N-C_{10}H_6-NH_2 + H^+$$

The solution is made slightly basic with sodium acetate in order to make this reaction complete.

Procedure

Prepare the Following Solutions:

(a) Dissolve about 0.8 g of sulfanilic acid (4-aminobenzenesulfonic acid) in 28 ml of glacial acetic acid and dilute the solution to about 100 ml with water.

(b) Dissolve about 0.5 g of 1-naphthylamine in 28 ml of glacial acetic acid and dilute the solution to about 100 ml with water.

(c) Dissolve about 14 g of sodium acetate trihydrate in water and dilute to about 50 ml.

(d) Weigh accurately 0.494 g of reagent-grade sodium nitrite, dissolve the salt in water, and dilute the solution to 1 liter in a volumetric flask. Pipet 10 ml of this solution into another 1-liter volumetric flask and dilute the solution to the mark. This solution now contains 0.0010 mg of nitrogen/ml.

Into seven 100-ml volumetric flasks, pipet 1-, 2-, 3-, 4-, 5-, 7-, and 10-ml portions of the standard nitrite solution. Secure two additional flasks, one for the blank and one for the unknown (Note 1). Pipet an aliquot (Note 2) of the unknown into one of these flasks. Then adjust the volume in each flask to about 50 ml with distilled water. Add to each flask 1 ml of the sulfanilic acid solution and allow the solutions to stand for 5 min. Then add to each flask 1 ml of the 1-naphthylamine solution and 1 ml of the sodium acetate. Finally, dilute each solution to the mark.

Using the blank as a reference and any one of the nitrite solutions prepared above, measure the absorbance at different wavelengths in the interval 400 to 600 nm. Take readings about 20 nm apart except in the region of maximum absorbance, where intervals of 5 nm are used. Plot the absorbance vs. wavelength and connect the points to form a smooth curve. Select the proper wavelength to use for the determination.

Using the selected wavelength, measure the absorbance of each of the standard solutions and the unknown. Plot the absorbance vs. the concentration of the standards and note whether Beer's law is obeyed. From the absorbance of the unknown solution, calculate the number of milligrams of nitrogen per liter (ppm) of the original unknown solution.

Notes

1. Prepared solutions may be used as unknowns. If a natural water is used, be sure that it is colorless and free of turbidity.
2. Consult the instructor concerning the volume of sample to be used.

EXPERIMENT 42. SPECTROPHOTOMETRIC DETERMINATION OF THE pK_a OF AN ACID-BASE INDICATOR[10]

In this experiment, spectrophotometry is employed to measure the pK_a of bromthymol blue, an acid-base indicator. The indicator (HIn) is a monoprotic acid and we can represent its dissociation as follows:

$$\text{HIn} \rightleftharpoons \text{H}^+ + \text{In}^-$$

As is shown in Chapter 5, the equilibrium expression for such a dissociation can be written

$$p\text{H} = pK_a - \log \frac{[\text{HIn}]}{[\text{In}^-]}$$

This can be rearranged to give

$$\log \frac{[\text{In}^-]}{[\text{HIn}]} = p\text{H} - pK_a$$

This is in the slope-intercept form of an equation for a straight line,

$$y = mx + b$$

where y is $\log[\text{In}^-]/[\text{HIn}]$, $m = 1$, and $b = -pK_a$. Hence, if the log term is plotted vs. pH, the slope is 1, the intercept is $-pK_a$, and the line should cross the pH axis at $p\text{H} = pK_a$ (Fig. 22.2). At the latter point $[\text{In}^-] = [\text{HIn}]$, and hence the log of the ratio of these terms is zero, making $p\text{H} = pK_a$.

The ratio $[\text{In}^-]/[\text{HIn}]$ can be determined spectrophotometrically. First, a solution of bromthymol blue is prepared in acidic solution (low pH) where essentially all of the indicator is in the HIn form. The absorption spectrum is then determined. Second, a solution is prepared in basic solution (high pH) where essentially all of the indicator is in the In^- form. The absorption spectrum of In^- is then determined. From the two absorption spectra the wavelengths of maximum absorbance of HIn and In^- are selected for further measurements.

[10] C. N. Reilley and D. T. Sawyer, *Experiments for Instrumental Methods*, McGraw-Hill Book Company, New York, 1961.

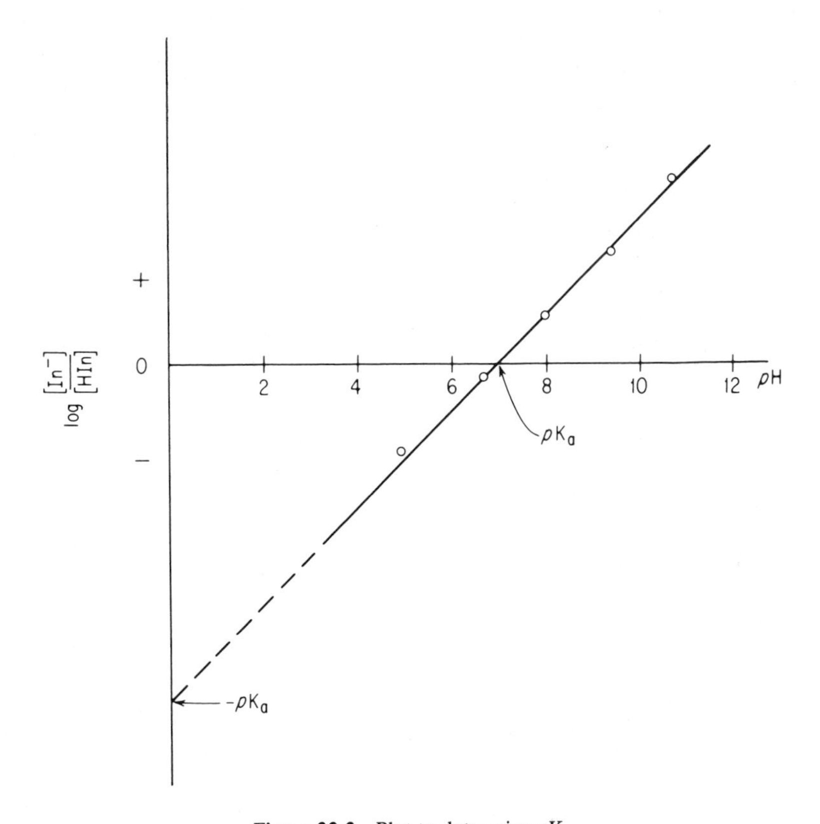

Figure 22.2 Plot to determine pK_a.

Buffered solutions with pH values on either side of the pK_a of bromthymol blue are then prepared and the absorbances measured at the selected wavelengths. The solutions contain the same total concentration of indicator, $\{[HIn] + [In^-]\}$, but the ratios vary with pH. Figure 22.3 shows a typical plot of absorbance vs. pH at the wavelength of maximum absorbance for the In^- species. The terms used in this figure are as follows:

$$A_a = \text{absorbance of HIn}$$

$$A_b = \text{absorbance of In}^-$$

$$A = \text{absorbance of mixture}$$

From the graph it is evident that

$$\frac{[In^-]}{[HIn]} = \frac{A - A_a}{A_b - A}$$

If the wavelength used is the one at which HIn shows maximum absorbance, the curve will be similar to that shown in Fig. 22.3, except that it will start at a high absorbance and curve down to a low absorbance value at high pH.

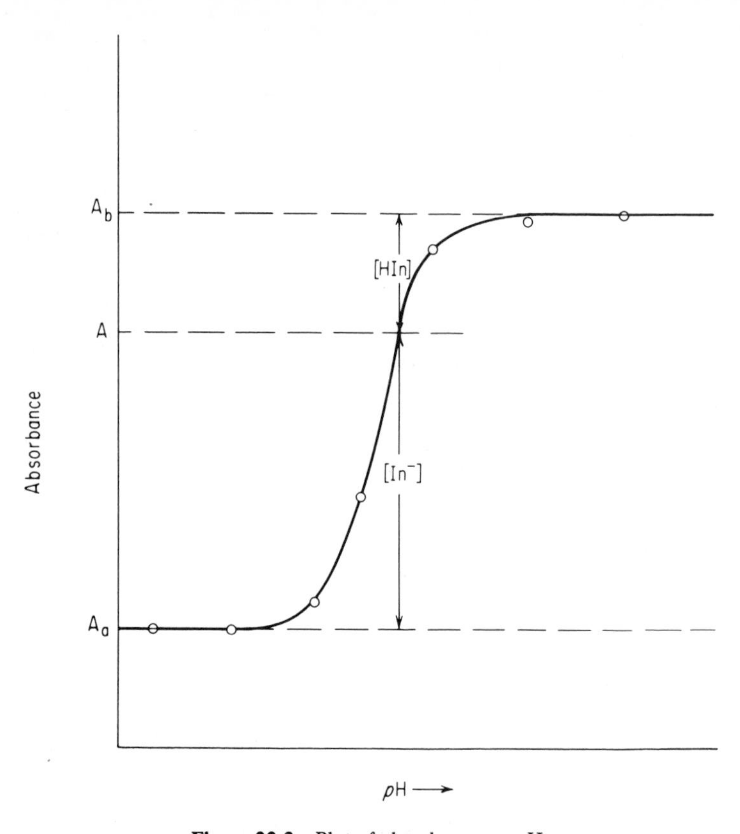

Figure 22.3 Plot of absorbance vs. pH.

Procedure

Prepare the Following Solutions:

(a) Dissolve 0.1 g of bromthymol blue in 100 ml of 20% ethanol.
(b) Dissolve 2.4 g of NaH_2PO_4 in water and dilute to 100 ml. The solution is 0.2 *M*.
(c) Dissolve 2.4 g of K_2HPO_4 in water and dilute to 100 ml. The solution of 0.2 *M*.
(d) Prepare a small amount of 3 *M* NaOH. Three grams of NaOH dissolved in 25 ml of solution will provide more than enough of this reagent.

Prepare a series of buffered solutions as follows: Secure seven clean 50-ml volumetric flasks (Note 1). Place 2.0 ml of the bromthymol blue solution in each flask, using a pipet. Then add the following volumes of phosphate solutions to these flasks. (These volumes can be measured with a graduated cylinder.)

Flask	KH_2PO_4, *ml*	Na_2HPO_4, *ml*
1	5	0
2	5	1
3	10	5
4	5	10
5	1	5
6	1	10
7	0	5

To flask 7 add 2 drops of the 3 *M* NaOH solution. Now dilute each solution to the mark and mix thoroughly. Measure and record the *p*H value of each solution using a *p*H meter.

To determine the absorption spectrum of bromthymol blue at low *p*H (the spectrum of HIn), measure the absorbance of solution 1 from 400 to 640 nm, using water as a reference (Note 2). Read the absorbance at 20-nm intervals except in the vicinity of the maximum in the spectrum, where readings should be taken every 5 nm. Plot the absorbance vs. wavelength.

To determine the absorption spectrum at high *p*H (the spectrum of In^-), measure the absorbance of solution 7 in the same manner as directed above. Plot the results on the same piece of graph paper.

Using the spectra obtained above, select two wavelengths at which further absorbance measurements will be made. Choose wavelengths at which HIn and In^- exhibit maximal differences in absorbance.

Now measure the absorbances of each of the seven solutions at the two wavelengths selected (Note 3). Prepare a graph of absorbance vs. *p*H for each of the two wavelengths. (See Fig. 22.3.) Determine the $[In^-]/[HIn]$ ratios in solutions 2 to 6 as explained above and shown in Fig. 22.3. Plot $\log [In^-]/[HIn]$ vs. *p*H (Fig. 22.2) and obtain a value of pK_a from the graph (Note 4).

Report the value of the pK_a of bromthymol blue in the manner desired by the instructor. Turn in all your graphs.

Notes

1. If it is more convenient to use 100-ml or 25-ml volumetric flasks, the amounts of reagents should be scaled up or down proportionately.
2. The *p*H of this solution is sufficiently low that most of the indicator is in the HIn form. If the absorbance reading at any wavelength is very high (above 0.8 to 1.0), the solutions should be diluted or less bromthymol blue used in each flask.
3. The instructor may wish measurements made at one wavelength only. Consult him before making the measurements.
4. Note that the value of pK_a can also be obtained from the graph in Fig. 22.3 by reading the *p*H at which $A_b - A = A_a - A$ (i.e., the absorbance halfway between A_a and A_b). Consult the instructor; he may wish you to determine the pK_a by several methods and report the average.

POTENTIOMETRIC TITRATIONS

It was pointed out in Chapter 12 that a potentiometric titration involves measurement of the difference in potential between an indicator electrode and a reference electrode during a titration. The difference in potential can be measured with a potentiometer or a *p*H meter. Generally, precise measurements of potential differences are made with a potentiometer. However, for the precision required in titrations a *p*H meter gives satisfactory results and is more convenient to use. A brief discussion of the *p*H meter is given below, followed by directions for performing several experiments.

*p*H Meters

An ordinary potentiometer cannot be employed with a glass electrode because of the high resistance, 1 to 100 megohms, of this electrode. The so-called *p*H meter is a voltage measuring device designed to be used with cells of high resistance. There are two common types available commercially, the potentiometric and the direct-reading. The former is basically a potentiometer, but since the off-balance currents are so small because of the high resistance of the cell, the current is amplified electronically so that it will affect a galvanometer or microammeter. The direct-reading instruments are electronic voltmeters of very high input resistance; the circuit is so arranged as to give a meter reading proportional to *p*H. The voltage of the glass-reference electrode pair is impressed across a very high resistance so that the current drawn is very low, of the order of 5×10^{-11} A. Since the resistance of the cell may be as large as 10^8 ohms, this means a voltage drop of 0.005 V:

$$E = I \times R = 5 \times 10^{-11} \times 10^{8} = 0.005$$

or an error of 0.5% at 1.000 V.

The direct-reading instruments are the most popular today, particularly since they can be line-operated. Several commercial models are available from such companies as Beckman, Leeds and Northrup, Coleman, Fisher, and Corning. The operating instructions are either printed directly on the instrument or are furnished in a pamphlet by the manufacturer.

EXPERIMENT 43. ACID-BASE TITRATIONS

The following experiments are chosen to illustrate the types of titration curves obtained with a strong acid, weak acid, and a polyprotic acid. The data can be used for standardizing a solution, analyzing an unknown, or determining the dissociation constant of a weak acid.

A typical experimental setup is shown in Fig. 22.4. The instructor will explain to you the operation of the *p*H meter and the precautions you should observe.

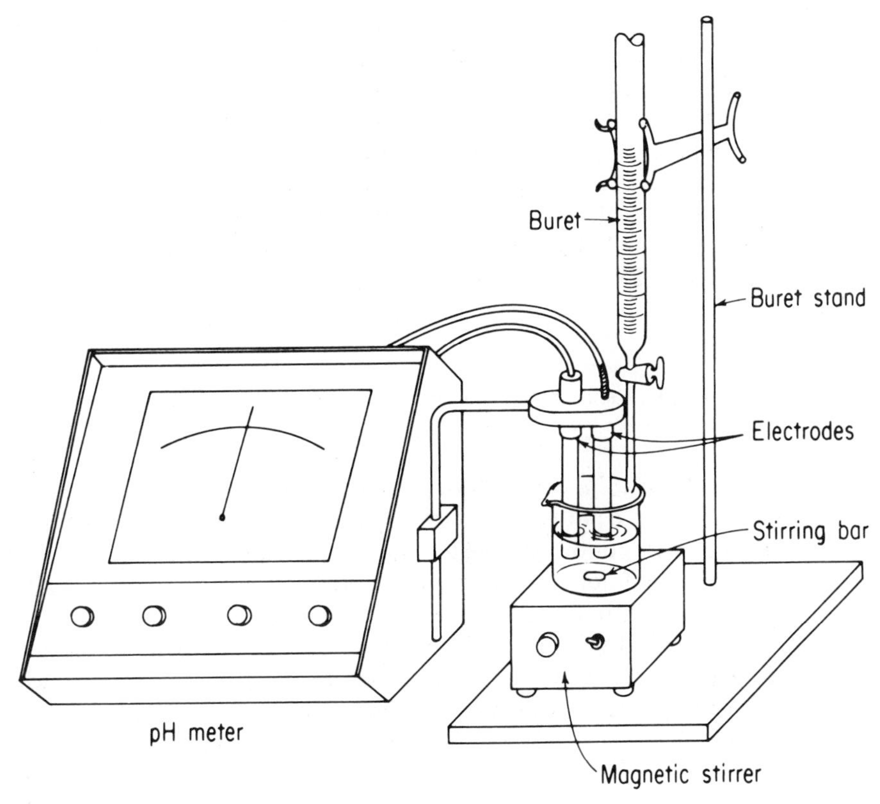

Figure 22.4 Potentiometric titration using *p*H meter.

Procedure

(a) Strong Acid–Strong Base. Prepare solutions of about 0.1 *N* hydrochloric acid and sodium hydroxide as directed in Experiment 1, page 564. Then pipet a 25.00-ml aliquot of the acid into a 250-ml beaker and dilute the solution to about 100 ml with distilled water. Insert the electrodes into the solution, being sure that they dip about $\frac{1}{2}$ inch below the surface. Adjust the mechanical or magnetic stirrer and set up a buret containing the base solution as shown in Fig. 22.4. Measure and record the *p*H of the solution before the addition of any titrant. Then add from the buret about 5 ml of the base solution and again measure the *p*H. Record this value as well as the buret reading at this point.

Proceed in this manner to record the *p*H and buret readings after the addition of about 10, 15, and 20 ml of titrant. Then add the titrant in about 1-ml intervals until the equivalence point is almost reached. (It may save time to run rapidly through the first titration to approximately locate the equivalence point. The second titration is then done carefully.) Then add the titrant in 0.1-ml intervals until the equivalence point is passed. It will be evident when this point is reached because of the large change in *p*H that occurs. Finally, record two additional readings at about 5 and 10 ml of excess titrant.

Make the following plots of the data: (1) pH vs. ml of NaOH; (2) $\Delta p\text{H}/\Delta V$ vs. ml of NaOH; and (3) $\Delta^2 p\text{H}/\Delta V^2$ vs. ml of NaOH. Determine the volume of base required by the acid from each plot and also calculate the volume by the analytical method described on page 318. Calculate the volume ratio of the two solutions.

Repeat the titration on two additional aliquots of acid (Note), adding to the first 2 drops of methyl red indicator and to the second 2 drops of phenolphthalein. Note the pH values at which these indicators change color and compare these with the data given in Table 6.3, page 140. Average the values obtained for the volume ratios and calculate the precision of the measurement.

Note

If standard solutions are already at hand, one titration is sufficient to illustrate the titration curves.

(b) Weak Acid–Strong Base. Weigh a sample of about 0.7 to 0.9 g of pure, dry potassium acid phthalate (Note) on the analytical balance. Dissolve the sample in about 100 ml of distilled water and titrate with the sodium hydroxide. Measure the pH at different increments of titrant as in part (a).

Plot the titration data in the same manner as above and compare the curves obtained for the weak acid with those obtained for the strong acid. Justify the selection of the indicator used in Experiment 3, page 567. Calculate the normality of the base and acid solutions. If further use is to be made of these solutions the standardization should be repeated (consult instructor).

Note

If the normalities of the solutions are already known, a sample of unknown purity can be titrated.

(c) pK of an Unknown Acid. Dissolve a sample of an unknown acid (consult instructor) in about 100 ml of distilled water in a 250-ml beaker and titrate the solution with standard sodium hydroxide as described in part (a). Plot the data as pH vs. ml of NaOH and determine the equivalence volume. Read from the curve the pH at half the volume required to reach the equivalence point. At this halfway point $p\text{H} = pK_a$, and the acid constant is thus determined. Report this value to your instructor and repeat the titration if he wishes this done. He may also wish to know the concentration of the unknown acid, if this is a solution, or the equivalent weight of the unknown if it is a solid. If the sample is a solid, it should be weighed on an analytical balance before it is dissolved.

(d) Titration of Phosphoric Acid. Pipet 25.00 ml of a phosphoric acid solution of unknown concentration into a 250-ml beaker. Dilute the solution to about 100 ml, insert the electrodes, and titrate with standard sodium hydroxide as directed in part (a). You should observe two breaks in the titration curve, one around pH 4 to 5 and the other around pH 9 to 10. Plot the titration curve as pH vs. ml of NaOH. Determine the following from the curve and report to the instructor: (a) the molarity of the acid solution and (b) the values of pK_{a_1} and pK_{a_2} for the first two dissociation constants of phosphoric acid (Note).

Note

For acids whose dissociation constants are greater than about 10^{-3} to 10^{-2}, an appreciable error is made by using the expression $pH = pK_a$ halfway to the equivalence point. The following expression can be used for such stronger acids:

$$pH = pK_a - \log \frac{C - [H^+]}{C + [H^+]}$$

The volume to the equivalence point, V_e, is first determined; then the pH at $\frac{1}{2}V_e$ is read and the value of $[H^+]$ calculated. The term C is the concentration of the anion halfway to the equivalence point and is given by

$$C = \frac{\frac{1}{2}V_e \times N}{\text{total volume}}$$

where the total volume is $\frac{1}{2}V_e$ + the starting volume. The latter volume should be known to ± 10 ml.

EXPERIMENT 44. REDOX TITRATIONS

There are many redox titrations that can be used to illustrate the potentiometric technique. In this experiment the titration of iron(II) with dichromate or cerium(IV) solution is carried out. The indicator electrode is platinum and the reference is a saturated calomel electrode.

Procedure

(a) Titration of Iron with Dichromate. Prepare a standard solution of potassium dichromate by weighing accurately about 1.25 g of the pure, dry salt. Dissolve the salt in water, transfer the solution to a 250-ml volumetric flask, and dilute the solution to the mark. Fill a buret with the solution.

Weigh accurately a sample of about 1.3 to 1.5 g of pure iron(II) ammonium sulfate. Dissolve the salt in about 100 ml of distilled water and add about 10 ml of concentrated sulfuric acid.

Insert a platinum and a saturated calomel electrode into the solution and adjust the stirrer if one is to be used. If a pH meter is to be employed, set the instrument to measure potential rather than pH. The platinum wire is the positive electrode. Add about 5 ml of the dichromate solution from the buret and then measure the potential. Continue the titration in the usual manner, recording the potential and volume of dichromate solution, until about 10 ml of excess titrant is added.

Make the following plots of the data: (1) potential vs. ml of titrant, (2) $\Delta E/\Delta V$ vs. ml of titrant, and (3) $\Delta^2 E/\Delta V^2$ vs. ml of titrant. Determine the volume of dichromate used, and from this volume and the weight of iron(II) ammonium sulfate, calculate the normality of the dichromate solution. Compare this value with the normality calculated from the weight of dichromate dissolved in 250 ml of solution. Repeat the titration if desired (consult instructor).

(b) Titration of Iron with Cerium (IV). Prepare 250 ml of a standard cerium(IV) solution from cerium(IV) ammonium nitrate as directed on page 593. Dissolve a sample of iron ore and reduce it with tin(II) chloride as directed in Experiment 25, page 590. Stop the procedure after adding excess tin(II) chloride; that is, do not remove the excess tin(II) ion with mercury(II) chloride. The titration curve will show two breaks: the tin reacting first, followed by the iron. The difference in volumes to the two end points gives the volume of titrant used by the iron sample.

Insert a platinum and a saturated calomel electrode into the iron solution in a 250-ml beaker. Adjust the volume to about 100 ml and set the *p*H meter to measure potential rather than *p*H. The platinum wire is the positive electrode. Add about 5 ml of the cerium(IV) solution from a buret and measure the potential. Continue the titration in the usual manner, recording the potential and volume of cerium(IV) solution, until about 10 ml of excess titrant is added.

Plot the potential vs. ml of titrant and measure the difference in volume between the first and second breaks. From this volume and the weight of the sample, calculate the percentage iron in the sample. Report this value to the instructor and repeat the determination if he wishes this.

Also plot $\Delta E/\Delta V$ vs. ml of titrant (for the titration of iron) and $\Delta^2 E/\Delta V^2$ vs. ml of titrant and compare these with the plots obtained in Experiment 43(a).

EXPERIMENT 45. PRECIPITATION TITRATIONS

A piece of polished silver wire serves as the indicator electrode for titrations involving silver ions. This electrode is available commercially for use with *p*H meters. A calomel electrode cannot dip directly into the solution being titrated since silver chloride would precipitate; instead, this electrode is placed in a saturated potassium nitrate solution in a separate vessel, and the two solutions are joined by a salt bridge containing potassium nitrate.[11] Directions are given for the titration of chloride with silver nitrate, and for the titration of a chloride–iodide mixture.

Procedure

(a) Titration of Chloride. Prepare a standard solution of silver nitrate by weighing accurately about 4.5 g of the pure salt. Dissolve the salt in water, transfer the solution to a 250-ml volumetric flask, and dilute the solution to the mark.

Accurately weigh a dried chloride sample of unknown purity (consult instructor as to the weight required) and dissolve the salt in about 100 ml of distilled

[11] Since the *p*H of the solution changes little during the precipitation titration, it is possible to use a glass electrode in place of calomel as the reference electrode, thereby eliminating the salt bridge.

water. Insert the silver electrode and the potassium nitrate salt bridge into the solution and adjust the stirrer. Set the *p*H meter to read potential. The silver wire is the positive electrode. Add about 5 ml of the silver nitrate solution from the buret and then measure the potential. Continue the titration in the usual manner, recording the potential and volume of silver nitrate, until about 10 ml of excess titrant is added.

Make the following plots of the data: (1) potential vs. ml of silver nitrate, (2) $\Delta E/\Delta V$ vs. ml of silver nitrate, and (3) $\Delta^2 E/\Delta V^2$ vs. ml of silver nitrate. Determine the volume of titrant, and from this volume and the molarity calculate the percentage of chloride in the sample. Repeat the titration if desired (consult instructor).

(b) Titration of a Chloride–Iodide Mixture. The standard solution of silver nitrate is prepared as in part (a). Secure from the instructor a solution containing both chloride and iodide ions of unknown concentration. Pipet a 25-ml aliquot into a 250-ml beaker and dilute the solution to about 100 ml with water. Insert the silver electrode and potassium nitrate salt bridge and adjust the stirrer. Set the *p*H meter to read potential. Titrate with silver nitrate in the usual manner, recording the potential and volume of titrant, noting carefully the volumes near the two equivalence points.

Plot the potential vs. ml of titrant and determine the volumes of titrant used by iodide and chloride. Calculate the molarities of these two ions and report the results to the instructor.

EXPERIMENT 46. IDENTIFICATION OF AN AMINO ACID[12]

Amino acids are important biological molecules which serve as building blocks for peptides and proteins. They have the general structure

$$\begin{array}{c} \mathrm{R} \\ | \\ \mathrm{H_2N{-}CH{-}CO_2H} \end{array}$$

where R is an organic group which is different in each amino acid. Note that the NH_2 group is attached to the carbon atom adjacent to the CO_2H group. For this reason these molecules are called "alpha" amino acids. There are about 20 different alpha amino acids that have been identified as units in the most important plant and animal proteins.

The acid-base equilibria of amino acids are indicated below for the simplest acid, glycine (R = H):

$$\underset{\text{Conjugate acid, A}}{\overset{+}{N}H_3CH_2CO_2H} \underset{H^+}{\overset{OH^-}{\rightleftharpoons}} \underset{\text{Neutral glycine, N}}{NH_2CH_2CO_2H} \rightleftharpoons \underset{\text{Dipolar ion, D}}{\overset{+}{N}H_3CH_2CO_2^-} \underset{H^+}{\overset{OH^-}{\rightleftharpoons}} \underset{\text{Conjugate base, B}}{NH_2CH_2CO_2^-}$$

[12] This experiment was designed by Ronald C. Johnson, who has kindly given us permission to use it here.

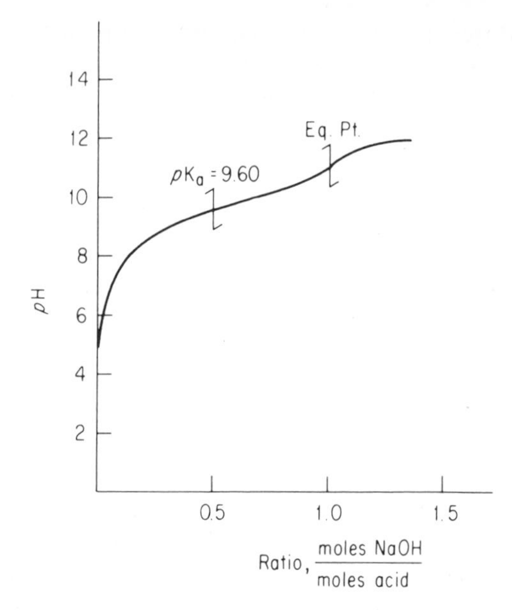

Figure 22.5 Titration curve of an amino acid such as glycine.

The neutral amino acid exists largely as the dipolar ion, D, and an aqueous solution of the acid is only slightly acidic. (A few amino acids have CO_2H or NH_2 groups in the R part of the molecule and hence are acidic or basic.) The $\overset{+}{N}H_3$ group in D is a weak acid, with a pK_a in the range 8 to 11. Hence when one titrates an amino acid such as glycine with NaOH, a curve such as that shown in Fig. 22.5 is obtained. The $\Delta pH/\Delta V$ in the titration curve is not a good one, and we would say that the titration is not feasible for the purposes of analysis. However, if a fairly concentrated solution of the acid is titrated, a large enough $\Delta pH/\Delta V$ can be obtained for the purpose of identifying the molecule.

Some amino acids can be obtained as the hydrochloride salts, that is, as the conjugate acid A (above). The conjugate acid is reasonably strong, with pK_a values in the range about 2 to 5. Hence the complete titration of such a compound gives two end points, as shown in Fig. 22.5. The first end point is good and can be used to help locate the second one which is not very sharp.

As mentioned above, some amino acids have a CO_2H or NH_2 group as part of R. For example, aspartic acid has the formula

$$\begin{array}{c} \qquad\quad CH_2CO_2H \\ \qquad\quad | \\ \overset{+}{N}H_3\text{—}CHCO_2^- \end{array}$$

Molecules containing the CO_2H group are acids with pK_a values in the range of about 3 to 6. The value in aspartic acid is 3.65. The pK_a of the $\overset{+}{N}H_3$ group is 9.60. Hence this molecule gives a titration curve with two end points similar to that shown in Fig. 22.6. Aspartic acid hydrochloride is a triprotic acid that gives a titration curve similar to that shown in Fig. 22.7.

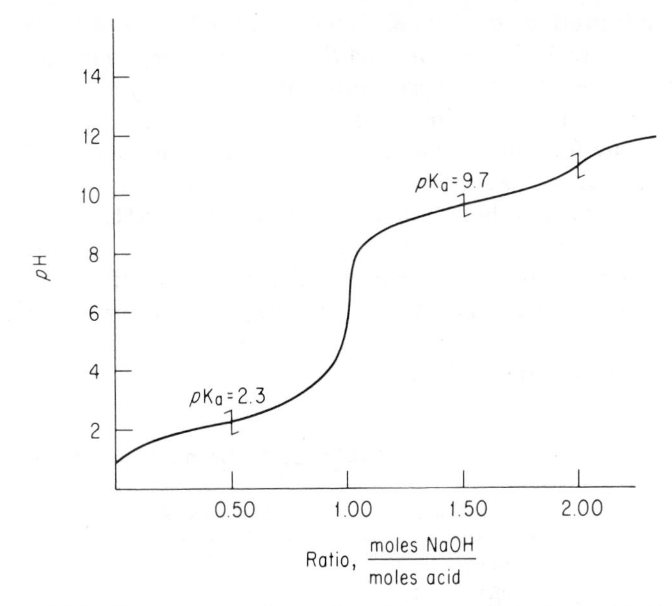

Figure 22.6 Titration curve of an amino acid such as alanine hydrochloride.

Lysine has the formula

$$\overset{+}{N}H_3\underset{}{CH}(CO_2^-)\text{—}(CH_2)_4NH_2$$

It can be obtained as the dihydrochloride in which both NH_2 groups as well as the CO_2^- group are protonated. As a consequence, lysine dihydrochloride titrates as

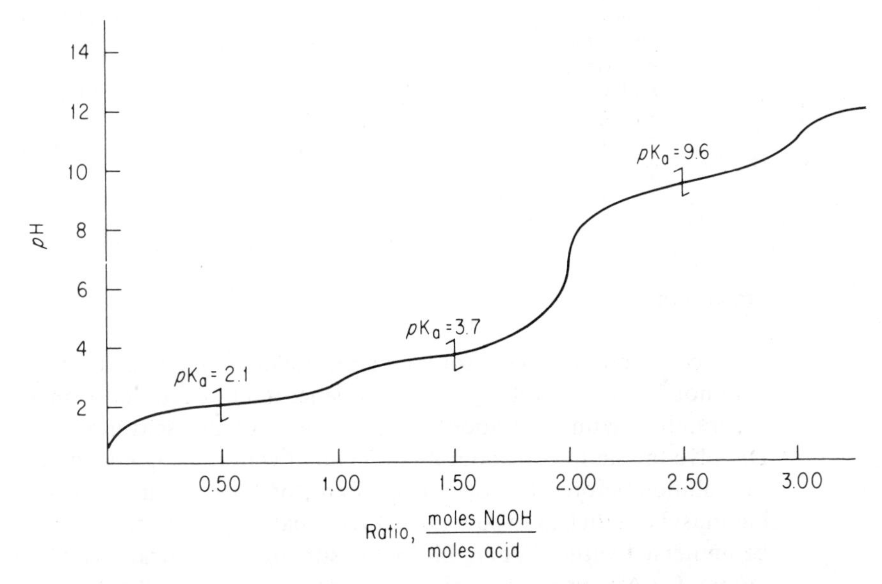

Figure 22.7 Titration curve of an amino acid such as aspartic hydrochloride.

a triprotic acid, with pK_a values of 2.20, 8.90, and 10.28. The first end point is a good one. The second and third are not very sharp and are not well separated.

The object of this experiment is to identify an amino acid by potentiometric titration with NaOH. From the data obtained, the molecular weight and pK_a value (or values) can be estimated. The number of possible unknowns will be limited to those shown in Table 22.1. Some of the compounds are more difficult to titrate than others. All have a weakly acidic $\overset{+}{N}H_3$ group and will give at least one end point similar to that shown in Fig. 22.4. In order to get an appreciable $\Delta pH/\Delta V$, relatively concentrated solutions are titrated. It should be kept in mind that one can locate the end point for the titration of a very weak acid group from the end point of a stronger acid group, provided, of course, that such a group is present in the molecule.

TABLE 22.1 Some Amino Acids

Name	Molecular Weight	pK_a
Alanine	89.1	9.69
Arginine · HCl	210.7	9.09
Asparagine	150	8.80
Aspartic acid	133.1	3.65, 9.60
Cysteine · HCl	176.6	1.71, 8.33, 10.78
Cystine	240.3	8.02, 8.71
Glutamic acid	147	4.25, 9.67
Glutamic acid · HCl	183.6	2.19, 4.25, 9.67
Glycine	75.1	9.60
Histidine	155.2	8.97
Histidine · HCl	209.6	5.97, 8.97
Leucine	131.2	9.60
Lysine · 2HCl	219.1	2.20, 8.90, 10.28
Methionine	149.2	9.21
Phenylalanine	166.2	9.13
Proline	115.1	10.60
Taurine	125.2	8.74
Tyrosine	181.2	9.11, 10.07
Valine	117.2	9.62

Procedure

Secure from the instructor your amino acid unknown and record its number in your notebook. Do not dry the sample since it may decompose on heating. Weigh accurately a sample of about 0.3 g (Note 1) of the acid into a 100-ml beaker. Dissolve the sample in about 20 to 30 ml of water, using heat if necessary. (Cool the solution before titration.) The volume of water should be kept at a minimum but must be sufficient to cover the stirring bar so that the tips of the electrodes can be immersed without being hit by the stirring bar. Titrate with standard NaOH (about 0.1 *N*), using 1.0-ml increments throughout the titration. If the *p*H changes by more than 0.3 unit/ml of base added, 0.5-ml increments would be appropriate. Continue the titration until the pH reaches a value of about 11.8.

Prepare a graph plotting pH vs. ml of NaOH. From the graph determine whether your acid is mono-, di-, or triprotic, determine its pK_a or pK_a's (Note 2), and its molecular weight. Identify the amino acid on the basis of your data and the information given in Table 22.1. Since many of the acids do not give sharp end points on titration and since the acids may not be completely pure, your molecular weight value may differ from the correct value by as much as 2%. The pK_a values may differ from those reported in Table 22.1 by 0.2 pK unit.

Turn in your graph to the instructor. Report the identity of your unknown and its molecular structure. (Consult a book on organic chemistry or biochemistry.) Finally, identify the acid group from which a proton is being removed in each titration step.

Notes

1. To the instructor: Compounds 1 to 5, 7 to 9, 12, 13, 15, and 19 are available and have been titrated successfully by beginning students. You may wish to confine your unknowns to this group of acids.

2. For acids with pK_a values of 3 or less the exact expression given on page 615 should be used.

ELECTROLYSIS

The principles of electrolysis are discussed in Chapter 13. Directions are given here for the determination of copper in a solution free of interfering metals and for the separation of copper and nickel by electrolysis.

EXPERIMENT 47. ELECTROLYTIC DETERMINATION OF COPPER

Apparatus

A schematic diagram of the necessary appartus is given in Fig. 13.1. The source of direct current is often a 6-V rectifier rather than a storage battery, and an ordinary rheostat of suitable size serves as the adjustable resistance. An inexpensive unit that can be easily assembled found wide usage.[13] It is desirable to provide mechanical stirring if the equipment is available. Commercial "electroanalyzers" which provide for rotation of one electrode to effect stirring can be purchased.

The electrodes are usually platinum, although copper gauze can be employed as the cathode for the deposition of copper. The cathode is usually a cylinder of platinum gauze, and the anode is a spiral of platinum wire or a small cylinder if the electrode is rotated.

Procedure

Clean the platinum electrodes by immersing them in warm 1 : 3 nitric acid for about 5 min. Then rinse them well with tap water and distilled water. Place the gauze cathode on a watch glass and dry it in an oven at about 105°C. Cool the cathode in a desiccator and then weigh it on the analytical balance. Avoid touching the gauze with the fingers, as this may leave grease on the surface and prevent copper from adhering.

The solution to be electrolyzed should contain about 2 ml of sulfuric acid and 1 ml of nitric acid (Note 1) per 100 ml. Obtain this solution or a solid unknown from the instructor (Note 2). Connect the electrodes properly to the apparatus, placing the spiral anode inside the gauze cylinder. Make sure that the electrodes do not touch. Raise the beaker (a tall-form one is convenient) around the electrodes and adjust the height so that the lower edge of the cathode almost touches the bottom of the beaker. About $\frac{1}{4}$ inch of the top of the cathode should not be covered by the solution. The solution can be diluted if necessary with distilled water. Cover the beaker with a split watch glass and adjust the rheostat to give its full resistance.

Turn on the stirrer and close the circuit to start the electrolysis. Only a small current should flow since the resistance is high. Gradually lower the resistance until the current is 2 to 4 A and the voltage is below 4 V (Note 3). Electrolyze at this voltage until the blue color of copper has disappeared (usually about 30 to 45 min). Add 0.5 g of urea (Note 4), continue the electrolysis for 15 min longer, and then add sufficient water to cover completely the top of the cathode (Note 5). Continue the electrolysis for an additional 15 min using 0.5 A current, and if no copper is deposited on the fresh cathode surface, the deposition is complete.

To stop the electrolysis, turn off the stirrer but do not interrupt the current at this time. Remove the support under the beaker, and slowly lower the beaker with one hand while washing the exposed portion of the cathode with a stream of water from the wash bottle. As soon as the cathode is completely out of the solution, cut off the current and raise a beaker of distilled water to cover the electrodes. Wash the electrodes with a second portion of distilled water and then disconnect the cathode. Dip the cathode into a beaker of acetone or alcohol and place it on a watch glass in the oven at about 105°C for 5 min (Note 6). Cool the electrode to room temperature and then weigh it accurately.

If a prepared solution was used, report the total weight of copper in your solution. If copper ore was employed, calculate the percentage of the metal in the sample. Duplicate samples should be run; consult the instructor as to his wishes.

The copper can be removed from the electrode by placing it in warm 1 : 3 nitric acid for a few minutes. The electrode is then rinsed well with tap water and distilled water.

Notes

1. If the acid has a yellow tinge indicating the presence of oxides of nitrogen, boil the solution until the oxides are expelled. Nitric acid improves the nature of the copper deposit by preventing the evolution of hydrogen at the cathode.

2. This solution can be prepared by dissolving pure copper foil in nitric acid according to procedure (b) of Experiment 34. It is then diluted in a volumetric flask and aliquot portions are given to the students. Alternatively, commercial unknowns of copper oxide which are readily dissolved in sulfuric acid can be used. If a brass sample is being analyzed, the filtrate from the lead determination by the sulfate method can be electrolyzed, after first adjusting the volume to about 100 ml and adding 1 ml of concentrated nitric acid.

3. If a mechanical or magnetic stirrer is not available, the electrolysis can be carried out without stirring by using a current of about 0.5 A. The electrolysis should then be allowed to run overnight.

4. Nitrite prevents complete deposition of copper and is removed by urea according to the equation

$$2NO_2^- + 2H^+ + (NH_2)_2CO \rightleftharpoons CO_2 + 2N_2 + 3H_2O$$

The nitrite is formed by the reaction

$$2H^+ + NO_3^- + 2e \rightleftharpoons H_2O + NO_2^-$$

5. If the solution is not to be used for further analysis, a few drops can be removed with a pipet and tested with concentrated ammonia. The deep blue of the copper-ammonia complex indicates the incomplete deposition of copper.

6. Do not heat the electrode any longer than this since the surface of the copper becomes oxidized easily.

EXPERIMENT 48. SEPARATION OF COPPER AND NICKEL BY ELECTROLYSIS

Copper and nickel are on opposite sides of hydrogen in the activity series and hence can be separated electrolytically by control of *p*H. Copper is deposited from an acid solution and the electrode weighed. The hydrogen ion concentration is lowered by addition of ammonia and then nickel can be deposited. The separation is useful in the determination of these two metals in coinage alloys of copper and nickel and in monel metal. These alloys usually contain a small amount of iron which must be separated before the electrolysis of nickel. Iron can be separated from nickel by precipitation of hydrous iron(III) oxide.

Procedure[14]

If the alloy (Note 1) is greasy, wash it with dilute ammonia, then with water, and finally with acetone. Dry the sample and weigh about 0.5 g into a 200-ml tall-form beaker. Add to the beaker 10 ml of water, 1 ml of concentrated sulfuric acid, and 2 ml of concentrated nitric acid. When the sample has dissolved, boil the solution to 100 ml and electrolyze the copper as directed in Experiment 47.

Wash the copper deposit thoroughly, taking care to save all the washings. Dry and weigh the electrode and return it to the apparatus without removing the copper (Notes 2 and 3).

[14] H. H. Willard, N. H. Furman, and C. E. Bricker, *Elements of Quantitative Analysis*, 4th ed., D. Van Nostrand Co., Inc., New York, 1956, p. 438.

Evaporate the filtrate slowly and carefully on a hot plate until fumes of sulfur trioxide appear (Note 4). Cool the residue and add to it carefully about 25 ml of water. Add about 10 ml of 1 : 1 ammonia to precipitate the iron. Filter off the hydrous oxide using a small filter paper and catch the filtrate in an electrolytic beaker. Wash the precipitate three times with small portions of water and then set the beaker containing the filtrate aside. Dissolve the precipitate by pouring over the filter paper a few milliliters of hot 3 *M* sulfuric acid, catching the filtrate in the original electrolytic beaker. Wash the paper with water and again precipitate the iron by adding 10 ml of 6 *M* ammonia. Filter the precipitate through the same paper and wash it with water, receiving the solution in the original beaker which contains the nickel solution.

The hydrous oxide can be ignited to Fe_2O_3 and weighed if desired (consult instructor). Otherwise, discard the precipitate.

To the filtrate containing the nickel, add 15 ml of concentrated ammonia and dilute the solution to about 100 ml. Electrolyze the nickel in the same manner as directed for copper. The completeness of deposition can be judged by the disappearance of the blue color of the nickel–ammonia complex. A few drops of the solution can be removed and tested for nickel with dimethylglyoxime.

When the deposition of nickel is complete, remove the solution and wash the electrode in the same manner as directed for copper. Rinse the electrode twice with distilled water and once with acetone or alcohol. Dry the electrode for 5 min in the oven, then cool and weigh it in the usual manner. Report the percentages of copper and nickel in the alloy.

Notes

1. The instructor may prefer to use prepared solutions of copper and nickel rather than the alloy. An aliquot portion is given each student and he reports the weight of each metal in his solution. About 2 ml of concentrated sulfuric acid is added, the solution is diluted to 100 ml, and copper is electrolyzed. No nitric acid is added, since it interferes with the deposition of nickel. The lengthy procedure required to remove iron is avoided if such solutions are used.
2. Nickel can be deposited on top of the copper.
3. If a copper–nickel solution is being analyzed, neutralize the solution at this point with concentrated ammonia (litmus), add an additional 15 ml of concentrated ammonia, and electrolyze the nickel as directed above.
4. This operation removes nitrates and oxidizes iron to the +3 state.

APPENDIX ONE

tables of equilibrium constants and standard potentials

TABLE 1 Dissociation Constants of Weak Acids and Bases (25°C)

Acid	Formula	K_a	pK_a	Conjugate base	K_b	pK_b
Acetic	CH_3COOH	1.8×10^{-5}	4.74	Acetate ion	5.6×10^{-10}	9.26
Ammonium ion	NH_4^+	5.6×10^{-10}	9.26	Ammonia	1.8×10^{-5}	4.74
Anilinium ion	$C_6H_5NH_3^+$	2.2×10^{-5}	4.66	Aniline	4.6×10^{-10}	9.34
Arsenic	H_3AsO_4	5.6×10^{-3}	2.26	Dihydrogen arsenate ion	1.8×10^{-12}	11.74
Dihydrogen arsenate ion	$H_2AsO_4^-$	1.7×10^{-7}	6.77	Monohydrogen arsenate ion	5.9×10^{-8}	7.23
Monohydrogen arsenate ion	$HAsO_4^{2-}$	3×10^{-12}	11.5	Arsenate ion	3×10^{-3}	2.5
Arsenious	H_3AsO_3	6×10^{-10}	9.2	Dihydrogen arsenite ion	1.7×10^{-5}	4.8
Dihydrogen arsenite ion	$H_2AsO_3^-$	3×10^{-14}	13.5	Monohydrogen arsenite ion	3×10^{-1}	0.5
Benzoic	$HC_7H_5O_2$	6.6×10^{-5}	4.18	Benzoate ion	1.5×10^{-10}	9.82
Boric	H_3BO_3	5.8×10^{-10}	9.24	Dihydrogen borate ion	1.7×10^{-5}	4.76
Carbonic	H_2CO_3	4.6×10^{-7}	6.34	Bicarbonate ion	2.2×10^{-8}	7.66
Bicarbonate ion	HCO_3^-	4.4×10^{-11}	10.36	Carbonate ion	2.3×10^{-4}	3.64
Chloroacetic, mono	$CH_2ClCOOH$	1.5×10^{-3}	2.82	Monochloroacetate ion	7×10^{-12}	11.18
Chloroacetic, di	$CHCl_2COOH$	5×10^{-2}	1.3	Dichloroacetate ion	2×10^{-13}	12.7
Chloroacetic, tri	CCl_3COOH	2×10^{-1}	0.7	Trichloroacetate ion	5×10^{-14}	13.3
Chromic	H_2CrO_4	1.8×10^{-1}	0.74	Bichromate ion	5.6×10^{-14}	13.26
Bichromate ion	$HCrO_4^-$	3.2×10^{-7}	6.49	Chromate ion	3.1×10^{-8}	7.51
Citric	$H_3C_6H_5O_7$	8.4×10^{-4}	3.08	Dihydrogen citrate ion	1.2×10^{-11}	10.92
Dihydrogen citrate ion	$H_2C_6H_5O_7^-$	1.8×10^{-5}	4.74	Monohydrogen citrate ion	5.6×10^{-10}	9.26
Monohydrogen citrate ion	$HC_6H_5O_7^{2-}$	4.0×10^{-6}	5.40	Citrate ion	2.5×10^{-9}	8.60
Cyanic	$HOCN$	2×10^{-4}	3.7	Cyanate ion	5×10^{-11}	10.3
Diethylammonium ion	$C_4H_{10}NH_2^+$	7.7×10^{-12}	11.11	Diethylamine	1.3×10^{-3}	2.89

Dimethylammonium ion	$C_2H_6NH_2^+$	1.9×10^{-11}	10.72	Dimethylamine	5.2×10^{-4}	3.28
Ethylammonium ion	$C_2H_5NH_3^+$	1.8×10^{-11}	10.75	Ethylamine	5.6×10^{-4}	3.25
Formic	$HCOOH$	1.8×10^{-4}	3.74	Formate ion	5.6×10^{-11}	10.26
Hydrazinium ion	$N_2H_5^+$	3.3×10^{-9}	8.48	Hydrazine	3.0×10^{-6}	5.52
Hydrocyanic	HCN	7.2×10^{-10}	9.14	Cyanide ion	1.4×10^{-5}	4.86
Hydrofluoric	HF	6×10^{-4}	3.22	Fluoride ion	1.7×10^{-11}	10.78
Hydrogen sulfide	H_2S	1×10^{-7}	7.0	Bisulfide ion	1.1×10^{-7}	7.0
Bisulfide ion	HS^-	1×10^{-15}	15.0	Sulfide ion	10	−1.0
Hypochlorous	$HOCl$	3×10^{-8}	7.5	Hypochlorite ion	3×10^{-7}	6.5
Methylammonium ion	$CH_3NH_3^+$	2×10^{-11}	10.7	Methylamine	5×10^{-4}	3.3
Nitrous	HNO_2	4.5×10^{-4}	3.35	Nitrite ion	2.2×10^{-11}	10.65
Oxalic	$H_2C_2O_4$	6.5×10^{-2}	1.19	Bioxalate ion	1.5×10^{-13}	12.81
Bioxalate ion	$HC_2O_4^-$	6.1×10^{-5}	4.21	Oxalate ion	1.6×10^{-10}	9.79
Phenol	C_6H_5OH	1.3×10^{-10}	9.89	Phenolate ion	7.7×10^{-5}	4.11
Phosphoric	H_3PO_4	7.5×10^{-3}	2.12	Dihydrogen phosphate ion	1.3×10^{-12}	11.88
Dihydrogen phosphate ion	$H_2PO_4^-$	6.2×10^{-8}	7.21	Monohydrogen phosphate ion	1.6×10^{-7}	6.79
Monohydrogen phosphate ion	HPO_4^{2-}	4.8×10^{-13}	12.32	Phosphate ion	2.1×10^{-2}	1.68
Phthalic	$C_6H_4(COOH)_2$	1.3×10^{-3}	2.89	Biphthalate ion	7.7×10^{-12}	11.11
Biphthalate ion	$C_6H_4C_2O_2H^-$	3.9×10^{-6}	5.41	Phthalate ion	2.6×10^{-9}	8.59
Pyridinium ion	$C_5H_5NH^+$	7.1×10^{-6}	5.15	Pyridine	1.4×10^{-9}	8.85
Sulfuric	H_2SO_4	Strong	—	Bisulfate ion	Weak	—
Bisulfate ion	HSO_4^-	1.2×10^{-2}	1.92	Sulfate ion	8.3×10^{-13}	12.08
Sulfurous	H_2SO_3	1.7×10^{-2}	1.77	Bisulfite ion	5.9×10^{-13}	12.23
Bisulfite ion	HSO_3^-	6.2×10^{-8}	7.21	Sulfite ion	1.6×10^{-7}	6.79
Tartaric	$H_2C_4H_4O_6$	9.4×10^{-4}	3.03	Bitartrate ion	1.1×10^{-11}	10.97
Bitartrate ion	$HC_4H_4O_6^-$	2.9×10^{-5}	4.54	Tartrate ion	3.4×10^{-10}	9.46

TABLE 2 Stepwise Formation Constants of Complex Ions*

Ligand	Cation	Ionic strength	Temp., °C	Logarithm of equilibrium constant K_1	K_2	K_3	K_4
Ammonia	Ag^+	1.0	25	3.37	3.78		
	Cd^{2+}	2.1	25	2.74	2.21	1.37	1.13
	Cu^{2+}	1.0	25	4.27	3.55	2.90	2.18
	Ni^{2+}	1.0	25	2.36	1.90	1.55	1.23
	Zn^{2+}	2.0	30	2.37	2.44	2.50	2.15
Chloride	Ag^+	0.2	25	2.85	1.87	0.32	0.86
	Fe^{3+}	1.0	25	0.62	0.11	−1.40	−1.92
	Hg^{2+}	0.5	25	6.74	6.48	0.85	1.00
	Pb^{2+}	1.0	25	0.88	0.61	−0.40	−0.15
Cyanide	Cd^{2+}	3.0	25	5.48	5.14	4.56	3.58
	Hg^{2+}	0.1	20	18.00	16.70	3.83	2.98
EDTA	Ag^+	0.1	20	7.32			
	Al^{3+}	0.1	20	16.13			
	Ba^{2+}	0.1	20	7.76			
	Ca^{2+}	0.1	20	10.70			
	Cd^{2+}	0.1	20	16.59			
	Co^{2+}	0.1	20	16.21			
	Cu^{2+}	0.1	20	18.79			
	Fe^{2+}	0.1	20	14.33			
	Fe^{3+}	0.1	20	25.1			
	Hg^{2+}	0.1	20	21.80			
	Mg^{2+}	0.1	20	8.69			
	Mn^{2+}	0.1	20	13.58			
	Ni^{2+}	0.1	20	18.56			
	Pb^{2+}	0.1	20	18.3			
	Sr^{2+}	0.1	20	8.63			
	Th^{4+}	0.1	20	23.2			
	TiO^{2+}	0.1	—	17.3			
	VO^{2+}	0.1	20	18.77			
	Zn^{2+}	0.1	20	16.26			
Thiocyanate	Ag^+	4.0	25	4.59	3.70	1.77	1.20
	Fe^{3+}	1.8	18	1.96	2.02	<-0.41	>-0.14
	Ni^{2+}	1.0	20	1.18	0.46	0.17	
Thiosulfate	Ag^+	0.2	20	10.00	3.36		

* Stepwise constants are defined as follows:

$$M^{4+} + X^- \rightleftharpoons MX^{3+} \qquad K_1 = \frac{[MX^{3+}]}{[M^{4+}][X^-]}$$

$$MX^{3+} + X^- \rightleftharpoons MX_2^{2+} \qquad K_2 = \frac{[MX_2^{2+}]}{[MX^{3+}][X^-]}$$

etc.

TABLE 3 Solubility Product Constants

Compound	Formula	Solubility product constant, K_{sp}
Aluminum hydroxide	$Al(OH)_3$	5×10^{-33}
Barium carbonate	$BaCO_3$	7×10^{-9}
Barium chromate	$BaCrO_4$	2×10^{-10}
Barium fluoride	BaF_2	3×10^{-6}
Barium iodate	$Ba(IO_3)_2$	6×10^{-10}
Barium oxalate	BaC_2O_4	2×10^{-7}
Barium sulfate	$BaSO_4$	1×10^{-10}
Cadmium carbonate	$CdCO_3$	3×10^{-14}
Cadmium oxalate	CdC_2O_4	1×10^{-8}
Cadmium sulfide	CdS	5×10^{-27}
Calcium carbonate	$CaCO_3$	5×10^{-9}
Calcium fluoride	CaF_2	4×10^{-11}
Calcium oxalate	CaC_2O_4	2×10^{-9}
Calcium sulfate	$CaSO_4$	6×10^{-5}
Copper(II) hydroxide	$Cu(OH)_2$	2×10^{-19}
Copper(II) iodate	$Cu(IO_3)_2$	1×10^{-7}
Copper(II) oxalate	CuC_2O_4	3×10^{-8}
Copper(II) sulfide	CuS	4×10^{-38}
Copper(I) bromide	$CuBr$	6×10^{-9}
Copper(I) chloride	$CuCl$	3×10^{-7}
Copper(I) iodide	CuI	1×10^{-12}
Copper(I) thiocyanate	$CuSCN$	4×10^{-14}
Iron(III) hydroxide	$Fe(OH)_3$	1×10^{-36}
Iron(II) hydroxide	$Fe(OH)_2$	2×10^{-14}
Iron(II) oxalate	FeC_2O_4	2×10^{-7}
Iron(II) sulfide	FeS	4×10^{-19}
Lead carbonate	$PbCO_3$	2×10^{-13}
Lead chloride	$PbCl_2$	1×10^{-4}
Lead chromate	$PbCrO_4$	2×10^{-14}
Lead fluoride	PbF_2	5×10^{-8}
Lead hydroxide	$Pb(OH)_2$	3×10^{-16}
Lead iodate	$Pb(IO_3)_2$	3×10^{-13}
Lead sulfate	$PbSO_4$	2×10^{-8}
Lead sulfide	PbS	3×10^{-28}

Compound	Formula	Solubility product constant, K_{sp}
Magnesium ammonium phosphate	$MgNH_4PO_4$	3×10^{-13}
Magnesium carbonate	$MgCO_3$	3×10^{-5}
Magnesium fluoride	MgF_2	7×10^{-9}
Magnesium hydroxide	$Mg(OH)_2$	1×10^{-11}
Magnesium oxalate	MgC_2O_4	9×10^{-5}
Manganese(II) hydroxide	$Mn(OH)_2$	4×10^{-14}
Manganese(II) sulfide	MnS	1×10^{-16}
Mercury(II) sulfide	HgS	3×10^{-52}
Mercury(I) bromide	Hg_2Br_2	3×10^{-23}
Mercury(I) chloride	Hg_2Cl_2	6×10^{-19}
Mercury(I) iodide	Hg_2I_2	7×10^{-29}
Nickel sulfide	NiS	1×10^{-25}
Silver arsenate	Ag_3AsO_4	1×10^{-22}
Silver bromate	$AgBrO_3$	6×10^{-5}
Silver bromide	$AgBr$	4×10^{-13}
Silver carbonate	Ag_2CO_3	8×10^{-12}
Silver chloride	$AgCl$	1×10^{-10}
Silver chromate	Ag_2CrO_4	2×10^{-12}
Silver cyanide	$Ag[Ag(CN)_2]$	2×10^{-12}
Silver hydroxide	$AgOH$	2×10^{-8}
Silver iodate	$AgIO_3$	3×10^{-8}
Silver iodide	AgI	1×10^{-16}
Silver oxalate	$Ag_2C_2O_4$	5×10^{-12}
Silver sulfide	Ag_2S	1×10^{-48}
Silver thiocyanate	$AgSCN$	1×10^{-12}
Strontium carbonate	$SrCO_3$	2×10^{-9}
Strontium fluoride	SrF_2	3×10^{-9}
Strontium oxalate	SrC_2O_4	6×10^{-8}
Strontium sulfate	$SrSO_4$	3×10^{-7}
Thallium(I) chloride	$TlCl$	2×10^{-4}
Thallium(I) sulfide	Tl_2S	1×10^{-22}
Zinc carbonate	$ZnCO_3$	3×10^{-8}
Zinc hydroxide	$Zn(OH)_2$	2×10^{-14}
Zinc oxalate	ZnC_2O_4	3×10^{-9}
Zinc sulfide	ZnS	1×10^{-24}

TABLE 4 Standard Potentials*

Redox couple	$E°$
$F_2 + 2H^+ + 2e \rightleftharpoons 2HF(aq)$	3.06
$F_2 + 2e \rightleftharpoons 2F^-$	2.87
$O_3 + 2H^+ + 2e \rightleftharpoons O_2 + H_2O$	2.07
$S_2O_8^{2-} + 2e \rightleftharpoons 2SO_4^{2-}$	2.01
$Co^{3+} + e \rightleftharpoons Co^{2+}$	1.82
$H_2O_2 + 2H^+ + 2e \rightleftharpoons 2H_2O$	1.77
$MnO_4^- + 4H^+ + 3e \rightleftharpoons MnO_2 + 2H_2O$	1.70
$PbO_2 + SO_4^{2-} + 4H^+ + 2e \rightleftharpoons PbSO_4 + 2H_2O$	1.69
$Au^+ + e \rightleftharpoons Au$	1.68
$HClO_2 + 2H^+ + 2e \rightleftharpoons HClO + H_2O$	1.64
$HClO + H^+ + e \rightleftharpoons \frac{1}{2}Cl_2 + H_2O$	1.63
$Ce^{4+} + e \rightleftharpoons Ce^{3+}$	1.61
$Bi_2O_4 + 4H^+ + 2e \rightleftharpoons 2BiO^+ + 2H_2O$	1.59
$BrO_3^- + 6H^+ + 5e \rightleftharpoons \frac{1}{2}Br_2 + 3H_2O$	1.52
$MnO_4^- + 8H^+ + 5e \rightleftharpoons Mn^{2+} + 4H_2O$	1.51
$PbO_2 + 4H^+ + 2e \rightleftharpoons Pb^{2+} + 2H_2O$	1.46
$Cl_2 + 2e \rightleftharpoons 2Cl^-$	1.36
$Cr_2O_7^{2-} + 14H^+ + 6e \rightleftharpoons 2Cr^{3+} + 7H_2O$	1.33
$MnO_2 + 4H^+ + 2e \rightleftharpoons Mn^{2+} + 2H_2O$	1.23
$O_2 + 4H^+ + 4e \rightleftharpoons 2H_2O$	1.23
$IO_3^- + 6H^+ + 5e \rightleftharpoons \frac{1}{2}I_2 + 3H_2O$	1.20
$ClO_4^- + 2H^+ + 2e \rightleftharpoons ClO_3^- + H_2O$	1.19
$Br_2(aq) + 2e \rightleftharpoons 2Br^-$	1.09
$Br_2(liq) + 2e \rightleftharpoons 2Br^-$	1.07
$Br_3^- + 2e \rightleftharpoons 3Br^-$	1.05
$VO_2^+ + 2H^+ + e \rightleftharpoons VO^{2+} + H_2O$	1.00
$AuCl_4^- + 3e \rightleftharpoons Au + 4Cl^-$	1.00
$NO_3^- + 4H^+ + 3e \rightleftharpoons NO + 2H_2O$	0.96
$NO_3^- + 3H^+ + 2e \rightleftharpoons HNO_2 + H_2O$	0.94
$2Hg^{2+} + 2e \rightleftharpoons Hg_2^{2+}$	0.92
$AuBr_4^- + 3e \rightleftharpoons Au + 4Br^-$	0.87
$Cu^{2+} + I^- + e \rightleftharpoons CuI$	0.86
$Hg^{2} + 2e \rightleftharpoons Hg$	0.85
$Ag^+ + e \rightleftharpoons Ag$	0.80
$Hg_2^{2+} + 2e \rightleftharpoons 2Hg$	0.79
$Fe^{3+} + e \rightleftharpoons Fe^{2+}$	0.77
$PtCl_4^{2-} + 2e \rightleftharpoons Pt + 4Cl^-$	0.73
$O + 2H^+ + 2e \rightleftharpoons H_2O$	0.70
$PtBr_4^{2-} + 2e \rightleftharpoons Pt + 4Br^-$	0.58
$MnO_4^- + e \rightleftharpoons MnO_4^{2-}$	0.56
$H_3AsO_4 + 2H^+ + 2e \rightleftharpoons HAsO_2 + 2H_2O$	0.56
$I_3^- + 2e \rightleftharpoons 3I^-$	0.54
$I_2(s) + 2e \rightleftharpoons 2I^-$	0.54
$Cu^+ + e \rightleftharpoons Cu$	0.52
$4H_2SO_3 + 4H^+ + 6e \rightleftharpoons S_4O_6^{2-} + 6H_2O$	0.51
$2H_2SO_3 + 2H^+ + 4e \rightleftharpoons S_2O_3^{2-} + 3H_2O$	0.40
$Fe(CN)_6^{3-} + e \rightleftharpoons Fe(CN)_6^{4-}$	0.36
$VO^{2+} + 2H^+ + e \rightleftharpoons V^{3+} + H_2O$	0.36
$Cu^{2+} + 2e \rightleftharpoons Cu$	0.34
$Hg_2Cl_2 + 2e \rightleftharpoons 2Hg + 2Cl^-$	0.28
$IO_3^- + 3H_2O + 6e \rightleftharpoons I^- + 6OH^-$	0.26
$AgCl + e \rightleftharpoons Ag + Cl^-$	0.22
$HgBr_4^{2-} + 2e \rightleftharpoons Hg + 4Br^-$	0.21
$Cu^{2+} + e \rightleftharpoons Cu^+$	0.15
$Sn^{4+} + 2e \rightleftharpoons Sn^{2+}$	0.15
$S + 2H^+ + 2e \rightleftharpoons H_2S$	0.14
$CuCl + e \rightleftharpoons Cu + Cl^-$	0.14
$AgBr + e \rightleftharpoons Ag + Br^-$	0.10
$S_4O_6^{2-} + 2e \rightleftharpoons 2S_2O_3^{2-}$	0.08
$CuBr + e \rightleftharpoons Cu + Br^-$	0.03
$2H^+ + 2e \rightleftharpoons H_2$	0.00
$HgI_4^{2-} + 2e \rightleftharpoons Hg + 4I^-$	−0.04
$Pb^{2+} + 2e \rightleftharpoons Pb$	−0.13
$CrO_4^{2-} + 4H_2O + 3e \rightleftharpoons Cr(OH)_3 + 5OH^-$	−0.13
$Sn^{2+} + 2e \rightleftharpoons Sn$	−0.14
$AgI + e \rightleftharpoons Ag + I^-$	−0.15
$CuI + e \rightleftharpoons Cu + I^-$	−0.19
$Ni^{2+} + 2e \rightleftharpoons Ni$	−0.25
$V^{3+} + e \rightleftharpoons V^{2+}$	−0.26
$PbCl_2 + 2e \rightleftharpoons Pb + 2Cl^-$	−0.27
$Co^{2+} + 2e \rightleftharpoons Co$	−0.28
$PbBr_2 + 2e \rightleftharpoons Pb + 2Br^-$	−0.28
$PbSO_4 + 2e \rightleftharpoons Pb + SO_4^{2-}$	−0.36
$PbI_2 + 2e \rightleftharpoons Pb + 2I^-$	−0.37
$Cd^{2+} + 2e \rightleftharpoons Cd$	−0.40
$Cr^{3+} + e \rightleftharpoons Cr^{2+}$	−0.41
$Fe^{2+} + 2e \rightleftharpoons Fe$	−0.44
$2CO_2(g) + 2H^+ + 2e \rightleftharpoons H_2C_2O_4(aq)$	−0.49
$Cr^{3+} + 3e \rightleftharpoons Cr$	−0.74
$Zn^{2+} + 2e \rightleftharpoons Zn$	−0.76
$H_2O + e \rightleftharpoons \frac{1}{2}H_2 + OH^-$	−0.83
$Mn^{2+} + 2e \rightleftharpoons Mn$	−1.18
$Al^{3+} + 3e \rightleftharpoons Al$	−1.66
$Mg^{2+} + 2e \rightleftharpoons Mg$	−2.37
$Na^+ + e \rightleftharpoons Na$	−2.71
$Ca^{2+} + 2e \rightleftharpoons Ca$	−2.87
$Sr^{2+} + 2e \rightleftharpoons Sr$	−2.89
$Ba^{2+} + 2e \rightleftharpoons Ba$	−2.90
$K^+ + e \rightleftharpoons K$	−2.93
$Li^+ + e \rightleftharpoons Li$	−3.05

*From W. M. Latimer *Oxidation Potentials*, 2nd ed., Prentice-Hall, Inc., Englewood Cliffs, N. J., 1952.

TABLE 5 Some Formal Potentials

Redox systems	Standard potential	Formal potential	Solution
$Ce^{4+} + e \rightleftharpoons Ce^{3+}$	—	1.23	$1M$ HCl
		1.44	$1M$ H_2SO_4
		1.61	$1M$ HNO_3
		1.7	$1M$ $HClO_4$
$Fe^{3+} + e \rightleftharpoons Fe^{2+}$	+0.771	0.68	$1M$ H_2SO_4
		0.700	$1M$ HCl
		0.732	$1M$ $HClO_4$
$Cr_2O_7^{2-} + 14H^+ + 6e \rightleftharpoons 2Cr^{3+} + 7H_2O$	+1.33	1.00	$1M$ HCl
		1.05	$2M$ HCl
		1.08	$3M$ HCl
		1.08	$0.5M$ H_2SO_4
		1.15	$4M$ H_2SO_4
		1.03	$1M$ $HClO_4$
$Fe(CN)_6^{3-} + e \rightleftharpoons Fe(CN)_6^{4-}$	+0.356	0.48	$0.01M$ HCl
		0.56	$0.1M$ HCl
		0.71	$1M$ HCl
		0.72	$1M$ H_2SO_4
		0.72	$1M$ $HClO_4$
$H_3AsO_4 + 2H^+ + 2e \rightleftharpoons H_3AsO_3 + H_2O$	+0.559	0.557	$1M$ HCl
		0.557	$1M$ $HClO_4$
$TiO^{2+} + 2H^+ + e \rightleftharpoons Ti^{3+} + H_2O$	+0.1	0.04	$1M$ H_2SO_4
$Pb^{2+} + 2e \rightleftharpoons Pb$	−0.126	−0.14	$1M$ $HClO_4$
$Sn^{2+} + 2e \rightleftharpoons Sn$	−0.136	−0.16	$1M$ $HClO_4$
$V^{3+} + e \rightleftharpoons V^{2+}$	−0.255	−0.21	$1M$ $HClO_4$

APPENDIX TWO

balancing oxidation-reduction equations

We shall describe briefly the oxidation number and the half-reaction methods for balancing redox equations. Both methods are presented in most general chemistry texts today.

Oxidation Number Method

In this method oxidation numbers are assigned to atoms in the reactants and products, and any changes in these numbers are attributed to a loss or gain of electrons. The number of electrons gained by the oxidizing agent is made equal to the number lost by the reducing agent by selecting appropriate coefficients for these two reactants.

Although the assignment of oxidation numbers can be made on an arbitrary basis, it is convenient to select the numbers as follows:

1. In ionic compounds the oxidation number is taken to be the same as the number of electrons gained or lost by the elements in forming the ion. For example, in sodium chloride, Na^+Cl^-. the oxidation number (also called oxidation state) of sodium is +1, that of chlorine −1. In zinc oxide, ZnO, zinc is taken as +2, oxygen as −2.
2. In covalent compounds of known structure, where electrons are shared by two atoms, the electrons are counted as belonging to the more electronegative of the two atoms. In HCl, for example, chlorine is assigned a number of −1, hydrogen a number of +1. In a covalent bond between like atoms, such as Cl_2, one electron is assigned to each atom, making the oxidation number of each Cl

atom zero. The oxidation number of any elemental substance, such as Na, H_2, O_3, P_4, etc., is zero.

In assigning oxidation numbers in compounds containing hydrogen, it is customary to start with hydrogen as +1. The only exception occurs in ionic hydrides, such as Na^+H^-, where it is taken as −1. Oxygen is next assigned a value of −2 except in peroxides, such as H_2O_2, where the value must be −1 if hydrogen is +1.

It should be noted that the algebraic sum of oxidation numbers in a neutral molecule is zero. Similarly, in a complex ion the sum of the oxidation numbers of the atoms must equal the charge of the ion. For example, in the ion HPO_4^{2-}, the oxidation numbers are H = +1, O = −2, and P = +5, giving the sum $+1 + 5 + 4(-2) = -2$.

The following examples illustrate the use of the oxidation number method to balance equations. Since the reactions occur in aqueous solution, we may need to add H^+, OH^-, or H_2O to balance oxygen and hydrogen atoms.

Example 1. Potassium permanganate, $KMnO_4$, oxidizes oxalic acid, $H_2C_2O_4$, in acid solution to form manganous ion, Mn^{2+}, and carbon dioxide, CO_2:

$$MnO_4^- + H_2C_2O_4 \longrightarrow Mn^{2+} + CO_2$$

Balance the equation adding H^+ and H_2O as needed.

Note that the oxidation number of manganese changes from +7 in MnO_4^- to +2 in Mn^{2+}. Hence each MnO_4^- ion gains five electrons. The oxidation number of carbon changes from +3 in $H_2C_2O_4$ to +4 in CO_2, a loss of one electron per carbon atom. Since each $H_2C_2O_4$ molecule contains two carbon atoms, the molecule loses two electrons. We can equate the electrons gained and lost by taking $2MnO_4^-$ ions for each $5H_2C_2O_4$ molecules. Hence we write

$$2MnO_4^- + 5H_2C_2O_4 \longrightarrow 2Mn^{2+} + 10CO_2 \qquad (1)$$

Next we balance the ionic charge by placing H^+ ions where needed. We need six plus charges on the left to make the charge +4 on each side of the arrow. The process is completed by adding eight molecules of water to the right side to balance the oxygen and hydrogen atoms.

$$2MnO_4^- + 5H_2C_2O_4 + 6H^+ \longrightarrow 2Mn^{2+} + 10CO_2 + 8H_2O \qquad (2)$$

Example 2. Chromate ion, CrO_4^{2-}, oxidizes sulfite ion, SO_3^{2-}, to sulfate, SO_4^{2-}, and is reduced to chromite, CrO_2^-, in basic solution:

$$CrO_4^{2-} + SO_3^{2-} \longrightarrow CrO_2^- + SO_4^{2-}$$

Balance the equation adding water and OH^- as needed.

Note that the oxidation number of chromium changes from +6 to +3, a gain of three electrons. Sulfur changes from +4 to +6, a loss of two electrons. Hence we write

$$2CrO_4^{2-} + 3SO_3^{2-} \longrightarrow 2CrO_2^- + 3SO_4^{2-}$$

Next balance the ionic charge by placing two hydroxide ions on the right side of the equation. The addition of one molecule of water to the left completes the balancing.

$$2CrO_4^{2-} + 3SO_3^{2-} + H_2O \longrightarrow 2CrO_2^- + 3SO_4^{2-} + 2OH^-$$

Half-Reaction Method

A redox equation can be balanced by writing separate reactions for the oxidizing and reducing agents. These "half-reactions" are balanced by adding electrons where needed. The numbers of electrons gained and lost in the two reactions are made the same, and the two equations are added to give the desired result.

Example 1. Balance the following equation for the reaction which occurs in aqueous acid.

$$MnO_4^- + C_2O_4^{2-} \longrightarrow Mn^{2+} + CO_2$$

Add H_2O and H^+ where needed.

First treat the oxidizing agent:

$$MnO_4^- \longrightarrow Mn^{2+}$$

Add water to the right to balance the oxygen:

$$MnO_4^- \longrightarrow Mn^{2+} + 4H_2O$$

Add H^+ to the left to balance hydrogen:

$$MnO_4^- + 8H^+ \longrightarrow Mn^{2+} + 4H_2O$$

The charge is balanced by placing five electrons on the left:

$$MnO_4^- + 8H^+ + 5e \longrightarrow Mn^{2+} + 4H_2O \qquad (1)$$

Second, treat the reducing agent:

$$C_2O_4^{2-} \longrightarrow 2CO_2$$

This is balanced except for the charge. Add two electrons to the right:

$$C_2O_4^{2-} \longrightarrow 2CO_2 + 2e \qquad (2)$$

Multiply equation (1) by 2 and equation (2) by 5, and add to give the final balanced equation:

$$2MnO_4^- + 5C_2O_4^{2-} + 16H^+ \longrightarrow 2Mn^{2+} + 10CO_2 + 8H_2O$$

Example 2. Balance the following reaction in basic solution:

$$Cr(OH)_3 + IO_3^- \longrightarrow I^- + CrO_4^{2-}$$

First treat the reducing agent:

$$Cr(OH)_3 \longrightarrow CrO_4^{2-}$$

Add H_2O to the left and H^+ to the right:

$$Cr(OH)_3 + H_2O \longrightarrow CrO_4^{2-} + 5H^+$$

Since we wish the equation to show OH^- ions rather than H^+ ions, we can add 5 OH^- to both sides. This yields

$$Cr(OH)_3 + H_2O + 5OH^- \longrightarrow CrO_4^{2-} + 5H_2O$$

or

$$Cr(OH)_3 + 5OH^- \longrightarrow CrO_4^{2-} + 4H_2O$$

Now balance the charge by adding three electrons to the right:

$$Cr(OH)_3 + 5OH^- \longrightarrow CrO_4^{2-} + 4H_2O + 3e \quad (1)$$

Next treat the oxidizing agent similarly, giving

$$IO_3^- + 6H^+ \longrightarrow I^- + 3H_2O$$

$$IO_3^- + 6H_2O \longrightarrow I^- + 3H_2O + 6OH^-$$

$$IO_3^- + 3H_2O + 6e \longrightarrow I^- + 6OH^- \quad (2)$$

Multiply equation (1) by 2 and add it to equation (2), giving

$$2Cr(OH)_3 + IO_3^- + 4OH^- \longrightarrow 2CrO_4^{2-} + I^- + 5H_2O$$

EXERCISES

1. Balance the following half reactions in aqueous acid:
   (a) $MnO_4^- \longrightarrow Mn^{3+}$
   (b) $Cr_2O_7^{2-} \longrightarrow Cr^{3+}$
   (c) $H_3AsO_3 \longrightarrow H_3AsO_4$
   (d) $PH_3 \longrightarrow HPO_3^{2-}$
   (e) $S_2O_3^{2-} \longrightarrow SO_4^{2-}$

2. Balance the following half reactions in aqueous base:
   (a) $MnO_4^- \longrightarrow MnO_2$
   (b) $O_2 \longrightarrow OH^-$
   (c) $Cu_2O \longrightarrow Cu(OH)_2$
   (d) $CrO_4^{2-} \longrightarrow Cr(OH)_3$
   (e) $Sn^{2+} \longrightarrow SnO_3^{2-}$

3. Balance the following equations, adding H_2O and H^+ where needed. Where basic medium is indicated, show OH^- ions rather than H^+ ions. Use either the oxidation number or half reaction method, as you prefer.
   (a) $H_2O_2 + MnO_4^- \longrightarrow Mn^{2+} + O_2$
   (b) $SO_3^{2-} + Br_2 \longrightarrow SO_4^{2-} + Br^-$
   (c) $I_2 \longrightarrow IO_3^- + I^-$ (basic)
   (d) $Mn^{2+} + BiO_3^- \longrightarrow MnO_4^- + Bi^{3+}$
   (e) $HO_2^- + CrO_2^- \longrightarrow CrO_4^{2-}$ (basic)
   (f) $Al + NO_3^- \longrightarrow AlO_2^- + NH_3$ (basic)
   (g) $MnO_4^- + VO^{2+} \longrightarrow VO_3^- + Mn^{2+}$
   (h) $C_7H_8O + Cr_2O_7^{2-} \longrightarrow C_7H_8O_2 + Cr^{3+}$
   (i) $SbH_3 + Cl_2O \longrightarrow H_4Sb_2O_7 + Cl^-$
   (j) $FeS + NO_3^- \longrightarrow Fe^{3+} + NO + S$
   (k) $MnO_4^- + CN^- \longrightarrow MnO_2 + CNO^-$ (basic)

(l) $UO_5^{2-} \longrightarrow UO_2^{2+} + O_2$
(m) $Zn + NO_3^- \longrightarrow Zn^{2+} + NH_4^+$
(n) $Fe + ClO_4^- \longrightarrow Fe^{3+} + Cl_2$
(o) $Pt + NO_3^- + Cl^- \longrightarrow PtCl_6^{2-} + NO_2$
(p) $VO_2^+ + V^{2+} \longrightarrow VO^{2+}$
(q) $S_2O_3^{2-} + I_2 \longrightarrow S_4O_6^{2-} + I^-$
(r) $I^- + Cr_2O_7^{2-} \longrightarrow I_2 + Cr^{3+}$
(s) $Cl^- + MnO_4^- \longrightarrow Cl_2 + Mn^{2+}$
(t) $MnO_4^- + Mn^{2+} \longrightarrow MnO_2$ (basic)

APPENDIX THREE

the literature of analytical chemistry

People have referred to an "information explosion" which threatens to inundate the scientists in a sea of paper. The volume of published research in chemistry is possibly more than doubling every 10 years, and wags are saying that it is easier to repeat work in the laboratory than to find it in the literature. Experts are considering various automated schemes for retrieving information, but it is impossible now to predict what library services may ultimately become available. Meanwhile, the chemist has to get along somehow, and there are ways of finding information in the literature that are more efficient than others. The student should certainly become familiar with the library during his undergraduate years.

The *primary* literature sources are published research papers in a host of journals, American and foreign. Some of these journals of a more general nature publish papers of broad interest, while others attract primarily a narrower readership, say, organic, physical, or analytical chemists. *Secondary* sources include brief abstracts of published papers, articles which review the literature in restricted areas, monographs in which experts describe at length the status of certain fields, treatises and other reference works, and textbooks. The beginning student will normally use mainly secondary sources, but as he pushes closer to the present boundary of knowledge, he should increasingly consult original research papers.

We can present here only a bare introduction to the literature of analytical chemistry. Further, it must be remembered that the research analytical chemist often confronts problems which have never been investigated before; it would be impossible to define ahead of time the fields in which he might read in order to find approaches to such problems, but it is not unusual to find organic reaction mechanisms, spectroscopy, electronics, and other such "nonanalytical" information brought to bear in analytical problems.

ANALYTICAL JOURNALS

Analyst, dating from 1877, is a publication of the Society for Analytical Chemistry in Great Britain.

Analytical Biochemistry, is a journal which presents applications of analytical chemistry to biochemistry.

Analytical Chemistry, published monthly by the American Chemical Society. Prior to 1947, this was called the Analytical Edition of *Industrial and Engineering Chemistry*, but the yearly volumes are numbered consecutively, without break, dating from 1929.

Analytica Chimica Acta, published by Elsevier Publishing Co. in The Netherlands since 1947. Papers are written in English, French, or German; each has a summary in all three languages.

Analytical Letters, is an international journal for rapid communication, published by Marcel Dekker, Inc., since 1968.

Chemical Instrumentation is a journal of experimental techniques in chemistry and biochemistry, published by Marcel Dekker, Inc., since 1969.

Chimie Analytique (now called *Analusis*) is the major French journal in the field.

CRC Critical Reviews in Analytical Chemistry presents informed, thorough, up-to-date, and critical reviews of important topics in analytical chemistry. It has been published since 1971 by The Chemical Rubber Company.

Journal of Chromatographic Science, published since 1963 by Preston Technical Abstracts Co., Evanston, Ill.

Journal of Chromatography, published since 1958 by Elsevier Publishing Co. in The Netherlands.

Journal of Electroanalytical Chemistry, since 1962, an Elsevier publication dealing with all aspects of theory and practice of electrical methods of analysis.

Separation Science, published since 1966, is an interdisciplinary journal of methods and processes of separation; a Marcel Dekker, Inc., publication.

Talanta, published since 1958 by Pergamon Press, Inc., contains research papers in English, French, and German, with summaries in all three languages.

Zeitschrift für analytische Chemie, dating from 1862, is the principal German analytical journal.

Zhurnal Analitischeskoi Khimii (Journal of Analytical Chemistry of the USSR) is the Russian analytical journal. It is available in English translation from Consultants Bureau, New York.

ABSTRACTS

Chemical Abstracts (*Chem. Abstr.* or *C.A.*), published since 1907 by the American Chemical Society, is the outstanding source of information for all fields of chemistry and certain related areas as well. *Chemical Abstracts* publishes brief abstracts of papers from nearly 10,000 periodicals, and it covers the patent literature as well. It appears biweekly. The abstracts are grouped into 74 sections, such as section 2, analytical chemistry, section 3, general physical chemistry, section 4, surface chemistry and colloids, and section 60, biochemical methods. Annual indexes provide access by subject, authors' names, and chemical formula. Decennial indexes covering 10-year periods (e.g., 1907–16, 1917–26) appeared through 1956; the last collective index covered a five-year period, 1967–71. The nomenclature and list of abbreviations used by *C.A.* are practically official for writers of American books and journal articles. A thorough search through *C.A.* is a tedious job, and workers frequently employ shortcuts based upon their experience or upon information obtained from other sources such as reference books, review articles, and monographs. But *C.A.* is the ultimate source, and information which cannot be found there can probably be best obtained by original work in the laboratory.

Chemisches Zentralblatt, the German abstract journal, dates back to 1830 and hence covers literature published before *C.A.* began. It is particularly valuable for this period, but has no advantage over *C.A.* for more recent information.

Analytical Abstracts, published by the Society for Analytical Chemistry, was formerly a section of the now-discontinued *British Abstracts*, appearing now in its own right. This is a short periodical, and a routine scan each month doesn't take much time.

Chemical Titles, published by the American Chemical Society, lists the titles of most papers in the world's chemical literature very promptly. Indexes are prepared on the basis of key words in the titles.

In addition to these more or less general abstract journals, abstracts sometimes appear catering to a particular special field. An example of these is *Gas Chromatography Abstracts*, published by the Institute of Petroleum in London. It appears annually, with subject and author indexes.

REVIEWS

Each April, *Analytical Chemistry* publishes along with its regular issue a separate Annual Review issue. In even years (e.g., 1964, 1966) fundamental topics are reviewed, while applied topics are covered in the alternate years. Typical fundamental reviews are gas chromatography, nucleonics, polarographic theory, emission spectrometry, nuclear magnetic resonance, and X-ray diffraction, while titles of applied reviews include clinical analysis, ferrous alloys, and paints and finishes. Each review will normally give hundreds of journal references, organized in some convenient fashion. These are the most comprehensive reviews of the analytical literature, but others are also published; for example, the *Annual Reports* on the progress of chemistry to the Chemical Society of London has an analytical section.

REFERENCE BOOKS AND ADVANCED TEXTS

Probably, at least for a few years, the best single reference work will be *Treatise on Analytical Chemistry*, edited by I. M. Kolthoff, P. J. Elving, and E. B. Sandell and published by Interscience Publishers, a division of John Wiley & Sons, Inc., New York. Publication began in 1959, and volumes are still appearing. The goal is "to present a concise, critical, comprehensive, and systematic, but not exhaustive, treatment of all aspects of classical and modern analytical chemistry." Publication is in three parts: Part I, Theory and Practice; Part II, Analytical Chemistry of the Elements; Part III, Analysis of Industrial Products.

A second general reference work with a similar goal, *Comprehensive Analytical Chemistry*, edited by C. L. Wilson and D. W. Wilson and published by Elsevier Publishing Co., Amsterdam, began to appear in 1959. Chapters written by experts are well referenced with regard to the original literature.

There is much information of analytical interest in the multivolume series *Technique of Organic Chemistry*, edited by A. Weissberger (John Wiley & Sons, Inc., New York). Volumes have been appearing since 1959.

The following are textbooks of a more or less general nature which are more advanced than the present one.

H. A. LAITINEN AND W. E. HARRIS, *Chemical Analysis*, 2nd ed., McGraw-Hill Book Company, New York, 1975.

S. SIGGIA, *Survey of Analytical Chemistry*, McGraw-Hill Book Company, New York, 1968.

T. B. SMITH, *Analytical Processes: A Physico-Chemical Interpretation*, 2nd ed., Edward Arnold Ltd., London, 1952.

H. F. WALTON, *Principles and Methods of Chemical Analysis*, 2nd ed., Prentice-Hall, Inc., Englewood Cliffs, N.J., 1964.

In addition, the elementary text *Quantitative Chemical Analysis*, 4th ed., by I. M. KOLTHOFF, E. B. SANDELL, E. J. MEEHAN, and S. BRUCKENSTEIN (Macmillan Publishing Co., Inc., New York, 1969) contains an unusual wealth of information for an elementary book.

SOME REPRESENTATIVE BOOKS ON SPECIAL TOPICS

R. N. ADAMS, *Electrochemistry at Solid Electrodes*, Marcel Dekker, Inc., New York, 1969.

M. R. F. ASHWORTH, *Titrimetric Organic Analysis*, Wiley–Interscience, New York, 1964–65.

A. J. BARD, ed., *Electroanalytical Chemistry*, Marcel Dekker, Inc., New York, 1966– (a continuing series).

R. G. BATES, *Determination of pH: Theory and Practice*, John Wiley & Sons, Inc., New York, 1964.

R. P. BAUMAN, *Absorption Spectroscopy*, John Wiley & Sons, Inc., New York, 1962.

E. W. BERG, *Physical and Chemical Methods of Separation*, McGraw-Hill Book Company, New York, 1963.

R. J. BLOCK and G. ZWEIG, *A Practical Manual of Paper Chromatography and Electrophoresis*, Academic Press, Inc., New York, 1962.

D. F. BOLTZ, ed., *Colorimetric Determination of Nonmetals*, John Wiley & Sons, Inc., New York, 1958.

S. J. CLARK, *Quantitative Methods of Organic Microanalysis*, Butterworth & Co. Ltd., London, 1956.

R. C. DENNY, *A Dictionary of Chromatography*, John Wiley & Sons, Inc., New York, 1976.

J. N. DONE and J. H. KNOX, *Applications of High-Speed Liquid Chromatography*, John Wiley & Sons, Inc., 1974.

R. A. DURST, ed., *Ion-Selective Electrodes, National Bureau of Standards Special Publication 314*, U.S. Government Printing Office, Washington, D.C., 1969.

C. DUVAL, *Inorganic Thermogravimetric Analysis*, 2nd ed., Elsevier Publishing Co., Inc., Amsterdam, 1963.

W. T. ELWELL and J. A. F. GIDLEY, *Atomic Absorption Spectrophotometry*, 2nd ed., Pergamon Press, Inc., New York, 1966.

H. A. FLASCHKA, *EDTA Titrations*, Pergamon Press, Inc., New York, 1959.

H. A. FLASCHKA and A. J. BARNARD, JR., *Chelates in Analytical Chemistry*, Vols. I–IV, Marcel Dekker, Inc., New York, 1967–72.

D. GLICK, ed., *Methods of Biochemical Analysis*, John Wiley & Sons, Inc., New York, 1954– (a continuing series).

E. HEFTMANN, *Chromatography*, Reinhold Publishing Corp., New York, 1961.

F. HELFFERICH, *Ion Exchange*, McGraw-Hill Book Company, New York, 1962.

W. F. HILLEBRAND, G. E. F. LUNDELL, H. A. BRIGHT, and J. I. HOFFMANN, *Applied Inorganic Analysis*, 2nd ed., John Wiley & Sons, Inc., New York, 1953.

W. HUBER, *Titrations in Nonaqueous Solvents*, Academic Press, Inc., New York, 1967.

B. L. KARGER, L. R. SNYDER, and C. HORVATH, *An Introduction to Separation Science*, John Wiley & Sons, Inc., New York, 1973.

J. J. KIRKLAND, ed., *Modern Practice of Liquid Chromatography*, Wiley–Interscience, New York, 1971.

I. M. KOLTHOFF and J. J. LINGANE, *Polarography*, 2nd ed., Interscience Publishers, Inc., New York, 1952.

I. M. KOLTHOFF and V. A. STENGER, *Volumetric Analysis*, 2nd ed., John Wiley & Sons, Inc., New York, 1942–57.

W. M. LATIMER, *The Oxidation States of the Elements and Their Potentials in Aqueous Solutions*, 2nd ed., Prentice-Hall, Inc., Englewood Cliffs, N.J., 1952.

J. J. LINGANE, *Electroanalytical Chemistry*, 2nd ed., Wiley–Interscience, New York, 1958.

J. J. LINGANE, *Analytical Chemistry of Selected Metallic Elements*, Reinhold Publishing Corporation, New York, 1966.

A. B. LITTLEWOOD, *Gas Chromatography: Principles, Techniques, and Applications*, 2nd ed., Academic Press, Inc., New York, 1970.

C. K. MANN, T. J. VICKERS, and W. M. GULICK, *Instrumental Analysis*, Harper & Row, Publishers, New York, 1974.

L. MEITES, *Polarographic Techniques*, 2nd ed., Wiley–Interscience, New York, 1965.

L. MEITES, ed., *Handbook of Analytical Chemistry*, McGraw-Hill Book Company, New York, 1963.

J. M. MILLER, *Separation Methods in Chemical Analysis*, John Wiley & Sons, Inc., New York, 1975.

J. MITCHELL, JR., I. M. KOLTHOFF, E. S. PROSKAUER, and A. WEISSBERGER, eds., *Organic Analysis*, John Wiley & Sons, Inc., New York, 1953– (a continuing series).

G. H. MORRISON and H. FREISER, *Solvent Extraction in Analytical Chemistry*, John Wiley & Sons, Inc., New York, 1957.

H. W. NURNBERG, ed., *Electroanalytical Chemistry*, John Wiley & Sons, Inc., New York, 1974.

J. P. PHILLIPS, *Automatic Titrators*, Academic Press, Inc., New York, 1959.

C. N. Reilley, ed., *Advances in Analytical Chemistry and Instrumentation*, Wiley–Interscience, New York, 1960– (a continuing series).

O. SAMUELSON, *Ion Exchange Separations in Analytical Chemistry*, John Wiley & Sons, Inc., New York, 1963.

E. B. SANDELL, *Colorimetric Determination of Traces of Metals*, 3rd ed., Wiley–Interscience, New York, 1959.

G. SCHWARZENBACH and H. FLASHKA, *Complexometric Titrations*, trans. by H. M. N. H. IRVING, Methuen & Co. Ltd., London, 1969.

S. SIGGIA and H. J. STOLTEN, *An Introduction to Modern Organic Analysis*, John Wiley & Sons, Inc., New York, 1956.

S. SIGGIA, *Quantitative Organic Analysis via Functional Groups*, 3rd ed., John Wiley & Sons, Inc, New York, 1963.

S. SIGGIA, ed., *Instrumental Methods of Organic Functional Group Analysis*, John Wiley & Sons, Inc., New York, 1972.

D. A. SKOOG and D. M. WEST, *Principles of Instrumental Analysis*, Holt, Rinehart and Winston, Inc., New York, 1971.

F. D. SNELL and F. M. BIFFEN, *Commercial Methods of Analysis*, Chemical Publishing Company, Inc., New York, 1964.

E. I. STERNS, *The Practice of Absorption Spectrophotometry*, John Wiley & Sons, Inc., New York, 1969.

H. A. STROBEL, *Chemical Instrumentation: A Systematic Approach to Instrumental Analysis*, 2nd ed., Addison-Wesley Publishing Co., Inc., Reading, Mass., 1973.

W. WAGNER and C. J. HULL, *Inorganic Titrimetric Analysis*, Marcel Dekker, Inc., New York, 1971.

H. F. WALTON, ed., *Ion Exchange Chromatography*, Dowden, Hutchinson and Ross, Inc., Stroudsburg, Pa., 1976.

APPENDIX FOUR

four-place table of logarithms

Four-place Table of Logarithms

No.	0	1	2	3	4	5	6	7	8	9
10	0000	0043	0086	0128	0170	0212	0253	0294	0334	0374
11	0414	0453	0492	0531	0569	0607	0645	0682	0719	0755
12	0792	0828	0864	0899	0934	0969	1004	1038	1072	1106
13	1139	1173	1206	1239	1271	1303	1335	1367	1399	1430
14	1461	1492	1523	1553	1584	1614	1644	1673	1703	1732
15	1761	1790	1818	1847	1875	1903	1931	1959	1987	2014
16	2041	2068	2095	2122	2148	2175	2201	2227	2253	2279
17	2304	2330	2355	2380	2405	2430	2455	2480	2504	2529
18	2553	2577	2601	2625	2648	2672	2695	2718	2742	2765
19	2788	2810	2833	2856	2878	2900	2923	2945	2967	2989
20	3010	3032	3054	3075	3096	3118	3139	3160	3181	3201
21	3222	3243	3263	3284	3304	3324	3345	3365	3385	3404
22	3424	3444	3464	3483	3502	3522	3541	3560	3579	3598
23	3617	3636	3655	3674	3692	3711	3729	3747	3766	3784
24	3802	3820	3838	3856	3874	3892	3909	3927	3945	3962
25	3979	3997	4014	4031	4048	4065	4082	4099	4116	4133
26	4150	4166	4183	4200	4216	4232	4249	4265	4281	4298
27	4314	4330	4346	4362	4378	4393	4409	4425	4440	4456
28	4472	4487	4502	4518	4533	4548	4564	4579	4594	4609
29	4624	4639	4654	4669	4683	4698	4713	4728	4742	4757
30	4771	4786	4800	4814	4829	4843	4857	4871	4886	4900
31	4914	4928	4942	4955	4969	4983	4997	5011	5024	5038
32	5051	5065	5079	5092	5105	5119	5132	5145	5159	5172
33	5185	5198	5211	5224	5237	5250	5263	5276	5289	5302
34	5315	5328	5340	5353	5366	5378	5391	5403	5416	5428
35	5441	5453	5465	5478	5490	5502	5514	5527	5539	5551
36	5563	5575	5587	5599	5611	5623	5635	5647	5658	5670
37	5682	5694	5705	5717	5729	5740	5752	5763	5775	5786
38	5798	5809	5821	5832	5843	5855	5866	5877	5888	5899
39	5911	5922	5933	5944	5955	5966	5977	5988	5999	6010
40	6021	6031	6042	6053	6064	6075	6085	6096	6107	6117
41	6128	6138	6149	6160	6170	6180	6191	6201	6212	6222
42	6232	6243	6253	6263	6274	6284	6294	6304	6314	6325
43	6335	6345	6355	6365	6375	6386	6395	6405	6415	6425
44	6435	6444	6454	6464	6474	6484	6493	6503	6513	6522
45	6532	6542	6551	6561	6571	6580	6590	6599	6609	6618
46	6628	6637	6646	6656	6665	6675	6684	6693	6702	6712
47	6721	6730	6739	6749	6758	6767	6776	6785	6794	6803
48	6812	6821	6830	6839	6848	6857	6866	6875	6884	6893
49	6902	6911	6920	6928	6937	6946	6955	6964	6972	6981
50	6990	6998	7007	7016	7024	7033	7042	7050	7059	7067
51	7076	7084	7093	7101	7110	7118	7126	7135	7143	7152
52	7160	7168	7177	7185	7193	7202	7210	7218	7226	7235
53	7243	7251	7259	7267	7275	7284	7292	7300	7308	7316
54	7324	7332	7340	7348	7356	7364	7372	7380	7388	7396
	0	1	2	3	4	5	6	7	8	9

Four-place Table of Logarithms (continued)

No.	0	1	2	3	4	5	6	7	8	9
55	7404	7412	7419	7427	7435	7443	7451	7459	7466	7474
56	7482	7490	7497	7505	7513	7520	7528	7536	7543	7551
57	7559	7566	7574	7582	7589	7597	7604	7612	7619	7627
58	7634	7642	7649	7657	7664	7672	7679	7686	7694	7701
59	7709	7716	7723	7731	7738	7745	7752	7760	7767	7774
60	7782	7789	7796	7803	7810	7818	7825	7832	7839	7846
61	7853	7860	7868	7875	7882	7889	7896	7903	7910	7917
62	7924	7931	7938	7945	7952	7959	7966	7973	7980	7987
63	7992	8000	8007	8014	8021	8028	8035	8041	8048	8055
64	8062	8069	8075	8082	8089	8096	8102	8109	8116	8122
65	8129	8136	8142	8149	8156	8162	8169	8176	8182	8189
66	8195	8202	8209	8215	8222	8228	8235	8241	8248	8254
67	8261	8267	8274	8280	8287	8293	8299	8306	8312	8319
68	8325	8331	8338	8344	8351	8357	8363	8370	8376	8382
69	8388	8395	8401	8407	8414	8420	8426	8432	8439	8445
70	8451	8457	8463	8470	8476	8482	8488	8494	8500	8506
71	8513	8519	8525	8531	8537	8543	8549	8555	8561	8567
72	8573	8579	8585	8591	8597	8603	8609	8615	8621	8627
73	8633	8639	8645	8651	8657	8663	8669	8675	8681	8686
74	8692	8698	8704	8710	8716	8722	8727	8733	8739	8745
75	8751	8756	8762	8768	8774	8779	8785	8791	8797	8802
76	8808	8814	8820	8825	8831	8837	8842	8848	8854	8859
77	8865	8871	8876	8882	8887	8893	8899	8904	8910	8915
78	8921	8927	8932	8938	8943	8949	8954	8960	8965	8971
79	8976	8982	8987	8993	8998	9004	9009	9015	9020	9025
80	9031	9036	9042	9047	9053	9058	9063	9069	9074	9079
81	9085	9090	9096	9101	9106	9112	9117	9122	9128	9133
82	9138	9143	9149	9154	9159	9165	9170	9175	9180	9186
83	9191	9196	9201	9206	9212	9217	9222	9227	9232	9238
84	9243	9248	9253	9258	9263	9269	9274	9279	9284	9289
85	9294	9299	9304	9309	9315	9320	9325	9330	9335	9340
86	9345	9350	9355	9360	9365	9370	9375	9380	9385	9390
87	9395	9400	9405	9410	9415	9420	9425	9430	9435	9440
88	9445	9450	9455	9460	9465	9469	9474	9479	9484	9489
89	9494	9499	9504	9509	9513	9518	9523	9528	9533	9538
90	9542	9547	9552	9557	9562	9566	9571	9576	9581	9586
91	9590	9595	9600	9605	9609	9614	9619	9624	9628	9633
92	9638	9643	9647	9652	9657	9661	9666	9671	9675	9680
93	9685	9689	9694	9699	9703	9708	9713	9717	9722	9727
94	9731	9736	9741	9745	9750	9754	9759	9763	9768	9773
95	9777	9782	9786	9791	9795	9800	9805	9809	9814	9818
96	9823	9827	9832	9836	9841	9845	9850	9854	9859	9863
97	9868	9872	9877	9881	9886	9890	9894	9899	9903	9908
98	9912	9917	9921	9926	9930	9934	9939	9943	9948	9952
99	9956	9961	9965	9969	9974	9978	9983	9987	9991	9996
	0	1	2	3	4	5	6	7	8	9

APPENDIX FIVE

answers to odd-numbered problems

CHAPTER 2. Errors and the Treatment of Analytical Data

1. (a) Iodine; (b) I, 0.00079 ppt; Co, 0.0017 ppt; Be, 0.0011 ppt
3. (a) $\bar{x}$ = 19.94; M = 19.94; R = 0.12; $\bar{d}$ = 0.04; $\bar{d}$ (rel) = 2.0 ppt; s = 0.046; v = 0.23%; (b) 19.94 ± 0.05 from s and R
5. (a) A: 0.09; 4.5 ppt; B: 0.11; 5.5 ppt (b) Work of A is more accurate and precise.
7. 0.0992; 5 ppt
9. 2 mg
11. 4 ppt
13. (a) 40; (b) 8; (c) 4
15. (a) 5; (b) 2; (c) 10; (d) 0.2
17. (a) 50; (b) 25.0; (c) 1.0×10^6; (d) 2000
19. (a) No; (b)(i) Yes; (ii) Yes; (iii) Yes
21. (a) No; (b) (i) No; (ii) No; (iii) No
23. (a) Yes, 0.1060; (b) 0.1030; 0.1021 – 0.1039
25. (a) 11.67; (b) 11.38
27. No
29. (a) 8.42 ± 0.19; (b) 8.42 ± 0.09
31. (a) 4; (b) 2
33. (a) 4.1 ± 0.1; (b) 20.98 ± 0.04
35. (a) 0.5; (b) 0.2; (c) 2; (d) 0.01; (e) 1
37. (a) 4.22×10^{-3}; (b) 9.6×10^7; (c) 42.16
39. (a) 6.670; (b) −4.595; (c) 2.82×10^3; (d) 1.8×10^6
41. (a) 0.26% high; (b) 6.5 ppt
43. 0.6 g

CHAPTER 3. Titrimetric Methods of Analysis

1. $2NaOH + H_3PO_4 \rightarrow Na_2HPO_4 + 2H_2O$
$Cr_2O_7^{2-} + 6Fe^{2+} + 14H^+ \rightarrow 2Cr^{3+} + 6Fe^{3+} + 7H_2O$
$Ag^+ + Br^- \rightarrow AgBr(s)$
$Hg^{2+} + 2Cl^- \rightarrow HgCl_2$
(a) 39.997; (b) 48.998; (c) 49.0307; (d) 151.90; (e) 169.873; (f) 119.002; (g) 162.30; (h) 74.551
3. (a) 0.200; (b) 0.0100; (c) 0.0665
5. (a) 0.200; (b) 0.0200; (c) 0.133
7. (a) 0.0253; (b) 0.0136; (c) 0.101

9. (a) 0.126; (b) 0.0816; (c) 0.101
11. (a) 0.105; 0.210; (b) 0.0883; 0.0883; (c) 0.768; 1.54 in reaction (ii); 0.384 in reaction (iv)
13. (a) 3.36; (b) 13.2; (c) 10.0; (d) 6.47
15. (a) 11.6; (b) 5.29
17. (a) 15.0; (b) 2.37×10^{-4}
19. (a) 0.0788; (b) 600 ml
21. 0.0880
23. 0.1251
25. (a) 0.1047; (b) 0.02094
27. (a) 1.091; (b) 0.1087
29. (a) 22.80; (b) 37.58
31. (a) 19.18; (b) 25.32
33. (a) 17.97; (b) 3.307
35. 24.5
37. 3.473
39. 0.271
41. 24.0
43. 960.0
45. (a) 33 g; (b) 0.33 g; (c) To require a convenient-size sample and minimize weighing errors.
47. 8.05

CHAPTER 4. Gravimetric Methods of Analysis

1. (a) 0.848002; (b) 0.33994; (c) 0.2351; (d) 0.966704; (e) 0.076732
3. 13.68
5. (a) 47.19; (b) 50.85
7. 19
9. 2.59 g
11. 0.400 g
13. 22.990
15. (a) 0.0701 g; (b) 0.0474 g; (c) 0.931 g; (d) 0.0601 g
17. 25.98
19. 93.47
21. 10.25% Na_2O; 35.56% K_2O
23. 1.27%
25. (a) 29.17; (b) 7.530
27. 6.821% NaCl; 35.95% NaBr

CHAPTER 5. Review of Chemical Equilibrium

1. (a) 2.22; (b) 1.00; (c) −1.00; (d) 9.40
3. (a) 2.0; (b) 0.50; (c) 1.8×10^{-5}; (d) 3.0×10^{-12}; (e) 4.0×10^{-15}
5. (a) 5.47; (b) 8.25; (c) 10.70; (d) 1.49
7. 4.0×10^{-5}
9. (a) 14.097; (b) 2.398; (c) 2.80; (d) 11.49
11. 4.9
13. (a) 3.92; (b) 9.26; (c) 1.51
15. (a) 10.96; (b) 5.17; (c) 7.00
17. 5.0×10^{-5}
19. (a) 1.26; (b) 1.30; (c) 7.00; (d) 13.65
21. (a) 1.15, 1.30; (b) 1.46, 1.54; (c) 1.91, 1.94; (d) 2.37, 2.38; (e) 2.87, 2.87
23. (a) 1×10^{-4}; (b) 3×10^{-3}; (c) 2
25. (a) Yes; (b) Yes; (c) No
27. (a) 3.5×10^{-5}; (b) 0.14
29. 0.1 *M*
31. 4.5×10^{-13}
33. 4.9×10^{-9}
35. (a) 4.1; (b) 3.6×10^2; (c) 9.8×10^3
37. 1.0×10^9

CHAPTER 6. Acid-Base Equilibria

1. (a) 1.36; (b) 1.51; (c) 1.74; (d) 2.54; (e) 4.46; (f) 7.00; (g) 9.54; (h) 11.60. Suitable indicators include bromcresol purple, bromthymol blue, and neutral red.
3. (a) 5.25; (b) 8.61; (c) 9.14; (d) 10.14; (e) 10.81; (f) 10.81; (g) 10.84; (h) 11.60
5. (a) 10.90; (b) 9.68; (c) 9.26; (d) 8.56; (e) 6.40; (f) 5.42; (g) 4.43; (h) 2.54
7. (a) 0.25; (b) 0.33; (c) 0.50; (d) 0.80; (e) 0.999
9. Derivation
11. (a) 8.81; (b) 10.78; (c) 8.62
13. 0.020
15. 95

17. 0.280

19. 4.62 ml of HOAc; 356 ml of NaOH

21. (a) 3.65; (b) (i) 3.35; (ii) 4.05; (iii) 8.41; (iv) 1.93; (v) 12.48; (c) ΔpH: (i) 0.30; (ii) 0.40; (iii) 4.80; (iv) 1.73; (v) 8.83

23. $[B^+] = 0.20$; $[BOH] = 0.40$

25. 5.15

27. (a) 9.70; (b) 9.70; (c) 9.70. Titration not feasible.

29. (a) 8.00, 8.80, 9.60; (b) 8.00, 8.85, 9.70; (c) 8.00, 8.91, 9.82

31. $[HA] = [OH^-] = 1 \times 10^{-6}$; $[Na^+] = [A^-] = 0.10$; $[H_3O^+] = 1 \times 10^{-8}$. Errors: 1% and 0.001%.

33. 7.9×10^3

CHAPTER 7. Acid-Base Equilibria in Complex Systems

1. (a) 1.0×10^{-5}, 5.00; 1.0×10^{-5}, 5.00; (b) 7.1×10^{-6}, 5.15; 1.0×10^{-5}, 5.00; (c) 1.6×10^{-5}, 4.79; 2.2×10^{-5}, 4.67; (d) 3.0×10^{-10}, 9.52; 1.7×10^{-10}, 9.76; (e) 4.5×10^{-9}, 8.35; 4.5×10^{-9}, 8.35; (f) 3.2×10^{-10}, 9.50; 1.0×10^{-11}, 11.00

3. (a) 1.80; (b) 2.12; (c) 4.67; (d) 9.77

5. (a) 1.62; 1.89; 2.26; 4.67; 6.84; 7.21; 9.77; 11.72; (b) Graph; (c) Methyl red is a suitable indicator for the first equivalence point, thymolphthalein for the second.

7. (a) 6.91; (b) 4.67

9. $H_2PO_4^-$ and HPO_4^{2-}; $[H_2PO_4^-]/[HPO_4^{2-}] = 0.65$

11.

pH	H_3C	H_2C^-	HC^{2-}	C^{3-}
1.0	100	—	—	—
2.0	0.92	0.08	—	—
3.0	0.54	0.45	0.008	3×10^{-5}
4.0	0.091	0.766	0.137	0.006
5.0	0.003	0.285	0.508	0.203
6.0	—	0.11	0.20	0.79
7.0	—	—	0.024	0.976
8.0	—	—	0.002	0.998

13. 18.46% Na_2CO_3; 9.37% NaOH

15. (a) 51.65; (b) 65.80; (c) 14.15; 46.16

17. (a) 43.96; (b) 44.92

19. (a) 11.08 ml; (b) 47.87

21. (a) 6.15; (b) 10.23

CHAPTER 8. Complex Formation Titrations

1. (a) 0.01135; (b) 4.224

3. 106.1 ppm $CaCO_3$; 41.55 ppm $MgCO_3$

5. 56.36% Bi; 37.71% Pb

7. 76.72

9. (a) 1.7×10^{13}; (b) 2.1×10^7

11. (a) 2.30; (b) 8.74; (c) 15.30

13. (a) 6.92; (b) 4.93; (c) 2.28

15. At pH 8: (a) 2.00; (b) 2.48; (c) 5.00; (d) 7.30; (e) 9.60; (f) 11.30. At pH 12: (a) 2.00; (b) 2.48; (c) 5.00; (d) 8.45; (e) 11.89; (f) 13.59. ΔpH: pH 8 = 4.60; pH 10 = 6.45; pH 12 = 6.89.

17. (a) 2.0×10^9; (b) 2.0×10^{10}

19. (a) 0.72; (b) 0.96; (c) 0.996; (d) 0.99998

21. (a) (i) M; (ii) M and N; (iii) M, N, and Q; (b) (i) 1.6×10^{10}; (ii) 0.10; (c) (i) 2.30; (ii) 2.30; (d) (i) 5.95; (ii) 7.25; (e) (i) 8.65; (ii) 3.00

CHAPTER 9. Solubility Equilibria

1. (a) 4×10^{-14}; (b) 2×10^{-19}; (c) 2×10^{-12}

3. (a) 5×10^{-9}; (b) 7×10^{-7}; (c) 7×10^{-6}

5. (a) 9×10^{-4}; (b) 4×10^{-4}; (c) 4×10^{-3}
7. (a) 2×10^{-8}; (b) $[Ag^+] = 0.010$; $[CrO_4^{2-}] = 2 \times 10^{-8}$; $[K^+] = 0.060$; $[NO_3^-] = 0.070$
9. (a) 1.70; (b) 2.30; (c) 2.22
11. $[Cl^-] = 0.030$; $[Ag^+] = 3 \times 10^{-9}$; $[Ca^{2+}] = 0.040$; $[NO_3^-] = 0.050$
13. (a) 1.00; (b) 1.48; (c) 4.00; (d) 6.20; (e) 8.40; (f) 10.36
15. Derivation
17. 1.00×10^{10}; 1.00×10^{-10}
19. (a) 100; (b) 0.4; (c) 1×10^{-4}
21. (a) 1×10^{-4}; (b) 0.2
23. 0.9
25. (a) CdS: −1.15, 1.35; ZnS: 0.00, 2.50; No (b) ZnS: 0.00, 2.50; FeS: 2.80, 5.30; Yes
27. $Al(OH)_3$: 3.57, 5.23; $Cu(OH)_2$: 5.15, 7.65; $Mg(OH)_2$: 9.00, 11.50. $Al(OH)_3$ from $Cu(OH)_2$, No; $Cu(OH)_2$ from $Mg(OH)_2$, Yes
29. (a) 2.5×10^{-3}; (b) 0.30; (c) Yes
31. 5.7×10^{-10}
33. (a) 4.50; (b) 0.089; (c) 4.69

CHAPTER 10. Oxidation-Reduction Equilibria

1. (a) +0.02 V, right, Cd positive; (b) −0.03 V, left, Cd negative; (c) −0.34 V, left, Pt negative; (d) +0.76 V (using 1.61 as $E°$ for cerium), right, Pt(Ce) positive
3. (a) −0.41 V; (b) −0.83 V; (c) +0.82 V; (d) +0.03 V
5. (a) 1.9×10^{37}; (b) 5.3×10^{-39}; (c) −0.79 V; −0.79 V
7. (a) 0.31; (b) left; (c) 2.9×10^3 joules; (d) -1.8×10^4 joules, right
9. Derivation
11. (a) 0.73 V; (b) 1.39 V; (c) 0.13 V; (d) 0.74 V
13. (a) 0.12 V; (b) 0.14 V; (c) 0.14 V; (d) 0.15 V; (e) 0.21 V; (f) 0.24 V
15. (a) 0.80 V; (b) 0.77 V; (c) 0.60 V; (d) 0.36 V; (e) 0.24 V; (f) 0.17 V
17. 1.7×10^{-7}
19. (a) −0.59 V; (b) +1.09 V
21. (a) 1×10^8; (b) 0.24 V
23. (a) $E = 0.25 + 0.059\,pH$; (b) 0.31 V; (c) 1.02 V
25. (a) 0.55 V; (b) 0.51 V
27. 2×10^{-16}
29. 0.52 V
31. (a) −0.76 V; (b) −0.34 V
33. (a) No; (b) 3.6×10^{-36}
35. (a) Yes; (b) 1.9×10^6; (c) 7.3×10^{-4}

CHAPTER 11. Applications of Oxidation-Reduction Titrations

1. (a) $2Fe^{2+} + H_2O_2 + 2H^+ \rightarrow 2Fe^{3+} + 2H_2O$
   (b) $Mn^{2+} + H_2O_2 + 2OH^- \rightarrow MnO_2 + 2H_2O$
   (c) $2Mn^{2+} + 5S_2O_8^{2-} + 8H_2O \rightarrow 2MnO_4^- + 10SO_4^{2-} + 16H^+$
   (d) $2Cr^{3+} + 3BiO_3^- + 4H^+ \rightarrow Cr_2O_7^{2-} + 3Bi^{3+} + 2H_2O$
   (e) $2Fe^{3+} + SO_2 + 2H_2O \rightarrow 2Fe^{2+} + SO_4^{2-} + 4H^+$
3. (a) $C_2H_6O_2 + 2MnO_4^- + 6H^+ \rightarrow 2CO_2 + 2Mn^{2+} + 6H_2O$
   EW = 6.0268
   (b) $C_2H_6O_2 + 6Ce^{4+} + 2H_2O \rightarrow 2H_2CO_2 + 6Ce^{3+} + 6H^+$
   EW = 10.345
   (c) $C_2H_6O_2 + HIO_4 \rightarrow 2H_2CO + IO_3^- + H^+ + H_2O$
   EW = 31.034
5. (a) 51.72; (b) 10.22; 8.920; 3.231
7. 5.6 ml
9. (a) $2MnO_4^- + 5SO_2 + 2H_2O \rightarrow 2Mn^{2+} + 4H^+ + 5SO_4^{2-}$
   (b) 36.00
11. 6.081
13. 0.200

15. (a) 0.101; 6.77 mg/ml
17. 3.685
19. Iron: 0.05952; chromium: 0.08608
21. (a) $3CHO_2^- + 2MnO_4^- + H_2O \rightarrow 2MnO_2 + 3CO_2 + 5OH^-$
(b) 36.7
23. (a) $IO_3^- + 5I^- + 6H^+ \rightarrow 3I_2 + 3H_2O$; 0.0200
(b) 60.00 ml
25. 16.8
27. (a) 0.21; (b) 5.2×10^{-6}; (c) Reaction slow at pH 5.

CHAPTER 12. Potentiometric Methods of Analysis

1.

Volume, ml	pH	*E*, V
0.00	1.40	0.33
10.0	1.56	0.34
20.0	1.78	0.36
39.0	3.14	0.44
39.9	4.15	0.49
40.0	7.00	0.66
40.1	9.85	0.83
50.0	11.82	0.95

3. Graphs [should be similar to Fig. 12.3(b) and (c)]
5. (a) 5.60; (b) 8.21
7. 30.17
9. 0.46
11. 0.62
13. 4.95
15. 40.0
17. $E = -0.42 + 0.059\,pH$
19. 0.53 V

CHAPTER 13. Other Electrical Methods of Analysis

1. (a) 1.23 V; (b) 1.23 V; (c) 1.23 V
3. 1.04 V
5. (a) 6.3×10^{-13}; (b) 4.8×10^{-40}
7. 0.16
9. (a) 14.58; (b) 23.05; (c) 30.17
11. −0.44 V
13. 1.9×10^{-7}
15. 3.2
17. (a) H_2 and AgCl; (b) 0.1471; (c) 0.24
19. (a) 0.349; (b) 1.47; (c) 27
21. (a) 5.8; (b) 5.0; (c) 4.5
23. 4.0×10^{-4} *M*
25. 0.0066 *M*
27. 7.7×10^{-6}

CHAPTER 14. Spectrophotometry

1. (a) 2, 0; (b) Infinite
3. (a) 0.0088; (b) 0.097; (c) 0.30; (d) 1.00; (e) 2.00
5. (a) 25; (b) 71; (c) 93
7. (a) 19; (b) 1.1×10^4
9. 0.045
11. 8×10^9
13. 215
15. 0.23
17. (a) −21.7%; (b) −4.3%; (c) −2.9%; (d) −5.4%; (e) −100%
19. 0.064 ml
21. (a) 0.60, 0.90, 1.20; (b) 50%, 25%, 13%, 6.3%; difference = 25, 12, 7; (c) 0.30, 0.60, 0.90; 100%, 50%, 25%, 13%; difference = 50, 25, 12
23. A = 2.77×10^{-4}; B = 3.15×10^{-4}
25. 1.8×10^9
27. MX_2
29. HIn, 0.00; In^-, 420; $pK_a = 6.63$
31. (a) 1.00; (b) 5.70

CHAPTER 16. Solvent Extraction

1. (a) 0.49; (b) 0.0002
3. 20.2
5. 50
7. 10
9. (a) 125 ml; (b) 80 ml
11. (a) 6.2×10^{-6}; (b) 3.1×10^{-4}; 6.2×10^{-3}; 6.2×10^{-2}; 0.31; 0.62
13. (a) $K = \dfrac{K_{DMT_2} K_f H_a^2}{K_{DHT}^2}$; (b) 2.00; 2.76; 3.00; 3.24; 4.50
15. $K = 1 \times 10^{-4}$; $n = 4$
17. 9×10^{-5}

CHAPTER 18. Liquid Chromatography

1. 3.29
3. 622.5

INDEX

A

B

C

D

E

F

G

H

I

J

K

L

M

N

O

P

Q

R

S

T

U

V

W

Z